Chemistry
For Degree Students

Organic Chemistry | Physical Chemistry | Laboratory Work

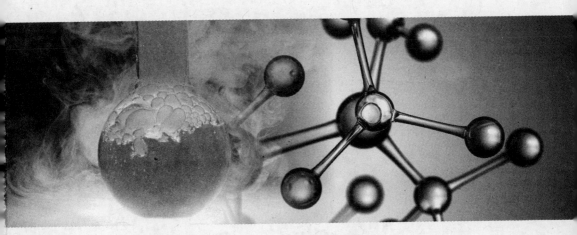

About the Author

Dr R L Madan has over 35 years experience of teaching chemistry at different levels. He has done extensive research work in the areas of adsorption, catalysis and surface chemistry at the Indian Institute of Technology, New Delhi. He has to his credit 15 research papers in the areas of adsorption, catalysis, polymers and superconductivity published in different national and international journals.

He started his teaching career at Government Postgraduate College, Gurgaon. He has held positions of Head of the Chemistry Department, Government Postgraduate College, Faridabad and Principal, Government College, Panchkula and Tigaon (Faridabad). He is a recipient of UNESCO and Swedish Institute Fellowships and has authored a number of books on Chemistry and Environmental Education.

Chemistry
For Degree Students

Organic Chemistry | Physical Chemistry | Laboratory Work

B.Sc. (Hons.) Semester II

As per UGC Choice Based Credit System (CBCS)

R L MADAN
M.Sc., Ph.D.
Former Head of the Chemistry Department
Government Postgraduate College, Faridabad
and *Principal*, Government College, Panchkula

S Chand And Company Limited
(ISO 9001 Certified Company)

S Chand And Company Limited
(ISO 9001 Certified Company)

Head Office: Block B-1, House No. D-1, Ground Floor, Mohan Co-operative Industrial Estate, New Delhi – 110 044 | Phone: 011-66672000

Registered Office: A-27, 2nd Floor, Mohan Co-operative Industrial Estate, New Delhi – 110 044 | Phone: 011-49731800

www.schandpublishing.com; e-mail: info@schandpublishing.com

Branches

Chennai	:	Ph: 23632120; chennai@schandpublishing.com
Guwahati	:	Ph: 2738811, 2735640; guwahati@schandpublishing.com
Hyderabad	:	Ph: 40186018; hyderabad@schandpublishing.com
Jalandhar	:	Ph: 4645630; jalandhar@schandpublishing.com
Kolkata	:	Ph: 23357458, 23353914; kolkata@schandpublishing.com
Lucknow	:	Ph: 4003633; lucknow@schandpublishing.com
Mumbai	:	Ph: 25000297; mumbai@schandpublishing.com
Patna	:	Ph: 2260011; patna@schandpublishing.com

© S Chand And Company Limited, 2022

All rights reserved. No part of this publication may be reproduced or copied in any material form (including photocopying or storing it in any medium in form of graphics, electronic or mechanical means and whether or not transient or incidental to some other use of this publication) without written permission of the copyright owner. Any breach of this will entail legal action and prosecution without further notice.

Jurisdiction: All disputes with respect to this publication shall be subject to the jurisdiction of the Courts, Tribunals and Forums of New Delhi, India only.

S. CHAND'S Seal of Trust

In our endeavour to protect you against counterfeit/fake books, we have pasted a hologram sticker over the cover of this book. The hologram displays the unique 3D multi-level, multi-colour effects of our logo from different angles when tilted or properly illuminated under a single source of light, such as 2D/3D depth effect, kinetic effect, gradient effect, trailing effect, emboss effect, glitter effect, randomly sparkling tiny dots, etc.

A fake hologram does not display all these effects.

First Edition 2022

ISBN: 978-93-550-1037-7   **Product Code:** H6CDSCHEM6810ENAA21O

PRINTED IN INDIA

By Vikas Publishing House Private Limited, Plot 20/4, Site-IV, Industrial Area Sahibabad, Ghaziabad – 201 010 and Published by S Chand And Company Limited, A-27, 2nd Floor, Mohan Co-operative Industrial Estate, New Delhi – 110 044.

Preface

It gives me immense pleasure to present this textbook to the teachers and students of B.Sc. Honours (Chemistry) as per Choice Based Credit System (CBCS) curriculum recommended by University Grants Commission and adopted by the Indian universities. Designed strictly as per B.Sc. Honours (Chemistry), Second Semester, this textbook comprehensively covers the following two branches of chemistry in two sections:

Organic Chemistry: This section familiarises the students with topics such as basics of organic chemistry, stereochemistry, chemistry of aliphatic hydrocarbons and aromatic hydrocarbons.

Physical Chemistry: This section acquaints the students with topics such as chemical thermodynamics, systems of variable composition, chemical equilibrium and solutions & colligative properties.

Additionally, a separate section, **Laboratory Work**, has been exclusively provided to help students achieve solid conceptual understanding and learn experimental procedures. While laboratory work on *Organic Chemistry* covers experiments on calibration of thermometer, purification of organic compounds, determination of melting point & boiling point and chromatography; *Physical Chemistry* covers experiments on thermochemistry.

A long span of interaction with students has guided me to give special attention to dark areas which students find difficult to grasp. Keeping that in view, this textbook has been written in a simple and lucid language with following special features to aid student's understanding of the subject matter:
- Well drawn illustrations, interesting examples and analogies from daily life have been provided to make the reading interesting and engaging
- A large number of problems which have appeared in previous years' examination papers of different universities have been provided with solution for the benefit of students
- Questions for practice have been provided at the end of each chapter as per the latest pattern of examination

I am thankful to the management and editorial team of S. Chand Publishing for the help and support in the publication of this textbook.

I believe that the textbook contains all that is needed to understand the subject in a systematic manner. However, I also believe that no work is perfect and there is always scope for improvement. I would welcome suggestions from the teaching fraternity and students for further improvement of the book.

R L Madan
Mob. 9971775666

Disclaimer : While the author of this book has made every effort to avoid any mistakes or omissions and has used his skill, expertise and knowledge to the best of his capacity to provide accurate and updated information, the author and S. Chand do not give any representation or warranty with respect to the accuracy or completeness of the contents of this publication and are selling this publication on the condition and understanding that they shall not be made liable in any manner whatsoever. S.Chand and the author expressly disclaim all and any liability/responsibility to any person, whether a purchaser or reader of this publication or not, in respect of anything and everything forming part of the contents of this publication. S. Chand shall not be responsible for any errors, omissions or damages arising out of the use of the information contained in this publication.
Further, the appearance of the personal name, location, place and incidence, if any; in the illustrations used herein is purely coincidental and work of imagination. Thus the same should in no manner be termed as defamatory to any individual.

Syllabus

CHEMISTRY-C III: ORGANIC CHEMISTRY- I

Basics of Organic Chemistry

Organic Compounds: Classification, and Nomenclature, Hybridization, Shapes of molecules, Influence of hybridization on bond properties.

Electronic Displacements: Inductive, electromeric, resonance and mesomeric effects, hyperconjugation and their applications; Dipole moment; Organic acids and bases; their relative strength.

Homolytic and Heterolytic fission with suitable examples. Curly arrow rules, formal charges; Electrophiles and Nucleophiles; Nucleophlicity and basicity; Types, shape and their relative stability of Carbocations, Carbanions, Free radicals and Carbenes.

Introduction to types of organic reactions and their mechanism: Addition, Elimination and Substitution reactions.

Stereochemistry

Fischer Projection, Newmann and Sawhorse Projection formulae and their interconversions; Geometrical isomerism: cis–trans and, syn-anti isomerism E/Z notations with C.I.P rules.

Optical Isomerism: Optical Activity, Specific Rotation, Chirality/Asymmetry, Enantiomers, Molecules with two or more chiral-centres, Distereoisomers, meso structures, Racemic mixture and resolution. Relative and absolute configuration: D/L and R/S designations.

Chemistry of Aliphatic Hydrocarbons

A. Carbon-Carbon sigma bonds

Chemistry of alkanes: Formation of alkanes, Wurtz Reaction, Wurtz-Fittig Reactions, Free radical substitutions: Halogenation -relative reactivity and selectivity.

B. Carbon-Carbon pi bonds

Formation of alkenes and alkynes by elimination reactions, Mechanism of E1, E2, E1cb reactions. Saytzeff and Hofmann eliminations.

Reactions of alkenes: Electrophilic additions their mechanisms (Markownikoff/ Anti Markownikoff addition), mechanism of oxymercuration-demercuration, hydroboration oxidation, ozonolysis, reduction (catalytic and chemical), syn and anti-hydroxylation (oxidation). 1, 2-and 1, 4-addition reactions in conjugated dienes and, Diels-Alder reaction; Allylic and benzylic bromination and mechanism, e.g. propene, 1-butene, toluene, ethyl benzene.

Reactions of alkynes: Acidity, Electrophilic and Nucleophilic additions. Hydration to form carbonyl compounds, Alkylation of terminal alkynes.

C. Cycloalkanes and Conformational Analysis

Types of cycloalkanes and their relative stability, Baeyer strain theory, Conformation analysis of alkanes: Relative stability: Energy diagrams of cyclohexane: Chair, Boat and Twist boat forms; Relative stability with energy diagrams.

Aromatic Hydrocarbons

Aromaticity: Hückel's rule, aromatic character of arenes, cyclic carbocations/carbanions and heterocyclic compounds with suitable examples. Electrophilic aromatic substitution: halogenation, nitration, sulphonation and Friedel-Craft's alkylation/acylation with their mechanism. Directing effects of the groups.

CHEMISTRY LAB-C II LAB

1. Checking the calibration of the thermometer
2. Purification of organic compounds by crystallization using the following solvents:
   (a) Water
   (b) Alcohol
   (c) Alcohol-Water
3. Determination of the melting points of above compounds and unknown organic compounds (Kjeldahl method and electrically heated melting point apparatus)
4. Effect of impurities on the melting point – mixed melting point of two unknown organic compounds
5. Determination of boiling point of liquid compounds. (boiling point lower than and more than 100 °C by distillation and capillary method)
6. Chromatography
   (a) Separation of a mixture of two amino acids by ascending and horizontal paper chromatography
   (b) Separation of a mixture of two sugars by ascending paper chromatography
   (c) Separation of a mixture of o-and p-nitrophenol or o-and p-aminophenol by thin layer chromatography (TLC)

CHEMISTRY -C IV: PHYSICAL CHEMISTRY- II

Chemical Thermodynamics

Intensive and extensive variables; state and path functions; isolated, closed and open systems; zeroth law of thermodynamics.

First law: Concept of heat, q, work, w, internal energy, U, and statement of first law; enthalpy, H, relation between heat capacities, calculations of q, w, U and H for reversible, irreversible and free expansion of gases (ideal and van der Waals) under isothermal and adiabatic conditions.

Thermochemistry: Heats of reactions: standard states; enthalpy of formation of molecules and ions and enthalpy of combustion and its applications; calculation of bond energy, bond dissociation energy and resonance energy from thermochemical data, effect of temperature (Kirchhoff's equations) and pressure on enthalpy of reactions. Adiabatic flame temperature, explosion temperature.

Second Law: Concept of entropy; thermodynamic scale of temperature, statement of the second law of thermodynamics; molecular and statistical interpretation of entropy. Calculation of entropy change for reversible and irreversible processes.

Third Law: Statement of third law, concept of residual entropy, calculation of absolute entropy of molecules.

Free Energy Functions: Gibbs and Helmholtz energy; variation of S, G, A with T, V, P; Free energy change and spontaneity. Relation between Joule-Thomson coefficient and other thermodynamic parameters; inversion temperature; Gibbs-Helmholtz equation; Maxwell relations; thermodynamic equation of state.

Systems of Variable Composition

Partial molar quantities, dependence of thermodynamic parameters on composition; Gibbs-Duhem equation, chemical potential of ideal mixtures, change in thermodynamic functions in mixing of ideal gases.

Chemical Equilibrium

Criteria of thermodynamic equilibrium, degree of advancement of reaction, chemical equilibria in ideal gases, concept of fugacity. Thermodynamic derivation of relation between Gibbs free energy of reaction and reaction quotient. Coupling of exoergic and endoergic reactions. Equilibrium constants and their quantitative dependence on temperature, pressure and concentration. Free energy of mixing and spontaneity; thermodynamic derivation of relations between the various equilibrium constants K_p, K_c and K_x. Le Chatelier principle (quantitative treatment); equilibrium between ideal gases and a pure condensed phase.

Solutions and Colligative Properties

Dilute solutions; lowering of vapour pressure, Raoult's and Henry's Laws and their applications. Excess thermodynamic functions.

Thermodynamic derivation using chemical potential to derive relations between the four colligative properties [(i) relative lowering of vapour pressure, (ii) elevation of boiling point, (iii) Depression of freezing point, (iv) osmotic pressure] and amount of solute. Applications in calculating molar masses of normal, dissociated and associated solutes in solution.

CHEMISTRY LAB- C IV LAB

Thermochemistry
  (a) Determination of heat capacity of a calorimeter for different volumes using change of enthalpy data of a known system (method of back calculation of heat capacity of calorimeter from known enthalpy of solution or enthalpy of neutralization).
  (b) Determination of heat capacity of the calorimeter and enthalpy of neutralization of hydrochloric acid with sodium hydroxide.
  (c) Calculation of the enthalpy of ionization of ethanoic acid.
  (d) Determination of heat capacity of the calorimeter and integral enthalpy (endothermic and exothermic) solution of salts.
  (e) Determination of basicity/proticity of a polyprotic acid by the thermochemical method in terms of the changes of temperatures observed in the graph of temperature versus time for different additions of a base. Also calculate the enthalpy of neutralization of the first step.
  (f) Determination of enthalpy of hydration of copper sulphate.
  (g) Study of the solubility of benzoic acid in water and determination of ΔH.

Any other experiment carried out in the class.

Contents

SECTION I: ORGANIC CHEMISTRY

1. Basics of Organic Chemistry I 2–29
- 1.1 Classification of Organic Compounds   3
- 1.2 Nomenclature of Organic Compounds   8
- 1.3 Nomenclature of Alicyclic Organic Compounds   12
- 1.4 Bond line Notation of Organic Compounds   13
- 1.5 Nomenclature of Aromatic Compounds   14
- 1.6 Nomenclature of Polycyclic Hydrocarbons   23
- *Key Terms*   25
- *Multiple Choice Questions*   25
- *Short Answer Questions*   26
- *Long Answer Questions*   26
- *Answers (Multiple Choice Questions)*   28
- *Suggested Readings*   30

2. Basics of Organic Chemistry II 30–95
- 2.1 Curved Arrow Notation   31
- 2.2 Drawing Electron Movements with Arrows   31
- 2.3 Electron Displacements in Organic Compounds   32
- 2.4 Inductive Effect   32
- 2.5 Electromeric Effect   37
- 2.6 Resonance or Mesomeric Effect   38
- 2.7 Comparison of Inductive, Electromeric and Resonance Effects   44
- 2.8 hyperconjugation or No-Bond Resonance   45
- 2.9 Organic Acids and Bases   50
- 2.10 Linear and Crossed Conjugation Systems   51
- 2.11 Hybridisation of Orbitals   52
- 2.12 Types of Hybridisation   53
- 2.13 Structures and Shapes of Certain Molecules   54
- 2.14 Some Important Bond Parameters   57
- 2.15 Localized and Delocalized Bonds   60
- 2.16 Homolytic and Heterolytic Bond Breaking (Fission)   62
- 2.17 Types of Organic Reactions   63
- 2.18 Formal Charge   65
- 2.19 Types of Reagents—Nucleophiles and Electrophiles   66
- 2.20 Nucleophilicity and Basicity   67
- 2.21 Reactive Intermediates   67
- 2.22 Carbocations   67
- 2.23 Carbanions   73
- 2.24 Free Radicals   76
- 2.25 Carbenes   80
- 2.26 Nitrenes   83
- 2.27 Arynes   85
- *Key Terms*   91
- *Multiple Choice Questions*   92
- *Short Answer Questions*   92

(xi)

Long Answer Questions 93
Answers (*Multiple Choice Questions*) 95
Suggested Readings 95

3. Stereochemistry 96–155

3.1　Introduction 97
3.2　Structural Isomerism 97
3.3　Stereoisomerism 99
3.4　Geometrical Isomerism 127
3.5　Conformational Isomerism (Analysis) 135
Key Terms 151
Multiple Choice Questions 151
Short Answer Questions 152
Long Answer Questions 153
Answers (*Multiple Choice Questions*) 155
Suggested Readings 155

4. Alkanes (Carbon-Carbon Sigma Bonds) 156–197

4.1　Alkanes 157
4.2　Structure and Nomenclature of Alkanes 157
4.3　Classification of the Carbon Atoms in Alkanes 158
4.4　Alkyl Groups 158
4.5　Structural Isomerism in Alkanes 159
4.6　Sources of Alkanes 164
4.7　General Methods of Preparation of Alkanes 164
4.8　Physical Properties of Alkanes 168
4.9　Chemical Properties of Alkanes 170
4.10　Mechanism of Halogenation of Alkane 174
4.11　Factors that influence the Orientation of Halogenation of Alkanes 177
4.12　Activation Energy 183
4.13　Relative Reactivities of Halogens in the Halogenation of Alkanes 185
4.14　Effect of Activation Energy on the Structure of Transition State in Halogenation of Alkanes 186
4.15　Reactivity and Selectivity of Halogens Towards Alkanes 187
Key Terms 193
Multiple Choice Questions 193
Short Answer Questions 195
Long Answer Questions 195
Answers (*Multiple Choice Questions*) 197
Suggested Readings 197

5. Alkenes (Carbon-Carbon Pi Bonds) 198–245

5.1　Nomenclature of Alkenes 199
5.2　Hindered Rotation Around Double Bond 199
5.3　Orbital Structure of Ethylene 200
5.4　Methods of Preparation of Alkenes 201
5.5　Physical Properties of Alkenes 211
5.6　Chemical Properties of Alkenes 212
5.7　Polymerisation 233
5.8　1, 2- and 1, 4 – Addition to Conjugated Dienes 235
5.9　Industrial Applications of Ethylene and Propene 236
Key Terms 242
Multiple Choice Questions 242

Short Answer Questions 243
Long Answer Questions 243
Answers (Multiple Choice Questions) 245
Suggested Readings 245

6. Alkynes 246–275

6.1 Nomenclature of Alkynes 247
6.2 Preparation of Alkynes 249
6.3 Physical Properties of Alkynes 253
6.4 Chemical Reactions of Alkynes 254
Key Terms 272
Multiple Choice Questions 272
Short Answer Questions 274
Long Answer Questions 274
Answers (Multiple Choice Questions) 275
Suggested Readings 275

7. Cycloalkanes 276–299

7.1 Cycloalkanes 277
7.2 Nomenclature of Substituted Cycloalkanes 277
7.3 Methods of Preparation of Cycloalkanes and its Derivatives 281
7.4 Blanc's Rule 287
7.5 Chemical Properties of Cycloalkanes 289
7.6 Baeyer Strain Theory 290
7.7 Sachse Mohr Theory of Strainless Rings and Stability of Cyclohexane 292
7.8 Difficulties in the Synthesis of Large-membered Rings 293
Key Terms 297
Multiple Choice Questions 297
Short Answer Questions 298
Long Answer Questions 298
Answers (Multiple Choice Questions) 299
Suggested Readings 299

8. Aromatic Hydrocarbons 300–348

8.1 Introduction 301
8.2 Arenes 301
8.3 Nomenclature of Benzene Derivatives 301
8.4 Structure of Benzene 304
8.5 Aromaticity 307
8.6 Huckel's Rule and Aromaticity 307
8.7 Annulenes 309
8.8 General Mechanism of Electrophilic Substitution in Benzene 313
8.9 σ- and π-Complexes 315
8.10 Effect of Substituents on Orientation and Reactivity of Benzene Ring 322
8.11 Theory of Reactivity in Aromatic Compounds on the Basis of
       Inductive and Resonance Effects 323
8.12 Ortho-Para Ratio 327
8.13 Substitution of Third Group in the Benzene Broup 329
8.14 Side Chain Halogenation 331
8.15 Side Chain Oxidation 333
8.16 Birch Reduction 335
8.17 Biphenyl or Diphenyl 336
Key Terms 343

(xiii)

Multiple Choice Questions 343
Short Answer Questions 345
Long Answer Questions 346
Answers (Multiple Choice Questions) 348
Suggested Readings 348

SECTION II: PHYSICAL CHEMISTRY

1. Chemical Thermodynamics – I (Zeroth and First Laws of Thermodynamics and Thermochemistry) 349–421

1.1 Thermodynamics 351
1.2 Definition of Certain Thermodynamics Terms 351
1.3 Internal Energy and Change in Internal Energy 354
1.4 Expression for Work of Expansion Against Constant Pressure 355
1.5 Heat 356
1.6 First Law of Thermodynamics 358
1.7 Enthalpy or Heat Content 359
1.8 Heat Capacity 361
1.9 Relationship of μ With Other Thermodynamic Quantities 366
1.10 Isothermal Reversible Expansion of a Gas 367
1.11 Values of $w, q, \Delta U$ and ΔH in Adiabatic Expansion of an Ideal Gas 372
1.12 Calculation of $w, q, \Delta U$ and ΔH in the Expansion of a Real (Van Der Walls) Gas in Isothermal Reversible Expansion 377
1.13 $q, w, \Delta E$ and ΔH in Adiabatic Reversible Expansion of a Real (or Van Der Waals) Gas 379
1.14 Thermochemistry 383
1.15 Heat of Reaction (or Enthalpy of Reaction) 387
1.16 Standard State 390
1.17 Different Kinds of Heat (or Enthalpy) of Reaction 395
1.18 Hess's Law of Constant Heat Summation 397
1.19 Bond Enthalpy or Bond Energy 409
1.20 Variation of Heat of Reaction with Temperature Kirchoff's Equation 414
Key Terms 418
Multiple Choice Questions 418
Short Answer Questions 419
Long Answer Questions 419
Answers (Multiple Choice Questions) 421
Suggested Readings 421

2. Chemical Thermodynamics – II (Second and Third Laws of Thermodynamics and Partial Molar Properties) 422–479

2.1 Introduction – Need for the Second Law 423
2.2 Carnot Cycle 424
2.3 Thermodynamic Scale of Temperature 428
2.4 Concept of Entropy 428
2.5 Entropy Change for an Ideal Gas Under Different Conditions 432
2.6 Entropy Change on Mixing of Ideal Gases 441
2.7 Helmholtz Function or Work Function 443
2.8 Gibb's Function or Gibb's Free Energy 443
2.9 Criteria For Feasibility or Spontaneity of a Process 448
2.10 Maxwell Relationships 450

2.11 Gibb's–Helmholtz Equation 453
2.12 Nernst Heat Theorem 457
2.13 Third Law of Thermodynamics 457
2.14 Residual Entropy 460
2.16 Partial Molar Quantities 464
2.17 Partial Molar Free Energy (Chemical Potential) 465
2.18 Gibbs-Duhem Equation 466
2.19 Chemical Potential of a Component in a System of Ideal Gases 469
2.20 Criteria for Phase Equilibrium for Multicomponent System 470
2.21 Clausius-Clapeyron Equation 471
Key Terms 476
Multiple Choice Questions 476
Short Answer Questions 477
Long Answer Questions 478
Answers (Multiple Choice Questions) 479
Suggested Readings 479

3. Chemical Equilibrium 480–513

3.1 Introduction 481
3.2 The State of Chemical Equilibrium 483
3.3 Law of Mass Action 483
3.4 Partial Molar Quantities 486
3.5 Free Energy Change as a Criterion of Spontaneity 488
3.6 Thermodynamic Derivation of the Law of Chemical Equilibrium 491
3.7 Van't Hoff Reaction Isotherm 494
3.8 Relation Between K_p and K_C 496
3.9 Van't Hoff Equation for the Temperature Dependence of Equilibrium Constant (Van't Hoff Reaction Isochore) 497
3.10 Coupling of Exoergic and Endoergic Reactions 500
3.11 Le-Chatelier's Principle 502
3.12 Physical Equilibria 508
Key Terms 510
Multiple Choice Questions 510
Short Answer Questions 511
Long Answer Questions 512
Answers (Multiple Choice Questions) 513
Suggested Readings 513

4. Solutions and Colligative Properties 514–570

4.1 Raoult's Law 515
4.2 Henry's Law 521
4.3 Excess Thermodynamic Functions 522
4.4 Colligative Properties 523
4.5 Determination of Vapour Pressure of a Liquid 524
4.6 Determination of Lowering of Vapour Pressure of the Solvent 524
4.7 Relation Between the Relative Lowering of Vapour Pressure and the Molecular Mass of the Solute (Raoult's Law) 525
4.8 Thermodynamic Derivation of the Expression for Relative Lowering of Vapour Pressure 526
4.9 Osmosis Phenomenon 529
4.10 Van't Hoff Relation Between the Osmotic Pressure of a Solution and Molecular Mass of the Solute 529
4.11 Van't Hoff Theory of Dilute Solution Taking the Example of Osmosis 533

4.12	Determination of Osmotic Pressure	534
4.13	Theories of Osmosis	539
4.14	Elevation in Boiling Point	541
4.15	Relation Between Elevation in Boiling Point of the Solution and the Molecular Weight of the Solute	544
4.16	Depression in Freezing Point	551
4.17	Relation Between Depression in Freezing Point and Molecular Mass of the Solute	552
4.18	Cause of Abnormal Molecular Masses of Solutes in Solutions	560
4.19	Van't Hoff Factor	561

Key Terms   567
Multiple Choice Questions   567
Short Answer Questions   568
Long Answer Questions   569
Answers (Multiple Choice Questions)   570
Suggested Readings   570

SECTION III: LABORATORY WORK

1. Calibration of a Thermometer 572–573
Viva-Voce Questions with Answers   573

2. Purification of Organic Compounds by Crystallization 574–578
2.1   Introduction   574
2.2   Selection of the Solvent   574
2.3   Fractional Crystallization   576
Viva-Voce Questions with Answers   577

3. Determination of Melting and Boiling Points 579–585
3.1   Determination of Melting Point   579
3.2   Determination of Boiling Point   582
Viva-Voce Questions with Answers   585

4. Paper Chromatography 586–592
4.1   Introduction   586
4.2   Circular Paper Chromatography or Radial Chromatography   587
Viva-Voce Questions with Answers   591

5. Thin Layer Chromatography 593–601
5.1   Introduction   593
5.2   Thin-Layer Chromatography (TLC)   593
5.3   Experimental Procedure For TLC   594
5.4 R_{st} Values   594
5.5   Adsorbents and Eluents   594
Viva-Voce Questions with Answers   600

6. Thermochemistry 602–616
6.1   Enthalpy of Solution   602
6.2   Enthalpy of Neutralization   603
Viva-Voce Questions with Answers   615

(xvi)

Section I

Organic Chemistry

- Basics of Organic Chemistry I
- Basics of Organic Chemistry II
- Stereochemistry
- Alkanes
- Alkenes
- Alkynes
- Cycloalkanes
- Aromatic Hydrocarbons

Chapter 1

Basics of Organic Chemistry I

LEARNING OBJECTIVES

After reading this chapter, you should be able to:
- classify organic compounds into different types
- know terms like word root, prefix and suffix
- know how to arrange prefix, word root and suffix
- to name a compound
- acquiant yourself about nomenclature rules
- understand the nomenclature of alicyclic compounds
- follow bond line notation
- learn nomenclature of aromatic compounds
- understand nomenclature of polycyclic compounds

1.1 CLASSIFICATION OF ORGANIC COMPOUNDS

There is a wide variety of ways in which carbon can bond with itself and other elements. This is the reason for the existence of a large number of organic compounds. According to an estimate, the number of organic compounds is nine million. To study these compounds systematically, we need to classify them into categories and sub-categories.

The two main categories of these compounds are:

1. Open chain compounds: These are aliphatic compounds and derive their name from Greek word meaning **Fatty**, because the earlier compounds of this category were obtained from fats. In such compounds, carbon atoms are joined to form open chains, straight or branched. A large **number of compounds of everyday use belong to this category. A few examples of this category are:**

$CH_3 - CH_2OH$ $CH_3 - COOH$ (Straight chain)
Ethyl alcohol   Acetic acid

$CH_3 - CH - CH_3$ $CH_3 - \underset{OH}{\overset{CH_3}{C}} - CH_3$ (Branched chain)
    $|$
    CH_3
Isobutane 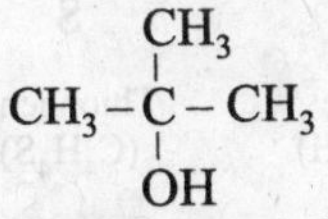   Tert.butyl alcohol

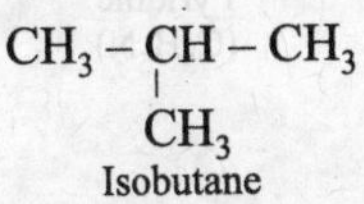

2. Closed chain or Cyclic compounds or Ring compounds: Organic compounds with closed chains of atoms are termed as **closed chains or cyclic compounds:** When a molecule contains two or more rings, it is called polycyclic. The cyclic compounds may be further sub-divided into two types viz., Homocyclic and Heterocyclic Compounds.

Homocyclic Compounds

The homocyclic ring compounds are made up of carbon atoms only. These include two types of compounds:

(i) Aromatic compounds: Compounds having benzene rings of six carbon atoms (*i.e.*, ring with single and double bonds at alternate positions) are called **aromatic compounds.** Aromatic compounds occur chiefly in plants and many of them are sweet smelling. Hence the name (aroma in Greek = sweet smell). Some examples of aromatic compounds are:

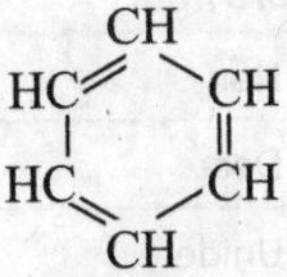

Benzene (C_6H_6) 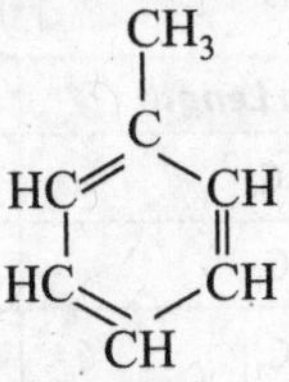   Toluene ($C_6H_5CH_3$) 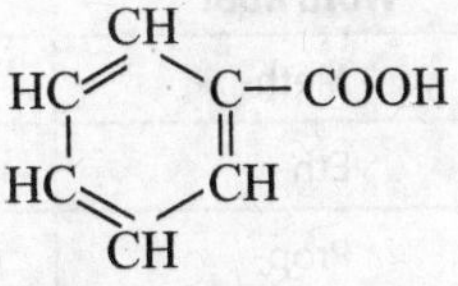   Benzoic acid (C_6H_5COOH) 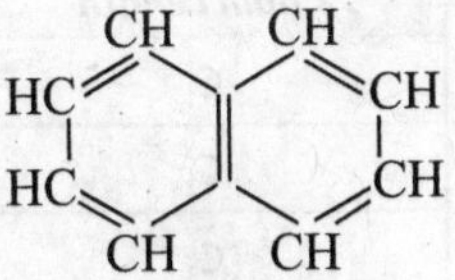   Naphthalene ($C_{10}H_8$)

(ii) Alicyclic compounds: These are homocyclic ring compounds which behave more like aliphatic than aromatic compounds. These have, therefore, been named as **aliphatic cyclic or alicyclic compounds.** These include polymethylenes such as cyclopropane, cyclobutane, etc. Such compounds contain no double bonds in their rings. Examples are:

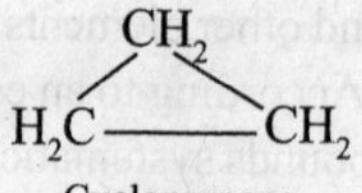

Cyclopropane (C_3H_6)

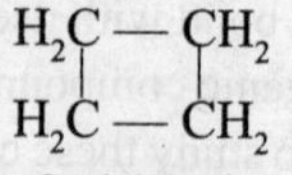

Cyclobutane (C_4H_8)

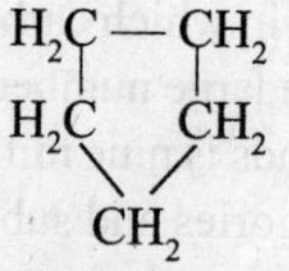

Cyclopentane (C_5H_{10})

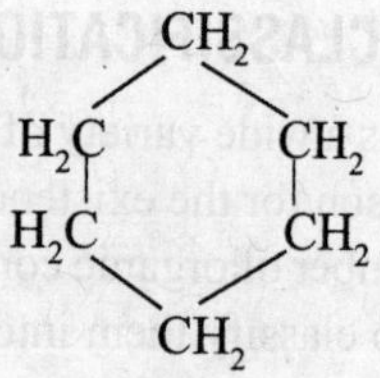

Cyclohexane (C_6H_{12})

Heterocyclic Compounds

Compounds having elements like O, N and S besides carbon at ring positions are called heterocyclic compounds. Examples are:

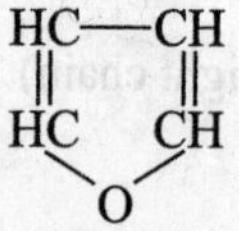

Furan (C_4H_4O)

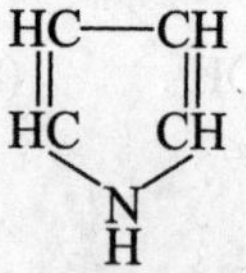

Pyrrole (C_4H_4NH)

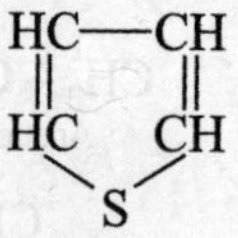

Thiophene (C_4H_4S)

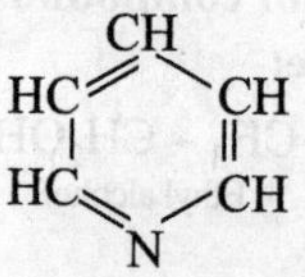

Pyridine (C_5H_5N)

1.1.1 Parts of the Name of a Compound

An organic compound used to be named after the plant or animal source from which it was obtained. As more and more compounds came to be identified, it became increasingly difficult to name them. International scientific body on chemistry IPUAC formulated guidelines to name the compounds using the following parts:

(i) **Word Root.** The word root represents the number of carbon atoms in the *parent chain*. The parent chain in the compound is selected by following certain rules as discussed later. For the chains upto four carbon atoms special word roots are used, but for the chains containing more than four carbon atoms, Greek numerals are used. The general word root for different aliphatic compounds is **ALK**. The word roots for carbon chains of different lengths are given below in Table 1.1.

TABLE 1.1 *Word Roots for Carbon Chains of Different Lengths*

Chain Length	Word Root	Chain Length	Word Root
C_1	Meth-	C_9	Non-
C_2	Eth-	C_{10}	Dec-
C_3	Prop-	C_{11}	Unidec-
C_4	But-	C_{12}	Dodec-
C_5	Pent-	C_{13}	Tridec-
C_6	Hex-	C_{18}	Octadec-
C_7	Hept-	C_{20}	Icos-
C_8	Oct-	C_{30}	Tricont-

(ii) **Primary Suffix.** Primary suffix is used to represent saturation or unsaturation in the carbon chain. While writing the name, primary suffix is added to the word root. Some of the primary suffixes are given below in Table 1.2.

Basics of Organic Chemistry I

TABLE 1.2 Some Primary Suffixes

Nature of Carbon Chain		Primary Suffix
Saturated Carbon Chain		ane
Unsaturated Carbon Chains:		
One	C = C bond	ene
two	C = C bond	adiene
three	C = C bond	atriene
one	C ≡ C bond	yne
two	C ≡ C bond	adiyne

(*iii*) **Secondary Suffix.** Secondary suffix is used to indicate the functional group in the organic compound. It is added to the primary suffix by dropping its terminal *e*. Secondary suffixes for various functional groups are given in Table 1.3.

TABLE 1.3 Some Organic Families and Secondary Suffixes

Class of Organic Compound	General Formula	Functional Group	Suffix	IUPAC Name of the Family (Word root + Prim. suffix + Sec. suffix)
Alcohols	R–OH	–OH	-ol	Alkanol
Thioalcohols	R–SH	–SH	-thiol	Alkanethiol
Amines	R–NH_2	–NH_2	-amine	Alkanamine
Aldehydes	R–CHO	–CHO	-al	Alkanal
Ketones	R–COR	>CO	-one	Alkanone
Carboxylic acids	R–COOH	–COOH	-oic acid	Alkanoic acid
Amides	R–$CONH_2$	–$CONH_2$	-amide	Alkanamide
Acid chlorides	R–COCl	–COCl	-oyl chloride	Alkanoyl chloride
Esters	R–COOR	–COOR	-oate	Alkyl alknoate
Nitriles	R–C ≡ N	–C ≡ N	-nitrile	Alkane nitrile

If the name of the secondary suffix begins with a consonant, then the terminal 'e' of the primary suffix is not dropped while adding secondary suffix to it.

The terminal 'e' of primary suffix is retained if some numerical prefix like di, tri, etc., is used before the secondary suffix.

(*iv*) **Prefix.** Prefix is a part of the name which appears before the word root. Prefixes are used to represent the names of alkyl groups or some functional groups as discussed below:

(*a*) *Alkyl groups* are formed by the removal of H atom from the alkanes. These are represented by the general formula C_nH_{2n+1} – or R –. Some alkyl groups along with their prefixes are given in Table 1.4.

TABLE 1.4 Some Alkyl Groups along with their Prefixes

Alkane	Alkyl Group	Prefix
CH_4	CH_3-	Methyl
C_2H_6	CH_3CH_2-	Ethyl
C_3H_8	$CH_3CH_2CH_2-$	n-Propyl
C_3H_8	CH_3-CH- $\quad\quad\quad\,|$ $\quad\quad\,CH_3$	isopropyl or (1-Methyl ethyl)

(b) *Some functional groups* are always indicated by the prefixes instead of secondary suffixes. These functional groups along with their prefixes are listed below in Table 1.5.

TABLE 1.5 Functional Groups which are always Represented by Prefixes

Functional Group	Prefix	Family	IUPAC Name
$-NO_2$	Nitro	$R-NO_2$	Nitroalkane
$-OR$	Alkoxy	$R-OR$	Alkoxyalkane
$-Cl$	Chloro	$R-Cl$	Chloroalkane
$-Br$	Bromo	$R-Br$	Bromoalkane
$-I$	Iodo	$R-I$	Iodoalkane
$-F$	Fluoro	$R-F$	Fluoroalkane
$-NO$	Nitroso	$R-NO$	Nitrosoalkane

(c) *In poly functional compound*, i.e., compound with more than one functional groups, one of the functional groups is treated as principal functional group and is indicated by the secondary suffix whereas other functional groups are treated as substituents and are indicated by the prefixes. The prefixes for various functional groups are given in Table 1.6.

TABLE 1.6 Prefixes for Functional Groups in Poly Functional Compound

Functional Group	Prefix	Functional Group	Prefix
$-OH$	Hydroxy	$-COOR$	Carbalkoxy
$-CN$	Cyano	$-COCl$	Chloroformyl
$-NC$	Isocyano	$-CONH_2$	Carbamoyl
$-CHO$	Formyl	$-NH_2$	Amino
$-SH$	Mercapto	$=NH$	Imino
$-SR$	Alkylthio	$>CO$	Keto or Oxo
$-COOH$	Carboxy		

1.1.2 Arrangement of Prefixes, Word Root and Suffixes in Naming an Organic Compound

The prefixes, word root and suffixes are arranged as follows while writing the name

Prefix(es) + Word root + prim. suffix + sec. suffix

The illustrated examples are:

$CH_3-CH_2-CH(CH_3)-CH_2-COOH$
(3-Methylpentanoic acid)

Methyl	pent	an	oic acid
Prefix	word Root	primary suffix	secondary suffix

$CH_3-CH=CH-CHO$ with CH_3 on carbon 3
(3-Chlorobut-2-enal)

Chloro	but	en	al
Prefix	word Root	primary suffix	secondary suffix

Note 1. *The name of organic compound may or may not contain prefix or secondary suffix but it always contains word root and primary suffix. For example, the names of straight chain alkanes, alkenes and alkynes do not contain any prefix and secondary suffix.*

$CH_3CH_2CH_2CH_3$
(Butane)

But	ane
Word root	primary suffix

$CH_3-CH_2CH=CHCH_3$
(Pent-2-ene)

Pent	ane
Word root	primary suffix

2. Numerical prefixes such as **di, tri**, etc., are used before the prefixes or secondary suffixes, if the compound contains more than one similar substituents or similar functional groups.

$CH_3-CH(CH_3)-CH(CH_3)-COOH$
(2,3-dimethylbutanoic acid)

Di	Methyl	but	An	oic acid
Numerical Prefix	prefix	word root	prim. suffix	Sec. suffix

While adding numerical prefix before the secondary suffix, the terminal 'e' of the primary suffix is not removed.

$CH_3-CH(OH)-CH_2OH$
(Propane 1,2-diol)

Prop	ane	di	ol
Word Root	prim. suffix	Numerical prefix	sec. suffix

$COOH-COOH$
(Ethane dioic acid)

Eth	ane	di	oic acid
word root	prim. suffix	numerical prefix	sec. suffix

3. *In case of* **alicyclic compounds**, *a separate secondary prefix cyclo is used immediately before the word root. Word root, here depends upon the number of carbon atoms in the ring.*

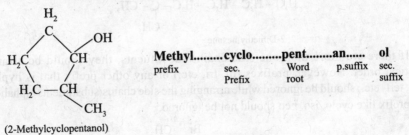

(2-Methylcyclopentanol)

Methyl	cyclo	pent	an	ol
prefix	sec. Prefix	Word root	p. suffix	sec. suffix

1.2 NOMENCLATURE OF ORGANIC COMPOUNDS

We describe below the rules for IUPAC nomenclature.

1. Longest possible chain rule. The longest possible continuous chain of carbon atoms containing the main functional group and also as many of the carbon-carbon multiple bond(s) is selected. This chain is taken as the parent chain. The name of the compound is derived from the alkane having the same number of carbon atoms as the parent chain.

For example,

Structure	Parent chain	Alkane from which name of the compound is derived
$\overset{1}{C}H_3 - \overset{2}{C}H_2 - \overset{3}{C}H - CH_3$ $\quad\quad\quad\quad\quad \overset{|}{C}H_2 - CH_2 - CH_3$ $\quad\quad\quad\quad\quad\quad 4 \quad\quad 5 \quad\quad 6$	Contains six carbon atoms	Hexane
$\quad\quad\quad CH_3$ $\overset{1}{C}H_2 = \overset{2|}{C} - \overset{3}{C}H_2 = \overset{4}{C}H_2$	Contains four carbon atoms	Butane

2. Lowest possible number rule for longest chain containing only substituents/side chains:

(a) The longest chain selected is numbered from that end which gives lowest number to the carbon bearing the substituents or side chains.

(b) The location of the substituent is indicated by a numeral which precedes the suffix. The numeral is the number of the carbon atom to which the substituents/side chain is attached. For example,

$$\overset{7}{H_3C} - \overset{6}{H_2C} - \overset{5}{H_2C} - \overset{4}{H_2C} - \overset{3}{H}\overset{|}{\underset{CH_3}{C}} - \overset{2}{H_2C} - \overset{1}{CH_3}$$
$$\quad 1 \quad\quad 2 \quad\quad 3 \quad\quad 4 \quad\quad 5 \quad\quad 6 \quad\quad 7$$

3-Methyl hepatne   (I)

The numbering is done from the right side as it gives lowest number (number 3) to the carbon bearing methyl group.

(c) If the same substituents/side chains occur more than once, prefixes di, tri, tetra, etc., are used to indicate their number. They are preceded by a number (one number for each group present) to indicate the carbon atom bearing them. For example,

$$\overset{6}{H_3C} - \overset{5}{H_2C} - \overset{4}{H_2C} - \overset{3}{H_2C} - \overset{2|}{\underset{\underset{CH_3}{|}}{\overset{\overset{CH_3}{|}}{C}}} - \overset{1}{CH_3}$$

2, 2-Dimethylhexane   (II)

(d) If there are more than one side chains/substituents, they should be arranged in alphabetical order. However, prefixes (di, tri, etc.) or any other prefix that is hyphenated (n-, sec-, tert- etc.) should be ignored while arranging the side chains/substituents alphabetically. But the prefix like cyclo, iso, neo should not be ignored.

$$\overset{6}{H_3C} - \overset{5}{CH_2} - \overset{4}{CH_2} - \overset{3}{\underset{Br}{\overset{|}{C}H}} - \overset{2}{\underset{CH_3}{\overset{|}{C}H}} - \overset{1}{CH_3}$$

3-bromo-2-methylhexane   (III)

The methyl is attached to carbon number 2 and bromine atom to carbon number 3. Since bromine comes first in alphabetical order, the exact name is 3-Bromo-2-methyl hexane and not 2-methyl-3-bromo hexane. Thus, the substituents are written in alphabetical order.

(e) If identical alkyl groups (side chains), are at equal distances from both the ends, the chain is numbered from the end where there are more side chains/substituents. For example,

$$\overset{1}{H_3C}-\overset{2}{HC}-\overset{3}{HC}-\overset{4}{H_2C}-\overset{5}{H_2C}-\overset{6}{HC}-\overset{7}{CH_3}$$
$$\qquad\quad |\quad\;\; |\qquad\qquad\quad\; |$$
$$\qquad\; CH_3\; CH_3\qquad\qquad CH_3$$
(IV)

2, 3, 6-Trimethylheptane
and not 2, 5, 6-Trimethylheptane

(f) If different alkyl groups are in equal position w.r.t. the ends of the chain, the chain is numbered from the end which gives the smaller number to the smaller alkyl substituents. For example,

$$\overset{7}{H_3C}-\overset{6}{CH_2}-\overset{5}{CH}-\overset{4}{CH_2}-\overset{3}{CH}-\overset{2}{CH_2}-\overset{1}{CH_3}$$
$$\qquad\qquad\quad |\qquad\qquad |$$
$$\qquad\qquad\; C_2H_5\qquad\; CH_3$$
(V)

5-Ethyl-3-methylheptane
and not 3-Ethyl-5-methylheptane

Note: A hyphen (-) is always put between the numeral and the side chain/substituent.

3. Lowest sum rule. The longest chain selected is numbered from that end which keeps the sum of numbers used to indicate the position of the side chains/substituents and functional group(s) as small as possible.

In example IV, the lowest sum = 2 + 3 + 6 = 11.

The name 2, 5, 6-Trimethyl heptane is incorrect as the sum of the number = 2 + 5 + 6 = 13.

Similarly in following example VI, the correct name is 2, 4, 4-Trimethyl hexane and not 3, 3, 5-Trimethyl hexane as the sum of the first set of numbers is 10 (2 + 4 + 4) and that of second set of numbers is 11 (3 + 3 + 5).

$$\qquad\qquad\qquad\qquad\qquad CH_3\quad\; CH_3$$
$$\qquad\qquad\quad\;\overset{6}{CH_3}-\overset{5}{H_2C}-\overset{|4\;\;3}{\underset{|3\;\;4}{C}}-\overset{}{CH_2}-\overset{|2\;\;1}{CH}-\overset{}{CH_3}$$
$$\qquad\qquad\qquad\qquad\qquad CH_3$$
(VI)

4. Lowest number for largest chain containing functional groups. When a compound has a functional group including a multiple bond and one or more side chains/subsituents, the lowest number should be given to the functional group even if it violates the lowest sum rule. For example,

$$\qquad\qquad\qquad\qquad\qquad\qquad\qquad\qquad\; Cl$$
$$\qquad\qquad\qquad\qquad\qquad\qquad\qquad\qquad\; |\;4$$
(i) $\qquad\qquad\qquad\overset{1}{H_2C}=\overset{2}{CH}-\overset{3}{CH_2}-\overset{}{CH_2}$ (VII)

4-Chlorobutene-1 or 4-Chloro-1-butene

The chain is numbered from left side so as to give lowest number to the double bond.

$$\qquad\qquad\quad\overset{1}{H_3C}-\overset{2}{HC}=\overset{3}{CH}-\overset{4}{CH}-\overset{5}{CH_3}$$
$$\qquad\qquad\qquad\qquad\qquad\qquad |$$
$$\qquad\qquad\qquad\qquad\qquad\; CH_3$$
(VIII)

4-Methylpent-2-ene

(ii) 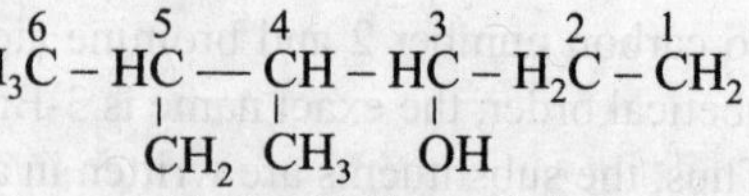 $\overset{6}{H_3C}-\overset{5}{HC}-\overset{4}{CH}-\overset{3}{HC}-\overset{2}{H_2C}-\overset{1}{CH_2}$
                  | | |
                 CH₂ CH₃ OH
4, 5-Dimethylhexan-3-ol   (IX)

(iii) 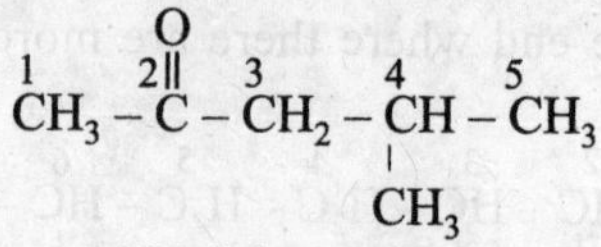 $\overset{1}{CH_3}-\overset{2}{\underset{\|}{C}}-\overset{3}{CH_2}-\overset{4}{CH}-\overset{5}{CH_3}$
with O double-bonded at C-2 and CH₃ on C-4
4-Methylpentan-2-one   (X)

If functional groups such as $-\underset{H}{\overset{|}{C}}=O$, $-\overset{O}{\underset{\|}{C}}-O-H$,

$-\overset{O}{\underset{\|}{C}}-NH_2$, $-\overset{O}{\underset{\|}{C}}-O-R$, $-C\equiv N$, $-\overset{O}{\underset{\|}{C}}-Cl$ etc.

are present in the molecule, the numbering of the parent chain in such cases must start from the carbon atom of the functional group and the number 1 given to the carbon atom of the above functional groups. For example,

(a) 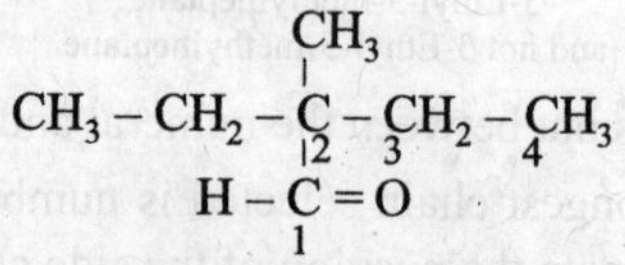 $CH_3-CH_2-\underset{\underset{\overset{|}{H-C=O}}{|}}{\overset{CH_3}{\underset{2}{C}}}-CH_2-CH_3$
                       1
2-Ethyl-2-methylbutanal   (XI)

(b) 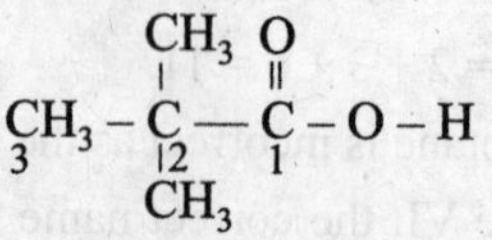 $\underset{3}{CH_3}-\underset{\underset{CH_3}{|}}{\overset{\overset{CH_3}{|}}{\underset{2}{C}}}-\overset{O}{\underset{\|}{\underset{1}{C}}}-O-H$
2, 2-Dimethylpropanoic acid   (XII)

(c) 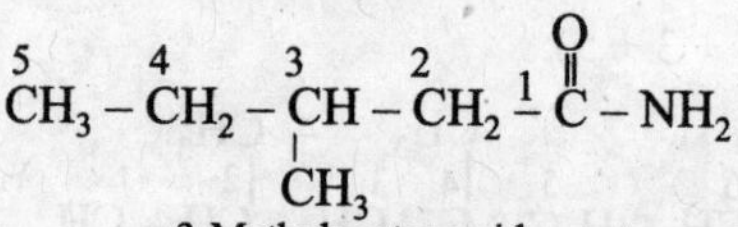 $\overset{5}{CH_3}-\overset{4}{CH_2}-\overset{3}{\underset{\underset{CH_3}{|}}{CH}}-\overset{2}{CH_2}-\overset{1}{\overset{O}{\underset{\|}{C}}}-NH_2$
3-Methylpentanamide   (XIII)

5. For longest chain containing two or more different functional groups (Poly functional compounds). When a compound contains two or more different functional groups, one of the functional groups is chosen as the **Principal functional group** and the remaining functional groups (secondary functional groups) are considered as the substituents. The following rules should be noted down:

When two or more functional groups are present in a compound, the principal functional group is chosen according to the following order of priority.

Sulphonic acid > carboxylic acid > acid anhydride > ester > acid chloride > amide > nitrile > isocyanide > aldehyde > ketone > alcohol > amine > alkene > alkyne > halo, nitro, alkoxy, alkyl

The prefixes for secondary functional groups are:

–OH	Hydroxy	–COOH	Carboxy
–OR	Alkoxy	–CN	Cyno
>C = O	Keto or oxo	–COOR	Carbalkoxy

(Contd...)

Basics of Organic Chemistry I

–CHO	Formyl	–CONH$_2$	Carboxamide
–NH$_2$	Amino	COCl	Chloroformyl
–NH	Imino		
–NHR	N-alkylamino		
–NR$_2$	N,N-dialkylamino		

Examples:

(i) H$_3$CO$\overset{2}{C}$H$_2\overset{1}{C}$H$_2$ – OH

Prefix : Methoxy
Word Root : eth
Pri. suffix : an
Sec. suffix : ol

2-Methoxyethanol

(ii) CH$_3$ – C(CH$_3$)(OH) —— C(CH$_3$)(OH) – CH$_3$

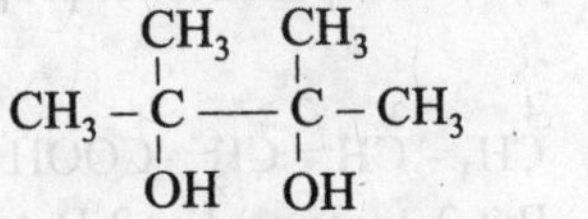

Numerical prefix : di
Prefix : methyl
Word root : but
Pri.suffix : ane
Numerical suffix : di
Suffix : ol

2, 3-Dimethylbutane-2, 3-diol

(iii) $\overset{6}{C}$H$_3$ – $\overset{5}{C}$H = $\overset{4}{C}$H – $\overset{3}{C}$H$_2$ – $\overset{2}{C}$(=O) – $\overset{1}{C}$H$_3$

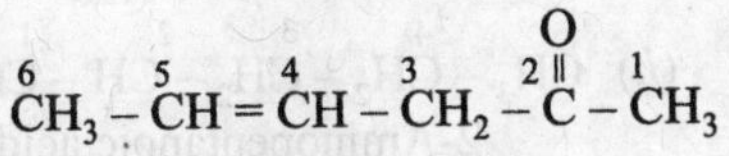

Word root : Hex
Pri. suffix : ene
Sec. suffix : one

(iv) HO$\overset{3}{C}$H$_2\overset{2}{C}$H$_2$ – $\overset{1}{C}$OOH

Prefix : Hydroxy
Word root : prop
Pri.suffix : an
Sec. suffix : oic acid

3-Hydroxypropanoic acid

(v) (CH$_3$)(CH$_3$)$\overset{8,7}{C}$ = $\overset{6}{C}$H – $\overset{5}{C} ≡ \overset{4}{C}$ – $\overset{3}{C}$H – $\overset{2}{C}$(OH)(CH$_3$) – $\overset{1}{C}$H$_3$

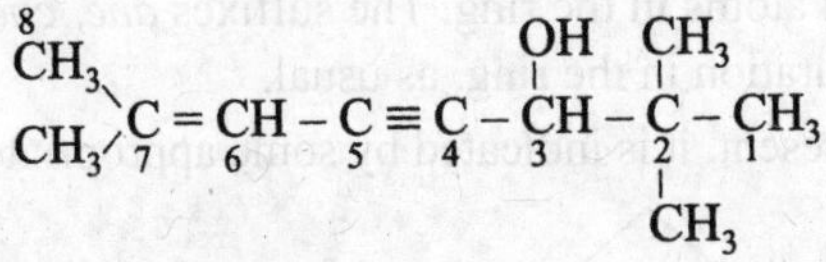

Prefix : Methyl
Word root : oct
Pri.suffixes : ene, yne
sec. suffix : ol

2, 2, 7-Trimethyloct-6-en-4yn-3-ol

(vi) H$\overset{4}{C} ≡ \overset{3}{C}$ – $\overset{2}{C}$H(OH) – $\overset{1}{C}$(=O) – O – H

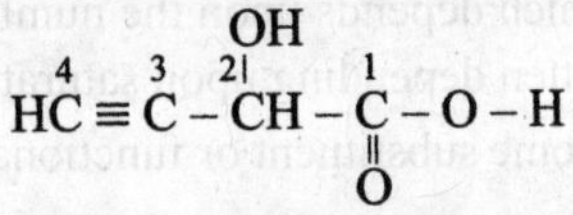

2-Hydroxybut-3-ynoic acid

(vii) H$_2\overset{4}{C}$ = $\overset{3}{C}$H – $\overset{2}{C}$H(NH$_2$) – $\overset{1}{C}$H$_2$OH

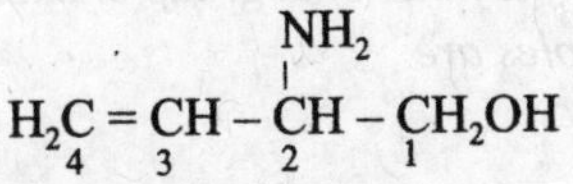

2-Amino but-3-en-1-ol

(viii) $\overset{3}{C}$H$_2$ = $\overset{2}{C}$H – $\overset{1}{C}$H$_2$OH

Prop-2-en-1-ol

6. If the chain of the carbon atom selected as the branched chain also contains multiple bonds or functional groups, the branched chain is separately numbered. This is done in such a way that the carbon of the branched chain which is attached to the parent chain is assigned no. 1. Also the names of such branched chains are written in brackets. For example:

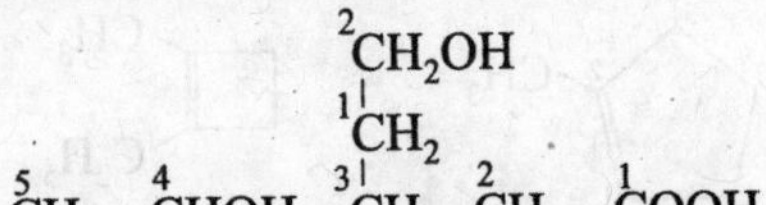

$\overset{5}{C}$H$_3$ – $\overset{4}{C}$HOH – $\overset{3}{C}$H($\overset{1}{C}$H$_2$–$\overset{2}{C}$H$_2$OH) – $\overset{2}{C}$H$_2$ – $\overset{1}{C}$OOH

4-Hydroxy-3 (2-hydroxy ethyl)-pentanoic acid

Select that chain of carbon atoms that includes the maximum number of functional groups.

It is numbered from that end which gives lowest number to the principal functional group.

Examples:

(a) $\overset{3}{C}H_2 - \overset{2}{C}H - \overset{1}{C}H_2$ with OH, OH, OH
1, 2, 3-Propanetriol or propane-1, 2 3-triol

(b) $\overset{5}{C}H_2 = \overset{4}{C}H - \overset{3}{C}H_2 - \overset{2}{C} - \overset{1}{C}H_3$ with =O on C2
Pent-4-en-2-one

(c) $\overset{4}{C}H_3 - \overset{3}{C}H = \overset{2}{C}H - \overset{1}{C}OOH$
But-2-enoic acid or 2-Butenoic acid

(d) $\overset{4}{C}H_3 - \overset{3}{C}H - \overset{2}{C}H_2 - \overset{1}{C}OOH$ with OH on C3
3-Hydroxybutanoic acid

(e) $\overset{4}{C}H_3 - \overset{3}{C}H - \overset{2}{C}H - \overset{1}{C}HO$ with OH on C3, CH3 on C2
3-Hydroxy-2-methylbutan-1-al

(f) $\overset{5}{C}H_3 - \overset{4}{C}H - \overset{3}{C}H = \overset{2}{C}H - \overset{1}{C}H_2OH$ with CH3 on C4
4-Methylpent-2-en-1-ol

(g) $\overset{4}{C}H_3 - \overset{3}{C}H - \overset{2}{C}H - \overset{1}{C}H_3$ with Cl on C3, OH on C2
3-Chlorobutan-2-ol

(h) $\overset{5}{C}H_3 - \overset{4}{C}H_2 - \overset{3}{C}H_2 - \overset{2}{C}H - \overset{1}{C}OOH$ with NH2 on C2
2-Aminopentanoic acid

1.3 NOMENCLATURE OF ALICYCLIC ORGANIC COMPOUNDS

Names of alicyclic compounds are derived by putting another prefix '*cyclo*' before the word root which depends upon the number of carbon atoms in the ring. The suffixes *ane, ene* or *yne* are written depending upon saturation or unsaturation in the ring, as usual.

If some substituent or functional group is present, it is indicated by some appropriate prefix

Cyclohexane   Cyclopentene

or suffix and its position is indicated by numbering the carbon atoms of the ring. *The numbering is done in such a way so as to assign least possible number to the functional group or substituent in accordance with the rules already discussed. Some examples are:*

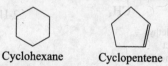

3-Methylcyclohexanol   1-Methyl-3-nitrocyclohexene

2, 3-Dimethylcyclopentene   1-Ethyl-2-methylcyclobutene

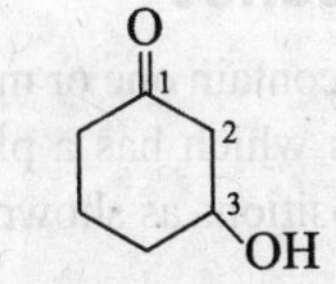

Basics of Organic Chemistry I   13

$$\overset{4}{C}H_3 - \overset{3}{C}H_3 - \overset{2}{C}H_3 = \overset{1}{C}H_2$$

3-Hydroxycyclohexanone 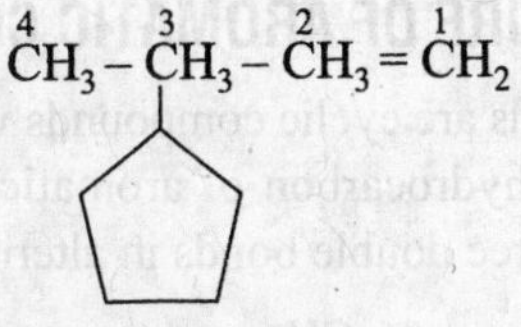   3-Cyclopentyl-1-butene

1.4 BOND LINE NOTATION OF ORGANIC COMPOUNDS

It is a simple, brief and convenient method of representing organic molecules. In these notations, the bonds between the carbon atoms are represented by lines. A single line (–) represents a single bond, two parallel lines (=) represent a double bond and three parallel lines (≡) represent a triple bond. The intersection of lines represents carbon atoms carrying appropriate number of H atoms. For example, 1, 3-butadiene ($CH_2 = CH - CH_2 = CH_2$) can be represented as follows:

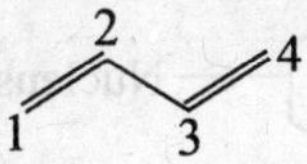

Some bond line structures along with their IUPAC names are given below:

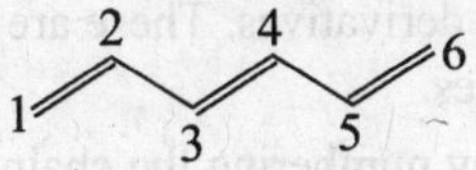

1,3,5-Hexatriene

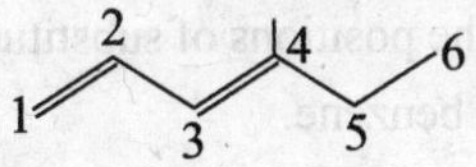

4-Methyl-1,3-pentadiene

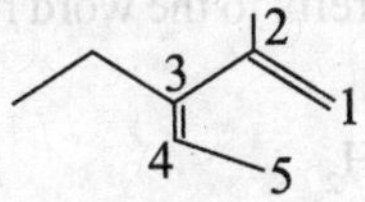

3-Ethyl-1,3-pentadiene

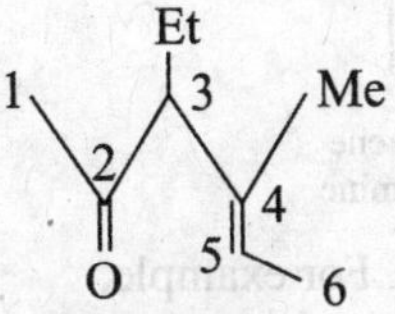

3-Ethyl-4 methylhex-4-en-2-one

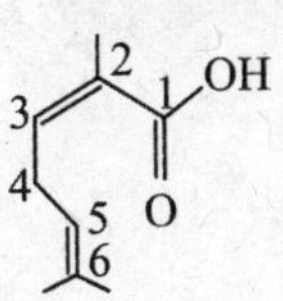

2, 6-Dimethyl-2, 5-heptadienoic acid

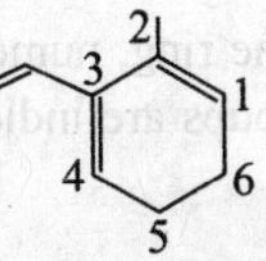

3-Ethenyl-2-methyl-1, 3-hexadiene

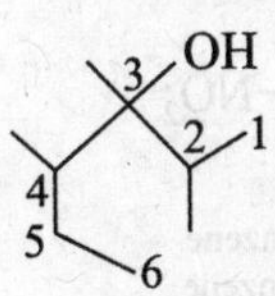

2, 3, 4-Trimethylhex-1-en-3-ol

1.5 NOMENCLATURE OF AROMATIC COMPOUNDS

Aromatic compounds are cyclic compounds which contain one or more benzene type rings. Benzene is a simplest hydrocarbon of aromatic series which has a planar cyclic ring of six carbon atoms having three double bonds in alternate positions as shown below:

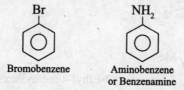

Benzene

The carbon atoms of benzene are numbered from 1 to 6 as shown above. The benzene ring is called the nucleus and alkyl groups attached to the ring are called **side chains**.

Benzene forms only one mono substituted derivative. However, it can form three disubstituted derivatives; namely 1, 2; 1, 3 and 1, 4 derivatives. These are respectively called *ortho* (or *o*-), *meta* (or *m*-) and *para* (or *p*-) derivatives.

Tri and poly substituted derivatives are named by numbering the chain in such a way that the parent group gets the lowest number and sum of the positions of substituents is the smallest.

Rule 1. The word root for benzene derivatives is benzene.

Rule 2. The name of substituent group is added as a prefix to the word root in case of mono-substituted benzenes. For example,

Bromobenzene   Aminobenzene
                  or Benzenamine

In some cases the name of the group is written as suffix. For example,

Benzene sulphonic acid

Rule 3. When two similar groups are attached to the benzene ring, numerical prefix *di* is placed before the name of the group, relative portion of the groups are indicated by suitable numbers or by the symbols *o*, *m* or *p*. For example,

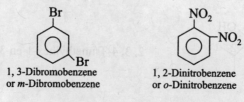

1, 3-Dibromobenzene   1, 2-Dinitrobenzene
or *m*-Dibromobenzene   or *o*-Dinitrobenzene

Basics of Organic Chemistry I

Rule 4. When two different groups are attached to the ring, the names of both groups are added as prefixes, in alphabetical order, to the word root and their relative positions are indicated. For example,

2-Bromoiodobenzene
or *o*-Bromoiodobenzene

If one of the groups a special name to the compound, than the name of the other group only is written as prefix. For example,

3-Chlorotoluene   4-Nitrophenol

Rule 5. In case of tri-substituted or higher-substituted derivatives, the positions of groups are indicated by numbers.

1, 3, 5-Trimethylbenzene   2, 4-Dichlorophenol

Names of Some Aromatic Compounds

1. Hydrocarbons (Arenes)

(a) *Hydrocarbons containing condensed rings*

Benzene   Naphthalene   Anthracene

(b) *Hydrocarbons containing one ring only*

Methyl benzene (Toluene)   Ethyl benzene   Phenylethene (styrene)   1, 2-Dimethyl benzene (*o*-Xylene)

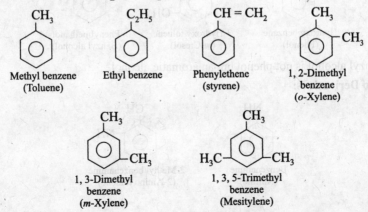

1, 3-Dimethyl benzene (*m*-Xylene)   1, 3, 5-Trimethyl benzene (Mesitylene)

2. Aromatic or Aryl radicals:

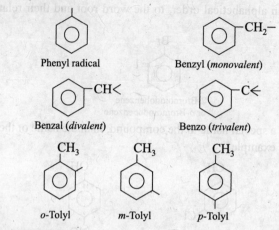

3. Halogen Derivatives:

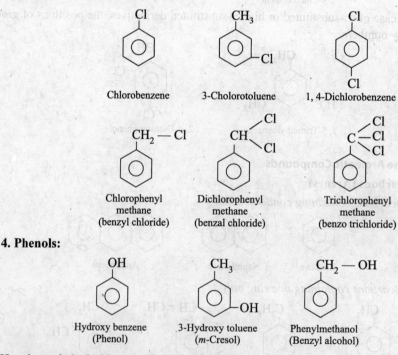

4. Phenols:

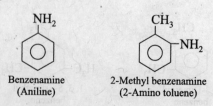

Here benzyl alcohol is not phenol but an aromatic alcohol.

5. Amino Derivatives:

Basics of Organic Chemistry I   17

N-Methylaniline   Diphenylamine

6. Ketones:

Methyl phenyl ketone
or
Acetophenone

Diphenyl ketone
or
Benzophenone

7. Aldeydes:

Benzaldehyde
(Benzenecarbaldehyde)

2-Hydroxy benzaldehyde
(Salicylaldehyde)

8. Carboxylic acids:

Benzoic acid

Phthalic acid
(1, 2-Benzene dicarboxylic acid)

9. Acid derivatives:

$O=C-Cl$ $O=C-NH_2$ $O=C-O-CH_3$

Benzoyl chloride   Benzamide   Methyl benzoate

$C \equiv N$

Benzonitrile
(phenyl cyanid)

Benzoic anhydride

10. Sulphonic acids:

SO_2OH CH_3

Benzene sulphonic acid SO_2OH
p-Toluene sulphonic acid (PTS)

11. Nitro derivatives:

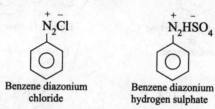

1, 3-Dinitro benzene

2, 4, 6-Trinitro toluene (T.N.T.)

2, 4, 6-Trinitro phenol (Picric acid)

12. Diazonium salts:

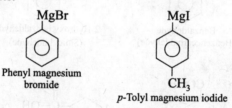

Benzene diazonium chloride

Benzene diazonium hydrogen sulphate

13. Grignard reagents:

Phenyl magnesium bromide

p-Tolyl magnesium iodide

14. Isocyanides:

Benzene carbylamine

o-Toluene carbylamine

Solved Examples

Example 1: Give the IUPAC name of the following compounds:

(i) $(CH_3CH_2CH_2)_4C$

(ii) $CH_2 = CH - CH_2 - CH = CH_2$

(iii) $CH_3 - \underset{CH_3}{CH} - \underset{CH_3}{CH} - \underset{CH_3}{CH} - CH_2CH_3$

(iv) $OHCH_2 - C \equiv C - CH_2OH$

(v) $(CH_3)_2 CH - CH_2 - COOH$

(vi) $CH_2 = CH - CH_2 - \underset{OH}{CH} - CH_2 - CH_3$

Solution:

(i) 4, 4-di *n*-propylheptane

(ii) 1, 4-pentadiene

(iii) 2, 3, 4-trimethylhexane

(iv) But-2-yne-1, 4-diol

(v) 3-methylbutanoic acid

(vi) Hex-5-en-3-ol

Example 2: Give IUPAC names of the following:

(i) $CH_3 - CH_2 - CH = CH - C \equiv CH$

(ii) $CH_3 - (CH_2)_2 - CH = CH - \underset{CH_3}{CH} - CHO$

(iii) $C_2H_5 - \underset{\underset{CH_2}{\|}}{C} - CH_2 - \underset{\underset{CH_3}{|}}{CH} - NH_2$ (iv) $CH_3 - CH = CH - \underset{\underset{O}{\|}}{C} - CH_3$

(v) $CH_3 - CO - CH_2 - CH_2 - COOH$ (vi) $CH_3 - \underset{\underset{OH}{|}}{CH} - CH = \underset{\underset{CH_3}{|}}{C} - CH_3$

(vii) $CH_3 - CH_2 - CH \diagup^{OH}_{\diagdown CH_3}$ (viii) $CH_3 - \underset{\underset{Br}{|}}{CH} - \underset{\underset{NO_2}{|}}{CH} - CH = CH_2$

(ix) $CH_3 - CH_2 - C \equiv C - CH_2 - \underset{\underset{O}{\|}}{C} - CH_2 - CH_3$

Solution:
- (i) Hex-3-ene-5-yne
- (ii) 2-Methylhept-3-enal
- (iii) 4-Amino-2-ethylpent-1-ene
- (iv) Pent-3-en-2-one
- (v) 4-Ketopentanoic acid
- (vi) 4-Methylpent-3-en-2-ol
- (vii) 1-Methylpropan-1-ol
- (viii) 4-Bromo-3-nitro pent-1-ene
- (ix) Oct-5-yn-3-one

Example 3: Write IUPAC names of the following:

(i) $HOOC - CH_2 - \underset{\underset{Br}{|}}{CH} - COOH$

(ii) $HC \equiv C - CH_2 - CH = CH_2$

Solution:
- (i) 2-Bromobutane-1, 4-dioic acid
- (ii) Pent-1-en-4-yne

Example 4: Write IUPAC names of the following:

(i) $(CH_3)_2 CH - N(CH_3)_2$

(ii) $O_2N - CH_2 - \underset{\underset{OCH_3}{|}}{CH} - CH_2 - COOCH_2 - CH_3$

(iii) $H - \underset{\underset{O}{\|}}{C} - \underset{\underset{CH_3}{|}}{CH_2} - CH_2 - \underset{\underset{O}{\|}}{C} - OH$

(iv) $H_3C - \underset{\underset{CH_2 - CH_2 - CH_3}{|}}{N} - CH_2 - CH_3$

Solution:
- (i) Isopropyl dimethylamine
- (ii) Ethyl-3-methoxy-4-nitrobutanoate
- (iii) 3-Formyl-3-methylpropanoic acid
- (iv) N-Ethyl-N-methyl-1-aminopropane

Example 5: Wrtie IUPAC names of the following:

(i) $CH_2 (COOH) CH (COOH) CH_2COOH$ (ii) benzene ring with two COOH groups

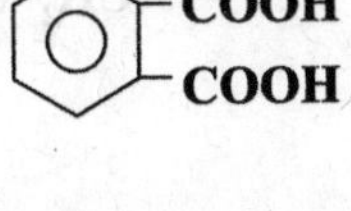

Solution:

(i) 3-Carboxypentane-1, 5-dicarboxylic acid

(ii) Benzene-1, 2-dicarboxylic acid

Example 6: Write IUPAC names of the following:

(i) Cl – CH$_2$ – CH = CH – CH – CHO
                                |
                                OCH$_3$

(ii) HO – CH – CH – CH – COOCH$_3$
            | | |
            CH$_3$ NH$_2$
                     Br (on middle CH)

(iii) CH$_3$ – CH – CH$_2$ – C – CH$_3$
              | |
              CH$_3$ – CH – CH$_3$ OCH$_3$
                                    (Cl on C)

(iv) Cyclobutane with 1-COOH, 2-OC$_2$H$_5$, 3-NH$_2$

Solution:

(i) 5-Chloro-2-methoxypent-3-enal

(ii) Methyl 2-amino-3-bromo-4-hydroxypentanoate

(iii) The compound can be written as

CH$_3$ – CH – CH – CH$_2$ – C – CH$_3$
         | | |
         CH$_3$ CH$_3$ Cl
                                OCH$_3$

2-Chloro-2-methoxy-4, 5 dimethylhexane

(iv) 3-Amino-2-ethoxycyclobutanoic acid

Example 7: Write the structural formulae of the following compounds:

(i) 2-Methyl-3-pentynoic acid
(ii) 3-Ethyl-5-hydroxy-3-hexenal
(iii) 2-Methyl-1, 3-butadiene
(iv) Hept-1-en-4-yne
(v) Pent-3-enoic acid
(vi) 1-Amino-4-methylpentan-2-one

Solution:

(i) $\overset{5}{C}H_3 – \overset{4}{C} \equiv \overset{3}{C} – \overset{2}{C}H – \overset{1}{C}OOH$
                                |
                                CH$_3$

(ii) $\overset{6}{C}H_3 – \overset{5}{C}H – \overset{4}{C}H = \overset{3}{C} – \overset{2}{C}H_2 – \overset{1}{C}HO$
              | |
              OH   C$_2$H$_5$

(iii) $\overset{4}{C}H_2 = \overset{3}{C} – \overset{2}{C}H = \overset{1}{C}H_2$
                    |
                    CH$_3$

(iv) $\overset{7}{C}H_3 – \overset{6}{C}H_2 – \overset{5}{C} = \overset{4}{C} – \overset{3}{C}H_2 – \overset{2}{C}H \equiv \overset{1}{C}H_2$

(v) $\overset{5}{C}H_3 – \overset{4}{C}H = \overset{3}{C}H – \overset{2}{C}H_2 – \overset{1}{C}OOH$

(vi) $\overset{5}{C}H_3 – \overset{4}{C}H – \overset{3}{C}H_2 – \overset{2}{C} – \overset{1}{C}H_2NH_2$
              | ‖
              CH$_3$ O

Example 8: Give the IUPAC names of:

(i) CH$_3$ – CH$_2$ – CH – CH$_2$ – CH – CH – CH$_3$
                    | | |
                    CH$_3$ CH$_3$
                                          |
                                          CH$_2$ – CH$_2$ – CH$_3$

(ii) $CH_3-\underset{\underset{CH_3}{|}}{\overset{\overset{CH_3}{|}}{C}}-CH-\underset{\underset{\underset{CH_3}{|}}{CH-CH_3}}{\overset{\overset{CH_3}{|}}{C}}=CH_2$

(iii) $CH_3-(CH_2)_3-C\equiv C-(CH_2)_3-CH_3$

(iv) $CH_2=CH-\underset{\underset{CH_3}{|}}{\overset{\overset{CH_3}{|}}{C}}-CH=CH-CH\overset{CH_3}{\underset{CH_3}{<}}$

(v) $CH_3-C\equiv C-\underset{\underset{CH_3}{|}}{CH}-CH_3$

Solution:

(i) $\overset{1}{C}H_3-\overset{2}{C}H_2-\overset{3}{C}H-\overset{4}{C}H_2-\overset{5}{C}H(-\overset{1'}{C}H-\overset{2'}{C}H_2) \longleftarrow$ Side chain
with CH_3 at 3 and CH_3 at 1', and chain $\overset{6}{C}H_2-\overset{7}{C}H_2-\overset{8}{C}H_3$

5-(1-methyl ethyl)-3-methyloctane

(ii) $\overset{5}{C}H_3-\overset{4}{\underset{\underset{CH_3}{|}}{\overset{\overset{CH_3}{|}}{C}}}-\overset{3}{C}H-\overset{2}{\underset{\underset{[CH-CH_2}{|}}{\overset{\overset{CH_3}{|}}{C}}}=\overset{1}{C}H_3$ Side chain
(with CH_3)

3-(1-methyl ethyl)-2, 4, 4 Trimethylpent-1-ene

(iii) 5-Decyne   (iv) 3, 3, 6-Trimethyl-1, 4-heptadiene

(v) 4-Methylpent-2-yne

Example 9: Write the structural formulae of the following compounds:

(i) 2-Methyl pent-2-ene-1-ol
(ii) Hex-1-en-4-yne
(iii) 4-Cyano-3-methoxybutanoic acid
(iv) 4-Amino-2-ethyl-2-pentenal
(v) 3, 5-Octadiene
(vi) 4-Chloro-2-isopropyl-3-methylcyclopentanone

Solution:

(i) $\overset{5}{C}H_3-\overset{4}{C}H_2-\overset{3}{C}H=\underset{\underset{CH_3}{|}}{\overset{2}{C}}-\overset{1}{C}H_2OH$

(ii) $\overset{6}{C}H_3-\overset{5}{C}\equiv\overset{4}{C}-\overset{3}{C}H_2-\overset{2}{C}H=\overset{1}{C}H_2$

(iii) $NC-\overset{4}{C}H_2-\underset{\underset{OCH_3}{|}}{\overset{3}{C}H}-\overset{2}{C}H_2\overset{1}{C}OOH$

(iv) $\overset{5}{C}H_3-\underset{\underset{NH_2}{|}}{\overset{4}{C}H}-\overset{3}{C}H=\underset{\underset{C_2H_5}{|}}{\overset{2}{C}}-\overset{1}{C}HO$

(v) $\overset{1}{C}H_3-\overset{2}{C}H_2-\overset{3}{C}H=\overset{4}{C}H-\overset{5}{C}H=\overset{6}{C}H-\overset{7}{C}H_2-\overset{8}{C}H_3$

(vi) [cyclopentanone ring with O (ketone), isopropyl (CH(CH_3)_2), CH_3, Cl, Cl substituents]

Example 10: Write IUPAC names of the following:

(i) 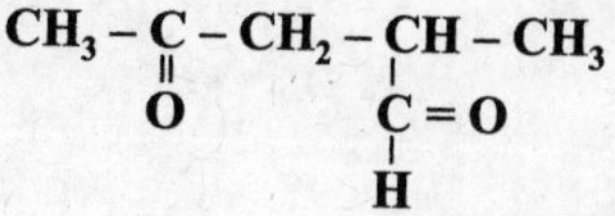 $CH_3-\underset{\underset{O}{\|}}{C}-CH_2-\underset{\underset{\underset{H}{|}}{C=O}}{CH}-CH_3$

(ii) 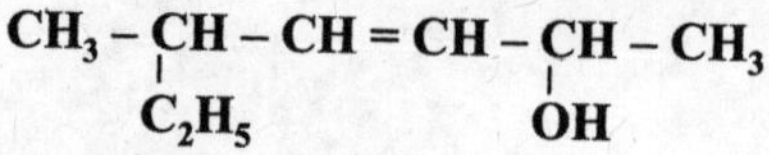 $CH_3-\underset{\underset{C_2H_5}{|}}{CH}-CH=CH-\underset{\underset{OH}{|}}{CH}-CH_3$

Solution:

(i) 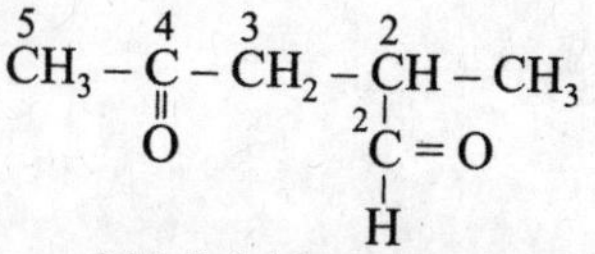 $\overset{5}{C}H_3-\underset{\underset{O}{\|}}{\overset{4}{C}}-\overset{3}{C}H_2-\underset{\underset{\underset{H}{|}}{\overset{2}{C}=O}}{\overset{2}{C}H}-CH_3$

2-Methyl-4-ketopentanal

(ii) 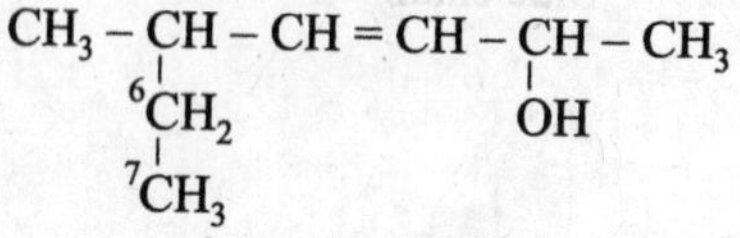 $CH_3-CH-CH=CH-\underset{\underset{OH}{|}}{CH}-CH_3$
$\underset{\underset{^7CH_3}{|}}{^6CH_2}$

5-methylhept-3-en-2-ol

Example 11: Write IUPAC names of the following:

(i) 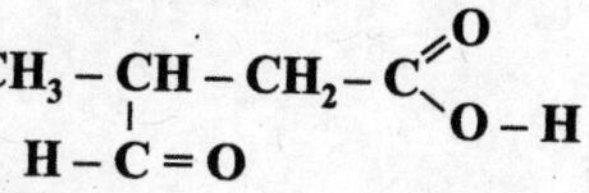 $CH_3-\underset{\underset{H-C=O}{|}}{CH}-CH_2-C\overset{\nearrow O}{\underset{\searrow O-H}{}}$

(ii) 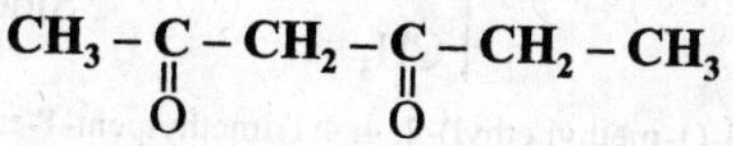 $CH_3-\underset{\underset{O}{\|}}{C}-CH_2-\underset{\underset{O}{\|}}{C}-CH_2-CH_3$

Solution:

(i) 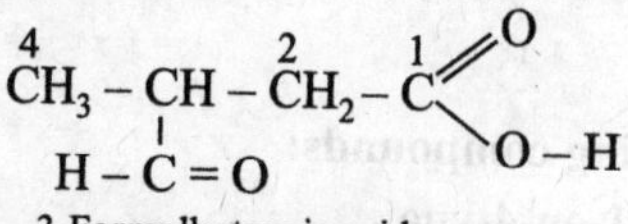 $\overset{4}{C}H_3-\underset{\underset{H-C=O}{|}}{CH}-\overset{2}{C}H_2-\overset{1}{C}\overset{\nearrow O}{\underset{\searrow O-H}{}}$

3-Formylbutanoic acid

(ii) 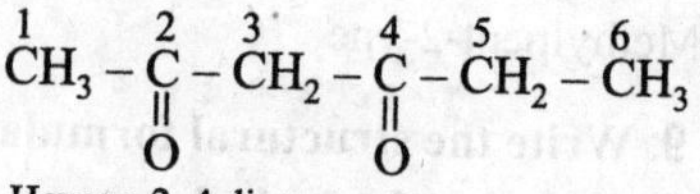 $\overset{1}{C}H_3-\underset{\underset{O}{\|}}{\overset{2}{C}}-\overset{3}{C}H_2-\underset{\underset{O}{\|}}{\overset{4}{C}}-\overset{5}{C}H_2-\overset{6}{C}H_3$

Hexane-2, 4 dione

Example 12: Write IUPAC names of the following:

(i) 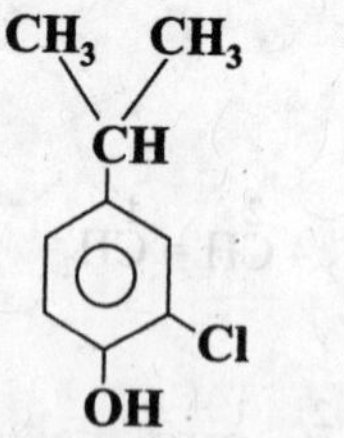 [benzene ring with CH(CH₃)₂ at top, Cl and OH positions]

(ii) 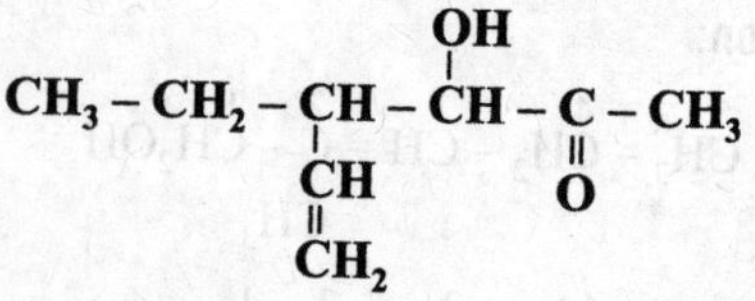 $CH_3-CH_2-CH-\underset{\underset{\underset{CH_2}{\|}}{CH}}{CH}-\underset{\underset{O}{\|}}{\overset{OH}{C}}-CH_3$

Solution:

(i) 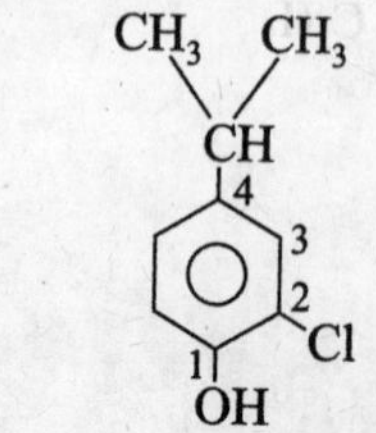 [benzene ring numbered, with isopropyl at 4, Cl at 2, OH at 1]

2-Chloro-4-isopropyl phenol

(ii) 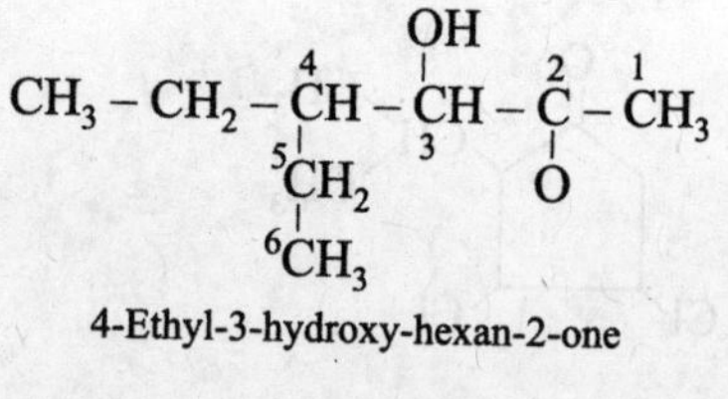 $CH_3-CH_2-\overset{4}{C}H-\overset{3}{\underset{\underset{^6CH_3}{|}}{\underset{^5CH_2}{|}}}{CH}-\underset{\underset{O}{\|}}{\overset{2}{C}}-\overset{1}{C}H_3$
$\overset{OH}{|}$

4-Ethyl-3-hydroxy-hexan-2-one

Example 13: Write IUPAC names of the following:

(i) $CH_3 - \underset{\underset{CH_3}{|}}{CH} - \underset{\underset{CH_2-CH_3}{|}}{C} = CH_2$

(ii) $CH_3 - \underset{\underset{H-C=O}{|}}{CH} - CH_2 - COOH$

Solution:

(i) $\overset{4}{C}H_3 - \underset{\underset{CH_3}{|}}{\overset{3}{C}H} - \underset{\underset{CH_2-CH_3}{|}}{\overset{2}{C}} = \overset{1}{C}H_2$

2 Ethyl-3-methylbut-2-ene

(ii) See Q. 11.

Example 14: Write IUPAC names of the following:

(i) $\underset{\underset{COOH}{|}}{CH_2} —— \underset{\underset{COOH}{|}}{CH} —— \underset{\underset{COOH}{|}}{CH_2}$

(ii) $CH_3 - \underset{\overset{|}{CH_3}}{C} = \underset{\overset{|}{CH_3}}{C} - COOC_2H_5$

Solution:

(i) $\underset{\underset{^5COOH}{|}}{\overset{4}{C}H_2} —— \underset{\underset{COOH}{|}}{\overset{3}{C}H} —— \underset{\underset{COOH}{|}}{\overset{2}{C}H_2}$

3-Carboxypentane-1, 5-dioic acid

(ii) $CH_3 - \underset{\overset{|}{CH_3}}{C} = \underset{\overset{|}{CH_3}}{C} - COOC_2H_5$

Ethyl-2-dimethylbut-2-enoate

1.6 NOMENCLATURE OF POLYCYCLIC HYDROCARBONS

We come across compounds, containing two or more rings fused together or having two or more carbon atoms in common. These common carbon atoms are called **bridgehead** carbons. Consider, for example, the following bicyclic compounds (containing two rings).

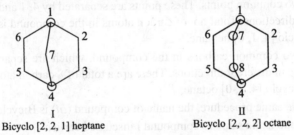

I
Bicyclo [2, 2, 1] heptane

II
Bicyclo [2, 2, 2] octane

Both these compounds have two bridge head carbons (common carbons) at positions 1 and 4. These common carbons have been shown as encircled.

1. We number various carbon atoms in such structures starting from any common carbon and go on numbering all carbon atoms in one direction till the cyclic structure is complete.

Bicyclo [1, 1, 0] butane

2. Then, we number the left out carbon atoms, in one direction. In compound I, there are 7 carbon atoms and in compound II, there are 8 carbon atoms. Thus, compound I is Bicycloheptane and compound II is Bicyclooctane.

3. To name these compounds, we find out, what are the numbers of carbon atoms separating the common carbons, when seen through different routes. Thus, in compound I, there are two carbons through one route, two carbons through the second route and one carbon through the third route, separating the common carbons. This compound will be

named as Bicyclo [2, 2, 1] heptane. Here, the numbers 2, 2, 1 represent the number of carbon atoms separating the common carbons in different directions.

These numbers in decreasing order are written in *square brackets* after the word bycyclo.

4. Coming to compound II, we find that there are 2 carbons in one of the directions, separating the common points. Consider the following compound.

There are 4 carbon atoms in all. Thus, it is bicyclobutane. No. 1 and 3 are common points. These common points are separated by one carbon in one direction, one carbon in second direction and no (zero) carbon in the third direction (direct link between position 1 and 2). Thus, the name of the compound is Bicyclo [1, 1, 0] butane.

Solved Examples

Example 15: Write names of the following bicyclic compounds.

Solution:
(i) There are two common points. These points are separated by 4, 1 and zero carbon atoms in different directions. Total no. of carbon atoms in the compound is 7. Thus, the name will be Bicyclo [4, 1, 0] heptane.
(ii) There are two common carbons in the compound, which are separated by 4, 2 and 0 carbon atoms in different directions. There are a total of 8 carbon atoms. Therefore, it is named as Bicyclo [4, 2, 0] octane.
(iii) Following the same procedure, the name of compound (iii) is Bicyclo [3, 3, 0] octane.
(iv) This compound is the same as compound I discussed earlier.

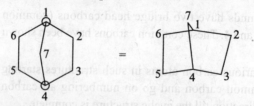

Positions 1 and 4 are common points. These are separated by 2, 2 and 1 carbon atoms in different directions. Therefore, the compound is named as Bicyclo [2, 2, 1] heptane.

Example 16: Name the following compounds.

Solution:

(i) There is a methyl group at position 2 and bromo group at position 3. Therefore, the compound can be named as 3-Bromo-2-methylbicyclo [2, 2, 1] heptane.

(ii) There is a methyl group at position 3 and ethyl group at position 2. Therefore, the name of the compound is : 2-Ethyl-3-methyl bicyclo [2, 2, 2] octane.

Key Terms

- Word root
- Preffix and suffix
- Lowest sum rule
- Bond notation
- Alicyclic compounds
- Polycyclic compounds

Evaluate Yourself

Multiple Choice Questions

1. Naphthalene has the molecular formula
   - (a) $C_{10}H_9$
   - (b) $C_{10}H_8$
   - (c) $C_{10}H_{10}$
   - (d) $C_{10}H_{11}$

2. Thioalcohols are compounds containing
   - (a) nitrogen
   - (b) phosphorus
   - (c) sulphur
   - (d) chlorine

3. The correct order of priority of function groups is
   - (a) Ester > Acid chloride > Amide > Nitrile
   - (b) Ester > Amide > Acid chloride > Nitrile
   - (c) Nitrile > Ester > Acid chloride > Amide
   - (d) Acid chloride > Ester > Amide > Nitrile

4. The primary suffix in $H_3COCH_2CH_2-OH$ is
   - (a) eth
   - (b) an
   - (c) ol
   - (d) none of the above

5. The correct IUPAC name of the compound $CH_3 - CHCl - CH(OH) - CH_3$ is
   - (a) 2-Chlorobutan–3–ol
   - (b) 3-Chlorobutan–2–ol
   - (c) Butan-2–ol chloride
   - (d) none of the above

6. The correct structure of benzene diazonium chloride is

(a)

(b)

(c)

(d) none of the above

Short Answer Questions

1. Write IUPAC names of the following:

   (i) $\underset{\underset{OH}{|}}{CH_2} - \underset{\underset{OH}{|}}{CH} - \underset{\underset{OH}{|}}{CH_2}$

   (ii) $CH_3 - \underset{\underset{OH}{|}}{CH} - \underset{\underset{CH_3}{|}}{CH} - CHO$

2. Give an idea of bond line notation of organic compounds.
3. Write the structures of the following:
   (i) Benzyl chloride
   (ii) Benzoyl chloride
   (iii) Benzotrichloride
4. Write the IUPAC name of the following compound:

   [cyclobutane ring with 1-COOH, 2-OC$_2$H$_5$, 3-NH$_2$]

5. Write the structure of bicyclo [1, 1, 0] butane.

Long Answer Questions

1. Write systematic names for the following:

   (i) $CH_3 - CH - COOH$ with phenyl group bearing Cl (para)

   (ii) $CH_2 - CH_2OH$ attached to benzene ring

   (iii) benzene ring with CH$_3$, two O$_2$N groups and NO$_2$

   (iv) $CH_3 - CH - CHBr - COOC_2H_5$ with phenyl group bearing NO$_2$

2. Write notes on:
   (i) Word root
   (ii) Primary suffix
   (iii) Secondary suffix
   (iv) Prefix
3. How are prefixes, word root and suffixes arranged in naming an organic compound? Illustrate in detail taking the examples of 2, 3-dimethylbutanoic acid and 2-methylcyclopentanol.
4. (a) Write IUPAC names of the following:

   (i) $Br-CH_2-CH=CH-\overset{\overset{O}{\|}}{C}-CH_3$

   (ii) $CH_2=CH-C\equiv CH$

   (b) Draw Structural formulae of the following compounds:

   (i) 4-Tert. butyl-5-isopropyl decane

   (ii) 2, 4-Dimethyl-3-ethylhexan-2-ol

5. (a) Give IUPAC names for the following:

   (i) $CH_3-\underset{\underset{O}{\|}}{C}-CH_2-CH_2-CHO$

   (ii) $CH_3-CH=CH-\underset{\underset{CH_3}{|}}{\overset{\overset{CH_3}{|}}{C}}-\underset{\underset{NH_2}{|}}{CH}-CH_3$

   (iii) $OHC-CH_2-CH_2-CH_2-CHO$

   (iv) $CH_3-\underset{\underset{CH_3}{|}}{\overset{\overset{CH_3}{|}}{C}}-O-CH_3$

6. Write structural formulae of the following compounds:
   (i) Hexanedioic acid
   (ii) 2-N, N-Dimethylaminobutane
   (iii) Pent-3-ynal
   (iv) 2-Ethylbut-1-ene

7. Write IUPAC names of the following:

   (a) $CH_2=CH-CH_2-\overset{\overset{O}{\|}}{C}-CH_3$

   (b) $CH_3-C\equiv C-\underset{}{\overset{\overset{CH_3}{|}}{CH}}-CH_3$

   (c) $CH_3-\overset{\overset{OH}{|}}{CH}-CH_2-\overset{\overset{O}{\|}}{C}-OH$

8. Write down IUPAC names of the following:

   (i) OHC – COOH

   (ii) – O – CH₃

   (iii) H₃C – C = CH – CH₃
              |
              C₂H₅

9. (a) Write IUPAC names of the following formulae:

   (i)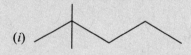

   (ii) [benzene ring] – CH₃

   (b) Write structural formulae of the following:

   (i) Pent-1, 4-diene

   (ii) 3-Hydroxypentanal

10. (a) Write IUPAC names of the following formulae:

    (i) [branched chain structure]

    (ii) – CH₂ – CH₂ – CH₃

    (b) Write structural formulae of the following:

    (i) Pent-1-en-4-yne

    (i) But-2-en-1-ol

11. Write the names of the following compounds:

    (i)   (ii)

    (iii)    (iv)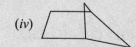

    (v)

Answers
Multiple Choice Questions

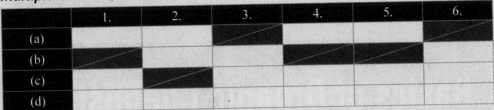

Suggested Readings

1. Morrison, R.N. & Boyd, R.N. Organic Chemistry, Dorling Kindersley (India) Pvt. Ltd.
2. Finar, I.L. Organic Chemistry (Volume I), Dorling Kindersley (India) Pvt. Ltd.

Chapter 2

Basics of Organic Chemistry II

LEARNING OBJECTIVES

After reading this chapter, you should be able to:
- derive the shaptes of molecules based upon the type of hybridisation
- learn the terms inductive effect, electromeric effect, resonance or mesomeric effect and hyperconjugation
- understand the effect of the above on the properties of different compounds
- follow the relative strength of organic acids and bases
- differentiate between homolytic and heterolytic fission
- learn curly arrows rules, formal charge, electrophiles and nucleophiles
- follow the types, shaptes and relative stability of carbocations, carbanions, free radicals and carbens
- learn the mechanisms of addition, elimination and substitution reations

2.1 CURVED ARROW NOTATION

In the study of organic reaction mechanisms, we come across movement of electrons from one atom to the other. This movement of electron is represented with the help of curved arrows. This method is known as curved arrow notation. For example, when the atom B is more electronegative than atom A in the following compound, then in the presence of a reagent, the flow of electrons that takes place may be represented with the help of **curved arrows**.

$$A = B \xrightarrow{\text{Reagent added}} \overset{\delta+}{A} - \overset{\delta-}{B}$$

Electronegativity plays an important role in knowing the direction of transfer of the shared pair of electron to one of the atoms.

(*i*) In a carbonyl group, $>C = O$, present in aldehydes and ketones, the displacement is towards the oxygen atom which is represented by a curved arrow as under

$$>C = \overset{\frown}{O} \longrightarrow >\overset{+}{C} - \overset{-}{O}$$

(*ii*) In propylene, the displacement of shared pair of electrons is towards the carbon atom which is away from the methyl group due to the reason that methyl group is electron repelling.

$$CH_3 - CH \overset{\frown}{=} CH_2 \longrightarrow CH_3 - \overset{+}{CH} - \overset{-}{CH_2}$$

2.2 DRAWING ELECTRON MOVEMENTS WITH ARROWS

There are instances in organic molecules where the separation of charges (complete +ve or –ve) does not take place. But, due to the presence of electron-attracting or electron-repelling groups, some kind of polarization or movement of electrons takes place. We represent it with the help of arrows along the bonds. For example, in the molecule of ethyl chloride, the arrows would be drawn as:

$$CH_3 \longrightarrow CH_2 \longrightarrow Cl$$

Presence of a halogen group in a carboxy acid helps in the release of protons. This happens because the electron movement takes place in the direction of the halogen group. Greater the number of halogen groups, greater will be the charge movement away from the carboxylic group and towards the halogen groups. This can be illustrated as under:

Chloroacetic acid

Dichloroacetic acid

Trichloracetic acid

2.2.1 Half-Headed and Double Headed Arrows

The curved arrows that we use to represent electron movement from one site to another can be classified as under:

(i) Half-headed arrows represented as ⌢ or ⌢

(ii) Double-headed arrows represented as ⌢

Half-headed or Single Headed Arrows

Half-headed or single headed arrows show the movement of a single electron. These are used in the discussion of radical chemistry mechanisms in particular. For example, we can depict the breaking of a bromine molecule into two bromine radicals as under:

$$Br—Br \longrightarrow \dot{B}r + \dot{B}r$$

Double-headed Arrows

Double-headed arrows show the movement of electron pair from one species to another. For example

$$H\ddot{O}^- \quad H^+ \longrightarrow H—O—H$$

The movement takes place from the tail of the arrow and goes to head.

Curly Arrow Rules

Golden Rules

1. Curly arrows are used to show the movement of pairs of electrons. Remember that a bond between two atoms is made up of two electrons.
2. Curly arrows always flow from electron-rich species to electron-poor species say from nucleophiles to electrophiles.
3. They start from lone pairs or bonds and end between a pair of atoms or on an atom.
4. The charges in any particular step should be balanced.
5. If electrons are taken out of a bond, that bond is broken.
6. If electrons are placed between two atoms, a bond is formed. The reaction $H_2O + H^+ \rightleftharpoons H_3O^+$ can be represented in terms of curly arrow rules as:

$$\overset{H}{\underset{H}{>}}\ddot{O}: \quad H^+ \rightleftharpoons \overset{H}{\underset{H}{>}}\overset{+}{O}—H$$

2.3 ELECTRON DISPLACEMENTS IN ORGANIC COMPOUNDS

The behaviour of an organic compound is influenced by the electron displacement taking place in its covalent bonds. These displacements may be permanent or temporary which take place in presence of another species in the molecule. The acidity and basicity of organic compounds, their stability, their reactivity towards other substances, etc. can be predicted based on such electronic displacements. These electron displacements are explained as under:

2.4 INDUCTIVE EFFECT

This is a permanent effect operating in polar covalent bonds. **The induction of a permanent dipole in a covalent bond between two unlike atoms of different electronegativities is**

called the inductive effect. The development of partial +ve and –ve charges is due to shift of the shared pair of electrons towards the more electronegative atom. This results in small fractional charges on the constituent atoms. When a carbon atom is bonded to a hydrogen (C – H) or another carbon (C – C) atom by a covalent bond as in alkanes, the sharing of electron-pair is symmetrical between them. Thus, no charges are induced on the atoms. However, when carbon is bonded to a halogen (X), charges are created.

$$—\overset{|}{\underset{|}{C}}—H \quad \text{or} \quad —\overset{|}{\underset{|}{C}}:H \qquad \text{(Symmetrical sharing of electron pair) Electronegativities of C and H are almost the same.}$$

$$—\overset{|}{\underset{|}{C}}—\overset{|}{\underset{|}{C}}— \quad \text{or} \quad —\overset{|}{\underset{|}{C}}:\overset{|}{\underset{|}{C}}— \qquad \text{(Symmetrical sharing of electron pair)}$$

$$—\overset{|}{\underset{|}{C}}:X \quad \text{or} \quad —\overset{\delta+|}{\underset{|}{C}}—X^{\delta-} \qquad \text{(Fractional charges due to greater electronegativity of X)}$$

The direction of displacement is shown by placing an arrow head midway along the line representing the sigma bond.

$$C_4 \rightarrow C_3 \rightarrow C_2 \rightarrow \overset{\delta+}{C_1} \rightarrow \overset{\delta-}{Cl}$$

The inductive effect of an atom or a group of atoms diminishes rapidly with distance. Infact, the inductive effect is almost negligible beyond two carbon atoms from the active atom or group.

Inductive effect does not involve actual transfer of electrons from one atom to another but simply helps in displacing them permanently.

Groups with – I effect:

$$NO_2 > CN > F > COOH > Cl > Br > I > OCH_3 > C_6H_5$$

Decreasing order of – I effect ⟶

Groups with + I effect:

$$(CH_3)_3\,C > (CH_3)_2\,CH > C_2H_5 > CH_3 > -H$$

Decreasing order of + I effect ⟶

Characteristics of inductive effect

1. Inductive effect arises due to displacement of σ-electrons only and occurs in molecules containing polar and single bonds only.
2. Inductive effect is permanent and irreversible.
3. Inductive effect diminishes progressively as we move away from electron-donating or electrons withdrawing group/atom along the molecule chain.
4. The electrons that are displaced do not leave their orbital. A little distortion of the orbital occurs which causes polarisation.
5. Depending upon the electrons-withdrawing or electron-donating strength, different atoms or groups polarize the covalent bonds to different extent.

Applications

(*i*) Reactivity of alkyl halides: The presence of halogen atoms in the molecule of alkyl halide creates a centre of low electron density on adjacent carbon which is readily attacked by the negatively charged reagents.

$$CH_3 \rightarrow Cl \quad \text{or} \quad \overset{\delta^+}{CH_3}\text{—}\overset{\delta^-}{Cl}$$
$$\uparrow$$
Site for the attack of electron rich reagent

(*ii*) Dipole moment: As the inductive effect increases, the dipole moment increases. This is because the dipole moment is the product of distance and charge.

$$CH_3 \rightarrow I \qquad CH_3 \rightarrow Br \qquad CH_3 \rightarrow Cl$$
$$1.64\ D \qquad\qquad 1.79\ D \qquad\qquad 1.83\ D$$

⎯⎯⎯⎯⎯⎯⎯⎯⎯⎯⎯⎯⎯⎯⎯⎯⎯⎯⎯⎯⎯→
Inductive effect increases

(*iii*) (*a*) Acidity of acetic acid and halogen substituted acetic acid: Electron-attracting groups increase the acid strength of the acid as such groups help in the removal of protons. Greater the number of such electron withdrawing group, greater is the acid strength of the acid. pK_a values of some acids are given below. It may be remembered that smaller the pK_a value, higher is the acid strength

$$Cl \leftarrow CH_2 \leftarrow \underset{\underset{O}{\|}}{C} \leftarrow O \leftarrow H \qquad\qquad H_3C\text{—}\underset{\underset{O}{\|}}{C}\text{—}O\text{—}H$$

Chloroacetic acid $\qquad\qquad\qquad$ Acetic acid
$pK_a = 2.86 \qquad\qquad\qquad\qquad pK_a = 4.76$

$$\underset{Cl}{\overset{Cl}{\diagdown}}CH \leftarrow \underset{\underset{O}{\|}}{C} \leftarrow O \leftarrow H \qquad\qquad \underset{Cl}{\overset{Cl}{\diagdown}}C\underset{Cl}{\diagup} \leftarrow \underset{\underset{O}{\|}}{C} \leftarrow O \leftarrow H$$

Dichloroacetic acid $\qquad\qquad\qquad$ Trichloroacetic acid
$pK_a = 1.25 \qquad\qquad\qquad\qquad pK_a = 0.65$

The decreasing order of acid strength

$$Cl_3CCOOH > Cl_2CHCOOH > ClCH_2COOH > CH_3COOH$$

(*b*) Relative acid strength of formic acid and acetic acid: Methyl group has an electron-releasing inductive effect (+*I* effect). It reduces the release of protons. Therefore acetic acid is a weaker acid than formic acid.

$$H_3C \rightarrow \underset{\underset{O}{\|}}{C} \rightarrow O \rightarrow H \qquad\qquad H\text{—}\underset{\underset{O}{\|}}{C}\text{—}O\text{—}H$$

$pK_a = 4.79 \qquad\qquad\qquad\qquad pK_a = 3.77$

(*c*) Relative acid strength of fluoroacetic acid, chloroacetic acid, bromoacetic acid and iodoacetic acid: Halogenated acids are much stronger acids than the parent acid and the acidity increases with the increase in electronegativity of the halogen present which helps in the release of protons. Thus, the strength of halogenated acids follows the order:

$$FCH_2COOH > ClCH_2COOH > BrCH_2COOH > ICH_2COOH$$

Inductive effect decreases with increase in distance of halogen atom from the carboxylic group, hence the strength of the acid is proportionally decreased. Thus,

$$CH_3CH_2CH(Cl)COOH > CH_3CH(Cl)CH_2COOH > CH_2ClCH_2CH_2COOH > CH_3CH_2CH_2COOH$$
α-Chlorobutyric acid β-Chlorobutyric acid γ-Chlorobutyric acid n-Butyric acid

(iv) (a) Relative reactivity of toluene and benzene in aromatic substitution reactions:
Aromatic substitution reactions are electrophilic in nature. Methyl has an electron-releasing inductive effect (+I effect). Therefore, toluene with higher electron density than benzene has greater reactivity in electrophilic substitution reactions.

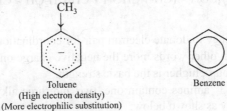

Toluene
(High electron density)
(More electrophilic substitution)

Benzene

(b) Relative reactivity of nitrobenzene and benzene in electrophilic aromatic substitution reactions

NO_2 group has –I inductive effect. It draws electrons from the benzene ring and decreases electron density on the ring
(Less reactive towards electrophilic substitution reactions)

Benzene

(v) Compared to water, phenol is more acidic (–I effect) but methyl alcohol is less acidic (+I effect):
This is because phenyl group has –I effect while methyl group has +I effect.

$$\langle \text{Ph} \rangle \leftarrow OH > H — OH > CH_3 \nrightarrow OH$$

Phenol Water Methyl alcohol

(vi) Inductive effect is dependent on state of hybridisation of the atoms linked by covalent bond:
Relative acidity of hydrocarbons may be given as

$$HC \equiv CH > CH_2 = CH_2 > CH_3 — CH_3$$
$$sp \qquad\qquad sp^2 \qquad\qquad sp^3$$

Electron density on s-orbital is more than on, p-orbital. Percentage of s character decrease from left to right.

Relative basicity of corresponding carbanions is as:

$$CH_3 — \bar{C}H_2 > CH_2 = \bar{C}H > CH \equiv \bar{C}$$

(vii) Strength of base: Base strength is defined as the tendency to donate an electron pair for sharing. Strength of a base can be explained on the basis of inductive effect.

(a) As compared to ammonia, methyl amine is more basic (+I effect) but aniline is less basic and diphenyl amine is a still weaker base (–I effect).

$$CH_3NH_2 > NH_3 > C_6H_5NH_2 > (C_6H_5)_2NH$$

(b) The decreasing order of base strength in alcohols from tertiary to primary is due to +I effect of alkyl groups.

$$(CH_3)_3COH > (CH_3)_2CHOH > CH_3CH_2OH > CH_3OH$$
$$\quad\quad 3° \quad\quad\quad\quad 2° \quad\quad\quad\quad 1° \quad\quad\quad methyl$$

(c) Greater the tendency to donate electron pair for coordination with proton, more is the basic nature, or in other words more the negative charge on nitrogen atom (due to +I effect of alkyl group), higher is the basic strength.

Aliphatic 1°, 2° and 3° amines contain one, two and three alkyl groups attached to the nitrogen atom respectively as shown below:

```
   H—N̈—H        R→N̈—H        R→N̈—H        R→N̈—R
     |              |              ↓              ↓
     H              H              R              R
  Ammonia      Primary amine   Secondary amine  Tertiary amine
```

On the basis of +I effect of alkyl groups alone, the expected order of basicity should be: 3° amine > 2° amine > 1° amine. This order is really followed by the amines in **gaseous state**. However, the observed order of basicity **in aqueous solution** is:

$$2° \text{ amine} > 1° \text{ amine} > 3° \text{ amine}$$

Explanation for the above observation is as under:

Basicity of an amine in aqueous solution does not depend only on the electron density on nitrogen in the compound but it also depends upon the stability of the conjugate acid obtained on accepting a proton which is in the order

$$1° \text{ amine} > 2° \text{ amine} > 3° \text{ amine}$$

The combined effect of electron density on N and the stability of the conjugate acid formed is that the observed order of basicity in aqueous solution is

secondary amine > primary amine > tertiary amine

(viii) Relative stabilities of carbocations: Greater the number of alkyl groups attached to positive carbon, greater the dispersal of charge and hence greater is the stability.

$$(CH_3)_3\overset{\oplus}{C} > (CH_3)_2\overset{\oplus}{CH} > (CH_3)\overset{\oplus}{CH_2} > \overset{\oplus}{CH_3}$$

(ix) +I group increases electron availability on carbonyl carbon: This therefore decreases the rate of nucleophilic addition.

On the other hand, electron-withdrawing –I group decreases electron availability on carbonyl carbon and therefore increases the rate of nucleophilic addition

$$CCl_3-\underset{\underset{}{\overset{\overset{O}{\|}}{}}}{C}-H > H-\underset{\underset{}{\overset{\overset{O}{\|}}{}}}{C}-H > H_3C-\underset{\underset{}{\overset{\overset{O}{\|}}{}}}{C}-H > CH_3-\underset{\underset{}{\overset{\overset{O}{\|}}{}}}{C}-CH_3$$

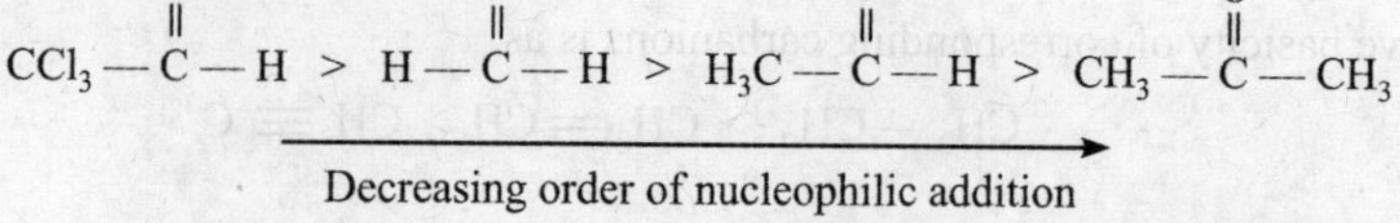

$$\overline{\quad\quad\quad\quad\quad\quad\quad\quad\quad\quad\quad\quad\quad\quad\quad\quad\quad\quad\quad\quad\quad\quad\quad\quad\quad\quad\quad}$$
Decreasing order of nucleophilic addition

2.5 ELECTROMERIC EFFECT

This is a temporary effect operating in unsaturated compounds only at the demand of a reagent. As soon as this attacking reagent is removed, the original condition is restored.

It involves the complete transfer of π-electrons of multiple bond, because π-bonds are loosely held and are easily polarisable.

$$>C=\ddot{\underset{..}{O}}: \underset{\text{Reagent withdrawn}}{\overset{\text{Reagent added}}{\rightleftarrows}} >\overset{+}{C}-\ddot{\underset{..}{O}}:^{\ominus}$$

The complete transfer of shared pair of π-electrons of a multiple bond to the more electronegative atom of the bonded atoms at the requirement of an attacking reagent is called electromeric effect (E-effect).

When the transfer of π-electrons takes place towards the attacking reagent (electrophile), the effect is called +E-effect.

For example;

$$>C=C< + H^+ \longrightarrow >\overset{+}{C}=C< \\ \quad\quad\quad\quad\quad\quad\quad\quad\quad | \\ \quad\quad\quad\quad\quad\quad\quad\quad\quad H$$

$$CH_3 \rightarrow CH=CH_2 + H^+ \longrightarrow CH_3-\overset{+}{CH}-CH_3$$
Propene

When the transfer of electrons takes place away from the attacking reagent, the effect is called –E effect. For example,

$$>C=O + CN^- \longrightarrow >C-O^- \\ \quad\quad\quad\quad\quad\quad\quad\quad\quad\quad | \\ \quad\quad\quad\quad\quad\quad\quad\quad\quad\quad CN$$

$$\left[>C=\ddot{\underset{..}{O}}: + :CN^{\ominus} \xrightarrow[\text{slow}]{\text{step (i)}} >C-\ddot{\underset{..}{O}}:^{\ominus} \xrightarrow[\text{fast}]{H^{\oplus} \text{ step (ii)}} >C-OH \\ \quad\quad\quad\quad\quad\quad\quad\quad\quad\quad\quad\quad\quad\quad\quad | \quad\quad\quad\quad\quad\quad\quad\quad\quad | \\ \quad\quad\quad\quad\quad\quad\quad\quad\quad\quad\quad\quad\quad\quad\quad CN \quad\quad\quad\quad\quad\quad\quad\quad CN \right]$$

When the *I*- and *E*-effect occur together in a molecule, they may be supporting or opposing each other. When they are opposing, the *E*-effect generally dominates over *I*-effect.

Applications

(*i*) Electrophilic addition reactions of unsaturated compounds involve electromeric effect or the polarisation of the carbon-carbon double bond in the presence of attacking electrophiles like H^+.

(*ii*) Nucleophilic addition reactions of carbonyl compounds involve electromeric effect or polarisation through –*E* electromeric effect of the carbon-oxygen double bond in the presence of a nucleophile.

(*iii*) Electrophilic substitution reactions of benzenoids involve polarisation through electromeric effect of the benzene ring when an electrophile approaches them. For example,

2.6 RESONANCE OR MESOMERIC EFFECT

If two double bonds in a molecule are separated by a single bond, they are said to be in **conjugation** and the molecule having such bonds are called conjugated molecules *e.g.*, 1, 3-butadiene.

$$H_2C = CH - CH = CH_2 \qquad H_2C = CH - \ddot{\underset{..}{C}l}:$$
1, 3-Butadiene   Vinyl chloride

Another example of conjugation is given by benzene which is a hexagonal ring of carbon atoms with three double bonds in the alternate positions. Alternatively, a **double** bond or a **triple** bond may also be in conjugation with a lone electron pair *e.g.*, in *vinyl chloride*.

The conjugated molecules do not exhibit the character of pure double bond or triple bond, for example, benzene is expected to be highly reactive since it has three double bonds in the ring but is actually quite stable. In order to explain the difference in the expected and the actual behaviour of the conjugated molecules, *Robinson* and *Ingold* stated that such compounds exist in two or more forms none of which can explain all the properties of the molecule under investigation. They called this concept as **mesomerism** or **mesomeric effect.** *Heisenberg* studied the same mathematically and named it as **resonance** or **resonance effect.** According to the concept,

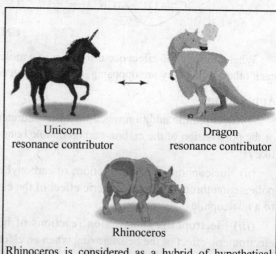

Unicorn resonance contributor

Dragon resonance contributor

Rhinoceros

Rhinoceros is considered as a hybrid of hypothetical unicorn and dragon because it possesses some properties of unicorn and some of the dragon.

If a compound having a certain molecular formula can be represented by different structural formulae which differ only in the arrangement of the electron pairs and not of the atoms, such like structures are called **resonating** *or* **contributing** *or* **canonical structures** *and the phenomenon is known as* **resonance**. *The compound cannot be represented completely by any of the resonating or contributing structures but by a mixture of all of them which is called* **resonance hybrid** *(cannot be actually represented). Resonance is indicated by the sign* ⟷ .

It may be noted that the various resonating or contributing structures do not actually exist i.e., they are all hypothetical structures. They have been given simply to explain certain properties of the compounds which cannot be otherwise explained.

Examples of Resonance

A common analogy of a resonance hybrid is that of a mule which is a hybrid of horse and donkey.

1. Monocarboxylic acid. The acidic character of monocarboxylic acid is explained with the help of resonance. The following contributing structures for the monocarboxylic acid are possible which differ in the position of the electron pairs.

$$R-C\overset{\overset{\ddot{\text{O}}:}{\|}}{\underset{}{}}\ddot{\text{O}}-H \longleftrightarrow R-C\overset{:\ddot{\text{O}}:}{\underset{}{|}}=\overset{+}{\ddot{\text{O}}}-H$$

As a result of resonance, the oxygen atom of the O — H group acquires a positive charge *i.e.*, it draws the electrons pair towards itself resulting in the release of protons.

2. Benzene. Benzene is a hybrid of two equivalent contributing structures (I and II) which differ in the position of the π-electron pairs representing double bonds. These were suggested by Kekule.

I ⟷ II

Three more contributing structures (III, IV and V) have been proposed by **Dewar**. Being less symmetrical they have small contribution (28% only) towards the hybrid. Benzene exists mainly in the form of **Kekule structures** (I and II). Benzene is, in fact, quite stable and its stability is explained

III ⟷ IV ⟷ V

with the help of resonance. As the π-electron charge is distributed over greater area *i.e.*, it gets delocalised.

3. 1, 3-Butadiene. The diene can be regarded as the hybrid of the following contributing structures:

$$\underset{(I)}{CH_2=CH-CH=CH_2} \longleftrightarrow \underset{(II)}{\overset{\oplus}{CH_2}-CH=CH-\overset{\ominus}{CH_2}} \longleftrightarrow \underset{(III)}{\overset{\ominus}{CH_2}-CH=CH-\overset{\oplus}{CH_2}}$$

The charged structures II and III make lesser contribution towards the hybrid. However, they do explain the 1, 4-addition in the conjugated dienes.

4. Carbon dioxide. The structural formula of carbon dioxide (CO_2) molecule is O = C = O. The standard C = O bond length is 122 pm but the bond length which is obtained for the molecule from spectroscopic studies is 115 pm. The difference in the bond length values can be explained by considering the following structures for carbon dioxide which are known as *resonating structures,* or *contributing structures* or *canonical structures.*

$$:\ddot{O}=C=\ddot{O}: \qquad :\overset{-}{\ddot{O}}-C\equiv O:\overset{+}{} \qquad :O\overset{+}{\equiv}C-\overset{-}{\ddot{O}}:$$
$$\text{(I)} \qquad\qquad \text{(II)} \qquad\qquad \text{(III)}$$

The carbon-oxygen bond length is the mean of all the bond length values. Carbon dioxide cannot be represented by any of the contributing structures but by an average of all. Carbon dioxide is said to be a resonance hybrid of these structures.

Conditions of resonance

Wheland has suggested the following conditions for resonance:

(*i*) The resonance or the contributing structures must differ only in the position of the electron pairs and not of the atomic nuclei.

(*ii*) The resonating structures must have the same number of paired and unpaired electrons.

(*iii*) The energies of the various resonating structures must be either same or nearly the same.

(*iv*) All the contributing or resonating structures do not contribute equally towards the hybrid. The equivalent structures have greater contribution. The contribution of any resonating structure towards the hybrid depends upon the following factors:

(*a*) Structures with more covalent bonds are more stable than the structures with less covalent bonds.

(*b*) The charged contributing structure is less stable as compared to the structure without any charge.

(*c*) Structure with negative charge on more electronegative atom is more stable than the structure with negative charge on less electronegative atom.

Effects of resonance

Some effects of the resonance are explained as under:

1. Stability. As a result of resonance, the energy of the hybrid decreases and its stability, therefore, increases. Greater the number of contributing structures, greater will be the stability of the hybrid. The stability also depends upon the equivalence of the contributing structures. Moreover, the charged structures have less contribution than the uncharged structures.

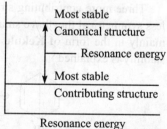

Resonance energy

2. Resonance energy. The relative stabilities of two different resonance hybrids may be compared in terms of **resonance energy.** *It may be defined as the difference in the energy content of the hybrid and its most stable contributing structure.* It may be expressed either as the difference in the heat of combustion of the contributing structures and the hybrid or in terms of the difference in the heat of hydrogenation values. Consider the following examples:

(*a*) **Heat of combustion value.** In benzene, the heat of combustion of the contributing structure is calculated by determining the values for different bonds mathematically.

Heat of combustion of 6 C — H bonds = 6 × 225.5 = 1355 kJ / mole
Heat of combustion of 3 C — C bonds = 3 × 206.0 = 618 kJ/mole
Heat of combustion of 3 C = C bonds = 3 × 491.0 = 1473 kJ/mole

3446 kJ/mole.

The value as determined experimentally for the hybrid is 3300 kJ/mole. Thus, the resonance energy for benzene is (3446 – 3300) kJ or **146 kJ/mole**.

Similarly, resonance energies for some typical molecules are as follows:

1, 3-Pentadiene (17.6 kJ / mole); 1, 3-Butadiene (14.6 kJ/mole) and 1, 3-Cyclohexadiene (7.5 kJ/mole).

(b) Heat of hydrogenation value. In benzene, the expected heat of hydrogenation value is 376 kJ/mole. But the value for the hybrid as determined experimentally is 230 kJ/mole *i.e.* the resonance energy is the same *i.e.* (376 – 230) kJ or (146 kJ/mole).

The relative stabilities of two hybrids can be compared in terms of the resonance energy value.

3. Bond length values. The resonance causes a change in bond length values. For example, the standard C — C bond length is 154 pm and C = C bond length is 134 pm. But the value as determined for benzene is 139 pm which does not coincide with any of the two values and is intermediate between the two values.

Types of resonance or mesomeric effect

Depending upon the nature of the functional group present adjacent to the multiple bond, the resonance effect (R) or mesomeric effect (M) is of two types depending upon the nature of the group which is present.

(i) Groups having electron-withdrawing resonance effect. Groups such as — NO_2, > C = O, — C ≡ N, — COOH etc. tend to withdraw the electrons from the multiple bond through resonance and are said to have – M or – R effect. For example, in the following compound, the aldehyde group (— CH = O) has – R effect.

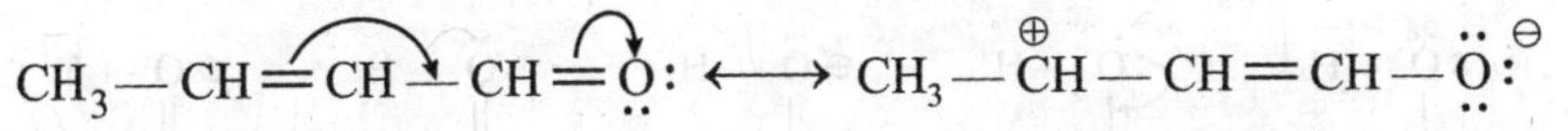

$$CH_3 — CH = CH — CH = \ddot{O}: \longleftrightarrow CH_3 — \overset{\oplus}{CH} — CH = CH — \ddot{\underset{\ominus}{O}}:$$

(ii) Groups having electron-releasing resonance effect. Groups such as — $\ddot{N}H_2$, — $\ddot{N}HR$, — $\ddot{N}R_2$, — $\ddot{O}$ — H, — $\ddot{O}$ — R etc. which can release electrons through resonance are said to have + M or + R effect. For example, the amine group (– NH_2) in the following compound has + R effect

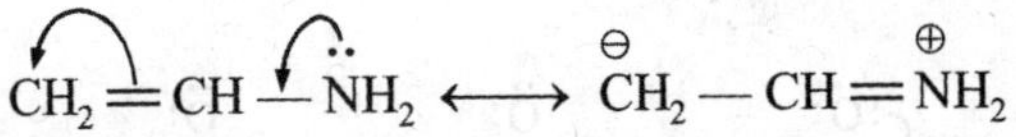

$$CH_2 = CH — \ddot{N}H_2 \longleftrightarrow \overset{\ominus}{CH_2} — CH = \overset{\oplus}{NH_2}$$

Applications of resonance effect

Resonance effect or mesomeric effect is quite useful in explaining many observations in organic compounds. A few typical applications are being listed as follows:

1. Explanation for the acidic character of carboxylic acids. The acidic character of carboxylic acids is due to release of H^+ ion in aqueous solution.

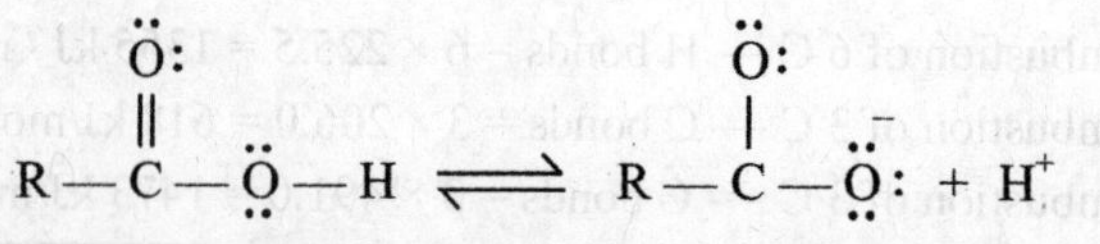

$$R-\overset{\overset{\ddot{O}:}{\|}}{C}-\ddot{\underset{\cdot\cdot}{O}}-H \rightleftharpoons R-\overset{\overset{\ddot{O}:}{|}}{C}-\ddot{\underset{\cdot\cdot}{O}}:^{-} + H^+$$

This is explained with the help of resonance. Both the acid and the carboxylate ion are resonance stabilised as shown below:

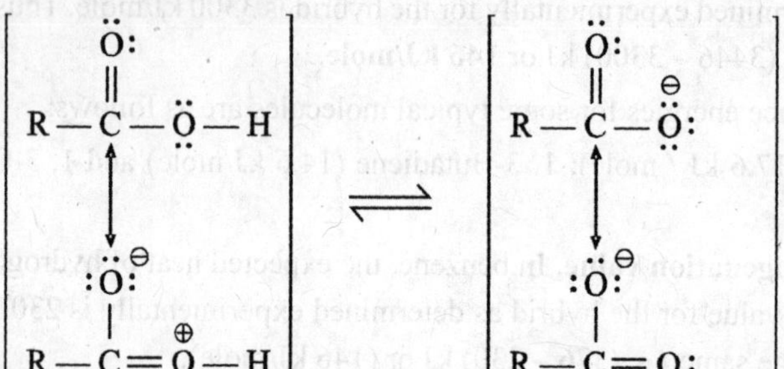

$$\left[R-\overset{\overset{\ddot{O}:}{\|}}{C}-\ddot{\underset{\cdot\cdot}{O}}-H \longleftrightarrow R-\overset{\overset{:\overset{\ominus}{\ddot{O}}:}{|}}{C}=\overset{\oplus}{\underset{\cdot\cdot}{O}}-H\right] \rightleftharpoons \left[R-\overset{\overset{\ddot{O}:}{\|}}{C}-\overset{\ominus}{\ddot{\underset{\cdot\cdot}{O}}}: \longleftrightarrow R-\overset{\overset{:\overset{\ominus}{\ddot{O}}:}{|}}{C}=\ddot{\underset{\cdot\cdot}{O}}:\right] + H^+$$

Acid   Carboxylate ion

While the contributing structures for the carboxylate ion are exactly equivalent, they are not so for the acid. *Therefore, hybrid for the ion is more stable than hybrid for the acid. The acid changes to a more stable ion by releasing a proton and this accounts for the acidic character of carboxylic acid.*

2. Explanation for acidic character of phenols. The acidic nature of phenols is due to the release of H^+ ion in aqueous solution.

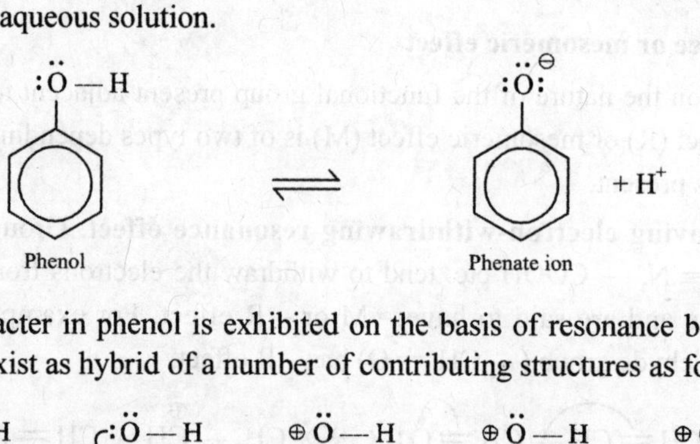

Phenol   ⇌   Phenate ion + H^+

The acidic character in phenol is exhibited on the basis of resonance both in phenol and phenate ion. They exist as hybrid of a number of contributing structures as follows:

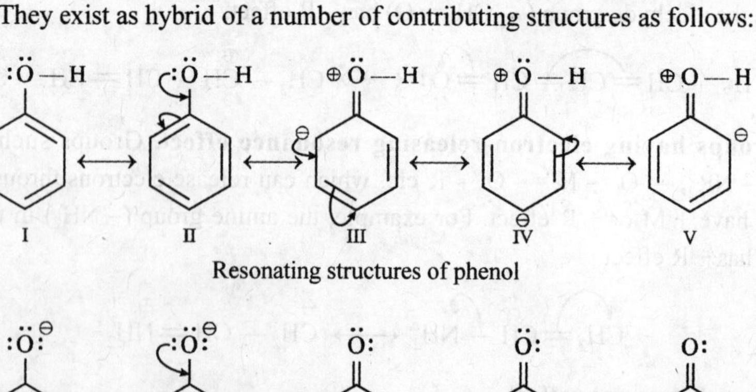

I   II   III   IV   V
Resonating structures of phenol

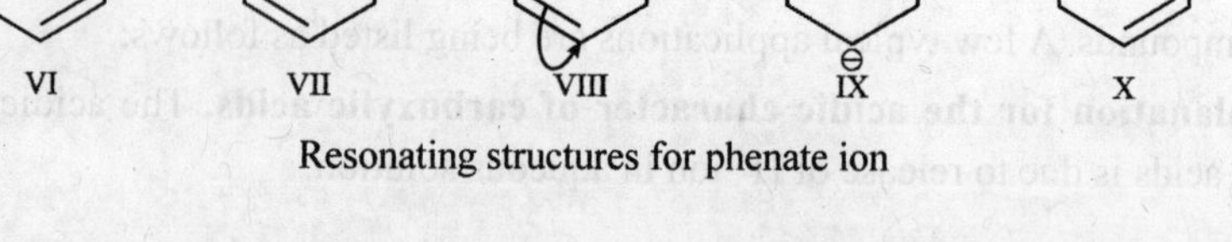

VI   VII   VIII   IX   X
Resonating structures for phenate ion

In case of phenol, contributing structures (III), (IV) and (V) involve charge separation *i.e.* they have both positive and negative charges. Energy is needed to separate the opposite charges. But no such structures are noticed in case of phenate ion. Therefore, the hybrid for the phenate ion is more stable and phenol changes to phenate ion by releasing proton. This accounts for the acidic character of phenol.

3. Comparison of relative basic strength of ethyl amine and acetamide. Ethyl amine is a stronger base than acetamide. In ethyl amine, the ethyl group with + I effect increases the electron density on the nitrogen atom. As a result, its electron releasing tendency is more. In acetamide the carbonyl group is an electron withdrawing group and takes away the electron pair from the nitrogen atom because of resonance as shown below:

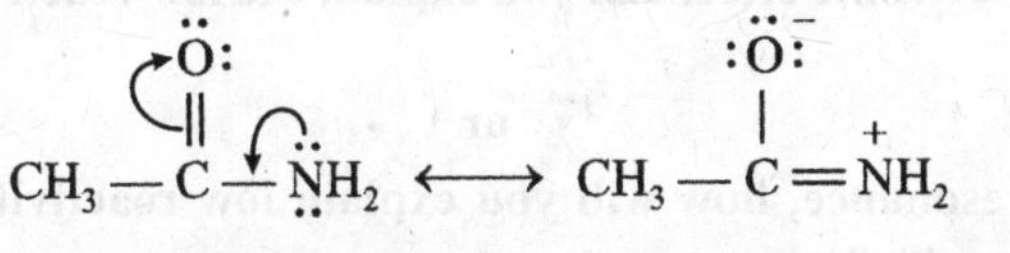

$$CH_3-\overset{\overset{\displaystyle :\ddot{O}:}{\|}}{C}-\ddot{N}H_2 \longleftrightarrow CH_3-\overset{\overset{\displaystyle :\ddot{O}:^-}{|}}{C}=\overset{+}{N}H_2$$

Resonance hybrids of acetamide

4. Comparison of relative basic strength of ethyl amine and aniline. Ethyl amine ($K_b = 5.6 \times 10^{-4}$) is much more basic than aniline ($K_b = 3.8 \times 10^{-10}$). In the former, the ethyl group with + I effect increases the electron density on the nitrogen atom and, thus, increases electron releasing tendency or basic character. In the latter, the phenyl group with − I effect tends to decrease the electron density on the nitrogen atom resulting in a decreased basic strength. Aniline is a hybrid of following contributing structures:

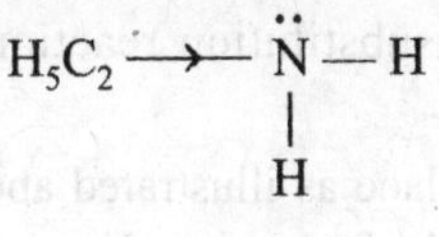

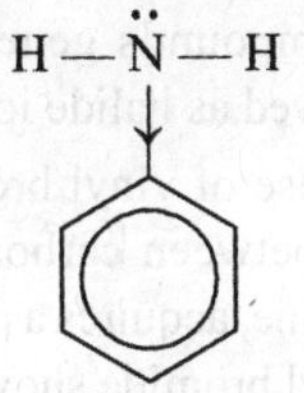

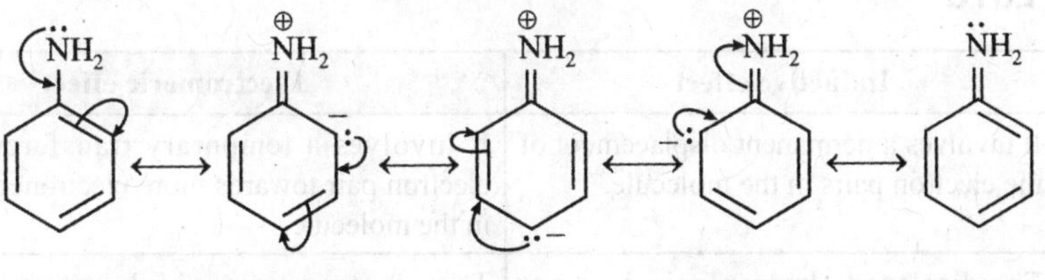

The electron density on the nitrogen atom decreases, consequently, the electron releasing tendency or basic strength decreases.

Solved Examples

Example 1: Which of the following canonical forms would contribute most towards resonance? Explain.

$$CH_2=CH-CH=\bar{C}H_2 \longleftrightarrow \bar{C}H_2-CH=CH-\overset{+}{C}H_2$$
$$(a) \qquad\qquad (b)$$

$$\longleftrightarrow \bar{C}H_2-\overset{+}{C}H-CH=CH_2$$
$$(c)$$

Solution: The resonance structures fulfilling the following conditions are more stable.
1. Greater number of covalent bonds
2. Less separation of charges
3. Negative charge on electronegative atom
4. Positive charge on electropositive atom
5. More dispersal of charges

Structure (*b*) takes care of dispersal of charge.

Hence, it is more stable and contributes more towards resonance.

Example 2: By what electronic effect can you explain the low reactivity of halogen atom in vinyl bromide.

or

On the basis of resonance, how will you explain low reactivity of vinyl bromide as compared to ethyl bromide?

Solution: Low reactivity of halogen in vinyl bromide can be explained on account of the phenomenon of resonance or mesomerism.

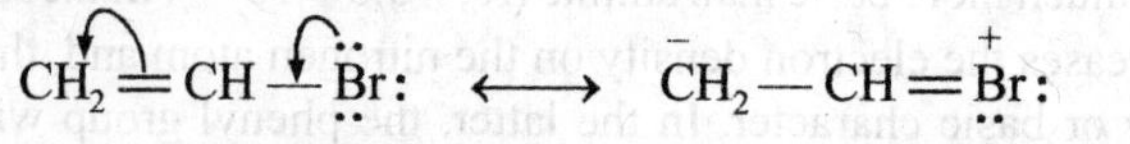

$$CH_2 = CH - \ddot{B}r: \longleftrightarrow \bar{C}H_2 - CH = \overset{+}{B}r:$$

Halogen compounds generally give nucleophilic substitution reactions, in which the halogen is removed as halide ion.

But in the case of vinyl bromide, resonance takes place as illustrated above. This creates a double bond between carbon and bromine. Removal of bromine thus becomes difficult. Moreover, bromine, acquires a positive charge and hence cannot be substituted by a nucleophile. That is why vinyl bromide shows low reactivity.

2.7 COMPARISON OF INDUCTIVE, ELECTROMERIC AND RESONANCE EFFECTS

Sl. Nos.	Inductive effect	Electromeric effect
1	It involves a permanent displacement of the electron pairs in the molecule.	It involves a temporary transference of the electron pair towards more electronegative atom in the molecule.
2	The displaced electron pair does not leave its molecular orbital. There is only a distortion in shape of the molecular orbital.	The electron pair which gets transferred, completely leaves its molecular orbital and takes up a new position.
3	The presence of the outside attacking reagent is not needed.	The presence of the outside attacking reagent is essential.
4	There is a partial separation of the charges.	There is a complete separation of the charges.
5	No ions are formed.	Ions are formed.
6	The presence of multiple bond is not essential.	The presence of multiple bond is essential.

Sl. Nos.	Resonance Effect	Inductive Effect
7	It occurs in conjugated systems.	It occurs in saturated compounds.
8	It involves the transference of the π-electrons.	It involves the displacement of the σ-electrons.
9	It involves delocalisation of electrons.	It does not involve any delocalisation of electrons.
10	It does not undergo any change in charge density with distance.	It decreases as we move away from the atom involved in the initial polar bond.

2.8 HYPERCONJUGATION OR NO-BOND RESONANCE

In the study of inductive effect, we have observed that the alkyl group has + I effect and the order of + I effects of different alkyl groups is as under:

$$-CH_3 < -C_2H_5 < -CH(CH_3)_2 < -C(CH_3)_3$$

But when the alkyl group is attached to an unsaturated system such as $-CH=CH_2$ group, or a benzene ring the order of the inductive effect simply gets reversed. This anomaly in the electron releasing tendency of the alkyl groups is explained on the basis of **hyperconjugation** which was developed by *Baker* and *Nathan*. It is also called *Baker-Nathan effect* or *no-bond resonance*.

Explanation. According to the concept, if an alkyl group carrying at least one hydrogen is attached to an unsaturated carbon atom, it releases the electrons of the C — H bond towards the multiple bond. For example, in propene ($CH_3 - CH = CH_2$), the following contributing structures are involved.

Hyperconjugation phenomenon

$$H-\underset{\underset{H}{\overset{\alpha|}{H}}}{C} - CH = CH_2 \longleftrightarrow H-\underset{H}{\overset{H^+}{C}} = CH - \bar{C}H_2$$

$$H-\underset{+H}{\overset{H}{C}} = CH - \bar{C}H_2 \longleftrightarrow H^+ \ \underset{H}{\overset{H}{C}} = CH - \bar{C}H_2$$

Since there is no bond between the α-carbon atom and one of the hydrogen atoms, the hyper conjugation is also called **no-bond resonance**. Now, hyperconjugation due to CH_3- group is expected to be more than due to CH_3CH_2- group since the latter has only two hydrogen atoms attached to the α-carbon atom directly linked to the double bonded carbon atom. Therefore, the number of contributing structures for ethyl group will be less. Similarly, the electron releasing tendency in $(CH_3)_2CH$ — group will be still less as it has only one hydrogen atom attached to the α-carbon atom and this tendency will be the least when the $(CH_3)_3C$ — group is attached to the $CH = CH_2$ group. Thus, the order of the electron releasing tendency or inductive effect of the alkyl groups get reversed because of the phenomenon of hyperconjugation.

Orbital concept of hyperconjugation

The orbital concept of hyperconjugation is illustrated with the help of propene $(H_3C — CH = CH_2)$. As mentioned earlier the electron pair of C — H bond (σ-bond) is involved in conjugation with the π-electron pair of the double bond. This may be shown with the help of Fig. 2.14.

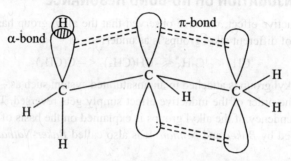

Fig. 2.1: Hyperconjugation in propene.

The σ — π conjugation is called hyperconjugation. Though hydrogen atom of the methyl group is not forming any real bond with the double bonded carbon atom but it is not actually free to leave its original position. From the orbital structure, it is evident that there is interaction between the electrons of the C — H bond and the π-electrons of the adjoining double bond. This will bring about a change in the bond lengths of C — H and C = C bonds. This has been supported by the experimental determination of the bond lengths.

Applications of hyperconjugation

The concept of hyperconjugation is quite useful in explaining relative stabilities of a number of species such as alkyl free radicals, carbocations and alkenes. These are briefly discussed as follows:

1. Relative Stabilities of Alkyl Free Radicals

The order of the relative stabilities of different alkyl free radicals is:

$(CH_3)_3C$ > $(CH_3)_2CH$ > CH_3CH_2 > CH_3
Tert-butyl   Isopropyl   Ethyl   Methyl
(3°)   (2°)   (1°)   (1°)

Ethyl free radical is regarded as a hybrid of four contributing structures.

```
      H       H              H    H              H    H
      |       |              |    |              |    |
  H — C — Ċ — C — H  ⟷  H — C = C — Ċ — H  ⟷  H — Ċ = C — C — H
      |       |              |    |              |    |
      H       H              H    H              H    H
        (I)                    (II)                    (III)
```

But there is no possibility of any hyperconjugation in methyl free radical. Therefore, ethyl free radical is more stable than the methyl free radical.

In isopropyl free radical, there are two methyl groups attached to the carbon atom which has single electron. As a result, six contributing structures in addition to normal structure are possible as shown below:

$$
\begin{array}{ccc}
\text{H}\overset{\text{H}}{\underset{\text{H}}{-\text{C}-}}\overset{\text{H}}{\underset{\text{H}}{\dot{\text{C}}-}}\overset{\text{H}}{\underset{\text{H}}{-\text{C}-}}\text{H} & \longleftrightarrow \text{H}\overset{\text{H}}{\underset{}{-\text{C}=}}\overset{\text{H}}{\underset{\text{H}}{\text{C}-}}\overset{\text{H}}{\underset{\text{H}}{-\text{C}-}}\text{H} & \longleftrightarrow \text{H}\overset{\text{H}}{\underset{\text{H}}{-\text{C}=}}\overset{}{\underset{\text{H}}{\text{C}-}}\overset{\text{H}}{\underset{\text{H}}{-\text{C}-}}\text{H} \\
\text{(I)} & \text{(II)} & \text{(III)}
\end{array}
$$

$$
\begin{array}{ccc}
\longleftrightarrow \text{H}\cdot \overset{\text{H}}{\underset{\text{H}}{\text{C}=}}\overset{\text{H}}{\underset{\text{H}}{\text{C}-}}\overset{\text{H}}{\underset{\text{H}}{-\text{C}-}}\text{H} & \longleftrightarrow \text{H}\overset{\text{H}}{\underset{\text{H}}{-\text{C}-}}\overset{}{\underset{\text{H}}{\text{C}=}}\overset{\text{H}}{\underset{\text{H}}{\text{C}-}}\text{H} & \longleftrightarrow \text{H}\overset{\text{H}}{\underset{\text{H}}{-\text{C}-}}\overset{\text{H}}{\underset{\text{H}}{\text{C}=}}\overset{\text{H}}{\underset{\text{H}}{\dot{\text{C}}\cdot}}\text{H} \\
\text{(IV)} & \text{(V)} & \text{(VI)}
\end{array}
$$

$$
\longleftrightarrow \text{H}\overset{\text{H}}{\underset{\text{H}}{-\text{C}-}}\overset{\text{H}}{\underset{\text{H}}{\text{C}=}}\overset{}{\underset{\dot{\text{H}}}{\text{C}-}}\text{H}
$$
(VII)

Similarly, *nine contributing structures* are possible for the tertiary butyl free radical in addition to its normal structure. Therefore, it is still more stable and thus, the order of the relative stabilities of the different alkyl free radicals can be explained.

2. Relative Stabilities of Alkyl Carbocations

The order of relative stabilities of the alkyl carbocations is

$$(CH_3)_3 C^\oplus > (CH_3)_2 CH^\oplus > CH_3 CH_2^\oplus > \overset{\oplus}{C}H_3$$

Tert-butyl   Isopropyl   Ethyl   Methyl

and the order is the same as in case of free radicals. This can also be explained with the help of hyperconjugation. The ethyl carbocation (primary) has three contributing structures in addition to its normal structure based upon hyperconjugation.

$$
\begin{array}{cccc}
\text{H}\overset{\text{H}}{\underset{\text{H}}{-\text{C}-}}\overset{\text{H}}{\underset{\text{H}}{\text{C}^+-}} & \longleftrightarrow \text{H}\overset{\text{H}^+}{\underset{\text{H}}{-\text{C}=}}\overset{\text{H}}{\underset{\text{H}}{\text{C}-}} & \longleftrightarrow \text{H}\overset{\text{H}}{\underset{^+\text{H}}{-\text{C}=}}\overset{\text{H}}{\underset{\text{H}}{\text{C}-}} & \longleftrightarrow \text{H}^+ \overset{\text{H}}{\underset{\text{H}}{\text{C}-}}\overset{\text{H}}{\underset{\text{H}}{\text{C}-}} \\
\text{(I)} & \text{(II)} & \text{(III)} & \text{(IV)}
\end{array}
$$

It is more stable than methyl carbocation where no resonance is possible. Similarly, the contributing structures in isopropyl carbocation and tertiary butyl carbocation are six and nine respectively in addition to the normal structure. Hence these are more stabilised and relative order of their stabilities is justified.

The order of the relative stabilities of the different alkyl carbocations can also be explained with the help of inductive effect (+*I* effect).

$$H_3C \rightarrow \underset{\underset{CH_3}{\uparrow}}{\overset{\overset{CH_3}{\downarrow}}{C\oplus}} \quad H_3C \rightarrow \underset{\underset{H}{\uparrow}}{\overset{\overset{CH_3}{\downarrow}}{C\oplus}} \quad H_3C \rightarrow \underset{\underset{H}{\uparrow}}{\overset{\overset{H}{\downarrow}}{C\oplus}} \quad H \rightarrow \underset{\underset{H}{\uparrow}}{\overset{\overset{H}{\downarrow}}{C\oplus}}$$

Tertiary butyl   Isopropyl   Ethyl   Methyl

Because of the +I effect, the alkyl group will tend to increase the electron density on positively charged carbon atom neutralising its +ve charge to some extent. As the reactivity of carbocation is due to the positive charge on it, the reactivity will tend to decrease and the stability will correspondingly increase. Thus, greater the number of the alkyl groups present, greater will be the stability of carbocation and lesser will be the reactivity. Thus, the relative stabilities of different alkyl carbocations can be explained.

3. Relative Stabilities of Alkenes

The order of the relative stabilities of different alkenes is:

$$R_2C = CR_2 > R_2C = CHR > RCH = CHR > RCH = CH_2 > CH_2 = CH_2$$

This can be explained on the basis of hyperconjugation. The presence of alkyl group (say methyl group) on the double bonded carbon atom is likely to increase the number of contributing structures due to hyperconjugation.

$$H - \underset{\underset{H}{|}}{\overset{\overset{H}{|}}{C}} - CH = CH_2 \quad\quad\quad H - \underset{\underset{H}{|}}{\overset{\overset{H}{|}}{C}} - HC = CH - \underset{\underset{H}{|}}{\overset{\overset{H}{|}}{C}} - H$$

Propene   2-Butene
(Three contributing structures)   (Six contributing structures)

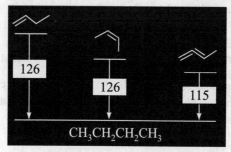

Stability of isomeric alkenes

Thus, greater the number of the contributing structures, greater will be the stability of the alkene.

4. Bond length

Hyperconjugation, like conjugation and resonance, also affects bond length.

$$H - \underset{\underset{H}{|}}{\overset{\overset{H\ 146\ pm}{|}}{C}} - CH = CH_2 \quad \longleftrightarrow \quad H - \underset{\underset{H}{|}}{\overset{\overset{H^+}{|}}{C}} = CH - \overset{\ominus}{C}H_2$$

135.3 pm

Bond length in propene is 1.46 pm in contrast to normal 1.54 pm (in propane). It is due to the partial double bond character acquired and hence a little shorter.

5. Dipole moment

Since hyperconjugation causes the separation of charges, it also affects the dipole moment in the molecule. Thus the dipole moment values of C — O bond in methanal and ethanal are quite different.

$$H-\underset{\mu = 2.27\ D}{\overset{H}{\underset{|}{C}}}=O \qquad H-\underset{\underset{H}{|}}{\overset{H}{\overset{|}{C}}}-CH=O \longleftrightarrow H-\underset{\underset{H}{|}}{\overset{H^+}{\overset{|}{C}}}=CH-O^-$$
$$\mu = 2.72\ D$$

6. Ortho-para directing property of methyl group in toluene is partly due to +I effect and partly due to hyperconjugation.

Types of Hyperconjugation

(a) Sacrafacial hyperconjugation

The essential condition is the attachment of alkyl group to double bond or triple bond.

$$H-\underset{\underset{H}{|}}{\overset{H}{\overset{|}{C}}}-C=C\!\!\!<\qquad \text{or}\qquad H-\underset{\underset{H}{|}}{\overset{H}{\overset{|}{C}}}-C\equiv C-$$

Carbon atom of alkyl group attached to double bond must contain atleast one hydrogen atom in hyperconjugation.

$$H-\underset{\underset{H}{|}}{\overset{H}{\overset{|}{C}}}-C=C- \longleftrightarrow H-\overset{H}{\overset{|}{C}}=C-\overset{\ominus}{C}-$$

$$H^+ \uparrow\downarrow \qquad\qquad H^+ \uparrow\downarrow$$

$$H-\overset{H}{\overset{|}{C}}=C-\overset{\ominus}{C}- \longleftrightarrow H^+ \quad \overset{H}{\overset{|}{C}}=C-\overset{\ominus}{C}-$$

It involves a sort of sacrifice of bond and hence the name sacrafacial.

(b) Isovalent hyperconjugation

This kind of hyperconjugation involves no sacrifice of bonds. Ethyl radicals have the same number of real bonds as the classical structure.

$$\underset{(I)}{H-\underset{\underset{H}{|}}{\overset{\overset{H}{|}}{C}}-\overset{\oplus}{\underset{\underset{H}{|}}{\overset{\overset{H}{|}}{C}}}} \longleftrightarrow \underset{(II)}{H-\underset{\underset{H}{|}}{\overset{\overset{H^{\oplus}}{|}}{C}}=\underset{\underset{H}{|}}{\overset{\overset{H}{|}}{C}}} \longleftrightarrow \underset{(III)}{H^{\oplus} \underset{\underset{H}{|}}{\overset{\overset{H}{|}}{C}}=\underset{\underset{H}{|}}{\overset{\overset{H}{|}}{C}}} \longleftrightarrow$$

2.9 ORGANIC ACIDS AND BASES

Acids

Organic acids are weak acids. They dissociate in water to a small extent compared to inorganic acids which dissociate almost completely in water. Strength of an organic acid is expressed in terms of dissociation constant (K_a). K_a indicates the extent to which the acid is ionised in aqueous solution. Let us take the example of acetic acid (CH_3COOH). It ionises in water as

$$CH_3COOH(aq) + H_2O(l) \rightleftharpoons CH_3COO^- + H_3O^+$$

Equilibrium constant of this reaction is written as

$$K_a = \frac{[CH_3COO^-][H_3O^+]}{[CH_3COOH]}$$

Ionization constants of some weak acids are given in Table 2.1 below:

TABLE 2.1 *Ionization constants of some weak acids*

Ionization Reaction	K_a at 25°C
$HSO_4^- + H_2O \rightleftharpoons H_3O^+ + SO_4^{2-}$	1.2×10^{-2}
$HF + H_2O \rightleftharpoons H_3O^+ + F^-$	3.5×10^{-4}
$HNCO + H_2O \rightleftharpoons H_3O^+ + NCO^-$	2×10^{-4}
$HCOOH + H_2O \rightleftharpoons H_3O^+ + HCOO^-$	1.8×10^{-4}
$CH_3COOH + H_2O \rightleftharpoons H_3O^+ + CH_3COO^-$	1.8×10^{-5}
$HClO + H_2O \rightleftharpoons H_3O^+ + ClO^-$	2.9×10^{-8}
$HBrO + H_2O \rightleftharpoons H_3O^+ + BrO^-$	2.8×10^{-9}
$HCN + H_2O \rightleftharpoons H_3O^+ CN^-$	4.9×10^{-10}

Bases

A base is a substance when added to water releases OH^- ions in solution. Organic bases are weak bases, they ionize to a small extent and produce a small concentration of OH^- ions. This is in contrast to inorganic bases like $NaOH$, KOH and $Ca(OH)_2$ which release a high concentration of OH^- ions. The strength of a base is measured in terms of K_b value which tells how much of it ionises in water to release OH^- ions. Take the case of an organic weak base trimethyl amine. In water solution, the following reaction takes place

$$(CH_3)_3 N\ (aq) + H_2O\ (l) \rightleftharpoons (CH_3)_3 NH^+\ (aq) + OH^-\ (aq)$$

The equilibrium constant for this reaction is expressed as

$$K_b = \frac{[(CH_3)_3 NH^+][OH^-]}{[(CH_3)_3 N]}$$

Ionization constants (K_b) for some weak bases are given in Table 2.2.

TABLE 2.2 K_b values of some weak bases

Ionization Reaction	K_b at 25°C
$(CH_3)_2NH + H_2O \rightleftharpoons (CH_3)_2 NH_2^+ + OH^-$	5.9×10^{-4}
$CH_3NH_2 + H_2O \rightleftharpoons CH_3NH_3^+ + OH^-$	4.4×10^{-4}
$(CH_3)_3N + H_2O \rightleftharpoons (CH_3)_3NH^+ + OH^-$	6.3×10^{-5}
$NH_3 + H_2O \rightleftharpoons NH_4^+ + OH^-$	1.8×10^{-5}
$C_6H_5NH_2 + H_2O \rightleftharpoons C_6H_5NH_3^+ + OH^-$	4.3×10^{-10}

2.10 LINEAR AND CROSSED CONJUGATION SYSTEMS

If the double bonds of a conjugated system lie along a length of the molecule in a straight line, it is called **a linear conjugated system** and if the double bonds do not lie along a straight line but contain a branched conjugated system, it is called **crossed conjugated system**. Look at the two structures given below:

$$\overset{1}{CH_2}=\overset{2}{CH}-\overset{3}{CH}=\overset{4}{CH}-\overset{5}{CH}=\overset{6}{CH_2}$$

Hexa-1, 3, 5-triene
(Linear conjugated triene)
I

$$\overset{1}{CH_2}=\overset{2}{C}-\overset{3}{C}\overset{\overset{CH_2}{\|}}{}-\overset{4}{CH}=\overset{5}{CH_2}$$

3-methylenepenta-1, 4-diene
(Crossed conjugted triene)
II

Compounds I and II are both conjugated trienes. The difference is that compound I contains all the double bonds in a straight line along the length of the molecule while the compound II does not. In compound II, the double bond at carbon-3 is not along the other two double bonds.

We can consider double bonds along with ketonic groups for conjugation, correspondingly, we shall have linear and crossed conjugated enones (en + one). Consider the following compounds containing the double bonds (enes) and the ketone group (one) in conjugation. Conjugation means in alternate positions.

$$\overset{6}{CH_2}=\overset{5}{CH}-\overset{4}{CH}=\overset{3}{CH}-\overset{2}{\underset{\|}{C}}\overset{O}{}-\overset{1}{CH_3}$$

Hexa-3, 5-dien-2-one
III
(Linear conjugated dienone)

$$\overset{1}{CH_2}=\overset{2}{CH}-\overset{3}{\underset{\|}{C}}\overset{O}{}-\overset{4}{CH}=\overset{5}{CH}-\overset{6}{CH_3}$$

Hexa-1, 4-dien-3-one
(IV)
(Crossed conjugated dienone)

It has been observed that out of the two trienes or dienones, the crossed conjugated moity has the lower resonance energy than the corresponding linear conjugated moity.

2.11 HYBRIDISATION OF ORBITALS

The electronic configuration of carbon is $1s^2\,2s^2\,2p_x^1\,2p_y^1$ as its At. No. is 6. As the bonds are formed by unpaired electrons, carbon should show a valency of 2. But, we observe that carbon shows tetravalency in its compounds like methane or carbon dioxide. This was explained by saying that the above-mentioned configuration is the configuration in the ground state. During the formation of bonds with other atoms, some of the electrons in lower energy levels are excited to higher levels. Thus, in the case of carbon, one electron from $2s$ level is excited to empty $2p_z$ level. (Fig. 2.2)

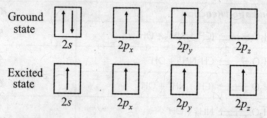

Fig. 2.2: *Ground and excited state configurations of carbon.*

Now, there are 4 electrons which are unpaired. These four electrons will form four bonds by overlapping with the orbitals of other atoms. For example, they combine with four hydrogen atoms to form methane (CH_4).

However, there is one thing which is surprising. We find that in methane, all the C—H bonds have the same bond energy and all the angles H—C—H are the same, *i.e.*, 109°– 28'. Looking back at the orbitals, which participate in bonding, *i.e.*, one *s*-orbital and three *p*-orbital which have different shapes and energies, we expect one of the bonds formed by *s*-orbital to be different from the three bonds formed by *p*-orbitals. Moreover, as the *p*-orbitals are inclined at an angle of 90° to each other, we expect the bond angles of 90° in methane also. But this actually is not so. To explain this anomaly, the concept of *hybridisation* was introduced.

It is assumed that the $2s$-orbital and three $2p$-orbitals which are associated with different shapes and energies mix with each other and produce four equivalent types of orbitals with the same shape and energy and are oriented symmetrically in space. These are called **hybrid** or **hybridised** orbitals and this phenomenon is called **hybridisation**. These hybridised orbitals then overlap with the orbitals of other atoms to form bonds.

Thus, hybridisation is the process of mixing of orbitals of different shapes and energies, to produce orbitals of equivalent shape and energy oriented symmetrically with respect to one another.

This phenomenon explains satisfactorily why the bond energies and bond angles are equal in the case of methane and many other molecules.

Necessary conditions for hybridisation

There are certain conditions which must be satisfied before the orbitals hybridise to produce equivalent type of orbitals. These are:

(*i*) The orbitals taking part in hybridisation should not differ too much in their energies. For example, $1s$ and $2p$ orbitals cannot hybridise. The participating orbitals must possess almost similar energies.

(*ii*) As many hybrid orbitals will be obtained as the number of orbitals combining together.

(*iii*) Whereas the paired orbitals do not take part in bonding, they can take part in hybridisation. In such a case, one of the positions in the geometrical shape is occupied by the lone pair of electrons. This happens in the case of ammonia and water molecules.

(*iv*) It is not essential for all the unpaired orbitals to participate in hybridisation. Those unpaired orbitals which do not take part in hybridisation form π-bonds by lateral overlapping.

2.12 TYPES OF HYBRIDISATION

Depending upon which atomic orbitals are combining together, we have different types of hybridisation, which are given below:

1. *sp* or diagonal hybridisation. This type of hybridisation occurs when one *s* orbital and one *p* orbital combine together to give two hybridised orbitals known as *sp* hybridised orbitals which have equal energies and they are oriented along the same line in opposite direction at an angle of 180°. Since, the orbitals are negatively charged, they repel each other. Hence, they orient themselves in such a way that they are maximum far apart from each other. Two orbitals can orient themselves along the same line in opposite directions to have minimum repulsive interactions. The shapes of the hybridised orbitals in *sp* hybridisation is given in Fig. 2.3.

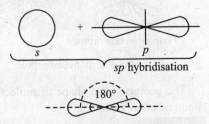

Fig. 2.3: *sp hybridisation*

2. *sp*² or trigonal hybridisation. This type of hybridisation takes place when one *s* orbital and two 2*p* orbitals combine together to give three *sp*² hybridised orbitals of equal energies. These hybrid orbitals are oriented at an angle of 120° in the same plane. The three orbitals point towards the corners of an equilateral triangle with the nucleus occupying the centre of the triangle. This is illustrated in Fig. 2.4.

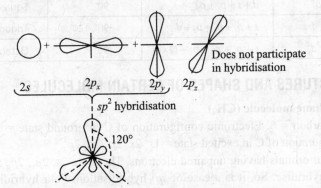

Fig. 2.4: *sp² hybridisation.*

3. *sp*³ or tetrahedral hybridisation. When one *s*-orbital and three *p*-orbitals combine together to produce equivalent type of orbitals, *sp*³ hybridisation takes place. Four *sp*³ hybridised orbitals are produced which have the same energy and shape and are oriented at an angle of 109°28' to one another. This is also known as *tetrahedral* hybridisation because the hybridised orbitals are pointing towards four corners of a tetrahedron with the nucleus occupying the centre of the tetrahedron. In this position, the hybridised orbitals experience minimum repulsive forces. This is illustrated in Fig. 2.5.

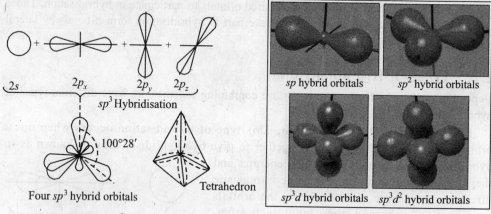

Fig. 2.5: sp^3 hybridisation.

The geometry and shape of molecules associated with different types of hybridisation is summarised in Table 2.3.

TABLE 2.3 *Hybridisation and shapes of molecules*

S. No.	Type of Hybridisation	Combination of Orbitals	Bond Angle	Geometry
1.	sp	$s + p_x$	180°	Linear
2.	sp^2	$s + p_x + p_y$	120°	Trigonal
3.	sp^3	$s + p_x + p_y + p_z$	109°28′	Tetrahedral
4.	dsp^2	$d + s + p_x + p_y$	90°	Square planer
5.	sp^3d	$s + p_x + p_y + p_z + d$	90°, 120°	Trigonal bipyramidal
6.	sp^3d^2	$s + p_x + p_y + p_z + (2×d)$	90°	Octahedral

2.13 STRUCTURES AND SHAPES OF CERTAIN MOLECULES

1. Shape of methane molecule (CH_4)

At. No. of carbon = 6. Electronic configuration of C in ground state = $1s^2\ 2s^2\ 2p_x^1\ 2p_y^1$. Electronic configuration of C in excited state = $1s^2\ 2s^1\ 2p_x^1\ 2p_y^1\ 2p_z^1$.

There are four orbitals having unpaired electrons. These are $2s$, $2p_x$, $2p_y$ and $2p_z$ orbitals. They will first hybridise. So, it is a case of sp^3 hybridisation. Four hybridised orbitals are oriented tetrahedrally as shown in Fig. 2.6.

All the four hybridised orbitals overlap with the $1s$-orbital of four hydrogen atoms. All these overlappings take place along the internuclear axes. Hence, four carbon-hydrogen σ-bonds are formed. The angle ∠HCH is 109°28′.

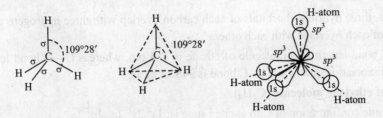

Fig. 2.6: *Shape of methane molecule.*

2. Shape of water molecule H₂O

Atomic number of oxygen = 8

Electronic configuration = $1s^2\, 2s^2\, 2p_x^2\, 2p_y^1\, 2p_z^1$

In the 2nd orbit, there are four orbitals, viz., $2s$, $2p_x$, $2p_y$ and $2p_z$. Out of these $2s$ and $2p_x$ cannot form the bonds as they are paired. But they can participate in hybridisation. Hence, it is a case of sp^3 hybridisation. The hybridised orbitals are oriented along the corners of a tetrahedron. Two of the positions on the tetrahedron are occupied by lone pair of electrons from $2s$ and $2p_x$ orbitals. The other two hybridised orbitals will overlap with the $1s$-orbitals of two hydrogen atoms to form a water molecule as shown in Fig. 2.7. We expect an angle of 109°28′ in HOH. Actually the angle is 105°. This is due to stronger repulsion of two lone pair of electrons with the shared pairs of electrons. Thus, it is of V shape.

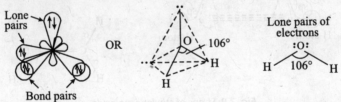

Fig. 2.7: *Shape of water molecule.*

3. Shape of ethane (C₂H₆) molecule

In the molecule of ethane two carbon atoms are linked to each other by a σ-bond and each carbon is linked to three hydrogens (Fig. 2.8)

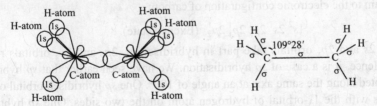

Fig. 2.8: *Shape of ethane molecule.*

Both the carbon atoms are sp^3 hybridised. Out of the four sp^3 hybridised orbitals of one carbon, one orbital overlaps with one of the hybridised orbitals of the second carbon making a C—C bond. The other three hybridised orbitals overlap with the $1s$-orbitals of three hydrogen atoms. Similarly, one hybridised orbital of the second carbon overlaps with one of the orbitals of the first carbon and three orbitals overlap with three $1s$-orbitals of three hydrogen atoms.

In other words, three hybridised orbitals of each carbon overlap with three hydrogens and the fourth orbital of each overlaps with each other.

The C—C bond length in a molecule of ethane is 154 pm, whereas C—H bond length is 110 pm. The dissociation energy of C—C bond is 348 kJ/mole.

4. Shape of ethylene molecule (C_2H_4)

The electronic configuration of C in excited state is: $1s^2\ 2s^1\ 2p_x^1\ 2p_y^1\ 2p_z^1$

Here, all the four unpaired orbitals do not participate in hybridization. Out of four, only three orbitals viz. $2s$, $2p_x$ and $2p_y$ hybridise. The fourth one $2p_z$ remains unhybridised. Hence, it is a case of sp^2 hybridisation. The hybridised orbitals are oriented in the same plane at an angle of 120°. We have two such carbons. One orbital of each carbon overlaps with each other to form a σ-bond. Two orbitals of each carbon overlap with two 1s-orbitals of hydrogen to form four C—H bonds. We are left with 2 unhybridised orbitals of two carbons which are oriented perpendicular to the plane of the rest of the molecule. They will overlap laterally giving rise to a weak π-bond between carbon-carbon. Hence, there will be a double bond, one σ-bond and one π-bond between carbon-carbon (Fig. 2.9).

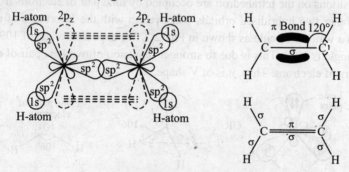

Fig. 2.9: *Shape of ethylene molecule.*

A double bond brings the carbon atoms together. Hence, the C = C bond length in C_2H_4 is shortened to 134 pm. The angle ∠HCH or ∠HCC is 120°, the trigonal angle. Bond dissociation energy of C = C bond is 614 kJ/mole.

5. Shape of acetylene molecule (C_2H_2)

Coming again to the electronic configuration of carbon:

$$1s^2\ 2s^1\ 2p_x^1\ 2p_y^1\ 2p_z^1\ \text{(Excited state)}.$$

Here, only $2s$ and $2p_x$ orbitals take part in hybridisation. $2p_y$ and $2p_z$ orbitals remain unhybridised. Hence, it is a case of sp hybridisation. We have seen earlier that sp hybridised orbital are oriented along the same axis at an angle of 180°. One sp hybridised orbital of each carbon overlaps with the 1s-orbital of hydrogen atom on the two sides. One sp hybridised orbital of each carbon overlaps with each other forming a C—C sigma bond. The unhybridised $2p_y$ orbital of one carbon overlaps laterally with the $2p_y$ orbital of the other carbon forming a π-bond. Similarly, $2p_z$ orbital of one carbon overlaps with $2p_z$ orbital of the other carbon laterally forming another π-bond. The molecule of acetylene has a linear shape (Fig. 2.10). There is a cylinder of negative charge around C—C bond.

A triple bond is formed between carbon-carbon. But of the three bonds, one is a σ-bond and two are π-bonds. Triple bond brings carbon atoms still nearer. The bond length C ≡ C is 1.20 pm.

Acetylene being a linear molecule, angle H — C — C is 180°. Bond dissociation energy of the $C \equiv C$ bond is 811 kJ/mole.

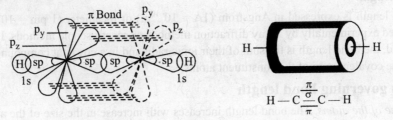

Fig. 2.10: *Shape of acetylene molecule (Linear shape).*

6. Shape of carbon dioxide molecule. At. No. of carbon is 6 and its electronic configuration in the excited state is:

$$1s^2\ 2s^1\ 2p_x^1\ 2p_y^1\ 2p_z^1$$

sp hybridisation takes place here. The hybridised orbitals are oriented along the same line at an angle of 180°. $2p_y$ and $2p_z$ orbitals do not participate in the hybridisation.

Coming to oxygen, its electronic configuration is:

$$1s^2\ 2s^2\ 2p_x^2\ 2p_y^1\ 2p_z^1$$

There are two such oxygen atoms in the molecule of CO_2. Two sp hybridised orbitals of carbon overlap with one unpaired 2p-orbital of each oxygen on the two sides forming two σ-bonds. And then, one unhybridised 2p-orbital of carbon overlaps with the other unpaired orbital of one oxygen laterally forming a π-bond. Similarly, the second unhybridised 2p-orbital of carbon overlaps with the other unpaired 2p-orbital of second oxygen atom (Fig. 2.11)

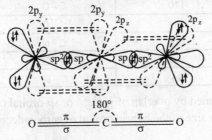

Fig. 2.11: *Shape of CO_2 molecule.*

The molecule of carbon dioxide has a linear shape.

2.14 SOME IMPORTANT BOND PARAMETERS

A number of factors including **hybridisation** govern bond properties. These are discussed as under:

1. Bond Length

When atoms are brought close to each other, attraction takes place between them and, therefore, the potential energy of the system keeps on decreasing till at a particular distance, the potential energy is minimum. If the atoms are further brought closer, the repulsion starts and the potential energy of the system begins to increase.

The average distance between the centres of the nuclei of the two bonded atoms is called its **bond length**.

Bond length is expressed in Angstrom (1Å = 10^{-10}m) or picometer (1 pm = 10^{-12}m). It is determined experimentally by X-ray diffraction methods or spectroscopic methods. In an ionic compound, the bond length is the sum of their ionic radii and in a covalent compound, it is the sum of the covalent radii of the constituent atoms.

Factors governing bond length

(i) Size of the atoms. The bond length increases with increase in the size of the atoms. For example, bond lengths of H—X are in the order: HI > HBr > HCl > HF.

(ii) Multiplicity of bond. The bond length decreases with the multiplicity of the bond. Thus, bond length of carbon-carbon bonds changes in the order: C ≡ C < C = C < C – C.

Multiplicity of the bond involves increasing overlapping of orbitals leading to decrease in bond length.

(iii) Type of hybridisation. An *s*-orbital is smaller in size. Greater the *s*-character, shorter is the hybrid orbital and, hence, shorter is the bond length. For example,

$$sp^3\ C-H > sp^2\ C-H > sp\ C-H$$

sp hybrid orbital has 50% *s* character (one *s* and one *p*). This percentage of *s* decreases as we move to sp^2 and sp^3.

The bond lengths of a few common bonds are given in Table 2.4.

TABLE 2.4 *Some bond lengths*

Bond	Bond length (pm)	Bond	Bond length (pm)
C – H	110	C – O	143
C – C	154	C = O	121
C = C	134	C – N	147
C ≡ C	120	C – Cl	177
O – H	96	C – Br	191

A C – H bond may be formed by overlap of sp^3, sp^2 or *sp*-orbital of carbon with an *s*-orbital of hydrogen. Therefore as the size of the hybrid orbital of carbon decreases, the length of C – H bond also decreases.

$$C(sp^3) - H\ >\ C(sp^2) - H\ >\ C(sp) - H$$
$$110\ \text{pm}\qquad\quad 107.6\ \text{pm}\qquad 106\ \text{pm}$$

2. Bond Energy

Energy is released when atoms come close together resulting in the formation of bond between them. Conversely, the energy is required to break the bond to get the separated atoms.

The amount of energy required to break one mole of bonds of a particular type so as to separate them into gaseous atoms is called **bond dissociation energy** *or simply* **bond energy**.

Bond energy is expressed in kJ mol^{-1}. Greater is the bond dissociation energy, stronger is the bond.

Since, a particular type of bond present in different molecules (*e.g.*, O – H bond in H – O – H) or even in the same molecule (*e.g.*, four C – H bonds in CH$_4$) do not possess the same bond energy, bond energies are usually the *average values*.

Table 2.5 gives the average bond energies of a few common bonds.

TABLE 2.5 *Some Bond Energies.*

Bond	Bond energy (kJ/mole)	Bond	Bond energy (kJ/mole)
C–H	414.2	C–O	359.8
C–C	347.3	C=O	748.9
C=C	610.9	C–N	305.4
C≡C	836.8	C–Cl	338.9
O–H	464.4	C–Br	284.5
N–H	389.1	C–I	213.4

Factors governing bond energy

(*i*) *Size of the atoms.* Bigger the size of the atoms, greater is the bond length and smaller is the bond energy.

(*ii*) *Multiplicity of bonds.* Greater the multiplicity of the bond, greater is the bond dissociation energy.

(*iii*) *Number of lone pairs of electrons present.* Greater the number of lone pairs of electrons present on the bonded atoms, greater is the repulsion between the atoms and, hence, smaller is the bond dissociation energy.

Bond	C–C	N–N	O–O	F–F
Lone pairs of electrons	0	1	2	3
Bond energy (kJ mol^{-1})	348	163	146	139

3. Bond Angle

A bond is formed by the overlap of atomic orbitals of different atoms. The direction of overlap gives direction of the bond.

The angle between the lines representing the directions of the bonds is called the **bond angle.**

It is expressed in degrees, minutes and seconds. H—C—H bond angle in CH$_4$, H—N—H bond angle in NH$_3$ and H—O—H bond angle in H$_2$O are shown in the Fig. 2.12.

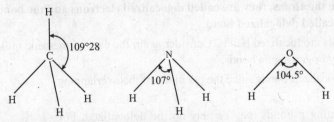

Fig. 2.12: *Bond angles in CH$_4$, NH$_3$ and H$_2$O*

Bond angle in a molecule depends upon the following factors:

(*i*) **Type of hybridisation.** The angles ∠HCH in methane, ethene and ethyne are expected to be 109° – 28′, 120° and 180° because of different orientations of hybridised orbitals resulting from sp^3, sp^2 and sp hybridisations respectively.

(*ii*) **Number of shared and lone pairs of electrons.** The hybridisation involved in the formation of CH_4, NH_3 and H_2O is the same viz sp^3. We expect a value of 109°–28' for ∠HCH, ∠HNH and ∠H–O–H. We find that the ∠HNH is 107° and the ∠HOH is 104.5°.

The variation in angles takes place because lone pair-bond pair repulsion is greater than bond pair-bond pair repulsion. Because of this effect, the ∠HNH is contracted to 107°. The molecule of water contains two lone pairs of electron. Thus the effect is two-fold and the ∠HOH is contracted to a much greater extent *i.e.* 104.5°.

(*iii*) **Unequal repulsion between σ and π bonds.** Coming back to the molecule of ethene, where we expict an angle of 120° for ∠HCH and ∠HCC, we find that ∠HCH is 116.6° while the angle ∠HCC is 121.7°

We explain it by saying that π-electrons being loosely bound repel the bond pair of C – H more than the bond pairs of C – H bonds repel each other.

2.15 LOCALIZED AND DELOCALIZED BONDS

A covalent bond is formed by the overlapping of atomic orbitals which may be hybridised or unhybridised. The electrons which form these bonds can be localized or delocalized.

If the electrons from the bond spend most of the thime in the space between the participating atom, they are called localized electrons and the bond formed between these atom is called localized bond.

But, if the electrons from the bond formed between two atoms are not confirmed to the space between the atoms, they are called delocalized electrons and the bond formed by such electrons is called delocalized bond.

Sigma(σ) bonds are localized bonds. Consider again the case of methane molecule. There are four carbon–hydrogen sigma bonds.

These are all localized bonds since the positions of the overlapping orbitals are fixed.

Electrons forming π-bonds may or may not be delocalized. For example, the electrons forming π bond in ehtene are restricted to the space between two carbon atoms. Therefore, these electrons are localized and the π bond in ethene is **localized.** However the situation is different in ethyne which contains two π-bonds in addition to a σ-bond. The two π-bonds are formed by the lateral overlap of *p*-orbitals of the two carbon atoms oriented at 90° to each other. Here, the electrons of

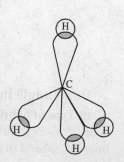

the two π-bonds will not be able to confine to their space and direction. The two will mix with each other forming a cylinder of electron cloud around C – C σ-bond. Thus, these electrons are **delocalized** and the two π-bonds formed are **delocalized bonds**. Delocalization occurs in order to provide greater stability to the molecule.

Now consider the case of 1, 3-Butadiene. This is a conjugated diene.

All four carbon atoms in 1, 3-butadiene are sp^2 hybridized. The sp^2 hybrid orbitals overlap with each other and with s orbitals of the hydrogen atoms to form C–C and C–H σ bonds. Since the bonds result from the overlap of trigonal sp^2 orbitals, all carbon and hydrogen atoms lie in one plane.

Also each carbon atom in 1, 3-butadiene possesses an unhybridized p orbital. The p orbitals are perpendicular to the plane of σ bonds. The p orbital on C-2 can overlap with the p orbitals on C-1 and C-3. The p orbital on C-3 can overlap with the p orbitals on C-2 and C-4. That is, all four p orbitals overlap to form a large π molecular orbital (Fig. 2.13). Each pair of π electrons is thus attracted, not by two, but all four carbons.

The overlap of p orbitals of C-2 and C-3 in both directions, which allows the π electrons to be spread over a larger area, is also an example of **delocalization**. This delocalization of π electrons is responsible for greater stability of 1, 3-butadiene.

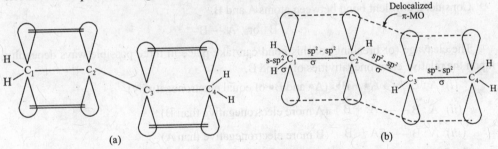

Fig. 2.13: *Orbital structure of 1, 3-Butadiene. (a) Without delocalization (b) With delocalization*

The C–C single bonds in 1, 3-butadiene are shorter (148 pm) than normal carbon–carbon single bonds. The C–C double bonds are longer (137 pm) than normal isolated carbon–carbon double bonds.

The delocalized molecular orbital contains four electrons. The central carbon-carbon single bond is shorter than the 154 pm bonds typical of alkanes because of its partial double bond character.

The electrons forming two π bonds in the molecule of 1, 3-butadiene are thus delocalized chemical bonds.

Another example of delocalized bonds is provided by the structure of benzene. It has a hexagonal structure in which the six carbon atoms occupy the corners of a regular hexagon. All the carbon atoms are linked to each other by sigma bonds. Also, each carbon is linked to one hydrogen by sigma bond. It may be noted that carbon atoms in benzene are sp^2 hybridised. Thus, all carbons lie in the same plane. There is one orbital in each carbon, which does not participate in hybridisation.

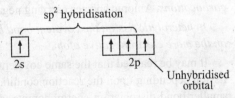

There are six such unhybridised orbitals these can overlap laterally in pairs to form three π bonds as shown in Fig. 2.14 (*a*) and (*b*). But the same situation exists as in 1, 3-butadiene. All the six orbitals overlap together forming a circular electron cloud above and below the benzene ring as shown in Fig. 2.14 (*c*).

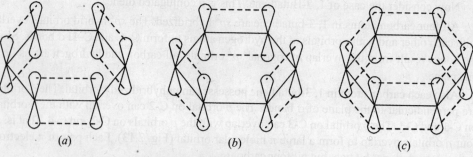

Fig. 2.14: *Bond delocalisation in benzene*

2.16 HOMOLYTIC AND HETEROLYTIC BOND BREAKING (FISSION)

Consider a covalent bond between atoms A and B.

$$A:B \quad \text{or} \quad A—B$$

The cleavage (or breaking) of this bond can take place in three possible ways depending upon the relative electronegativities of A and B.

(*i*) $A:B \longrightarrow A\bullet + B\bullet$ (A• and B• of equal electronegativity)

(*ii*) $A:B \longrightarrow \overset{-}{A}: + \overset{+}{B}$ (A more electronegative than B)

(*iii*) $A:B \longrightarrow \overset{+}{A} + :\overset{-}{B}$ (B more electronegative than A)

The first type of cleavage is called *homolytic fission* or *homolysis* and leads to the formation of very reactive species called *'free radicals'* (atoms or groups of atoms containing odd or unpaired electrons).

In homolytic fission the covalent bond breaks in such a way that each fragment carries one unpaired electron.

Homolytic fission usually occurs in non-polar bonds. High temperature, presence of ultraviolet radiations and radical initiators such as peroxide favour homolytic fission.

The second and third types of cleavage is called *heterolytic fission* and leads to the formation of ionic species. These ionic species are also very reactive and carry charges on carbon. Cationic species carrying positive charge on a carbon atom are called *carbonium ions* or *carbocations*. Anionic species carrying negative charge on carbon atom are called *carbanions*.

In heterolytic fission the covalent bond breaks in such a way that the pair of electrons stays on the more electronegative atom.

It may be realised that the same covalent bond can undergo homolytic fission or heterolytic fission depending upon the reaction conditions. Homolytic fission is generally associated with smaller bond dissociation energy compared to heterolytic fission. This can be understood by the fact, that in hetrolytic fission, both the shared electrons are going to be shifted to one atom which requires a greater amount of energy.

$$CH_3 \overset{\frown}{-} \overset{\frown}{Cl} \longrightarrow \dot{C}H_3 + \dot{C}l; \quad \Delta H = +352 \text{ kJmol}^{-1} \quad \text{Homolytic fission}$$

$$CH_3 \overset{\frown}{-} Cl \longrightarrow \overset{+}{C}H_3 + :\bar{C}l; \quad \Delta H = +952 \text{ kJmol}^{-1} \quad \text{Heterolytic fission}$$

2.17 TYPES OF ORGANIC REACTIONS

Organic reactions may be classified into four main types:

(a) Substitution reactions
(b) Addition reactions
(c) Elimination reactions
(d) Rearrangements

These are separately described as under.

(a) Substitution reactions. *A substitution reaction is one in which a part of one molecule is replaced by other atom or group without causing a change in the rest of the molecule.* Following are some examples of substitution reactions.

(i) $CH_4 + Cl_2 \longrightarrow CH_3Cl + HCl$ (Free-radical substitution reaction)
   Methane   Methyl chloride

(ii) $C_2H_5Br + KOH \longrightarrow C_2H_5OH + KBr$ (Nucleophilic substitution reaction)

(iii) $C_6H_6 + Cl_2 \xrightarrow{FeCl_3} C_6H_5Cl + HCl$ (Electrophilic substitution reactions)
    Benzene   Chlorobenzene

The substitution reactions may be brought about by free-radicals, nucleophilic or electrophilic reagents. Thus, there are free radical substitution reactions, nucleophilic substitution reactions and electrophilic substitution reactions.

(b) Addition reactions. *When two molecules of same or different substances combine together giving rising to a new product, it is an addition reaction.* Examples of addition reactions are:

(i) $CH_2 = CH_2 + Br_2 \longrightarrow CH_2BrCH_2Br$ (Electrophilic addition reactions)
   Ethane   Ethylene dibromide

(ii) $CH_3CHO + HCN \longrightarrow CH_3CH(OH)CN$ (Nucleophilic addition reaction)
    Acetaldehyde   Acetaldehyde cyanohydrin

(iii) $CH_3-CH=CH_2 + HBr \xrightarrow{Peroxides} CH_3-CH_2-CH_2Br$
     Propene   n-Propyl bromide

Addition reactions could be brought about by free-radical, electrophilic or nucleophilic reagents.

As in the case of substitution, there are free radical, electrophilic and nucleophilic substitution reactions.

(c) Elimination reactions. *These reactions involve the removal of atoms or groups from a molecule to form a new compounds containing multiple bonds.* Dehydrohalogenation of alkyl halides is a common example of this type reaction.

$$CH_3-CH_2CH_2Cl + KOH \xrightarrow{Alc.} CH_3CH=CH_2 + KCl + H_2O$$
n–Propyl chloride　　　　　　　　Propene

Elimination reactions are of two types. Bimolecular eliminations or E_2 reactions and monomolecular or E_1 elimination reactions.

In E_2 reactions, rate determining step involves two reacting species. Dehydrohalogenation of tertiary alkyl halides with alc. KOH proceeds by E_2 mechanism. In E_1 reactions, the rate determining step involves only one reacting species. Dehydrohalogenation of primary alkyl halides with alc. KOH usually proceeds through E_1 mechanism.

Elimination reactions can also be classified as under:

(i) **α-Elmination reactions.** Loss of two atoms/groups occurs from the *same atoms* of the substrate molecule. Base catalysed dehydrohalogenation of chloroform to form dichlorocarbene is an example of such reaction

$$HO^- + H-CCl_2-Cl \longrightarrow :CCl_2 + H_2O + Cl^-$$
　　　　　　　　　　　　　Dichlorocarbene

(ii) **β-Elimination reactions.** Loss of two atoms/groups occurs from *adjacent atoms* of the substrate molecule. For example

$$H-CH_2-CH_2-OH \xrightarrow{H^+} CH_2=CH_2 + H_2O \quad \text{(Acid-catalysed dehydration)}$$
　　　　　　　　　　　　Ethene

$$HO^- \;\; H-CH_2-CH_2-Br \xrightarrow[\Delta]{Alc\;KOH} CH_2=CH_2 + KBr + H_2O \quad \text{(Base-catalysed dehydrohalogenation of alkyl halides)}$$
　　　　　　　　　　　　　　　　Ethene

(ii) **γ-Elimination.** Loss of two atoms or groups occurs from α- and γ-positions *i.e.* from the two positions intervened by another position, leading to the formation of three-membered ring. For example

$$\underset{\underset{Br\quad\quad Br}{}}{H_2C-CH_2-CH_2} \xrightarrow[\Delta]{Zn\;dust} \underset{Cyclopropane}{H_2C-CH_2-CH_2\text{ (ring)}} + ZnBr_2$$

(d) Rearrangement reactions. *Rearrangement reactions involve the migration of an atom or a group from one atom to the other within the same molecule.*

It is interesting to note that the first organic compound, *i.e.*, urea synthesized in the laboratory by Wohler actually involved a rearrangement reaction.

$$NH_4CNO \xrightarrow{\Delta} NH_2-CO-NH_2$$
Amm. cyanate　　　　　Urea

Another important example of such reactions is *Hoffmann bromamide reaction*. This reaction involves the migration of an alkyl groups from the carbon to the nitrogen atom of an amide with the simultaneous elimination of CO as carbonate ion under the influence of Br_2/KOH.

$$CH_3-CONH_2 + Br_2 + 4KOH \longrightarrow CH_3-NH_2 + K_2CO_3 + 2KBr + 2H_2O$$
Acetamide　　　　　　　　　　　　　Methylamine

Similarly maleic acid, when heated in a sealed tube, is converted into fumaric acid.

$$\begin{array}{c} H-C-COOH \\ \| \\ H-C-COOH \end{array} \xrightarrow{\Delta} \begin{array}{c} H-C-COOH \\ \| \\ HOOC-C-H \end{array}$$

Maleic acid → Fumaric acid

2.18 FORMAL CHARGE

A formal charge is the charge assigned to an atom in a molecule, assuming that electrons in all chemical bonds are shared equally between atoms, regardless of electronegativity difference. In determining the best Lewis or resonance structure, the formal charge on each of the atoms should be as close to zero as possible.

Formal charge on an atom in a molecule can be obtained from the following equation

Formal charge $= V - N - \dfrac{B}{2}$

V = No. of valence electrons of the neutral atom in ground state.

N = No. of non-bonding valence electrons on this atom.

B = Total number of electrons shared in bonds with other atoms in the molecule.

Consider the case of CO_2, it is a neutral molecules with 16 total valence electrons.

There are three possible structures of CO_2 as given below:

(i) $:\ddot{O}-\ddot{C}-\ddot{C}:$

(ii) $:\ddot{O}-\dot{C}=\ddot{O}:$

(iii) $:\ddot{O}=C=\ddot{O}:$

How to decide about the actual structure of CO_2. As per the rule that structure should be chosen which gives the formal charge on each of the three atoms as close to zero as possible.

We find by calculation that although sum of the formal charges of the three atoms in all the three structures is zero, it is only structure (iii) where the sum of formal charges of all the atoms as well as formal charge of each atom is zero. Therefore we choose structure (iii) as the correct structure of CO_2. Let us take the case of ozone. Its structure is written as :

$$\ddot{O}=\overset{+}{\overset{..}{O}}-\underset{..}{\overset{..}{O}}:$$

Formal charge on L.H.S. oxygen $= V - N - \dfrac{B}{2} = 6 - 4 - \dfrac{4}{2} = 0$

Formal charge on middle oxygen $= V - N - \dfrac{B}{2} = 6 - 2 - \dfrac{6}{2} = +1$

Formal charge R.H.S. oxygen $= V - N - \dfrac{B}{2} = 6 - 6 - \dfrac{2}{1} = -1$

There is no other possible linear structure for ozone which would give the formal charge on each oxygen equal to zero. A cyclic structure would be highly unstable. Hence we would accept this as the structure for ozone. Thus, the concept of formal charge lets us decide, the most probable stable structure of a molecule.

2.19 TYPES OF REAGENTS—NUCLEOPHILES AND ELECTROPHILES

Nucleophilic reagents or nucleophiles

A nucleophilic reagent is a reagent with an atom having an unshared or lone pair of electrons. Such a reagent is in search of a point where it can share these electrons to form a bond. Nucleophiles are of two types:

(*i*) **Neutral nucleophiles.** These are the nucleophiles which are neutral in charge. But they carry some unshared electrons which they like to share with some positive centre or electron deficient centre. Ammonia $\ddot{N}H_3$, water $H_2\ddot{O}$ and alcohols $R — \ddot{O} — H$ are examples of neutral nucleophiles.

(*ii*) **Negative nucleophiles.** These are the nucleophiles which carry negative charge. Examples of this type of nucleophiles are hydroxyl ions (OH^-), halide ion (X^-), alkoxide ion (RO^-) and cyanide ion (CN^-). Carbanions also come in the category of negative nucleophiles.

Both neutral and negatively charged nucleophiles contain at least one lone pair of electrons which is donated to electron deficient species. Thus they behave as *Lewis bases*. It may be noted that nucleophiles attack the substrate molecule at a site which has the *least electron density*.

Electrophilic reagents or electrophiles

An electrophile is a reagent containing electron deficient atoms. Such species have a tendency to attach themselves to centres of high electron density. There are two types of electrophiles:

(*i*) **Neutral electrophiles.** These electrophile do not carry any net charge. Lewis acids like $AlCl_3$, $FeCl_3$ and BF_3 belong to this category of electrophiles. Sulphonium ion (SO_3) carries no net charge, but it acts as an electrophile for sulphonation in benzene rings. This is because of its structure.

$$\overset{-}{O} — \overset{++}{\underset{\underset{O}{\|}}{S}} — \overset{-}{O}$$

As the positive charge is concentrated and the negative charge is scattered, it acts as an electrophile. Substances like $SnCl_4$ which have vacant *d*-orbitals would like to accommodate electrons in them. Thus such substances also act as electrophiles.

(*ii*) **Positive electrophiles.** These electrophiles carry a net positive charge. Examples of this category of electrophiles are hydrogen ion (H^+), hydronium ion (H_3O^+), nitronium ion (NO_2^+), and chloronium ion (Cl^+). In the halogenation and nitration of aromatic systems, these electrophiles are involved.

Both positive and neutral electrophiles are short of one pair of electrons. Therefore, they have a tendency to seek electrons from other species. Thus, they behave as *Lewis acids*. They would attach to other molecules at the site of *highest electron density*.

Example 3: Classify the following as nucleophiles and electrophiles: H_3O^+, NH_3, $AlCl_3$, ROH, BF_3, CN^-, SO_3.

Solution:

Nucleophiles	Electrophiles
NH_3	H_3O^+
ROH	$AlCl_3$
CN^-	BF_3
	SO_3 (Sulphonium ion)

Example 4: Pick up from the following, the electrophiles and nucleophiles:

$$AlCl_3, PH_3, HNO_3, R_3N$$

Solution: Electrophiles: $AlCl_3$, HNO_3.

Nucleophiles: PH_3, R_3N.

Example 5: Pick up from the following, the electrophiles and nucleophiles:

$$BF_3, NH_3, ROH, SnCl_4$$

Solution: Electrophiles: BF_3, $SnCl_4$.

Nucleophiles: NH_3, ROH.

2.20 NUCLEOPHILICITY AND BASICITY

Nucleophilicity is the ability of a group with a lone pair of electrons to push another such group out of a carbon centre. Basicity is the ability of group to combine with a proton or to denote a lone pair of electrons.

Thus nucleophilicity involves donation of electrons to carbon while basicity involves donation to hydrogen.

Let us consider the nucleophilicity of halogen anions. These anions are solvated by means of hydrogen bonding. Fluorine is the smallest anion and has the largest charge density and therefore is solvated most heavily leading to the largest solution radius followed by chlorine, bromine and iodine.

For nucleophilic attack, this solvent cage has to be broken to enable to reaction to take place. Breaking the cage requires energy and that is shown up as activation energy and therefore iodide is most reactive because it is solvated the least. Then comes bromide, chloride and fluoride.

However, if we move to aprotic solvent like DMS, the situation is different. There is no hydrogen bonding, the anions are 'naked', no cage is to be broken. In such case smaller anions become better nucleophiles.

To sum up, basicity is a subset of nucleophilicity. All nucleophiles are Lewis bases, they donate a lone pair of electrons. A base is just the name we give to nucleophile when it is forming a bond to a proton (H^+). Nucleophile attacks carbon while a base attacks a proton.

2.21 REACTIVE INTERMEDIATES

In organic reactions, reactants do not change into products in one step. The change normally takes place via an intermediate product, which is short-lived. From the intermediate product, the reaction passes on to the products. Various reaction intermediates that we come across in the study of organic reactions are:

(*i*) Carbonium ion (or carbocation) (*ii*) Carbanion
(*iii*) Free radical (*iv*) Carbene
(*v*) Nitrene (*vi*) Arynes (benzyne)

2.22 CARBOCATIONS

These are defined as the species in which the positive charge is carried by the carbon atom with six electrons in its valence shell. These are formed by the heterolytic fission in which an atom or group along with its pair of electrons leaves the carbon. In heterolytic fission, the shared pair of electrons between two atoms goes to one atom only.

$$R-CH_2-X \xrightarrow[\text{Fission}]{\text{Heterolytic}} R\overset{\oplus}{C}H_2 + \overset{\ominus}{\ddot{X}}$$

Stability of Carbocations

1. The order of stability among simple alkyl carbocations is: tertiary > secondary > primary. In most of the reactions, primary and secondary carbocations get rearranged to tertiary carbocations. Both *n*-propyl fluoride and isopropyl fluoride form the same isopropyl cation (2° carbocation). Similarly all the four butyl fluorides *viz.*, *n*-, iso-, sec. and tertiary butyl fluorides form the same tert-butyl cation. There are two factors which determine the stability:

   (*i*) *Hyperconjugation or resonance.*
   (*ii*) *Field effect or inductive effect of groups.*

Hyperconjugation

A large number of canonical forms can be written for tertiary carbocation compared to those for primary carbocation. Consider hyperconjugation in primary and tert-carbocation.

$$R-\underset{H}{\overset{H}{\underset{|}{C}}}-\overset{H}{\underset{H}{\overset{|}{C}}}{}^{\oplus} \longleftrightarrow H^{\oplus}\ R-\underset{H}{\overset{|}{C}}=\underset{H}{\overset{|}{C}} \longleftrightarrow R-\underset{H}{\overset{H}{\underset{|}{C}}}=\underset{\overset{\oplus}{H}}{\overset{H}{\underset{|}{C}}}$$

Hyperconjugation in primary carbocation

$$R-\underset{H}{\overset{H}{\underset{|}{C}}}-\overset{CH_2R}{\underset{CH_2R}{\overset{|}{C}}}{}^{\oplus} \longleftrightarrow H^{\oplus}\ R-\underset{H}{\overset{CH_2-R}{\underset{|}{C}}}=\underset{CH_2-R}{\overset{|}{C}} \longleftrightarrow H^{\oplus}\ R-\underset{H}{\overset{C-R}{\underset{|}{CH}}}-\overset{\|}{\underset{CH_2-R}{C}} \longleftrightarrow \text{etc.}$$

Hyperconjugation in tertiary cation

Greater the number of canonical forms, greater is the stability.

Field effect

Electron donating effect of alkyl groups increases the electron density around the +vely charged carbon. This results in reducing the magnitude of positive charge on it and thus the charge is delocalised on α-carbon. Dispersal of positive charge increases the stability. Of carbocation of all the simple cation, *tert*-butyl cation being most stable. Even *tert*-pentyl and *tert*-hexyl cations produce *tert*-butyl cation at high temperature. Lower alkanes like methane, ethane and propane when treated with **super acid*** also yield *tert*-butyl cation as the main product. Salts of *tert*-butyl cation, like $(CH_3)_3 C^+ SbF_3^-$ have been prepared from super acid solutions.

---

*A mixture of SbF_5 + HF or SbF_5 + FSO_3H dissolved in liquid SO_2 is one of the strongest acid solutions and are commonly called super acids.

Basics of Organic Chemistry II 69

2. The stability of carbocation containing a conjugated double bond is usually greater due to increased delocalisation by resonance. In such carbocations, the positive charge is dispersed on at least two carbon atoms. Consider the following carbocation.

3. Allylic type cabocations have been prepared from the solution of conjugated diene in conc. sulphuric acid and are found to be most stable.

4. Carbocations can be stabilised through aromatisation also. 1-Bromocyclohepta-2, 4, 6-triene tropylium bromide is a crystalline solid. It is highly soluble in water and forms bromide ions in solution. Clearly, the compound is not covalent in nature. The reason for such a behaviour is the stability of tropylium cation which follows Huckel's rule (6 π electrons) for aromaticitiy.

1, Bromo-2, 4, 6-Triene tropylium bromide

Tropylium cation

Its stability is also explained in terms of canonical structures which stabilise the tropylium cation.

5. Substituted cyclopropenyl cation possesses even more aromatic stabilisation ($n = 0$).

In the above tripropyl substituted cation, all the carbons are sp^2 hybridised. Thus all the p-orbitals including the vacant p-orbital of the positively charged carbon overlap forming a delocalised π molecular orbital, leading to stabilization.

6. Benzyl cation can be written in the following canonical forms:

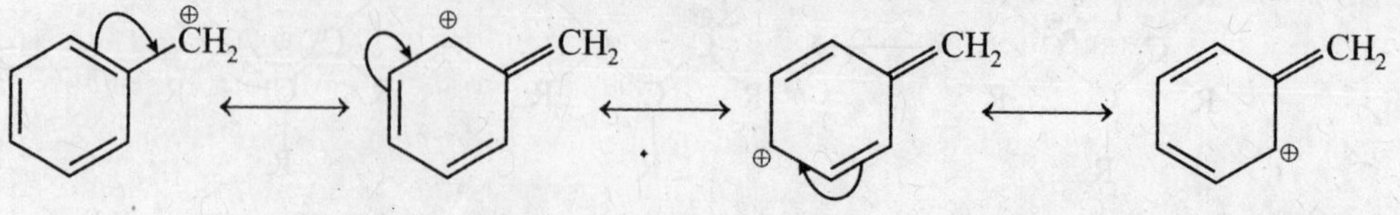

Due to larger number of canonical forms and greater dispersal of positive change, benzyl carbocation is still more stable.

7. Triphenyl chloromethane ionises in SO_2 as

$$(C_6H_5)_3 C \cdot Cl \rightleftharpoons (C_6H_5)_3 C^\oplus + Cl^-$$

Triphenyl methyl cation has been isolated as solid salt as $(C_6H_5)_3 C^+ \cdot BF_4^-$. The stability of triarylmethyl cation is further increased if the rings are substituted at *ortho* and *para* positions by electron donating groups.

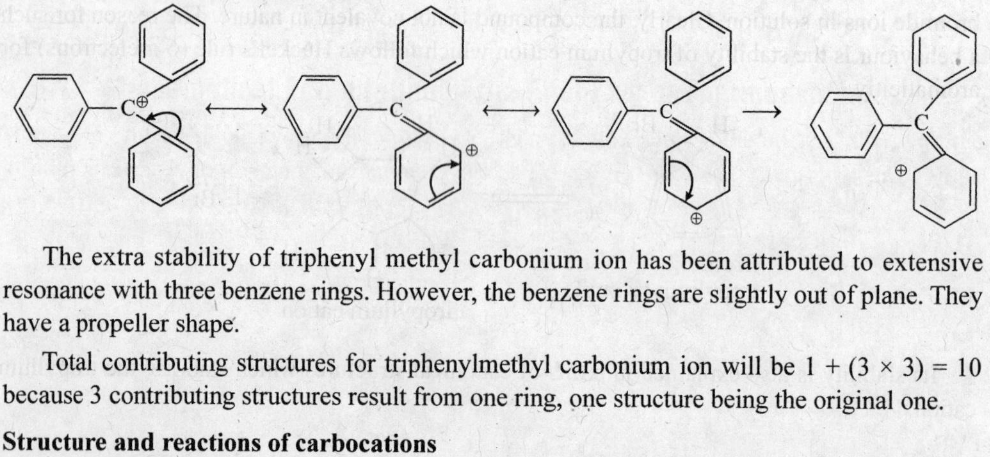

The extra stability of triphenyl methyl carbonium ion has been attributed to extensive resonance with three benzene rings. However, the benzene rings are slightly out of plane. They have a propeller shape.

Total contributing structures for triphenylmethyl carbonium ion will be 1 + (3 × 3) = 10 because 3 contributing structures result from one ring, one structure being the original one.

Structure and reactions of carbocations

Structure. The carbon atom bearing the positive charge in a carbocation is sp^2 hybridised. The three sp^2 hybrid orbitals are utilised in the formation of sigma bonds with three atoms or groups. The third unused p-orbital remains vacant. Thus, the carbocation is a flat species having all the three bonds in one plane with a bond angle of 120° between them as shown below:

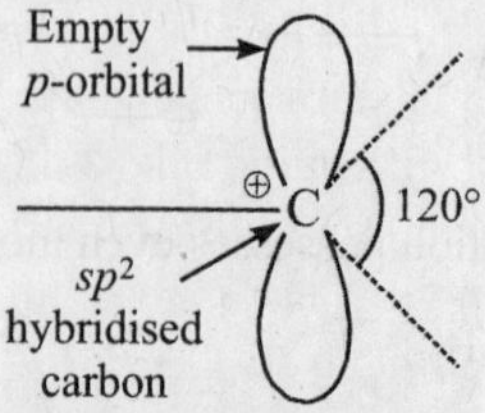

Reactions

Carbocations, which are shortlived species and very reactive give the following reactions:

1. Proton loss. Carbocation may lose a proton to form an alkene. An ethyl carbocation loses a proton to form ethene.

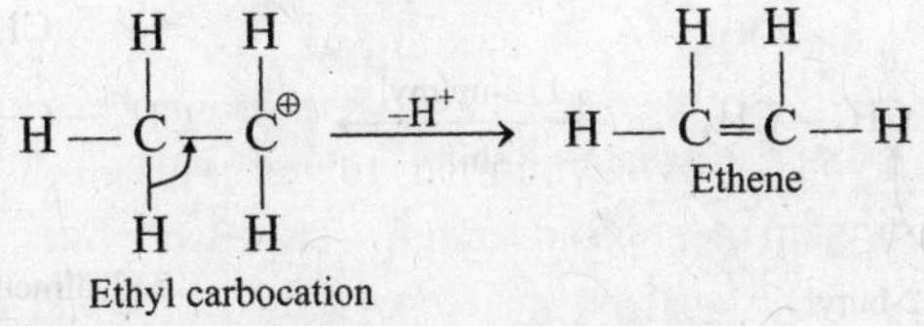

$$H-\underset{\underset{H}{|}}{\overset{\overset{H}{|}}{C}}-\overset{\overset{H}{|}}{\underset{\underset{H}{|}}{C}}^{\oplus} \xrightarrow{-H^+} H-\underset{\underset{H}{|}}{\overset{\overset{H}{|}}{C}}=\underset{|}{\overset{|}{C}}-H$$
Ethyl carbocation   Ethene

2. Combination with nucleophiles. Carbocations combine with nucleophiles to acquire a pair of electrons. For example, a highly reactive methyl carbocation with hydroxyl ion to form methyl alcohol.

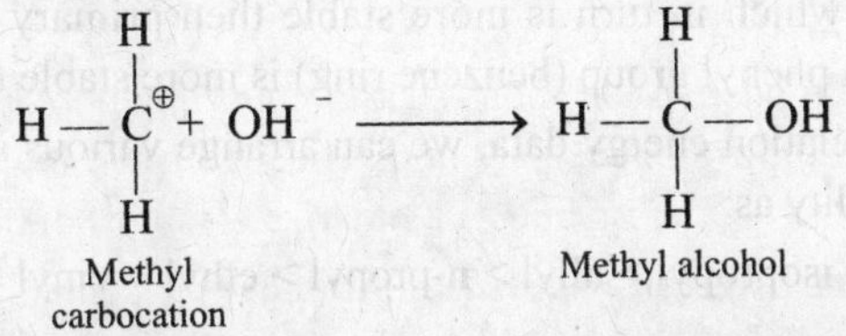

$$H-\overset{\overset{H}{|}}{\underset{\underset{H}{|}}{C}}^{\oplus} + OH^- \longrightarrow H-\overset{\overset{H}{|}}{\underset{\underset{H}{|}}{C}}-OH$$
Methyl carbocation   Methyl alcohol

3. Addition to alkene. A carbocation may add to an alkene to produce another carbocation possessing higher molecular weight.

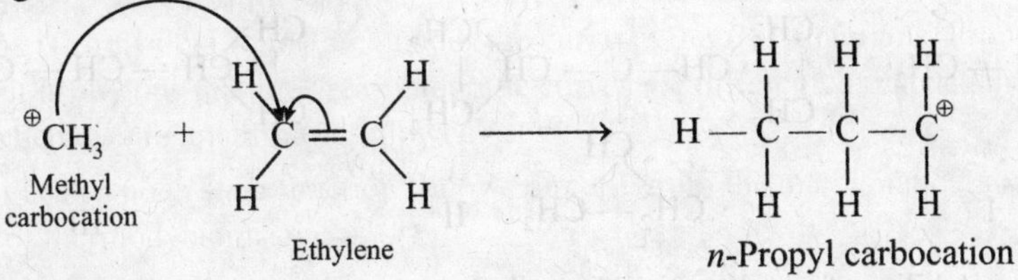

$$^{\oplus}CH_3 + \underset{\underset{H}{|}}{\overset{\overset{H}{|}}{C}}=\underset{\underset{H}{|}}{\overset{\overset{H}{|}}{C}} \longrightarrow H-\overset{\overset{H}{|}}{\underset{\underset{H}{|}}{C}}-\overset{\overset{H}{|}}{\underset{\underset{H}{|}}{C}}-\overset{\overset{H}{|}}{\underset{\underset{H}{|}}{C}}^{\oplus}$$
Methyl carbocation   Ethylene   n-Propyl carbocation

4. Rearrangement. The migration of alkyl or aryl or hydrogen along with an electron pair to the positive centre takes place which results in the formation of more stable carbocation. Some examples are given below:

1, 2-Hydride shift

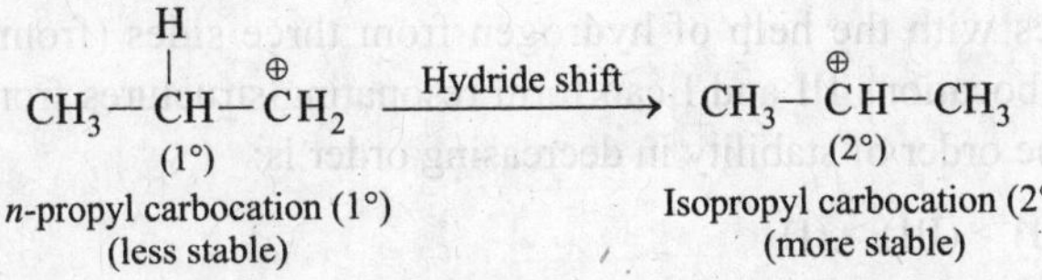

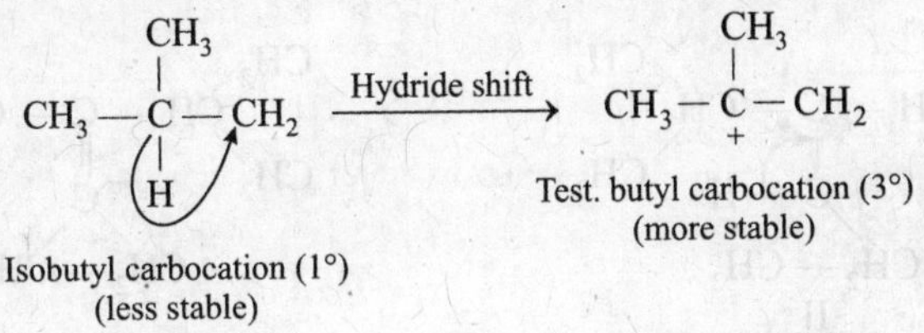

$$CH_3-\overset{\overset{H}{|}}{\underset{|}{C}H}-\overset{\oplus}{C}H_2 \xrightarrow{\text{Hydride shift}} CH_3-\overset{\oplus}{C}H-CH_3$$
(1°)   (2°)
n-propyl carbocation (1°)   Isopropyl carbocation (2°)
(less stable)   (more stable)

$$CH_3-\underset{\underset{H}{|}}{\overset{\overset{CH_3}{|}}{C}}-\overset{+}{C}H_2 \xrightarrow{\text{Hydride shift}} CH_3-\underset{+}{\overset{\overset{CH_3}{|}}{C}}-CH_2$$
Isobutyl carbocation (1°)   Test. butyl carbocation (3°)
(less stable)   (more stable)

1, 2-Methyl shift

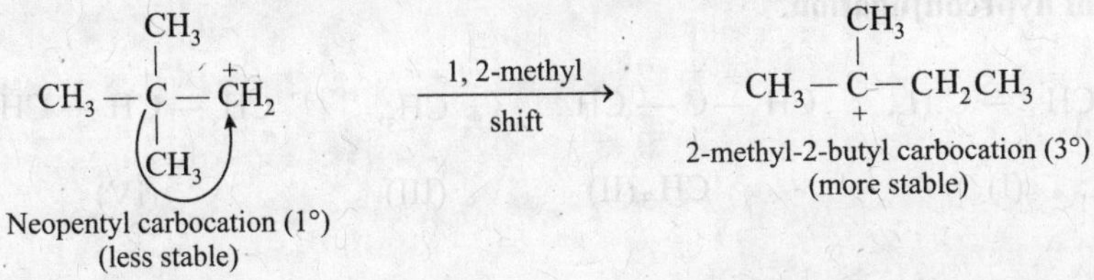

$$CH_3-\underset{\underset{CH_3}{|}}{\overset{\overset{CH_3}{|}}{C}}-\overset{+}{C}H_2 \xrightarrow{\text{1, 2-methyl shift}} CH_3-\underset{+}{\overset{\overset{CH_3}{|}}{C}}-CH_2CH_3$$
Neopentyl carbocation (1°)   2-methyl-2-butyl carbocation (3°)
(less stable)   (more stable)

$$CH_3-\underset{\underset{CH_3}{|}}{\overset{\overset{CH_3}{|}}{C}}-\overset{+}{C}H-CH_3 \xrightarrow{\text{1, 2-methyl shift}} CH_3-\underset{|}{\overset{\overset{CH_3}{|}}{C}}\overset{+}{-}CH-CH_3$$

3; 3-dimethyl-2-butyl   2, 3-dimethyl-2-butyl
carbocation (2°)   carbocation (3°)
(less stable)   (more stable)

Such a rearrangement takes place because a tertiary carbocation is more stable than secondary carbocation, which in turn is more stable then primary cabocation. Similarly, a carbocation attached to a phenyl group (benzene ring) is more stable than others.

From the bond dissociation energy data, we can arrange various alkyl/aryl carbocations in order of decreasing stability as:

tert-Butyl > benzyl > isopropyl > allyl > n-propyl > ethyl > vinyl > methyl > phenyl

Example 6: Arrange the following in order of decreasing stability and give suitable explanation for it.

[Structures I, II, III showing cyclopropyl-substituted carbocations]

I   II   III

Solution: The stability of a carbocation depends upon two factors:

(i) Hyperconjugation

(ii) Inductive effect of groups attached.

In this problem, it is hyperconjugation or resonance which is important. Structure II can form maximum number of resonating structures and hence this is most stable. It can form resonating structures with the help of hydrogen from three sides (from three cyclopropyl groups) whereas carbocations III and I can form resonating structures from two and one side respectively. Thus the order of stability in decreasing order is:

(II) > (III) > (I)

[Structure showing II with –H⁺ arrow leading to product structure]

Example 7: Arrange the following in order of increasing stability and explain the order on the basis of hyprconjugation.

$$CH_3-\overset{+}{C}H_2, \quad CH_3-\underset{\underset{CH_3}{|}}{\overset{+}{C}}-CH_3, \quad \overset{+}{C}H_3, \quad CH_3-\overset{+}{C}H-CH_3$$

(I)   (II)   (III)   (IV)

Solution: Greater the number of resonating structures, greater is the stability attached to a carbocation. Carbocation (II) which is a tertiary carbocation will give the maximum number of resonating structures involving hydride shift, like

$$\overset{+}{C}H_2 - CH - CH_3 \quad\quad CH_3 - \overset{+}{C}H - CH_2 \quad\quad CH_3 - CH - CH_3$$
$$|||$$
$$CH_3 CH_3 {}^+CH_2$$

Greater the number of hydrogens on carbon atoms in the immediate neighbourhood of carbocation, greater the number of resonating structures and hence greater the stability.

This is followed by secondary carbocation IV, primary carbocation I and methyl carbocation (III). Hence the order of stability in increasing order is:

$$\overset{+}{C}H_3 < \overset{+}{C}H_3 - CH_2 < CH_3 - \overset{+}{C}H - CH_3 < CH_3 - \overset{+}{C} - CH_3$$
$$|$$
$$CH_3$$

Example 8: Order of stability of carbocation is as under. Explain.

$$(C_6H_5)_3 \overset{+}{C} > (C_6H_5)_2 \overset{+}{C}H > C_6H_5 \overset{+}{C}H_2$$
$$(I)(II)(III)$$

Solution: Carbocation (I) gives the maximum number of resonating structures, as this is linked to three benzene rings, hence this will have maximum stability.

For details, see Art. 2.8.

The order of stability in decreasing order is:

$$I > II > III$$

Example 9: Why tropyllium bromide gives precipitate with $AgNO_3$ and CH_3Br does not?

Solution: This is because tropyllium bromide behaves like an ionic compound whereas methyl bromide does not. Ionic nature of tropyllium ion is because of extra stability of tropyllium ion. For details refer to Art. 2.9.

2.23 CARBANIONS

These are defined as the *species with an unshared pair of electrons and a negative charge on the central carbon atom*. A carbanion may be formed in one of the following ways:

(*i*) An atom or group leaves carbon without the electron pair (Heterolytic fission).

$$R - \underset{\underset{H}{|}}{\overset{\overset{H}{|}}{C}} - Y \xrightarrow{\text{Heterolytic fission}} R - \underset{\underset{H}{|}}{\overset{\overset{H}{|}}{C}}{}^{\ominus} : + Y^+$$
$$\text{Carbanion}$$

(*ii*) An anion adds to a carbon-carbon double or triple bond

$$-\underset{|}{\overset{|}{C}} = \underset{|}{\overset{|}{C}} - + :Y^\ominus \longrightarrow -\underset{|}{\overset{\ominus}{\overset{|}{C}}} - \underset{|}{\overset{|}{C}} - Y$$
$$\text{Carbanion}$$

(*iii*) A strong base abstracts a proton from a carbonyl or cyano compound

$$C_2H_5O^- \;H\!-\!CH_2\!-\!\overset{\displaystyle O}{\overset{\|}{C}}\!-\!H \longrightarrow \left[CH_2\!-\!\overset{\displaystyle O}{\overset{\|}{C}}\!-\!H \;\longleftrightarrow\; \underset{\displaystyle H}{\underset{|}{CH_2\!=\!C}}\!-\!\overset{\displaystyle O^-}{} \right] + C_2H_5OH$$

Resonance stabilized

(*iv*) A strong base abstracts hydrogen of terminal acetylenes

$$H_2N^- + H\!-\!C\!\equiv\!C\!-\!H \longrightarrow \;:C\!\equiv\!CH + NH_3$$

Every carbanion possesses unshared pair of electrons and is therefore a base. When the carbanion accepts a proton, it gives a conjugate acid. The stability of the carbanion is directly related with the strength of the conjugated acid.

$$\underset{\text{(Base)}}{\text{Carbanion}} + H^+ \longrightarrow \text{Conjugate acid}$$

Stability of carbanions

1. It is important to note that weaker the acid, stronger is the base and hence lower is the stability of the carbanion. Clearly, the order of stability of carbanion can be determined from the order of the strength of conjugate acids. Carbanions are highly unstable in solution as compared to carbocations.

2. The order of stability of carbanions is:

Vinyl > Phenyl > Cyclopropyl > Ethyl > *n*-propyl > Isobutyl

Also it has been found that the stability of carbanions decreases in the order methyl > prim-carbanion > sec-carbanion.

$$\underset{\displaystyle H}{\underset{|}{\overset{\displaystyle H}{\overset{|}{H\!-\!C\!-\!}}}} > CH_3 \longrightarrow CH_2 > \overset{\displaystyle CH_3}{\overset{|}{\underset{\displaystyle CH_3}{\underset{|}{CH}}}}$$

This stability order can be explained simply by the field effects. The presence of electron donating alkyl groups in secondary (Isopropyl) carbanion results in greater localisation of negative charge on the central carbon atom and hence the stability falls. Cyclopropyl carbanion has greater stability due to greater *s*-character on the carbanionic carbon.

3. Vinyl and phenyl carbanions are more stable due to resonance. In cases where a double or triple bond is located at α-position to the carbanionic carbon, the ion is stabilised by resonance in which the unshared pair overlaps with π-electrons of the double bond. The stability of allylic and benzylic carbanions can be illustrated as under.

$$R\!-\!CH\!=\!CH\!-\!\overset{\ominus}{C}H_2 \;\longleftrightarrow\; R\!-\!\overset{\ominus}{C}H\!-\!CH\!=\!CH_2$$

4. Diphenyl methyl and triphenyl methyl carbanions are more stable then even benzylic and can be kept in solution for a longer time if water is completely excluded.

5. When the carbanionic carbon is conjugated with a carbon-oxygen or carbon-nitrogen multiple bond, then the stability of carbanion is increased. The reason is that the presence of electronegative atoms helps in the dispersal of negative charge and thus such carbanions are better capable of bearing the negative charge. Thus,

$$R-\overset{\ominus}{C}H-\overset{\parallel}{\underset{O}{C}}-R \longleftrightarrow R-CH=\underset{:\overset{..}{O}:^{\ominus}}{C}-R$$

6. The increase in the *s*-character at the central carbon increases the stability of carbon. Thus, the order of stability is expressed as:

$$R-C\equiv\overset{\ominus}{\overset{..}{C}} > R_2C=\overset{\ominus}{\overset{..}{C}}H > \overset{\ominus}{Ar} > R_3C-\overset{\ominus}{C}H_2$$

Acetylenic carbon contains 50% *s*-character being *sp*-hybridised.

7. **Stabilisation by Sulphur or Phosphorus:** The bonding of sulphur or phosphorus atom with carbanionic carbon increases the stability. It is probably due to the overlap of unshared pair with an empty *d*-orbital. Consider a carbanion containing sulphur atom:

$$R-\underset{\underset{O}{\parallel}}{\overset{\overset{O}{\parallel}}{S}}-\overset{R}{\underset{R}{\overset{|}{C^{\ominus}}:}} \longleftrightarrow R-\underset{\underset{:O:}{\parallel}}{\overset{\overset{O}{\parallel}}{S}}-\overset{R}{\underset{R}{\overset{|}{C}}} \longleftrightarrow \text{etc.}$$

8. **Aromatic nature.** Some carbanions are stable as they are aromatic. *e.g.*, cyclopentadienyl anion is a stable carbanion.

Huckel's $(4n + 2)$ rule is satisfied here with 6π electrons in the ring.

Structure and reactions of carbanions

Structure. The carbanions, being unstable have not been isolated. Therefore, their structure is not known with certainty. Most likely, the central carbon atom with an unshared pair of electrons (and negative charge) is sp^3 hybridised. The three sp^3 hybrid orbitals are used in forming three sigma bonds with the other atoms. The fourth sp^3-orbital is occupied by a lone pair of electron. In fact, the structure of carbanion is quite similar to that of ammonia as shown in the figure.

Carbanions which get stabilized by resonance involving the lone pair of electrons and the π electrons of multiple bonds, should be planar as this is the necessary condition for resonance to take place.

Carbanions of $R_1R_2R_3C^-$ are optically inactive. Why?

If the sp^3 orbital containing the lone pair of electrons is considered as fourth substituent, then carbanions of the type $R_1 R_2 R_3 C^-$ should be optically active as the carbon is linked to four different groups (chiral) and it should be possible to obtain two optical forms. But in actual practice, these forms have not been obtained. We explain it by saying that the pyramid structure of the carbonion is not rigid and it undergoes rapid inversion as shown below:

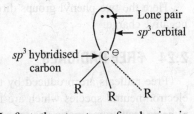

Consequently, one optical form (say structure I) gets converted to structure II and a racemic mixture is obtained which is not easily resolvable.

Reactions

Some important reactions of carbanions are described below:

1. Reaction with positive species or electrophiles. A carbanion reacts with a proton or with another action.

$$R-\underset{R}{\overset{R}{C}}{:}^{\ominus} + Y^{\oplus} \longrightarrow R-\underset{R}{\overset{R}{C}}-Y$$

2. Reactions involving the displacement of an atom or a group. Such reactions are nucleophilic substitution (SN^2) reactions observed in alkyl halides.

$$\overset{\ominus}{R} + \overset{|}{\underset{|}{C}}-X \longrightarrow R-\overset{|}{\underset{|}{C}} + \overset{\ominus}{\underset{..}{\overset{..}{X}}}$$

3. Addition to carbonyl compounds. Carbanions attack an electron deficient carbon in the carbonyl compounds. Such reactions are called nucleophilic addition reactions.

$$\overset{\ominus}{R} + \underset{|}{\overset{|}{C}}=O \longrightarrow -\underset{|}{\overset{R}{\underset{|}{C}}}-\overset{..}{\underset{..}{O}}{:}^{\ominus}$$

4. Rearrangement. In some cases, carbanions may rearrange to form more stable species. Consider the rearrangement in triphenylmethyl carbanion.

$$(C_6H_5)_3\,C-\overset{\ominus}{\underset{..}{C}}H_2 \longrightarrow (C_6H_5)_2\,\overset{\ominus}{\underset{..}{C}}-CH_2-C_6H_5$$

Here the two phenyl groups directly attached to the central carbon help in dispersing the negative charge.

2.24 FREE RADICALS

Free radicals are produced by the homolytic fission of a covalent bond. These are odd electron neutral species which are formed by the homolytic fission of a covalent bond. Free radicals are paramagnetic due to the presence of unpaired electron. Formation of free radicals is favoured by the presence of UV light, heat and organic peroxides. Reactions involving radicals widely occur in the gas phase. Such reactions also occur in solutions, particularly if carried in non-polar solvents. An important characteristics of free radical reactions is that, once initiated, they proceed very fast. The free radicals can be detected by magnetic susceptibility measurements.

$$R-R \xrightarrow{\text{Homolytic Fission}} 2R\bullet \text{ (Free radical)}$$

$$Cl-Cl \xrightarrow{h\nu} 2Cl\bullet \text{ (Free radical)}$$

$$\underset{\text{Acylperoxide}}{RCO-O-OCOR} \xrightarrow{\text{Thermal}} 2RCOO\bullet \text{ (Free radical)}$$

$$(CH_3)_4Pb \xrightarrow{\Delta} Pb + 4\overset{\bullet}{C}H_3$$
Tetramethyl lead → Methyl free radical

Stability of free radicals

1. Simplest alkyl free radicals are highly reactive like carbocations and carbanions. Their life time is extremely short in solution. The relative stability of simple alkyl radicals has the order:

$$R_3\overset{\bullet}{C} > R_2\overset{\bullet}{C}H > R\overset{\bullet}{C}H_2 > \overset{\bullet}{C}H_3$$
$$\quad 3° \quad\quad 2° \quad\quad 1° \quad\quad Methyl$$

It can be explained on the basic of hyperconjugation similar to that in carbocations.

$$R-\underset{H}{\overset{H}{C}}-\underset{H}{\overset{H}{C}}\bullet \longleftrightarrow R-\underset{H}{\overset{H^\bullet}{C}}=\underset{H}{\overset{H}{C}} \longleftrightarrow R-\underset{H}{\overset{H}{C}}=\underset{H\bullet}{\overset{H}{C}}-H$$

Greater the number of the resonating structures, greater is the stability of the free radical.

2. Allylic and benzylic type of free radicals are more stable and comparatively less reactive than simple alkyl radicals. The reason is the delocalisation of the unpaired electron over the π-orbital system.

$$R-CH=CH-\overset{\bullet}{C}H_2 \longleftrightarrow R-\overset{\bullet}{C}H-CH=CH_2 \equiv R-CH\cdots CH\cdots CH_2$$
Allylic radical

Benzylic radical

3. Triphenyl methyl and triarylmethyl radicals are much more stable in solution at room temperature. The stability of such radicals is due to resonance.

$$(C_6H_5)_3\overset{\bullet}{C} \longleftrightarrow \text{[ring]}=C(C_6H_5)_2 \longleftrightarrow \text{[ring]}=C(C_6H_5)_2 \longleftrightarrow etc.$$

Steric hindrance to dimerisation is probably the major cause of their stability. If each aromatic nucleus in the radical has a bulky *p*-substitutent, then irrespective of any substitution at the *o*-positions, dimerisation will be greatly inhibited and hence radical stability increases.

Structure and reactions of free radicals

Structure. The state of hybridisation of carbon atom having the unpaired electron is not clearly established. However, it is believed to be either pyramidal (like ammonia)

Structure of a free radical

or planar. But experimental evidence suggests strongly that the alkyl radicals such as methyl radical are actually planar with sp^2 hybridisation. The three coplanar hybrid orbitals are used in the formation of three sigma bonds with other atoms. The unhybridised orbital which lies in a plane at right angles to the plane of the hybrid orbitals, carries the unpaired electron. ESR spectra of $\overset{\bullet}{C}H_3$ and other simple alkyl radicals indicate that these radicals have planar structures. This is also in accordance with the fact that optical activity is lost when a free radical is generated at an asymmetric carbon. As a general rule, we can say that simple alkyl free radicals prefer a planar or near planar shape. However, the free radicals in which the carbon is connected to atoms of high electronegativity prefer a pyramidal shape. The increase in electronegativity causes the deviation from the planar geometry.

Reactions

Some important reactions of free radicals are described below:

1. Halogenation of aliphatic hydrocarbons

$$Cl-Cl \xrightarrow{h\nu} 2Cl\bullet$$
$$CH_4 + Cl\bullet \longrightarrow CH_3\bullet + HCl$$
$$CH_3 + Cl_2 \longrightarrow CH_3Cl + Cl$$

In the presence of sunlight, chlorination of methane gives chloromethane, dichloromethane, chloroform and carbon tetrachloride.

2. Addition. Halogen addition to alkenes takes place with free radical mechanism. The addition of chlorine to tetrachloroethene is photochemically catalysed. A molecule of chlorine undergoes homolytic fission giving two chlorine radicals. Each radical is capable of initiating a reaction chain.

$$Cl-Cl \underset{}{\overset{h\nu}{\rightleftharpoons}} C\bullet-Cl\bullet$$

$$Cl_2C=C.Cl_2 + Cl\bullet \longrightarrow Cl_2\overset{\bullet}{C}-CCl_3 \xrightarrow{Cl-Cl} \overset{\bullet}{C}l + Cl_3C-CCl_3$$

$\overset{\bullet}{C}l$ radical further adds to tetrachloroethene. This continues till the whole of tetrachloroethene has been converted into hexachloroethane. Chain termination occurs through radical-radical collision. The radical reactions are inhibited by the presence of oxygen. The reason is that the molecule of oxygen has two unpaired electrons and behaves as a biradical. This biradical combines with highly reactive radical intermediate and converts it to less reactive peroxy radical which is unable to propagate the chain reaction.

3. Consider the addition of chlorine to benzne in the presence of light. The reaction proceeds by radical intermediates and gives benzene hexachloride. But the attack of chlorine radical on toluene results in preferential hydrogen abstraction giving substitution in CH_2 group. Clearly, it is because of the greater stability of benzyl radical due to delocalisation.

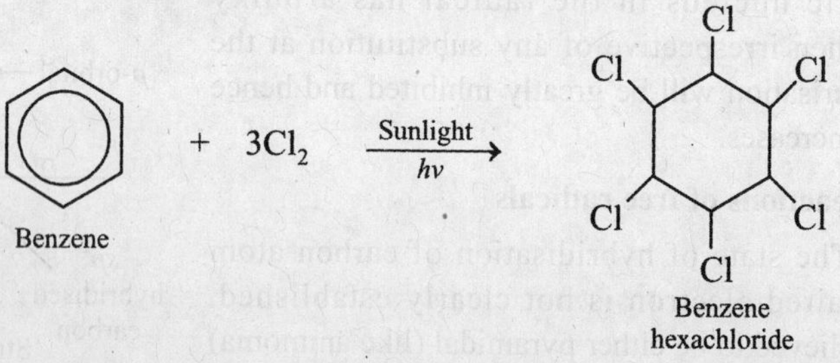

Benzene + 3Cl$_2$ $\xrightarrow{\text{Sunlight}\atop h\nu}$ Benzene hexachloride

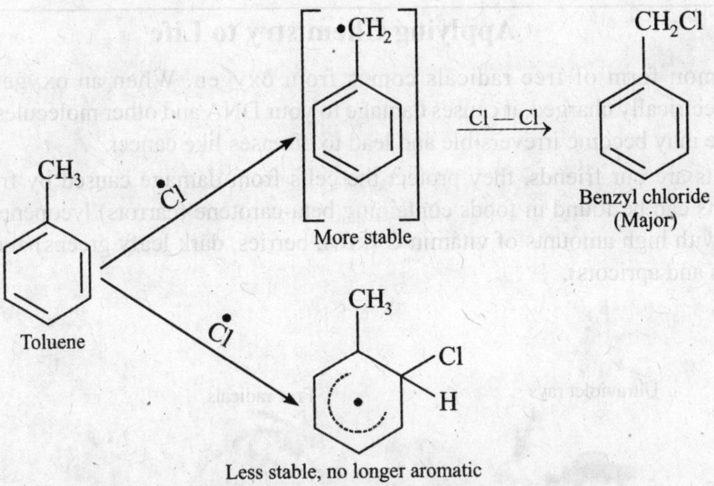

Basics of Organic Chemistry II

[Diagram: Toluene + Cl• → benzyl radical (More stable) with resonance on ring; reacts with Cl—Cl to give Benzyl chloride (Major, CH_2Cl on ring). Alternative pathway: Cl• adds to ring giving CH_3-substituted cyclohexadienyl radical with Cl and H — Less stable, no longer aromatic.]

4. The addition of HBr to propene in presence of peroxides yields 1-Bromopropane. The reaction involves the formation of radical intermediates giving anti-Markownikoff addition. In this reaction, $\overset{\bullet}{Br}$ radical attacks propene to form a more stable secondary free radical.

$$ROOR \longrightarrow 2R\overset{\bullet}{O}$$

$$R\overset{\bullet}{O} + H-Br \longrightarrow RO-H + \overset{\bullet}{Br}$$

$$CH_3CH=CH_2 + \overset{\bullet}{Br} \longrightarrow CH_3\overset{\bullet}{C}H-CH_2BR$$

$$\downarrow HBr$$

$$\overset{\bullet}{Br} + CH_3-CH_2-CH_2BR$$
$$\text{1-Bromopropane}$$

5. Vinyl polymerisation. Radical reactions also produce polymers of great importance. Like other radical reactions, polymerisation reactions also constitute three step viz.; initiation, propagation and termination. In initiation step, a free radical is formed under the influence of peroxide. The radical formed then propagates the chain reaction.

$$\overset{\bullet}{R} + CH_2=CH \longrightarrow RCH_2-\overset{\bullet}{CH_2} \xrightarrow{CH_2=CH_2} RCH_2-CH_2-CH_2-\overset{\bullet}{CH_2}$$

and so on.

In the final step, the collisions between the radicals terminate the reaction

$$R(CH_2)_n^{\bullet} + \overset{\bullet}{R} \longrightarrow R(CH_2)_n R$$

$$R(CH_2^{\bullet})_n + R(\overset{\bullet}{CH_2})_n \longrightarrow R(CH_2)_n R$$

Polythene, PVC, teflon etc. are the polymers which are formed from the respective monomers by radical pathways.

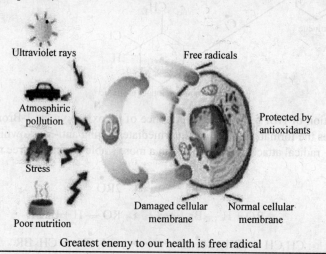

Applying Chemistry to Life

Most common form of free radicals comes from oxygen. When an oxygen molecule becomes electrically charged, it causes damage to your DNA and other molecules. Over time this damage may become irreversible and lead to diseases like cancer.

Antioxidants are our friends, they protect the cells from damage caused by free radicals. Antioxidants can be found in foods containing beta-carotene (carrots) lycopene (tomatoes) and foods with high amounts of vitamin C (kiwi, berries, dark leafy greens) and vitamin E (seeds, nuts and apricots).

Greatest enemy to our health is free radical

Example 10: Arrange the following free-radicals in increasing order of stability and explain your answer.

$$\overset{\bullet}{C_2H_5}, \quad (C_2H_5)_2 \overset{\bullet}{C}(C_2H_5), \quad (C_2H_5)_2 \overset{\bullet}{CH}, \quad \overset{\bullet}{CH_3}$$

Solution: The order of stability can be given considering the hyperconjugation phenomenon.

(i) Ethyl radical is a hybrid of four resonating structures.

$$\underset{I}{H-\underset{\underset{H}{|}}{\overset{\overset{H}{|}}{C}}-\overset{\bullet}{\underset{\underset{H}{|}}{\overset{\overset{H}{|}}{C}}}-H} \longleftrightarrow \underset{II}{H-\underset{\underset{H}{|}}{\overset{\overset{H}{|}}{C}}=\underset{\underset{H}{|}}{C}-H} \longleftrightarrow \underset{III}{\overset{\bullet}{H}\underset{\underset{H}{|}}{C}=\underset{\underset{H}{|}}{C}-H} \longleftrightarrow \underset{IV}{H-\underset{\underset{\overset{\bullet}{H}}{|}}{\overset{\overset{H}{|}}{C}}=\underset{\underset{H}{|}}{C}-H}$$

2.25 CARBENES

These are defined as the neutral organic species containing a divalent carbon atom having a set of electrons but no charge on it. For example,

(a) (i) : CH_2
     Methyl carbene

(ii) : CCl_2
    Dichlorocarbene

Formation of carbenes

They can be generated by the following methods:

Basics of Organic Chemistry II 81

(b) (i) By the action of UV light on diazomethane

$$CH_2N_2 \longrightarrow :CH_2 + N_2$$
$$\text{Methylene}$$
$$\text{(Carbene)}$$

(ii) By the action of sodium ethoxide on chloroform

$$\underset{\text{Chloroform}}{CHCl_3} + C_2H_5ONa \longrightarrow \underset{\substack{\text{Dichloro}\\\text{carbene}}}{:CCl_2} + C_2H_5OH + NaCl$$

(iii) By thermal or photochemical decomposition of cyclopropanes and oxiranes

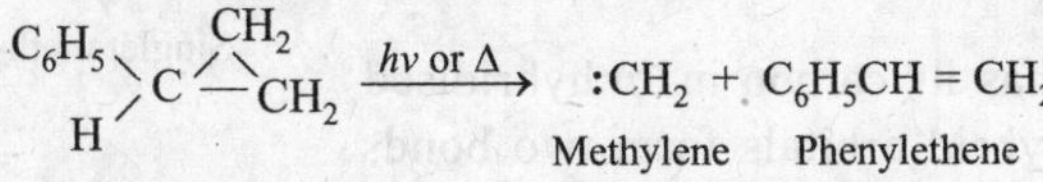

$$\underset{H}{\overset{C_6H_5}{>}}C\underset{CH_2}{\overset{CH_2}{\triangle}} \xrightarrow{h\nu \text{ or } \Delta} \underset{\text{Methylene}}{:CH_2} + \underset{\text{Phenylethene}}{C_6H_5CH=CH_2}$$

These are highly reactive species.

Stability

1. The two unbonded electrons of a carbene may be either paired or unpaired. When the two electrons on the carbon atom are paired, then it is called *singlet* carbene. In the *triplet* state, it is called a biradical as the two electrons on carbon atom are unpaired.

$$\underset{\text{Singlet methylene}}{H—\ddot{C}—H \quad \text{or} \quad \uparrow\downarrow CH_2} \qquad \underset{\text{Triplet methylene}}{H—\dot{C}—H \quad \text{or} \quad \uparrow\uparrow CH_2}$$

In the triplet state, carbene is relatively more stable as it has lower energy content.

2. The singlet and the triplet species can be distinguished by the common addition reaction of carbene to double bond to form cyclopropane derivatives. The addition of singlet species to *cis*-Butene forms a *cis* product. The reason is that the movement of two pairs of electrons occur either simultaneously or with one rapidly succeeding the other.

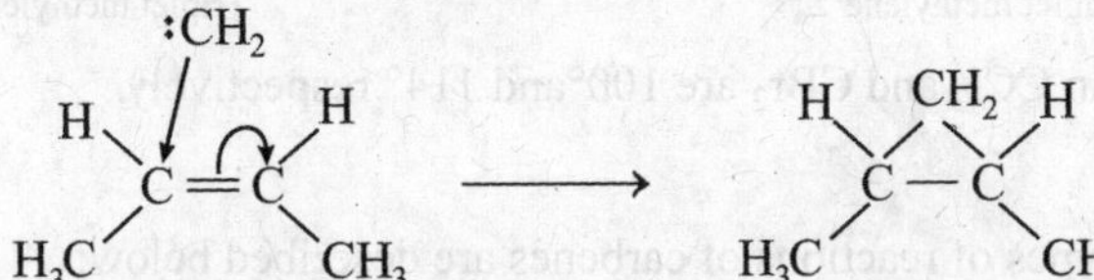

3. In case of triplet carbene, the two unpaired electrons cannot form a new covalent bond simultaneously as they have parallel spins. Thus, one of the unpaired electrons will form a bond with the electron from the double bond that has opposite spin. Now, two unpaired electrons with the same spin are left and thus no bond is formed. The bond formation is possible by collision process when one of the electrons can reverse its spin. This is illustrated in the diagram below:

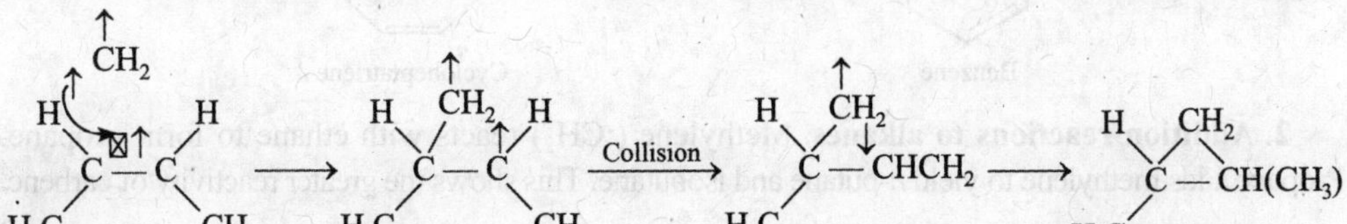

During this time of collision, there is a free rotation around C—C single bond and the product formed is a mixture of *cis* and *trans* isomers. It has been found that the difference in energy between the singlet and the triplet methylene is about 37–46 kJ/mole.

Structure and reactions of carbenes
Structure

The structure of two types of carbenes are different.: Singlet carbenes have their carbon in sp^2 hybridised state. Of the three sp^2 hybrid orbitals, two are used in forming two single bonds with monovalent atoms or groups attached to carbon. The unshared pair of electrons is present in the third sp^2 hybrid orbital and the unhybridised p-orbital is empty.

Triplet carbene has its carbon in sp hybridised state. The two sp hybrid orbitals form two bonds with monovalent atoms or groups attached to carbon. The two unhybridised p-orbitals contain one electron each.

The structure of triplet methylene is revealed by ESR measurements since these species are biradicals. Its geometry is found to be bent with bond angle of about 136°. The electronic spectra of singlet methylene formed in flash photolysis of diazomethane tell that it is also bent and the bond angle is about 103°.

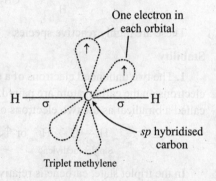

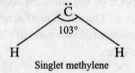

The bond angles in CCl_2 and CBr_2 are 100° and 114° respectively.

Reactions

Some important types of reactions of carbenes are described below:

1. Addition to carbon-carbon double bonds. The reactions of singlet and triplet methylene to 2-Butene has been discussed above. Carbenes also add to aromatic systems and the intermediate products rearrange with ring enlargement reaction. Carbenes also show addition reactions to C = N bonds.

Benzene $+ :CH_2 \longrightarrow$ Cycloheptatriene

2. Addition reactions to alkanes. Methylene ($:CH_2$) reacts with ethane to form propane. Propane adds methylene to yield n-butane and isobutane. This shows the greater reactivity of carbene.

$$CH_3-CH_2-CH_3 \xrightarrow{:CH_2} CH_3-CH_2-CH_2-CH_3 + CH_3 + CH_3-CH-CH_3$$
$$\text{Propane} \qquad\qquad n\text{-Butane} \qquad\qquad\qquad\qquad |$$
$$\qquad\qquad\qquad\qquad\qquad\qquad\qquad\qquad\qquad\qquad CH_3$$
$$\qquad\qquad\qquad\qquad\qquad\qquad\qquad\qquad\qquad \text{Isobutane}$$

The reaction of carbenes with higher alkanes yield a number of possible products. It shows that the addition of carbenes is not all selective. Dichlorocarbenes do not give reactions involving insertion.

3. An important reaction of carbenes is dimerisation.

$$R_2\ddot{C} + R'_2\ddot{C} \quad\quad R_2C = CR'_2$$

As the carbenes are highly reactive, the dimer formed carries so much energy that it dissociates again.

4. We also come across rearrangement reactions in carbenes with the migration of alkyl group or hydrogen atom. These rearrangements are much more rapid. Thus, the addition and insertion reactions of carbenes are seldom encountered with alkyl or dialkyl carbenes. Many rearrangements of carbenes directly give stable molecules. Some rearrangements involved in carbenes are:

$$CH_3 - CH_2 - CH - \ddot{C}H \longleftrightarrow CH_3 - CH_2 - CH = CH_2$$
Carbene $\quad\quad\quad\quad\quad\quad\quad\quad\quad$ 1-Butene

$$R - \underset{\underset{O}{\|}}{C} - \ddot{C}H \longrightarrow O = C = CHR \quad (\text{Wolff rearrangement})$$
Acyl Carbene $\quad\quad\quad\quad$ Ketene

$$\triangle\!\!-\!\ddot{C}H \longrightarrow \square$$

2.26 NITRENES

Organic species having general formula as R — N: are called nitrenes. These are analogous to carbenes and are sometimes referred to as azocarbenes. The structure R — N: shows that nitrogen is bonded to one atom or group only and there are two non-bonded electron pairs. Thus, there is only a set of electrons around nitrogen. This explains why nitrenes are electron deficient and reactive species.

The most common method of nitrene generation involves photolytic or thermolytic decomposition of azides, acyl azides and isocyanates

$$HN_3 \xrightarrow{h\nu \text{ or } \Delta} H - \ddot{N}{:} + N_2 \uparrow$$

$$CH_3 - N_3 \xrightarrow{h\nu} CH_3 - \ddot{N}{:} + N_2 \uparrow$$

$$C_6H_5 - N_3 \xrightarrow{h\nu} C_6H_5 - \ddot{N}{:} + N_2 \uparrow$$

$$RCON_3 \xrightarrow{h\nu} N_2 + RCO\ddot{N}$$
$\quad\quad\quad\quad\quad\quad\quad\quad\quad\quad$ Alkyl nitrene

$$R - N = C = O \xrightarrow{h\nu} CO + R - \ddot{N}{:}$$

Stability of nitrenes

Nitrenes are highly reactive species and cannot be isolated under ordinary conditions. Alkyl nitrenes have been isolated by trapping in matrices at 4 K while aryl nitrenes (less reactive) can be trapped at 77 K. Nitrenes can be generated both in the singlet and the triple states but its ground state is probably the triplet state.

$$R-\ddot{N}\uparrow\downarrow \text{ or } R-\ddot{N}:$$
Singlet

$$R-\ddot{N}\uparrow \text{ or } R-\ddot{N}\cdot$$
Triplet

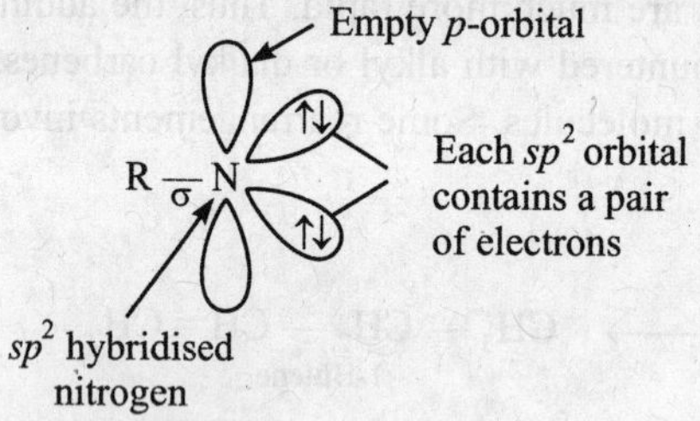

- Empty *p*-orbital
- Each sp^2 orbital contains a pair of electrons
- sp^2 hybridised nitrogen

Singlet nitrene

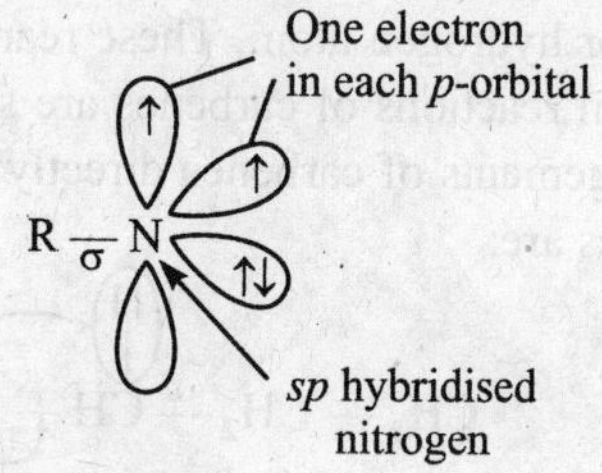

- One electron in each *p*-orbital
- *sp* hybridised nitrogen

Triplet nitrene

Reactions of Nitrenes

Nitrenes show the following reaction:

1. Addition to C = C bonds. Nitrenes add to alkenes to form cyclic product. Such reactions are common for acyl nitrenes.

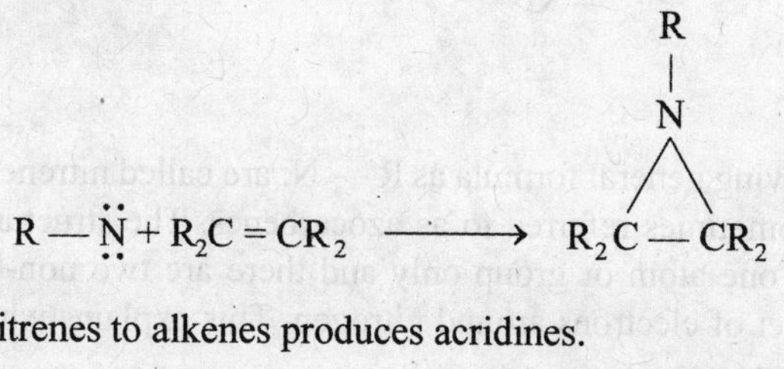

$$R-\ddot{N} + R_2C=CR_2 \longrightarrow R_2C-CR_2$$
$$\overset{R}{\underset{N}{|}}$$

Cycloaddition of nitrenes to alkenes produces acridines.

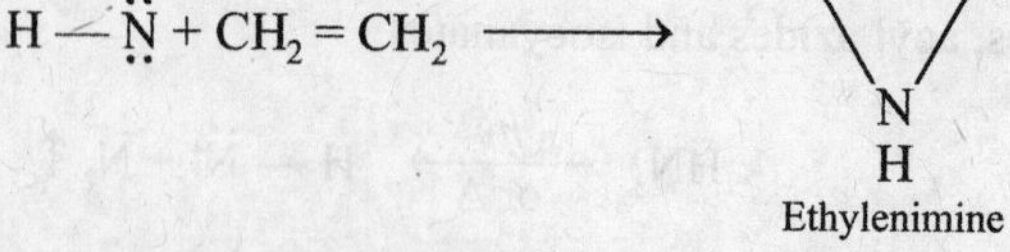

$$H-\ddot{N} + CH_2=CH_2 \longrightarrow \underset{H}{\underset{N}{\triangle}}$$

Ethylenimine

$$\underset{H}{\overset{CH_3}{\underset{}{>}}}C=C\underset{H}{\overset{CH_3}{<}} + :N-\ddot{H} \longrightarrow \underset{\underset{H}{\underset{|}{N}}}{\overset{H_3C}{\underset{}{>}}HC-CH\overset{CH_3}{<}}$$

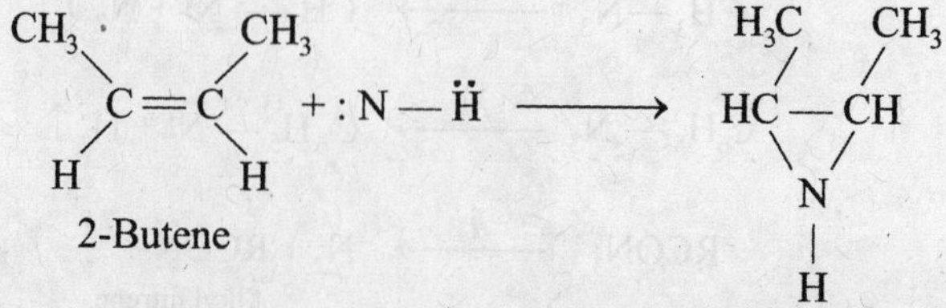

2-Butene

Acridine

Basics of Organic Chemistry II

2. Insertion. Nitrenes, especially acyl and sulphonyl nitrenes can insert into C — H and certain other bonds.

$$R'-\overset{O}{\overset{\|}{C}}-\ddot{N}: + R_3C-H \longrightarrow R'\overset{O}{\overset{\|}{C}}-\overset{H}{\overset{|}{N}}-CR_3$$

3. Abstraction. Nitrenes are also capable of extracting hydrogen atom form an alkane to form an alkyl radical.

$$R-\ddot{N} + R-H \longrightarrow R\dot{N}-H + \dot{R}$$

4. Rearrangements. Nitrenes undergo rearrangements also due to the migration of a group from adjacent atom to the electron deficient nitrogen. Insertion and addition to C = C bond reactions do not proceed in alkyl nitrenes as the rearrangement is much more faster.

$$R-CH-\ddot{N} \longrightarrow RCH=NH$$
$$\overset{|}{H}$$

5. Dimerisation. Nitrenes also dimerise to form a di-imide. Aryl nitrenes yield azobenzenes.

$$2C_6H_5-\ddot{N} \longrightarrow C_6H_5-N=N-C_6H_5$$

6. Ring enlargement. Azepines are prepared by the ring enlargement of phenyl azide.

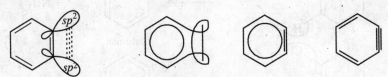

2.27 ARYNES

These are highly reactive species which contain an additional bond between two carbon atoms of the benzene rings. Taking benzyne as an example of aryne, the new bond of benzyne is formed by the sideways overlap of sp^2 orbitals belonging to two neighbouring carbon atoms. It is depicted as:

Representation of benzyne

It is evident that the new extra bond orbital lies along the side of the benzene ring and it has little interaction with the π-electron cloud of the benzene ring. Since the sideways overlapping is not very effective, the new bond is a weak one and hence benzyne is highly reactive molecule.

Generation of arynes

By the action of sodamide in liquid ammonia in aryl halides. It is obtained as an intermediate in nucleophilic substitution of aryl halides to form aniline

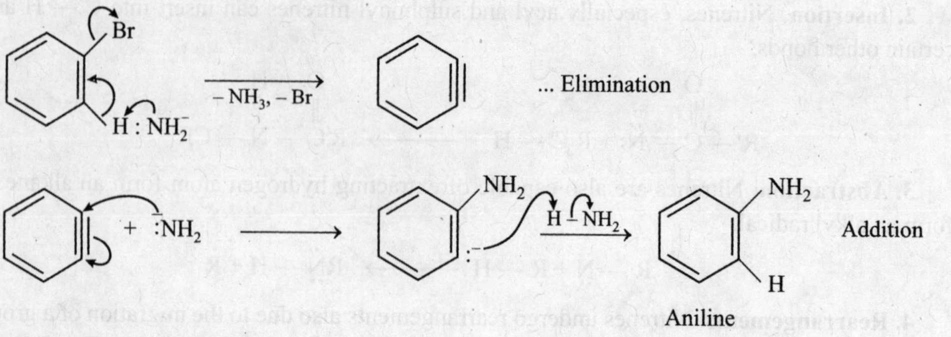

[Elimination scheme showing bromobenzene with ortho H being abstracted by NH₂⁻, losing NH₃ and Br⁻ to form benzyne ... **Elimination**]

[Addition scheme: benzyne + :NH₂⁻ → phenyl carbanion → abstracts H from NH₂ to give Aniline ... **Addition**]

Aryne (Benzyne) Mechanism

It has been found that electron-withdrawing groups activate the aryl halides towards bimolecular nucleophilic substitution reactions. In the absence of such activation, aryl halides can also be made to undergo nucleophilic substitution in the presence of strong nucleophiles. Thus, when chlorobenzene is treated with very strong nucleophile, viz. amide ion, it is converted into aniline.

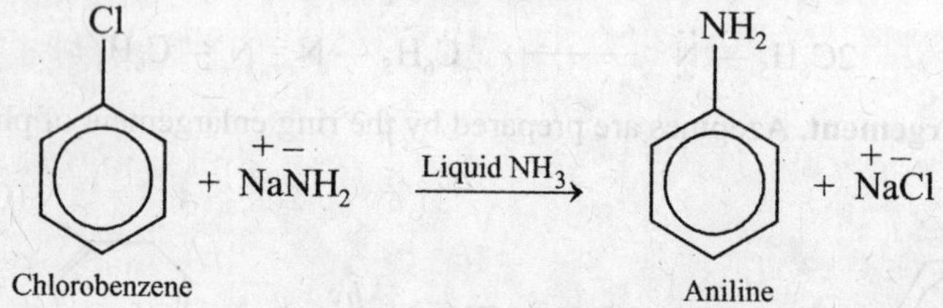

$$\text{Chlorobenzene} + \text{NaNH}_2 \xrightarrow{\text{Liquid NH}_3} \text{Aniline} + \text{NaCl}$$

The above type of nucleophilic substitution reaction occurs by benzyne mechanism. It involves both **elimination** and **addition** as represented below:

(I) Elimination. In the elimination step, the amide ion abstracts a proton from one of the ortho-positions with respect to the halogen. The resulting carbanion loses the halides ion to form the benzyne, as illustrated below:

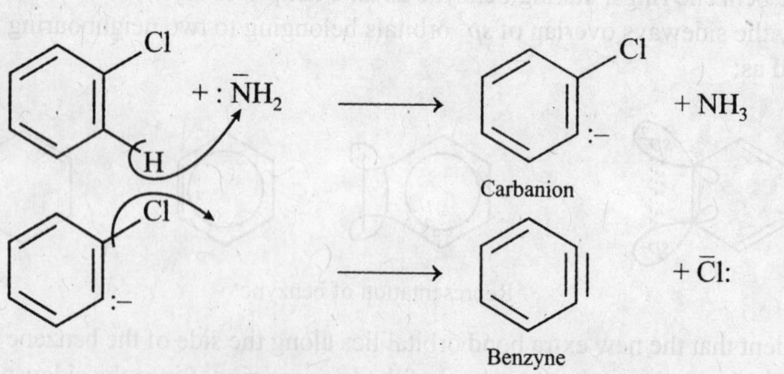

[Scheme: o-H of chlorobenzene abstracted by :NH₂⁻ → Carbanion (with Cl and negative charge) + NH₃; then carbanion loses Cl⁻ → Benzyne + Cl⁻]

(II) Addition. In the addition step, the amide ion attacks the benzyne molecule to form the carbanion which abstracts a proton from solvent ammonia to yield the final substituted product. These reactions are given below:

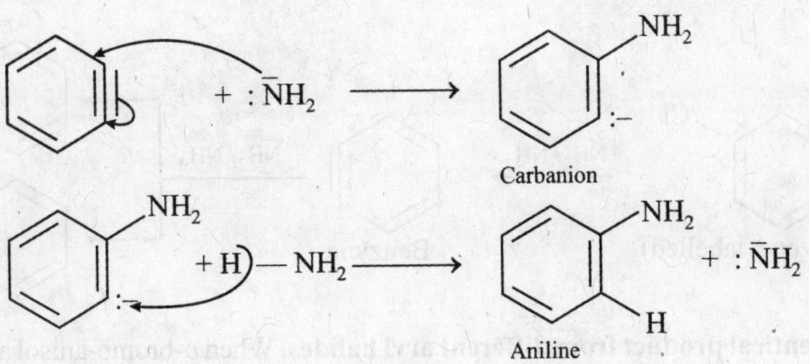

Evidence in support of benzyne mechanism

There is ample experimental evidence to prove the truth of benzyne mechanism of aromatic substitution.

(i) Isotope effect. The amination of o-deuteriobromobezene is slower than that of bromobenzene. This observation indicates that the cleavage of *ortho* hydrogen is involved in the rate determining step during the elimination stage. This can further be illustrated as follows:

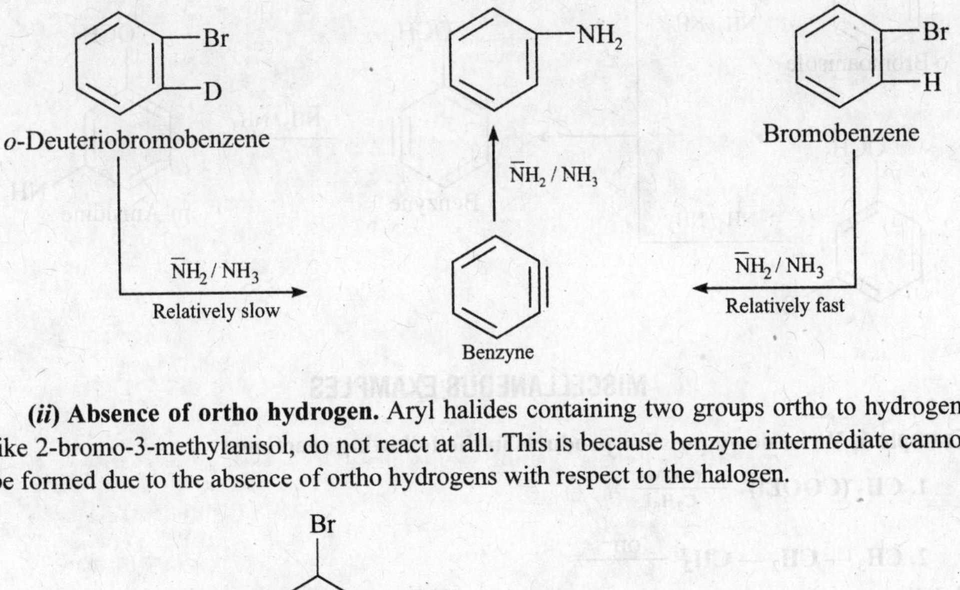

(ii) Absence of ortho hydrogen. Aryl halides containing two groups ortho to hydrogen, like 2-bromo-3-methylanisol, do not react at all. This is because benzyne intermediate cannot be formed due to the absence of ortho hydrogens with respect to the halogen.

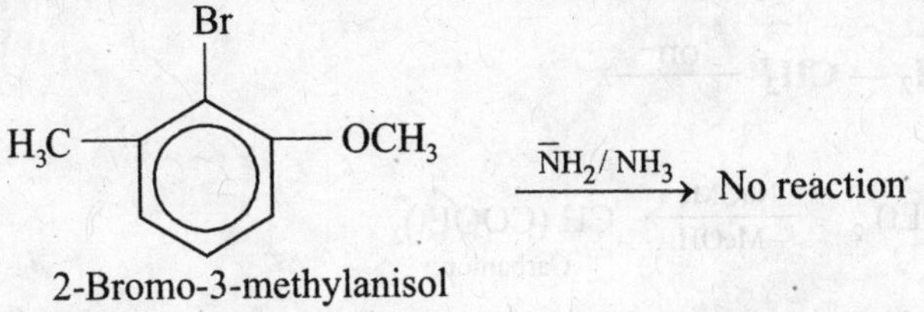

(iii) Tracer studies. When labelled chlorobenzene in which chlorine is linked to C^{14} isotope (C^*) is allowed to react with amide ion, two types of aniline are formed. In one type, the —NH_2 group is bonded to C^{14} while in the other it is attached to carbon *ortho* to the labelled carbon. This can be explained as follows:

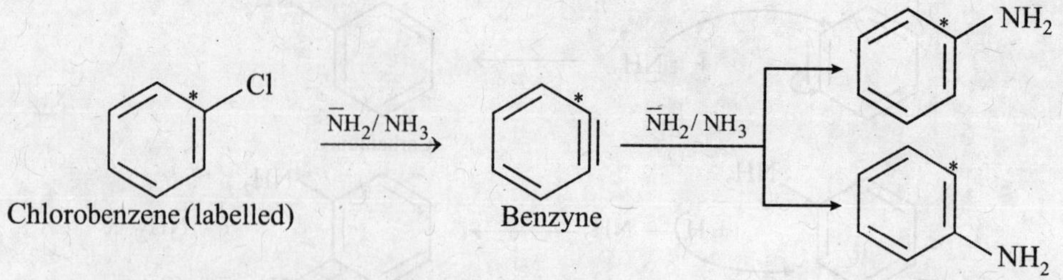

(iv) Identical product from different aryl halides. When *o*-bromo-anisol and *m*-bromo-anisol are allowed to react separately with ion in liquid NH$_3$, the same product, *i.e.*, *m*-anisidine is formed. The formation of the same product is due to the formation of the same intermediate benzyne as shown below. The amide ion attacks the benzyne species at *meta* position and not *ortho* position probably because of steric factor.

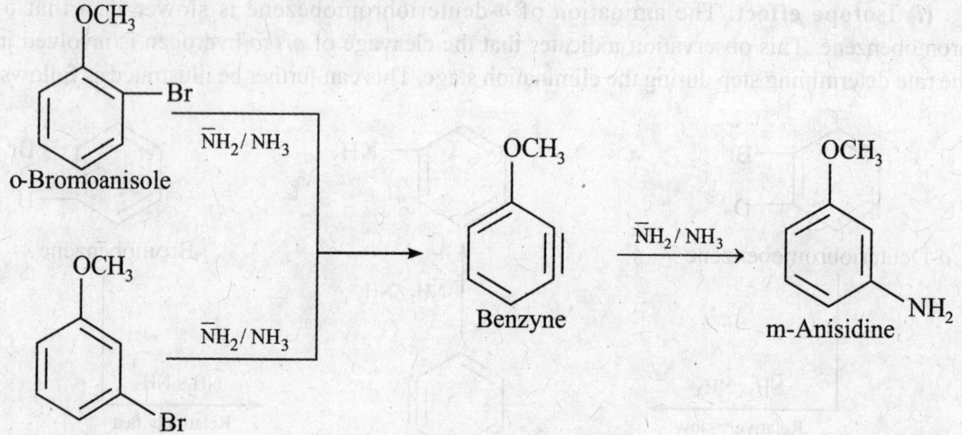

MISCELLANEOUS EXAMPLES

Example 1: Write down possible products in the following reactions:

1. $CH_2(COOEt)_2 \xrightarrow[C_2H_5I]{NaOMe}$

2. $CH_3-CH_2-CH_2^+ \xrightarrow{OH^-}$

Solution.

(1) $CH_2(COOEt)_2 \xrightarrow[-MeOH]{NaOMe} \bar{C}H(COOEt)_2$ (Carbanion)

$\xrightarrow{C_2H_5I} CH(COOEt)_2 \xrightarrow{Hydrolysis} C_2H_5CH_2COOH$ (Butanoic acid)
   $\quad\quad\quad\quad\; |$
   $\quad\quad\quad\; C_2H_5$

(2) $CH-CH_2-\overset{+}{C}H_2 \longrightarrow CH_3-\overset{+}{C}H-CH_3 \xrightarrow{OH^-} CH_3-CH(OH)CH_3$
(*n*-Propyl carbocation) (more stable carbocation) (Propanol-2)

Basics of Organic Chemistry II   89

Example 2: How do you explain that
  (i) Benzyl carbocation is much more stable than phenylethyl carbocation?
  (ii) Triphenyl methyl carbocation is a very stable carbocation?

Solution:

(i)   [C₆H₅–$\overset{+}{C}H_2$ Benzyl carbocation]   [C₆H₅–CH₂–$\overset{+}{C}H_2$ Phenylethyl carbocation]

Benzyl carbocation exists in a number of canonical forms and hence is more stable than phenyl ethyl carbocation in which there is no possibility of resonance.

[Resonance structures of benzyl carbocation shown]

(ii) Triphenylmethyl carbocation is a very stable carbocation. See Art. 2.8.

Example 3: What happens when 2, 3-Dimethyl butan-2, 3-diol is treated with dil. H_2SO_4?

Solution:

$$CH_3-\underset{\underset{CH_3}{|}}{\overset{\overset{OH}{|}}{C}}-\underset{\underset{CH_3}{|}}{\overset{\overset{OH}{|}}{C}}-CH_3 \xrightarrow[\text{Step I}]{H_2SO_4} CH_3-\underset{\underset{CH_3}{|}}{\overset{\overset{}{}}{\overset{+}{C}}}-\underset{\underset{CH_3}{|}}{\overset{\overset{O-H}{|}}{C}}-CH_3$$

2, 3-Dimethyl butan-2, 3 diol (Pinacol)   Carbocation

$$\xrightarrow{\text{Step II}} \left[CH_3-\underset{\underset{CH_3}{|}}{\overset{\overset{CH_3}{|}}{C}}-\overset{\overset{:O-H}{|}}{\underset{+}{C}}-CH_3 \rightleftharpoons CH_3-\underset{\underset{CH_3}{|}}{\overset{\overset{CH_3}{|}}{C}}-\overset{\overset{\overset{+}{O}H}{|}}{C}-CH_3 \right] \xrightarrow{\text{Step III}} CH_3-\underset{\underset{CH_3}{|}}{\overset{\overset{CH_3}{|}}{C}}-\overset{\overset{}{\underset{O}{||}}}{C}-CH_3$$

Pinacolone

Migration of methyl group in step II is facilitated by the resonance possible in the product.

Example 4: Arrange the following carbocations in the order of increasing stability:

$$C_6H_5\overset{+}{C}H_2, \quad (C_6H_5)_2\overset{+}{C}H, \quad (CH_3)_3\overset{+}{C}, \quad (CH_3)_3\overset{+}{C}H$$

Solution: $(C_6H_5\overset{+}{C}H_2) < (C_6H_5)_2\overset{+}{C}H < (CH_3)_2\overset{+}{C}H < (CH_3)\overset{+}{C}$

Example 5: Which of the two, single and triplet carbene is more stable and why?

Solution: Triplet carbenes are more stable than singlet carbenes because two unhybridised p-orbitals contain one electron each while unshared pair of electrons is present in the third sp^2 hybrid orbital and the unhybridised p-orbital is empty.

Example 6: Mention the methods by which you can infer that a particular reaction involves a free radical intermediate or not.

Solution: Methods:

   1. Formation of free radicals is favoured by the presence of u.v. light, heat and organic peroxides.

2. Free radicals are paramagnetic due to the presence of unpaired electrons.
3. Reactions involving free radicals widely occur in the gas phase.

Example 7: Explain why benzyl carbocation is more stable than cyclohexyl carbocation.

Solution: Benzyl carbocation is able to stabilise itself through resonance as shown below:

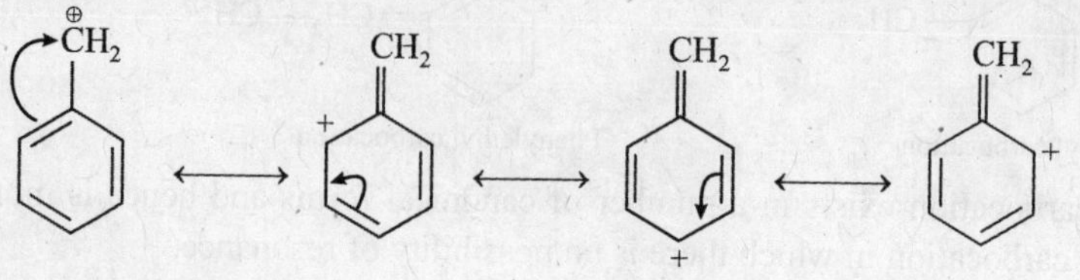

There is no such possibility in cyclohexyl carbocation

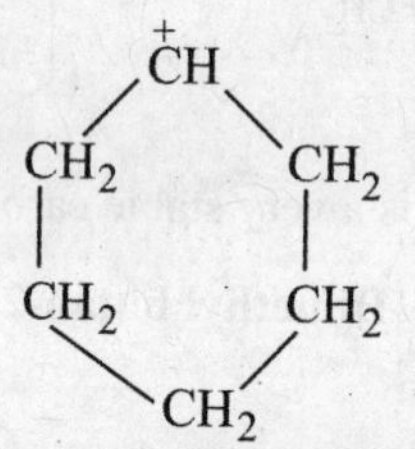

Cyclohexyl carbocation

Example 8: Give the orbital structure of $\overline{CH_3}$. Name its hybridisation and its shape.

Solution: The central carbon atom in CH_3 with a pair of unshared electrons is sp^3 hybridised. The three sp^3 hybrid orbitals are used in forming three sigma bonds with hydrogen. The fourth sp^3 hybrid orbital is occupied by a lone pair of electrons. The structure is similar to that of ammonia in which one hybrid orbital of nitrogen is occupied by a lone pair of electrons. The structure may be depicted as:

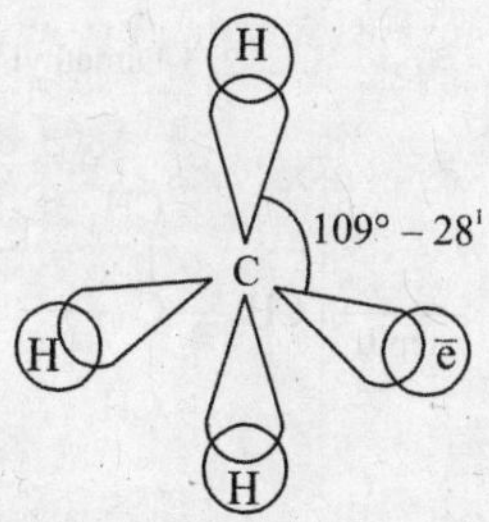

Thus it has a tetrahedral geometry. Three hydrogens are directed towards three corners of the tetrahedron. The fourth position is occupied by unshared pair of electrons.

Example 9: The carbocation F_3C^+ is more stable than carbocation $F_3C - \overset{+}{C}$.

Solution: In $F_3C - C^+$, the group F_3C is strongly electron-withdrawing. It draws the electrons towards itself and intensifies the positive charge on carbon. Thus dispersal of positive charge does not take place which is essential for the stability of a carbocation.

The carbocation F_3C^+ is also unstable if we consider the inductive effect of F atoms. But it gets stabilized due to another reason. The lone pair of electrons on each of the three F-atoms overlap with the empty p-orbital of carbon atom carrying the positive charge. This disperses the positive charge on carbon and stabilizes the F_3C^+ carbocation. Out of the inductive effect and overlapping effect, the latter dominates and the carbocation stabilizes.

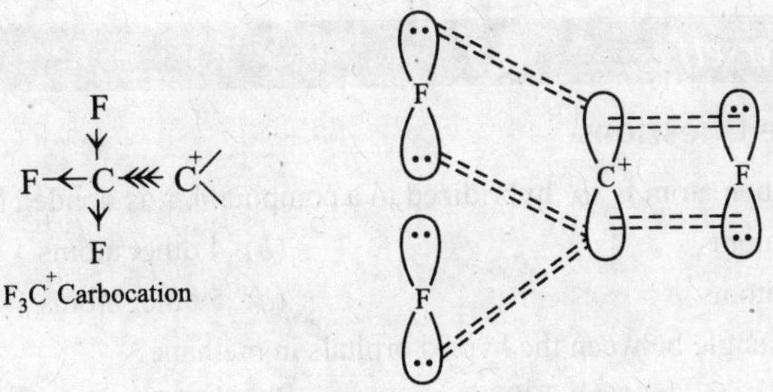

F_3C^+ Carbocation

Example 10: Explain why a less stable carbocation undergoes rearrangement to a more stable carbocation, but a less stable free radical does not.

Solution: Rearrangement to a more stable state takes place through the formation of a transition state.

In the case of *carbocation*, the central carbon atom is sp^2 hybridised and the unhybridised p-orbital is empty (as shown in the Fig.). During the rearrangement to more stable form, the two electrons of migrating σ-bond can be accommodated into the empty p-orbital. Thus rearrangement to more stable carbocation is facilitated.

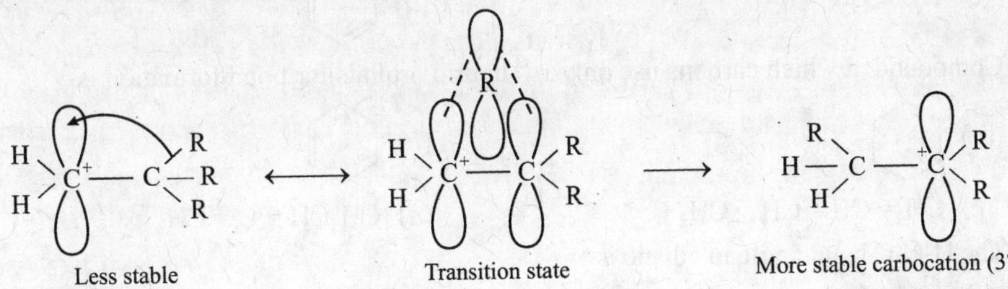

Less stable carbocation (1°)   Transition state   More stable carbocation (3°)

In the case of *free radical*, the central atom is again sp^2 hybridised but the unhybridised orbital contains one electron. Now the two electrons of the migrating σ-bond cannot move into p-orbital which already has one electron. This will be in contravention of *Pauli exclusion principle*. Therefore free radical rearrangement does not occur since these are energetically unfavourable.

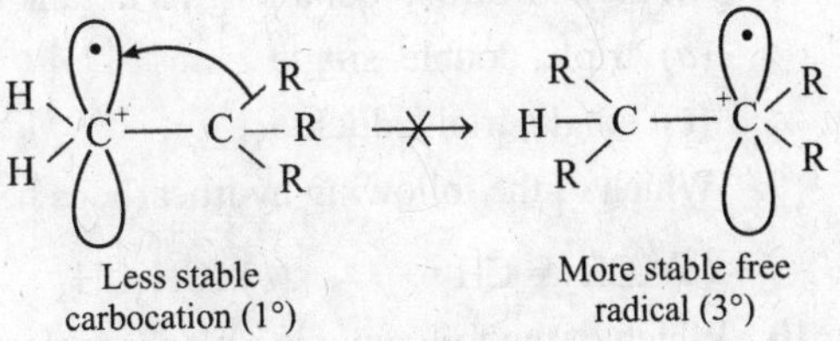

Less stable carbocation (1°)   More stable free radical (3°)

Key Terms

- Mesomeric effect
- Curly arrow rules
- Free radicals and carbenes
- Hyperconjugation
- Carbocations and carbanions

Evaluate Yourself

Multiple Choice Questions

1. When the carbon atom is sp^2 hybridized in a compound, it is bonded to
   (a) 2 other atoms
   (b) 4 other atoms
   (c) 3 other atoms
   (d) 5 other atoms
2. What is bond angle between the hybrid orbitals in methane?
   (a) 180°
   (b) 120°
   (c) 109.5°
   (d) 115.5°
3. What is the bond length of a carbon-carbon double bond?
   (a) 120 pm
   (b) 134 pm
   (c) 154 pm
   (d) 168 pm
4. Which of the following is the correct order of bond lengths:
   (a) $C-C < C=C < C\equiv C$
   (b) $C-C > C\equiv C > C=C$
   (c) $C\equiv C > C-C > C=C$
   (d) $C\equiv C < C-C > C=C$
5. The carbon-carbon bond length is maximum in
   (a) $CH_2=CH_2$
   (b) CH_3CH_3
   (c) $HC\equiv CH$
   (d) benzene
6. Compound in which carbons use only sp^2 hybrid orbitals for bond formation is
   (a) cyclohexane
   (b) benzene
   (c) $CH_2=CH-CH=CH_2$
   (d) $CH_3CH=C=CH_2$
7. The H-C-C bond angle in ethane is
   (a) 60°
   (b) 109°28′
   (c) 120°
   (d) 118°28′
8. The carbon-carbon bond lengths in rank of increasing bond length is:
   (a) triple, double, single
   (b) single, double, triple
   (c) single, triple, double
   (d) triple, single, double
9. Which of the following hydrocarbons has the shortest C-C bond length?
   (a) $CH_2=CH_2$
   (b) CH_3CH_3
   (c) $HC=CH$
   (d) benzene
10. Which of the following is a planar molecule?
    (a) Formaldehyde
    (b) Acetone
    (c) Formic acid
    (d) Acetic acid

Short Answer Questions

1. Ammonia and water molecules involve sp^3 hybridisation. But the $\angle HNH$ and $\angle HOH$ are not 109° − 28′. Explain.
2. What is heat of hydrogenation? How does it help in determining the resonance energy of benzene?
3. With the help of a diagram, explain the concept of hyperconjugation.
4. Comment on the unusual stability of the following:
   (i) cyclopentadienyl anion and
   (ii) triphenylmethyl anion

5. What is meant by curved arrow notation? Illustrate with an example.
6. Where do we use half headed and double headed arrows? Give examples.
7. Why is phenol a stronger acid than aliphatic alcohol?
8. Show how the carbanion $\overset{\ominus}{C}H_2-\underset{\underset{\displaystyle O}{\|}}{C}-CH_3$ is stabilised due to resonance?
9. Give the decreasing order of basicity of the following carbanions:
$$\overset{\ominus}{C_2H_5} > \overset{\ominus}{HC}=CH_2 > \overset{\ominus}{C}\equiv CH$$
10. Give the bond angles and shapes of molecules involving sp, sp^2 and sp^3 hybridisation.

Long Answer Questions

1. (a) Explain what do you understand by inductive effect.
   (b) How does inductive effect help in explaining.
      (i) Relative acidic strengths of acetic acid and formic acid?
      (ii) Relative basic strengths of primary, secondary and tertiary amines?
2. What is meant by inductive effect? Describe with the help of inductive effect that dichloro acetic acid is a stronger acid than acetic acid.
3. What is hyperconjugation? With its help, explain which of the following free radicals is the most stable:
$$CH_3-\overset{\bullet}{C}H_2,\ \overset{\bullet}{C}H_3,\ (CH_3)_3\overset{\bullet}{C},\ (CH_3)_2\overset{\bullet}{C}H$$
4. Indicate the type of hybridization of each carbon atom in the following structures:
   (a) CH_3CH_3
   (b) $CH_3C\equiv CCH_3$
   (c) $CH_2=C=CH_2$
   (d) $CH_3C\equiv N$

   Ans. (a) CH_3-CH_3 both sp^3

   (b) $CH_3-C\equiv C-CH_3$: sp^3, sp, sp, sp^3

   (c) $H_2C=C=CH_2$: sp^3, sp

   (d) $CH_3-C\equiv N$: sp^3, sp

5. (a) Explain the relative stabilities of various alkyl free radicals on the basis of hyperconjugation.
   (b) Describe the two types of mesomeric effects.
   (c) Differentiate between inductive and electromeric effects.
6. Indicate the type of hybridization of each atom in the following structures:
   (a) $H-\underset{\underset{\displaystyle }{\|}}{\overset{\overset{\displaystyle O}{\|}}{C}}-H$
   (b) CH_3-O-CH_3
   (c) $H-C\equiv N$
   (d) CH_3NH_2

Ans. (a)
$$H-\underset{sp^2}{\overset{\overset{\displaystyle O}{\|}}{C}}-H$$

(b) $CH_3 - \overset{..}{\underset{..}{O}} - CH_3$ sp^3

(c) $H - C \equiv \overset{..}{N}$ sp

(d) $CH_3 - \overset{..}{N}H_2$ sp^3

7. Classify the following into electrophiles and nucleophiles and give reasons?
   (i) Water   (ii) Silver cation   (iii) Methylene carbene

8. Which of the following will generate the most stable carbocation and why?
$$CH_3CH_2OH, \; CH_3CHOHCH_3, \; C_6H_5CHOHCH_3$$

9. List the factors that determine the stability of a carbocation. Give an example where field effect controls the stability.

10. Name the different methods for the determination of mechanism of a reaction. Explain one of them.

11. Explain why ethyl carbocation ($CH_3 - \overset{+}{C}H_2$) is more stable than n-propyl carbocation ($CH_3 - CH_2 - \overset{+}{C}H_2$)?

   Hint: Greater the number of hyperconjugative structures of a carbocation, greater is the stability.

$$\underset{(I)}{H-\underset{H}{\overset{H}{|}}{\underset{|}{C}}-\overset{+}{C}H_2} \longleftrightarrow \underset{(II)}{H-\underset{H}{\overset{H^+}{|}}{\underset{|}{C}}=CH_2} \longleftrightarrow \underset{(III)}{\overset{+}{H}\underset{H}{\overset{H}{|}}{\underset{|}{C}}=CH_2} \longleftrightarrow \underset{(IV)}{H-\underset{H^+}{\overset{H}{|}}{\underset{|}{C}}=CH_2}$$

$$\underset{(I)}{CH_3-\underset{H}{\overset{H}{|}}{\underset{|}{C}}-\overset{+}{C}H_2} \longleftrightarrow \underset{(II)}{CH_3-\underset{H}{\overset{H^+}{|}}{\underset{|}{C}}=CH_2} \longleftrightarrow \underset{(III)}{CH_3-\underset{H^+}{\overset{H}{|}}{\underset{|}{C}}=CH_2}$$

We find that ethyl carbocation has four hyperconjugation structures while n-propyl carbocation has three. Thus, ethyl carbocation is more stable.

12. Hydrolysis of $(CH_3)_2 \, C = CH - CH_2 \, Cl$ and $(CH_3)_2 - \underset{\underset{Cl}{|}}{C} - CH = CH_2$ yields the same products in the same ratio. How do we explain this?

13. Give a list of curly arrow rules.

14. What is meant by formal charge? What is its utility?

Answers
Multiple Choice Questions

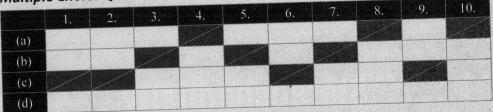

Suggested Readings
1. Kalsi, P.S. Steriochemistry. Conformation and Mechanism. New Age International 2005.
2. Eiliel, E.L. & Wilen, S.H. Stereochemistry of Organic Compounds. Wiley, London.

Chapter 3

Stereochemistry

LEARNING OBJECTIVES

After reading this chapter, you should be able to:
- draw Fischer, Newmann and Sawhorse projection formulae
- learn the interconversion of these projection formulae
- learn geometrical isomerism, cis-trans and syn-anti isomerism with E/Z notation with C.I.P. rules
- understand terms like optical activity, specific rotation and chirality
- study enantiomers with molecules having two or more chiral centres
- learn the terms like distreoisomers, meso compounds, racemic mixtures and resolution
- learn how to arrive at absolute configuration with D/L and R/S designations

3.1 INTRODUCTION

One of the striking features of organic compounds is that they exhibit isomerism. Compounds having the same molecular formula and molecular mass but having different characteristics are called isomers. This phenomenon is called isomerism. The relative position of different atoms or groups in these isomers is different and that causes the difference in properties. Isomerism may be classified into two types:

1. Structural isomerism. In this type of isomerism, the molecules differ in the structural arrangement of the atoms or groups.

2. Stereoisomerism. Here the isomers possess the same structural arrangement but differ with respect to the arrangement of atoms or groups in space.

The two types of isomerisms are discussed in detail as under:

3.2 STRUCTURAL ISOMERISM

If two or more compounds differ in the relative arrangement of atoms in the molecule, they are said to be structural isomers and this phenomenon is known as **structural isomerism.** There are different kinds of structural isomerisms as under:

(*i*) Chain isomerism or Nuclear isomerism.

(*ii*) Position isomerism.

(*iii*) Functional isomerism.

(*iv*) Metamerism.

(*v*) Tautomerism.

Each one of these is discussed as under.

Chain Isomerism or Nuclear Isomerism

If different compounds of the same class of organic compounds, having the same molecular formula, differ in the structure of carbon chain, they are called chain isomers. Examples of this type of isomerism are:

(*a*) *n*-Butane and Isobutane (Mol Formula = C_4H_{10})

```
    H   H   H   H                    H       H       H
    |   |   |   |                    |       |       |
H — C — C — C — C — H    and    H — C ——— C ——— C — H
    |   |   |   |                    |       |       |
    H   H   H   H                    H    H—C—H     H
                                          |
         n-Butane                         H
                                      Isobutane
```

(*b*) *n*-Pentane, isopentane and neo pentane (Mol. Formula = C_5H_{12})

```
    H   H   H   H   H                     H       H       H   H
    |   |   |   |   |                     |       |       |   |
H — C — C — C — C — C — H    ;      H — C ——— C ——— C — C — H
    |   |   |   |   |                     |       |       |   |
    H   H   H   H   H                     H    H—C—H     H   H
                                               |
         n-Pentane                             H
                                          Isopentane
```

$$\text{Neo-pentane structure:}$$

```
           H
           |
     H  H—C—H  H
     |  |     |
 H—C———C———C—H
     |  |     |
     H  H—C—H  H
           |
           H
```
Neo-pentane

Position Isomerism

If different compounds, belonging to some homologous series, with same molecular formula have same carbon skeleton but differ in the position of substituent or functional group; these are known as *position isomers*. Examples of this type of isomers are:

(a) 1-Propanol and 2-propanol (Mol. Formula = C_3H_8O)

$$\underset{\text{1-Propanol}}{CH_3-CH_2-CH_2OH} \quad \text{and} \quad \underset{\text{2-Propanol}}{CH_3-\underset{\underset{\text{OH}}{|}}{CH}-CH_3}$$

(b) 1-Butene and 2-Butene (Mol. Formula = C_4H_8)

$$\underset{\text{1-Butene}}{CH_2=CH_2-CH_2-CH_3} \quad \text{and} \quad \underset{\text{2-Butene}}{CH_3-CH=CH-CH_3}$$

Functional Isomerism

Different compounds, with same molecular formula but different functional groups are known as *functional isomers*. For example:

(a) Ethyl alcohol and Dimethyl ether (Mol. Formula = C_2H_6O)

$$\underset{\text{Ethyl alcohol}}{CH_3CH_2OH} \quad \text{and} \quad \underset{\text{Dimethyl ether}}{CH_3-O-CH_3}$$

(b) Propionaldehyde and acetone (Mol. Formula = C_3H_6O)

$$\underset{\text{Propionaldehyde}}{CH_3CH_2CHO} \quad \text{and} \quad \underset{\text{Acetone}}{CH_3-CO-CH_3}$$

Metamerism

This is a special kind of structural isomerism in which different compounds, with same molecular formula, belong to same homologous series but differ in the distribution of alkyl groups around a central atom. Examples are:

(a) Diethyl ether and methyl propyl ether (Mol. Formula = $C_4H_{10}O$)

$$\underset{\text{Diethyl ether}}{C_2H_5-O-C_2H_5} \quad \text{and} \quad \underset{\text{Methyl propyl ether}}{CH_3-O-C_3H_7}$$

(b) Diethyl ketone and methyl propyl ketone (Mol. Formula = $C_5H_{10}O$)

$$\underset{\text{Diethyl ketone}}{C_2H_5-CO-C_2H_5} \quad \text{and} \quad \underset{\text{Methyl propyl ketone}}{CH_3-CO-C_3H_7}$$

This type of isomers are known as *metamers* and the phenomenon is known as *metamerism*.

Stereochemistry

Tautomerism

Compounds whose structures differ in the arrangement of atoms but which exist simultaneously in dynamic equilibrium with each other, are called *tautomers*. This phenomenon is called *tautomerism*.

In most of the cases tautomerism is due to shifting of a hydrogen atoms from one carbon (or oxygen or nitrogen) to another with the rearrangement of single or double bonds. For example,

$$CH_3-\underset{\underset{\text{Ethyl acetoacetate (keto form)}}{}}{\overset{\overset{O}{\|}}{C}}-CH_2-COOC_5H_5 \rightleftharpoons CH_3-\underset{\underset{\text{Ethyl acetoacetate (Enolic form)}}{}}{\overset{\overset{OH}{|}}{C}}=CH-COOC_5H_5$$

$$CH_3-CH_2-\underset{\underset{\text{Nitroethane}}{}}{N\underset{O}{\overset{O}{\diagup\!\!\!\!\diagdown}}} \rightleftharpoons CH_3-CH=\underset{\underset{\text{Isonitroethane}}{}}{N\underset{O}{\overset{OH}{\diagup\!\!\!\!\diagdown}}}$$

Example: Write the structures of all possible isomers of C_3H_8O giving their IUPAC names.

Solution: Position isomers

$$\underset{\text{Propanol–1}}{CH_3CH_2CH_2OH} \qquad \underset{\text{Propanol–2}}{CH_3-CHOHCH_3}$$

Functional Isomers

$$\underset{\text{Propanol–1}}{CH_3CH_2CH_2OH} \qquad \underset{\text{Methoxyethane}}{CH_3OCH_2CH_3}$$

3.3 STEREOISOMERISM

Compounds having different three-dimensional relative arrangement of atoms in space are called stereoisomers. This phenomenon is called *stereoisomerism*. These compounds are said to have different configurations. Stereoisomerism is of the following different kinds:

(*i*) Optical isomerism

(*ii*) Geometrical isomerism

(*iii*) Conformational isomerism.

Optical Isomerism or Enantiomerism

Before taking up optical isomerism, let us understand the terms: plane polarised light, optical activity and specific rotation.

Plane Polarised Light

Ordinary light has vibrations taking place at right angles to the direction of propagation of light spread in all the possible planes. If we pass ordinary light through **nicol**. prism vibrations in all planes except one are cut off. Thus light coming out of nicol prism has vibrations only in one plane. Such a light is called *plane polarised light*.

Optical Activity

Behaviour of certain substances is strange. When a plane polarised light is passed through the solution of such substances, the light coming out of the solution is found to be in a different plane. The plane of polarised light is rotated. *Such substances, which rotate the plane of plane polarised light when placed in its path are known as optically active substances* and the phenomenon is known as *optical activity*. The angle of rotation (α) of plane polarised light is known as *Optical rotation*. The substances which rotate the plane of polarised light to the clockwise or right direction are known as dextrorotatory or having positive (+) rotation and those which rotate the plane polarised light to the anticlockwise or left direction are known as *laevorotatory* or having negative (−) rotation. Substances which do not rotate the plane of polarised light are said to be optically inactive.

The instrument used for measuring optical rotation is called polarimeter. It consists of a light source, two nicol prisms and in between a tube to hold the solution of organic substance. The schematic representation of a polarimeter is given in Fig. 3.1.

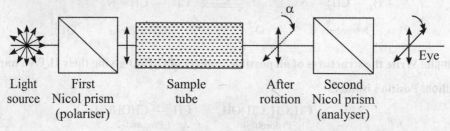

Fig. 3.1: *Measurement of optical rotation*

Specific Rotation

The angle of rotation of plane-polarised light or optical rotation (α) of an organic substance depends not only on the kind of molecules but also varies considerably with the number of molecules that light encounters in its path which in turn depends on the concentration of the solution used and the length of polarimeter tube containing it. Besides this, it depends on temperature, wavelength of light and nature of solvent used.

The optical activity of a substance is expressed in terms of *specific rotation*. $[\alpha]_\lambda^t$ which is a constant quantity, characteristic of a particular substance,

$$[\alpha]_\lambda^t = \frac{\alpha}{l \times c}$$

where, α = observed rotation in degrees
l = length of polarimeter tube in decimeter
c = concentration of substance in gm per ml of solution
t and λ signify the temperature and wavelength of light used.
When $l = 1$ and $c = 1$, $[\alpha]_\lambda^t = \alpha$

Specific rotation is thus defined as the optical rotation produced by a compound when plane polarised light passes through one decimeter length of the solution having concentration one gram per millilitre. Usually the monochromatic light used is D line of sodium ($\lambda = 589$ nm). Thus specific rotation of cane sugar can be expressed as

$$[\alpha]_D^{20°C} = +66.5° \text{ (water)}$$

In this expression D stands for D line of sodium, 20°C is temperature of measurement, + sign shows the dextrorotation and water is the solvent used.

Louis Pasteur, while studying the crystallography of salts of tartaric acid made a peculiar observation. He observed that optically inactive sodium ammonium tartarate existed as a mixture of two different types of crystals which were mirror images of each other. With the help of a hand lens and a pair of forceps, he carefully separated the mixture into two different types of crystals. These crystals were mirror images of each other and were called enantiomorphs and the phenomenon as *enantiomorphism*. Although the original mixture was optically inactive; each type of crystals when dissolved in water, were found to be optically active. Moreover the specific rotations of the two solutions were exactly equal but of opposite sign *i.e.* one solution rotated the plane polarised light to the right or clockwise while the other to the left or anticlockwise and to the same extent. Two types of crystals or solutions were identical in all other physical and chemical properties. *Isomers which are non-superimposable mirror images of each other are called enantiomers.*

van't Hoff (1852-1911)
van't Hoff was a Dutch Chemist. His work on three dimensional molecular properties is widely acclaimed. He received the first Nobel prize in chemistry in 1901 for his work on rates of reactions, chemical equilibrium and osmotic pressure.

According to La Bell and van't Hoff the four valencies of a carbon atom are directed towards the four corners of a regular tetrahedron at the centre of which lies the carbon atom. Consider a compound of formula C_{LMNO} having four different groups L, M, N and O attached to a carbon atom. This compound can be represented by two models which look like mirror images of each other.

Enantiomerism in real life

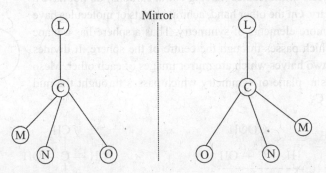

It is important to note here that these two molecules cannot be superimposed on each other *i.e.* they will not coincide in all their parts. We may turn them in as many ways as we like but we find that though two groups of each may coincide, the other two do not. Hence these must represent two isomers of formula C_{LMNO}. Lactic acid $CH_3CHOHCOOH$ and sec-Butyl chloride $C_2H_5CHClCH_3$, exist as two optically active isomers which are enantiomers *i.e.* mirror images of each other. Mirror images of the two compounds are represented as above.

COOH   COOH C_2H_5 C_2H_5

Enantiomers of Lactic Acid   Enantiomers of Sec. Butyl Chloride

The carbon atom to which four different groups are attached, is known as asymmetric or chiral carbon atom or stereogenic centre.

If two of the groups attached to carbon are same, we shall observe that it is possible to superimpose the mirror images on each other. Such a compound will not show optical isomerism or *enantiomerism.*

Hence non-superimposability of the mirror images is responsible and essential for the type of stereoisomerism known as enantiomerism.

The term optical isomerism is used for the existence of stereoisomers which differ in their behaviour towards the plane polarised light. Thus enantiomeric molecules are always non-superimposable mirror images of each other. The non-superimposability of mirror images arises due to **chiral** or **asymmetric** nature of molecule. A molecule is said to be chiral if it has no plane of symmetry and is therefore non-superimposable on its mirror image.

It may be concluded with the remarks that chirality is the fundamental condition of enantiomerism or optical isomerism.

Elements of Symmetry

(i) Plane of symmetry

It is defined as an imaginary plane which divides the molecule into two equal halves such that one half is the mirror image of the other half. It may pass through atoms, between the atoms or both. Chiral objects or molecules (having chiral carbons) do not have a plane of symmetry. On the other hand, achiral objects or molecules have at least one or more elements of symmetry. Thus a sphere has a plane of symmetry which passes through the centre of the sphere. It divides the sphere into two halves which are mirror images of each other. Meso tartaric acid has a plane of symmetry which passes throught the mid point of C_2 and C_3.

$\overset{1}{C}OOH$
$H - \overset{2}{C} - OH$

$H - \overset{3}{C} - OH$
$\overset{4}{C}OOH$

Meso tartaric acid

CH_3
$H - C - OH$
COOH

Lactic acid

Therefore one half of the meso tartaric acid is mirror image of the other half. Lactic acid, $CH_3CH(OH)$ COOH on the other hand has no plane of symmetry which can divide into two equal halves.

(ii) Centre of symmetry

A centre of symmetry is defined as a point from which lines when drawn on one side and extended an equal distance on the other side will meet identical points (or atoms or groups). For example *trans* –1, 3-dimethylcyclobutane has a centre of symmetry. Three dimensional (and not two dimensional) models need to be drawn to understand this point.

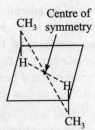
1, 3-Dimethylcyclobutane

(iii) Alternating axis of symmetry

A molecule is said to possess an n-fold alternating axis of symmetry if rotation through an angle of 360/n about this axis and then followed by reflection in a plane perpendicular to the axis gives a molecule which is indistinguishable from the original molecule. Consider Z^+ and Z^- as chiral group (+) – CH(CH_3) C_2H_5 and (–) –$CH(CH_3)$ C_2H_5 respectively. Rotation of molecule (*a*) through 90° about AB which passes through the centre of the ring and perpendicular to the plane gives (*b*). Reflection of (*b*) gives the original molecule (*a*). Thus it has 4-fold 360/90=4 alternating axis of symmetry and hence is achiral (Fig 3.2)

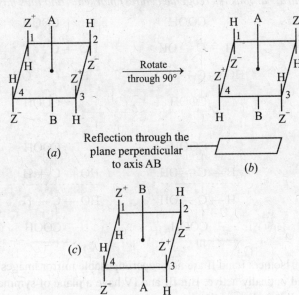

Fig. 3.2: *Four-fold alternating axis of symmetry*

Therefore, absence of an alternating axis of symmetry is the necessary and sufficient condition for a molecule to exhibit optical activity.

Chirality and Dissymmetry

An object which is non-superimposable on its mirror image is called **chiral** object. And an object which is superimposable on its mirror image is called **achiral** object. Examples of chiral objects including letters from English alphabets are: A pair of hands, shoes, gloves, letters P, F, J etc.

Examples of achiral objects and letters are: Ball, sphere, letters A, O, M, etc.

Achiral objects or molecules possess a plane of symmetry that can divide the molecule into two identical halves. Chiral objects or molecules do not possess a plane of symmetry that can divide the molecule into two identical halves. The **chirality** is also known as **dissymetry** while **achirality** is also known as **symmetry**.

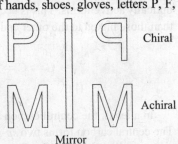
Chiral and achiral objects

Thus chirality or dissymmetry is the property associated with certain compounds which do not have a plane of symmetry and are thus non-superimposable on their mirror images. Dissymmetry is an essential condition for optical activity. In the study of enantiomers containing one chiral carbon atom, we find that they do not have a plane of symmetry and hence are dissymmetric in nature, for example, lactic acid, sec-butyl chloride etc. We can thus say that compounds with a chiral carbon atom are optically active in nature.

This is not true, however, for compounds with two or more chiral carbon atoms. Here, one or more isomers may be optically inactive in spite of the presence of chiral carbon atoms. Consider *for example, the case of tartaric acid, CH(OH) COOHCH (OH) COOH. It has two chiral carbons (stereogenic centres) marked * and thus has four isomers as shown below:*

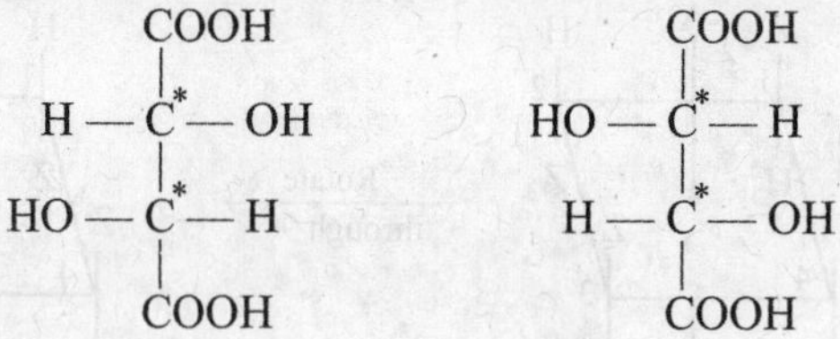

```
        COOH                    COOH
         |                       |
    H — C*— OH              HO — C*— H
         |                       |              I and II are optically
    HO — C*— H               H — C*— OH         active due to the absence
         |                       |              of plane of symmetry
        COOH                    COOH
         I                       II


        COOH                    COOH
         |                       |
    H — C*— OH              HO — C*— H
    -----|-----             -----|-----         III and IV are optically
    H — C*— OH              HO — C*— H          inactive due to the presence
         |                       |              of plane of symmetry
        COOH                    COOH            (shown by dotted line)
         III                     IV
```

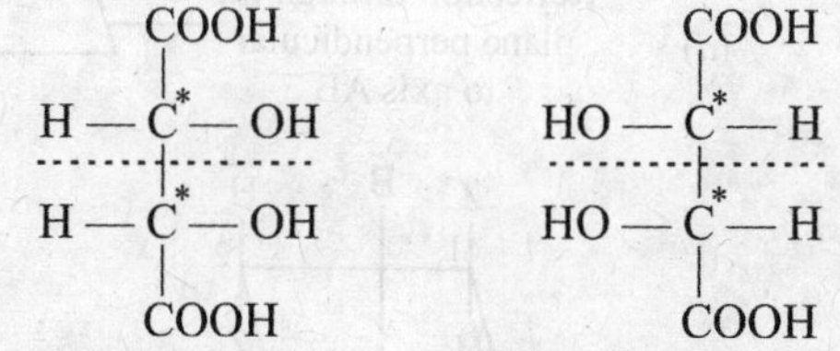

Isomers I and II are non-superimposable mirror images of each other, they are dissymetric and optically active. But III and IV have a plane of symmetry. Therefore, they are symmetric and optically inactive in nature.

In other words, optical activity in organic compounds is due to the presence of dissymmetry and not due to chiral carbon atoms alone.

No. of optical isomers for a compound is given by 2^n where n is the number of chiral carbon atoms.

It may be noted that some organic compounds are optically active even though they do not contain chiral atom. For example,

Substituted allenes. Dienes with double bonds in adjacent positions are called allenes. Substituted allenes may be represented as RCH = C = CHR. Pentane –2, 3-diene exists in two optically active forms. This is because the groups linked to one end carbon are in different plane than those linked to the other end carbon.

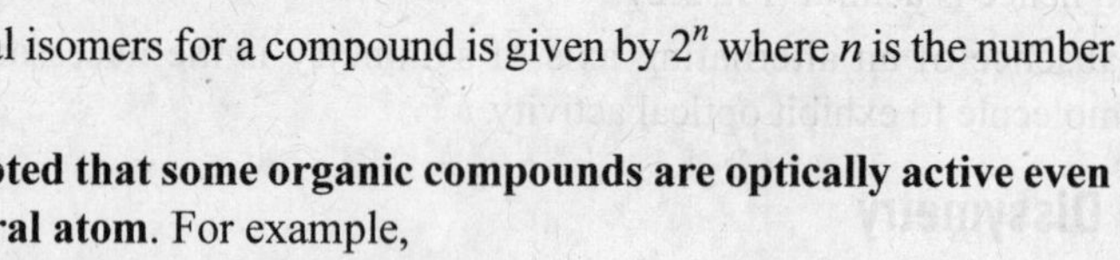

$$\begin{array}{c} H \\ \diagdown \\ H_3C \end{array} C=C=C \begin{array}{c} H \\ \diagup \\ \diagdown CH_3 \end{array} \qquad \begin{array}{c} H \\ \diagdown \\ CH_3 \end{array} C=C=C \begin{array}{c} H \\ \diagup \\ \diagdown CH_3 \end{array}$$

In allenes, the central carbon is *sp* hybridized, and the terminal carbons are sp^2 hybridized. The central carbon forms two sp-sp^2 σ bonds. The central carbon also has two *p* orbitals which are mutually perpendicular. These forms π bonds with the *p* orbitals on the other carbon atoms.

Stereochemistry

As a result, the substituent at one end of the molecule are in a plane which is perpendicular to that of the substituents at the other end, so that the compound exists in two forms which are non-superimposable mirror images and are optically active.

Substituted biphenyl. Benzene rings in substituted biphenyl lie in different planes. Hence the compound exhibits optical activity.

We can conclude from the above discussion that the most essential condition for a compound to show optical activity is the presence of dissymmetry and not chirality. However, compounds containing only one chiral carbon are always dissymmetric since they do not have a plane of symmetry.

The compounds, with two or more similar chiral carbon atoms which are optically inactive due to the presence of plane of symmetry, are called **meso** compounds.

Consider the following isomer of tartaric acid

```
        COOH
         |
    H — C — OH
         |           ← Plane of
    H — C — OH          symmetry
         |
        COOH
```

It has a plane of symmetry denoted by the denoted line. The optical rotation of upper half of the molecule is neutralized by that of the lower half as the two rotations are in opposite directions. The net result is that the molecule is optically inactive. **Such compounds are known as meso compounds**.

Similarly, we come across meso structures in 2, 3-Dichlorobutane. This compound also possesses two asymmetric carbon atoms (stereogenic centres) marked with *.

$$CH_3 - \overset{*}{C}H - \overset{*}{C}H - CH_3$$
$$||$$
$$ClCl$$

The different isomers can be written as

```
        CH3                    CH3
         |                      |
    H — C*— Cl              Cl — C*— H       I and II show optical
         |                      |            activity due to absence
    Cl — C*— H              H — C*— Cl       of plane of symmetry
         |                      |            (Dissymmetry)
        CH3                    CH3
         I                      II


        CH3                    CH3
         |                      |
    H — C*— Cl              Cl — C*— H       III and IV are optically
    ─────|──────            ─────|──────     inactive because of the
    H — C*— Cl              Cl — C*— H       presence of plane of
         |                      |            symmetry
        CH3                    CH3
         III                    IV
```

Structures III and IV are meso compounds because the optical activity of the upper half of the molecule is neutralised by that of the lower half.

Properties of enantiomers

(i) They have identical physical properties but differ in direction of rotation of plane polarised light. Though the two enantiomers rotate the plane polarised light in opposite direction, the extent of rotation is the same.

(ii) They have identical chemical properties except towards optically active reagents. The rates of reaction of optically active reagents with two enantiomers differ and sometimes one of the enantiomers does not react at all.

(iii) In biological system (–) or *l*-glucose is neither metabolised by animals nor fermented by yeast whereas (+) or *d*-glucose undergoes both these processes and plays an important role in animal metabolism and fermentation. Similarly mould *penicillium glaucum* consumes only *d*-tartaric acid when fed with a mixture of equal quantities of *d*- and *l*-tartaric acid.

(iv) When equal amounts of enantiomers are mixed together an optically inactive racemic modification denoted by (±) or *dl* is obtained.

Solved Examples

Example 1: Given below are the structural formulae of all alkanes with molecular formula C_6H_{14}. Which of these exhibit enantiomerism?

(i) $CH_3(CH_2)_4CH_3$

(ii) $CH_3(CH_2)_2 - CH - CH_3$
              |
              CH_3

(iii) $CH_3CH - CH - CH_3$
          | |
         CH_3 CH_3

(iv) $(CH_3)_3C - CH_2CH_3$

Solution: There is no carbon in any of the compounds above which is chiral *i.e.* attached to four different groups.

Chirality is the necessary condition for a molecule to exhibit enantiomerism. Hence none of the compounds above shows enantiomerism.

Example 2: Which of the following compounds exhibit enantiomerism?

(i) $CH_2OH\ CHOH\ CHO$

(ii) $\ CH_3$
          $\,|$
     $CH_3 - CH - CHCl - CH_3$

(iii) $CH_3 - CH_2 - CH - CH_3$
                      |
                     CH_3

(iv) $CH_3CHOHCH_3$

(v) CH_2NH_2COOH

Solution: Compounds (i) and (ii) have chiral carbons marked with asterisks and hence these two compounds show enantiomerism.

$CH_2OH\overset{*}{C}HOHCHO$

CH_3
$|$
$CH_3 - CH - \overset{*}{C}HCl - CH_3$

Compounds (*iii*), (*iv*) and (*v*) have no chiral carbon atom in the molecule.
Hence they do not show enantiomerism.

Example 3: Point out the optically active compounds out of the following:

(*i*) $CH_2OH — CHOH — CHO$ (*ii*) $CH_3 — CHOH — CH_2OH$

(*iii*)
```
        CHOH
         |
  H  —  C  —  OH
         |
  HO —  C  —  H
         |
        COOH
```

Solution: Compounds (*i*) and (*ii*) have one chiral carbon (the middle one) and hence both these compounds are optically active compounds. (*iii*) has two chiral carbons (the middle ones) and therefore this compound is also optically active.

Flying Wedge representation

Spatial arrangement of atoms or groups in a molecule is called its configuration. The three dimensional (3D) structure of an organic molecule may be represented by flying wedge model. We use three types of lines in this model. A solid wedge (—) represents a bond lying above the plane of the paper projecting towards the observer, an ordinary line (—) represents a bond lying in the plane of the paper and a dotted wedge (|||||) represents a bond lying below the plane of paper projecting away from observer. For convenience, we can use a dotted line (----) in place of dotted wedge. For example the flying wedge representation of CHPQR is given alongside

Fischer's projection formula for planar representation of three dimensional molecules.

Emil Fischer in 1891 introduced a simple method for representing three dimensional molecules in one plane. It is known as *Fischer projection formula*. Following points are to be observed for this purpose:

(*i*) The chiral molecule is imagined in such a way that two groups point towards the observer and two away from the observer. The groups pointing towards the observer are written along the horizontal line (shown by thick wedge-like bonds) and those pointing away are written along the vertical line. The central carbon is present at the crossing of the horizontal and vertical lines.

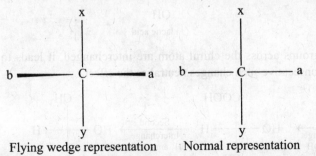

Flying wedge representation   Normal representation

Emil Fischer (1852–1919)
Emil Fischer was a German Chemist and recipient of 1902 Nobel Prize in chemistry. He developed the Fischer Projection, a symbolic way of drawing asymmetric compounds.

Thus if a, b, x and y are four groups attached to a carbon, the molecule will be represented by the projection formula as above. Here a and b groups point towards the observer (or above the plane) and groups x and y are away from the observer (or below the plane).

(*ii*) The longest chain of carbon atoms in the molecule (with the most oxidised carbon) should be represented along the vertical line. Lactic acid, therefore, according to the above conventions will be represented as

$$\begin{array}{c} \text{COOH} \\ | \\ \text{H} - \text{C} - \text{OH} \\ | \\ \text{CH}_3 \end{array} \qquad \text{or} \qquad \begin{array}{c} \text{COOH} \\ \text{H} - \!\!\!\!\!\!\!\!\!\!\mid\!\!\!\!\!\!\!\!\!\!-\text{OH} \\ \text{CH}_3 \end{array}$$

– COOH, – CHO, – CH$_2$OH, etc. groups should be located on the vertical line.

(*iii*) We can avoid writing carbon at the crossing of the vertical and horizontal lines. A crossing automatically means the presence of a carbon.

(*iv*) If necessary, planar formula may be imagined to be rotated from end to end without lifting it from the plane of the paper. Rotation by 180° in the plane of the paper does not create any change in the configuration of the molecule.

$$\begin{array}{c} \text{COOH} \\ \text{H}-\!\!\!\!\!\!\mid\!\!\!\!\!\!-\text{OH} \\ \text{CH}_3 \end{array} \xrightarrow{\text{Rotation through 180°}} \begin{array}{c} \text{CH}_3 \\ \text{HO}-\!\!\!\!\!\!\mid\!\!\!\!\!\!-\text{H} \\ \text{COOH} \end{array}$$

I (–) lactic acid II (–) lactic acid

In the above rotation by 180°, II has been obtained from I. There has been no change in configuration of the molecule. I and II are in fact the same thing.

(*v*) Rotation by 90° or 270° brings about a change in configuration of the molecule. Consider the following rotation.

$$\begin{array}{c} \text{COOH} \\ \text{H}-\!\!\!\!\!\!\mid\!\!\!\!\!\!-\text{OH} \\ \text{CH}_3 \end{array} \xrightarrow{\text{Rotation 90°}} \begin{array}{c} \text{H} \\ \text{CH}_3-\!\!\!\!\!\!\mid\!\!\!\!\!\!-\text{COOH} \\ \text{OH} \end{array}$$

(–) lactic acid (+) lactic acid

(*vi*) If the positions of two groups across the chiral atom are interchanged, it leads to inversion of configuration. Two consecutive such changes neutralise the effect.

$$\begin{array}{c} \text{COOH} \\ \text{H}-\!\!\!\!\!\!\mid\!\!\!\!\!\!-\text{OH} \\ \text{CH}_3 \end{array} \xrightarrow[\text{H and OH}]{\text{First Interchange}} \begin{array}{c} \text{COOH} \\ \text{HO}-\!\!\!\!\!\!\mid\!\!\!\!\!\!-\text{H} \\ \text{CH}_3 \end{array} \xrightarrow[\text{CH}_3 \text{ and COOH}]{\text{Second Interchange}} \begin{array}{c} \text{CH}_3 \\ \text{HO}-\!\!\!\!\!\!\mid\!\!\!\!\!\!-\text{H} \\ \text{COOH} \end{array}$$

(–) lactic acid (+) lactic acid (–) lactic acid
I II III

Structures III and I are the same because as per rule (iv) above, III on rotation through 180° will give I.

Absolute configuration of optical isomers

In the earlier days, as the modern techniques of finding out configuration were not available, Fischer assigned the following configurations to the (+) and (−) enantiomers of glyceraldehyde arbitrarily and denoted them by capital letters D and L respectively. Small letters *d* and *l* represent sign of rotation, while capital letters D and L represent configuration.

```
                    MIRROR
       CHO           |           CHO
        |            |            |
  H ————OH           |      HO————H
        |            |            |
       CH₂OH         |           CH₂OH
  D (+) glyceraldehyde       L (−) glyceraldehyde
```

The relative configurations of a number of other optically active compounds have been established by correlating them with D(+) or L(−) glyceraldehyde. All those optically active compounds, which are obtained from D(+) glyceraldehyde through a sequence of reactions *without breaking the bonds of asymmetric carbon atom*, are designated as D configuration irrespective of their sign of rotation and the other enantiomer as L configuration.

For example,

```
       CHO                        COOH
        |          Br₂/H₂O         |
  H————OH         ————————→   H————OH
        |                          |
       CH₂OH                      CH₂OH
  D(+) Glyceraldehyde         D (−) Glyceric acid
                                   │
                                   │ PBr₃
                                   ↓
       COOH                       COOH
        |          Zn, H⁺          |
  H————OH         ←————————   H————OH
        |                          |
       CH₃                        CH₂Br
  D(−) Lactic acid           D (−) 3-Bromo-
                             2-hydroxypropanoic acid
```

In all the D configurations, —OH attached to asymmetric carbon atom is written on the right hand side of Fischer projection formula. Similarly, in all the L-configurations, —OH attached to the lowest asymmetric carbon atom is written on the left hand side in the Fisher's projection formula.

R and S specification for the configuration of an optically active compound

Cahn, Ingold and Prelog developed a method which can be used to designate the configuration of all the molecules containing asymmetric carbon atom (chiral centre). This system is known as *Cahn-Ingold-Prelog system* or R and S system and involves two steps.

Step I. The four different atoms or groups of atoms attached to chiral carbon atom are assigned a sequence of priority according to the following set of sequence rules.

Sequence Rule 1. *If the four atoms, directly attached to asymmetric carbon atom, are all different, the priority depends on their atomic number. The atom of higher atomic number gets higher priority.* For example, in chloroiodomethane sulphonic acid the priority sequence is I, Cl, SO₃H, H

$$\underset{(4)}{\underset{|}{\text{H}}}\underset{(1)}{\text{I}} - \underset{|}{\overset{\overset{\text{SO}_3\text{H (3)}}{|}}{\text{C}}} - \underset{(2)}{\text{Cl}}$$

We consider the atom of the group which is directly linked to the central carbon.

Sequence Rule 2. *If Rule 1 fails to decide the relative priority of two groups it is determined by similar comparison of next atoms in the group and so on.* In other words, if two atoms directly attached to chiral centre are same, the next atoms attached to each of these atoms, are compared. For example in 2-butanol two of the atoms directly attached to chiral centre are carbon themselves. To decide the priority between the two groups —CH₃ and —CH₂CH₃, we proceed like this. Methyl carbon is further linked to H, H and H. The sum of atomic numbers of three H is 3. The methylene carbon of the ethyl group is linked to two hydrogens and one carbon directly. The sum of at. no. of two H and one C is 8. Thus ethyl group gets the priority over methyl. Hence the priority sequence is OH, C₂H₅, CH₃, H.

$$\underset{(2)}{\text{CH}_3 - \text{CH}_2} - \underset{\underset{\text{H (4)}}{|}}{\overset{\overset{\text{OH (1)}}{|}}{\text{C}}} - \underset{(3)}{\text{CH}_3}$$

2-Butanol

In 2-methyl-3-pentanol, the C, C, H of isopropyl gets priority over the C, H, H of ethyl, so the priority sequence is OH, isopropyl, ethyl, H.

$$\text{CH}_3 - \underset{(2)}{\overset{\overset{\text{CH}_3}{|}}{\text{CH}}} - \underset{\underset{(1)}{\underset{|}{\text{OH}}}}{\overset{\overset{\text{H (4)}}{|}}{\text{C}}} - \underset{(3)}{\text{CH}_2} - \text{CH}_3$$

2-Methyl-3-pentanol

In 1, 2-dichloro-3-methylbutane the Cl, H, H of CH₂Cl gets priority over the C, C, H of isopropyl due to atomic number of Cl being higher than that of C. So the priority sequence is Cl, CH₂Cl, isopropyl, H.

$$\text{CH}_3 - \underset{(3)}{\overset{\overset{\text{CH}_3}{|}}{\text{CH}}} - \underset{\underset{(1)}{\underset{|}{\text{Cl}}}}{\overset{\overset{\text{H (4)}}{|}}{\text{C}}} - \underset{(2)}{\text{CH}_2\text{Cl}}$$

1, 2-dichloro-3-methylbutane

Sequence Rule 3. *A doubly or triply bonded atom is considered equivalent to two or three such atoms; but two or three atoms, if attached actually, get priority over doubly or triply* bonded atom. In glyceraldehyde, O, O, H of — CHO gets priority over the O, H, H of — CH$_2$OH; so the priority sequence is — OH, — CHO — CH$_2$OH — H.

$$\begin{array}{c} (2)\ \text{CHO} \\ | \\ (4)\ \text{H} - \text{C} - \text{OH}\ (1) \\ | \\ \text{CH}_2\text{OH} \\ (3) \end{array}$$

Glyceraldehyde

Step II. After deciding the sequence of priority for four atoms or groups attached to asymmetric carbon atom; the molecule is visualised in such a way that the atom or group of lowest or last (*i.e.* fourth) priority is directed away from us, while the remaining three atoms or groups are pointing towards us. Now if on looking at these three groups (pointing towards us) in the order of their decreasing priority, our eye moves in clockwise direction, the configuration is specified as R (from Latin word *rectus* meaning right) and on the other hand if our eye moves in anticlockwise direction the configuration is specified as S (from Latin word *sinister* meaning left).

The following examples illustrate the above method for specification of configuration as R and S to molecules of compounds containing an asymmetric or chiral carbon atom.

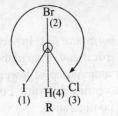 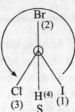 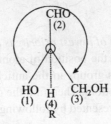 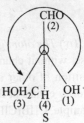

Configuration on the basis of projection formula

When a compound is represented by the Fischer projection formula, the configuration can be easily determined without constructing the model. To determine whether the eye travels clockwise or anticlockwise, we have to place the group or atom of the lowest priority at the bottom of the Fischer projection formula. The following four situations arise:

(*i*) *The atom/group of lowest priority is at the bottom*. In such a case, simply rotate the eye in the order of decreasing priorities. The configuration is R if the eye travels in clockwise direction and S if the eye travels in *anticlockwise* direction.

For example, Glyceraldehyde, represented by the following projection formula has R configuration:

$$\begin{array}{c} (1)\ \text{OH} \\ (3)\quad |\quad (2) \\ \text{HOH}_2\text{C} - \text{C} - \text{CHO} \\ | \\ \text{H} \end{array}$$

R—Glyceraldehyde

(*ii*) *The atom/group of lowest priority is at the top*. In such case, rotate the molecule by 180° so as to bring atom/group of lowest priority at the bottom. This can be done by reversing the position of all the atoms or groups. Then find the direction in the order of decreasing priorities. For example, the compound CHBrClI, represented by following projection formula, has S configuration:

$$\underset{\underset{Br}{|}}{\overset{\overset{H}{|}}{I-C-Cl}} \equiv \underset{\underset{H}{|}}{\overset{\overset{(2)\ Br}{|}}{\overset{(3)}{Cl}-\overset{(1)}{C}-I}}$$

S-Bromochloroiodomethane

(iii) *The atom/group of lowest priority is at the right hand side of the horizontal line.* In such case, change the position of atoms or groups in clockwise direction so that atom/group of the lowest priority comes at the bottom but do not change the position of the atom/group at the top of the vertical end. Then find the direction in the order of decreasing priorities. For example, CHBrClI, represented by following projection formula, has S configuration.

$$\underset{\underset{Cl}{|}}{\overset{\overset{Br}{|}}{I-C-H}} \equiv \underset{\underset{H}{|}}{\overset{\overset{(2)\ Br}{|}}{\overset{(3)}{Cl}-\overset{(1)}{C}-I}}$$

S-Bromochloroiodomethane

(iv) *The atom/group of lowest priority is at the left hand side of the horizontal line.* In such case, without changing the position of atom/group at the top of the vertical end, change the position of other atoms/groups in the anticlockwise direction so that atom/group of lowest priority comes at the bottom. Then find the direction in the order of decreasing priorities. For example, CHBrClI, represented by following projection formula, has R configuration.

$$\underset{\underset{Cl}{|}}{\overset{\overset{Br}{|}}{H-C-I}} \equiv \underset{\underset{H}{|}}{\overset{\overset{(2)\ Br}{|}}{\overset{(1)}{I}-\overset{(3)}{C}-Cl}}$$

R-Bromochloroiodomethane

Solved Examples

Example 4: Assign R and S configuration to the following:

(i) $\underset{\underset{CH_3}{|}}{\overset{\overset{Cl}{|}}{H-C}} - \underset{\underset{H}{|}}{\overset{\overset{CH_3}{|}}{C-H}}$

(ii) $\underset{\underset{Cl}{|}}{\overset{\overset{NH_2}{|}}{HO-C-CH_3}}$

(iii) $\underset{\underset{COOH}{|}}{\overset{\overset{H}{|}}{HO-C-CH_3}}$

(iv) $\underset{\underset{CHO}{|}}{\overset{\overset{C_2H_5}{|}}{CH_3-C-COOH}}$

Solution: (*i*) 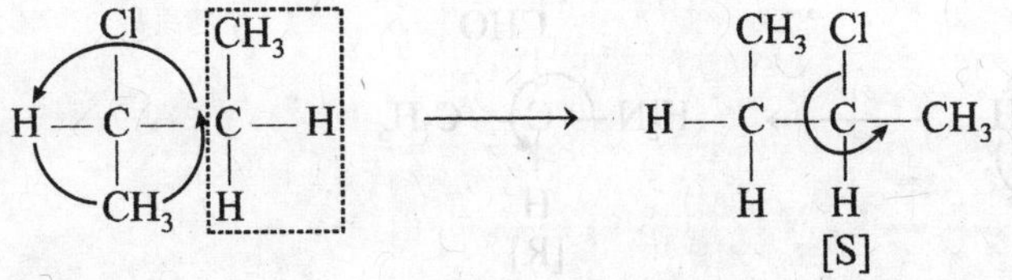

(*ii*) 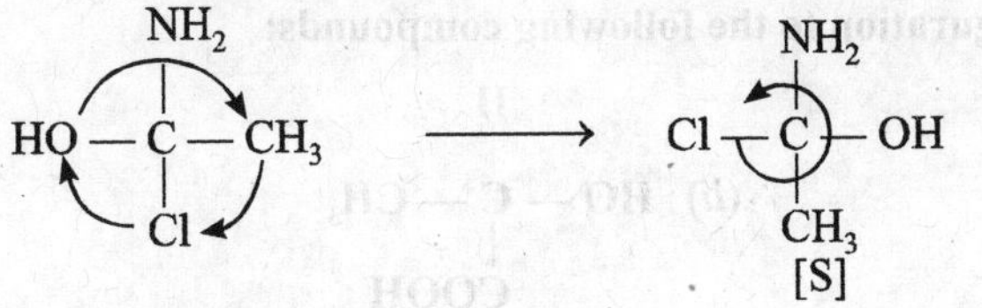

(*iii*) 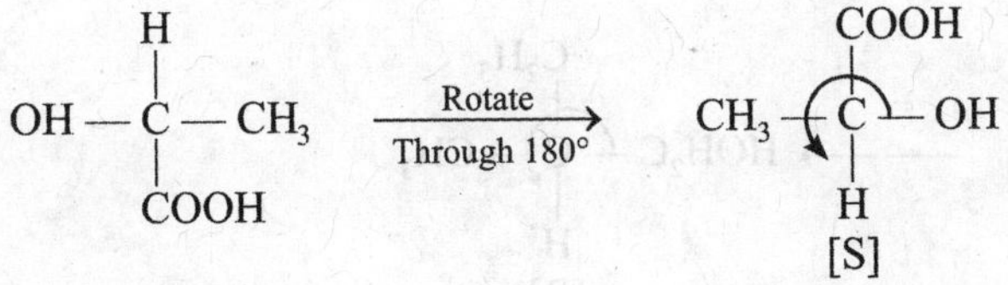

(*iv*) 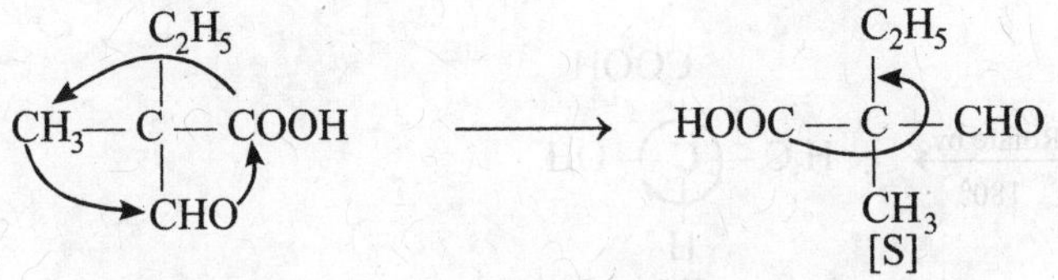

Example 5: Assign R and S configuration to the following:

(*i*) 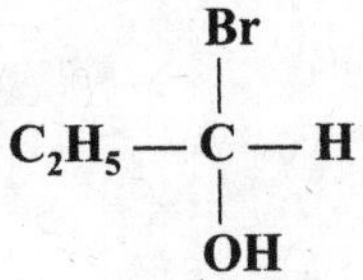

(*ii*) 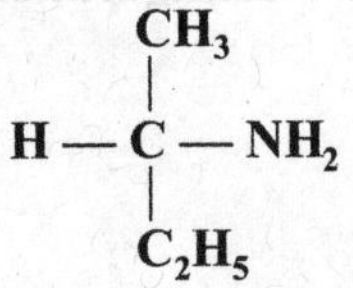

(*iii*) 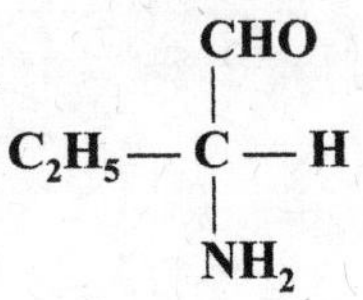

Solution: (*i*) 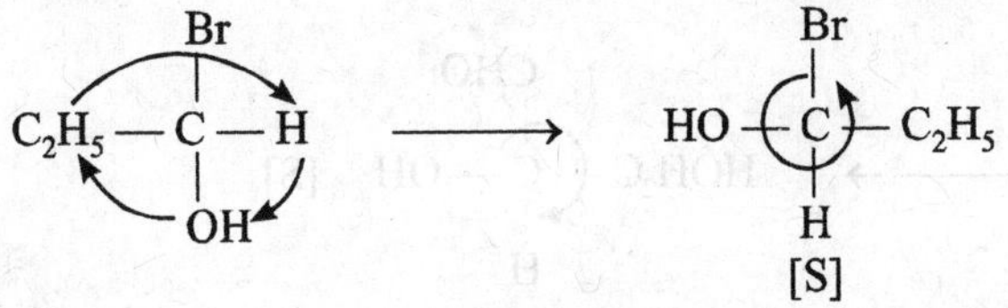

(*ii*) 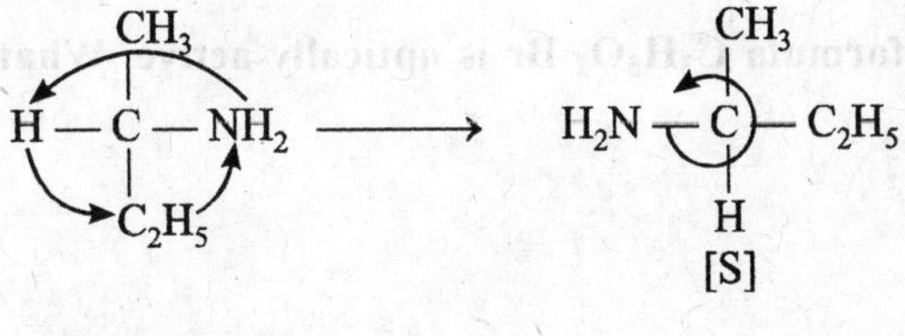

(iii)
$$C_2H_5-\underset{NH_2}{\overset{CHO}{\underset{|}{C}}}-H \longrightarrow H_2N-\underset{H}{\overset{CHO}{\underset{|}{C}}}-C_2H_5 \quad [R]$$

Example 6: Assign R and S configuration to the following compounds:

(i) $CH_3-\underset{CH_2OH}{\overset{C_2H_5}{\underset{|}{C}}}-H$

(ii) $HO-\underset{COOH}{\overset{H}{\underset{|}{C}}}-CH_3$

Solution: (i)
$$CH_3-\underset{CH_2OH}{\overset{C_2H_5}{\underset{|}{C}}}-H \longrightarrow HOH_2C-\underset{H}{\overset{C_2H_5}{\underset{|}{C}}}-CH_3 \quad [R]$$

(ii)
$$OH-\underset{COOH}{\overset{H}{\underset{|}{C}}}-CH_3 \xrightarrow{\text{Rotate by } 180°} H_3C-\underset{H}{\overset{COOH}{\underset{|}{C}}}-OH \quad [S]$$

Example 7: Assign R and S configuration to the following:

(i) $H-\underset{CH_3}{\overset{COOH}{\underset{|}{C}}}-NH_2$

(ii) $OH-\underset{CH_2OH}{\overset{CHO}{\underset{|}{C}}}-H$

Solution: (i)
$$H-\underset{CH_3}{\overset{COOH}{\underset{|}{C}}}-NH_2 \longrightarrow H_2N-\underset{H}{\overset{COOH}{\underset{|}{C}}}-CH_3 \quad [R]$$

(ii)
$$HO-\underset{CH_2OH}{\overset{CHO}{\underset{|}{C}}}-H \longrightarrow HOH_2C-\underset{H}{\overset{CHO}{\underset{|}{C}}}-OH \quad [S]$$

Example 8: A carboxy acid of the formula $C_3H_5O_2$ Br is optically active. What is its structure?

Solution: Two structures are possible for a carboxy acid with the above molecular formula

$$CH_3 - \underset{\underset{Br}{|}}{\overset{\overset{H}{|}}{C}} - COOH \qquad BrCH_2 - \underset{\underset{H}{|}}{\overset{\overset{H}{|}}{C}} - COOH$$

$$\text{I} \qquad\qquad\qquad \text{II}$$

Of the two structures, structure I has a chiral (or asymmetric) carbon atom whereas structure II has none. Therefore the structure showing optical activity is I.

Molecules with Two or More Chiral Centres

In such a case, firstly the configuration about each of the chiral carbon is specified and then with the help of numbers, the specification pertaining to the carbon atom of that number is written. Thus the configurations of isomers of 2, 3, 4-Trihydroxybutanal are:

$$\overset{1}{C}HO \qquad\qquad\qquad \overset{1}{C}HO$$
$$H - \overset{2}{C} - OH \qquad\qquad HO - \overset{2}{C} - H$$
$$H - \overset{3}{C} - OH \qquad\qquad HO - \overset{3}{C} - H$$
$$\overset{4}{C}H_2OH \qquad\qquad\qquad \overset{4}{C}H_2OH$$

(2R, 3R) -2, 3, 4-Trihydroxy butanal (2S, 3S) -2, 3, 4-Trihydroxy butanal
(A)   (B)

The configurations are explained as follows:

In the compound A above, sequence of groups attached to C_2 is OH, CHO, CHOH — CH_2OH and H.

Now in order to fix specification, first we consider C_2 and ignore C_3

$$\begin{array}{c} CHO \\ | \\ H - C_2 - OH \\ | \\ CHOHCH_2OH \end{array} \quad \xrightarrow{\text{Interchanges}} \quad \begin{array}{c} CHO \\ | \\ HO - C_2 - CHOHCH_2OH \\ | \\ H \end{array}$$

Ignoring (C_3)   (2R)

Similarly we consider C_3 and we ignore C_2. Thus

$$\begin{array}{c} CHOHCHO \\ | \\ H - C_3 - OH \\ | \\ CH_2OH \end{array} \quad \xrightarrow{\text{Interchanges}} \quad \begin{array}{c} CHOHCHO \\ | \\ HO - C_3 - CH_2OH \\ | \\ H \end{array}$$

Ignoring (C_2)   (3R)

(Here the sequence of groups attached to C_3 is OH, CHOHCHO, CH_2OH and H)

Hence the configuration of compound A is (2R, 3R) – 2, 3, 4-Trihydroxy butanal. Similarly the configuration of compund B can be derived as explained below:

For C$_2$:

$$\underset{\text{Ignoring (C}_3)}{\overset{\text{CHO}}{\underset{\text{CHOHCH}_2\text{OH}}{\text{HO}-\text{C}_2-\text{H}}}} \xrightarrow{\text{Interchanges}} \underset{(2S)}{\overset{(2)\,\text{CHO}}{\underset{\text{H}}{\overset{(3)}{\text{HOH}_2\text{CHOHC}}-\text{C}_2-\overset{(1)}{\text{OH}}}}}$$

For C$_3$:

$$\underset{\text{Ignoring (C}_2)}{\overset{\text{CHOHCHO}}{\underset{\text{CH}_2\text{OH}}{\text{HO}-\text{C}_3-\text{H}}}} \xrightarrow{\text{Interchanges}} \underset{(3S)}{\overset{(2)\,\text{CHOHCHO}}{\underset{\text{H}}{\overset{(3)}{\text{HOH}_2\text{C}}-\text{C}-\overset{(1)}{\text{OH}}}}}$$

Hence configuration of compund B is (2S, 3S) – 2, 3, 4-Trihydroxybutanal.

Number of stereoisomers (optical forms and meso forms)

Different cases are discussed as under:

1. When the molecule is unsymmetrical and contains n stereogenic or chiral carbon atoms.

Number of optical isomers (OI) or d/l or (+)/(–) forms = 2^n

Number of meso isomers (MI) = 0

Total No. of isomers (OI + MI) = $2^n + 0 = 2^n$

Consider for example the molecule $CH_3 - \overset{*}{C}HBr - \overset{*}{C}HBr - COOH$. This molecule is unsymmetrical and has two dissimilar chiral carbon atoms. Total no. of isomers for this compound is 2^2 as shown below (structure I, II, III, & IV)

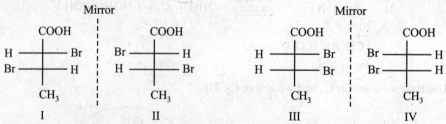

2. When the molecule is symmetrical and has even no. of stereogenic centres or chiral carbon atoms.

Number of optical isomers (OI) = 2^{n-1}

No. of meso isomers (MI) = $2^{(n/2-1)}$

Total No. of isomers (OI + MI) = $2^{(n-1)} + 2^{(n/2-1)}$

Consider for example tortaric acid $HOOC - \overset{*}{C}HOH - \overset{*}{C}HOH - COOH$. It is symmetrical and has two chiral carbon atoms shown by asterisks. It has a total of 3 isomers

No. of optical isomers (OI) = $2^{2-1} = 2$

Number of meso isomers (MI) = $2^{(2/2-1)} = 2^0 = 1$

Total number of isomers (OI + MI) = $2 + 1 = 3$

3. When the molecule is symmetrical and has odd number of chiral atoms.

No of optical isomers (OI) = $2^{(n-1)} - 2^{(n/2-1/2)}$

No. of meso isomers (MI) = $2^{(n/2-1/2)}$

Total No. of isomers (OI + MI) = $2^{n-1} - 2^{(n/2-1/2)} + 2^{(n/2-1/2)} = 2^{n-1}$

Consider the molecule HOOC–$\overset{*}{C}$HOH–$\overset{*}{C}$HOH–$\overset{*}{C}$HOH–COOH. It has three chiral carbon atoms.

	Mirror		Plane of symmetry		
COOH	|	COOH	COOH	COOH	No. of optical isomers = 2
HO—┼—H	|	H—┼—OH	H—┼—OH	H—┼—OH	
H—┼—OH	|	HO—┼—H	H—┼—OH	HO—┼—H	No. of meso isomers = 2
H—┼—OH	|	HO—┼—H	H—┼—OH	H—┼—OH	
COOH	|	COOH	COOH	COOH	Total no. of isomers = 2 + 2 = 4
I		II	III	IV	

Solved Examples

Example 9: Prove that the presence or absence of chiral carbon atom in a molecule is not the necessary and sufficient criterion for existence of optical activity.

Solution: Optical activity is a property which is related to dissymmetry in the molecule. Dissymmetry occurs normally in compounds with chiral carbon atoms. But sometimes there are deviations. For example,

(i) Consider the case of meso tartaric acid

$$\begin{array}{c} \text{COOH} \\ | \\ \text{H—}\overset{*}{\text{C}}\text{—OH} \\ | \\ \text{H—}\overset{*}{\text{C}}\text{—OH} \\ | \\ \text{COOH} \end{array}$$

Although there are two chiral carbons marked with asterisks, still the compound does not show optical activity. This is because the molecule has a plane of symmetry and is thus non-dissymmetric.

(ii) There are molecules which do not contain a chiral carbon but still show optical activity. Consider the optically active substance

$$RCH = C = CHR$$

Optical activity of this molecule is explained on the basis of dissymmetry in the molecule. Thus it can be remarked that presence or absence of chiral C-atom is not the necessary and sufficient criterion from the existence of optical activity.

In fact it is the dissymmetry criterion which is responsible for the same.

A dissymmetric molecule is that which has no plane of symmetry.

Example 10: Assign R and S configuration to the following:

(i)
$$HO-\underset{|}{\overset{COOH}{\underset{|}{C^2}}}-H$$
$$H-\underset{|}{\overset{|}{\underset{COOH}{C^3}}}-OH$$

(ii)
$$H-\underset{|}{\overset{NH_2}{\underset{COOC_2H_5}{C}}}-COOH$$

(iii)
$$ClCH_2-\underset{|}{\overset{Cl}{\underset{CH_3}{C}}}-CH(CH_3)_2$$

Solution: *(i)* **First consider C_2.** The groups in order of priority attached to C_2 are —OH, —COOH, —CH(OH) COOH, H

$$HO-\underset{CH(OH)COOH}{\overset{COOH}{C^2}}-H \equiv HOOC(OH)CH-\underset{H}{\overset{COOH}{C}}-OH \quad \text{Thus the configuration around } C_2 \text{ is S}$$

Now consider C_3. The groups in order of priority attached to C_3 are —OH, —COOH, —CH(OH)COOH, H

$$H-\underset{COOH}{\overset{CH(OH)COOH}{C^3}}-OH \equiv HO-\underset{H}{\overset{CH(OH)COOH}{C^3}}-COOH \quad \text{Thus the configuration around } C_3 \text{ is S}$$

Thus this structure has the configuration 2S, 3S

(ii)
$$H-\underset{|}{\overset{NH_2}{\underset{COOC_2H_5}{C}}}-COOH$$

The groups attached to central carbon, in order of priority are —NH_2, —$COOC_2H_5$, —COOH and H

$$H-\underset{COOC_2H_5}{\overset{NH_2}{C}}-COOH \equiv HOOC-\underset{H}{\overset{NH_2}{C}}-COOC_2H_5. \quad \text{It has R configuration}$$

(iii)
$$ClCH_2-\underset{|}{\overset{Cl}{\underset{CH_3}{C}}}-CH(CH_3)_2$$

The groups attached, in order of priority, to the central carbon are —Cl, —CH_2Cl, —$CH(CH_3)_2$ and CH_3.

$$\text{ClCH}_2-\overset{\text{Cl}}{\underset{\text{CH}_3}{\text{C}}}-\text{CH}(\text{CH}_3)_2 \qquad \text{The compound has thus S configuration}$$

Example 11: Assign R and S configuration to the following Fischer projection.

(i)
```
      Cl
      |
H ---- OH
      |
H ---- OH
      |
      Cl
```

(ii)
```
         C₂H₅
         |
H₃C ---- OH
         |
H  ----  CH₃
         |
         OH
```

Solution. (i) The groups attached to C-2, in order of priority are —Cl, —OH, —CH(OH) Cl and —H

$$H-\overset{\text{Cl}}{\underset{\text{CH(OH)Cl}}{\overset{2}{\text{C}}}}-OH \equiv HO-\overset{\text{Cl}}{\underset{\text{H}}{\overset{2}{\text{C}}}}-CH(OH)Cl \qquad \text{Configuration around C-2 is S}$$

Now consider configuration around C-3.

The groups around C-3 in order of priority are —Cl, —OH, —CH(OH)Cl and H

$$H-\overset{\text{CH(OH)Cl}}{\underset{\text{Cl}}{\overset{3}{\text{C}}}}-OH \equiv HO-\overset{\text{CH(OH)Cl}}{\underset{\text{H}}{\overset{3}{\text{C}}}}-Cl \qquad \text{The configuration around C-2 is } R$$

Thus complete configuration of the compound is 2S, 3R

(ii)
```
         C₂H₅
         |
H₃C ---- OH
         |
H  ----  CH₃
         |
         OH
```

The groups around C_2, in order of priority are —OH, —CH(OH) CH₃, C₂H₅ and CH₃

$$H_3C-\overset{\text{C}_2\text{H}_5}{\underset{\text{CH(OH)CH}_3}{\overset{2}{\text{C}}}}-OH \equiv HO-\overset{\text{C}_2\text{H}_5}{\underset{\text{CH}_3}{\text{C}}}-CH(OH)CH_3 \qquad \text{The configuration around C-2 is S}$$

Configuration around C-3

$$H-\overset{\text{C(OH) (CH}_3\text{) C}_2\text{H}_5}{\underset{\text{OH}}{\overset{3}{\text{C}}}}-CH_3 \qquad \text{The groups around C-3, in order of priority are —OH, —C(OH) (CH}_3\text{) C}_2\text{H}_5\text{, —CH}_3 \text{ and —H}$$

$$\underset{\underset{\text{OH}}{|}}{\overset{\overset{\text{C(OH)(CH}_3\text{)C}_2\text{H}_5}{|}}{\text{H}-\overset{3}{\text{C}}-\text{CH}_3}} \equiv\equiv\equiv \underset{\underset{\text{H}}{|}}{\overset{\overset{\text{C(OH)(CH}_3\text{)C}_2\text{H}_5}{|}}{\text{H}_3\text{C}-\overset{3}{\text{C}}-\text{OH}}}$$

The configuration around C-3 is S

Thus complete configuration of the compound is 2S, 3S.

Example 12: A compound $C_4H_{10}O$ shows optical activity. Identify the compound and write the possible stereoisomers.

Solution: With the molecular formula $C_4H_{10}O$, a no. of alcohols and ethers are possible. But we are interested in a compound with a chiral carbon atom so as to give optical activity. Such a compound with the formula $C_4H_{10}O$ is 2-Butanol. It exists in two enantiomeric forms.

$$\underset{\underset{\text{H}}{|}}{\overset{\overset{\text{CH}_3}{|}}{\text{HO}-\text{C}-\text{C}_2\text{H}_5}} \qquad \underset{\underset{\text{H}}{|}}{\overset{\overset{\text{CH}_3}{|}}{\text{C}_2\text{H}_5-\text{C}-\text{OH}}}$$

$\qquad\qquad$ S-2-Butanol $\qquad\qquad\qquad$ R-2-Butanol

Racemic modifications

Racemic modification is the term used for a mixture of equal amounts of enantiomers. A racemic mixture is optically inactive because of external compensation. The optical activity caused by one enantiomer is neutralised by the activity of the other enantiomer. The notation for a racemic modification or mixture is ± or *dl*. A racemic mixture may also be denoted by the letters R and S. For examples RS-sec. butyl chloride.

When a chiral compound is synthesised from an achiral reactant, a racemic variety of products is obtained. For example, when propionic acid is borminated, α-bromo propionic acid (a chiral product) is obtained. The two enantiomers (+) and (−) α-bromopropionic acids are formed in equal quantities and the product is a racemic mixture. It is optically inactive.

Optical isomerism (enantiomerism) in compounds having two dissimilar chiral carbon atoms (Distereoisomers)

Two asymmetric carbon atoms are said to be dissimilar when atoms or groups attached to one asymmetric carbon atom are different from those attached to the other. Compounds of this type exist in 2^2 i.e., 4 stereoisomers. For example 3-chloro-2-butanol is a compound with two dissimilar chiral carbons marked 1 and 2 and exists in the following four forms:

$$\underset{\underset{\text{CH}_3}{|}}{\overset{\overset{\text{CH}_3}{|}}{\text{H}-\text{C}-\text{OH}\atop\text{Cl}-\text{C}-\text{H}}} \qquad \underset{\underset{\text{CH}_3}{|}}{\overset{\overset{\text{CH}_3}{|}}{\text{HO}-\text{C}-\text{H}\atop\text{H}-\text{C}-\text{Cl}}} \qquad \underset{\underset{\text{CH}_3}{|}}{\overset{\overset{\text{CH}_3}{|}}{\text{H}-\text{C}-\text{OH}\atop\text{H}-\text{C}-\text{Cl}}} \qquad \underset{\underset{\text{CH}_3}{|}}{\overset{\overset{\text{CH}_3}{|}}{\text{HO}-\text{C}-\text{H}\atop\text{Cl}-\text{C}-\text{H}}}$$

$\qquad$ (I) $\qquad\qquad\qquad$ (II) $\qquad\qquad\qquad$ (III) $\qquad\qquad\qquad$ (IV)

The structures I and II are non-superimposable mirror images, so these are a pair of enantiomers. Similarly structures III and IV are another pair of enantiomers. These two pairs of enantiomers give rise to two racemic modifications.

The structures I and III are neither enantiomers nor superimposable. Such type of stereoisomers are called *diastereomers*. Similarly I and IV, II and III, and II and IV, are the other pairs of diastereomers. *Diastereomers can be defined as those stereoisomers which are not mirror images of each other.*

Unlike enantiomers, diastereomers have different physical properties and may rotate the plane of polarised light in the same or different directions and to different extent.

Another example of this type is 2, 3-dichloropentane.

```
     C₂H₅              C₂H₅              C₂H₅              C₂H₅
      |                 |                 |                 |
 H — C — Cl        Cl — C — H        H — C — Cl        Cl — C — H
      |                 |                 |                 |
 Cl — C — H        H — C — Cl        H — C — Cl        Cl — C — H
      |                 |                 |                 |
     CH₃               CH₃               CH₃               CH₃
      (I)               (II)              (III)             (IV)
```
|_____| |_____|
    A pair of enantiomers   Second pair of enantiomers

Properties of Diastereomers

1. They show similar, but not identical, chemical properties (as they contain the same functional groups). Rates of reactions of diastereomers with a given reagent are generally different.
2. They have different physical properties like m.p. and b.p., densities, refractive indices, specific rotations, solubilities etc. in a given solvent.
3. They can be separated by techniques like fractional crystallisation, fractional distillation and chromatography.

Example 13: Write four configurations of tartaric acid and select the pairs forming (*i*) Enantiomers, (*ii*) Diastereomers. Which of them are optically active and which of them are not. Why?

Solution. Tartaric acid is an example of compounds with two similar chiral carbon atoms. Here the four groups attached to both carbons (C-2 and C-3) are the same. Various configurations of tartaric acid are:

```
    COOH             COOH             COOH             COOH
     |                |                |                |
 H — C — OH      HO — C — H       H — C — OH      HO — C — H
     |                |                |                |
OH — C — H       H — C — OH       H — C — OH      HO — C — H
     |                |                |                |
    COOH             COOH             COOH             COOH
     (I)              (II)             (III)            (IV)
```

If we rotate configuration IV by an angle of 180°, we get configuration III. Hence structures III and IV represent the same configuration. Thus there are only three configurations of tartaric acid viz., I, II & III.

(*i*) Structures I and II are enantiomers, they are mirror images of each other. There is no plane of symmetry in either of them. Therefore both I and II are optically active but display the optical activity in opposite directions.

(*ii*) Structures I and III are not mirror images of each other. Hence they are not enantiomers. Such pairs of compounds having the identical molecular formula and identical groups are called diastereomers.

Similarly structures II and III also form a pair of diastereomers.

Compound III has two chiral carbons, but still it optically inactive. This is because there is a plane of symmetry as indicated by dotted line. There being no dissymmetry, the compound is not optically active. Such a compound is caused meso comp. Thus III and IV represent meso tartaric acid.

Retention and inversion of configuration

A reactant with a chiral molecule in a reaction gives a product which may have the same configuration as the reactant or an opposite configuration. *If the configuration of the reactant and product is the same, the reaction is said to proceed with retention of configuration-And if the reactant and the product have opposite configurations, the reaction is said to occur with inversion of configuration consider the following examples.*

S_N^2 *reaction of alkyl halide with* OH^- *ions*. This reaction takes place with *inversion* of configuration as shown below

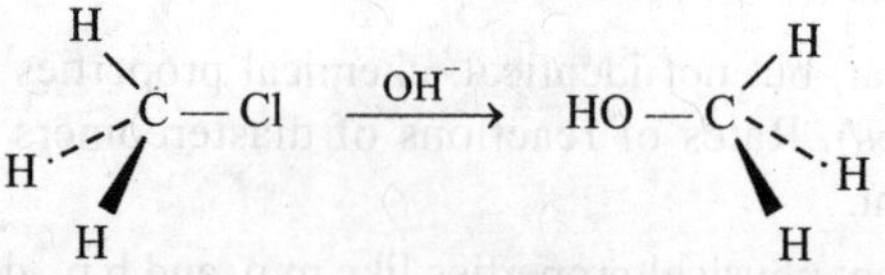

$$\text{H}_3\text{C—Cl} \xrightarrow{OH^-} \text{HO—CH}_3$$

Reaction of a secondary alcohol with tosyl chloride (p-toluenesulphonyl chloride) in the presence of pyridine.

This reaction takes place with retention of configuration and the product is alkyl tosylate

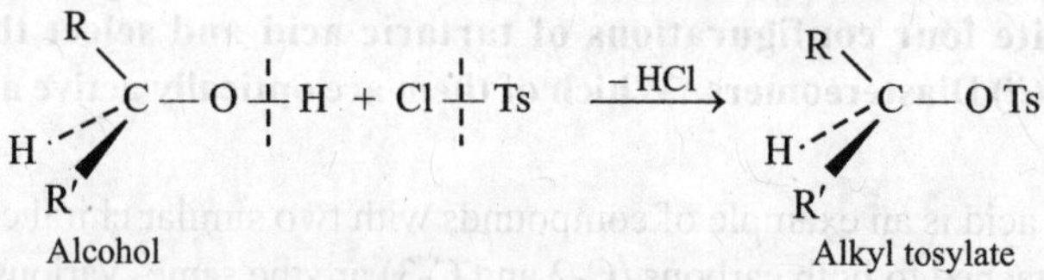

$$\underset{\text{Alcohol}}{R\text{—C—O}{\mid}H} + Cl{\mid}Ts \xrightarrow{-HCl} \underset{\text{Alkyl tosylate}}{R\text{—C—OTs}}$$

Racemisation

Under suitable conditions, most of the optically active compounds can lose their optical activity without undergoing any change in their structure *i.e.*, the two enantiomers forms are convertible into each so that the final result is racemic modification. The transformation of an optically active enantiomer into the optically inactive racemic modification under the influence of heat, light or chemical reagents is known as racemisation. Thus if the starting material is the (+) form, then after treatment, half will be converted into (−) form. If the starting material is (−) form, half will be converted into (+) form. For example, (+) or (−) lactic acid on warming with sodium hydroxide gets converted into (±) lactic acid.

Though different mechanisms have been presented for the racemisation of different types of compounds in most of the cases it occurs via the formation of an intermediate which is no longer dissymmetric or chiral. For example, a ketone in which chiral carbon atom is joined to a hydrogen atom can undergo racemisation via the formation of enolic form by tautomeric change.

$$\underset{\text{(+) form}}{\begin{array}{c} C_2H_5 \\ | \\ CH_3-C-H \\ | \\ C=O \\ | \\ C_6H_5 \end{array}} \rightleftarrows \underset{\text{loss of chirality}}{\begin{array}{c} CH_3 \quad C_2H_5 \\ \diagdown \diagup \\ C \\ || \\ C \\ | \\ C_6H_5 \end{array}} \rightleftarrows \underset{\text{(−) form}}{\begin{array}{c} C_2H_5 \\ | \\ H-C-CH_3 \\ | \\ C=O \\ | \\ C_6H_5 \end{array}}$$

The intermediate enolic form, which is no longer chiral when reverts to the stable keto form, it is equally likely to produce (+) or (−) forms and thus racemisation takes place.

Example 14: Giving examples distinguish between the following: Meso and Racemic forms.

Solution: Meso form of a compound is optically inactive form, in spite of the presence of asymmetric carbon atoms in it. This is because there is a plane of symmetry in the molecule. The activity of one part is neutralised by the activity of the other part. An example is meso tartaric acid.

Racemic form means a mixture of equal amounts of d and l forms of a compound, like mixture of equal amounts of d and l lactic acid or ± tartaric acid.

$$\begin{array}{c} COOH \\ | \\ H-C-OH \\ | \\ H-C-OH \\ | \\ COOH \end{array} \text{meso tartaric acid}$$

Such a mixture does not show any optical activity.

It may be mentioned that optical inactivity of the meso form is due to **internal compensation** as the activity of one part of the molecule is neutralised by that of the other part.

Optical inactivity of the racemic form is due to external compensation as the two forms d- and l-neutralise the optical activity of each other.

Threo and Erythro Notations

Diastereomers are optical isomers of a compound which are not mirror images of each other and hence are not enantiomers. For example d-tartaric acid is a diastereomer of mesotartaric acid. But the two are not enantiomers.

As for as the absolute configuration of a compound is concerned we derive it from the configuration of glyceraldehyde which has been taken as arbitrary standard.

$$\underset{\text{D (+) glyceraldehyde}}{\begin{array}{c} CHO \\ | \\ H-\!\!\!-\!\!\!-\!\!\!-OH \\ | \\ CH_2OH \end{array}} \qquad \underset{\text{L (−) glyceraldehyde}}{\begin{array}{c} CHO \\ | \\ HO-\!\!\!-\!\!\!-\!\!\!-H \\ | \\ CH_2OH \end{array}}$$

D and L stand for the configuration while (+) and (−) signs denote the actual direction of rotation of plane polarised light. If H and OH are on the LHS and RHS of central carbon, it donates D configuration. If OH and H are on the LHS and RHS of central carbon it denotes L configuration.

Any compound that can be obtained from or converted into D (+) glyceraldehyde has D configuration. Similarly any compound that can be obtained from or converted into L (−) glyceraldehyde has L configuration.

Thus, when we have to decide the configuration of compounds (particularly sugars) containing more than three carbons, we shall check the configuration of lower two carbons as shown below.

$$X \quad \boxed{\begin{array}{c} | \\ H-C-OH \\ | \\ CH_2OH \end{array}} \qquad Y \quad \boxed{\begin{array}{c} | \\ HO-C-H \\ | \\ CH_2OH \end{array}}$$

D-series   L-series

If the arrangement corresponds to X, the compound has D configuration. If the arrangement corresponds to Y, the compound belongs to L configuration.

Let us take the example of CHO $\overset{*}{C}$H OH $\overset{*}{C}$HOH CH$_2$OH. This contains two asymmetric carbon atoms marked with asterisks. There are four optical isomers possible, all of which are known.

If the H on the third carbon atom (from the bottom) is on the left hand side, the compound is *erythro*, while if the H on the third carbon atom is on the right hand side, the compound is *threo*. The structures of D-erythrose and D-threose which may be assumed to be obtained from D-glyceraldehyde are given below

$$\begin{array}{c} CHO \\ H \!-\!\!\!\!\!\!\!\!-\!\!\!\!\!\!\!\!-\!\!\!\!\!\!\!\!- OH \\ H \!-\!\!\!\!\!\!\!\!-\!\!\!\!\!\!\!\!-\!\!\!\!\!\!\!\!- OH \\ CH_2OH \\ \text{D (–) erythrose} \\ \text{I} \end{array} \longleftarrow \begin{array}{c} CHO \\ H \!-\!\!\!\!\!\!\!\!-\!\!\!\!\!\!\!\!-\!\!\!\!\!\!\!\!- OH \\ CH_2OH \\ \text{D (+) Glyceraldehyde} \end{array} \longrightarrow \begin{array}{c} CHO \\ HO \!-\!\!\!\!\!\!\!\!-\!\!\!\!\!\!\!\!-\!\!\!\!\!\!\!\!- H \\ H \!-\!\!\!\!\!\!\!\!-\!\!\!\!\!\!\!\!-\!\!\!\!\!\!\!\!- OH \\ CH_2OH \\ \text{D (–) Threose} \\ \text{II} \end{array}$$

Compounds I and II are diastereomers of each other from the D-series. Similarly there will be another pair of diastereomers from the L-series. It must be noted that erythro compound (I) on oxidation gives mesotartaric acid.

$$\begin{array}{c} CHO \\ H \!-\!\!\!\!\!\!\!\!-\!\!\!\!\!\!\!\!-\!\!\!\!\!\!\!\!- OH \\ H \!-\!\!\!\!\!\!\!\!-\!\!\!\!\!\!\!\!-\!\!\!\!\!\!\!\!- OH \\ CH_2OH \\ \text{I} \end{array} \xrightarrow{[O]} \begin{array}{c} COOH \\ H \!-\!\!\!\!\!\!\!\!-\!\!\!\!\!\!\!\!-\!\!\!\!\!\!\!\!- OH \\ H \!-\!\!\!\!\!\!\!\!-\!\!\!\!\!\!\!\!-\!\!\!\!\!\!\!\!- OH \\ COOH \\ \text{meso tartaric acid} \end{array}$$

Similarly the erythro compound from L-series on oxidation will also produce meso tartaric acid.

Example 15: **Predict whether 3-chlorohexane will be optically active or not? Give reasons for your answer.**

Solution:

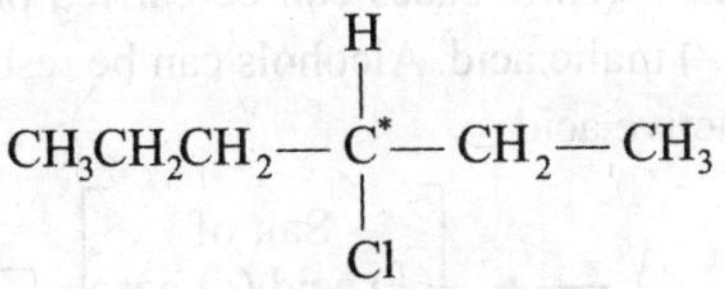

$$CH_3CH_2CH_2 - \overset{H}{\underset{Cl}{\overset{|}{C^*}}} - CH_2 - CH_3$$

3-chlorohexane

The compound contains are asymmetric carbon atom (marked with *). The four different groups attached are
1. — H
2. — Cl
3. — CH_2 — CH_3
4. — CH_3 — CH_2 — CH_3

Hence the compund shows optical activity.

Methods for the resolution of racemic mixtures

The separation of racemic modification into enantiomers is called resolution. Since the two enantiomers in a racemic mixture have identical physical and chemical properties, these cannot be separated by usual methods of fractional distillation or fractional crystallisation. Special methods are adopted for their separation as given below:

(i) Mechanical separation

This method was first adopted by Pasteur for separating the enantiomers of ammonium tartarate. When racemic modification is crytallized from a solution, two types of crystals are obained. These are mirror images of each other consisting of (+) and (–) forms which can be separated by hand picking with the help of a pair of tweezers and a powerful lens. This is a very laborious method and can be applied only to those compounds which give well defined distinguishable crystals of enantiomers.

(ii) Biochemical method

Certain micro-organisms grow in a racemic mixture, consuming only one of the enantiomers while leaving the other unaffected. Thus *penicillium glaucum* when placed in (±) tartaric acid, consumes only (+) tartaric acid and leaves (–) tartaric acid unused. The major disadvantage of this method is that one of the enantiomers get destroyed.

(iii) Chemical method

This method is mostly used for the resolution of racemic modification. In this method the racemic modification is treated with an optically active reagent to get a pair of *diastereomers*. Since diastereomers differ in their physical properties, it is possible to separate them by physical methods such as fractional crystallisation, fractional distillation etc. The pure diastereomers are then decomposed, into a mixture of optically active reagent and corresponding enantiomer, which can be separated.

Suppose the racemic modification is an (±) acid. When it is treated with an optically acitve, say (–) base, it gives a mixture of two salts, one of (+) acid (–) base, the other of (–) acid (–) base.

These salts are neither superimposable nor mirror images; so these are diastereomers having different physical properties and can be separated by fractional crystallisation. After separation the optically active acids can be recovered in pure forms by adding a mineral acid.

The commonly used optically active bases for the purpose are naturally occurring alkaloids such as (–) brucine, (–) quinine, (–) strychnine and (+) cinchonine.

Similarly, the resolution of racemic bases can be carried out using a naturally occurring optically active acid such as (–) malic acid. Alcohols can be resolved in a similar way by ester formation using an optically active acid.

$$\begin{bmatrix}(+)\text{ acid}\\(-)\text{ acid}\end{bmatrix} + (-)\text{ base} \longrightarrow \begin{matrix}\begin{bmatrix}\text{Salt of}\\(+)\text{ acid }(-)\text{ base}\end{bmatrix} \xrightarrow{H^+} (+)\text{ acid}\\ \\ \begin{bmatrix}\text{Salt of}\\(-)\text{ acid }(-)\text{ base}\end{bmatrix} \xrightarrow{H^+} (-)\text{ acid}\end{matrix}$$

  Racemic   Diastereomers easily   Enantiomers
 modification   separable   in pure form

Asymmetric Synthesis

The preparation of an optically active dissymmetric (chiral) compound from non-dissymmetric molecules under the influence of an optically active substance is known as asymmetric synthesis.

Whenever a dissymmetric product is synthesised from a non-dissymmetric reactants, it is always an optically inactive racemic modification (as described earlier). However, by the use of an optically active reagent, one of the enantiomers can be obtained in excess so that the resulting product is optically active.

For example pyruvic acid on direct reduction yields the optically inactive racemic lactic acid, but pyruvic acid, pre-esterified with an optically active alcohol say (–) menthol, on reduction and subsequent hydrolysis yields predominantly (–) lactic acid.

$$CH_3-\underset{\underset{\text{Pyruvic acid}}{}}{\overset{\overset{O}{\|}}{C}}-COOH \xrightarrow[\text{reduction}]{2H} \underset{(\pm)\text{ Lactic acid}}{CH_3-CHOH-COOH}$$

$$\underset{\text{Pyruvic acid}}{CH_3-\overset{\overset{O}{\|}}{C}-COOH} + \underset{(-)\text{ Menthol}}{C_{10}H_{19}OH} \longrightarrow \underset{(-)\text{ Menthyl pyruvate}}{CH_3-\overset{\overset{O}{\|}}{C}-COOC_{10}H_{19}}$$

$$\Big\downarrow \begin{matrix}\text{reduction}\\2H\end{matrix}$$

$$\underset{(-)\text{ Menthol}}{C_{10}H_{19}OH} + \underset{\substack{(-)\text{ Lactic acid}\\\text{(predominantly)}}}{CH_3-CHOH-COOH} \xleftarrow{H_2O,\ H^+} \underset{(-)\text{ Menthyl lactate}}{CH_3-CHOH-COOC_{10}H_{19}}$$

Walden Inversion

In 1893, Walden was able to convert an optically active compound into its enantiomer by a series of replacement reactions. For example, conversion of (–) malic acid into (+) malic acid and vice-versa can be achieved as under:

$$\begin{array}{c}\text{CHOH—CO}_2\text{H}\\|\\\text{CH}_2\text{—CO}_2\text{H}\end{array} \underset{\text{KOH}}{\overset{\text{PCl}_5}{\rightleftharpoons}} \begin{array}{c}\text{CHCl—CO}_2\text{H}\\|\\\text{CH}_2\text{—CO}_2\text{H}\end{array}$$

(−) Malic acid   (+)-Chlorosuccinic acid

(I)   (II)

↑ AgOH   ↓ AgOH

$$\begin{array}{c}\text{CHCl—CO}_2\text{H}\\|\\\text{CH}_2\text{—CO}_2\text{H}\end{array} \underset{\text{KOH}}{\overset{\text{PCl}_5}{\rightleftharpoons}} \begin{array}{c}\text{CHOH—CO}_2\text{H}\\|\\\text{CH}_2\text{—CO}_2\text{H}\end{array}$$

(−) chlorosuccinic acid   (+)- malic acid

(IV)   (V)

During the change of (−) malic acid (I) or (+) malic acid (III) there must be a change in configuration in one of the two steps. If the configuration of II and III is same, the change must have taken place between I and II. Any single reaction in which change of configuration takes place, is termed as Walden Inversion.

In other words if an atom or group of atoms directly attached to chiral carbon atoms is replaced by another atom or group of atoms the configuration of product is sometimes found to be different from that of starting compound. Such phenomenon which involves inversion of configuration during a reaction is called *Walden Inversion or Optical Inversion*.

For example, in SN^2 hydrolysis of 2-bromooctane inversion of configuration takes place.

$$\begin{array}{c}C_6H_{13}\\|\\H—C—Br\\|\\CH_3\end{array} \xrightarrow{OH^-} \begin{array}{c}C_6H_{13}\\|\\HO—C—H\\|\\CH_3\end{array}$$

(−) 2-Bromooctane   (+) 2-Octanol

It must be kept in mind here that inversion of configuration may or may not lead to a change in direction of rotation. The change, if any, is only a matter of chance as in the example given above.

3.4 GEOMETRICAL ISOMERISM

Two carbon atoms joined by a single bond (σ bond) are capable of free rotation around each other, but this rotation is hindered in case of compounds containing carbon-carbon double bond. According to molecular orbital theory, carbon atoms involved in double bond formation are sp^2 hybridised so that each carbon atom has three planar sp^2 hybridised orbitals and fourth p orbital having its lobes at right angles to the plane of sp^2 orbitals. The formation of π bond involves the overlapping of p orbitals. With the formation of a π bond between C — C along with a σ bond which is already existing, there remains no possibility of rotation along C — C axis. Neither of the two doubly bonded carbon atoms can be rotated about double bond without destroying π orbital which requires large amount of energy. Thus at ordinary temperature, the rotation about a carbon-carbon double bond is restricted or hindered and gives rise to a kind of stereoisomerism known as *Geometrical Isomerism*.

Geometrical Isomerism, also known as cis-trans isomerism takes place in compounds containing carbon-carbon double bond in which each of the two doubly bonded carbon atoms is attached to two different atoms or groups. All the compound with general formula of the type $C_{AB} = C_{DE}$ or $C_{AB} = C_{AB}$ show geometric isomerism. If either of the two carbon atoms carries two identical groups as in $C_{AB} = C_{AA}$ or $C_{AB} = C_{DD}$, the isomerism does not exist. *This isomerism is due to difference in the relative spatial arrangement of the atoms or groups about the doubly bonded carbon atoms.*

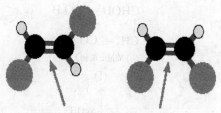

No rotation about the double bond

Conditions to be fulfilled by a compound to exhibit geometrical isomerism

For a compound to show geometrical isomerism the following conditions are necessary:

(*i*) The molecule must contain a carbon-carbon double bond about which there is no free rotation.

(*ii*) Each of the fouble bonded carbon atoms must be attached to two different atoms or groups.

In case of compounds with formula of the type $C_{AB} = C_{AB}$; if two similar groups are on the same side of double bond, the isomer is known a *cis-* and if two similar groups are on the opposite sides of the double bond the isomer is known as *trans-* such as:

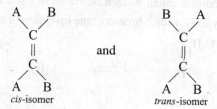

For example,

2-butene exists in two isomeric forms.

cis-2-butene   *trans*-2-butene

Similarly butene-dioic acid exists in two isomeric forms; *cis-* form is called maleic acid and *trans-* form is called fumaric acid.

Maleic acid (*cis*-isomer)   Fumaric acid (*trans*-isomer)

Determination of configuration of a geometrical isomer

Different methods available for determination of configuration of a geometrical isomers are described below:

(i) From Dipole moments. Generally *cis*-isomer has greater dipole moment as compared to *trans*-isomer. In case of cis- the similar groups being on the same side, the electronic effects are additive; while in case of *trans*-isomer, the similar groups being on opposite side, the electronic effects cancel each other.

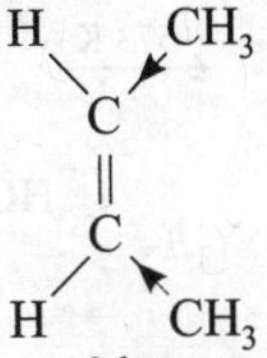

cis-2-butene
$\mu = 0.4$

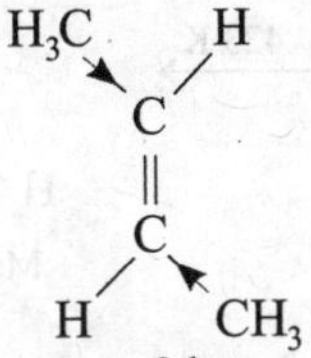

trans-2-butene
$\mu = 0$

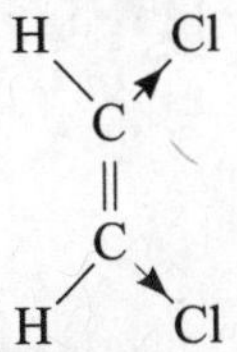

cis-2-Dichlorethene
$\mu = 1.85$

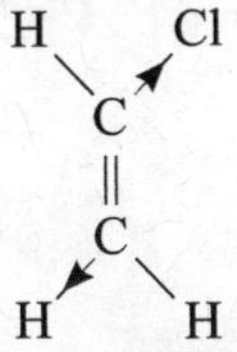

trans-1, 2-dichloroethen
$\mu = 0$

(ii) From boiling point. Generally speaking, a *cis* isomer has a higher boiling point compared to the *trans* isomer. This is because of higher dipole moment and higher polarity in the molecule which acts as the binding force and is responsible for higher b.p. of the *cis* isomer. Boiling point of *cis*-2-butene is 277 K while the *trans*-2-butene boils at 274 K.

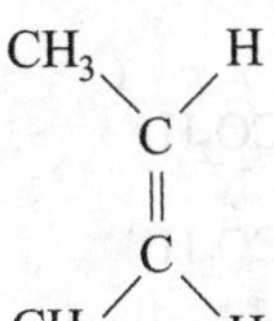

cis-2-butene
b.p. 277 K

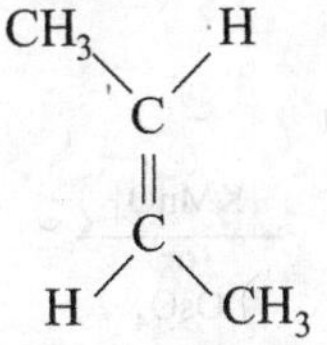

trans-2-butene
b.p. 274 K

(iii) From melting point. The isomers show a reverse trand here. A *cis* isomer has a lower melting point compared to the *trans* isomer. Here the factor that is important is the size of the molecule. It can be realised that a *cis* isomer will occupy a smaller volume compared to the *trans* isomer as illustrated by dotted lines. Thus maleic acid melts at 403 K whereas fumaric acid melts at 575 K.

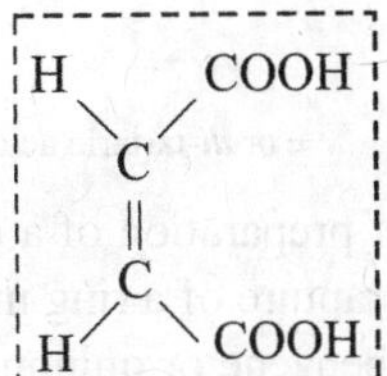

Maleic acid (*cis*)
m.p. 403 K

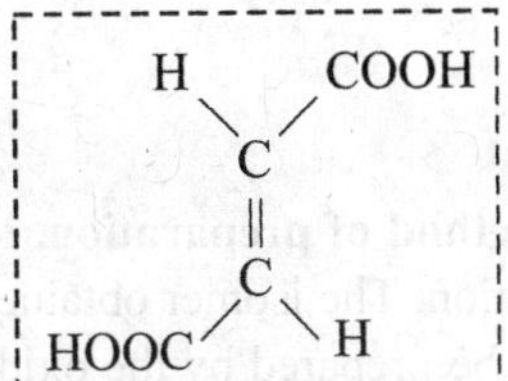

Fumaric acid (*trans*)
m.p. 575 K

(*iv*) **From the formation of cyclic compounds.** Two geometric isomers (*cis* and *trans*) can be distinguished through reactions that lead to formation of ring. *Cis*-isomer undergoes ring closure more readily than the *trans*-isomer. For example maleic acid readily loses water when heated to about 423 K, to give an anhydride; while fumaric acid does not give anhydride at this temperature. Rather it must be heated to 573 K to get the same anhydride. Hydrolysis of anhydride yields only maleic acid.

$$\underset{\substack{\text{Maleic acid}\\(\textit{cis}\text{-2-isomer})}}{\underset{H}{\overset{H}{\diagdown}}\underset{\diagup}{\overset{C}{\underset{\|}{C}}}\underset{CO_2H}{\overset{CO_2H}{\diagup}}} \xrightarrow{423\text{ K}} \underset{\text{Maleic anhydride}}{\underset{H}{\overset{H}{\diagdown}}\underset{\diagup}{\overset{C}{\underset{\|}{C}}}\underset{CO}{\overset{CO}{\diagup}}O} \xleftarrow{573\text{ K}} \underset{\substack{\text{Fumaric acid}\\(\textit{trans}\text{-2-isomer})}}{\underset{HO_2C}{\overset{H}{\diagdown}}\underset{\diagup}{\overset{C}{\underset{\|}{C}}}\underset{H}{\overset{CO_2H}{\diagup}}}$$

↓ hydrolysis

only maleic acid and not fumaric acid

(*v*) **From the formation of the type of optical isomer.** Maleic acid and fumaric acid, both on treatment with KMnO$_4$ or OsO$_4$ yield optically inactive variety of tartaric acid. Maleic acid yields meso tartaric acid while fumaric acid yields racemic (± or *dl*) tartaric acid.

Maleic acid (*cis*-2-isomer) $\xrightarrow{\text{KMnO}_4 \text{ or OsO}_4}$ Meso-tartaric acid

Fumaric acid (*trans*-isomer) $\xrightarrow{\text{KMnO}_4 \text{ or OsO}_4}$ ± or *dl*-tartaric acid

(*vi*) **From the method of preparation.** Method of preparation of a compound sometimes leads to its configuration. The isomer obtained by the rapture of a ring must be the *cis*-isomer, *e.g.*, maleic acid can be prepared by the oxidation of benzene or quinone, so it must be a *cis*-

Stereochemistry **131**

```
Benzene    or    Quinone    —[O]→    Maleic acid (cis-2-isomer)
```

isomer.14

E and Z designations of geometrical isomers (Cahn-Ingold-Prelog or C.I.P. Rules)

The *cis* and *trans*- designated can be used only for the compounds in which two doubly bonded carbon atoms are having similar atoms or groups e.g., of the type $C_{AB} = C_{AB}$. But, when the two doubly bonded carbon atoms are having different atoms or groups attached to them e.g., of the type $C_{AB} = C_{DE}$; it is not possible to assign them *cis* or *trans* configurations. To overcome this difficulty, a more general system for designating the configuration of geometric isomers has been adopted. This system developed by *Cahn*, Ingold and Prelog originally for the absolute configuration of optical isomers is known as *E* and *Z system* and is based on priority of attached groups. The atoms or groups attached to each carbon of the double bond, are assigned first and second priority. If the atoms or groups having higher priority attached to two carbons are on the same side of double bond the configuration is designated as Z (derived from German word *Zusammen* meaning together) and if the atoms or groups of higher priority are on the opposite side of the double bond, the configuration is designated as E (derived from German word *-entgegen* meaning across or opposite).

Priorities of atoms or groups are determined in the same way as for R & S configurations of optical isomers. At. weights or atomic numbers of atoms directly linked with ethylenic carbon atoms are taken into consideration.

Let us consider an example in which two doubly bonded atoms are attached to four different halogens such as $C_{BrF} = C_{ICl}$. Since Br is having higher priority over F and I is having priority over Cl (due to their higher atomic numbers). The isomer in which Br and I are on the same side of double bond will be called Z and the isomer in which Br and I are on the opposite sides of double bond will be called E.

```
   F      Br           F      Br
    \    /              \    /
      C                   C
      ‖                   ‖
      C                   C
    /    \              /    \
   Cl     I            I      Cl
   Z-isomer            E-isomer
```

In the same way *cis* and *trans* isomers of 2-butene can be called Z and E-2-butenes respectively.

```
   H      CH₃          H      CH₃
    \    /              \    /
      C                   C
      ‖                   ‖
      C                   C
    /    \              /    \
   H      CH₃          H₃C    H
```

Similarly maleic acid can be specified as Z-isomer and fumaric acid as E-isomer.

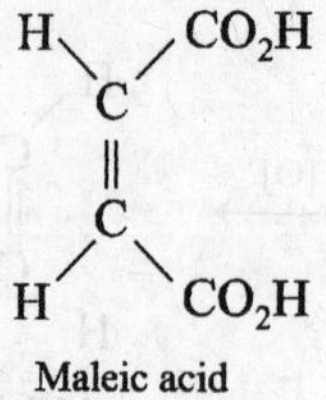

132 Chemistry for Degree Students—B.Sc. (Hons.) Semester II

Maleic acid (Z-isomer)

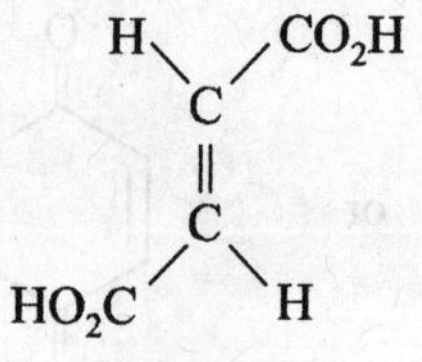

Fumaric acid (E-isomer)

In determining the configuration, we have to select the group of higher priority on one carbon. Similarly we select the group of higher priority on the other carbon atom. If these two groups are on the same side of double bond, the configuration is Z, otherwise it is E.

Some more examples are:

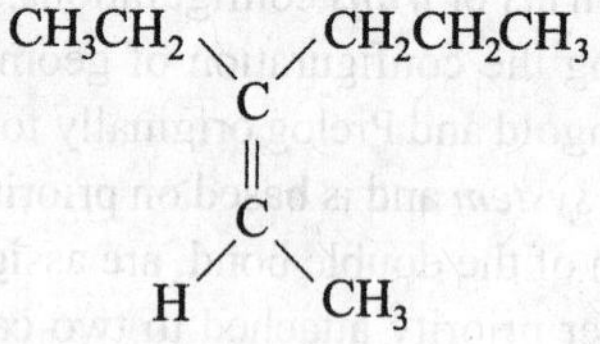

Z-3-Ethyl-2-hexene and E-3-Ethyl-2hexene

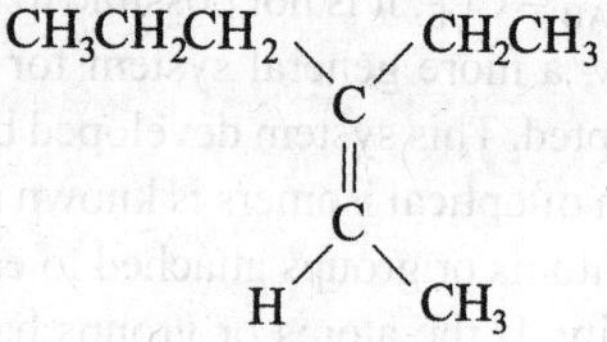

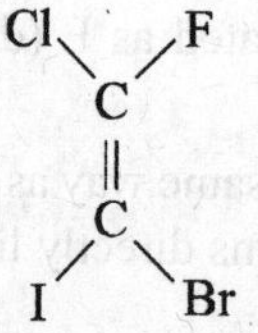

Z-1-Bromo-2-chloro 2-fluoro-1-iodoethene and E-1-Bromo-2-chloro 2-fluoro-1-iodoethene

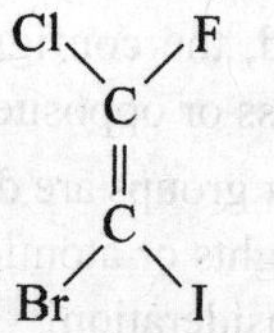

Solved Examples

Example 16: Assign E and Z configurations to the following compounds.

(i) CH_3, H / C=C / Br, Cl

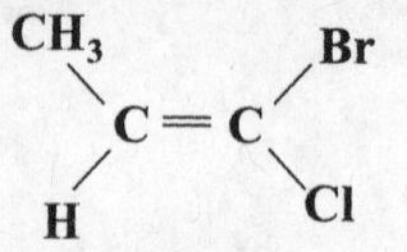

(ii) CH_3, H / C=C / CH_2CH_3, $CH(CH_3)_2$

(iii) H, $(CH_3)_2CH$ / C=C / C_2H_5, C_6H_5

(iv) CH_3, H / C=C / H, CHO

(v) CH_3, H / C=C / H, $COOH$

(vi) I, Br / C=C / C_6H_5, $NHCH_3$

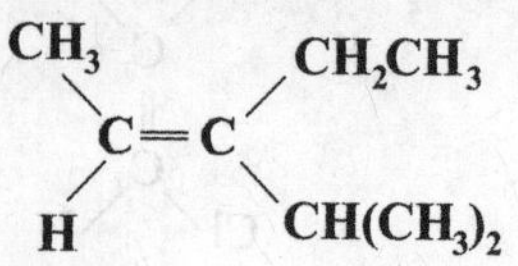

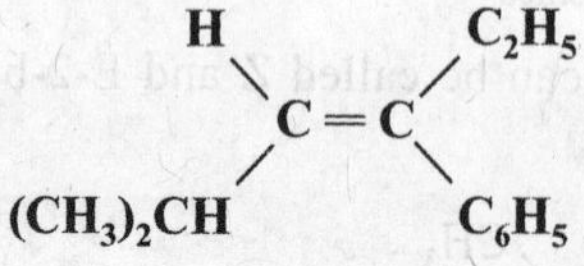

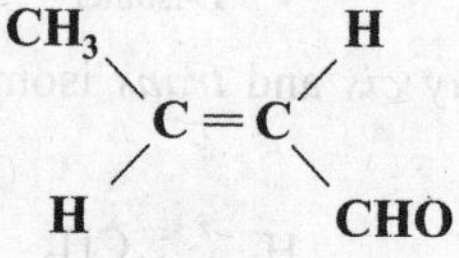

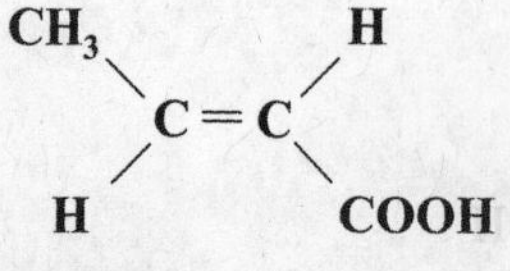

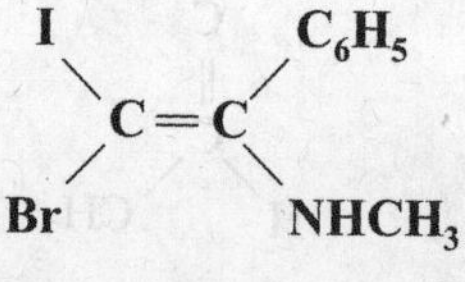

Solution: (*i*) Out of —CH_3 and —H, —CH_3, has higher priority, out of —Br and —Cl, —Br has higher priority. Groups of higher priority lie on the same side. Therefore the configuration is Z.

(*ii*)
$$CH_3CH_2CH_3$$
$$\diagdown\diagup$$
$$C=C$$
$$\diagup\diagdown$$
$$HCH(CH_3)_2$$

Out of — CH_3 and — H, — CH_3 has higher priority. Out of two groups on the second carbon, — $CH(CH_3)_2$ has higher priority. The groups of higher priority lie on opposite sides of the double bond. Therefore the configuration is E.

Following the same analogy, the students can see the configurations of compounds (*iii*), (*iv*), (*v*) and (*vi*) are Z, E, E and E respectively.

Example 17: Assign E and Z configurations to the following:

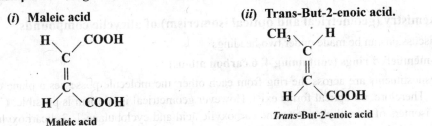

Solution: In compound I, the groups of higher priority on the two carbons linking double bond are CH_3— and Cl. Since these are on the same side of the double bond it has Z configuration.

In compound II, the groups of higher priority on the two carbons linking the double bond are CH_3— and D-(Deuterium) since these are on opposite sides of the double bond, it has E configuration.

Example 18: Assign E and Z configuration to the following:

(*i*) **Maleic acid**

$$HCOOH$$
$$\diagdown\diagup$$
$$C$$
$$\|$$
$$C$$
$$\diagup\diagdown$$
$$HCOOH$$

Maleic acid

(*ii*) **Trans-But-2-enoic acid.**

$$CH_3H$$
$$\diagdown\diagup$$
$$C$$
$$\|$$
$$C$$
$$\diagup\diagdown$$
$$HCOOH$$

Trans-But-2-enoic acid

Solution: In maleic acid, groups of higher priority on the two carbon atoms are — COOH and — COOH. These two groups are on the same side of the double bond. Hence configuration is Z.

In *trans*-but-2-enoic acid, the groups of higher priority on the two carbons are — CH_3 and — COOH, since these are on the opposite sides of the double bond, it has E configuration.

Geometrical isomerism in oximes (Syn-Anti Isomerism)

Oximes of aldehydes having the general formula R—CH=N—OH are capable of exhibiting geometrical isomerism as — H and — OH groups may be present on the same side or opposite sides of the double bond. The two stereoisomers thus obtained are named as *syn* (equivalent of *cis*) and *anti* (equivalent of *trans*). Two geometrical isomers of an aldoxime may be represented as:

syn-aldoxime   anti-aldoxime

Geometrical isomers of benzaldoxime have been actually isolated.

syn-banzaldoxime has m.p. 35°C while *anti*-benzaldoxime has m.p. 130°C. The two compounds may be shown as under:

syn-benzaldoxime   anti-benzaldoxime

Oximes of ketones ($RR'C = NOH$) can also show geometrical isomerism provide R and R' are not the same. Thus benzophenone oxime does not exhibit geometrical isomerism because the two groups attached to carbonyl carbon are the same. Phenyl tolyl ketoxime is known to exist in two geometrical forms

syn-phenyl tolyl oxime   anti-phenyl tolyl oxime

Stereochemistry (geometrical and optical isomerism) of alicyclic compounds

The discussion can be made under two headings

Even-membered rings (containing 4, 6 carbon atoms)

If the substituents are across the ring from each other, the molecule possesses a plane of symmetry. Therefore, no optical forms exist. However geometrical isomerism is possible. *Cis* and *trans* isomers of 3-methylcyclobutane carboxylic acid and cyclobutane-1, 2-dicarboxylic acid are shown below

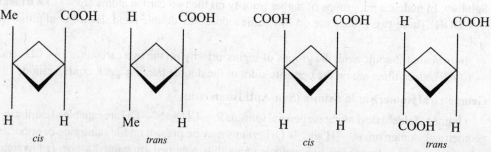

3-methylcyclobutane carboxylic acids   Cyclobutane-1, 2-dicarboxylic acids

Stereochemistry **135**

Odd membered rings (containing 3, 5 carbon atoms)

A disubstituted (on different carbons) odd membered ring has two asymmetric carbon atoms. Therefore two diastereomeric pairs of enantiomers are possible. Take the case of 2-methylcyclopropanecarboxylic acid. Geometrical and optical isomers have been represented as under

One pair of enantiomers (cis form / Mirror / cis form)

Second pair of enantiomers (trans form / Mirror / trans form)

If however, the two substituents are the same, there will be one pair of enantiomers and a meso form. Take the case of cyclopropane-1, 2-dicarboxylic acid. *trans* form exists as a pair of enantiomers while the *cis* form as meso compound

Pair of enantiomers (trans / Mirror / trans) — Meso cyclopropane-1, 2-dicarboxylic acid (cis)

3.5 CONFORMATIONAL ISOMERISM (ANALYSIS)

Carbon-carbon single bond in alkanes is a sigma bond formed by the overlapping of sp^3 hybrid atomic orbitals along the inter-nuclear axis. The electron distribution in such a bond is symmetrical around internuclear axis; so that free rotation of one carbon against the other is possible without breaking sigma or single covalent bond. Consequently such compound can have different arrangements of atoms in space, which can be converted into one another simply by rotation around single bond, without breaking it. These different arrangements are known as *Conformational Isomers or Rotational Isomers or Conformers*. Since the potential energy barrier for their inter-coversion is very low, it is not possible to isolate them at room temperature. At least 60-85 kJ/mole must be the energy difference between two conformers to make them isolatable at room temperature.

Hence, conformations can be defined as different arrangements of the atoms which can be converted into one another by rotation around single bonds.

Newman and Sawhorse representations for the conformations of ethane

Alkanes can have an infinite number of conformations by rotation around carbon-carbon single bonds. In ethane two carbon atoms are linked by a single bond and each carbon atom is further linked with three hydrogen atoms. If one of the carbon atoms is allowed to rotate about carbon-carbon single bond keeping the other carbon stationary, an infinite number of arrangements of the hydrogens of one carbon, with respect to those of the other, are obtained. All these arrangements are called conformations (Bond angles and bond lengths remain the same).

Newman representation

This can be easily understood with the help of Newman Projection formulae. The molecule is viewed from front to back in the direction of carbon-carbon single bond. The carbon nearer to the eye is represented by a point and three hydrogen attached to it are shown by three lines at an angle of 120° to one another. The carbon atom away from the eye is represented by a circle and three hydrogens attached to it are shown by shorter lines at an angle of 120° to each other.

Out of infinite number of conformations, Newman Projection formulae for two extreme cases are as shown below:

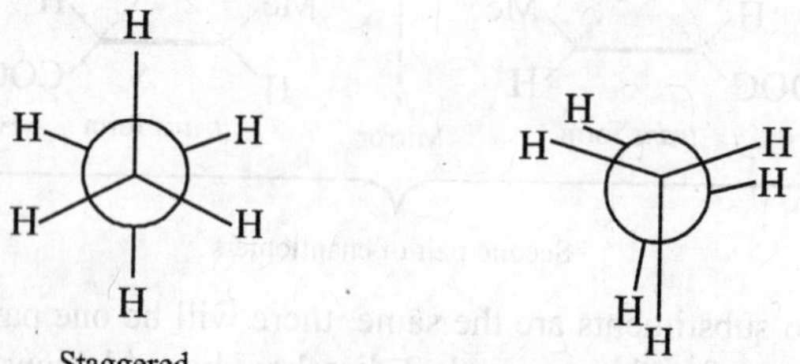

Newman Projection Formulae for conformations of ethane.

The conformation in which the H atoms of two carbons are as far apart as possible, is known as *Staggered conformation* and the conformation in which the H atoms of back carbon are just behind those of the front carbon is known a *Eclipsed conformation*. These are converted into one another by rotation of one carbon against the other through 60°. The other conformations, in between these two, are known as *skew conformations*.

Sawhorse representation

In this representation, the molecule is visualised slightly from above and from the right and then projected on the paper. The bond between two carbons is drawn diagonally and is a bit longer for the sake of clarity. The lower left hand carbon is taken as front carbon and the upper right hand carbon is taken as back carbon. The Sawhorse representation of staggered and eclipsed conformations of ethane are given below:

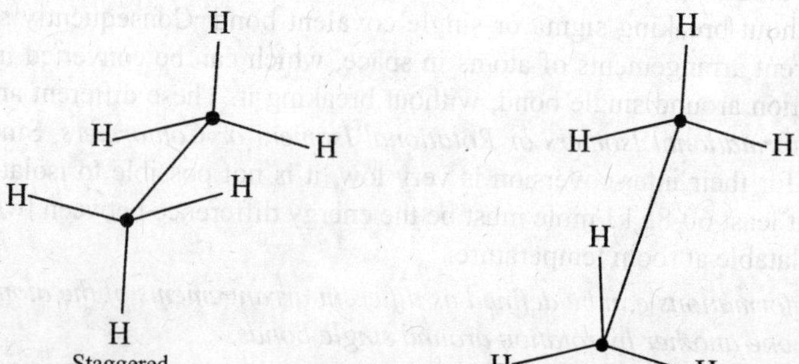

Sawhorse representation for conformations of ehtane

Flying wedge representation

In this representation, the bonds connecting the two carbon atoms are drawn horizontally in the plane of paper. Wedges (—) represent the bonds lying above the plane of the paper. Ordinary lines denote the bonds lying in the plane of the planar and dotted lines (----) represent bonds lying below the plane of the paper. Flying edge representation for staggered and eclipsed conformations of ethane are shown in Fig.

Fig: *Flying wedge representations for ethane*

Pitzer in 1936 found that the rotation is not completely free. Rather there exists a potential energy barrier which restricts the free rotations. It means that the molecule spends most of its time in the most stable staggered conformation and it spends least time in the least stable eclipsed conformation; the energy difference being 12 kJ/mole in the case of ethane.

The energy required to rotate the molecule about carbon-carbon bond is called rotational or torsional energy.

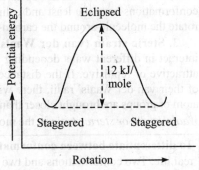

Rotational or torsional energy of ethane

Difference between conformation and configuration

Conformation

Structures containing different arrangement of atoms of a molecule in space which can arise by rotation about a single bond are called conformers. For example ethane exists in different conformations called staggered, eclipsed and skew conformations. The energy difference between different conformers is rather small. This phenomenon is called conformation. The property of conformation is also exhibited by *n*-butane, cyclohexane, stilbene dichloride etc.

Configuration

Structures of a compound differing in the arrangement of atoms or groups around a particular atom in space are called configurations. Enantiomers, distereomers and geometrical isomers come under this category. For example *d*-and *l*-lactic acids are configurations of lactic acid.

$$
\begin{array}{cc}
\text{CH}_3 & \text{CH}_3 \\
| & | \\
\text{HO—C—H} & \text{H—C—OH} \\
| & | \\
\text{COOH} & \text{COOH} \\
d\text{-lactic acid} & l\text{-lactic acid}
\end{array}
$$

cis and *trans* butenes are configuration of butene

CH₃\C=C/CH₃ over H/ \H   H\C=C/CH₃ over CH₃/ \H

cis-2-butene   *trans*-2-butene

Factors affecting relative stability of conformations

The following factors play a vital role in the stability of conformations.

1. Angle strain. Every atom has the tendency to have the bond angles that match those of its bonding orbitals. If there is any deviation from this normal bond angle, the molecule suffers from *angle strain*. Conformations suffering from angle strain are found to be less stable.

2. Torsional strain. There is a tendency on the part of two carbons linked to each other to have their bonds staggered. That is why the staggered form of any molecule like ethane, *n*-butane is most stable. As the bonds of two connected carbons move towards eclipsed state, a *torsional strain* is set up in the molecule thus raising its energy. Thus the staggered conformations have the least and eclipsed the highest torsional strain. The energy required to rotate the molecule around the carbon-carbon bond is called *torsional energy*.

3. Steric strain (van der Waals' Strain). Groups attached to two linked carbons can interact in different ways depending upon their size and polarity. These interactions can be attractive or repulsive. If the distance between the groups or atoms is just equal to the sum of their van der Waals' radii, there will be attractive interactions between them. And if these atom or groups are brought closer than this distance, there will be repulsions leading to *van der Waals' strain or steric strain* in the molecule.

> **To differentiate between conformation and configuration**, we are taking an analogy from real life. Two conformations and two configurations of a dog are considered here.
> - Conformations are different spatial arrangements of the same compound (for example, anti and gauche conformers; They cannot be separated. Some conformations are more stable than others.
>
> **Different Conformations**
>
>
>
> Stable   Unstable
>
> - Compounds with different configurations are different compounds. They can be separated from each other. Bonds have to be broken to interconvert compound with different configurations.
>
> **Different Configurations**
>
>
>
> Stable   Unstable

4. Dipole-dipole Interactions. Atoms or groups attached to bonded carbons orient or position themselves to have favourable dipole-dipole interactions. It will be their tendency to have maximum dipole-dipole attractions. Hydrogen bond is a particular case of powerful dipole-dipole attractions.

The stability of a conformer is determined by the net effect of all the above factors.

Conformations and change in dipole-moment of 1, 2-dibromoethane with temperature

The conformations of 1, 2-dibromoethane have been extensively studied by dipole-moment measurement. The conformations are depicted as under:

In the liquid state, the percentage of anti forms is 65 corresponding to conformational free energy of 3.5 kJ mol^{-1} in favour of *anti*.

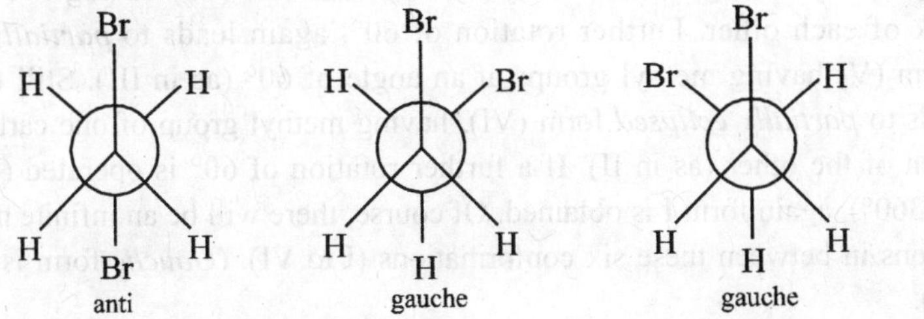

anti   gauche   gauche

stability of *anti*-form is due to combined effects of steric factor and dipole-dipole interactions. It goes in favour of formation of *anti*- conformer. Dipole movement increases with increase of temperature.

Conformations of butane

n-butane is an alkane with four carbon atoms, which can be considered to be derived from ethane by replacing one hydrogen on each carbon with a methyl group. If we consider the rotation about the central carbon-carbon bond ($C_2 - C_2$), the situation is somewhat similar to ethane; but *n*-butane has more than one staggered and eclipsed conformations (unlike ethane which has only one staggered and one eclipsed conformation). Newman Projection formulae for various staggered and eclipsed conformations of *n*-butane are as given below:

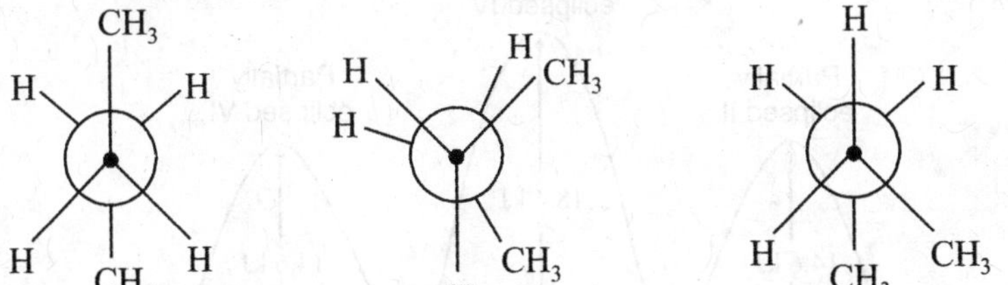

Completely staggered or anti form
I

Partially eclipsed form
II

Partially staggered or gauche form or skew form
III

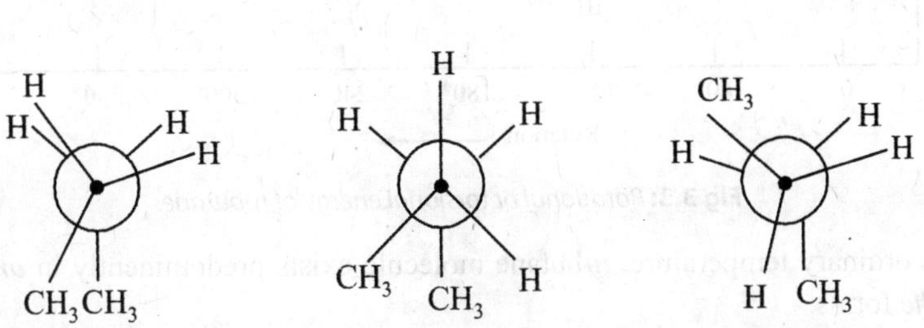

Completely eclipsed form
IV

Partially staggered or gauche
V

Partially eclipsed form
VI

The *completely staggered* conformation, (I) also known as *anti* form, is having the methyl goups as far apart at possible. Let us see how these forms have been obtained. Let us start from structure I. Holding the back carbon (represented by circle) fixed along with its groups, — H, — H and — CH_3, rotate the front carbon (shown by a dot) in clockwise direction by an angle of 60°. Groups attached to it will also move. *Partially eclipsed* form (II) is obtained. In this conformation (II), methyl group of one carbon is at the back of hydrogen of the other carbon. Further rotation of 60° leads of a *partially staggered* conformation (III), also known as *gauche* form, in which the two methyl groups are at an angle of 60°. Rotation by another 60° gives rise to a *fully eclipsed* form (IV) having two methyl groups at the back of each other. Further rotation of 60°, again leads to *partially staggered or gauche* form (V) having methyl groups at an angle of 60° (as in III). Still further rotation of 60° leads to *partially eclipsed* form (VI), having methyl group of one carbon at the back of hydrogen of the other (as in II). If a further rotation of 60° is operated (completing the rotation of 360°), again form I is obtained. Of course, there will be an infinite number of other conformations in between these six conformations (I to VI). (*Gauche* form is also known as skew form).

Out of these six conformations, the *completely staggered* or *anti conformation* (I) is most stable and *partially staggered* or *gauche* conformation (III or V) is slightly less stable: the energy difference being only 3.8 kJ/mole. On the other hand the *completely eclipsed* conformation (IV) is least stable and *partially eclipsed* conformation (II or VI) is slightly more stable, again the energy difference being 3.8 kJ/mole. (This is due to presence of steric strain between two methyl groups). The energy difference between most stable conformation (I) and least stable conformation (IV) is about 18.4 kJ/mole while that between I and II (or VI) is about 14.6 kJ/mole. (Fig 3.3)

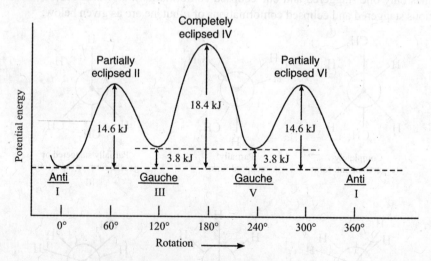

Fig 3.3: *Rotational or torsional energy of n-butane.*

Thus at ordinary temperature, *n*-butane molecule exists predominantly in *anti* form with some *guache* forms.

Conformational enantiomers of 2, 3-Dimethylbutane and 2, 2, 3-Trimethylbutane

Conformational isomers of 2,3-Dimethylbutane

[Newman projections I, II, III showing conformers of 2,3-dimethylbutane]

Structure I is supposed to have maximum stability out of these three isomers, II and III are conformational enantiomers (mirror-images) also. This can be visualized after rotation of III as a whole.

Conformational isomers of 2, 2, 3-Trimethylbutane

[Newman projections I, II, III showing conformers of 2,2,3-trimethylbutane]

All these structures have the same stability and the compound may be assumed to occur in one conformation only.

Conformations of cyclohexane

Chair conformation

Chair conformation of cyclohexane is represented below:

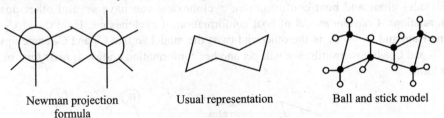

Newman projection formula   Usual representation   Ball and stick model

Chair conformation of *n*-butane.

This is the most stable conformation of cyclohexane. In this conformation, all the bond angles are tetrahedral and the C — H bonds on adjacent carbons are in a staggered position. This conformation has no strain and has minimum energy.

Boat Conformation

There is no angle strain in the molecule as all the angles are tetrahedral. But hydrogens on four carbon atoms (C_2 and C_3, C_5 and C_6) are eclipsed. As a result, there is considerable torsional strain. Also, two hydrogens pointing towards each other at C_1 and C_4 (called flagpole hydrogens) are very close to each other. This brings in van der Waals strain in the molecule. Due to these reasons boat conformation is less stable than chair conformation by an amount 28.8 kJ/mole.

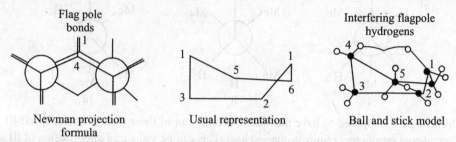

Boat conformation of cyclohexane

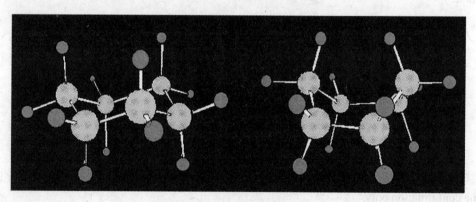

Conformations of cycloalkanes

Twist conformation of cyclohexane. Sequence of changes in going from chair form to boat form (Interconversions)

Besides chair and boat conformation, cyclohexane can have several other possible conformations. Consider model of boat conformation of cyclohexane. Hold $C_2 - C_3$ bond in one hand and $C_5 - C_6$ in the other and twist the model so that C_2 and C_5 come upwards and C_3 and C_6 go downwards. We will get another conformation known as *twist form* or *skew boat* form.

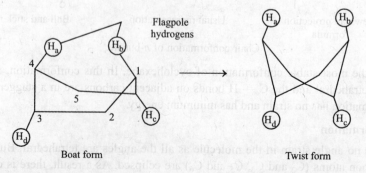

During this twisting, the flagpole hydrogens (H_a and H_b) move apart while the hydrogens H_c and H_d move closer. If this motion is continued another boat form will be obtained in which H_c and H_d become the *flagpole hydrogens*.

In twist forms the distance between H_a and H_b is equal to that of H_c and H_d and the steric strain is minimum; also the torsional strain of C_2 — C_3 and C_2 — C_6 (due to their being eclipsed) is partly relieved. Thus the twist form is more stable than boat form by about 5.4 kJ/mole, but it is much less stable than chair form by 23.4 kJ/mole.

If we want to convert the chair form into boat form it will have to pass through a half chair form having considerable angle strain and torsional strain. The energy difference between chair form and half chair form being about 46 kJ/mole, half chair form is quite unstable.

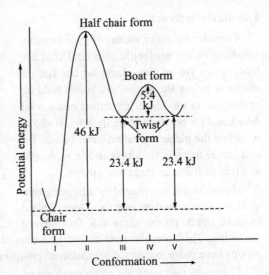

Fig. 3.4: *Potential energy diagram of the conformers of cyclohexane*

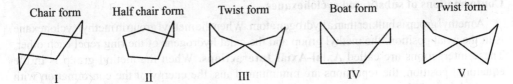

Equatorial and axial bonds in cyclohexane

Consider the structure of chair form of cyclohexane as given below:

Although, the cyclohexane ring is not planar completely, but for approximation, we can take it as planar. Consider the position of various hydrogens in the chair conformation. There are two distinct kinds of hydrogens, Six of the hydrogens which are marked H_e are almost oriented within the plane of cyclohexane ring. These are called **Equatorial hydrogen atoms**. The bonds by which these are held to the ring are called **Equatorial bonds**.

Equatorial and axial bonds in cyclohexane

We again observe that six hydrogen atoms which are shown as H_a in the figure above are oriented perpendicular to the cyclohexane ring. These are called **Axial hydrogen atoms** and the bonds by which they are held to the ring are called **Axial bonds**.

It may be noted that there is one axial and one equatorial hydrogen on each carbon in the chair conformation of cyclohexane.

1, 3-diaxial interaction

Consider the chair model of cyclohexane. Looking at the molecule, we find that six hydrogens lie in the plane while six lie above or below the plane. Six bonds holding hydrogens in the plane are called *equatorial bonds* and six bonds holding hydrogens above or below the plane are called *axial bonds*. By and large, there is no stress in the molecule and it is as stable as staggered ethane.

If a hydrogen is replaced by a larger group or atom, than crowding occurs. Atoms linked by axial bonds on the same side face severe crowding. This interaction is called 1, 3-diaxial interaction. Generally speaking, atoms or groups have more free space in equatorial position than in axial position.

There are two possible chair conformations of methyl cyclohexane one with $-CH_3$ in equatorial position and the other with $-CH_3$ in axial position.

It is observed that $-CH_3$ in equatorial position faces less crowding by hydrogens compared to $-CH_3$ in the axial position. Methyl group in the axial position is approached more closely by axial hydrogens on $C-3$ and $C-5$. This is called *1, 3-diaxial interaction*.

Conformations of substituted cyclohexanes

A methyl group is bulkier than a hydrogen atom. When the methyl group in methylcyclohexane is in the axial position, the methyl group and the axial hydrogens of the ring repel each other. These interactions are called **Axial-Axial Interactions.** When the methyl group is in the equatorial position, the repulsions are minimum. Thus, the energy of the conformation with equatorial methyl group is lower. At room temperature, about 95% methylcyclohexane molecules are in the conformation in which the methyl group is equatorial.

The bulkier the group, the greater is the energy difference between equatorial and axial conformations. In other words, a cyclohexane ring with a bulky substituent (e.g., *t*-Butyl group) is more likely to have that group in the equatorial position.

Conversion of Axial Bonds to Equatorial Bonds And Vice Versa

Each of the six carbon atoms of cyclohexane has one equatorial one axial hydrogen atom. Thus, there are six equatorial hydrogens and six axial hydrogens. In the flipping and reflipping between conformations, axial hydrogens become equatorial hydrogen while equatorial hydrogens become axial. This is represented in Fig. below:

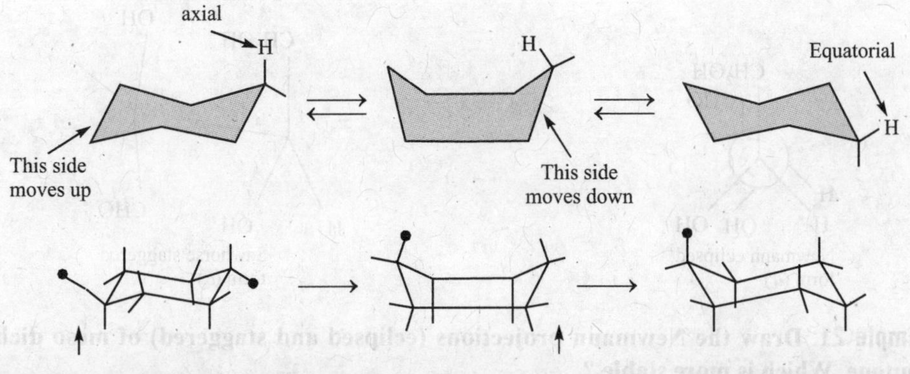

Solved Examples

Example 19: Out of the following two conformational forms of 1, 2-dibromo ethane which will have higher dipole moment?

(a) [Newman projection with Br at top-front, H's on sides and bottom-front; back: H top, H sides, Br bottom]

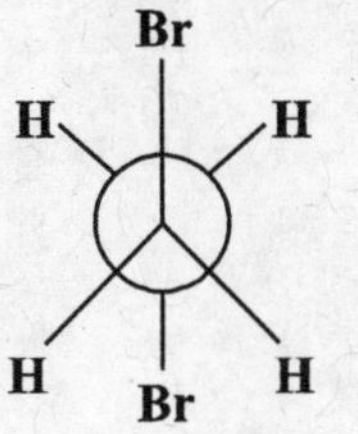

(b) [Newman projection with Br and Br gauche]

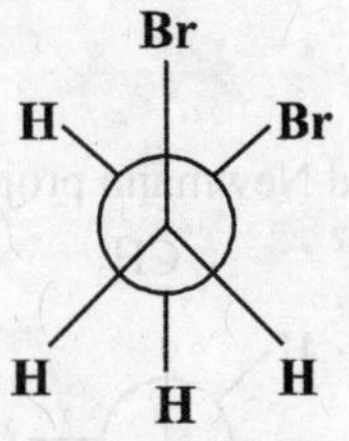

Solution: In structure (b) two Br groups are close to each other. This will result in repulsion between them leading to separation of charge and thus creation of higher dipole moment.

Example 20: Convert the Fischer's projection of D (–) erythrose sugar (a) to the following:

(a)
```
        CHO
   H ——|—— OH
        |
        |—— OH
        |
       CH₂OH
```
D (–) Erythrose

→ Sawhorse eclipsed form (b)
→ Sawhorse staggered form (c)
→ Newmann eclipsed form (d)
→ Newmann staggered form (e)

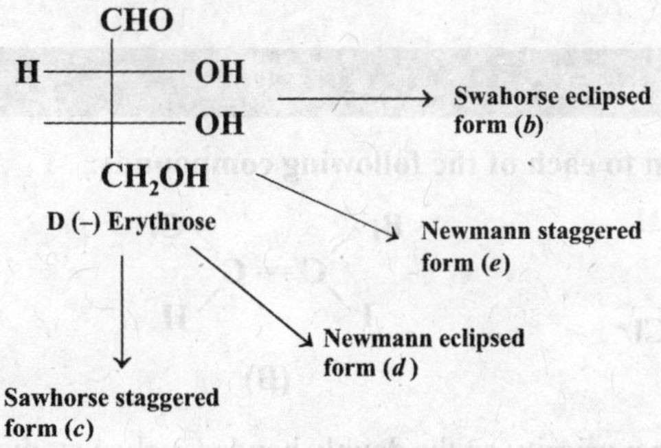

Solution:

Newmann staggered form (e)

D (–) Erythrose

Sawhorse eclipsed form (b)

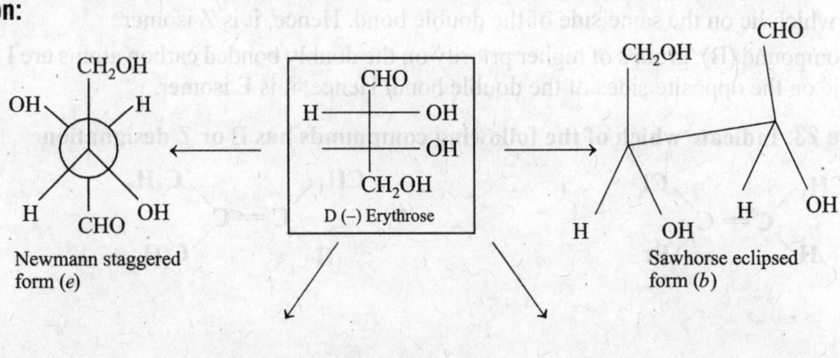

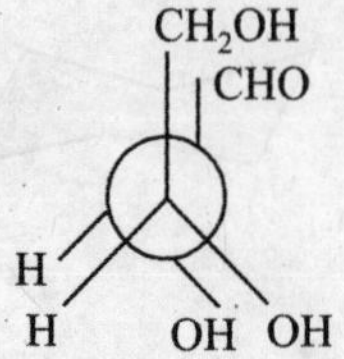

Newmann eclipsed form (d)

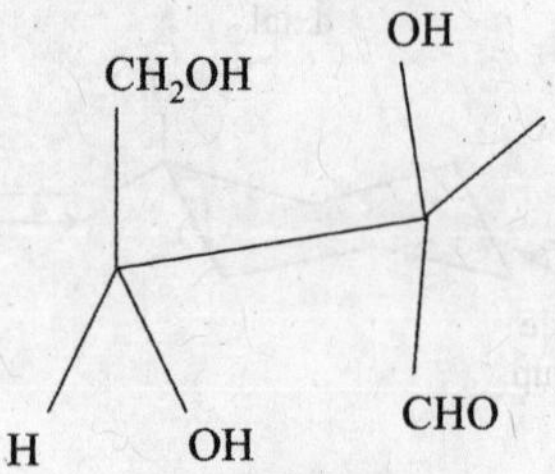

Sawhorse staggered form (c)

Example 21: Draw the Newmann projections (eclipsed and staggered) of meso dichlorobutane. Which is more stable ?

Solution: Meso dichlorobutane may be written as:

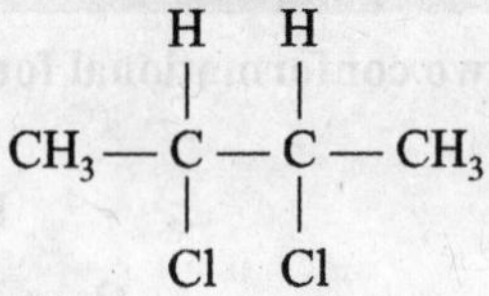

$$CH_3 - \underset{\underset{Cl}{|}}{\overset{\overset{H}{|}}{C}} - \underset{\underset{Cl}{|}}{\overset{\overset{H}{|}}{C}} - CH_3$$

Eclipsed and staggered Newmann projections may be written as:

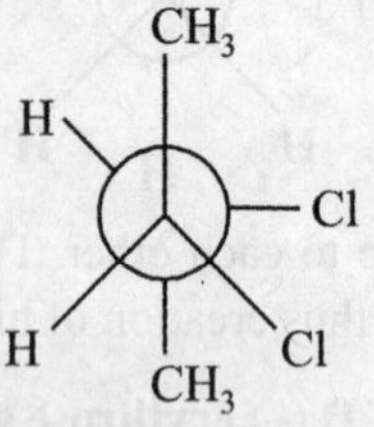

(Staggered) 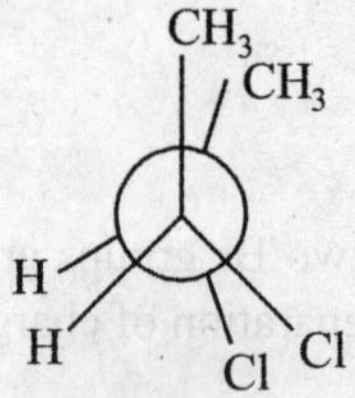   (Eclipsed)

Solved Examples—Isomerism

Example 22: Assign E or Z configuration to each of the following compounds:

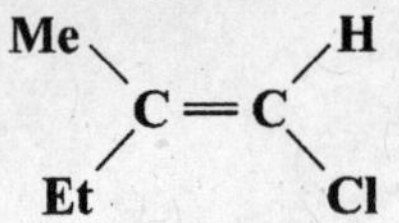

(A) 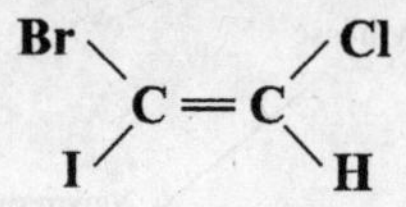   (B)

Solution: In compound (A), groups of higher priority on the doubly bonded carbon atoms are Et and Cl, which lie on the same side of the double bond. Hence, it is Z isomer

In compound (B), groups of higher priority on the doubly bonded carbon atoms are I and Cl, which lie on the opposite sides of the double bond. Hence, it is E isomer.

Example 23: Indicate which of the following compounds has E or Z designation

(i)   (ii)

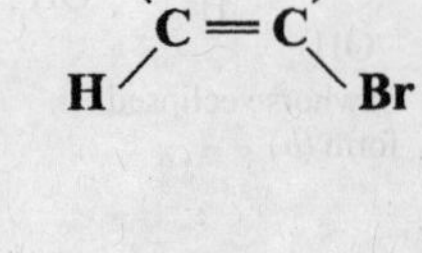

(iii) [structure: (CH3CH2)(CH3)C=CH2 type]

(iv) H\C=C/COOH with HOOC and H (cis arrangement shown)

Solution:

(i) CH₃\C=C/Cl with H and Br — **E-isomer**

(ii) CH₃\C=C/C₃H₇ with H and C₂H₅ — **Z-isomer**

(iii) **E-isomer**

(iv) H\C=C/COOH with HOOC\ /H — **E-isomer**

Example 24: Assign E or Z configuration to the following:

(a) CH₃\C=C/H with CH₃—CH₂ and CH₂—C(=O)—H

(b) HC≡C\C=C/D with H and CH=CH₂

Solution:

(a) Groups of higher priority on the two doubly bonded carbon atoms are on the same side of the double bond. Hence, its configuration is Z.

(b) Groups of higher priority on the two doubly bonded carbon atoms are on the opposite sides of the double bond (D stands for heavy hydrogen). Hence, its configuration is E.

Example 25: Assign R or S configuration to the following:

(a) H on top, HO—C—CH₂OH, CHO at bottom

(b) OH, CHO, H, CH₂OH arrangement

(c) Cl, OH, C, CH₃, CH₂CH₃ arrangement

Solution: We need to bring the group of lowest priority *i.e.* H at the bottom.

(Follow the rules given in the chapter)

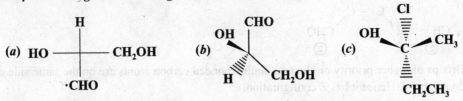

(a) HO—C(H)(CH₂OH)(CHO) ≡ HOH₂C—C—OH with H down, CHO up [S]

We move in the anticlockwise direction in moving from group of highest priority to the group of lowest priority (1 → 2 → 3). Hence it has S configuration.

148 Chemistry for Degree Students—B.Sc. (Hons.) Semester II

(b)

$$\underset{H}{\overset{OH}{\diagdown}}\underset{CH_2OH}{\overset{CHO}{\diagup}} \equiv \begin{array}{c} CHO \\ H \text{———} OH \\ CH_2OH \end{array} \equiv \underset{H}{\overset{\overset{②}{CHO}}{HO—\overset{①}{C}\curvearrowright\overset{③}{—CH_2OH}}}$$

We move in the clockwise direction in moving from group of highest priority to the group of lowest priority, (leaving H). Hence, it has R-configuration.

(c)

$$\underset{CH_2CH_3}{\overset{H}{OH\diagdown\!\!\!\!\diagup C\diagdown\!\!\!\!\diagup CH_3}} \equiv \begin{array}{c} H \\ HO \text{———} CH_3 \\ CH_2CH_3 \end{array} \equiv \underset{H}{\overset{\overset{②}{CH_2CH_3}}{\underset{③}{CH_3}\text{—}\overset{①}{C}\curvearrowleft\text{—OH}}}$$

We follow anticlockwise direction in moving from group of highest priority to the group of lowest priority. Hence, it has S configuration.

Example 26: Assign A or Z rotations to the following compounds showing priorities for various groups:

(i) $\underset{CH_2=CH}{\overset{(CH_3)_3C}{\diagdown}}C=C\underset{CHO}{\overset{COCH_3}{\diagup}}$ (ii) $\underset{H}{\overset{D}{\diagdown}}C=C\underset{Br}{\overset{NH_2}{\diagup}}$

Solution:

(i) $\underset{\underset{②}{CH_2=CH}}{\overset{\overset{①}{(CH_3)_3C}}{\diagdown}}C=C\underset{\underset{②}{CHO}}{\overset{\overset{①}{COCH_3}}{\diagup}}$

Groups of higher priority on the two doubly bonded carbon atoms are on the same side of the double bond. Hence it has Z configuration.

(ii) $\underset{\underset{②}{H}}{\overset{\overset{①}{D}}{\diagdown}}C=C\underset{\underset{①}{Br}}{\overset{\overset{②}{NH_2}}{\diagup}}$

Here, D stands for deuterium *i.e.* heavy hydrogen. Groups of higher priorities on the two doubly bonded carbon atoms are on the opposite sides of the double bond. Hence, it is E configuration.

Example 27: Assign R or S configuration to the following compounds showing priorities for various groups.

Stereochemistry

(i)
```
         H
         |
H₃COC ———+——— CO₂H
         |
       CH₂OH
```

(ii)
```
         C≡CH
         |
H₃C ———+——— H
         |
       CH=CH₂
```

Solution: To assign the configuration to the compounds, we need to bring the group of lowest priority *i.e.* H at the bottom as per the rules explained in the chapter

(i)
```
         H
         |
CH₃OC ———+——— CO₂H    ≡     HO₂C ——(↶)—— COCH₃
         |                         |
       CH₂OH                       H
```
with priorities ① HO₂C, ② COCH₃, ③ CH₂OH, ④ H

We have to trace anti-clockwise direction in moving from the group of lowest priority to the group of highest priority. Hence, it has S configuration.

(ii)
```
         C≡CH                        C≡CH
         |                           |
H₃C ———+——— H        ≡     CH₂=CH ——(↶)—— CH₃
         |                           |
       CH=CH₂                        H
```
with priorities ① C≡CH, ② CH₂=CH, ③ CH₃, ④ H

We have to trace anti-clockwise direction in moving from group of highest priority to group of lowest priority (leaving H). Hence, it has S configuration.

Example 28: Discuss the optical isomerism exhibited by diphenyl compounds.

Solution: Substituted biphenyls show optical isomerism when substituents in the 2-positions are large enough to prevent rotation about the bond joining the two benzene rings. For example, biphenyl-2, 2'-disulphonic acid exists in two forms.

Biphenyl-2, 2'-disulphonic acid
(a biphenyl)

These two forms are non-superimposable mirror images. They do not interconvert at room temperature because the energy required to twist one ring through 180° relative to the other is too high. This is because, during the twisting process, the two-SO_3H group must come into very close proximity when the two benzene rings become coplaner and strong repulsive forces are introduced.

Example 29: Draw the Fischer projection formulas for all possible stereoisomers of 2, 3, 4- trihydroxyglutaric acid. Comment on the stereogenicity of C-3 centre in the active and the meso isomers.

Solution: Following are the stereoisomers of 2, 3, 4-trihydroxyglutaric acid:

```
    COOH              COOH              COOH
    |                 |                 |
HO—C—H            H—C—OH            H—C—OH
    |                 |                 |
 H—C—OH           HO—C—H             H—C—OH
    |                 |                 |
 CH(OH)COOH       CH(OH)COOH        CH(OH)COOH
     I                II                III
```

The Fischer projection formula of compound I is:

```
        COOH
         |
  HO ——|—— H
         |
   CH(OH) — CH(OH)COOH
```

The Fischer projection formula of compound II is:

```
        COOH
         |
   H ——|—— OH
         |
   CH(OH) — CH(OH)COOH
```

The following points are to be observed:
(*i*) The central carbon is present at the crossing of horizontal and vertical lines.
(*ii*) The longest chain of carbon atoms is present along the vertical line.

Example 30: How many asymmetric carbon atoms are created during the complete reduction of benzil PhCOCOPh with LiAlH$_4$? Also write the number of possible stereo-isomers of the product?

Solution: Ph—$\overset{O}{\overset{\|}{C}}$—$\overset{O}{\overset{\|}{C}}$—Ph $\xrightarrow{\text{LiAlH}_4}$ Ph—$\overset{*}{C}$HOH—$\overset{*}{C}$HOH—Ph
             Benzil   1   2
                                      1, 2-Diphenylethane-1, 2-diol

The product contains two asymmetric carbon atoms. The three stereomeric forms of the product are given below.

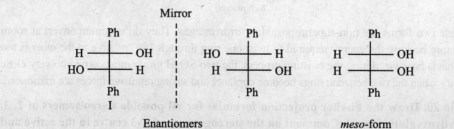

                        Enantiomers   *meso*-form

Example 31: Give proper reasons, state whether the following statements are true or false.
(a) An achiral compound can have chiral centres
(b) When an achiral molecule reacts to give a chiral molecule, the product is always racemic
(c) A molecule with R-configuration is always dextrorotatory
(d) In chemical reactions, if S-configuration of a stereoisomer having one chiral centre is converted to R-configuration, it always means that inversion of configuration has occurred.
(e) An optically inactive substance must be achiral

Solution:
(a) The statement is true. Consider the molecule of *meso*-tartaric acid. It has two chiral centres, yet it is optically inactive (achiral)
(b) This statement is false, if an optically active solvent or a catalyst such as an enzyme is used, the product is likely to be optically active
(c) This statement is false, sign of rotation has nothing to do with the configuration. For example both *dextro* and *laevo* glyceraldehyde have R-configuration
(d) This is a false statement, configurations R and S depend upon the priority sequence of the groups attached and not upon their absolute configurations.
(e) This statement is false. A racemic mixture because of the presence of equal amounts of two enantiomers is optically active but still it contains a chiral compound.

Key Terms

- Fischer projection formula
- Newmann projection formula
- Sawhorse projection formula
- Distereoisomers

Evaluate Yourself

Multiple Choice Questions

1. The isomers of a substance must have
   (a) same chemical properties
   (b) same molecular weight
   (c) same structural formula
   (d) same functional groups

2. Ethanol (CH_3CH_2OH) and dimethyl ether (CH_3OCH_3) are best considered as
   (a) structural isomers
   (b) stereoisomers
   (c) enantiomers
   (d) diastereomers

3. Which of the following statements is false about tautomers ?
   (a) Tautomers are structural isomers
   (b) Tautomers are structural isomers which exist in dynamic equilibrium

(c) Tautomerism involves movement of atoms
(d) Tautomers have independent existence
4. How many structural isomers are possible for C_4H_9Br ?
   (a) 2
   (b) 3
   (c) 4
   (d) 5
5. Alkenes show geometrical isomerism due to:
   (a) asymmetry
   (b) rotation around a single bond
   (c) resonance
   (d) restricted rotation around a double bond
6. Two compounds have the same composition and also have the same atoms attached to the same atoms, although with different orientations in space. These compounds are
   (a) identical
   (b) position isomers
   (c) structural isomers
   (d) stereoisomers
7. Compounds with the same molecular formula but different structural formulas are called
   (a) alkoxides
   (b) iso compounds
   (c) isomers
   (d) ortho compounds
8. How many isomeric aromatic hydrocarbons are possible for C_8H_{10} ?
   (a) 3
   (b) 4
   (c) 5
   (d) 6
9. How many isomers are possible for the compound with molecular formula C_4H_8 ?
   (a) 2
   (b) 4
   (c) 6
   (d) 8
10. The compounds $CH_3CH_2OCH_2CH_3$ and $CH_3OCH_2CH_2CH_3$ are
    (a) enantiomers
    (b) conformational isomers
    (c) metamers
    (d) optical isomers

Short Answer Questions

1. Draw a diagram of nicol prism and show how a plane polarised light is obtained from it.
2. Name different elements of symmetry. Describe briefly one of them.
3. Pentene −2, 3 − diene is optically active although it does not contain chiral carbon atoms. Explain.
4. How will you give flying edge representation of the molecule CHClBrI?
5. What are distereomers? Give its three properties.
6. What is meant by inversion and retention of configuration?
7. What are enantiomers? Write their characteristics?
8. What is chiral carbon atom?
9. What is the most essential condition of compound to be optically active in nature?
10. Assign R and S configurations to the following Fischer projection formulae:

```
        CH=CH₂                           Cl
         |                                |
   H ────┼──── COOH            H₅C₂ ─────┼───── H
         |                                |
        C₂H₅                             Br
```

Long Answer Questions

1. Which of the following can exhibit geometrical isomerism:
    (*i*) 2-Hexene
    (*ii*) 2, 3-Dimethyl-2-butene
    (*iii*) 4-Methyl-2-pentene
    (*iv*) 2, 3-Dimethyl-1-butene.
2. Draw the structures of each of the following compounds showing its stereochemistry:
    (*i*) (2Z, 4Z)-2, 4-hexadiene
    (*ii*) (2R, 3S)-Tartaric acid.
3. What do you understand by the E and Z notations for representing geometrical isomers? Assign E and Z configurations to the following:

```
    CH₃CH₂CH₂     CH₂CH₃            F       Br
           \    /                    \     /
            C                         C
            ‖                         ‖
            C                         C
           / \                       / \
          H   CH₃                   I   Cl
```

4. Assign R and S configurations to the following compounds:

```
            CHO                               H
             |                                |
       HO — C — H                       I — C — Cl
   (i)       |                   (ii)         |
       HO — C — H                             Br
             |
           CH₂OH


            NH₂                              NH₂
             |                                |
   (iii) HO — C — H                 (iv) H — C — CH₃
             |                                |
            CH₃                             COOH
```

5. Distinguish between meso and racemic forms of tartaric acid.
6. Assign R and S configurations to the following:

```
           COOH                             CHO
            |                                |
   (i) H — C — NH₂              (ii) HO — C — H
            |                                |
           CH₃                             CH₂OH
```

CHCOOH
‖
7. Write down the configuration of CHCOOH according to E and Z sequence rules.
8. Explain biological method for the resolution of the racemic modification.
9. Predict whether 3-chlorohexane will be optically active or not. Give reasons for your answer.
10. Which of the following compounds exhibit geometrical isomerism ? Write *cis* and *trans* isomers:
    (*i*) 2-Butene
    (*ii*) 2-Pentene
    (*iii*) 1, 1-Dibromo-1-butene
    (*iv*) 1, 1-Dichloro-1-propene
    (*v*) 1-Pentene
    (*vi*) 2-Methyl-2-butene.
11. What do you understand by the E and Z notations for representing geometrical isomers. Assign E and Z configurations to the following:

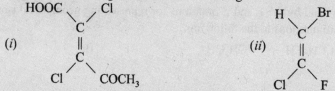

12. Optical activity is linked with the presence of asymmetric carbon atom. Justify the statement with a suitable example.
13. Assign R and S configurations to the following compounds:

```
          NH₂                         NH₂
          |                           |
(i) H — C — CH₃             (ii) HO — C — H
          |                           |
          COOH                        CH₃

          OH                          H
          |                           |
(iii) H₃C — C — C₆H₅         (iv) H₃C — C — C₆H₅
          |                           |
          H                           NH₂
```

14. (*a*) What do the symbols E and Z stand for ? Illustrate briefly the E and Z system of naming a pair geometrical isomers. What are the advantages of E and Z system over conventional *cis-trans* system.

(*b*) Using E and Z notations write the IUPAC names of

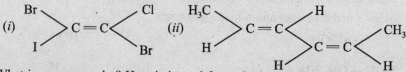

15. (*a*) What is sequence rule ? How is it used for assigning configuration ? Discuss with suitable example.

(*b*) When a compound is expected to be optically active and why?

Answers

Multiple Choice Questions

	1.	2.	3.	4.	5.	6.	7.	8.	9.	10.
(a)		■								
(b)	■							■		
(c)				■		■			■	■
(d)			■		■		■			

Suggested Readings

1. Stereochemistry and the chemistry of natural produces. Dorling Kindersley (India) Pvt. Ltd.
2. Kalsi, P.S. Steriochemistry. Conformation and Mechanism. New Age International 2005.

Chapter 4

Alkanes (Carbon-Carbon Sigma Bonds)

LEARNING OBJECTIVES

After reading this chapter, you should be able to:
- learn methods of preparation of alkanes
- learn Wurtz reaction and Wurtz-Fittig reactions
- learn free radical substitution reactions
- formulate the mechanism of free radical substitution
- understand relative reactivity and selectivity in halogenation reactions of alkanes

4.1 ALKANES

Alkanes which are the hydrocarbons, represent the parent family from which all other families have been derived by replacing one or more hydrogen atoms with suitable functional groups. These are also known as the saturated hydrocarbons or paraffins (*para*-little, *affin*-reactive). These have the general formula C_nH_{2n+2} and are open chain in nature. The bonds present in alkane molecules are single covalent bonds and are sigma (σ) bonds in nature. Moreover, the carbon atoms involved in these molecules have their normal tetrahedral bond angles. Alkanes have low reactivity because of their stable structure.

4.2 STRUCTURE AND NOMENCLATURE OF ALKANES

The first member of the alkane family is methane (CH_4). In this molecule, the carbon atom is located in the centre of a regular tetrahedron [Fig. 4.1(*a*)] and the four H atoms joined to it by C — H bonds, are directed towards its four corners. The H — C — H bond angle is 109° 28'. The spatial formula of methane can also be indicated by *wedge projection* [Fig. 4.1(*b*)] where the thick line indicates the C — H bond that is directed towards the eye of observer. The normal line represents the bond that lies in the plane of the paper while the two C — H bonds indicated by dotted lines project behind the plane of the paper.

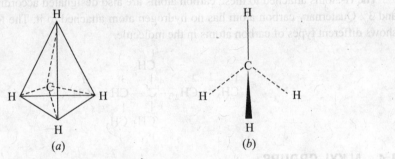

Fig. 4.1: *Spatial and planar representation of methane.*

The next member of the family is ethane (C_2H_6) and can be obtained by replacing on H atom in methane molecule with a methyl (CH_3) group.

$$CH_4 \xrightarrow[+CH_3]{-H} CH_3CH_3$$

Methane   Ethane

Similarly, propane (C_3H_8), the third member of the family is formed in a similar way by replacing one H–atom in ethane with CH_3 group.

$$CH_3CH_3 \xrightarrow[+CH_3]{-H} CH_3CH_2CH_3$$

Ethane   Propane

The IUPAC and common names of the first six members of the alkane family are given in the following table.

Value of n	Alkane	IUPAC name	Common name
1	CH_4	Methane	Methane
2	$CH_3 \cdot CH_3$	Ethane	Ethane

3	$CH_3 \cdot CH_2 \cdot CH_3$	Propane	Propane
4	$CH_3 \cdot CH_2 \cdot CH_2 \cdot CH_3$	Butane	n-Butane
5	$CH_3 \cdot (CH_2)_3 \cdot CH_3$	Pentane	n-Pentane
6	$CH_3 \cdot (CH_2)_4 \cdot CH_3$	Hexane	n-Hexane

The prefix *n*- is used in the common names of the compounds with four or more carbon atoms. This indicates a straight or normal chain of carbon atoms and no branching is involved.

4.3 CLASSIFICATION OF THE CARBON ATOMS IN ALKANES

The carbon atoms in an alkane molecule are classified as *primary* (1°), *secondary* (2°) *tertiary* (3°) and *quaternary* (4°).

(*i*) A primary carbon is linked to one or no carbon atom.

(*ii*) A secondary carbon is linked to two carbon atoms.

(*iii*) A tertiary carbon is linked to three carbon atoms.

(*iv*) A quaternary carbon is linked to four carbon atoms.

The H-atoms attached to these carbon atoms are also designated accordingly *i.e.*, as 1°, 2° and 3°. Quaternary carbon atom has no hydrogen atom attached to it. The following example shows different types of carbon atoms in the molecule.

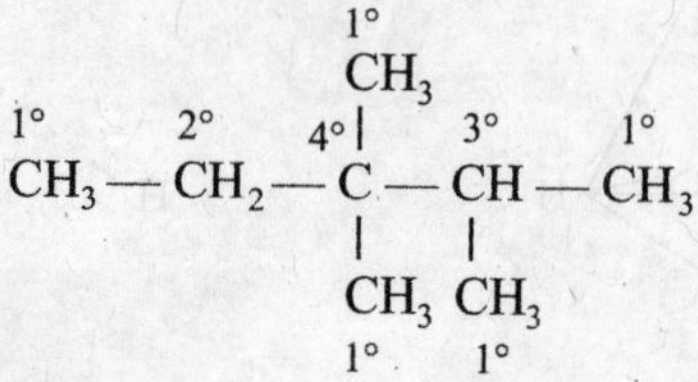

$$\overset{1°}{CH_3} - \overset{2°}{CH_2} - \overset{4°}{\underset{\underset{1°}{CH_3}}{\overset{\overset{1°}{CH_3}}{C}}} - \overset{3°}{\underset{\underset{1°}{CH_3}}{CH}} - \overset{1°}{CH_3}$$

4.4 ALKYL GROUPS

Removal of the H-atom from a carbon atom in alkane molecule forms a group called *alkyl group* with formula $C_nH_{(2n+1)}$. It may be noted that the number of the alkyl groups which are formed from a particular alkane molecule depends upon the total number of different carbon atoms present. For example, methane and ethane molecules have only primary carbon atom and give only one alkyl group while propane, and *n*-butane with both primary and secondary carbon atoms yield two different alkyl groups as shown below:

$$\underset{\text{Ethane}}{\overset{1°}{CH_3} - \overset{1°}{CH_3}} \xrightarrow{-H(1°)} \underset{\text{Ethyl group}}{CH_3 - CH_2 -}$$

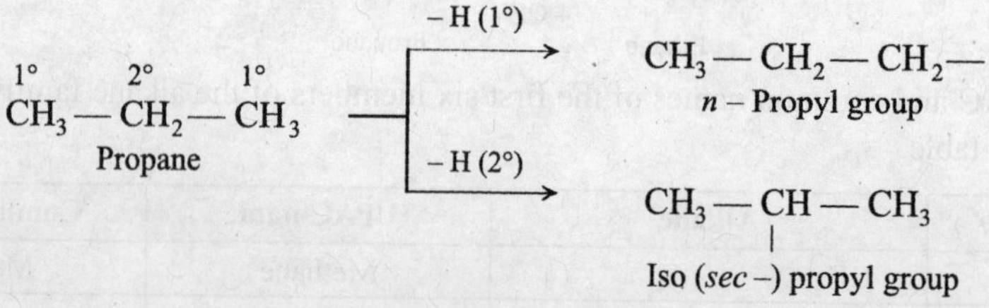

$$\underset{\text{Propane}}{\overset{1°}{CH_3} - \overset{2°}{CH_2} - \overset{1°}{CH_3}} \begin{array}{c} \xrightarrow{-H(1°)} \underset{n-\text{Propyl group}}{CH_3 - CH_2 - CH_2 -} \\ \\ \xrightarrow{-H(2°)} \underset{\text{Iso }(sec-)\text{ propyl group}}{CH_3 - \underset{|}{CH} - CH_3} \end{array}$$

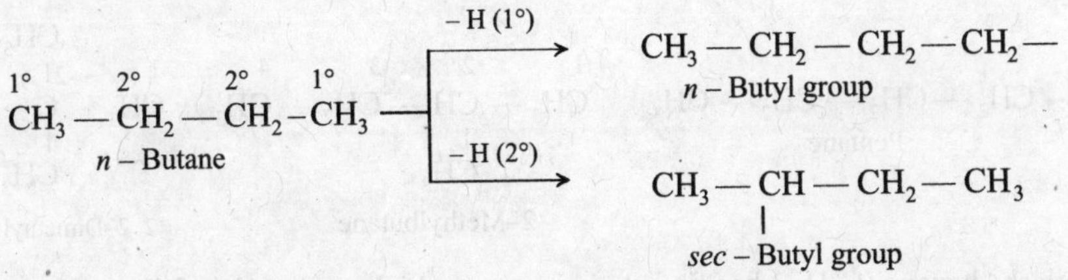

Alkanes (Carbon-Carbon Sigma Bonds) 159

$$\underset{n-\text{Butane}}{\overset{1°\quad 2°\quad 2°\quad 1°}{CH_3-CH_2-CH_2-CH_3}} \begin{array}{c} \xrightarrow{-H(1°)} \underset{n-\text{Butyl group}}{CH_3-CH_2-CH_2-CH_2-} \\ \\ \xrightarrow{-H(2°)} \underset{sec-\text{Butyl group}}{CH_3-\underset{|}{CH}-CH_2-CH_3} \end{array}$$

Similarly, isobutane having primary & tertiary carbon atoms also gives two alkyl groups, as under:

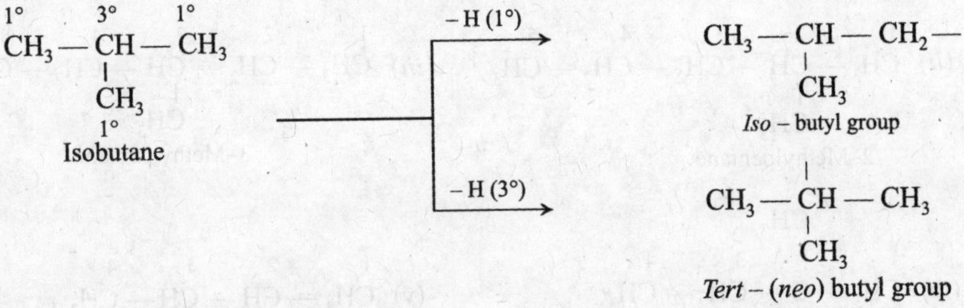

$$\underset{\text{Isobutane}}{\overset{1°\quad 3°\quad 1°}{CH_3-\underset{\underset{1°}{\overset{|}{CH_3}}}{CH}-CH_3}} \begin{array}{c} \xrightarrow{-H(1°)} \underset{Iso-\text{butyl group}}{CH_3-\underset{\underset{CH_3}{|}}{CH}-CH_2-} \\ \\ \xrightarrow{-H(3°)} \underset{Tert-(neo)\text{ butyl group}}{CH_3-\underset{\underset{CH_3}{|}}{\overset{|}{C}}-CH_3} \end{array}$$

iso represents branching by two similar groups (*iso*-propyl and *iso*-butyl), *neo* indicates branching by *three* such groups (*neo*-butyl). A particular alkyl group will be secondary in case the H-atom attached to secondary carbon atom has been replaced. Thus, *iso* and *sec* prefixes for these groups should not be confused with each other.

4.5 STRUCTURAL ISOMERISM IN ALKANES

Alkanes exhibit only **chain isomerism** which is discussed as follows:

In the first three members of the alkane family, the carbon atoms are linked to each other to form straight chains (C — C or C — C — C). However, in the next member butane (C_4H_{10}), there is a scope for branching *i.e.* there may be a straight chain or branched chain of carbon atoms as shown below:

$$\underset{\text{(Straight chain)}}{C-C-C-C} \qquad \underset{\text{(Branched chain)}}{\overset{\overset{}{C-C-C}}{\underset{\underset{C}{|}}{}}}$$

The alkanes corresponding to these structures are called butane (*n*-butane) and 2-Methylpropane (isobutane).

$$\underset{n\text{-butane}}{H_3C-CH_2-CH_2-CH_3} \qquad \underset{iso\text{-butane}}{CH_3-\underset{\underset{CH_3}{|}}{CH}-CH_3}$$

These are the examples of the **chain** isomers which may be defined as *the compounds having the same molecular formula but differing in the nature of the chains of the carbon atoms (straight or branched)*.

It may be noted that as the number of the carbon atoms in a molecule increases, the extent of branching also increases accordingly. Thus, pentane (C_5H_{12}) has three chain isomers given below:

$$CH_3-CH_2-CH_2-CH_2-CH_3$$
Pentane

$$\overset{1}{C}H_3-\overset{2}{C}H-\overset{3}{C}H_2-\overset{4}{C}H_3$$
$$\underset{CH_3}{|}$$
2–Methylbutane

$$\overset{1}{C}H_3-\overset{2}{\underset{CH_3}{\overset{CH_3}{|}}}{C}-\overset{3}{C}H_3$$
2,2–Dimethylpropane

Similarly, hexane (C_6H_{14}) has five chain isomers which are listed as follows:

(i) $CH_3-CH_2-CH_2-CH_2-CH_2-CH_3$
Hexane

(ii) $\overset{1}{C}H_3-\overset{2}{C}H-\overset{3}{C}H_2-\overset{4}{C}H_2-\overset{5}{C}H_3$
      $\quad\quad\quad |$
      $\quad\quad\ CH_3$
2–Methylpentane

(iii) $\overset{1}{C}H_3-\overset{2}{C}H_2-\overset{3}{C}H-\overset{4}{C}H_2-\overset{5}{C}H_3$
      $\quad\quad\quad\quad\quad\ |$
      $\quad\quad\quad\quad\ CH_3$
3–Methylpentane

(iv) $\overset{1}{C}H_3-\overset{2}{\underset{CH_3}{\overset{CH_3}{|}}}{C}-\overset{3}{C}H_2-\overset{4}{C}H_3$
2,2–Dimethylbutane

(v) $\overset{1}{C}H_3-\overset{2}{C}H-\overset{3}{C}H-\overset{4}{C}H_3$
     $\quad\quad\quad |\quad\ |$
     $\quad\quad CH_3\ CH_3$
2,3–Dimethylbutane

For stereoisomerism and conformational isomerism in alkanes refer to chapter 3.

IUPAC Nomenclature

(i) The longest carbon chain in the molecule is identified. The compound is named as a derivative of hydrocarbon corresponding to the longest chain. For example, if there are five carbon atoms in the longest chain, the hydrocarbon will be named as a derivative of pentane and so on.

(ii) The position of different alkyl groups is noted. The numbering is done in such a way that the sum of the numbers corresponding to the position of alkyl groups (called locant) is minimum. This is called *lowest sum rule*. This is illustrated with the help of examples given below:

(a) $\overset{5}{C}H_3-\overset{4}{C}H_2-\overset{3}{\underset{CH_3}{\overset{CH_2-CH_3}{|}}}{C}-\overset{2}{\underset{CH_3}{\overset{|}{C}H}}-\overset{1}{C}H_3$
3-Ethyl-2,3-dimethylpentane

(b) $\overset{1}{C}H_3-\overset{2}{\underset{CH_3^{\ 4}CH_2}{\overset{CH_3}{|}}}{C}-\overset{3}{\underset{|}{\overset{CH_3}{|}}}{C}-CH_2-CH_3$
$\quad\quad\quad\quad\quad\quad\quad\quad\quad\quad |$
$\quad\quad\quad\quad\quad\quad\quad\quad\ ^5CH_2-{}^6CH_3$
3-Ethyl-2,2,3-trimethylhexane

The position of side chain alkyl groups in the longest chain is called **locant**. We shall be using this term here on in the chapter.

(iii) If *two different chains of equal length* are possible, select the one with larger number of side chains or alkyl groups. Consider the following compound

$$\overset{1}{CH_3}-\overset{2}{CH}-\overset{3}{CH}-\overset{4}{CH_2}-\overset{5}{CH_2}-\overset{6}{CH_3} \equiv \overset{4}{CH_3}-\overset{3}{CH}-\overset{2}{CH}-\overset{1}{CH_2}-CH_2-CH_3$$
$$\underset{}{|}\underset{}{|}\underset{}{|}\underset{}{|}$$
$$CH_3CH_2CH_3CH_3\underset{5}{CH_2}\underset{6}{CH_3}$$

We can identify two six carbon longest chains in the same compound as shown by bold lines. The longest chain in the L.H.S. is to be considered because it has two side chains at positions 2 and 3. The longest chain in R.H.S. contains only one side chain at position 4.

(iv) When two or more simple substituents are present on the parent chain, each alkyl group is arranged in *alphabetical* order before the name of the parent alkane. In deciding the alphabetical order of various groups, the prefixes *iso* and *neo* are considered to be part of the name of the alkyl group while the prefixes *sec-* and *tert-* are not.

(v) When two different substituents are present at equivalent positions, the numbering of the chain is done in such a way that the substituent which comes first in the alphabetical order gets the lower number (locant).

(vi) When the same substituent occurs more than once on the parent chain at different positions, the locant of each substituent is separated by commas and the related numerical prefixes such as *di*(for two), *tri* (for three), *tetra* (for four) etc. are attached to the name of the substituent alkyl group. However, the prefixes *di, tri*, etc, themselves are not considered while deciding the alphabetical order of the alkyl groups.

(vii) In case the substituent on the parent alkane is complex having a branched chain, it is named as substituted alkyl group by separate numbering of the carbon atoms of this group. The name of such a complex substituent is always enclosed in brackets to avoid confusion with numbering in the parent chain. Consider the following example.

$$\begin{array}{c} \overset{3}{CH_3} \leftarrow \text{Complex substituent} \\ | \\ CH_3 - \underset{1}{\overset{2}{C}} - CH_3 \\ | \\ CH_2 \end{array}$$
$$\overset{1}{CH_3}-\overset{2}{CH_2}-\overset{3}{CH_2}-\overset{4}{CH_2}-\overset{5}{CH}-\overset{6}{CH_2}-\overset{7}{CH_2}-\overset{8}{CH_2}-\overset{9}{CH_3}$$

We have a complex substituent on the parent chain at position 5. The numbering on the main chain and side chain is shown as above. The name of the compound would be 5 – (2, 2-Dimethylpropyl) nonane.

(viii) If the chain length of two complex substituents is of equal length, then the complex substituent with greater number of alkyl groups forms part of the longest carbon chain while the other one is taken as complex substituent. For example, out of the two possible complex chains in the following compound, the one bordered with a dotted line is accepted

$$\overset{10}{CH_3}-\overset{9}{CH_2}-\overset{8}{CH_2}-\overset{7}{CH_2}-\overset{6}{CH_2}-\overset{5}{CH}-\overset{1}{CH_2}-\overset{2}{CH_2}-\overset{3}{CH_2}-\overset{4}{CH_3}$$
$$\overset{4}{CH_2}$$
$$CH_3-\overset{3}{C}-CH_3$$
$$\underset{2}{CH_2}-\underset{1}{CH_3}$$

with CH_2-CH_3 branch at position 5 shown in dotted box.

The name of the compound would be 5 – (2-Ethylbuytl) – 3, 3-dimethyldecane

(ix) While deciding the alphabetical order of various substituents, the name of the complex susbtituent is considered to begin with the first letter of its complete name. For example, if the complex substituent is 1, 2–dimethylbutyl, then the letter **d** and not **m** will be considered for the purpose of such a compound.

(x) When the names of two or more complex substituents are identical the priority of naming is given to the substituent which has the lowest locant (position) at the first cited point of difference within the complex substituent. For example, if the complex susbstituents are 1-methylbutyl and 2-methylbutyl, then priority of naming will be given to 1-methylbutyl (because 1 comes before 2).

The examples that follow will illustrate some of the points:

Solved Examples

Example 1: Give IUPAC names of the following compounds:

(i)
$$CH_3-CH-CH-CH_3$$
$$\quad\quad |\quad\quad\ |$$
$$\quad\quad CH_2\ \ CH_2$$
$$\quad\quad |\quad\quad\ |$$
$$\quad\quad CH_3\ \ CH_3$$

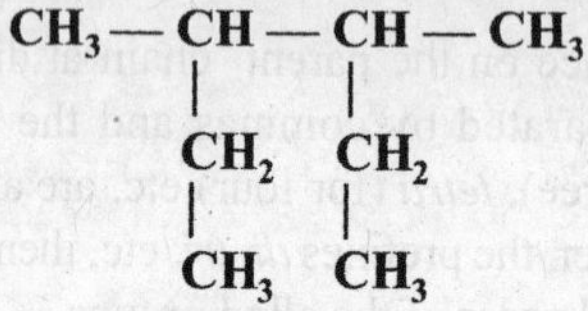

(ii)
$$\quad\quad\quad\quad CH_3\ \ CH_2-CH_3$$
$$\quad\quad\quad\quad\ |\quad\quad\ |$$
$$CH_3-CH-CH-CH_2-CH_2-CH_3$$

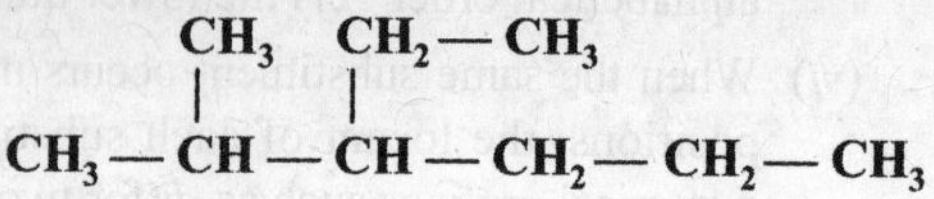

(iii)
$$\quad\quad\quad CH_3$$
$$\quad\quad\quad\ |$$
$$CH_3-C-CH_2-CH-CH_3$$
$$\quad\quad\quad\ |\quad\quad\quad\ |$$
$$\quad\quad\quad CH_3\quad\quad CH_3$$

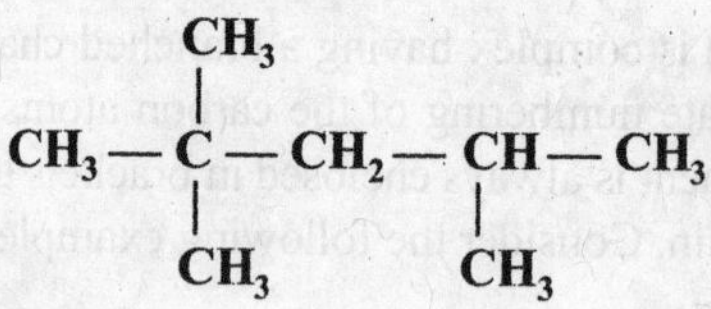

(iv)
$$\quad\quad\quad\quad\quad\quad\quad\quad\quad\quad CH_3$$
$$\quad\quad\quad\quad\quad\quad\quad\quad\quad\quad\ |$$
$$CH_3-CH-CH_2-CH-CH_2-CH-CH_2CH_3$$
$$\quad\quad\ |\quad\quad\quad\quad\ |$$
$$\quad\quad CH_3\quad\quad CH(CH_3)_2$$

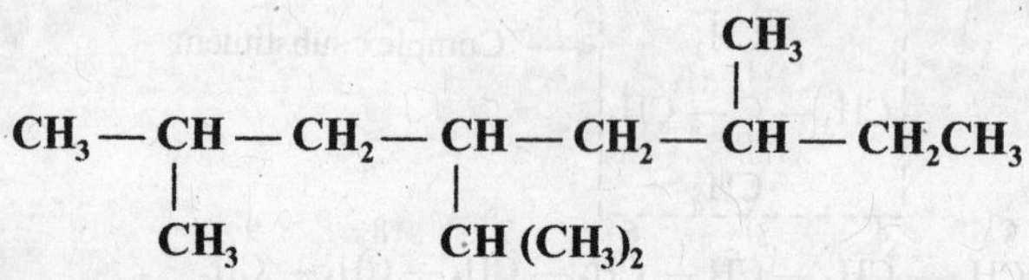

Solution:

(i) 3, 4 – Dimethylhexane.

(ii) 3-Ethyl-2 – methylhexane.

(iii) 2, 2, 4-Trimethylpentane.

(iv) 2, 6-Dimethyl-4 – isopropyloctane.

Example 2: Write the structural formulae and IUPAC names of various compounds possible with the molecular formula C_6H_{14}.

Solution: Various structures possible are given below:

(i) $CH_3CH_2CH_2CH_2CH_2CH_3$ — *n*-Hexane

(ii) 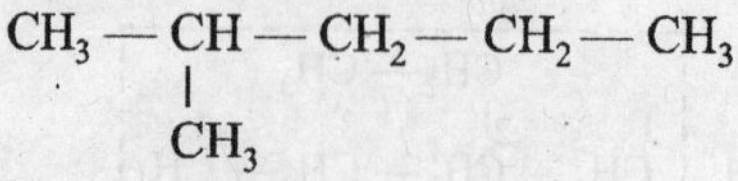
$$CH_3-CH-CH_2-CH_2-CH_3$$
$$\quad\quad\ |$$
$$\quad\quad CH_3$$
2-Methylpentane

(iii) 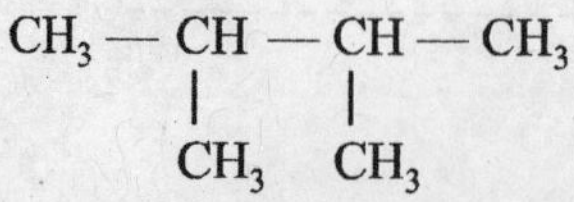
$$CH_3-CH-CH-CH_3$$
$$\quad\quad\ |\quad\ |$$
$$\quad\quad CH_3\ CH_3$$
2, 3-Dimethylbutane

(iv) CH₃—C(CH₃)(CH₃)—CH₂—CH₃   2, 2-Dimethylbutane

Problems for Practice

Give IUPAC names to the following compounds:

(i) CH₃—CH(CH₃)—CH₂—C(CH₃)(CH₃)—CH₃

(ii) CH₃CH₂CH(CH₃)—CH(CH₂CH₃)—CH₂CH₃

(iii) CH₃CH₂—C(CH₃)(CH₂CH₃)—CH₂—CH(CH₃)—CH₃

(iv) (CH₃)₃C—C(CH₃)(CH₂CH₂CH₃)—CH₂CH₃

(v) CH₃CH₂—C(CH₃)(CH₂CH₂CH₃)—CH(C₂H₅)—C₂H₅

(vi) (CH₃)₃CCH₂CH₂CH(CH₃)CH₂CH₃

(vii) CH₃CH₂CH₂CH[C(CH₃)₃]CH₂CH₃

(viii) CH₃—CH(CH₃)—CH(CH₃)—CH₂—CH₃ ; CH₃—CH(CH₃)—CH₃

(ix) CH₃CH₂—CH(C₂H₅)—C(CH₃)[C(CH₃)₃]—CH₂CH₂CH(CH₃)CH₂CH₃

(x) CH₃CH(CH₃)CH₂CH₂CH[CH₂CH(CH(CH₃)₂)]CH(CH₃)₂

(xi) CH₃CH₂CH₂—CH(CH₃)—CH(CH₂CH₃)—CH₂CH₂CH₃

(xii) (CH₃CH₂CH₂)₂CH—CH[CH(CH₃)₂]—CH₂CH₂CH₃

(xiii) CH₃—[CH₂]₅—CH(CH₃)—CH₂—CH(CH₂CH₃)—CH₂CH₂CH₂CH₃

(xiv) CH₃CH₂CH₂—CH[CH(CH₃)₂]—CH(CH₂CH₃)—C[CH(CH₃)₂](CH₃)—CH₂CH₃

(xv) [CH₃(CH₂)₃]₂CH₂—CH(CH₃)—CH(CH₃)₂

Answers

(i) 2, 2, 4-Trimethylpentane (ii) 3-Ethyl-4-methylhexane

(iii) 4-Ethyl-2, 4-dimethylhexane (iv) 3-Ethyl-2, 2, 3-trimethylhexane

(v) 3, 4-Diethyl-4-methylheptane (vi) 2, 2, 5-Trimethylheptane
(vii) 4-(1, 1-Dimethylethyl) heptane (viii) 3-Ethyl-2, 4, 5-trimethylheptane
(ix) 4-(1, 1-Dimethylethyl)-3-ethyl-4, 7-dimethyldecane
(x) 2, 8-Dimethyl-5-(2-methylpropyl) nonane
(xi) 4-Ethyl-5-methyloctane
(xii) 4-(1-Methylethyl)-5-propyloctane
(xiii) 7 – (1, 2-Dimethylpentyl)-5-ethyltridecane
(xiv) 3, 3-Diethyl-4 methyl-5-(1-methylethyl) octane
(xv) 4-Butyl-2, 3-dimethyloctane

4.6 SOURCES OF ALKANES

Two main sources of alkanes are natural gas and petroleum. Both these substances are generally found together in underground deposits. Natural gas contains about 80% methane, 10% ethane, the rest being mixture of higher hydrocarbons. Petroleum contains hydrocarbons containing upto 40 carbon atoms. Wood turpentine and some pine trees contain n-heptane and cabbage leaves contain n-nonacosane ($C_{29}H_{60}$).

4.7 GENERAL METHODS OF PREPARATION OF ALKANES

Different methods of preparation of alkanes are given below:

(1) Hydrogenation of unsaturated hydrocarbons (Sabatier-Senderen's reaction). A mixture of unsaturated hydrocarbons and hydrogen is passed our finely divided platinum, palladium or nickel at 523-573 K. Alkenes and alkynes are reduced to alkanes.

$$\begin{matrix} CH \\ \| \\ CH_2 \end{matrix} + H_2 \xrightarrow[523-573\,K]{Pt,\,Pd\,or\,Ni} \begin{matrix} CH_3 \\ | \\ CH_3 \end{matrix}$$

$$\underset{\text{Propene}}{CH_3-CH=CH_2} + H_2 \xrightarrow[523-573\,K]{Pt} \underset{\text{Propane}}{CH_3-CH_2-CH_3}$$

$$\underset{\text{Acetylene}}{CH\equiv CH} + 2H_2 \xrightarrow[523-573\,K]{Pt} \underset{\text{Ethane}}{CH_3-CH_3}$$

(2) Reduction of alkyl halides. This may be done in different ways:

(i) Chemical reducing agents like zinc and acetic acid or hydrochloric acid, zinc-copper couple and alcohol; magnesium-amalgam and water etc., convert alkyl halides into alkanes.

$$\underset{\text{Alkyl halide}}{R-X} + 2[H] \xrightarrow[\text{Zn / Acetic acid}]{Zn-Cu\,/\,alcohol} \underset{\text{Alkane}}{R-H} + H-X$$

$$C_2H_5-Br + 2[H] \xrightarrow[\text{Zn / Acetic acid}]{Zn-Cu\,/\,alcohol} \underset{\text{Ethane}}{C_2H_6} + HBr$$

(*ii*) Alkyl iodides are easily reduced by heating with concentrated HI at about 423 K in a sealed tube.

$$R-I + HI \xrightarrow[423\ K]{P} R-H + I_2$$

$$C_2H_5I + HI \xrightarrow[423\ K]{P} C_2H_6 + I_2$$

(3) Wurtz reaction. This reaction involves the interaction of two molecules of alkyl halides and sodium metal in dry ether.

$$R-\overline{\underline{X + 2Na + X}}-R \xrightarrow{\text{Dry ether}} R-R + 2\ NaX$$

$$CH_3-\overline{\underline{Br + 2Na + Br}}-CH_3 \xrightarrow{\text{Dry ether}} CH_3-CH_3 + 2NaBr$$

$$C_2H_5-\overline{\underline{Br + 2Na + Br}}-C_2H_5 \xrightarrow{\text{Dry ether}} CH_3-CH_2-CH_2-CH_3 + 2NaBr$$
n-Butane

When different alkyl halides are used, a mixture of three alkenes is obtained as shown below:

$$\underset{\text{Methyl bromide}}{CH_3-Br} + 2Na + \underset{\text{Ethyl bromide}}{Br-C_2H_5} \longrightarrow \underset{\text{Propane}}{CH_3-CH_2-CH_3} + 2NaBr$$

$$CH_3-Br + 2Na + Br-CH_3 \longrightarrow \underset{\text{Ethane}}{CH_3-CH_3} + 2NaBr$$

$$C_2H_5-Br + 2Na + Br-C_2H_5 \longrightarrow CH_3-CH_2-CH_2-CH_3 + 2NaBr$$

Wurtz reaction is suitable for the preparation of symmetrical alkanes. With secondary and tertiary alkyl halides, many side products are formed.

Mechanism of Reaction

Two mechanisms have been suggested for the reaction.

Mechanism I

The reaction takes place in two steps as shown below:

$$RX + 2Na \longrightarrow R-Na + NaX$$

$$RNa + XR \longrightarrow R-R + NaX$$

Mechanism II

Again the reaction takes place in two steps as under:

$$RX + Na \longrightarrow \overset{\bullet}{R} + NaX$$

$$\overset{\bullet}{R} + \overset{\bullet}{R} \longrightarrow \underset{\text{Hydrocarbon}}{R-R}$$

(4) Wurtz-Fittig reaction. If we take a mixture of an alkyl halide and an aryl halide and treat them with sodium metal in dry ether, then a substituted benzene compound is obtained. This reaction is known as Wurtz-Fittig reaction.

$$CH_3-Br + 2Na + Br-C_6H_5 \xrightarrow{\text{Dry ether}} CH_3-C_6H_5 + 2NaBr$$
$$\text{Toluene}$$

(5) Corey-House Synthesis. It is a better method than Wurtz reaction. An alkyl halide and a lithium dialkyl copper are reacted to give a higher hydrocarbon.

$$\underset{\substack{\text{Alkyl}\\\text{halide}}}{R'-X} + \underset{\substack{\text{Lithium}\\\text{dialkyl copper}}}{R_2CuLi} \longrightarrow \underset{\text{Alkane}}{R-R'} + R-Cu + LiX$$

(R and R' may be same or different)

For example,

$$\underset{n\text{-Propyl bromide}}{CH_3CH_2CH_2Br} + \underset{\substack{\text{Lithium dimethyl}\\\text{copper}}}{(CH_3)_2CuLi} \longrightarrow \underset{n\text{-Butane}}{CH_3CH_2CH_2CH_3} + \underset{\text{Methyl copper}}{CH_3Cu} + LiBr$$

A notable feature of this reaction is that it can be used for preparing symmetrical, unsymmetrical, straight-chain or branched-chain alkanes.

For a better yield of the product, the alkyl halide used should be primary whereas lithium dialkyl copper may be primary, secondary or tertiary.

(6) Indirect reduction of alkyl halides. Alkyl halides react with magnesium in dry ether to form alkyl magnesium halides (Grignard reagent). Decomposition of the Grignard reagent with water yield alkanes.

$$\underset{\text{Alkyl halide}}{R-X} + Mg \xrightarrow{\text{Dry ether}} \underset{\text{Grignard reagent}}{R-MgX}$$

$$R-MgX + H-OH \longrightarrow \underset{\text{Alkane}}{R-H} + Mg\begin{matrix}OH\\X\end{matrix}$$

$$\underset{\substack{\text{Methyl magnesium}\\\text{bromide}}}{CH_3-MgBr} + H-OH \longrightarrow CH_4 + Mg\begin{matrix}OH\\X\end{matrix}$$

(7) Decarboxylation of Carboxylic Acid Salts. Sodium salts of a fatty acid are heated with soda lime (NaOH + CaO) to form alkanes.

$$R-\boxed{COONa + NaO}-H \xrightarrow{CaO} R-H + Na_2CO_3$$

$$\underset{\text{Sodium acetate}}{CH_3-COONa} + NaOH \xrightarrow{CaO} CH_4 + Na_2CO_3$$

A probable mechanism for this reaction is

$$HO\overset{O^-}{\underset{\underset{O}{\parallel}}{C}}-R \rightleftharpoons H-O-\overset{O^-}{\underset{\underset{O}{\parallel}}{C}}-R \rightleftharpoons R^- + H-O-\overset{O^-}{\underset{\underset{O}{\parallel}}{C}} \longrightarrow \underset{\text{Alkane}}{R-H} + {}^-O-C\begin{matrix}O^-\\\parallel\\O\end{matrix}$$

(8) Kolbe's Electrolytic Process. By the electrolysis of conc. solution of sodium or potassium salts of fatty acids, alkanes are obtained.

$$2R-COONa \longrightarrow 2RCOO^- + 2Na^-$$

$$2R-COO^- \xrightarrow{-2e^-} \begin{array}{c} RCOO \\ | \\ RCOO \\ \text{(Unstable)} \end{array} \to R-R + 2CO_2 \quad \text{[At anode]}$$

$$2Na^+ + 2e^- \longrightarrow 2Na + 2H_2O$$
$$\downarrow$$
$$2NaOH + H_2 \quad \text{[At cathode]}$$

$$2CH_3COONa \xrightarrow{\text{Electrolysis}} \underbrace{CH_3-CH_3 + 2CO_2}_{\text{[At anode]}} + \underbrace{2Na + 2H_2O}_{\text{[At cathode]}}$$

A probable mechanism for this reaction involves free radicals.

$$CH_3-\overset{O}{\underset{\|}{C}}-\ddot{\underset{..}{O}}:^- \longrightarrow CH_3-\overset{O}{\underset{\|}{C}}-\dot{O}: + e^-$$
Acetate free radical

$$CH_3 \overset{\frown}{:}\overset{O}{\underset{\|}{C}}\overset{\frown}{\smile}\ddot{O}: \longrightarrow \dot{C}H_3 + \overset{O}{\underset{\|}{C}}{=}O$$
Methyl
free radical

$$\dot{C}H_3 + \dot{C}H_3 \longrightarrow CH_3-CH_3$$
Ethane

(9) By Reduction of Alcohols, Aldehydes, Ketones and Acids. The reduction is carried out with the help of HI and P.

(i) $\underset{\text{Alcohol}}{R-OH} + 2HI \xrightarrow[423\,K]{P} R-H + H_2O + I_2$

$\underset{\text{Ethyl alcohol}}{C_2H_5-OH} + 2HI \xrightarrow[423\,K]{P} \underset{\text{Ethane}}{C_2H_6} + H_2O + I_2$

(ii) $\underset{\text{Aldehyde}}{R-\overset{O}{\underset{\|}{C}}-H} + 4HI \xrightarrow[423\,K]{P} \underset{\text{Alkane}}{R-CH_3} + H_2O + 2I_2$

$\underset{\text{Acetaldehyde}}{CH_3-\overset{O}{\underset{\|}{C}}-H} + 4HI \xrightarrow[423\,K]{P} \underset{\text{Ethane}}{CH_3-CH_3} + H_2O + 2I_2$

(iii) $\underset{\text{Ketone}}{R-\overset{O}{\underset{\|}{C}}-R} + 4HI \xrightarrow[423\,K]{P} \underset{\text{Alkane}}{R-CH_2-R} + H_2O + 2I_2$

$\underset{\text{Acetone}}{CH_3-\overset{O}{\underset{\|}{C}}-CH_3} + 4HI \xrightarrow[423\,K]{P} \underset{\text{Alkane}}{CH_3-CH_2-CH_3} + H_2O + 2I_2$

(iv) $\underset{\text{Acid}}{R-\overset{O}{\overset{\|}{C}}-OH} + 6HI \xrightarrow[423 K]{P} \underset{\text{Alkane}}{R-CH_3} - 2H_2O + 3I_2$

$\underset{\text{Formic acid}}{H-COOH} + 6HI \xrightarrow[473 K]{P} CH_4 + 2H_2O + 3I_2$

$CH_3-COOH + 6HI \xrightarrow[473 K]{P} CH_3-CH_3 + 2H_2O + 3I_2$

4.8 PHYSICAL PROPERTIES OF ALKANES

1. Solubility. Alkanes being non-polar in nature are insoluble in water but soluble in non-polar solvents like ether, benzene, etc. Solubility decreases with increase in molecular size.

2. Boiling Points. As alkanes are non-polar molecules, they have only weak van der Waals' forces operating between them. van der Waals' forces of attraction depend upon the molecular size and increase with increase in molecular size. Among the chain isomers of a compound, the branched chain isomer has lower boiling point compared to straight-chain isomer. For example, the boiling points of three isomeric pentanes are given below:

$CH_3CH_2CH_2CH_2CH_3$
n-Pentane
(b.p = 309 K)

$CH_3-CH-CH_2-CH_3$
$\quad\quad |$
$\quad\quad CH_3$
Isopentane
(b.p = 301 K)

$\quad\quad CH_3$
$\quad\quad |$
CH_3-C-CH_3
$\quad\quad |$
$\quad\quad CH_3$
Neopentane
(b.p = 281 K)

Explanation: The boiling point of a substance depends upon the force of attraction between the molecules. Force of attraction between the molecules depends upon the area of contact between them. Area of contact and force of attraction will be maximum for straight chain compound. As branching increases, the area of contact decreases. Consequently, the force of attraction and boiling point will decrease.

3. Melting Point. Alkanes show lower melting point because of absence of charge. Melting point is expected to increase with increase in molecular size. This is found to be true but the increase is not uniform.

Alkane	Melting Point (K)
C_3H_8	85.9
n-C_4H_{10}	138
n-C_5H_{12}	143.3
n-C_6H_{14}	179

We find that increase in melting point is relatively more in going from odd-carbon to even-carbon alkane compared to when we move from even-carbon to odd-carbon alkane.

Explanation. It may be realised that in even-carbon alkanes (*e.g.* C_6H_{14}) the packing of molecules can take place to a greater extent, resulting in greater attraction forces and hence greater melting points. This is not so with odd-carbon alkanes.

Representation of two chain of *n*-C_6H_{14} (greater attractive forces)

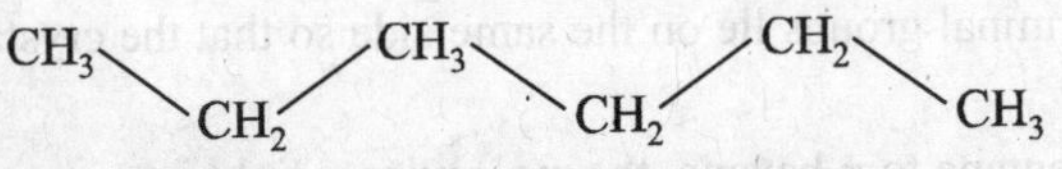

Solved Examples

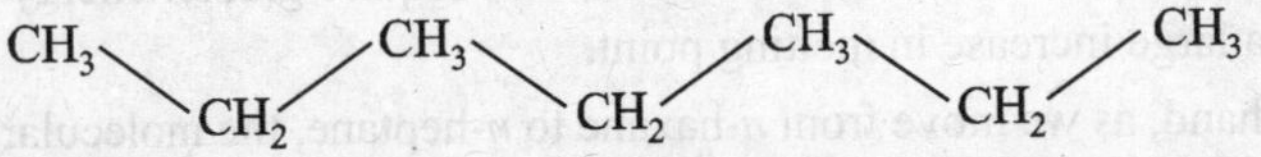

Example 3: Write the structural formula of *n*-pentane, isopentane and neopentane. Which of these has highest boiling point and why?

Solution:

Structural formulae of pentanes

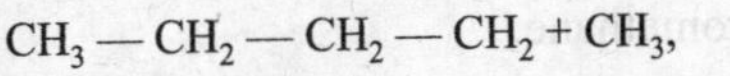

$CH_3 — CH_2 — CH_2 — CH_2 + CH_3$
n-Pentane

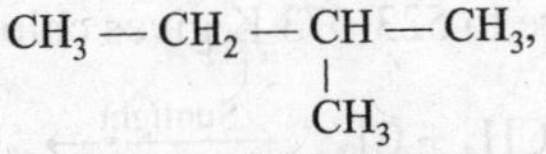

$CH_3 — CH_2 — CH — CH_3$
                       |
                       CH_3
Isopentane

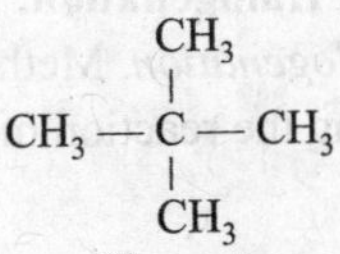

$CH_3 — \underset{\underset{CH_3}{|}}{\overset{\overset{CH_3}{|}}{C}} — CH_3$
Neopentane

Highest boiling point. Of the three pentanes, *n*-pentane has the highest boiling point.

Explanation: All the three pentanes are non-polar compounds having only weak intermolecular forces of attraction (van der Waals' forces). The molecular weights of the three isomers are naturally the same but they have different surface areas. As the surface area of *n*-pentane is larger than those of the other two, it has strongest intermolecular forces of attraction. Therefore it has highest boiling point followed by iso- and neopentanes.

Example 4: As we move up from *n*-pentane to *n*-hexane, there is an increase of about 35° in the melting point but on moving from *n*-hexane to *n*-heptane the increase is only about 4°. How can you account for these observations?

Solution: Disproportionate variation in melting points of the three alkanes can be explained on the basis of their molecular weights and their ability to fit in the crystal structures as follows.

In alkanes, the carbon atoms form zig-zag chains as depicted below for *n*-pentane, *n*-hexane and *n*-heptane.

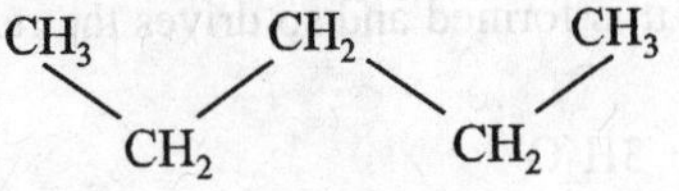

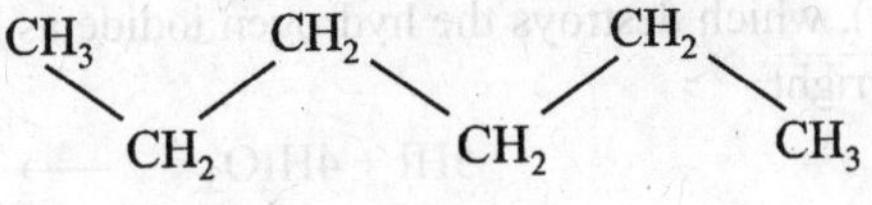

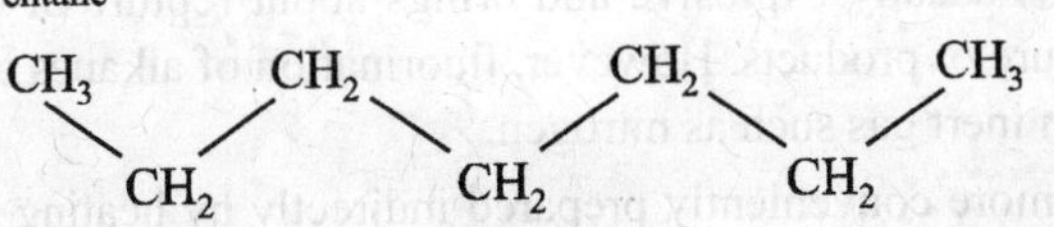

It is apparent that in *n*-hexane, the two terminal methyl groups lie on opposite sides of the zig-zig chain which enables this chain to fit in closely in the crystal structure. But in *n*-pentane and *n*-heptane, the two terminal groups lie on the same side so that the crystal structure is not so closely packed.

As we move from *n*-pentane to *n*-hexane, the molecular weight increases, also the packing arrangement becomes more closed. Both these factors require greater energy to melt the solid. As such there is a large increase in melting point.

On the other hand, as we move from *n*-haxane to *n*-heptane, the molecular weight increases but the packing arrangement does not remain so close. While the melting point would tend to increase due to increase in molecular weight, the decrease in close packing would tend to have the opposite effect. The net result is that there is only a small increase in melting point in this case.

4.9 CHEMICAL PROPERTIES OF ALKANES

Chemical properties of alkanes are described as under:

1. Halogenation. Displacement or substitution of hydrogen atoms by halogens is known as *halogenation*. Methane on treatment with chlorine in the presence of diffused sunlight or by heating the reaction mixture to 523-673 K gives chloromethane.

$$CH_4 + Cl_2 \xrightarrow[\text{or } 523-673K]{\text{Sunlight}} \underset{\text{Chloromethane}}{CH_3Cl} + HCl$$

The reaction does not stop here. Chloromethane is further chlorinated to dichloromethane. Dichloromethane is chlorinated to trichloromethane and trichloromethane is finally chlorinated to tetrachloromethane. This sequence of reactions is given below:

$$CH_3Cl + Cl_2 \longrightarrow \underset{\text{Dichloromethane}}{CH_2Cl_2} + HCl$$

$$CH_2Cl_2 + Cl_2 \longrightarrow \underset{\text{Trichloromethane}}{CHCl_3} + HCl$$

$$CHCl_3 + Cl_2 \longrightarrow \underset{\text{Tetrachloromethane}}{CCl_4} + HCl$$

Bromination of alkanes takes place in similar manner but less readily. Direct iodination of alkanes is not possible as the reaction is reversible.

$$R-H + I_2 \rightleftharpoons R-I + HI$$

However, iodination can be carried out in the presence of oxidising agent such as iodic acid (HIO_3), which destroys the hydrogen iodide as soon as it is formed and so drives the reaction to the right.

$$5HI + 4HIO_3 \longrightarrow 3I_2 + 3H_2O$$

Direct fluorination is usually explosive and brings about repture of C — C and C — H bonds leading to a mixture of products. However, fluorination of alkanes can be carried out by diluting fluorine with an inert gas such as nitrogen.

Alkyl fluorides are more conveniently prepared indirectly by heating alkyl chlorides with inorganic fluorides such as AsF_3, SbF_3, AgF, Hg_2F_2 etc.

$$2\underset{\text{Ethyl chloride}}{CH_3CH_2Cl} + Hg_2F_2 \longrightarrow 2\underset{\text{Ethyl fluoride}}{CH_3CH_2F} + Hg_2Cl_2$$

2. Nitration.
It is a process in which hydrogen atom of alkane is replaced by nitro group ($-NO_2$). Alkanes undergo nitration when treated with fuming HNO_3 in the vapour phase between 423–748 K. For example,

$$CH_3 - [H + HO] - NO_2 \longrightarrow CH_3 - NO_2 + H_2O$$
$$\text{Nitro methane}$$

Higher alkanes on nitration give a mixture of all possible mono derivatives. For example,

$$CH_3 - CH_2 - CH_3 + HNO_3 \xrightarrow{673 K} CH_3 - CH_2 - CH_2 - NO_2 +$$
$$\text{1-Nitropropane}$$

$$\underset{\text{2-Nitropropane}}{CN_3 - \underset{\underset{NO_2}{|}}{CH} - CH_3} + \underset{\text{Nitroethane}}{CN_3 - CH - NO_2} + \underset{\text{Nitromethane}}{CH_3 - NO_2}$$

Mechanism. The reaction takes place by the following *free-radical* mechanism.

$$H - O - NO_2 \xrightarrow[\text{Homolytic fission}]{423-673 K} H\dot{O} + \dot{N}O_2$$

$$R - H + \dot{O}H \longrightarrow R\cdot + H_2O$$

$$R\cdot + \cdot NO_2 \longrightarrow R - NO_2$$

3. Sulphonation.
It is a process in which hydrogen atom of alkane is replaced by sulphonic acid group ($-SO_3H$). Sulphonation of hexane and higher members may be carried out by treating the alkane with oleum (fuming sulphuric acid). For example,

$$\underset{n\text{-Hexane}}{C_6H_{13} - H + HO - SO_3H} \xrightarrow{\Delta} \underset{\text{Hexane sulphonic acid}}{C_6H_{13} - SO_3H} + H_2O$$

Mechanism Sulphomation occurs by *free-radical* mechanism

$$HO - SO_3H \xrightarrow[\text{Homolytic fission}]{423-673 K} H\dot{O} + \dot{S}O_3H$$

$$R - H + \dot{O}H \longrightarrow R\cdot + H_2O$$

$$R\cdot + \cdot NO_2 \longrightarrow \underset{\text{Alkanesulphonic acid}}{R - \dot{S}O_3H}$$

The ease of replacement of hydrogen atom is

Tertiary hydrogen > Secondary hydrogen > Primary hydrogen

4. Oxidation.
(*i*) *Combustion*. Alkanes burn readily in excess of air or oxygen to form CO_2 and H_2O. This process is called combustion and is accompanied by the evolution of a large amount of heat. For example,

$$C_nH_{2n+2} + O_2 \xrightarrow{\Delta} nCO_2 + (n+1) H_2O + \text{Heat}$$

$$CH_4 + 2O_2 \xrightarrow{\Delta} CO_2 + 2H_2O + 890 \text{ kJ}$$

Alkanes are, therefore, used as fuels in the form of kerosene oil, gasoline, LPG.

(*ii*) *Catalytic oxidation.* When alkanes are heated in a limited supply of air or oxygen at high pressure and in the presence of suitable catalyst, they are oxidised to alcohols, aldehydes or fatty acids. For example,

$$2CH_4 + O_2 \xrightarrow[\text{Cu Tube}]{\text{523 K, 100 Atm}} 2CH_3OH$$

$$CH_4 + O_2 \xrightarrow[\text{Mo}_2O_3]{\text{673 K, 200 Atm}} HCHO + H_2O$$

(*iii*) Alkanes containing tertiary hydrogens can be oxidised by such reagents to form alcohols. For example,

$$\underset{\text{Iso-butane}}{CH_3-\underset{\underset{CH_3}{|}}{\overset{\overset{CH_3}{|}}{C}}-H} + O \xrightarrow{KMnO_4} \underset{\text{tert-Butyl alcohol}}{CH_3-\underset{\underset{CH_3}{|}}{\overset{\overset{CH_3}{|}}{C}}-OH}$$

5. Aromatisation. When alkanes containing six or more carbon atoms are heated to high temperature (773 K) under high pressure and in the presence of a catalyst (oxides of chromium, vanadium, molybdenum supported on alumina), they are converted into aromatic hydrocarbons. The process involves cyclisation, isomerisation and dehydrogenation and is known as *aromatisation*. For example,

n–hexane $\xrightarrow[\text{Cyclisation}]{-H_2}$ Cyclohexane $\xrightarrow[-3H_2]{\text{Dehydrogenation}}$ Benzene

Similarly

n–heptane $\xrightarrow[\text{2–Dehydrogenation}]{\text{1–Cyclisation}}$ Toluene + $4H_2$

6. Thermal decomposition or pyrolysis or cracking. The process of decomposition of bigger molecules into a number of simpler molecules by the action of heat is called *pyrolysis or cracking*. This process involves the breaking of C — C and C — H bonds leading to the formation of lower alkanes and alkenes. For example,

$$CH_3-CH_2-CH_3 \xrightarrow{733 \text{ K}} \underset{\text{Ethylene}}{CH_2=CH_2} + \underset{\text{Methane}}{CH_4} \quad \text{or} \quad CH_3-CH=CH_2 + H_2 \quad \text{Propene}$$

7. Alkylation. Alkanes containing a tertiary carbon atom react with alkenes on heating in the presence of conc H_2SO_4 or HF to produce larger branched chain alkanes: For example,

Alkanes (Carbon-Carbon Sigma Bonds)   173

$$CH_3-\underset{\underset{CH_3}{|}}{\overset{\overset{CH_3}{|}}{C}}-H + CH_2=\underset{\underset{CH_3}{|}}{C}-CH_3 \xrightarrow[\Delta]{Conc.\ H_2SO_4} CH_3-\underset{\underset{CH_3}{|}}{\overset{\overset{CH_3}{|}}{C}}-CH_2-\underset{\underset{CH_3}{|}}{\overset{\overset{H}{|}}{C}}-CH_3$$

      Isobutane   Isobutylene   2, 2, 4-Trimethylpentane
                                                                                   (Isooctane)

Mechanism

    The reaction takes place through carbocation intermediates.

(i) $CH_3-\underset{\underset{CH_3}{|}}{C}=CH_2 + H^+ \longrightarrow CH_3-\underset{\underset{CH_3}{|}}{\overset{+}{C}}-CH_3$

                                      3° Butyl carbocation

(ii) $CH_3-\underset{\underset{CH_3}{|}}{\overset{+}{C}}-CH_3 + CH_2=\underset{\underset{CH_3}{|}}{C}-CH_3 \longrightarrow CH_3-\underset{\underset{CH_3}{|}}{\overset{\overset{CH_3}{|}}{C}}-CH_2-\underset{\underset{CH_3}{|}}{\overset{+}{C}}-CH_3$

                                Isobutylene

(iii) $CH_3-\underset{\underset{CH_3}{|}}{\overset{\overset{CH_3}{|}}{C}}-CH_2-\overset{+}{\underset{\underset{CH_3}{|}}{C}}-CH_3 + H-\underset{\underset{CH_3}{|}}{\overset{\overset{CH_3}{|}}{C}}-CH_3 \xrightarrow{Hydride\ Transfer}$

$$CH_3-\underset{\underset{CH_3}{|}}{\overset{\overset{CH_3}{|}}{C}}-CH_2-\underset{\underset{CH_3}{|}}{\overset{\overset{H}{|}}{C}}-CH_3 + CH_3-\overset{\overset{CH_3}{|}}{\underset{\underset{CH_3}{|}}{\overset{+}{C}}}$$

                                  2, 2, 4-Trimethyl pentane

step (ii) and (iii) are repeated over and over again.

8. Isomerization. On heating with anhy. $AlCl_3$ and HCl, n-alkanes are converted into branched chain alkanes. This process is called *isomerization*.

$$CH_3-CH_2-CH_2-CH_3 \xrightarrow[\Delta]{AlCl_3/HCl} CH_3-\underset{\underset{Isobutane}{}}{\overset{\overset{CH_3}{|}}{CH}}-CH_3$$

Mechanism of the reaction

    The reaction involves formation of carbocation, followed by 1, 2-methyl shift and 1, 2-hydride shift to form more stable 3°carbocation.

$$CH_3CH_2CH_2CH_3 \xrightarrow{Dehydrogenation} CH_3CH=CH-CH_3 + H_2$$
           n-Butane   2-Butene

$$AlCl_3 + HCl \rightleftharpoons H\overset{+}{A}lCl_4^-$$

$$CH_3CH=CHCH_3 + \overset{+}{H}\ AlCl_4^- \longrightarrow CH_3\overset{+}{C}H-CH_2CH_3 + AlCl_4^-$$
                                                          2° Butyl carbocation

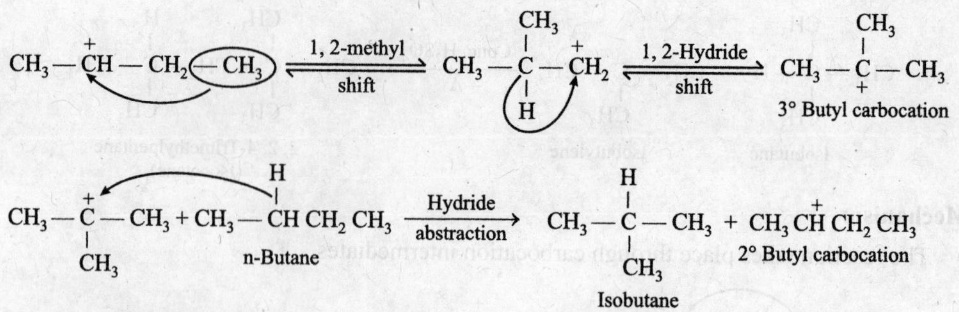

$$CH_3-\overset{+}{C}H-CH_2-CH_3 \xrightleftharpoons[\text{shift}]{1,2\text{-methyl}} CH_3-\underset{H}{\overset{CH_3}{\underset{|}{C}}}-\overset{+}{C}H_2 \xrightleftharpoons[\text{shift}]{1,2\text{-Hydride}} CH_3-\underset{+}{\overset{CH_3}{\underset{|}{C}}}-CH_3$$
3° Butyl carbocation

$$CH_3-\underset{CH_3}{\overset{+}{\underset{|}{C}}}-CH_3 + CH_3-\overset{H}{\underset{|}{C}H}-CH_2\,CH_3 \xrightarrow{\text{Hydride abstraction}} CH_3-\underset{CH_3}{\overset{H}{\underset{|}{C}}}-CH_3 + CH_3\,\overset{+}{C}H\,CH_2\,CH_3$$

n-Butane   Isobutane   2° Butyl carbocation

4.10 MECHANISM OF HALOGENATION OF ALKANE

The halogenation of alkanes occurs by a free radical mechanism. It involves three steps: (a) Chain Initiation, (b) Chain propagation, (c) Chain termination. Mechanism of halogenation is explained by considering the chlorination of methane.

(a) **Chain Initiation.** When a mixture of CH_4 and Cl_2 is heated or subjected to diffused sunlight, Cl_2 absorbs energy and undergoes homolytic fission producing chlorine free radicals. One molecule gives rise to two radicals.

Step (1) $:\!\overset{..}{\underset{..}{Cl}}\!\cdot\cdot\!\overset{..}{\underset{..}{Cl}}\!: \xrightarrow{h\nu \text{ or } \Delta} :\!\overset{..}{\underset{..}{Cl}}\!\cdot + \cdot\overset{..}{\underset{..}{Cl}}\!:$
Chlorine free radicals

(b) **Chain propagation.** The chlorine free radical produced above collides with a molecule of methane forming hydrogen chloride and a methyl free-radical. The methyl free-radical in turn reacts with a molecule of chlorine forming methyl chloride and chlorine free-radical. The newly formed chlorine radical can react with another molecule of methane as in step (2) generating methyl free-radical and hydrogen chloride. The methyl free-radical can again repeat step (3) and so on. Thus, the sequence of reactions in steps (2) and (3) is repeated over and over again and thus the chain is propagated. In other words, a single photon of light initially absorbed by chlorine can bring about the conversion of a large number of molecules of methane into methyl chloride.

Step (2) $H-\underset{H}{\overset{H}{\underset{|}{\overset{|}{C}}}}-H + Cl\cdot \longrightarrow H-\underset{H}{\overset{H}{\underset{|}{\overset{|}{C}}}}\cdot + HCl$
Methyl free-radical

Step (3) $H-\underset{H}{\overset{H}{\underset{|}{\overset{|}{C}}}}\cdot + Cl-Cl \longrightarrow H-\underset{H}{\overset{H}{\underset{|}{\overset{|}{C}}}}-Cl + Cl\cdot$

However, the above reaction does not stop at methyl chloride stage but proceeds further till all the H-atoms of methane are replaced by chlorine atoms giving a mixture of mono-, di-, tri- and tetra-chloromethane.

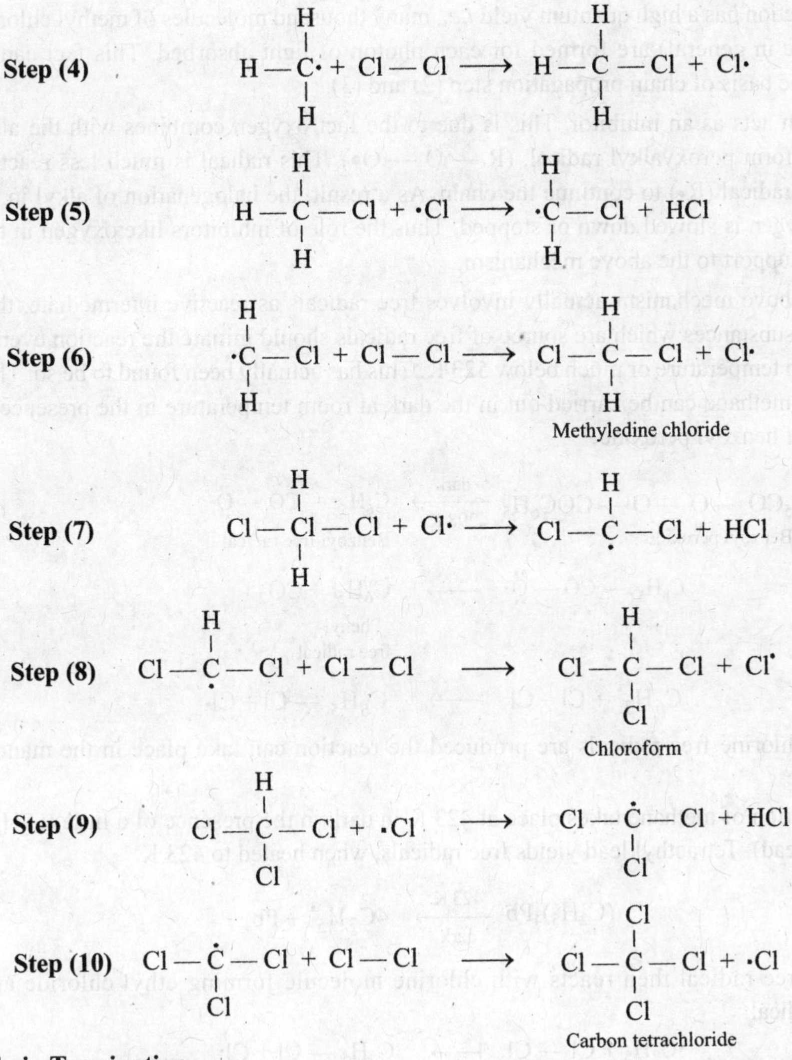

Step (4) $H_3C\cdot + Cl-Cl \longrightarrow H_3C-Cl + Cl\cdot$

(shown as CH₃ with H's drawn out)

Step (5) $CH_3-Cl + \cdot Cl \longrightarrow \cdot CH_2-Cl + HCl$

Step (6) $\cdot CH_2-Cl + Cl-Cl \longrightarrow Cl-CH_2-Cl + Cl\cdot$

Methyledine chloride

Step (7) $Cl-CH_2-Cl + Cl\cdot \longrightarrow Cl-\overset{\cdot}{C}H-Cl + HCl$

Step (8) $Cl-\overset{\cdot}{C}H-Cl + Cl-Cl \longrightarrow Cl-CHCl-Cl + Cl\cdot$

(product: CHCl₃) Chloroform

Step (9) $CHCl_3 + \cdot Cl \longrightarrow \cdot CCl_3 + HCl$

Step (10) $\cdot CCl_3 + Cl-Cl \longrightarrow CCl_4 + \cdot Cl$

Carbon tetrachloride

Chain Termination

The chain reactions mentioned above, however, come to an end if the free radicals combine amongst themselves to form neutral molecules. Some of the chain terminating steps are:

$$\cdot Cl + \cdot Cl \longrightarrow Cl-Cl$$

$$\cdot CH_3 + \cdot Cl \longrightarrow CH_3-Cl$$

$$\cdot CH_3 + \cdot CH_3 \longrightarrow CH_3-CH_3$$

Evidence in support of free radical mechanism of halogenation of alkanes

The following points support the free radical mechanism:

(*i*) Reaction does not take place in dark at room temperature but requires energy in the form of heat or light. This is due to the fact that the chain initiation step (1) is endothermic and hence needs a large amount of energy to break the Cl — Cl bond into radicals.

(*ii*) The reaction has a high quantum yield *i.e.*, many thousand molecules of methyl chloride (or alkyl halide in general) are formed for each photon of light absorbed. This fact can be explained on the basis of chain propagation step (2) and (3).

(*iii*) Oxygen acts as an inhibitor. This is due to the fact oxygen combines with the alkyl free radical to form peroxyalkyl radical, (R — O — O•). This radical is much less reactive than alkyl free radical (R•) to continue the chain. As a result, the halogenation of alkyl in the presence of oxygen is slowed down or stopped. Thus the role of inhibitors like oxygen in this reaction gives support to the above mechanism.

(*iv*) If the above mechanism actually involves free radicals as reactive intermediate, then the addition of substances which are source of free radicals should initiate the reaction even in the dark at room temperature or much below 523 K. This has actually been found to be so. Thus chlorination of methane can be carried out in the dark at room temperature in the presence of small amount of benzoyl peroxide.

$$C_6H_5CO-O-O-COC_6H_5 \xrightarrow[298\,K]{dark} C_6H_5-CO-\ddot{O}\cdot$$
$$\text{Benzoyl peroxide} \qquad \qquad \text{Benzoyl free radical}$$

$$C_6H_5-CO-\ddot{O}\cdot \longrightarrow C_6H_5\cdot + CO_2$$
$$\qquad \qquad \qquad \text{Phenyl free radical}$$

$$C_6H_5\cdot + Cl-Cl \longrightarrow C_6H_5-Cl + Cl\cdot$$

Once, the chlorine free radicals are produced the reaction can take place in the manner explained above.

(*v*) Chlorination of methane takes place at 423 K in dark in the presence of a little $(C_2H_5)_4Pb$ (Tetraethyl lead). Tetraethyl lead yields free radicals, when heated to 423 K.

$$(C_2H_5)_4Pb \xrightarrow[dark]{423\,K} 4C_2H_5\cdot + Pb$$

The ethyl free radical then reacts with chlorine molecule forming ethyl chloride and chlorine free radical.

$$\cdot C_2H_5 + Cl-Cl \longrightarrow C_2H_5-Cl + Cl\cdot$$

The chlorine free radical thus formed brings about chlorination of methane as explained above.

Chlorination and bromination of propane, butane and isobutane

Halogenation of an alkane containing more than one type (primary, secondary, tertiary) of hydrogens, gives a mixture of isomeric products. For example, chlorination of propane, butane and isobutane give the following products in the proportion given below each.

(*i*) $CH_3-CH_2-CH_3 \xrightarrow[298K]{Cl_2\ light} \underset{\substack{n-\text{Propyl chloride}\\(45\%)}}{CH_3-CH_2-CH_2-Cl} + \underset{\substack{\text{Isopropyl chloride}\\(55\%)}}{CH_3-\underset{Cl}{\overset{|}{CH}}-CH_3}$

(*ii*) $CH_3-CH_2-CH_2-CH_3 \xrightarrow[298K]{Cl_2\ light} \underset{\substack{n-\text{Butyl chloride}\\(28\%)}}{CH_3-CH_2-CH_2-CH_2-Cl}$

$$+ \ CH_3 - CH_2 - \underset{\underset{Cl}{|}}{CH} - CH_3$$
sec–Butyl chloride
(72%)

(iii) $CH_3 - \underset{\underset{CH_3}{|}}{CH} - CH_3 \xrightarrow{Cl_2} CH_3 - \underset{\underset{CH_3}{|}}{CH} - CH_2 - Cl + CH_3 - \underset{\underset{CH_3}{|}}{\overset{\overset{CH_3}{|}}{C}} - Cl$

Isobutane   Isobutyl chloride   tert–Butyl chloride
(64%)   (36%)

Bromination gives the corresponding bromides but in different proportions:

(iv) $CH_3 - CH_2 - CH_3 \xrightarrow[\substack{\text{light} \\ \text{400 K}}]{Br_2} CH_3 - CH_2 - CH_2 - Br + CH_3 - \underset{\underset{Br}{|}}{CH} - CH_3$

Propane   n-propyl bromide   Isopropyl bromide
(3%)   (97%)

(v) $CH_3 - CH_2 - CH_2 - CH_3 \xrightarrow[\substack{\text{light} \\ \text{400 K}}]{Br_2} CH_3 - CH_2 - CH_2 - Br$

n–Butane   n–Butyl bromide
(2%)

$$+ \ CH_3 - CH_2 - \underset{\underset{Br}{|}}{CH} - CH_3$$
sec–Butyl bromide
(98%)

(vi) $CH_3 - \underset{\underset{Br}{|}}{CH} - CH_3 \xrightarrow[\substack{\text{light} \\ \text{400 K}}]{Br_2} CH_3 - \underset{\underset{CH_3}{|}}{CH} - CH_2 - Br + CH_3 - \underset{\underset{CH_3}{|}}{\overset{\overset{CH_3}{|}}{C}} - Br$

iso–Butane   Isobutyl bromide   Tert-Butyl bromide
(1%)   (99%)

The result given above shows that the relative amounts of the different isomeric products differ largely depending upon the halogen used. It is also important to note that the bromination, in constrast to chlorination, leads to the formation of only one of the possible isomeric products. This is reflected in the percentages like 97%, 98% and 99% for one of the products in each reaction. Thus bromine atom is more selective in the site of attack than chlorine.

4.11 FACTORS THAT INFLUENCE THE ORIENTATION OF HALOGENATION OF ALKANES

Factors that influence the halogenation process are:
   (i) Collision frequency
  (ii) Probability factor
 (iii) Energy factor
Let us take the example of chlorination of propane.

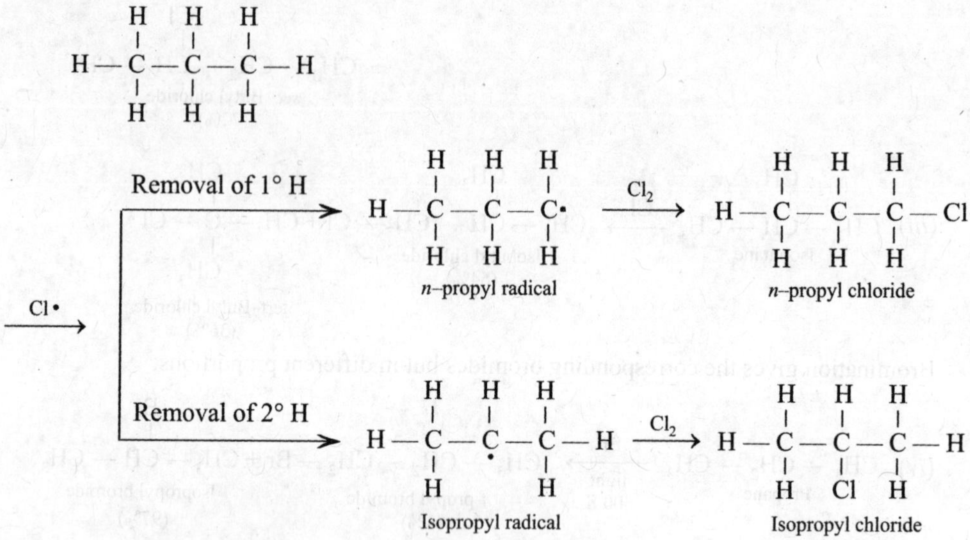

```
        H   H   H
        |   |   |
    H — C — C — C — H
        |   |   |
        H   H   H
```

Removal of 1° H → n-propyl radical →[Cl₂] n-propyl chloride

Cl•

Removal of 2° H → Isopropyl radical →[Cl₂] Isopropyl chloride

The relative amounts of *n*-propyl chloride and isopropyl chloride depend upon the relative rates at which the intermediate *n*-propyl and isopropyl radicals are formed as the rate-determining step is the formation of *n*-propyl and isopropyl radicals by the attack of chlorine radical on propane at the proper site.

The rates at which the two propyl radicals are formed depend upon the above factors.

***(i)* Collision frequency.** The collision frequency is same for the two reactions because both involve the collisions of the same particles viz., propane and chlorine atom.

***(ii)* Probability factor.** Since propane has six primary hydrogens and two secondary hydrogens, the probability of removal of primary hydrogens as compared to the secondary hydrogens should be in the ratio of 6 : 2 or 3 : 1.

If we consider the probability factor and collision frequency only, the chlorination of propane would be expected to yield *n*-propyl chloride and isopropyl chloride in the ratio of 3 : 1. But in actual practice, the two chloride are formed roughly in equal amounts *i.e.*, the ratio is 1 : 1 or 3 : 3 (see the experimental data).

***(iii)* Energy factor.** The probability of formation of isopropyl chloride is about three times as large as that of formation of *n*-propyl chloride. This means that collisions with secondary hydrogens are about three times more effective than the collisions with primary hydrogens. This implies that isopropyl radicals are formed more easily than *n*-propyl radicals in the above reactions. We know that isopropyl radical is secondary radical while *n*-propyl radical is a primary radical and the decreasing order of stability is:

Tert-radical > sec-radical > primary radical

It is, therefore, clear that isopropyl chloride would be more stable (and hence easily formed) than *n*-propyl chloride. Now the stability or the ease of formation of an alkyl radical is related to the E_{act} (energy of activation) of the reaction leading to the formation of that radical. Hence E_{act} of the reaction leading to the formation of isopropyl radical is smaller than E_{act} of the reaction leading to the formation of *n*-propyl radical. In other words, E_{act} for the abstraction of a secondary hydrogen is less than E_{act} for the abstraction of a primary hydrogen. Hence the orientation in chlorination of propane is determined by the energy factor, *i.e.*, the E_{act} for the abstraction of secondary and primary hydrogens in propane.

Chlorination of isobutane presents a similar problem. Here there are nine primary hydrogens and one tertiary hydrogen. The abstraction of one of the nine primary hydrogens leads to the formation of isobutyl chloride while the abstraction of single tertiary hydrogen leads to the formation of tert-butyl chloride. We would, therefore, expect that the probability factor favours the formation of *iso*-butyl chloride (where primary hydrogen is replaced by chlorine) by the ratio of 9 : 1. But isobutyl chloride and tert-butyl chloride are formed roughly in the ratio of 2 : 1 or 9 : 4.5. Naturally, about 4.5 times as many collisions with tertiary hydrogens are effective as with primary hydrogens.

$$
\begin{array}{c}
CH_3 \\
| \\
CH_3-CH-CH_3
\end{array}
\xrightarrow{Cl\cdot}
\begin{cases}
\text{Removal of } 1°\,H \longrightarrow CH_3-\overset{CH_3}{\underset{|}{CH}}-\overset{\bullet}{CH_2} \xrightarrow{Cl_2} CH_3-\overset{CH_3}{\underset{|}{CH}}-CH_2Cl \\
\quad\quad\quad\quad\quad\quad\quad \text{Isobutyl radical} \quad\quad\quad\quad\quad \text{Isobutyl chloride} \\
\\
\text{Removal of } 3°\,H \longrightarrow CH_3-\overset{CH_3}{\underset{\bullet}{C}}-CH_3 \xrightarrow{Cl_2} CH_3-\overset{CH_3}{\underset{\underset{Cl}{|}}{C}}-CH_3 \\
\quad\quad\quad\quad\quad\quad\quad \text{tert–Butyl radical} \quad\quad\quad\quad\quad \text{tert–Butyl chloride}
\end{cases}
$$

This means that E_{act} for the abstraction of a tertiary hydrogen (3° H) is less than E_{act} for the abstraction of a primary hydrogen (1° H) and, in fact, even less than for the abstraction of secondary hydrogen (2° H).

Study of chlorination of many alkanes have shown that at room temperature, the relative rates of abstraction of tertiary, secondary and primary hydrogens are in the ratio 5.0 : 3.8 : 1.0. With the help of these values, we can predict fairly well the ratio of isomeric chlorination products obtained from a given alkane. The expected ratio of *n*-propyl chloride and isopropyl chloride in the chlorination of propane can be calculated as follows:

$$\frac{\text{yield of } n\text{--propyl chloride}}{\text{yield of isopropyl chloride}} = \frac{\text{No. of } 1°\,H}{\text{No. of } 2°\,H} \times \frac{\text{Reactivity of } 1°\,H}{\text{Reactivity of } 2°\,H} = \frac{6}{2} \times \frac{1}{3.8} = \frac{6}{7.6}$$

or

$$\frac{\text{yield of isopropyl chloride}}{\text{yield of } n\text{--propyl chloride}} = \frac{7.6}{6} = 1.26$$

Thus, yield is isopropyl chloride = 1.26 × yield of *n*-propyl chloride.

$$\frac{\text{Actual yield of isopropyl chloride}}{\text{Actual yield of } n\text{--propyl chloride}} = \frac{55}{45} = 1.22$$

Similarly, in case of isobutane, we get the following result:

$$\frac{\text{yield of isobutyl chloride}}{\text{yield of tert-butyl chloride}} = \frac{\text{No. of } 1°\,H}{\text{No. of } 3°\,H} \times \frac{\text{Reactivity of } 1°\,H}{\text{Reactivity of } 3°\,H} = \frac{9}{1} \times \frac{1}{5} = \frac{9}{5}$$

$$\frac{\text{yield of isobutyl chloride}}{\text{yield of tert-butyl chloride}} = \frac{9}{5} = 1.8$$

Thus, yield of isobutyl chloride = 1.80 × yield of tert-butyl chloride

$$\frac{\text{Actual yield of isobutyl chloride}}{\text{Actual yield of tert-butyl chloride}} = \frac{16}{9} = 1.78$$

In bromination of alkane, the same sequence of reactivity *i.e.*, 3° > 2° > 1° is found, but the relative reactivity rates are much larger. At 400 K, the relative rates for abstraction of tertiary. Secodary and primary hydrogens are 1600 : 82 : 1. The observed of *n*-propyl bromide and isopropyl bromide can be calculated as follows:

$$\frac{\text{yield of } n\text{-propyl bromide}}{\text{yield of isopropyl bromide}} = \frac{\text{No. of } 1°\text{ H}}{\text{No. of } 2°\text{ H}} \times \frac{\text{Reactivity of } 1°\text{ H}}{\text{Reactivity of } 2°\text{ H}} = \frac{6}{2} \times \frac{1}{82} = \frac{3}{82}$$

or $\dfrac{\text{yield of } n\text{-propyl bromide}}{\text{yield of iso-propyl bromide}} = \dfrac{3}{82}$

Similarly,

$$\frac{\text{yield of iso-butyl bromide}}{\text{yield tert-butyl bromide}} = \frac{\text{No. of } 1°\text{ H}}{\text{No. of } 3°\text{ H}} \times \frac{\text{Reactivity of } 1°\text{ H}}{\text{Reactivity of } 3°\text{ H}}$$

$$= \frac{9}{1} \times \frac{1}{1600} = \frac{9}{1600}.$$

Solved Examples

Example 5: An alkane with molecular weight 72 formed only one monochloro substitution product. Suggest a structure for the alkane.

Solution: Molecular weight 72 suggests that the alkane has the molecular formula C_5H_{12}. Out of three isomeric pentanes viz., *n*-pentane, isopentane and neopentane the last one is the answer of the question. This is because only neopentane has one type of hydrogen (primary) and hence would give one monochloro substitutional product.

$$CH_3-\underset{\underset{CH_3}{|}}{\overset{\overset{CH_3}{|}}{C}}-CH_3 \xrightarrow[-HCl]{Cl_2} CH_3-\underset{\underset{CH_3}{|}}{\overset{\overset{CH_3}{|}}{C}}-CH_2Cl$$

Neopentane   Neopentyl chloride
(2, 2-Dimethylpropane)

Example 6: Give the structure of pentane that would be expected to produce the largest number of isomeric monochloro derivatives.

Solution:

There are three isomers of pentanes.

$CH_3-CH_2-CH_2-CH_2-CH_3$ $CH_3-\underset{\underset{CH_3}{|}}{CH}-CH_2-CH_3$
n–pentane   Isopentane

$H_3C-\underset{\underset{CH_3}{|}}{\overset{\overset{CH_3}{|}}{C}}-CH_3$ (neopentane)

Out of the three isomers, isopentane would give largest number of isomeric derivatives as it contains three types of replaceable hydrogen (primary, secondary and tertiary).

Example 7: Calculate the percentage of expected isomers during monobromination of n-butane. The relative rates of substitution per 3°, 2° and 1° hydrogen are 1600 : 82 : 1.

Solution:

$$CH_3 — CH_2 — CH_2 — CH_3 \xrightarrow{Br_2}$$
$$n\text{-butane}$$

$$CH_3 — CH_2 — CH_2 — CH_2Br + CH_3 CH_2 CHBr CH_3$$
$$n\text{-butyl bromide} \qquad\qquad sec\text{-butyl biomide}$$

$$\frac{\text{Yield of } n\text{-butyl bromide}}{\text{Yield of } sec\text{-butyl bromide}} = \frac{\text{No. of 1° H} \times \text{Reactivity of 1° H}}{\text{No. of 2° H} \times \text{Reactivity of 2° H}} = \frac{6 \times 1}{4 \times 82} = \frac{6}{328}$$

$$\% \text{ of } n\text{-butyl bromide} = \frac{6}{334} \times 100 = 1.79\%$$

$$\% \text{ of sec-butyl bromide} = 100 - 1.79 = 98.21\%$$

Example 8: Predict the proportions of isomeric products from chlorination at room temperature of (i) n-Butane (ii) 2, 3-dimethyl butane.

Solution: (i) n-Butane

$$CH_3 — CH_2 — CH_2 — CH_3 \xrightarrow{Cl_2} CH_3 — CH_2 — CH_2 — CH_2Cl + CH_3 — CH_2 — \underset{\underset{Cl}{|}}{CH} — CH_3$$
$$n\text{–Butyl chloride} \qquad\qquad sec\text{–Butyl chloride}$$

$$\frac{\text{yield of } n\text{-Butyl chloride}}{\text{yield of sec-Butyl chloride}} = \frac{\text{No. of 1° H}}{\text{No. of 2° H}} \times \frac{\text{Reactivity of 1° H}}{\text{Reactivity of 2° H}}$$

$$= \frac{6}{4} \times \frac{1}{3.8} = \frac{3}{7.6} = \frac{28\%}{72\%}$$

(ii) 2, 3-dimethylbutane

$$H_3C — \underset{\underset{CH_3}{|}}{HC} — \underset{\underset{CH_3}{|}}{HC} — CH_3 + Cl_2 \longrightarrow CH_3 — \underset{\underset{CH_3}{|}}{CH} — \underset{\underset{CH_3}{|}}{CH} — CH_2Cl \qquad\qquad (I)$$
$$\text{1-Chloro-2, 3-dimethylbutane}$$

$$+ CH_3 — \underset{\underset{CH_3}{|}}{CH} — \underset{\underset{Cl}{|}}{CH} — CH_3 \qquad\qquad (II)$$
$$\text{2-Chloro-3-methylbutane}$$

$$\frac{\text{yield of I (Primary)}}{\text{yield of II (Sec.)}} = \frac{\text{No. of 1° H}}{\text{No. of 3° H}} \times \frac{\text{Reactivity of 1° H}}{\text{Reactivity of 3° H}}$$

$$= \frac{12}{2} \times \frac{1}{5} = \frac{6}{5} = \frac{55\%}{45\%}$$

Example 9: Chlorination reaction of certain higher alkanes can be used for laboratory preparations. For example, we can prepare neopentyl chloride from neopentane and cyclopentyl chloride from cyclopentane. How do you account for this fact?

Solution:

Chlorination of neopentane $\begin{bmatrix} \text{CH}_3 \\ | \\ \text{CH}_3-\text{C}-\text{CH}_3 \\ | \\ \text{CH}_3 \end{bmatrix}$ and cyclopentane [pentagon] can be carried out because all the hydrogens of each compound are equivalent and replacement of any one yields the same product, *i.e.* neopentyl chloride $\begin{bmatrix} \text{CH}_3 \\ | \\ \text{CH}_3-\text{C}-\text{CH}_2\text{Cl} \\ | \\ \text{CH}_3 \end{bmatrix}$ and cyclopentyl chloride [pentagon with Cl] respectively.

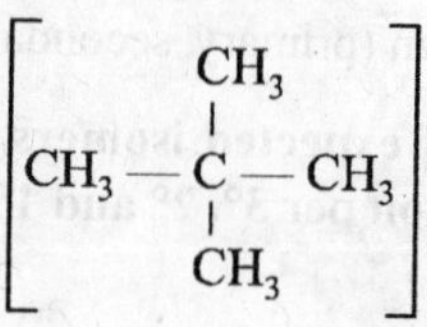

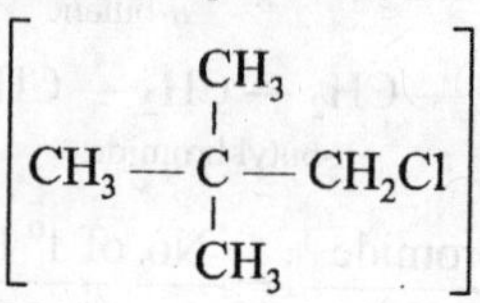

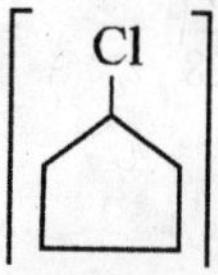

Example 10: Monochlorination of an equimolar mixture of methane and ethane does not yield a mixture of methyl chloride and ethyl chloride in equimolar proportions. How do you account for this observation?

Solution: Monochlorination of an equimolar mixture of methane and ethane yields a mixture of methyl chloride and ethyl chloride in which ethyl chloride is formed about 400 times as much as methyl chloride. This can be attributed to two factors as discussed below:

(1) Probability factor. Ethane contains six hydrogen atoms each of which can be replaced by chlorine while methane contains only four hydrogen atoms. Therefore the probabilities of formation of ethyl chloride and methyl chloride will be in the ratio 6 : 4 or 3 : 2.

(2) Relative reactivities of different classes of hydrogen atoms. This is the most important factor which determines the relative amount of products in the halogenation of methane and ethane (or other alkanes). This in turn depends upon the relative stabilities of alkyl radicals formed in the step which controls the overall process of halogenation. In case of methane and ethane, this step may be written as:

$$CH_4 + \dot{C}l \longrightarrow \dot{C}H_3 + HCl$$
$$CH_3-CH_3 + \dot{C}l \longrightarrow CH_3-\dot{C}H_2 + HCl$$

Now the stability of ethyl radical (relative to ethane) is more than the stability of methyl radical (relative to methane). Therefore the transition state involved in the formation of ethyl radical from ethane would be more stable than the transition state involved in the formation of methyl radical from methane. As such transition state of ethyl radical requires lesser E_{act} than transition state of methyl radical and therefore the former is formed more readily than the latter (Fig. 4.2). This means that carbon hydrogen bond in ethane breaks more readily than carbon hydrogen bond in methane or the primary hydrogen atoms in ethane are more reactive than hydrogen atoms in methane. As a consequence ethane undergoes chlorination much more readily than methane so that the mixture formed contains a very high proportion of ethyl chloride as compared to methyl chloride.

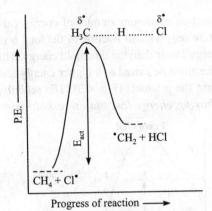
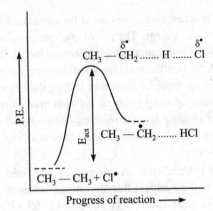

Greater E_{act} of transition state, lesser ease of formation of free radical, lesser reactivity of hydrogens.

Lesser E_{act} of transition state, greater ease of formation of free radical, greater reactivity of hydrogens.

Fig. 4.2: *Illustration of relative reactivities of hydrogens of methane and primary hydrogens of ethane.*

Example 11: On chlorination, an equimolar mixture of neopentane and ethane yields neopentyl chloride and ethyl chloride in the ratio of 2.3 : 1. How does the reactivity of a primary hydrogen in neopentane compare with that of primary hydrogen in ethane?

Solution: No. of primary hydrogen atoms in ethane = 6

No. of primary hydrogen atoms in neopentane

$$\begin{bmatrix} CH_3 \\ | \\ CH_3 - C - CH_3 \\ | \\ CH_3 \end{bmatrix} = 12$$

$$\frac{\text{yield of neopentyl chloride}}{\text{yield of ethyl chloride}} = \frac{\text{No. of } 1^\circ \text{ H} \times \text{Reactivity of } 1^\circ \text{ H in neopentane}}{\text{No. of } 1^\circ \text{ H} \times \text{Reactivity of } 1^\circ \text{ H in ethane}}$$

$$\frac{2.3}{1} = \frac{12 \times \text{Reactivity in neopentane}}{6 \times \text{Reactivity in ethane}}$$

or $\quad \dfrac{\text{Reactivity in neopentane}}{\text{Reactivity in ethane}} = \dfrac{2.3 \times 6}{12} = \dfrac{1.15}{1}$ 1.15 : 1.

4.12 ACTIVATION ENERGY

Consider the reaction

$$F^\bullet + CH_3 - H \longrightarrow \bullet CH_3 + H - F \quad \Delta H = -134 \text{ kJ}$$
$$\Delta H^\circ = 435 \text{ kJ} \qquad\qquad 569 \text{ kJ} \quad E_{act} = 5 \text{ kJ}$$

For a reaction to occur, collision between reactant molecules is a must. Thus the above mentioned reaction takes place only if the collision between a chlorine atom and a methane molecule takes place. But all the collisions are not effective. For a collision to be effective, the reactant molecules must be associated with a certain minimum amount of energy called *threshold energy*. The no. of such molecules constitute only a small fraction of the total no. of molecules or most of the molecules have kinetic energy lower than the threshold energy. Thus, in order that the reaction should occur, the molecules must be raised to a higher energy level, *i.e.*, they must be activated before they react to form the products (Fig. 4.3). *The additional amount of energy which the reactant molecules having energy less than threshold energy must acquire so that their collisions result in chemical reaction is called activation energy.*

In the above mentioned exothermic reaction we expect that the energy released in the formation of H—F bond (569 kJ/mole) appears to be sufficient to break the weaker C—H bond of methane (435 kJ/mole). However, it is not so. Clearly, the bond breaking and bond-making are not perfectly synchronized, and the energy released through bond formation is not completely available for bond breaking. In other words, certain minimum amount of energy (5 kJ/mole as shown experimentally) must be supplied to initiate the reaction. This is the *activation energy* for the reaction.

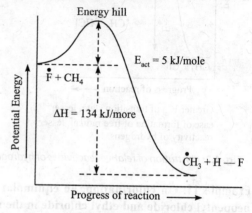

Fig. 4.3: *Concept of activation energy*

There is another aspect which must be kept in mind. In addition to the adequate energy (kinetic energy) which the colliding molecules possess at the time of collision, they must be properly oriented before the collision can be really effective in casuing the reaction. The number of properly oriented collisions in a reaction is quite small as compared to the total number of collisions.

Thus in the above mentioned example, the methane molecule must be oriented in such a way that it presents a hydrogen atom to the full force of the impact. In this case, only about one collision in eight is properly oriented.

In general, in collisions, sufficient energy (E_{act}) and proper orientation are essential for a reaction to take place.

As the reaction proceeds, changes in the potential energy takes place and these energy changes are shown in Fig. 4.3.

The potential energy of methane molecule and fluorine atom is shown on the left and that of the methyl free radical (CH_3) and hydrogen fluoride on the right (Fig. 4.3). As the reaction begins, the kinetic energy of the reactants (due to their motion) is changed into potential energy.

With the increase in potential energy, we move up the energy hill and move down the other side. During the descent, the potential is changed back into kinetic energy till we reach the level of the products ($CH_3\bullet$ and HF). As the potential energy of the product is less than that of the reactants; there must be a corresponding increase in kinetic energy. In other words, the molecules of the products ($\bullet CH_3$ and HF) move faster than the molecules of the reactants (CH_4 and $\bullet F$) so there would be a rise in temperature or heat will be evolved. The height of the hill top or energy hill (in kJ/mole) above the level of the reactant is the energy of activation (5 kJ/mole).

4.13 RELATIVE REACTIVITIES OF HALOGENS IN THE HALOGENATION OF ALKANES

The magnitude of E_{act} determines the rate of a reaction. Reactions with low E_{act} proceed at faster rate than the one with a high E_{act} at the same temperature. This relationship between the rate of a reaction and the magnitude of the energy of activation can be explained with the help of halogenation of methane as follows:

$$CH_4 + X_2 \xrightarrow{(X_2 = F_2, Cl_2, Br_2, I_2)} CH_3-X + HX$$

The rate determining step here is the abstraction of hydrogen by halogen atom *i.e.*, how fast methyl radical is formed.

$$CH_4 + X\bullet \longrightarrow CH_3\bullet + HX$$

Formation of CH_3 radical in chain propagation step is difficult but once formed it gets readily converted into the alkyl halide.

Thus, the overall rate of halogenation of methane depends upon how fast the methyl radical is formed which, in turn, depends on the E_{act} for the formation of methyl radical. Thus,

$$\begin{array}{llll}
F\bullet + CH_3-H \longrightarrow H-F + \bullet CH_3 & \Delta H = -134 \text{ kJ} \\
435\text{ kJ}569\text{ kJ} & E_{act} = 5 \text{ kJ} \\
Cl\bullet + CH_3-H \longrightarrow H-Cl + \bullet CH_3 & \Delta H = +4 \text{ kJ} \\
435\text{ kJ}431\text{ kJ} & E_{act} = 17 \text{ kJ} \\
Br\bullet + CH_3-H \longrightarrow H-Br + \bullet CH_3 & \Delta H = +67 \text{ kJ} \\
435\text{ kJ}368\text{ kJ} & E_{act} = 78 \text{ kJ} \\
I\bullet + CH_3-H \longrightarrow H-I + \bullet CH_3 & \Delta H = +138 \text{ kJ} \\
435\text{ kJ}297\text{ kJ} & E_{act} = 138 \text{ kJ}
\end{array}$$

Potential energy diagrams for the above mentioned reactions are given in Fig. 4.4

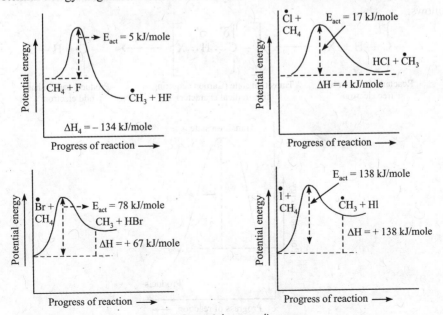

Fig. 4.4: *Potential energy diagrams.*

It is clear the decreasing order of E_{act} for chain propagation step is:
Iodination > Bromination > Chlorination > Fluorination
Hence order of halogenation should be:
Fluorination > Chlorination > Bromination > Iodination

This order is in conformity with the observed order of reactivity of F_2, Cl_2, Br_2 and I_2 in halogenation of methane in particular and alkanes in general.

4.14 EFFECT OF ACTIVATION ENERGY ON THE STRUCTURE OF TRANSITION STATE IN HALOGENATION OF ALKANES

For a collision to be effective collision, the reactant molecules must be associated with a certain amount of energy called threshold energy. Thus, they also have to cross an energy barrier (hill top) between the reactants and products before reaction can be accomplished. The top of the energy hill (energy barrier) represents an intermediate structure called the *transition state*. The transition state corresponds to a particular arrangement of atoms of reacting species as they are converted from reactants into product. It represents a configuration in which old bonds are partially broken and new bonds are partially formed. *The activation energy* (E_{act}) *is the difference in energy content between the transition state and reactants,* as shown in Fig. 4.5.

Hence the transition state of a reaction is the state of highest potential energy acquired by the reactants during their change into the products. The transition state is unstable due to its high energy content and cannot be isolated.

Let us see how the concept of transition state helps to explain the difference in the rates of halogenation of alkanes. We have already seen that the difference in reactivity of alkanes towards halogen atoms are due mainly to the difference in E_{act}; a more stable radical is associated with low activation energy for its formation. This is turn, implies that the transition state leading to its formation would be more stable.

The transition state for the abstraction of hydrogen atom from an alkane may be represented as follows:

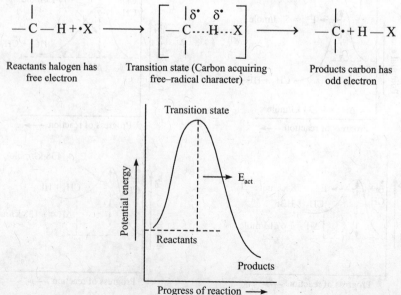

Fig. 4.5: *Transition state*

Alkanes (Carbon-Carbon Sigma Bonds) **187**

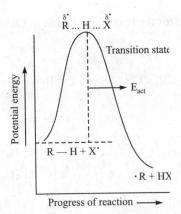

Fig. 4.6: *Transition state in halogenation of alkane.*

The C—H bond is partly broken and H—X bond is partly formed (Fig. 4.6). Depending upon the extent to which the C—H bond is broken, the alkyl group is associated with the character of free radical it will form. Factors that stabilise the resulting free radical (delocalisation of the odd electron) tend to stabilize the nascent free radical in the transition state.

4.15 REACTIVITY AND SELECTIVITY OF HALOGENS TOWARDS ALKANES

Bromine is less reactive towards alkanes than chlorine but bromine is more selective than chlorine. The relative rates of substitution by bromine per hydrogen atom (3°, 2°, 1° hydrogen) are 1600 : 82 : 1 as compared to 5.0 : 3.8 : 1 in case of chlorine.

Thus the reaction of isobutane and bromine, for example, gives mainly tert-butyl bromide.

$$CH_3-\underset{\underset{CH_3}{|}}{\overset{\overset{CH_3}{|}}{C}}-H + Br_2 \xrightarrow[400\ K]{light} CH_3-\underset{\underset{CH_3}{|}}{\overset{\overset{CH_3}{|}}{C}}-Br + CH_3-\underset{\underset{H}{|}}{\overset{\overset{CH_3}{|}}{C}}-CH_2Br$$
$$(99\%) \qquad (1\%)$$

Further, bromine atom is less reactive than chlorine atom. Thus, selectivity is related to reactivity and it can be generalised that the less reactive the reagent the more selective it is in its attack under a set of similar reaction. The greater selectivity of bromine can be explained in terms of transition state theory. According to a postulate made by Professor G.S. Hammond **the structure of transition state of endothermic step (with high value of E_{act}) of a reaction resembles the products of that step more than it does the reactants. The structure of transition state of exothermic state with low value of E_{act} of a reaction resembles the reactant of that step more than it does the products.**

This principle can be explained with the help of potential energy diagrams for the chlorination and bromination of isobutane.

The abstraction of hydrogen by the highly reactive chlorine atom is exothermic and has low E_{act}. According to the above postulate, the transition state resembles the reactants more than it does the products. In other words, the transition state is reached early and the carbon-hydrogen bond is slightly stretched. Atoms and electrons are distributed almost in the same way as in the reactants; carbon is still tetrahedral. The alkyl radical has developed very little free radical character. The transition state for the hydrogen (1° and 3°) abstraction steps may be shown as follows:

$$CH_3-\underset{\underset{CH_3}{|}}{CH}-CH_3 + Cl\bullet \longrightarrow \underset{\text{Reactant like structure}}{CH_3-\underset{\underset{CH_3}{|}}{CH}-CH_2......H......Cl}$$

$$\downarrow$$

$$Cl + CH_3-\underset{\underset{CH_3}{|}}{CH}-CH_2Cl \xleftarrow{Cl_2} CH_3-\underset{\underset{CH_3}{|}}{CH}-CH_2 + HCl$$

$$CH_3-\underset{\underset{CH_3}{|}}{\overset{\overset{CH_3}{|}}{C}}-H + {}^\bullet Cl \longrightarrow \underset{\text{Product like structure}}{CH_3-\underset{\underset{CH_3}{|}}{\overset{\overset{CH_3}{|}}{C}}\overset{\delta\bullet}{.....}H\overset{\delta\bullet}{......}Br}$$

$$\downarrow$$

$$Cl\bullet + CH_3-\underset{\underset{CH_3}{|}}{\overset{\overset{CH_3}{|}}{C}}-Cl \xleftarrow{Cl_2} CH_3-\underset{\underset{CH_3}{|}}{\overset{\overset{CH_3}{|}}{C}}\bullet + HCl$$

The transition state in both cases resembles the reactants. Since the reactants in both the cases are the same and the same type of C—H bonds are broken (primary or tertiary), it has a relatively small influence on the relative rates of the reactions. The two reactions proceed with similar (but not identical) rates because their respective activation energies are similar (See Fig. 4.7).

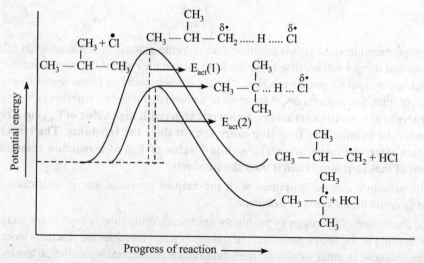

Fig. 4.7: *Activation energies in exothermic reactions.*

Alkanes (Carbon-Carbon Sigma Bonds) **189**

In the above diagram it isclear that E_{act} are similar, but because tertiary C — H bond is broken more easily than primary C—H bond, the reaction (2) has low E_{act} and proceeds comparatively at faster rate.

On the other hand, the abstraction of hydrogen by the less reactive bromine atom is endothermic and has a very high E_{act}. The transition state resembles the products more than it does the reactants. In other words, the transition state is reached late and the carbon-hydrogen bond is broken to a considerable extent. Atoms and electrons are distributed almost in the same way as in the products; the carbon is almost trigonal. The alkyl radical has developed considerable free radical character.

The transition states may be shown in the following ways:

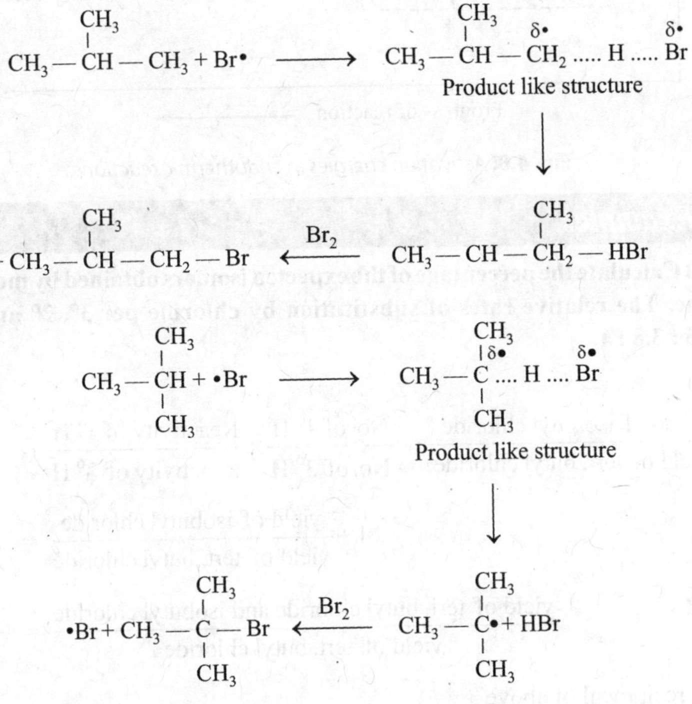

$$CH_3-\underset{\underset{CH_3}{|}}{CH}-CH_3 + Br^\bullet \longrightarrow CH_3-\underset{\underset{CH_3}{|}}{CH}-CH_2\overset{\delta\bullet}{.....}H\overset{\delta\bullet}{.....}Br$$
Product like structure

$$\downarrow$$

$$^\bullet Br + CH_3-\underset{\underset{CH_3}{|}}{CH}-CH_2-Br \xleftarrow{Br_2} CH_3-\underset{\underset{CH_3}{|}}{CH}-CH_2-HBr$$

$$CH_3-\underset{\underset{CH_3}{|}}{CH} + {}^\bullet Br \longrightarrow CH_3-\underset{\underset{CH_3}{|}}{\overset{\overset{CH_3}{|}}{C}}\overset{\delta\bullet}{....}H\overset{\delta\bullet}{....}Br$$
Product like structure

$$\downarrow$$

$$^\bullet Br + CH_3-\underset{\underset{CH_3}{|}}{\overset{\overset{CH_3}{|}}{C}}-Br \xleftarrow{Br_2} CH_3-\underset{\underset{CH_3}{|}}{\overset{\overset{CH_3}{|}}{C}}^\bullet + HBr$$

The transition states in both steps resemble the products in energy and structure. Since the products in both the cases are different, the type of C—H bond being broken has a lot of influence on the relative rates of reactions. They proceed with different rates. Abstraction of tertiary hydrogen takes place much faster. Bromine is, therefore, more selective in its attack. (See Fig. 4.8)

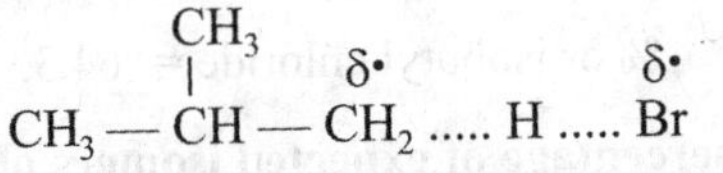

$$CH_3-\underset{\underset{CH_3}{|}}{CH}-CH_2\overset{\delta\bullet}{.....}H\overset{\delta\bullet}{.....}Br$$

It is clear from the diagram that transition state for reaction (1) resembles a less stable primary radical while the transition state for reaction (2) resembles a more stable radical. The E_{act} for reaction (2) is much lower than that for reaction (1). The product contains mostly tert-butyl bromide.

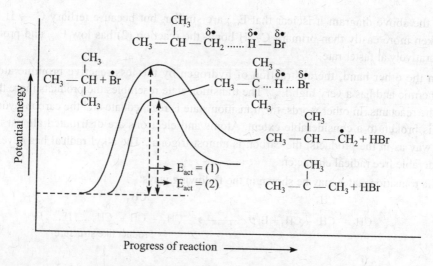

Fig. 4.8: *Activation energies in endothermic reactions.*

Solved Examples

Example 12: Calculate the percentage of the expected isomers obtained by monochlorination of isobutane. The relative rates of substitution by chlorine per 3°, 2° and 1° hydrogen atoms are 5 : 3.8 : 1.

Solution:

$$\frac{\text{yield of isobutyl chloride}}{\text{yield of tert. butyl chloride}} = \frac{\text{No. of 1° H}}{\text{No. of 3° H}} \times \frac{\text{Reactivity of 1° H}}{\text{Reactivity of 3° H}} = \frac{9}{1} \times \frac{1}{5} = \frac{9}{5}$$

$$1 + \frac{\text{yield of isobutyl chloride}}{\text{yield of tert. butyl chloride}} = 1 + \frac{9}{5}$$

$$\frac{\text{yield of tert. butyl chloride and isobutyl chloride}}{\text{yield of tert. butyl chloride}} = \frac{14}{5}$$

Take the reciprocal of above

$$\frac{\text{yield of tert. butyl chloride}}{\text{Total yield}} = \frac{5}{14}$$

$$\text{\% of tert. butyl chloride} = \frac{5}{14} \times 100 = 35.7$$

% or isobutyl chloride = 64.3.

Example 13: Calculate the percentage of expected isomers obtained by monobromination of propane. The relative rates of substitution by bromine per 3°, 2° and 1° hydrogen atoms are 1600 : 82 : 1.

Solution:

$$\frac{\text{yield of }n\text{-propyl bromide}}{\text{yield of iso-propyl bromide}} = \frac{3}{82}$$

Add 1 to both sides

$$1 + \frac{\text{yield of } n\text{-propyl bromide}}{\text{yield of iso-propyl bromide}} = 1 + \frac{3}{82}$$

$$\frac{\text{Total yield}}{\text{yield of iso-propyl bromide}} = \frac{85}{82}$$

or

$$\frac{\text{yield of iso-propyl bromide}}{\text{Total yield}} = \frac{82}{85}$$

% of iso-propyl bromide = 96.5

% of n-propyl bromide = 3.5

Example 14: Fluorocarbon (C_5F_{12}) has lower boiling point than pentane (C_5H_{12}) even though it has a far higher molecular weight. Explain.

Solution: Molecular weight is not the only criterion that affects the boiling point of a compound. There is a wide gap between the electronegativities of C and F. This results in separation of charges and creation of dipoles on the molecules of C_5F_{12}. This causes repulsions between the molecules of C_5F_{12}.

C_5H_{12} on the other hand does not exhibit this types of repulsive interactions between the molecules because of comparable electronegativities of C and H.

Example 15: Which would have a higher boiling point and why?

$$CH_3CH_2CH_2CH_2CH_3 \quad \text{or} \quad H_3C-\underset{\underset{CH_3}{|}}{\overset{\overset{CH_3}{|}}{C}}-CH_3$$

Solution: Boiling point of an alkane depends upon the magnitude of van der Waals forces between the molecules which in turn depends upon the area of contact or surface area of the molecule. Branched alkanes tend to be spherical and thus possess smaller area of contact between the molecules compared to straight chain molecules. Higher the van der Walls force of attraction between the molecules, higher will be the boiling point. Thus n-pentane has a greater b.p. than neopentane.

Example 16: An equimolar mixture of neopentane and ethane yields neopentyl chloride and ethyl chloride in the ratio 2.3 : 1. How does the reactivity of a primary hydrogen in neopentane compare with that of a primary hydrogen in ethane?

Solution: The reaction involved may be written as

$$\underset{\text{Ethane}}{CH_3-CH_3} + Cl_2 \longrightarrow \underset{\text{Chloroethane}}{CH_3\,CH_2\,Cl} + HCl$$

$$\underset{\text{Neopentane}}{CH_3-\underset{\underset{CH_3}{|}}{\overset{\overset{CH_3}{|}}{C}}-CH_3} + Cl_2 \longrightarrow \underset{\text{Neopentyl chloride}}{CH_3-\underset{\underset{CH_3}{|}}{\overset{\overset{CH_3}{|}}{C}}-CH_2Cl} + HCl$$

No. of 1° H-atoms in ethane = 6, No. of 1° H-atoms in neopentane = 12

It is given that

$$\frac{\text{Amount of Neopentyl Chloride}}{\text{Amount of Ethyl Chloride}} = \frac{2.3}{1}$$

The reactivity of their 1° H-atoms can be compared as

$$\frac{\text{No. of 1° H-atoms in neopentane}}{\text{No. of 1° H-atoms in ethane}} \times \frac{\text{Reactivity of 1° H in neopentane}}{\text{Reactivity of 1° H in ethane}} = \frac{2.3}{1}$$

or

$$\frac{12 \times \text{Reactivity of 1° H in neopentane}}{6 \times \text{Reactivity of 1° H in ethane}} = \frac{2.3}{1}$$

or

$$\frac{\text{Reactivity of 1° H in neopentane}}{\text{Reactivity of 1° H in ethane}} = \frac{2.3}{1} \times \frac{6}{12} = 1.15 : 1.$$

Example 17: Electrolysis of an aqueous solution of sodium propionate gives butane, ethane, ethene and ethyl propanoate. Propose a mechanism to explain the formation of these products.

Solution: The following mechanism explains the formation of the products

$$C_2H_5 - \underset{\underset{\text{Propanoate ion}}{}}{\overset{O}{\underset{\|}{C}}} - \ddot{\underset{..}{O}}{:}^- \xrightarrow{-e^-} C_2H_5 - \overset{O}{\underset{\|}{C}} - \ddot{\underset{..}{O}}{:} \longrightarrow \dot{C}_2H_5 + \overset{O}{\underset{\|}{C}} = O$$

Other products are formed from, ethyl radical as under:

(i) $\dot{C}_2H_5 + \dot{C}_2H_5 \longrightarrow C_2H_5 - C_2H_5$
    Butane

(ii) $CH_3 - \dot{C}H_2 + H - CH_2 - \dot{C}H_2 \xrightarrow{\text{Disproportionation}} \underset{\text{Ethane}}{CH_3 - CH_3} + \underset{\text{Ethene}}{CH_2 = CH_2}$

(iii) $C_2H_5 - \underset{\underset{\text{Propanoate free radical}}{}}{\overset{O}{\underset{\|}{C}}} - \dot{O} + \dot{C}_2H_5 \longrightarrow \underset{\text{Ethyl propanoate}}{C_2H_5 - \overset{O}{\underset{\|}{C}} - OC_2H_5}$

Example 18: Bromination and not chlorination is a usual synthetic method for the preparation of alkyl halides. Explain.

Solution: Bromination of alkanes takes place more selectively than chlorination. 1-Bromobutane and 2-bromobutane are obtained in the ratio of 1:49 while under the similar conditions, 1-chlorobutane and 2-chlorobutane are obtained in the ratio of 7:18 upon chlorination of butane. Thus there is lack of selectivity on chlorination. 2-Bromobutane is obtained almost in pure state while 2-chlorobutane is obtained in only 78% of the mixture. That is why bromination is preferred to chlorination for preparation of alkyl halides.

Example 19: What are the different synthetic routes for the synthesis of 2-methylpentane from three-carbon compounds? Compare their efficiencies.

Solution: The two synthetic routes are given below

(i) $CH_3CH_2CH_2Br \xrightarrow[-LiBr]{Li/Ether} CH_3CH_2CH_2-Li \xrightarrow{CuI} (CH_3CH_2CH_2)_2CuLi$

Lithium di-n-propylcuprate

$\xrightarrow[S_N2]{(CH_3)_2CH-Br} CH_3CH_2CH_2-\underset{\underset{CH_3}{|}}{CH}-CH_3$

2-methylpentane

(ii) $CH_3-\underset{\underset{CH_3}{|}}{CH}-Br \xrightarrow[-LiBr]{Li/ether} CH_3-\underset{\underset{CH_3}{|}}{CH}-Li \xrightarrow{CuI} \left[CH_3-\underset{\underset{CH_3}{|}}{CH}\right]_2 CuLi$

Isopropyl bromide

Lithium diisopropylcuprate

$\xrightarrow[S_N2]{CH_3CH_2CH_2Br} CH_3-\underset{\underset{CH_3}{|}}{CH}-CH_2CH_2CH_3$

2-methylpentane

S_N2 reactions take place more readily with 1° than with 2° alkyl halides. Therefore, the method (i) is more efficient.

APPLYING CHEMISTRY TO LIFE

Petrol is a mixture of higher alkanes. Petrol is a major source of energy. With the development of civilization and a rise in the standard of living in the form of automobiles and household gadgets, our energy needs are constantly moving up. Worldwide, there is a severe scarcity of energy or power. Efforts are on in the scientific research to develop techniques to synthesise petrol as the fossil fuels are constantly dwindling and may reduce to nil someday. Fischer Tropsch synthesis is one such endeavor in this direction.

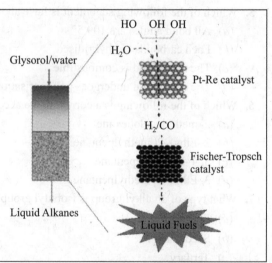

Key Terms

- Wurtz reaction
- Wurtz-Fittig reaction
- Free-radical substitution
- Reactivity and selectivity

Evaluate Yourself

Multiple Choice Questions

1. Hydrocarbons are
   (a) insoluble in water
   (b) composed of carbon and hydrogen
   (c) both (a) and (b)
   (d) None of these

2. A tertiary carbon is bonded directly to:
   (a) 2 hydrogens
   (b) 3 carbons
   (c) 2 carbons
   (d) 4 carbons
3. Which molecular formula indicates 2-methylpentane?
   (a) C_5H_{12}
   (b) C_6H_{14}
   (c) C_5H_{10}
   (d) C_6H_{12}
4. Hydrocarbons are
   (a) Composed of carbon and hydrogen
   (b) Composed of carbon, hydrogen, and oxygen
   (c) Composed of carbon and oxygen
   (d) Composed of carbon and nitrogen
5. Which of the following statement is false about propane?
   (a) All bond angles are 109.5°
   (b) Each carbon is sp^3 hybridised
   (c) The compound is combustible
   (d) The compound undergoes polymerisation to give polypropylene.
6. Which of the following is a correct name according to the IUPAC rules?
   (a) 2–methylcyclohexane
   (b) 2–Ethyl–2–methylpentane
   (c) 3, 4–Dimethylpentane
   (d) 3–Ethyl–2–methylpentane
7. What type of an alkyl group is isobutyl group?
   (a) Primary
   (b) Secondary
   (c) Tertiary
   (d) None of these
8. How many monochlorinated isomers would result from the reaction of chlorine with n–butane in the presence of UV light?
   (a) 2
   (b) 4
   (c) 6
   (d) 8
9. A compound of formula C_3H_8 does not react with bromine in Cl_4 in the dark. The compound could be
   (a) alkane
   (b) cycloalkane

(c) alkene
   (d) cycloalkene
10. Which halogen does not react appreciably with methane in a free–radical substitution reaction?
    (a) Chlorine
    (b) Bromine
    (c) Iodine
    (d) Fluorine

Short Answer Questions

1. Write down the various structured isomers of an alkane with molecular mass 72. Which of them has the minimum boiling point? Give reasons.
2. How can you synthesize n-butane from the following compounds?
    (i) n-Butyl bromide
    (ii) Pentanoic acid
3. Describe the mechanism of Kolbe's reaction.
4. Justify that the order of reactivity of halogenation of alkanes is:
$$F_2 > Cl_2 > Br_2 > I_2$$
5. Chlorination of alkanes is an exothermic reaction and yet it requires either high temperature or exposure to light for its success.
6. Explain primary, secondary, tertiary and quaternary carbon atom giving examples.
7. How will you convert propene into propane and acetylene into ethane by Sabatier and Senderen's method?
8. Give in brief the steps involved in the mechanism of Wurtz reaction.
9. Increase in melting point of alkanes is more in going from odd-carbon to even-carbon alkane compared to when we move from even-carbon to odd-carbon alkane. Explain.
10. Give two evidences in support of free radical mechanism in the halogenation of alkanes.

Long Answer Questions

1. Justify the statements:
    (a) The halogenation of alkanes in the presence of tetraethyl lead proceeds at a lower temperature than when it is carried in its absence.
    (b) The chlorination of alkanes is an exothermic reaction yet it requires ultra-violet light for its success.
    (c) Fluorination of methane is faster as compared to its chlorination.
2. Write notes on:
    (i) Relative reactivity of alkanes in halogenation.
    (ii) Oxidation of alkanes.

3. (a) What do you understand by conformation? Draw the potential energy diagram for the conformations of *n*-butane and arrange them in increasing order of stability.

   (b) In the halogenation of methane with Cl and Br free radicals, the values of ΔH under similar conditions are:

   $$\dot{Cl} + CH_4 \longrightarrow HCl + \dot{CH_3} \quad \Delta H = 1 \text{ k cal/mole}$$
   $$E_{act} = 4 \text{ k cal/mole}$$

   $$\dot{Br} + CH_4 \longrightarrow HBr + \dot{CH_3} \quad \Delta H = +16 \text{ k cal/mole}$$
   $$E_{act} = \text{Not given}$$

   Give the relative rates of the above reactions giving reasons for your answer.

4. What is the order of reactivity of the different halogens towards methane? Give a suitable explanation in support of your answer.

5. Write down the various structural isomers of an alkane with molecular mass 72. Which of them has minimum boiling point? Give reasons.

6. Calculate the percentage of isomers expected during the monochlorination of butane at room temperature. The relative reactivities for 3°, 2° and 1° hydrogens are 5.0, 3.8 and 1.0 respectively.

7. (a) How are alkanes prepared by:
      (i) Corey-house reaction
      (ii) Wurtz reaction

   (b) Give the structure of pentane which would be of expected to give largest number of isomeric monochloro derivative.

   (c) Discuss the trends in mp and bp in alkanes.

   (d) The relative rates of substitution of bromine per 2° and 1° hydrogen atoms are 82 : 1. Calculate the percentage of expected isomers obtained during monobromination of propane at 400 K.

8. (a) Write notes on:
      (i) Wurtz reaction
      (ii) Corey-House synthesis

   (b) Discuss the mechanism of halogenation of alkanes. Also give evidences in support of the above mechanism.

9. (a) Explain why chlorination of *n*-butane in the presence of light gives a mixture of 72% of 2-chloropropane and 28% of 1-chlorobutane.

   (b) Explain why?
      (i) The boiling points of *n*-alkanes increase as the molecular mass increases.
      (ii) The boiling points of isomeric alkanes decrease with increase in the branching of the chain.

   (c) Give two methods for the preparation of alkanes.

10. (a) Discuss the mechanism of chlorination of methane giving potential energy diagram.

    (b) Justify that the order of relative reactivity of halogenation of alkanes is
    $$F_2 > Cl_2 > Br_2 > I_2$$

Answers

Multiple Choice Questions

	1.	2.	3.	4.	5.	6.	7.	8.	9.	10.
(a)				■			■	■	■	
(b)		■	■							
(c)	■									■
(d)					■	■				

Suggested Readings

1. Morrison, R.N. & Boyd, R.N. Organic Chemistry, Dorling Kindersley (India) Pvt. Ltd.
2. Finar, I.L. Organic Chemistry (Volume I), Dorling Kindersley (India) Pvt. Ltd.

Chapter 5

Alkenes
(Carbon-Carbon Pi Bonds)

LEARNING OBJECTIVES

After reading this chapter, you should be able to:
- learn the methods of preparation of alkenes
- learn the methods of preparation of alkynes
- study the properties of alkenes and alkynes
- formulate the mechanism of E1, E2 and E1cb reactions
- learn Saytzelf and Hofmann eliminations
- understand the mechanism of Markownikoff and anti-Markownikoff reactions, oxymercuration-deoxymercuration, hydroboration oxidation, ozonolysis, syn and anti-hydroxylation reactions
- learn 1,2 and 1,4-addition reactions, Diels-Alder reactions

Alkenes (Carbon-Carbon Pi Bonds)

5.1 NOMENCLATURE OF ALKENES

Alkenes are unsaturated hydrocarbons having the general formula C_nH_{2n}. They contain two hydrogen atoms less than required to form alkanes with the same number of carbon atoms. This is made possible by introducing a double bond between two carbon atoms. This symbolises unsaturation in the compound. Example of alkenes are ethylene and propylene.

Nomenclature

Nomenclature of alkenes is explained in the form of table below:

S.No.	Structure	Common name	IUPAC name
1.	$CH_2 = CH_2$	Ethylene	Ethene
2.	$CH_3CH = CH_2$	Propylene	Propene
3.	$CH_3CH_2CH = CH_2$	β-butylene	But-1-ene
4.	$CH_3CH = CH - CH_3$	α-butylene	But-2-ene
5.	$CH_3 - \underset{\underset{CH_3}{\mid}}{C} = CH_2$	Isobutylene	2-MethylPropene

Solved Examples

Example 1: Write the IUPAC names of isomers having the molecule formula C_5H_{10}.

Solution: Names and structure of five isomers having the molecules formula C_5H_{10} are given below.

1. $CH_3 - CH_2 - CH_2 - CH = CH_2$ pent-1-ene

2. $\underset{H}{\overset{CH_3-CH_2}{\diagdown}} C = C \underset{H}{\overset{CH_3}{\diagup}}$ cis-pent-2-ene

3. $\underset{H}{\overset{CH_3-CH_2}{\diagdown}} C = C \underset{CH_3}{\overset{H}{\diagup}}$ trans-pent-2-ene

4. $\overset{4}{C}H_3 - \overset{3}{C}H = \overset{2}{\underset{\underset{CH_3}{\mid}}{C}} - \overset{1}{C}H_3$ 2-Methylbut-2-ene

5.2 HINDERED ROTATION AROUND DOUBLE BOND

The lack of freedom of rotation of the bonded atoms around a carbon-carbon double bond is known as hindered rotation.

Explanation: A double bond is constituted of a σ bond and a π bond between two carbon atoms.

$$\diagup C \underset{\sigma}{\overset{\pi}{=}} C \diagdown$$

It is thus associated with greater bond dissociation energy compared to a single bond. Bond length is also shorter compared to a single bond. It is therefore much more difficult to break a double bond.

Rotation round a double bond would involve breaking the double bond which requires a large amount of energy. Hence rotation around the double bonds is difficult.

5.3 ORBITAL STRUCTURE OF ETHYLENE

In the molecule of ethylene (C_2H_4), the two carbon atoms are lin/ked to each other through a double bond while each carbon atom is separately linked to two hydrogen atoms through single bonds.

Since hybridised orbitals lie in the same plane all the carbon and hydrogen atoms of ethylene also lie in the same plane.

The unhybridised 2π orbitals of the two carbon atoms overlap in a plane perpendicular to the plane of carbon and hydrogen atoms, to form carbons-carbon π bonds as shown below.

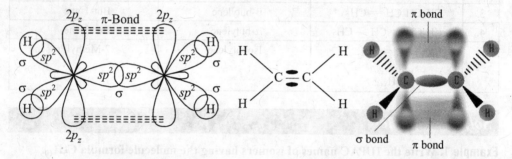

The carbon-carbon double bond length in ethylene is smaller than the carbon-carbon single bond length in ethane.

The carbon-carbon double bond length in ethylene is somewhat smaller than the carbon-carbon single bond length in ethane because of the following reasons:

(i) In ethylene, the two carbon atoms are sp^2 hybridised while in ethane the carbon atoms are sp^3 hybridised. The size of sp^2 hybrid orbitals is slightly smaller than that of sp^3 orbitals. Therefore, the length of the bond formed by the mutual overlapping of sp^2 orbitals would be smaller than bond formed by the mutual overlapping of sp^3 orbitals.

(ii) Carbon-carbon double bond also involves the additional sidewise overlap of unhybridised $2p_z$ orbitals. As a result, the carbon atoms are brought still closer and the bond length decreases.

Problems for Practice

Q 1. Write IUPAC names of the following compounds

(i) $CH_3-\underset{\underset{CH_3}{|}}{C}=\underset{\underset{CH_3}{|}}{C}-CH_3$

(ii) $CH_3-CH_2-\underset{\underset{CH_3}{|}}{CH}-\underset{\underset{CH_3}{|}}{C}=CH_2$

(iii) $CH_3-CH_2-CH=\underset{\underset{CH_2CH_3}{|}}{C}-CH_3$

(iv) $CH_3-\underset{\underset{CH_3}{|}}{\overset{\overset{CH_3}{|}}{C}}-\overset{\overset{CH_3}{|}}{CH}-CH=CH_2$

Alkenes (Carbon-Carbon Pi Bonds)

(v) $(CH_3)_2CHCH = CH\ CH\ (CH_3)_2$

(vi) $CH_3CH_2CH_2 - \underset{|}{CH} - \underset{|}{CH} - CH_2CH_2CH_3$
    positions: $CH_2=CH$ and CH_2CH_3

(vii) $CH_3CH_2CH_2CH_2 - \underset{|}{C} = CHCH_3$
     $CH_2CH_2CH_3$

Q 2. Draw the structures corresponding to the following IUPAC names

(i) 3, 4-Diisopyl – 2, 5-dimethylhex-3-ene

(ii) 2, 4, 4-Trimethyl-3-(1–methylethyl) pent-1-ene

Answers

Q 1. (i) 2, 3-Dimethylbut-2-ene   (ii) 3-Ethyl-2-methylpent-1-ene

(iii) 3-Methylhex-3-ene   (iv) 3, 3, 4-Trimethylpent-1-ene

(v) 2, 5-Dimethylhex-3-ene   (vi) 4-Ethyl-3-propylhep-1-ene

(vii) 3-Propyl hept-2-ene

Q 2. (i) $(CH_3)_2CH - \underset{|}{C} = \underset{|}{C} - CH(CH_3)_2$
         $(CH_3)_2CH\quad CH(CH_3)_2$

(ii) $CH_3 - \underset{\underset{CH_3}{|}}{\overset{\overset{CH_3}{|}}{C}} - \underset{\underset{CH(CH_3)_2}{|}}{CH} - \overset{\overset{CH_3}{|}}{C} = CH_2$

5.4 METHODS OF PREPARATION OF ALKENES

By Dehydration of Alcohols

An alcohol is converted into an alkene by dehydration. Dehydration is brought about by the use of an acid say conc. H_2SO_4 and application of heat. It involves heating the alcohol with H_2SO_4 or by passing the vapours of alcohol over alumina Al_2O_3 at 623-673 K.

$$-\underset{\underset{H}{|}}{\overset{|}{C}} - \underset{\underset{O-H}{|}}{\overset{|}{C}} - \xrightarrow[\Delta]{Acid} -\overset{|}{C} = \overset{|}{C} - + H_2O$$

1. $CH_3CH_2OH \xrightarrow[433\text{-}443\ K]{H_2SO_4} CH_2 = CH_2 + H_2O$
   Ethylene

2. $CH_3CH_2 - CH_2 - CH_2OH \xrightarrow[413\ K]{H_2SO_4} CH_3 - CH = CH - CH_3$
   Butan-1-ol   But-2-ene [Major product]

3. $CH_3 - \underset{\underset{^3CH_3}{|}}{\overset{\overset{^1CH_3}{|}}{^2C}} - OH \xrightarrow[363\ K]{20\%\ H_2SO_4} \underset{CH_3}{\overset{CH_3}{>}}C = CH_2 + H_2O$
   2-Methyl propan-2-ol   2-Methyl prop-1-ene

4. $\underset{CH_3}{\overset{CH_3}{>}}CHOH \xrightarrow[623\ K]{Al_2O_3} CH_3 - CH = CH_2$
   Isopropyl alcohol   Propylene

Reactivity and Orientation (Regioselectivity)

From the above examples it is observed that the ease of dehydration is usually the greatest in 3° alcohols, that is why the 3° alcohols require milder conditions for dehydration while simple 1° alcohols are dehydrated only with difficulty. Therefore, the order of reactivity and the ease of dehydration of alcohols is as follows:

$$3° > 2° > 1°$$

When isomeric alkenes can be formed, one of the isomers has the tendency to dominate [$CH_3-CH(OH)-CH_2-CH_3$, butan-2-ol] upon dehydration is likely to form but-1-ene and but-2-ene, but actually yields almost exclusively but-2-ene. This is in accordance with Saytzeff rule.

Saytzeff rule states that in the dehydration of alcohols, hydrogen is preferentially eliminated from the carbon atom which contains fewer number of hydrogens attached to it. This is because more highly alkylated olefins are more stable than others.

Since dehydration is an E_1 elimination the Saytzeff rule [Elimination yielding the most highly alkylated olefins] applies to those cases where more than one olefins can be formed.

$$CH_3-\underset{\underset{CH_3}{|}}{\overset{\overset{O-H}{|}}{C}}-CH_2-CH_3 \xrightarrow{\text{Conc. } H_2SO_4} CH_3-\underset{\underset{CH_3}{|}}{C}=CH-CH_3 + H_2C=\underset{\underset{CH_3}{|}}{C}-CH_2-CH_3$$

2-Methylbutan-2-ol   2-Methylbut-2-ene (Major product)   2-Methylbut-1-ene (Minor)

Mechanism of dehydration of alcohols. The mechanism of dehydration of alcohols is given schematically below:

1. *Attachment of the proton to alcoholic oxygen*

$$-\overset{|}{\underset{|}{C}}-\overset{|}{\underset{H}{C}}-\ddot{\underset{..}{O}}-H + \overset{+}{H} \rightleftharpoons -\overset{|}{\underset{|}{C}}-\overset{|}{\underset{H}{C}}-\overset{+}{\underset{H}{O}}-H$$

Alcohol   Protonated alcohol

2. *Removal of water molecule*

$$-\overset{|}{\underset{|}{C}}-\overset{|}{\underset{H}{C}}\overset{+}{\underset{H}{O}}-H \rightleftharpoons -\overset{|}{\underset{|}{C}}-\overset{|}{\underset{H}{C^+}} + H_2O$$

Carbonium ion

3. *Removal of proton*

$$-\overset{|}{\underset{|}{\underset{H}{C}}}-\overset{+}{\underset{|}{C}} \rightleftharpoons \;\;\rangle C = C\langle \; + H^+$$

*For those alcohols that yield volatile olefins, the reaction can always be driven to completion by distilling off the olefin as it is formed.

Alkyl shift and hydrogen shift

Sometimes it is found that the alkene obtained by dehydration of alcohols does not fit in the mechanism. For example,

$$\underset{\text{3, 3-Dimethylbutan-2-ol.}}{CH_3-\underset{\underset{CH_3}{|}}{\overset{\overset{CH_3}{|}}{C}}-\underset{\underset{OH}{|}}{CH}-CH_3} \xrightarrow{\text{Conc. } H_2SO_4} \underset{\substack{\text{2, 3-Dimethylbut-2-ene}\\ \text{(Major product)}}}{CH_3-\underset{\underset{CH_3}{|}}{C}=\underset{\underset{CH_3}{|}}{C}-CH_3}$$

This can be explained by considering a rearrangement of carbonium ions. A carbonium ion can arrange to form a more stable carbonium ion. The mechanism of rearrangement is given below:

$$CH_3-\underset{\underset{CH_3:\overset{..}{\underset{..}{O}}-H}{|}}{\overset{\overset{CH_3}{|}}{C}}-CH-CH_3 + \overset{+}{H} \rightleftharpoons CH_3-\underset{\underset{CH_3\ \overset{+}{O}-H}{|}}{\overset{\overset{CH_3}{|}}{C}}-CH-CH_3$$

$$\Big\Updownarrow -H_2O$$

$$\underset{\text{3° Carbonium ion (more stable)}}{CH_3-\underset{\underset{CH_3\ H}{|}}{\overset{\overset{CH_3}{|}}{\overset{+}{C}}}-CH-CH_3} \underset{\text{Rearrangement}}{\overset{\text{1, 2 Alkyl shift}}{\rightleftharpoons}} \underset{\text{2° Carbonium ion (less stable)}}{CH_3-\underset{\underset{CH_3}{|}}{\overset{\overset{CH_3}{|}}{C}}-\overset{+}{CH}-CH_3 + H_2O}$$

$$\Big\Updownarrow -H^+$$

$$\underset{\substack{\text{2, 3-Dimethylbut-2-ene}\\ \text{(major product)}}}{CH_3-\underset{\underset{CH_3}{|}}{C}=\underset{\underset{CH_3}{|}}{C}-CH_3}$$

Thus, we see that the carbonium ion which is the intermediate product in the dehydration of alcohol changes to a more stable carbonium ion by the shifting of alkyl group from one position to another. If the alkyl group shifts to the neighbouring carbon atom, it is called **1, 2 alkyl shift**.

Somethings, a hydride ion shifts from the neighbouring carbon to the carbonium ion, thereby producing a more stable carbonium ion. This is called **1, 2 hydride shift**.

Solved Examples

Example 2: Predict the major product obtained on dehydration of the following alcohols:

(i) $(CH_3)_2 C(OH) CH_2 - CH_3$

(ii) $CH_3-\underset{\underset{CH_3}{|}}{\overset{\overset{OH}{|}}{C}}-\underset{\underset{CH_3}{|}}{CH}-CH_3$ (iii) $CH_3-\underset{\underset{CH_3}{|}}{CH}-CH_2-CH_2OH$

Solution:

(i) $(CH_3)_2-\underset{\underset{OH}{|}}{C}-CH_2-CH_3 \xrightarrow{H^+} (CH_3)_2C-CH_2-CH_3$
$\phantom{(i) (CH_3)_2-C-CH_2-CH_3 \xrightarrow{H^+} (CH_3)_2C}\overset{|}{\underset{}{H^{\oplus}OH}}$

$\downarrow -H_2O$

$(CH_3)_2C=CH_2-CH_3 \xleftarrow{-H^+} (CH_3)_2C^{\oplus}-CH_2-CH_3$
$\underset{H}{|}$
Tert. carbonium ion

(ii) $CH_3-\underset{\underset{CH_3}{|}}{\overset{\overset{OH}{|}}{C}}-\underset{\underset{CH_3}{|}}{CH}-CH_3 \xrightarrow{H^+} CH-\underset{\underset{CH_3}{|}}{\overset{\overset{\overset{\oplus}{HOH}}{|}}{C}}-\underset{\underset{CH_3}{|}}{CH}-CH_3$

$\downarrow -H_2O$

$CH_3-\underset{\underset{CH_3}{|}}{C}=\underset{\underset{CH_3}{|}}{C}-CH_3 \xleftarrow{-H^+} CH_3-\underset{\underset{CH_3}{|}}{\overset{+}{C}}-\underset{\underset{CH_3}{|}}{\overset{\overset{H}{|}}{C}}-CH_3$
2, 3 Dimethylbutene-2

(iii) $CH_3-\underset{\underset{CH_3}{|}}{CH}-CH_2-CH_2-OH \xrightarrow{H^+} CH_3-\underset{\underset{CH_3}{|}}{CH}-CH_2-CH_2-\underset{\underset{H}{|}}{\overset{\oplus}{OH}}$

$\downarrow -H_2O$

1,2 Hydride shift $ CH_3-\underset{\underset{CH_3}{|}}{\overset{\overset{H}{|}}{C}}-\overset{\oplus}{CH}-CH_3 \xleftarrow{1,2\text{-Hydride shift}} CH_3-\underset{\underset{CH_3}{|}}{CH}-\underset{\underset{H}{|}}{CH}-\overset{\oplus}{CH_2}$
Sec. carbonium ion Primary carbonium ion

$CH_3-\underset{\underset{CH_3H}{||}}{\overset{\oplus}{C}}-CH-CH_3 \longrightarrow CH_3-\underset{\underset{CH_3}{|}}{C}=CH-CH_3$
Tert. carbonium ion 2-Methylbutene-2

Preparation of alkenes by dehydrohalogenation of alkyl halides

Elimination Reaction

Alkyl halides can form alkenes by the loss of molecule of hydrogen halide under the influence of a base catalyst (alcoholic potash). The reaction is known as *dehydrohalogenation of alkyl halides*. It involves the removal of the halogen atom together with the hydrogen atom from the adjacent carbon atom.

$$CH_3 - CH_2X \xrightarrow{OH^-} CH_2 = CH_2 + H_2O + X^-$$

Mechanism of reaction (E_2 mechanism). Elimination bimolecular mechanisms)

$$\underset{\text{Alkyl halide}}{-\underset{H}{\underset{|}{C}}-\underset{X}{\underset{|}{C}}-} \xrightarrow{-HX} \underset{\text{Alkene}}{-C=C-}$$

Orientation in dehydrohalogenation. In some cases, this reaction yields a single product (alkene) and in other cases yields a mixture of alkenes. For example, out of 1-chlorobutane and 2-chlorobutane the former yields only but-1-ene while the latter gives but-1-ene and but-2-ene. Thus, the preferred product is the alkene that has the greater number of alkyl groups attached to the doubly bonded carbon atoms. In other words, the ease of formation of alkene has the following order:

$$R_2C = CR_2 > R_2C = CHR > R_2C = CH_2 > RCH = CHR > RCH = CH_2 > CH_2 = CH_2$$

This order is based on heat of hydrogenation values.

Hence in dehydrohalogenation, the more stable the alkene more easily it is formed.

In the dehydrohalogenation of alkyl halide to form alkene, the preferred product is the alkene which is more highly alkylated at the doubly bound carbon atoms. This is another statement of Saytzeff Rule.

This is also clear from the transition state formed in the dehydrohalogenation of alkyl halides as discussed below. The double bond is partly formed and the transition state acquires alkene character. Factors that stabilise an alkene also stabilise an *incipient* alkene in the transition state.

$$\underset{X}{\underset{|}{-\underset{|}{\overset{H}{C}}-\underset{|}{C}-}} \xrightarrow{OH^-} \left[\underset{X^{\delta-}}{\underset{\vdots}{-\underset{|}{\overset{\overset{\delta-}{HO}\cdots H}{C}}-\underset{|}{C}-}} \right]$$

$$\longrightarrow\ >C=C< \ + \ X^- + H_2O$$

Reactivity of alkyl halides in dehydrohalogenation. The decreasing order of reactivity of alkyl halides in this reaction is:

Tert-Alkyl halides > Sec-Alkyl halides > Primary Alkyl halides
(3°) (2°) (1°)

As we move from primary to secondary and from secondary to tertiary halides, the structure becomes more branched at carbon atom bearing the halogen. This increased branching has two results. (i) the number of hydrogen atoms available for attack by a base is more and thus there is greater probability towards elimination, (ii) it leads to the formation of more highly branched (hence more stable) alkene hence more stable transition state and low E_{act}.

As a result of these two factors, the decreasing order of reactivity of alkyl halides for dehydrohalogenation is:

$$3° > 2° > 1°$$

The mechanism discussed above is called E_2 type of mechanism of dehydrohalogenation.

Mechanism of Dehydrohalogenation (E_1 mechanism,) Elimination unimolecular mechanism)

Some secondary and tertiary alkyl halides undergo dehydrogenation in a solution of low base concentration by a different mechanism known as E_1 mechanism (E for elimination; 1 for unimolecular). *This mechanism operates through a two-step process in which the rate detemining step involves only one molecule.* It is believed that in the first step, which is a slow and rate-determining step, the alkyl halide dissociates into halide ion and carbocation. In the second step which is a fast one, the carbocation loses a proton to the OH^- ion to form the alkene.

The complete mechanism is as shown below:

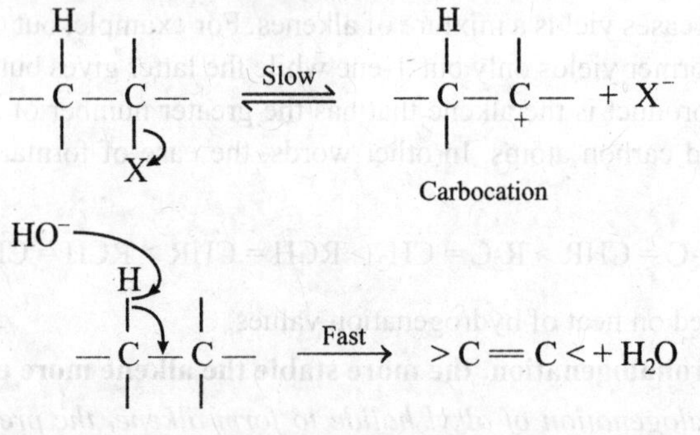

$$-\underset{|}{\overset{H}{C}}-\underset{X}{\overset{|}{C}}- \;\rightleftharpoons\; \underset{\text{Slow}}{} \;\; -\underset{|}{\overset{H}{C}}-\underset{+}{\overset{|}{C}}- + X^-$$

Carbocation

$$HO^- \;\; -\underset{|}{\overset{H}{C}}-\underset{|}{\overset{|}{C}}- \;\xrightarrow{\text{Fast}}\; >C=C< + H_2O$$

In certain alkyl halides (say 3°), there can be slight variation in the mechanistic approach as the carbon cation initially formed undergoes rearrangement to form more stable carbocation and thus yields highly branched alkene as the chief product of the reaction.

For example, action of alcoholic potash on 2-chloro-3, 3-dimethylbutane yields chiefly 2, 3-dimethylbut-2-ene as the carbocation initially formed undergoes rearrangement as shown below:

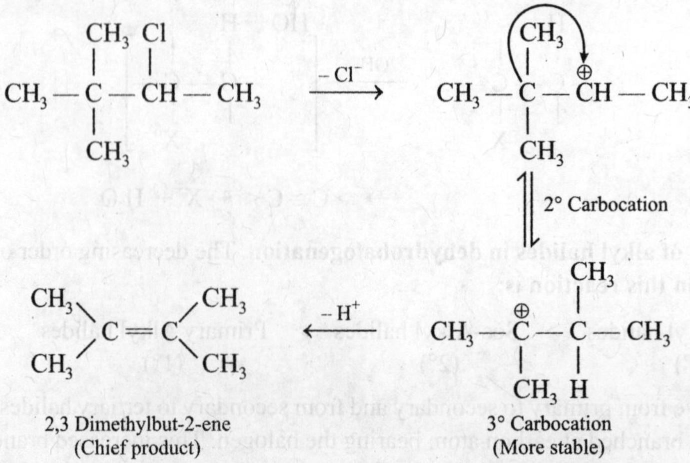

$$CH_3-\underset{\underset{CH_3}{|}}{\overset{\overset{CH_3}{|}}{C}}-\overset{Cl}{\underset{|}{CH}}-CH_3 \;\xrightarrow{-Cl^-}\; CH_3-\underset{\underset{CH_3}{|}}{\overset{\overset{CH_3}{|}}{C}}-\overset{\oplus}{CH}-CH_3$$

2° Carbocation

$$\underset{CH_3}{\overset{CH_3}{}}{>}C=C{<}\underset{CH_3}{\overset{CH_3}{}} \;\xleftarrow{-H^+}\; CH_3-\overset{\oplus}{\underset{|}{C}}-\underset{\underset{H}{|}}{\overset{\overset{CH_3}{|}}{C}}-CH_3$$

2,3 Dimethylbut-2-ene
(Chief product)

3° Carbocation
(More stable)

This is an example of E_1 type of mechanism of dehydrohalogenation.

For a given alkyl group, the order of reactivity of different alkyl halides is RI > RBr > RCl > RF. For the same halogen atom, the order of reactivity is: Tertiary alkyl halide > sec. alkyl halide > Primary alkyl halide. We thus observe that the order of reactivity of alkyl halides in both E_1 and E_2 types of reactions is the same. It is tertiary > secondary > primary.

E1CB Reaction Mechanism

An elimination reaction that occurs when a compound bearing a poor bearing group and an acidic hydrogen is treated with a base as shown below

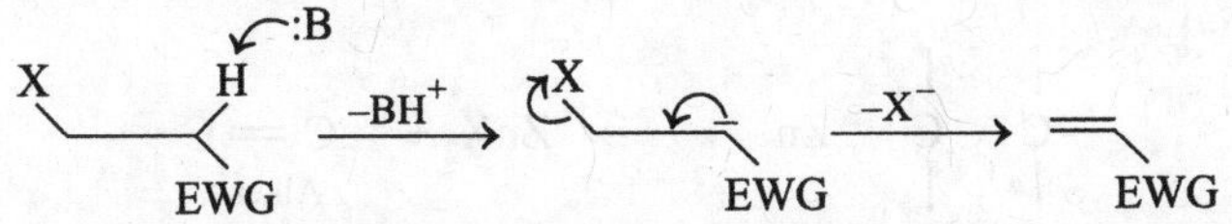

ElCB stands for Elimination Unimolecular Conjugate Base and
EWG stands for Electron Withdrawing Group

The reaction is unimolecular from the conjugate base of the starting compound which in turn is formed by deprotonation of the starting compound by a suitable base

Electron withdrawing group can be one of the following:

A carbonyl group (keto, aldehyde, ester)

A nitro group

An electron deficient aromatic group

A common example of E1CB reaction is dehydration of aldol. This is explained as under.

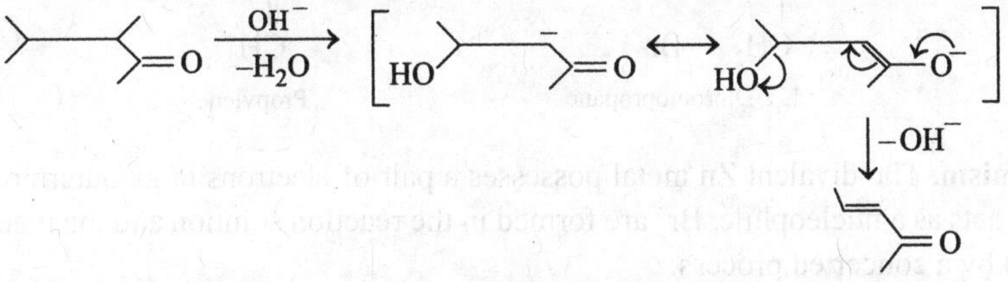

Example 3: For a given halogen, what is the order of reactivity of various alkyl halides? Explain.

Solution: For a given halogen, the order of reactivity of alkyl halides is:

tertiary > secondary > primary

This is because in a tertiary alkyl halide, there is maximum number of hydrogens on the neighbouring carbon atom. So the chances of attack by hydroxyl ion (E_2 mechanism) are maximum leading to maximum reactivity. This is followed by secondary and primary alkyl halides.

Even if the reaction takes place by E_1 mechanism, tertiary alkyl halide has greatest reactivity because the carbonium ion formed in the first step has the maximum stability leading to completion of reaction. A tertiary carbonium ion (carbocation) has maximum stability followed by secondary and followed by primary carbonium ion.

Example 4: For a given alkyl group, what is the order of reactivity with different halogens?

Solution: Order of reactivity with different halogens, with a given alkyl group is

I > Br > Cl > F

Alkyl iodide has the maximum reactivity because C — I bond can be broken most easily leading to the formation of products, C — I bond has the minimum bond dissociation energy. Reactivity of alkyl halide decreases as we go to bromide, chloride and fluoride.

Preparation of alkenes by dehalogenation of dihalides

Dehalogenation of 1, 2 dihaloalkanes (*vic*-dihalides) involves the treatment to the vic-dihalide with reactive metals like Zn in acetic acid and it yields the corresponding alkenes.

$$\underset{\text{1, 2 dihaloalkane}}{-\underset{\underset{X}{|}}{C}-\underset{\underset{X}{|}}{C}-} + Zn \longrightarrow ZnX_2 + \underset{\text{Alkene}}{>C=C<}$$

For example:

$$\underset{\text{1,2 Dibromo ethane}}{\underset{|}{CH_2-Br} \atop \underset{|}{CH_2-Br}} + Zn \longrightarrow ZnBr_2 + \underset{\text{Ethylene}}{\underset{CH_2}{\overset{CH_2}{\|}}}$$

$$\underset{\text{1, 2 Dibromopropane}}{\underset{|}{CH_3} \atop {CH-Br} \atop \underset{|}{CH_2-Br}} + Zn \longrightarrow ZnBr_2 + \underset{\text{Propylene}}{\underset{CH_2}{\overset{CH_3}{\underset{\|}{CH}}}}$$

Mechanism. The divalent Zn metal possesses a pair of electrons in its outermost shell and therefore it acts as a nucleophile, Br⁻ are formed in the reaction solution and the reaction can be represented by a concerted process.

$$\underset{Br}{\underset{|}{-C}}\overset{Zn : Br}{\underset{|}{-C-}} \longrightarrow >C=C< + :Br^- + [ZnBr]^+ \quad \text{[Trans elimination process]}$$

$$[ZnBr]^+ \longrightarrow Zn^{2+} + Br^-$$

Preparation of alkenes by (*i*) reduction of alkynes and (*ii*) dehydrogenation of alkanes

(*i*) Alkenes can be prepared by the partial reduction of alkynes using metallic *sodium* or *lithium* in liquid ammonia (**Birch reduction**), or by using a calculated quantity of hydrogen over palladium catalyst.

Reduction with sodium in liquid ammonia yields *trans* alkene, whereas *cis* alkene results from H_2 in the presence of Pd.

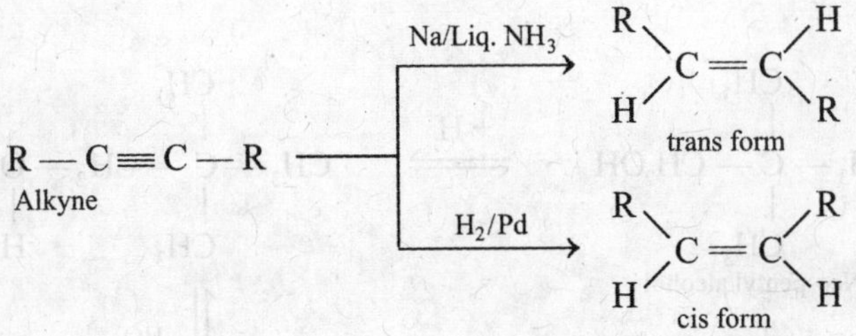

$$R-C\equiv C-R \xrightarrow{\text{Na/Liq. NH}_3} \underset{\text{trans form}}{\overset{R}{\underset{H}{>}}C=C\overset{H}{\underset{R}{<}}}$$

$$R-C\equiv C-R \xrightarrow{\text{H}_2/\text{Pd}} \underset{\text{cis form}}{\overset{R}{\underset{H}{>}}C=C\overset{R}{\underset{H}{<}}}$$

Alkyne

(*ii*) **Dehydrogenation of alkanes.** It involves the passing of the alkane in the vapour phase over a bed of catalysts (oxides of chromium and aluminium).

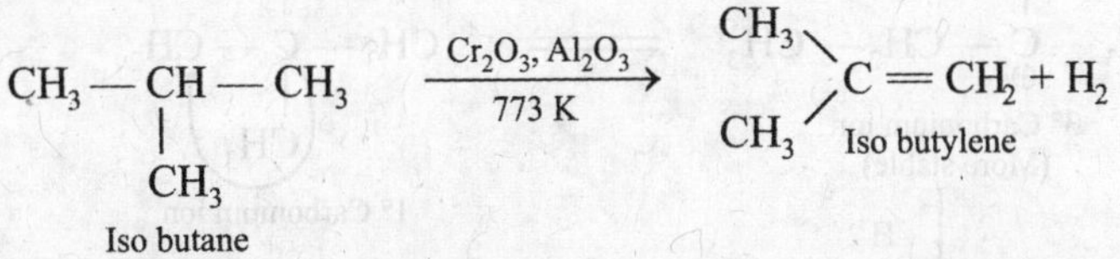

$$\underset{\text{Iso butane}}{CH_3-\underset{\underset{CH_3}{|}}{CH}-CH_3} \xrightarrow[773\text{ K}]{Cr_2O_3,\,Al_2O_3} \underset{\text{Iso butylene}}{\overset{CH_3}{\underset{CH_3}{>}}C=CH_2}+H_2$$

Example 5: Which alcohol of each pair would you expect to be more easily dehydrated? Also name the main products of dehydration.

1. $CH_3-CH_2-CH_2-CH_2OH$, $CH_3CH_2\underset{\underset{OH}{|}}{CH}-CH_3$

2. $(CH_3)_2CH-\underset{\underset{OH}{|}}{\overset{\overset{CH_3}{|}}{C}}-CH_3$, $(CH_3)_2CH-CH(CH_3)CH_2OH$

Solution:

	Alcohol	Main product of dehydration
1.	$CH_3CH_2\underset{\underset{OH}{\|}}{CH}-CH_3$ Butan-2-ol	$CH_3-CH=CH-CH_3$ But-2-ene
2.	$CH_3-\underset{\underset{CH_3}{\|}}{CH}-\underset{\underset{O-H}{\|}}{\overset{\overset{CH_3}{\|}}{C}}-CH_3$ 2,3-Dimethylbutan-2-ol	$CH_3-\underset{\underset{CH_3}{\|}}{C}=\overset{\overset{CH_3}{\|}}{C}-CH_3$ 2,3-dimethylbut-2-ene

Example 6: When neopentyl alcohol $(CH_3)_3.C.CH_2OH$ is heated with acid, it is slowly converted into a mixture of two isomeric alkenes of the formula C_5H_{10}. Name these alkenes according to IUPAC system of nomenclature and also show how they are formed. Which one of them is the major product and why?

Solution:

$$CH_3-\underset{\underset{CH_3}{|}}{\overset{\overset{CH_3}{|}}{C}}-CH_2OH \rightleftharpoons \overset{+H^+}{} CH_3-\underset{\underset{CH_3}{|}}{\overset{\overset{CH_3}{|}}{C}}-CH_2-\overset{+}{\underset{\underset{H}{|}}{O}}-H$$

Neo-pentyl alcohol

$$\Big\updownarrow -HO_2$$

$$CH_3-\underset{\oplus}{\overset{\overset{CH_3}{|}}{C}}-CH_2-CH_3 \rightleftharpoons CH_3-\underset{\underset{CH_3}{|}}{\overset{\overset{CH_3}{|}}{C}}-\overset{+}{C}H_2$$

3° Carbonium ion
(More stable)   1° Carbonium ion

$$\Big| -H^+$$

$$CH_3-\underset{\underset{CH_3}{|}}{\overset{\overset{CH_3}{|}}{C}}=CH-CH_3 \qquad CH_2=\underset{\underset{CH_3}{|}}{\overset{}{C}}-CH_2-CH_3$$

2-Methyl but-2-ene   2-Methyl but-1-ene
(Major product)   (Minor)

Preparation of alkenes by Hofmann elimination

Hofman elimination reaction is given as under:

$$\left[R-CH_2-CH_2-\underset{\underset{CH_3}{|}}{\overset{\overset{CH_3}{|}}{N^+}}-CH_3 \right] OH^- \xrightarrow{\Delta} R-CH=CH_2 + (CH_3)_3N + H_2O$$

Quaternary amm. hydroxide   Alkene   Trimethyl amine

This reaction is used to *eliminate* nitrogen from an organic basic compound by exhaustive *methylation*. Quaternary ammonium hydroxide is obtained from the corresponding quaternary ammonium halide by treatment with aqueous Ag_2O as under.

$$2RCH_2CH_2\overset{+}{N}(CH_3)_3 X^- + Ag_2O + H_2O \longrightarrow 2RCH_2CH_2\overset{+}{N}(CH_3)_3 OH^- + 2AgX$$

Mechanism

The reaction follows E_2 mechanism where the hydroxide ion removes a proton and the positively charged nitrogen acts as the leaving group

$$\overset{HO^-}{\frown}\overset{H}{\underset{\underset{\overset{+}{N}(CH_3)_3}{|}}{\overset{|}{-C-CH_2}}} \xrightarrow{E_2 \text{ reaction}} >C=CH_2 + (CH_3)_3N + H_2O$$

5.5 PHYSICAL PROPERTIES OF ALKENES

(*i*) The first three members of alkenes i.e. ethene, propane and butenes are colourless gases; the next eleven members ($C_5 – C_{15}$) are liquids while the higher alkenes are solids. Ethene has a pleasant smell but other members are odourless.

(*ii*) Alkenes are insoluble in water but quite soluble in non-polar solvents like benzene, CCl_4.

(*iii*) Densities of alkenes increase with increase in molecular mass. Maximum density of alkenes is 0.89 g cm^{-3}. Thus all alkenes are lighter than water.

(*iv*) **Dipole moment.** The *trans* isomer has generally smaller dipole moment than *cis* isomer. For example:

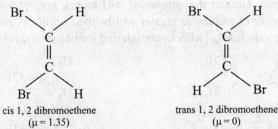

cis 1, 2 dibromoethene
($\mu = 1.35$)

trans 1, 2 dibromoethene
($\mu = 0$)

This is due to the fact that in *trans* isomer, similar groups are on opposite sides of the molecule and their bond polarities tend to cancel each other. But in *cis* isomers the bond polarities are not cancelled out.

(*v*) **Boiling points.** Generally speaking *cis* isomers have higher boiling points than *trans* isomers. This is due to the fact that *cis* isomers have greater intermolecular forces of attraction on account of higher dipole moments of their molecules. For example, boiling point of *cis*-2-butene is 277 K while that of *trans*-2-butene is 274 K.

(*vi*) **Melting points.** Out of *cis* and *trans* isomers, the *trans* compound has higher melting point. This is because *trans* compounds are more symmetrical than their *cis* isomers and are, therefore, more closely packed in the crystal lattice. For example, melting point of maleic acid (*cis* isomer) is 403 K while fumaric acid (*trans* isomer) is 503 K.

Maleic acid
(cis isomer, less symmetrical)

Fumaric acid
(trans isomer, more symmetrical)

Relative Stability of alkenes

Stability of an alkene can be estimated by measuring its heat of hydrogenation. *Heat of hydrogenation of an alkene may be defined as the amount of heat evolved when one mole of the alkene is hydrogenated.* Heat of hydrogenation is a measure of the stability of the alkene. The lower the heat of hydrogenation, the more stable the alkene is. We observe from heat of hydrogenation studies that (*i*) an unsubstituted alkene such as ethene has the highest heat of hydrogenation.

(*ii*) Greater the number of alkyl groups attached to the doubly bonded carbon atoms, more stable is the alkene. The order of stability of alkenes is given as under

$R_2C = CR_2 > R_2C = CHR > RCH = CHR > R_2C = CH_2 > RCH = CH_2 > CH_2 = CH_2$

(*iii*) The *trans*-isomer is slightly more stable than the corresponding *cis*-isomer.

Explanation of relative stability of alkenes.

We can explain the relative stabilities based on the concept of hyperconjugation which utilises σ-H atoms. Number of σ-H atoms and heat of hydrogenation (Δ_H) of some alkenes are given below. We expect the Δ_H values inversely proportional to the number of σ-H atoms. This is found to be so. Greater the number of σ-H atoms, greater will be the number of hyperconjugation structures and hence greater will be the stability. A greater stability of the alkene means smaller value of Δ_H (without considering the negative sign)

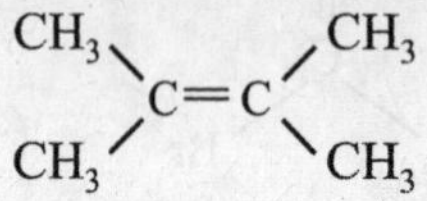

2, 3-Dimethyl but-2-ene (I)
σ-H = 12, Δ_H = – 112.2 kJmol^{-1}

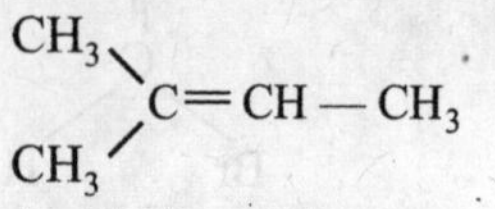

2-Methyl but-2-ene (II)
σ-H = 9, Δ_H = – 112.5 kJmol^{-1}

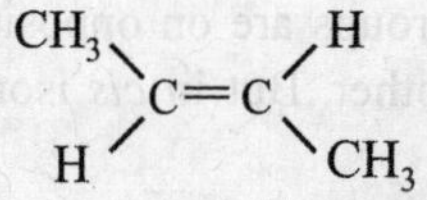

trans-But-2-ene (III)
σ-H = 6, Δ_H = – 115.4 kJmol^{-1}

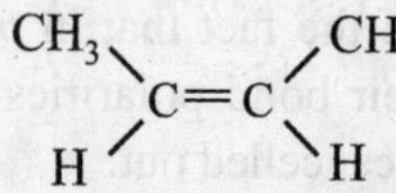

cis-But-2-ene (IV)
σ-H = 6, Δ_H = – 119.6 kJmol^{-1}

Propene (V)
σ-H = 3, Δ_H = – 125.8 kJmol^{-1}

Ethene (VI)
σ-H = zero, Δ_H = – 137.1 kJmol^{-1}

Among the *cis*-trans isomers, *trans* is more stable than the *cis* isomer. Consider for example compounds (III) and (IV) above. In compound (IV), the two methyl groups lie on the same and cause steric hindrance. This makes the structure slightly unstable and hence we above greater heat of hydrogenation.

5.6 CHEMICAL PROPERTIES OF ALKENES

Addition Reactions

(*i*) Addition of hydrogen (Hydrogenation)

Alkene reacts with hydrogen gas in the presence of Pt, Pd or Raney Ni to form alkanes. This is an exothermic process.

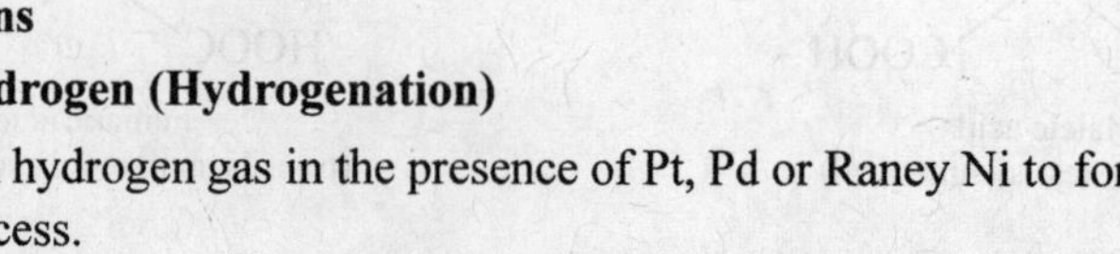

$$\ce{>C=C< + H_2 ->[\text{Pt, Pd or Ni}] -C-C- + Heat}$$

Mechanism

Hydrogenation of alkenes is a *cis-addition* process. Catalyst has a big role to play in the reaction as illustrated below.

Hydrogen molecule gets absorbed on the surface of the catalyst using the vacant *d*-orbitals of the catalyst metal. Then the alkene molecule is adsorbed through π-electrons. Then the adsorbed H-atom is transferred to the same face of unsaturated compound giving *cis*-addition product. The product is desorbed from the catalyst surface to make way for the addition of more alkene molecules on the surface. This process continues till the hydrogenation process is complete.

(ii) Addition of Halogens

Alkenes add a molecule of halogen to form alkyl halides.

$$CH_2 = CH_2 + Br_2 \longrightarrow CH_2Br - CH_2Br$$

Brown colour → 1, 2 Dibromoethane (colourless)

On adding bromine water to a compound containing a double bond, brown colour due to bromine disappears.

Mechanism of halogenation

Halogenation of an alkene takes place by electrophilic addition mechanism. A halogen molecule (Cl_2, Br_2 or I_2) is non-polar in nature. But when this molecule approaches the alkene molecule, the π bonds have the effect of polarising the halogen molecule. The positive end of the polarised halogen molecule is attached to one ethylenic carbon atom with the simultaneous attachment of the negative end to the second ethylenic carbon atom forming a halonium ion, with the release of halide ion. The halide ion then attacks the halonium ion from the back side to avoid steric hindrance, thus giving rise to a dihalide. The steps of the mechanism are shown as under, taking the example of bromination.

$$Br - Br \xrightarrow{\pi \text{ Electron cloud}} \overset{\delta^+}{Br} - \overset{\delta^-}{Br}$$

Polarised bromine molecule

Step 1. Alkene + Polarised Br—Br ⇌ π-Complex ⟶ Bromonium ion + Br^-

Step 2.

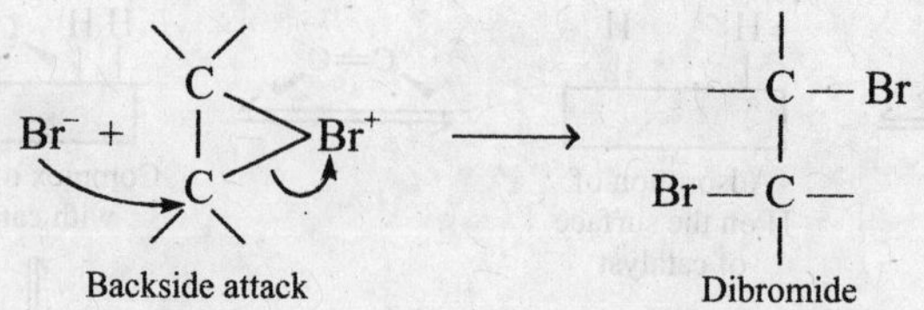

$$Br^- + \underset{\text{Backside attack}}{\overset{|}{\underset{|}{C}}\!\!\diagdown\!\!\overset{|}{\underset{|}{C}}\!\!\diagup\!\! Br^+} \longrightarrow \underset{\text{Dibromide}}{\begin{array}{c} -\!\!C\!-\!Br \\ | \\ Br\!-\!C\!- \end{array}}$$

In the π-complex, there is only electrostatic attraction between the polarised bromine molecule and the alkene, there is no actual bonding.

Evidence to prove that addition of halogen to an alkene proceeds through a halonium ion and not through carbonium ion (carbocation).

Addition of halogen to an alkene proceeds through the formation of a halonium ion and not through a carbonium ion. This can be proved like this:

It is observed that halogenation of alkenes gives rise to products that are optically active. This phenomenon can be explained by halonium ion mechanism and not by carbonium ion mechanism.

In halonium ion mechanism, the halide (in the second step) attacks from the back of halonium ion, giving rise to *trans* dihalide product.

In carbonium ion mechanism, the halide can attack from both sides of the carbonium ion (as it is flat in shape), giving rise to both *cis* and *trans* isomers.

Trans product obtained from halonium ion mechanism explains optical activity of the actual product obtained.

If we assume the carbonium ion mechanism, the mixture of *cis* and *trans* products will give us a racemic modification which will have no net optical activity. This is against the observations. Practically, we get an optically active compound.

This is schematically shown as under taking the example of bromination of cyclohexene.

Carbonium ion mechanism

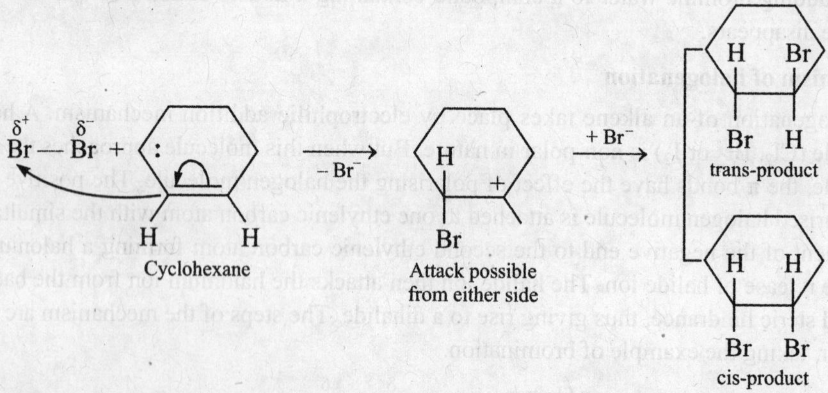

Halonium ion mechanism

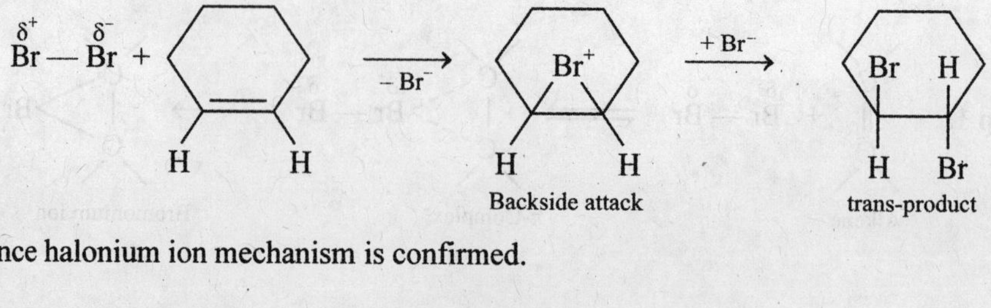

Hence halonium ion mechanism is confirmed.

(iii) Reaction of Alkenes with Halogens in the Presence of Water

Alkenes react with halogens in the presence of water to form halohydrins. For example:

$$\text{>C=C<} + Cl_2 + H_2O \longrightarrow -\overset{|}{\underset{Cl}{C}}-\overset{|}{\underset{OH}{C}}- + HCl$$

Chlorohydrin

Mechanism of the Reaction

The reaction takes place through the intermediate formation of a halonium ion as shown below:

$$\overset{C}{\underset{C}{\|}} + \overset{\delta^+}{Cl}-\overset{\delta^-}{Cl} \rightleftharpoons \begin{array}{c} C \\ | \\ C \end{array} \overset{\delta^+}{\underset{}{>}}\overset{\delta^-}{Cl}-Cl \longrightarrow \begin{array}{c} C \\ | \\ C \end{array} > Cl^+ + Cl^-$$

$$H_2\ddot{O}: + \begin{array}{c} C \\ | \\ C \end{array} > Cl^+ \longrightarrow \begin{array}{c} -C-Cl \\ | \\ H-O^+-C- \\ | \\ H \end{array} \longrightarrow \begin{array}{c} -C-Cl \\ | \\ HO-C- \\ | \end{array} + H^+$$

Addition of an unsymmetrical molecule over a double bond. Markownikoff's rule

When olefins are treated with hydrogen halides either in the gas phase or in an inert non-ionizing solvent (pentane), addition occurs to form alkylhalides.

$$H_2C = CH_2 + HX \longrightarrow H_3C-CH_2X$$
Ethylene   Ethylhalide

With propylene ($CH_3-CH=CH_2$), the next higher homologue, the addition of HX yields two products (1-halopropane and 2-halopropane).

Vladimir Markownikoff (1837–1904)
Vladimir Markownikoff was a Russian Chemist. He is best known for the rule after his name describing the addition of halogen halide to alkenes

$$CH_3-CH=CH_2 + H-X \longrightarrow \begin{cases} CH_3-CH_2-CH_2X \\ \text{1-Halopropane} \\ \\ CH_3-CH-CH_3 \\ \quad\quad\quad | \\ \quad\quad\quad X \\ \text{2-Halopropane} \end{cases}$$

However, in actual practice 2-halopropane is exclusively obtained.

The exclusive formation of the above product is in accordance with Markownikoff Rule which governs the addition of unsymmetrical reagents to unsymmetrical alkenes. *It states that the ionic addition of unsymmetrical reagents to unsymmetrical olefins proceeds in such a way that the more positive part of the reagent, becomes attached to the carbon atom with larger number of hydrogen atoms.*

When the hydrogen halides or water is added, hydrogen atom constitutes the more positive part or in other words, the hydrogen atom becomes attached to the olefinic carbon atom which carries the larger number of hydrogen atoms, for example:

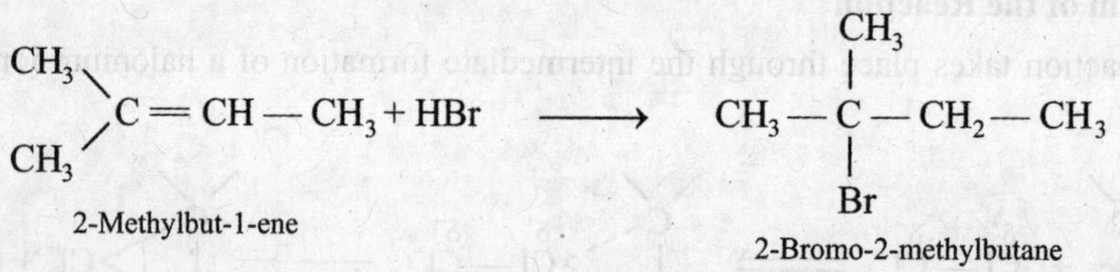

$$(CH_3)_2C=CH-CH_3 + HBr \longrightarrow CH_3-\underset{\underset{Br}{|}}{\overset{\overset{CH_3}{|}}{C}}-CH_2-CH_3$$

2-Methylbut-1-ene　　　　　　　　2-Bromo-2-methylbutane

Theoretical explanation of the orientation of addition to olefins is related to the relative stability of carbonium ions. We know that relative stability of carbonium ions is: tertiary > secondary > primary > CH_3^+. If there is the probability of more then one carbonium ions being formed, the addition of electrophile yields the more stable one.

That is why **2 bromopropane** is exclusively formed by the addition of HBr to propylene (an unsymmetrical alkene) and not **1-bromopropane** because secondary isopropyl carbonium ion is more stable than primary *n*-propyl carbonium ion.

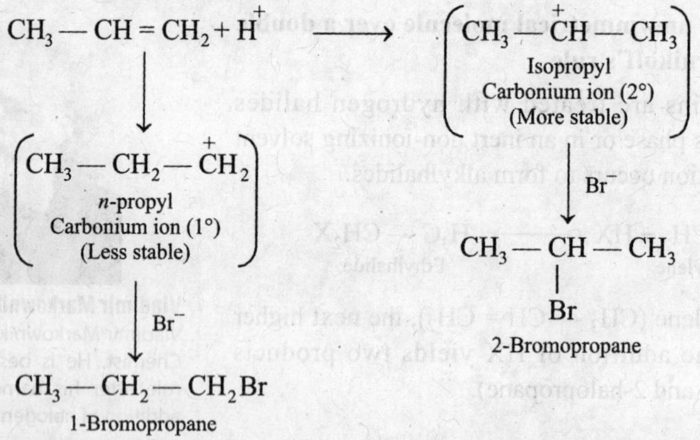

$$CH_3-CH=CH_2 + H^+ \longrightarrow \begin{pmatrix} CH_3-\overset{+}{C}H-CH_3 \\ \text{Isopropyl} \\ \text{Carbonium ion (2°)} \\ \text{(More stable)} \end{pmatrix}$$

$$\downarrow \qquad\qquad\qquad\qquad \downarrow Br^-$$

$$\begin{pmatrix} CH_3-CH_2-\overset{+}{C}H_2 \\ n\text{-propyl} \\ \text{Carbonium ion (1°)} \\ \text{(Less stable)} \end{pmatrix} \qquad CH_3-\underset{\underset{Br}{|}}{C}H-CH_3$$

$$\downarrow Br^- \qquad\qquad\qquad\qquad \text{2-Bromopropane}$$

$$CH_3-CH_2-CH_2Br$$
1-Bromopropane

Considering the electronic effect of methyl group in propylene, the polarization of the double bond due to the (+I) electron-repelling inductive effect of the methyl group is depicted below.

$$CH_3 \rightarrowtail \overset{\delta^+}{C}H=\overset{\delta^-}{C}H_2$$

Thus, the +ve and –ve ends of the dipole (H — X) will add to the –ve and +ve ends of the double bonds, respectively yielding **2-halopropane**.

Mechanism

$$CH_3 - \overset{\delta^+}{CH} = \overset{\delta^-}{CH_2} \quad \overset{\delta^+}{H} - \overset{\delta^-}{X} \xrightarrow{\text{Electrophilic attack}} CH_3 - \overset{+}{CH} - CH_3$$

$$\downarrow :X^- \quad \text{Intermediate 2°-carbonium ion (Nucleophilic attack by } :X^-)$$

$$CH_3 - \underset{|}{\overset{X}{CH}} - CH_3$$
$$\text{2-Halopropane}$$

Thus the modern statement of Markownikoff's rule is:

In the ionic addition of an unsymmetrical reagent to a double bond, the positive portion of the adding reagent becomes attached to the carbon atom of the double bond so as to yield the more stable carbonium ion.

Anti-Markownikoff's addition peroxide effect (Kharasch Effect)

If has been observed that the addition of H — Br to propylene in the presence of peroxides, yields predominantly 1-bromopropane, that is, the reagent adds onto the olefins under these conditions in a manner contrary to Markownikoff's rule thus suggesting a change in the mechanism.

This change in the orientation of addition due to the presence of peroxides is known as the peroxide effect.

Here the change in the mode of addition of the reagent is due to a change from an ionic *mechanism to a free-radical mechanism*. Markownikoff's addition requires the initiation by H^+. Anti-Markownikoff's addition requires initiation by $(\overset{\bullet}{Br})$. Each species attacks the olefin molecule at the centre of highest electron density to yield the most stable intermediate carbonium ion or free radical.

Kharasch in his laboratory
Morris Selig Kharasch (1895-1957) was a pioneering organic chemist best known for his work on free radicals and polymerizations

Mechanism of addition of HBr to propylene in the presence of peroxides

It involves the following steps:

(i) Chain initiation step. The reaction is initiated by the alkoxy radical produced by the homolytic fission of peroxides, which abstracts an atom of hydrogen from HBr generating bromine free radicals $(\overset{\bullet}{Br})$.

$$R - O - O - R \longrightarrow 2R\overset{\bullet}{O}$$
Peroxides   Alkoxy radical

$$R - \overset{\bullet}{O} + H - Br \longrightarrow R - OH + \overset{\bullet}{Br}$$
(Chain initiation)

(ii) Chain propagation step. The $\overset{\bullet}{Br}$ then attacks the propylene molecule to give a more stable secondary radical addition species:

$$CH_3-CH=CH_2 + \dot{B}r \longrightarrow CH_3-\dot{C}H-CH_2Br + CH_3-CH-CH_2^{\cdot}$$
$$\text{2° free radical} \qquad\qquad\qquad |$$
$$\text{(More stable)} \qquad\qquad\qquad Br$$
$$\text{1° free radical}$$
$$\text{(Less stable)}$$

The secondary radical then reacts with another H — Br molecule to yield the product, and another Br which can further propagate the reaction:

$$CH_3-\dot{C}H-CH_2Br + H-Br \longrightarrow CH_3-CH_2-CH_2Br + \dot{B}r$$
$$\text{(Chain propagation)} \qquad\qquad\qquad \text{1-Bromopropane}$$

This mechanism is supported by the fact that small amount of peroxide can influence addition to a large number of molecules of an alkene and a small amount of an inhibitor such as hydroquinone or diphenyl amine can prevent this change.

Peroxide oxide effect or anti-Markownikoff's rule (or Kharasch effect) is observed in the addition of H — Br and not H — Cl or H — I.

Explanation

In case of HCl, it is probably due to the fact that H — Cl bond (430 kJ/mole) is stronger than H — Br bond (368 kJ/mole) and is not broken homolytically by the free radicals generated by peroxides. As such free radical addition of HCl to alkenes is not possible. In case of HI, H — I bond (297 kJ/mole) is no doubt weaker than H — Br bond and can be broken easily. But the iodine atoms (I•) thus formed readily combine amongst themselves to form iodine molecules rather than add to the olefins.

Addition of (*i*) Sulphuric acid (*ii*) water to an alkene

Sulphuric acid adds to an alkene to form alkyl hydrogen sulphate white water adds in the presence of an acid to form alcohol.

(*i*) Mechanism of addition of sulphuric acid to alkene. It is a two step electrophilic addition reaction which takes place as follows:

$$\underset{\text{Alkene}}{\overset{\displaystyle\diagdown C}{\underset{\displaystyle\diagup C}{\|}}} \overset{\delta^+ \quad \delta^-}{H-OSO_3H} \longrightarrow \underset{\text{Carbocation}}{\begin{array}{c}|\\-C-H\\|\\-C^+\\|\end{array}} + \bar{O}SO_3H$$

$$\begin{array}{c}|\\-C-OH\\|\\-C^+\\|\end{array} \quad +\bar{O}SO_3H \longrightarrow \underset{\text{Alkyl hydrogen sulphate}}{\begin{array}{c}|\\-C-H\\|\\-C-OSO_3H\\|\end{array}}$$

(*ii*) Mechanism of addition of water to an alkene. Alkenes react on water in the presence of an acid to form alcohols.

$$\underset{\text{Alkene}}{\diagup}C=C\diagdown + H_2O \xrightarrow{H_3O^+} \underset{\text{Alcohol}}{\begin{array}{c}|\quad|\\-C-C-\\|\quad|\\H\quad OH\end{array}}$$

Alkenes (Carbon-Carbon Pi Bonds) 219

[Mechanism scheme: Alkene + H_3O^+ ⇌ Carbocation + H_2O; Carbocation + $:\ddot{O}-H$ with H ⇌ Protonated alcohol; Protonated alcohol + H_2O ⇌ Alcohol + H_3O^+]

Ozonolysis

Ozone undergoes a reaction with olefins in an inert solvent at low temperatures to yield unstable addition compounds called 'ozonides'. These ozonides are not easily isolated and yield carbonyl compounds on further treatment either with boiling water and Zn or zinc and acetic acid.

This overall reaction is called as *Ozonolysis*.

Ozonolysis is a very useful reaction for illucidating the structures of olefinic compounds. Ozone has a dipolar resonance structure with a bond angle of about 120°.

[Resonance structures of ozone]

The central atom (oxygen) is relatively electron deficient and thus can act as an electrophile for attack on the π electrons of the olefins. The final structure of the ozonide has been shown to have a structure resulting from the complete rupture of the carbon-carbon double bond.

[Scheme: Alkene + O_3 → Addition intermediate → Rearrangement → Ozonoide]

[Scheme: Ozonoide → with H_2O/Zn → two carbonyl compounds + H_2O_2; with Zn, AcOH → two carbonyl compounds + H_2O]

The important thing to note about ozonides is that the *carbon-carbon bond has been broken.* For example:

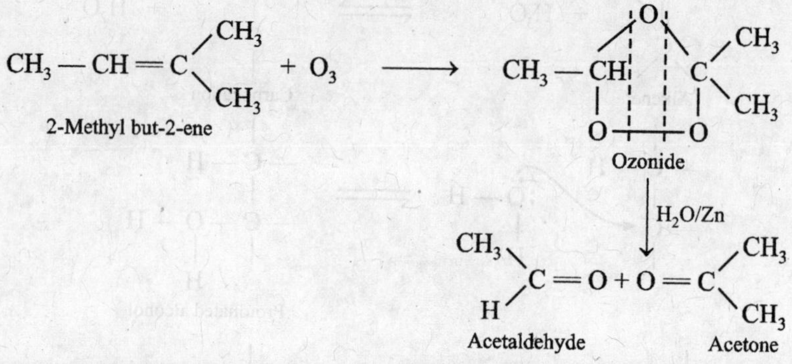

$$CH_3-CH=C(CH_3)(CH_3) + O_3 \longrightarrow$$
2-Methyl but-2-ene

$$CH_3-CH \underset{O-O}{\overset{O}{\overline{}}} C(CH_3)(CH_3)$$
Ozonide

$$\xrightarrow{H_2O/Zn}$$

$$\underset{H}{\overset{CH_3}{>}}C=O + O=C\underset{CH_3}{\overset{CH_3}{<}}$$
Acetaldehyde   Acetone

By identifying the products of ozonolysis, it is possible to locate the double bond in alkene.

Solved Examples

Example 7: On reductive ozonolysis, an unsaturated hydrocarbon gave the following compounds:
(*i*) Ethanediol   (*ii*) Propanone   (*iii*) Ethanol

Write the structural formula of the hydrocarbon and write its IUPAC name.

Solution: The hydrocarbon is

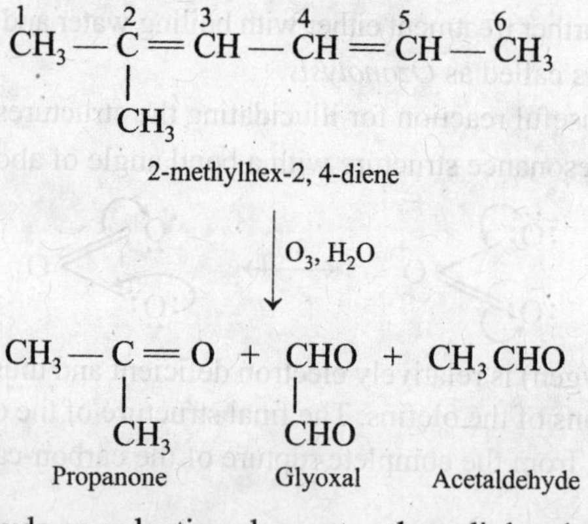

$$\overset{1}{CH_3}-\underset{\underset{CH_3}{|}}{\overset{2}{C}}=\overset{3}{CH}-\overset{4}{CH}=\overset{5}{CH}-\overset{6}{CH_3}$$

2-methylhex-2, 4-diene

$$\downarrow O_3, H_2O$$

$$CH_3-\underset{\underset{CH_3}{|}}{C}=O + \underset{\underset{CHO}{|}}{CHO} + CH_3CHO$$

Propanone   Glyoxal   Acetaldehyde

Glyoxal and acetaldehyde on reduction change to ethanediol and ethyl alcohol respectively.

Example 8: Complete the following reaction

$$CH_3CH=CHCH_3 + O_3 \longrightarrow ? \xrightarrow{Zn/H_2O} ?$$

Solution: $CH_3CH=CHCH_3 \xrightarrow{O_3} CH_3CH\underset{O}{\overset{O-O}{||}}CH-CH_3 \xrightarrow{Zn/H_2O} 2CH_3CHO + H_2O_2$

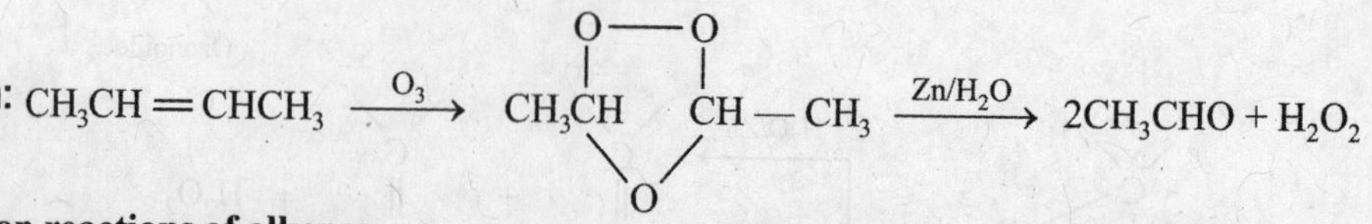

Oxidation reactions of alkenes
(a) Syn and anti Hydroxylation

Conversion of alkene into 1, 2-diols is known as 1, 2 hydroxylation reaction; an — OH group is added on to each carbon atom of the double bond. The most commonly used reagents

Alkenes (Carbon-Carbon Pi Bonds) 221

are potassium permanganate and osmium tetroxide (OsO_4). These reagents give rise to syn products.

$$\underset{\text{cis-2-butene}}{{}^{CH_3}_{H}}C=C{}^{CH_3}_{H} \xrightarrow[\text{or } OsO_4]{KMnO_4} \underset{\text{syn-Dihydroxy product}}{H-\underset{\underset{OH}{|}}{\overset{\overset{CH_3}{|}}{C}}-\underset{\underset{OH}{|}}{\overset{\overset{CH_3}{|}}{C}}-H} \quad (1,2\text{ diol})$$

The two hydroxy groups have been added on the same side of double bond.

The two reagents ($KMnO_4$ and OsO_4) are known to react by similar mechanism which involves cyclic intermediates of the following type which are subsequently hydrolyzed in aqueous solutions to glycol and the reduced form of the reagent.

[cyclic Mn epoxide intermediate] and [cyclic Os epoxide intermediate]

[reaction of Mn cyclic intermediate → syn product + Mn byproduct]

[Cyclic osmium ester + 2H₂O → syn product + Osmic acid]

The permanganate reaction (Baeyer's test) is usually carried out in cold, dilute aqueous alkaline solution.

$$\underset{}{\overset{CH_2}{\underset{CH_2}{\|}}} + H_2O + O \text{ (from } KMnO_4) \longrightarrow \underset{\text{syn Glycol}}{CH_2OH-CH_2OH}$$

$$\underset{\text{Propylene}}{CH_3-CH=CH_2} + H_2O + O \text{ (from } KMnO_4) \longrightarrow \underset{\text{syn (Propylene glycol)}}{CH_3-\underset{\underset{OH}{|}}{CH}-\underset{\underset{OH}{|}}{CH_2}}$$

Both OsO_4 and $KMnO_4$ give syn hydroxylation products. However, if we treat an alkene with a peroxy acid (say metachloroperbenzoic acid) and hydrolyse the epoxide in the presence of an acid or base, we obtain antiDihydroxy product (anti glycol). This will be a case of antihydroxylation. This can be illustrated as under:

$$\underset{\text{trans-2-butene}}{\underset{H}{\overset{CH_3}{>}}C=C\underset{CH_3}{\overset{H}{<}}} \xrightarrow{\text{metachloro per benzoic acid}} \underset{\text{epoxide}}{\underset{H}{\overset{CH_3}{>}}C-C\underset{CH_3}{\overset{H}{<}}} \xrightarrow{H^+/H_2O} \underset{\text{anti Dihydroxy product}}{\begin{array}{c} CH_3 \\ | \\ H-C-OH \\ | \\ HO-C-H \\ | \\ CH_3 \end{array}}$$

This is a case of antihydroxylation.

(b) Oxidation with hot KMnO₄ solution

An alkene on heating with alkaline $KMnO_4$ solution undergoes cleavage of carbon-carbon double bond. Carboxylic acids, ketones and carbon dioxide are formed depending upon the nature of the alkene. With *terminal alkenes* (double bond at the end of the chain), one of the products is methanoic acid which on further oxidation gives $CO_2 + H_2O$.

$$\underset{\text{Ethene}}{CH_2=CH_2} + 4[O] \xrightarrow[373-378 \text{ K}]{KMnO_4, KOH} 2H-\overset{O}{\overset{\|}{C}}-OH \xrightarrow{[O]} 2CO_2 + 2H_2O$$

$$\underset{\text{Propene}}{CH_3-CH=CH_2} + 4[O] \xrightarrow[373-378 \text{ K}]{KMnO_4, KOH} CH_3\overset{O}{\overset{\|}{C}}-OH + HCOOH \xrightarrow{[O]} CO_2 + H_2O$$

$$\underset{}{CH_3-\underset{\underset{CH_3}{|}}{C}=CH_2} + 4[O] \xrightarrow{KMnO_4, KOH} \underset{\text{Propenone}}{CH_3-\underset{\underset{CH_3}{|}}{C}=O} + HCOOH \xrightarrow{[O]} CO_2 + H_2O$$

With *non-terminal alkene* (double bond not at the end of the chain), carboxylic acids or ketones or both are obtained depending upon the structure

$$\underset{\text{But-2-ene}}{CH_3-CH=CH-CH_3} + 4[O] \xrightarrow[373-378 \text{ K}]{KMnO_4 + KOH} \underset{\text{Ethanoic acid}}{CH_3COOH + CH_3COOH}$$

$$\underset{\text{Pent-2-ene}}{CH_3CH_2-CH=CH-CH_3} + 4[O] \xrightarrow[373-378 \text{ K}]{KMnO_4 + KOH} \underset{\text{Propanoic acid \quad Ethanoic acid}}{CH_3CH_2COOH + CH_3COOH}$$

$$\underset{\text{2, 2-Dimethyl but-2-ene}}{\underset{CH_3}{\overset{CH_3}{>}}C=C\underset{CH_3}{\overset{CH_3}{<}}} + 2[O] \xrightarrow[373-378 \text{ K}]{KMnO_4 + KOH} \underset{}{\underset{CH_3}{\overset{CH_3}{>}}C=O} + \underset{\text{Propanone}}{\underset{CH_3}{\overset{CH_3}{>}}C=O}$$

Hydroboration of alkenes

On treatment with diborane $(BH_3)_2$, alkenes form alkyl boranes which are further converted into di and tri alkylboranes. For example:

$$\underset{\text{Propylene}}{2CH_3-CH=CH_2} + (BH_3)_2 \longrightarrow \underset{n\text{-Propylborane}}{2CH_3-CH_2-CH_2BH_2}$$

$$CH_3-CH_2-CH_2BH_2 + CH_3-CH=CH_2 \longrightarrow \underset{\text{Di-}n\text{-propylborane}}{(CH_3-CH_2-CH_2)_2 BH}$$

$$(CH_3-CH_2-CH_2)_2 BH + CH_3-CH=CH_2 \longrightarrow \underset{\text{Tri-}n\text{-propylborane}}{(CH_3-CH_2-CH_2)_3 B}$$

Alkyboranes in turn are readily oxidised by alkaline solution of H_2O_2 to form alcohols. For example:

$$(CH_3CH_2CH_2)_3 B + 3H_2O \xrightarrow{H_2O_2/OH^-} 3CH_3-CH_2-CH_2OH + H_3BO_3$$

This reactions for the preparation of alcohol from alkene is called **hydroboration**.

Mechanism. Diborane participates in the reaction in its monomeric form *i.e.*, BH_3. It behaves as an electrophile since the boron atom in the molecule is electron deficient as it has only six valence electrons.

Since BH_3 is an electrophile, hydroboration of alkanes also involves an electrophilic addition mechanism. But unlike other addition reactions of alkenes, hydroboration takes place in a single step through a transition state as shown below by considering the hydroboration of propylene.

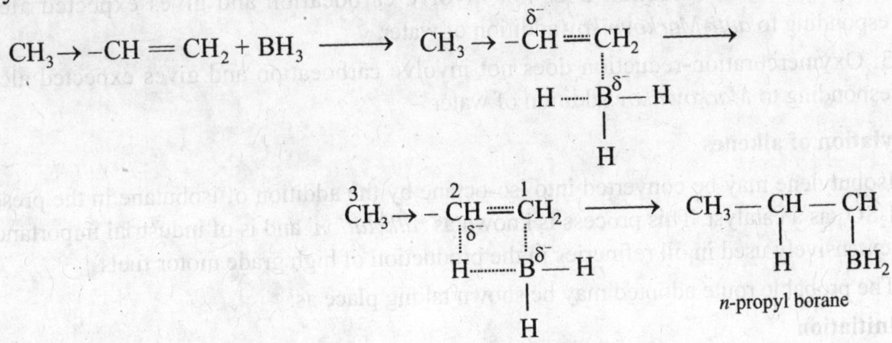

$$CH_3 \rightarrow -CH=CH_2 + BH_3 \longrightarrow CH_3 \rightarrow -\overset{\delta-}{CH}=CH_2 \longrightarrow$$
$$H\cdots\overset{|}{\underset{|}{B^{\delta-}}}-H$$
$$H$$

$$\overset{3}{CH_3} \rightarrow -\overset{2}{\underset{\delta-}{CH}}=\overset{1}{CH_2} \longrightarrow CH_3-CH-CH$$
$$H\cdots\overset{|}{\underset{|}{B^{\delta-}}}-H \qquad\qquad \overset{|}{H}\quad \overset{|}{BH_2}$$
$$H \qquad\qquad\qquad\qquad\qquad n\text{-propyl borane}$$

Oxymercuration-reduction

Alkenes react with mercuric acetate, $(CH_3COO)_2 Hg$ abbreviated as $Hg(OAC)_2$ in aqueous tetrahydrofuran (THF) to form the intermediate which on reduction with $NaBH_4$ in alkaline medium give alcohols.

$$\underset{\text{But-1-ene}}{CH_3CH_2CH=CH_2} \xrightarrow[\text{THF}-H_2O]{(CH_3COO)_2Hg} CH_3CH_2-\underset{\underset{OH}{|}}{CH}-\underset{\underset{HgOCOCH_3}{|}}{CH_2} \xrightarrow[OH]{NaBH_4} \underset{\text{2-Butanol}}{CH_3CH_2-\underset{\underset{OH}{|}}{CH}-CH_3}$$

This process is called *oxymercuration-reduction* or *oxymercuration-demercuration* and may be used for the preparation of alcohol corresponding to Markovnikov addition of water to alkenes.

Mechanism

The reaction takes place through electrophilic addition reaction and resembles halohydrin formation. An intermediate mercurinium is formed in the reaction.

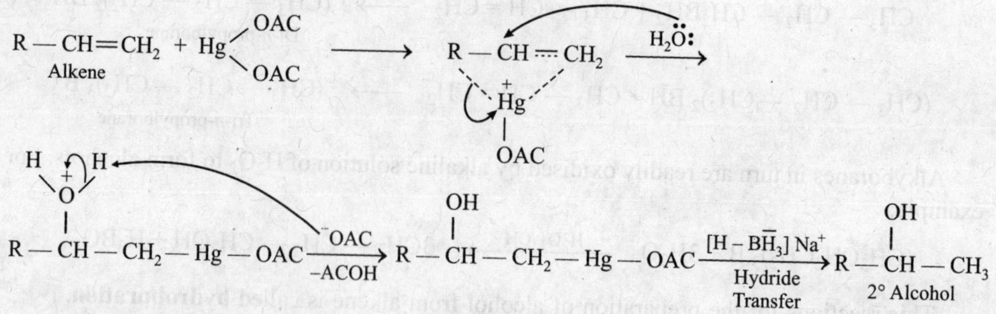

$$R-CH=CH_2 + Hg\begin{smallmatrix}OAC\\OAC\end{smallmatrix} \longrightarrow R-CH \cdots CH_2 \xrightarrow{H_2\ddot{O}:}$$
$$\text{Alkene} \qquad\qquad\qquad\qquad \underset{\underset{OAC}{|}}{\overset{+}{Hg}}$$

$$R-\underset{\underset{O}{|}}{CH}-CH_2-Hg-OAC \xrightarrow[-ACOH]{OAC} R-\underset{\underset{OH}{|}}{CH}-CH_2-Hg-OAC \xrightarrow[\text{Hydride Transfer}]{[H-B\bar{H}_3]\,Na^+} R-\underset{\underset{OH}{|}}{CH}-CH_3$$
$$\overset{H}{\underset{}{\diagdown}}\overset{+}{\underset{}{\diagup}}\overset{H}{\underset{}{}} \qquad\qquad\qquad\qquad\qquad\qquad\qquad\qquad\qquad\qquad\qquad\qquad 2°\text{ Alcohol}$$

Hydration, Hydroboration-oxidation and oxymercuration-reduction reactions are all used for the preparation of alcohols. A *comparative study* of the three processes for the preparation of alcohols may be made as under.

1. Hydration of alcohols gives alcohols corresponding to Markovnikov's addition of water to alkenes. As carbocation is the intermediate, it may *rearrange* to a more stable carbocation. As a result we may get rearranged (unexpected) alcohol

2. Hydrocarbon-oxidation does not involve carbocation and gives expected alcohol corresponding to *anti-Markovnikov* addition of water.

3. Oxymercuration-reduction does not involve carbocation and gives expected alcohol corresponding to *Markovnikov* addition of water.

Alkylation of alkenes

Isobutylene may be converted into iso-octane by the addition of isobutane in the presence of H_2SO_4 as a catalyst. This process is known as *'alkylation'* and is of industrial importance as it is extensively used in oil refineries in the production of high grade motor fuels.

The probable route adopted may be shown taking place as:

Initiation

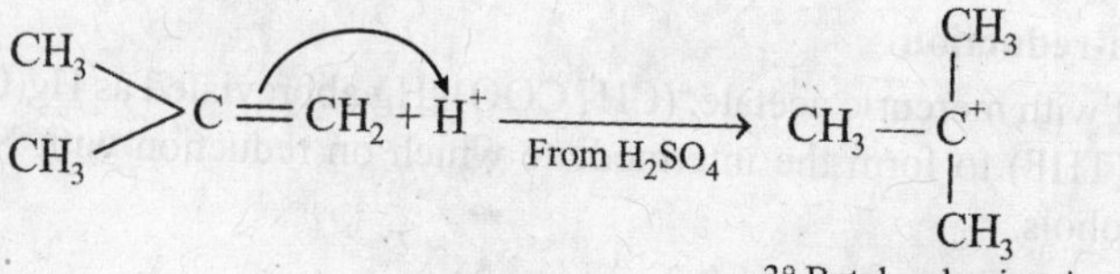

$$\underset{CH_3}{\overset{CH_3}{\diagdown}}C=CH_2 + H^+ \xrightarrow{\text{From }H_2SO_4} CH_3-\underset{\underset{CH_3}{|}}{\overset{\overset{CH_3}{|}}{C^+}}$$
$$\text{3° Butyl carbonium ion}$$

Propagation. There are two propagation steps. in the first step; addition of 3° butyl carbonium ion with isobutylene to form a new carbonium ion ($^+C_8H_{17}$) takes place.

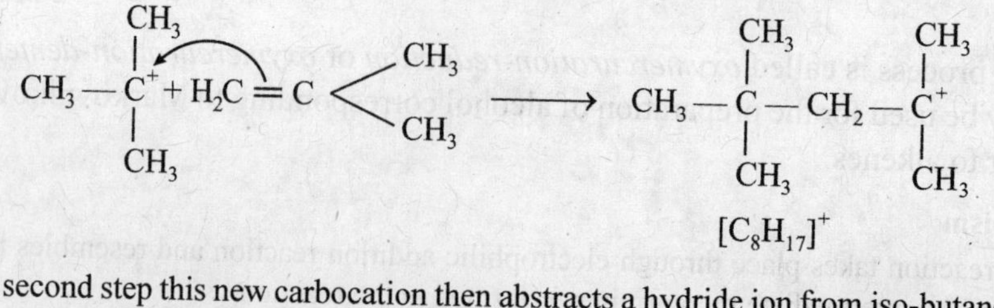

$$CH_3-\underset{\underset{CH_3}{|}}{\overset{\overset{CH_3}{|}}{C^+}} + H_2C=C\underset{CH_3}{\overset{CH_3}{\diagup}} \longrightarrow CH_3-\underset{\underset{CH_3}{|}}{\overset{\overset{CH_3}{|}}{C}}-CH_2-\underset{\underset{CH_3}{|}}{\overset{\overset{CH_3}{|}}{C^+}}$$
$$[C_8H_{17}]^+$$

In the second step this new carbocation then abstracts a hydride ion from iso-butane to form isooctane and a new 3° butyl carbonium ion.

Alkenes (Carbon-Carbon Pi Bonds)   225

$$CH_3-\underset{\underset{CH_3}{|}}{\overset{\overset{CH_3}{|}}{C}}-CH_2-\overset{\overset{CH_3}{|}}{\underset{\underset{CH_3}{|}}{C^+}} + CH_3-\overset{\overset{H}{|}}{\underset{\underset{CH_3}{|}}{C}}-CH_3 \longrightarrow CH_3-\overset{\overset{CH_3}{|}}{\underset{\underset{CH_3}{|}}{C^+}} + CH_3-\underset{\underset{CH_3}{|}}{\overset{\overset{CH_3}{|}}{C}}-CH_2-\overset{\overset{CH_3}{|}}{\underset{\underset{CH_3}{|}}{C}}-H$$

Isobutane   Iso-octane
(2, 2, 4 tri-methyl pentane)

Termination. It involves the loss of a proton from the carbonium ions to form an alkene.

$$CH_3-\underset{\underset{CH_3}{|}}{\overset{\overset{CH_3}{|}}{C}}-\overset{\overset{H}{|}}{CH}-\overset{\overset{CH_3}{|}}{\underset{\underset{CH_3}{|}}{C^+}} \xrightarrow{-H^+} \overset{5}{CH_3}-\underset{\underset{CH_3}{|}}{\overset{\overset{CH_3}{|}}{\overset{4}{C}}}-\overset{3}{CH}=\overset{2}{C}\underset{CH_3}{\overset{CH_3}{<}}$$

2, 4, 4-Tri methylpent-2-ene
(Higher %)
(A)
Or

$$CH_3-\underset{\underset{CH_3}{|}}{\overset{\overset{CH_3}{|}}{C}}-CH_2-\underset{\underset{H-CH_2}{|}}{\overset{\overset{CH_3}{|}}{C^+}} \xrightarrow{-H^+} \overset{5}{CH_3}-\underset{\underset{CH_3}{|}}{\overset{\overset{CH_3}{|}}{\overset{4}{C}}}-\overset{3}{CH_2}-\overset{\overset{CH_3}{|}}{\overset{2}{C}}=\overset{1}{CH_2}$$

2, 4, 4-Tri methylpent-1-ene
(B)

Reaction of alkenes with sulphur monochloride

Ethylene and sulphur monochloride produce the powerful blistering agent (Vesicant) known as mustard gas used during World War I. Though it is a liquid (b.pt. 490 K), small quantities of its vapours in air are highly toxic.

$$\begin{array}{c}CH_2=CH_2\\H_2C=CH_2\end{array} + \begin{array}{c}S_2Cl_2\\\text{Sulphur}\\\text{monochloride}\end{array} \longrightarrow \begin{array}{c}CH_2Cl\\|\\CH_2\end{array}-S-\begin{array}{c}CH_2Cl\\|\\CH_2\end{array} + S$$

(Ethylene)   Mustard gas
[β, β'-Dichlorodiethyl sulphide]

Action of carbenes on alkenes

When treated with diazomethane (CH_2N_2) or ketene ($CH_2=C=O$) in the presence of light, alkenes add on carbene, a methylene, to form substituted cyclopropane.

$$>C=C< + CH_2N_2 \xrightarrow{\text{Light}} >C-C< + N_2$$
$$\underset{CH_2}{\underset{|}{\vee}}$$

Alkene   Diazomethane   Substituted cyclopropane

$$CH_3-CH=CH-CH_3 + CH_2N_2 \xrightarrow{\text{Light}} CH_3-CH-CH-CH_3 + N_2$$
$$\underset{CH_2}{\underset{\vee}{}}$$

Alkene   Diazomethane   1, 2-Dimethyl cyclopropane

It is of theoretical interest that this reaction is stereospecific with singlet methylene occurring exclusively in one spatial direction, that is, addition of carbene to *cis*-but-2-ene yields *cis* −1, 2-dimethyl cyclopropane whereas the reaction with *trans*-but-2-ene yields *trans* −1, 2-dimethyl cyclopropane.

Addition of carbenes on alkenes

Alkenes undergo addition reactions with carbenes, both in the singlet and triplet forms*, to form cyclopropanes. Since a ring is generated as a result of such an addition reaction, the reaction is known as *cycloaddition*. For example, when treated with diazomethane ketone, (CH_2N_2) or ($CH_2 = C = O$) which produce carbene in the presence of light, alkenes add on methylene to from cyclopropanes.

$$>C=C< + :CH_2 \longrightarrow >C-C<$$
$$\text{Carbene} \qquad\qquad\qquad \overset{\displaystyle\vee}{\underset{H_2}{C}}$$
$$\text{(Singlet or triplet)} \qquad\qquad \text{Cyclopropane}$$

Addition of singlet methylene takes place in a *sterospecific* manner to yield product having the same stereochemistry as the alkene. For example, singlet methylene reacts with *cis*-2 butene to form *cis*-1, 2-dimethylcyclopropane. Similarly, it reacts with *trans*-2 butene to form *trans*-1, 2-dimethylcyclopropane.

[Reaction scheme: *cis*-2-Butene + :CH₂ (Singlet methylene, obtained from CH₂N₂ in liquid *cis*-2-butene) → *cis*-1, 2 Dimethyl-cyclopropane]

[Reaction scheme: *trans*-2-Butene + :CH₂ (Singlet methylene, obtained from CH₂N₂ in liquid *trans*-2-butene) → *trans*-1, 2 Dimethyl-cyclopropane]

Addition of **triplet methylene** is *non-stereospecific* and produces a mixture of products. For example, addition of triplet methylene to either *cis*-2-butene or *trans*-2 butene produces a *mixture of cis and trans-1, 2-dimethylcyclopropanes.*

[Reaction scheme: $CH_3-CH=CH-CH_3 + \cdot\dot{C}H_2$ (2-Butene, cis or trans) + Triplet methylene (obtained from CH₂N₂ in gaseous butene) → Mixture of *cis* and *trans* 1, 2-dimethyl cyclopropanes]

---

* For difference between singlet an triplet forms of carbenes, see chapter 2, under carbenes.

Different mechanistic routes in the two cases are responsible for giving different results.

Mechanism of addition of singlet methylene

Addition of singlet methylene involves single step *electrophilic addition*. Being electron deficient in nature, singlet methylene seeks electrons from carbon-carbon double bond. As a result singlet methylene gets simultaneously attached to both the doubly bond carbon atoms. Therefore, stereochemistry of the reactant remains preserved in the products, *cis*-alkenes yield *cis*-dialkyl cyclopropanes while *trans*-alkenes yield *trans*-dialkylcyclopropanes. The complete mechanism may be depicted as:

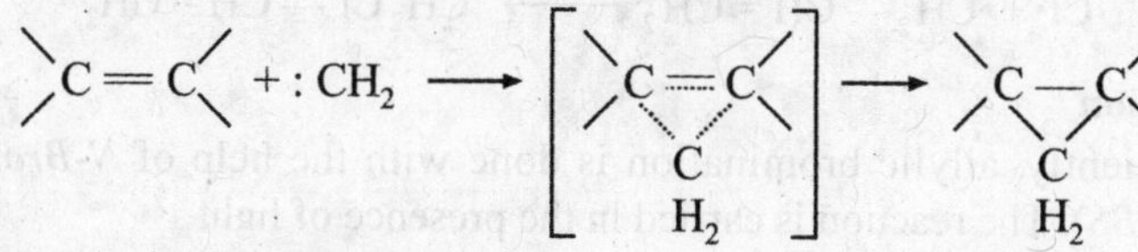

$$\text{>C=C<} + :CH_2 \longrightarrow \left[\text{>C=C<} \atop CH_2 \right] \longrightarrow \text{>C—C<} \atop CH_2$$

Mechanism of addition of triplet methylene

Triplet methylene reacts with alkenes in a two step free radical mechanism. Triplet methylene is a diradical and it attacks the alkene to form an intermediate diradical having the conformation A. This diradical has a life time long enough for rotation of groups to take place around central carbon-carbon bond. This gives rise to conformation B. Ring closures of A and B lead to the formation of both *cis* and *trans* products as shown below:

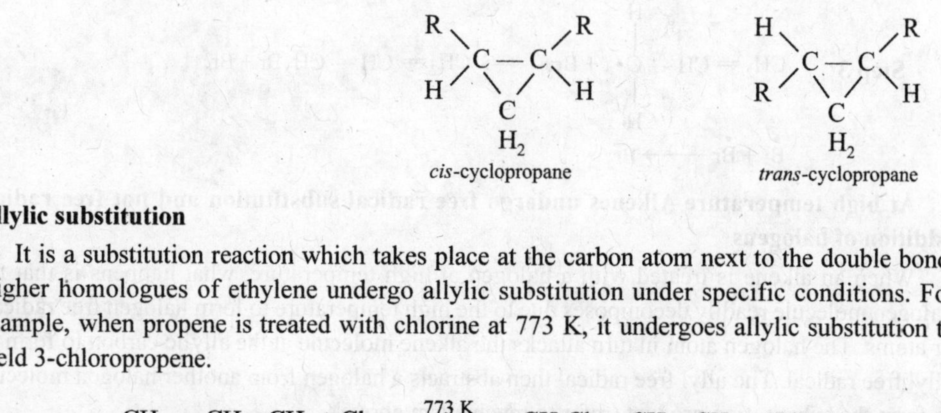

[Reaction scheme: *cis*-alkene + •CH_2 (Triplet Methylene) → intermediate A → *cis*-cyclopropane; A undergoes Rotation → B → *trans*-cyclopropane]

Allylic substitution

It is a substitution reaction which takes place at the carbon atom next to the double bond. Higher homologues of ethylene undergo allylic substitution under specific conditions. For example, when propene is treated with chlorine at 773 K, it undergoes allylic substitution to yield 3-chloropropene.

$$CH_3—CH=CH_2 + Cl_2 \xrightarrow{773\ K} \underset{\text{3-Chloropropene}}{CH_2Cl—CH=CH_2} + HCl$$

Mechanism of the reaction. Allylic substitution takes place by free radical chain mechanism as given below:

1. Chain initiating step

$$Cl—Cl \xrightarrow{773\ K} 2Cl\bullet$$

2. Chain propagating step

$$Cl\bullet + CH_3-CH=CH_2 \longrightarrow HCl + [\bullet CH_2-CH=CH_2 \longleftrightarrow CH_2=CH-\overset{\bullet}{C}H_2]$$
Allyl radical

$$Cl-Cl + \bullet CH_2-CH=CH_2 \longrightarrow CH_2Cl-CH=CH_2 + Cl\bullet$$

3. Possible chain terminating steps

$$Cl\bullet + Cl\bullet \longrightarrow Cl_2$$

$$Cl\bullet + \bullet CH_2-CH=CH_2 \longrightarrow CH_2Cl-CH=CH_2$$

Allylic bromination

More conveniently, allylic bromination is done with the help of *N-Bromosuccinimide* (abbreviated as *NBS*). The reaction is carried in the presence of light

$$CH_3CH=CH_2 + \underset{NBS}{\begin{matrix}CH_2-CO\\ |\\ CH_2-CO\end{matrix}}\!\!\!\!\!>\!\!NBr \xrightarrow{h\nu} \underset{Propene}{BrCH_2CH=CH_2} + \underset{Succinimide}{\begin{matrix}CH_2CO\\ |\\ CH_2CO\end{matrix}}\!\!\!\!\!>\!\!NH$$

Propene

Mechanism

Step 1

$$\begin{matrix}CH_2-CO\\ |\\ CH_2-CO\end{matrix}\!\!\!\!\!>\!\!N\frown Br \longrightarrow \begin{matrix}CH_2-CO\\ |\\ CH_2-CO\end{matrix}\!\!\!\!\!>\!\!\overset{\bullet}{N:} + \overset{\bullet}{Br}$$
Bromine radical

Step 2

$$CH_2=CH-\underset{H}{\overset{H}{\underset{|}{\overset{|}{C}}}}\!-\!\frown\!H+\overset{\bullet}{Br} \longrightarrow CH_2=CH-\underset{H}{\overset{H}{\underset{|}{\overset{|}{C}}}}\!\bullet + HBr$$

Step 3

$$CH_2=CH-\underset{H}{\overset{H}{\underset{|}{\overset{|}{C}}}}\!\bullet + Br_2 \longrightarrow CH_2=CH-CH_2Br + \overset{\bullet}{Br}$$

$$\overset{\bullet}{Br} + \overset{\bullet}{Br} \longrightarrow Br_2$$

At high temperature Alkenes undergo free radical substitution and not free radical addition of halogens

When an alkene is treated with a halogen at high temperature, what happens is that the halogen molecule readily decomposes due to the high temperature to form halogen free radicals or atoms. The halogen atom in turn attacks the alkene molecule at the allylic carbon to form an allyl free radical. The allyl free radical then abstracts a halogen from another halogen molecule to form the substitution product (refer to mechanism above).

The question is why the halogen atom does not attack the alkene at the double bond to form the free radical and then add another halogen atom to form an addition product. The answer is that halogen atom does attack the alkene even at the double bond to form a free radical. But due to very high temperature, this free radical readily eliminates the halogen atom before another halogen atom could be added to form the addition product.

Alkenes (Carbon-Carbon Pi Bonds)

That is:

$$Cl\bullet + CH_3-CH=CH_2 \underset{\text{Elimination of Cl}}{\overset{\text{Addition of Cl}\bullet}{\rightleftharpoons}} CH_3-\overset{\bullet}{C}H-CH_2Cl$$

Thus, the formation of addition product is prevented.

Benzylic bromination

Toluene on treatment with N-bromosuccinimide gives benzyl bromide. We call this benzylic bromination.

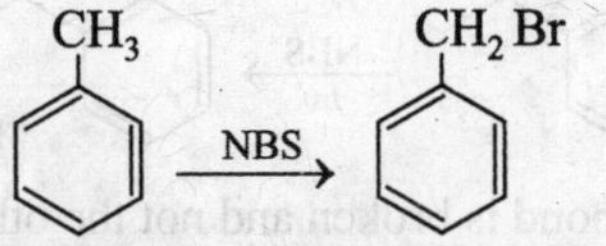

Toluene $\xrightarrow{\text{NBS}}$ Benzyl bromide

Benzylic C–H bonds have bond strength of about 90 kcal/mol which is less than normal primary C–H bond strength of 100 kcal/mol. That means it is quite easy to form the benzyl radical.

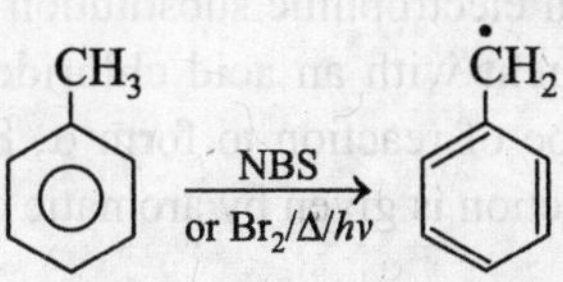

$$\text{PhCH}_3 \xrightarrow[\text{or Br}_2/\Delta/h\nu]{\text{NBS}} \text{Benzyl radical}$$

Benzyl radical is quite stable because of resonance structures

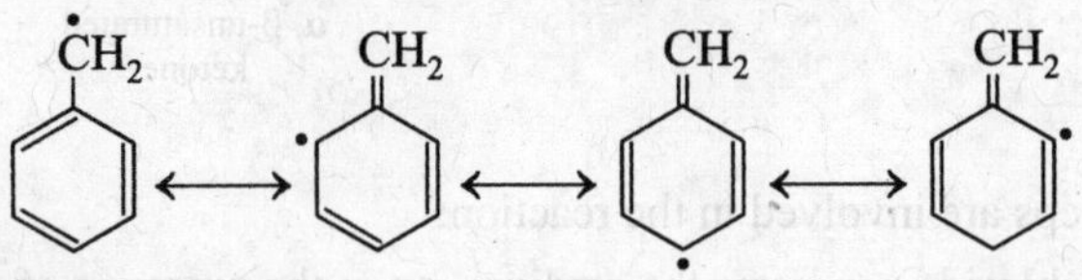

[resonance structures of benzyl radical]

Thus the ease of formation of benzyl bromide can be explained on the ease of formation and stability of benzyl radical

Mechanism of reaction

Step 1

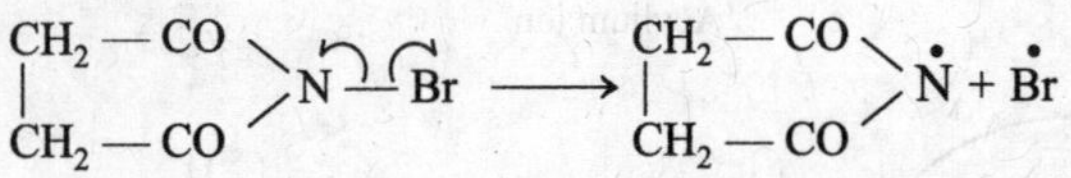

$$\begin{matrix} CH_2-CO \\ | \\ CH_2-CO \end{matrix}\!\!\!\!>\!\!N\!-\!Br \longrightarrow \begin{matrix} CH_2-CO \\ | \\ CH_2-CO \end{matrix}\!\!\!\!>\!\!\overset{\bullet}{N} + \overset{\bullet}{Br}$$

Step 2

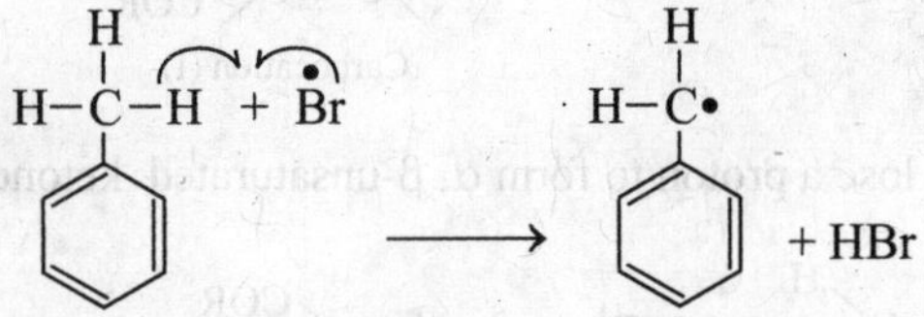

$$\text{PhCH}_2\text{-H} + \overset{\bullet}{Br} \longrightarrow \text{PhCH}_2\bullet + HBr$$

Step 3

$$\text{PhCH}_2\bullet + Br-Br \longrightarrow \text{PhCH}_2\text{-Br} + \overset{\bullet}{Br}$$

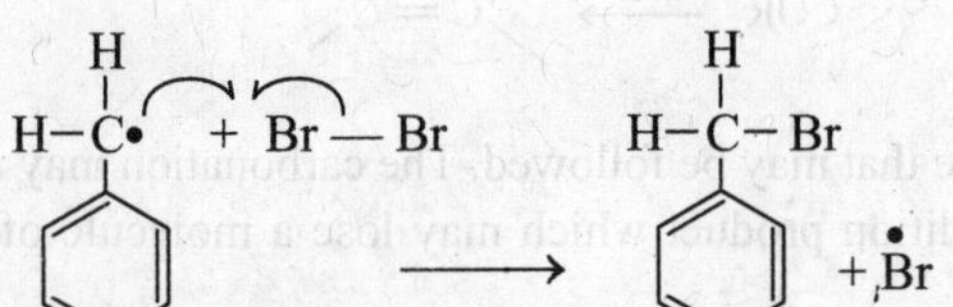

$$\overset{\bullet}{Br} + \overset{\bullet}{Br} \longrightarrow Br_2$$

For benzylic bromination, we can also take Br_2 instead of NBS because aromatic rings are not easily affected by Br_2. Let us compare NBS bromination of toluene and isopropylbenzene.

$$Ph-CH_3 \xrightarrow[h\nu]{NBS} Ph-CH_2Br$$

$$Ph-CH(CH_3)_2 \xrightarrow[h\nu]{NBS} Ph-CBr(CH_3)_2$$

Thus only the benzylic C–H bond is broken and not the other alkyl bonds by NBS.

Vinylic substitution

Alkenes undergo electrophilic addition reactions with hydrogen, halogens, halogen acids, etc, but they also take part in electrophilic substitution at the unsaturated carbon atom. For example, an alkene on treatment with an acid chloride in the presence of anhydrous $AlCl_3$ undergoes Friedel Crafts type of reaction to form α, β-unsaturated ketones. It maybe remembered that Friedel-Crafts reaction is given by aromatic compounds normally.

$$\underset{}{\overset{}{>}C=C\overset{H}{<}} + ROCl \xrightarrow{Anhy. AlCl_3} \underset{\substack{\alpha,\ \beta\text{-unsaturated} \\ \text{ketone}}}{\overset{}{>}C=C\overset{COR}{<_H}}$$

Mechanism

The following steps are involved in the reaction:

Step 1 The acid chloride generates the acylium ion in the presence of anhy. $AlCl_3$

$$R-\overset{\overset{O}{\|}}{C}-Cl + AlCl_3 \rightleftharpoons [R-\overset{+}{C}=\overset{..}{\overset{..}{O}}: \longleftrightarrow R-C\equiv \overset{+}{\overset{..}{O}}:] + AlCl_4^-$$

Acylium ion

$$>C=C\overset{H}{<} + O=\overset{+}{C}-R \longrightarrow \overset{+}{>C}-C\overset{H}{<_{COR}}$$

Carbocation (I)

Step 2 The carboration may lose a proton to form α, β-unsaturated ketone

$$\underset{I}{\overset{+}{>C}-C\overset{H}{<_{COR}}} \xrightarrow{-H^+} >C=C\overset{COR}{<}$$

There is an alternative course that may be followed. The carbonation may accept a chloride ion from $AlCl_4^-$ to form an addition product which may lose a molecule of HCl to form α, β-unsaturated ketone

Epoxidation reaction

When an alkene is treated with peroxybenzoic acid, an epoxide is formed. This reaction is called **epoxidation** reaction

$$\underset{\text{Alkene}}{>C=C<} + \underset{\text{Peroxybenzoic acid}}{C_6H_5-\overset{O}{\underset{\|}{C}}-O-OH} \longrightarrow \underset{\text{Epoxide}}{-\overset{|}{\underset{|}{C}}-\overset{|}{\underset{|}{C}}-} + \underset{\text{Benzoic acid}}{C_6H_5-\overset{O}{\underset{\|}{C}}-OH}$$

Mechanism of the reaction

$$C_6H_5\overset{O}{\underset{\|}{C}}-O-O-H \underset{\text{Slow}}{\rightleftharpoons} R-\overset{O}{\underset{\|}{C}}-O^- + \overset{+}{OH}$$
Electrophile

$$>C=C< + \overset{+}{OH} \longrightarrow >\overset{+}{C}-\overset{|}{C}- \longrightarrow >C-C< \xrightarrow{-H^+} \underset{\text{(Epoxide)}}{>C-C<}$$

Solved Examples

Example 9: What happens when:

(i) 2, 3-Dimethyl-2-pentene is subjected to ozonolysis.

(ii) when propylene is treated with sodium borohydride and boron trifluoride followed by the treatment of product thus formed with alkaline hydrogen peroxide.

Solution: (i) When 2, 3-dimethyl-2-pentene is subjected to ozonolysis, a mixture of acetone and ethyl methyl ketone is obtained.

$$\underset{\text{2, 3 Dimethyl-2-pentene}}{CH_3-CH_2-\underset{\underset{CH_2}{|}}{C}=\underset{\underset{CH_3}{|}}{C}-CH_3} + O_3 \longrightarrow CH_3-CH_2-\underset{\underset{O\;---\;O}{|}}{\overset{\overset{H_3C}{|}}{C}}\overset{O}{\diagdown}\underset{\underset{CH_3}{|}}{\overset{\overset{CH_3}{|}}{C}}-CH_3$$

$$\xrightarrow{Zn/H_2O} \underset{\text{Ethyl methyl ketone}}{CH_3-CH_2-\underset{\underset{CH_3}{|}}{C}=O} + \underset{\text{Acetone}}{CH_3-\underset{\underset{CH_3}{|}}{C}=O}$$

(*ii*) On treatment with sodium borohyride and boron tri-fluoride, propylene yields tripropyl borane which on treatment with alkaline hydrogen peroxide yields *n*-propyl alcohol.

$$3NaBH_4 + 4BF_3 \xrightarrow{273K} 2(BH_3)_2 + 3NaBF_4$$

$$2CH_3CH=CH_2 + (BH_3)_2 \longrightarrow 2CH_3-CH_2-CH_2BH_2$$

$$CH_3-CH_2-CH_2BH_2 \xrightarrow{CH_3CH=CH_2} (CH_3-CH_2-CH_2)_2BH$$

$$\xrightarrow{CH_3-CH=CH_2} \underset{\text{Tri } n\text{-propyl borane}}{(CH_3-CH_2-CH_2)_3B}$$

$$(CH_3-CH_2-CH_2)_3B + 3H_2O_2 \xrightarrow{NaOH} \underset{n\text{-propyl alcohol}}{3CH_3CH_2OH + H_3BO_3}$$

Example 10: Write equations for the reactions and name the product formed when isobutylene reacts with:

(*i*) Cl_2 and water
(*ii*) HBr
(*iii*) HBr in the presence of peroxide
(*iv*) H_2O in the presence of H^+
(*v*) HI in the presence of peroxide
(*vi*) H_2SO_4

Solution:

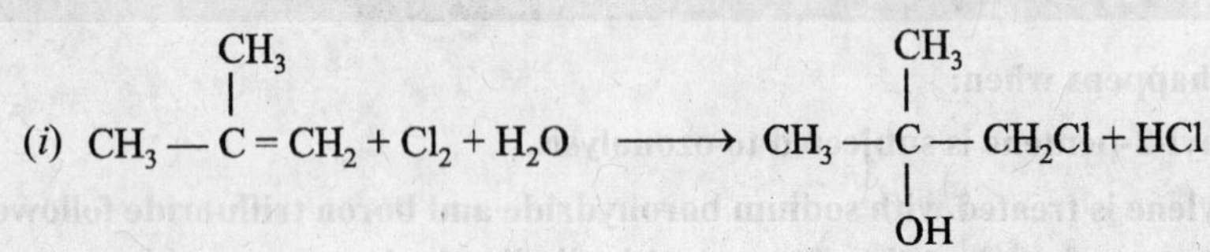

(*i*)
$$CH_3-\underset{\underset{}{|}}{\overset{\overset{CH_3}{|}}{C}}=CH_2 + Cl_2 + H_2O \longrightarrow CH_3-\underset{\underset{OH}{|}}{\overset{\overset{CH_3}{|}}{C}}-CH_2Cl + HCl$$
1-Chloro-2-methyl-2-propanol

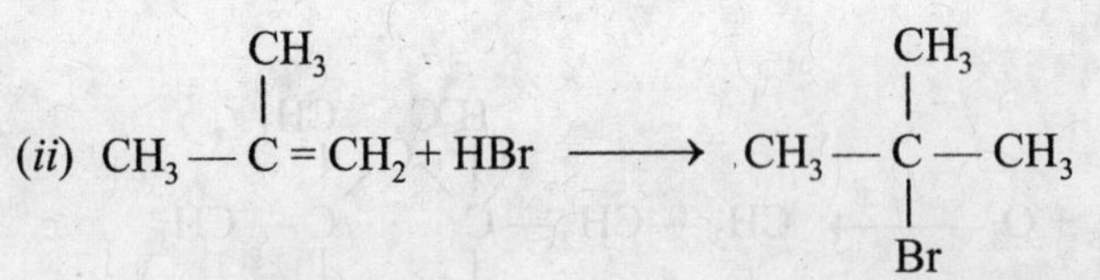

(*ii*)
$$CH_3-\overset{\overset{CH_3}{|}}{C}=CH_2 + HBr \longrightarrow CH_3-\underset{\underset{Br}{|}}{\overset{\overset{CH_3}{|}}{C}}-CH_3$$
2-Bromo-2-methylpropane

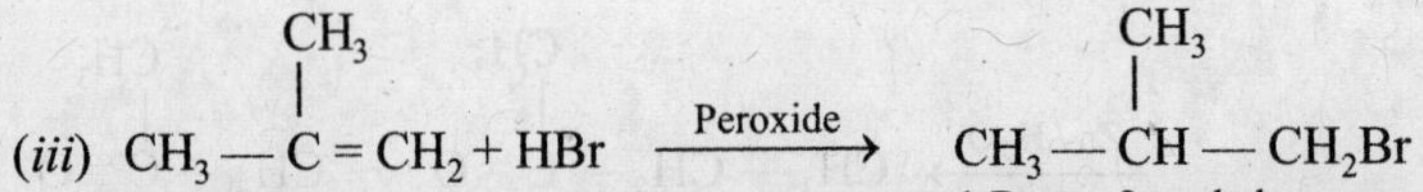

(*iii*)
$$CH_3-\overset{\overset{CH_3}{|}}{C}=CH_2 + HBr \xrightarrow{\text{Peroxide}} CH_3-\overset{\overset{CH_3}{|}}{CH}-CH_2Br$$
1-Bromo-2-methylpropane

Alkenes (Carbon-Carbon Pi Bonds)   233

(iv) $CH_3-\underset{\underset{CH_3}{|}}{C}=CH_2 + H_2O \xrightarrow{H^+} CH_3-\underset{\underset{OH}{|}}{\overset{\overset{CH_3}{|}}{C}}-CH_3$

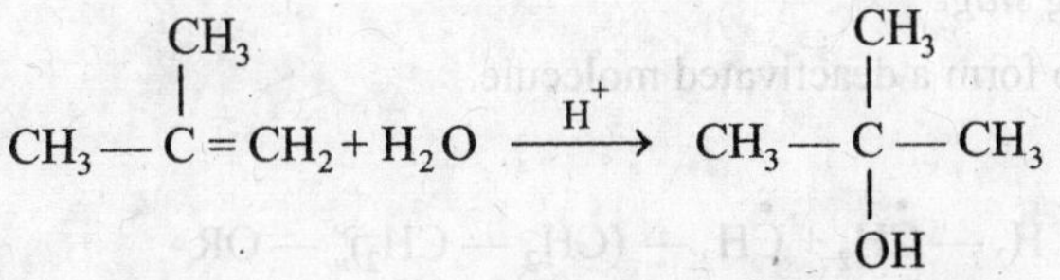

2-methyl-2-propanol

(v) $CH_3-\underset{\underset{CH_3}{|}}{C}=CH_2 + HI \xrightarrow{Peroxide} CH_3-\underset{\underset{I}{|}}{\overset{\overset{CH_3}{|}}{C}}-CH_3$

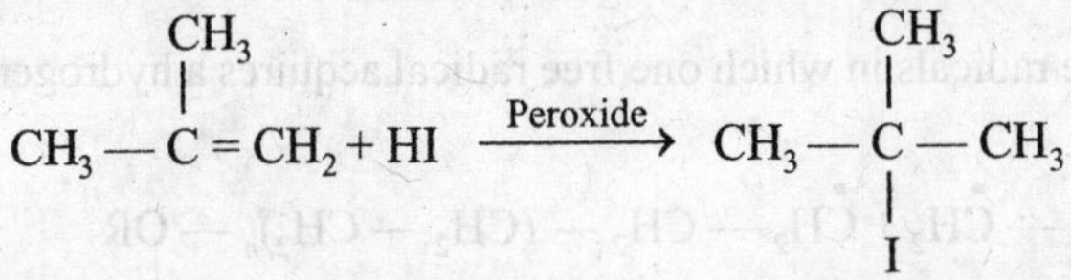

2-Iodo-2-methylpropane

(vi) $CH_3-\underset{\underset{CH_3}{|}}{C}=CH_2 + H_2SO_4 \longrightarrow CH_3-\underset{\underset{OSO_3H}{|}}{\overset{\overset{CH_3}{|}}{C}}-CH_3$

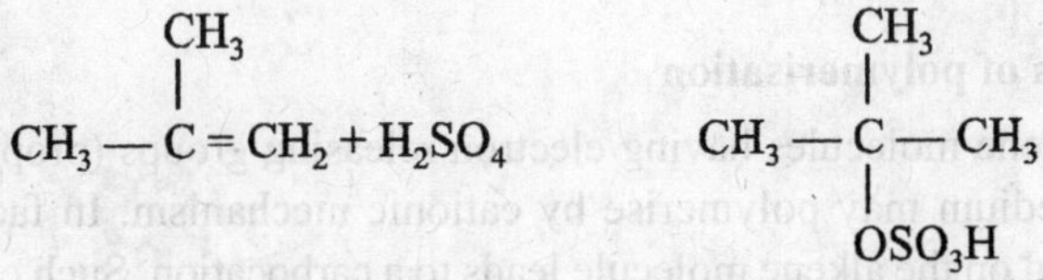

Tert. butyl hydrogen sulphate

5.7 POLYMERISATION

It is the process of formation of a large molecule from small molecules. For example, when heated at a high temperature and pressure in the presence of traces of oxygen, ethylene undergoes polymerisation to form polythene.

$$nCH_2=CH_2 \xrightarrow[\text{Traces of } O_2]{\text{High temp. and pressure}} (-CH_2-CH_2-)_n$$

Free radical chain mechanism of addition polymerisation

Many alkenes and substituted alkenes undergo polymerisation by free radical chain mechanism. Such reactions usually occur at high temperature under pressure or in the presence of catalysts such as peroxides and salts of peracids. The catalyst provides the free radical to initiate the chain process which is carried on by chain propagating steps. The mechanism is illustrated below with the help of polymerisation of ethylene in the presence of peroxide.

1. *Chain initiating state*

$$\underset{\text{Peroxide}}{R-O-O-R} \xrightarrow{h\nu} \underset{\text{Free radical}}{2\overset{\bullet}{R}O}$$

2. *Chain propagating stage*

$$\overset{\bullet}{R}O + CH_2=CH_2 \longrightarrow RO-CH_2-\overset{\bullet}{C}H_2$$

$$RO-CH_2-\overset{\bullet}{C}H_2 + CH_2=CH_2 \longrightarrow RO-CH_2-CH_2-CH_2-\overset{\bullet}{C}H_2$$

The free radical formed in each step adds to a fresh molecule of alkene and so the chain progresses till a large molecule of the polymer is obtained and the chain is terminated.

3. *Possible chain terminating stage*

(i) Coupling of free radicals to form a deactivated molecule.
For example:

$$RO-(CH_2-CH_2)_n-CH_2-\overset{\bullet}{C}H_2 + \overset{\bullet}{C}H_2-(CH_2-CH_2)_n-OR$$
$$\longrightarrow RO-(CH_2-CH_2)_n-CH_2-CH_2-CH_2-(CH_2-CH_2)_n-OR$$

(ii) Disproportionation of free radicals in which one free radical acquires a hydrogen from another and both get deactivated.

$$RO-(CH_2-CH_2)_n-CH_2-\overset{\bullet}{C}H_2 + \overset{\bullet}{C}H_2-CH_2-(CH_2-CH_2)_n-OR$$
$$\longrightarrow RO-(CH_2-CH_2)_n-CH=CH_2 + CH_3-CH_2-(CH_2-CH_2)_n-OR$$

Cationic and anionic mechanism of polymerisation

(a) Cationic mechanism. Alkene molecules having electron releasing groups (propylene, isobutylene etc.) in the acidic medium may polymerise by cationic mechanism. In fact, the electrophile attack of proton of acid on the alkene molecule leads to a carbocation. Such cations may further take part in the chemical combination leading to the formation of polyalkene. The cationic polymerisation is illustrated with the example of isobutyne.

$$\underset{\text{Isobutylene}}{CH_3-\underset{\underset{CH_3}{|}}{C}=CH_2} + H^{\oplus} \underset{\text{Acid}}{\longrightarrow} \underset{\text{Carbocation}}{CH_3-\underset{\underset{\oplus}{|}}{C}-CH_3}$$

$$CH_3-\underset{\underset{CH_3}{|}}{\overset{\overset{CH_3}{|}}{C\oplus}} + CH_2=\underset{\underset{CH_3}{|}}{\overset{\overset{CH_3}{|}}{C}}-CH_3 \longrightarrow CH_3-\underset{\underset{CH_3}{|}}{\overset{\overset{CH_3}{|}}{C}}-CH_2-\underset{\oplus}{\overset{\overset{CH_3}{|}}{C}}-CH_3$$

$$\xrightarrow{\text{Repetition}} CH_3-\underset{\underset{CH_3}{|}}{\overset{\overset{CH_3}{|}}{C}}-(-CH_2-\underset{\underset{CH_3}{|}}{\overset{\overset{CH_3}{|}}{C}}-)_n-CH_2-\underset{\oplus}{\overset{\overset{CH_3}{|}}{C}}-CH_3$$

$$\xrightarrow{-H^{\oplus}} CH_3-\underset{\underset{CH_3}{|}}{\overset{\overset{CH_3}{|}}{C}}-(-CH_2-\underset{\underset{CH_3}{|}}{\overset{\overset{CH_3}{|}}{C}}-)_n-CH=\overset{\overset{CH_3}{|}}{C}-CH_3$$
$$\text{Polyisobutylene}$$

(b) Anionic mechanism. The anionic mechanism is noticed in alkene molecules having some electron withdrawing groups present in them *e.g.* vinyl chloride, etc. The anionic polymerisation is carried in the presence of a suitable base such as sodamide ($NaNH_2$), *n*-butyl lithium (*n*-C_4H_9Li) etc.

$$B: + CH_2=CH-Cl \longrightarrow B-CH_2-\overset{\ominus}{CH}-Cl$$
$$\text{Base} \qquad\qquad\qquad\qquad \text{Carbanion}$$

$$B-CH_2-\overset{\ominus}{CH}-Cl + CH_2=CH-Cl \longrightarrow B-CH_2-CH(Cl)-CH_2-\overset{\ominus}{CH}-Cl$$

$$\downarrow \text{Repetition}$$

$$\text{Polyvinyl chloride (P.V.C.)}$$

Polymerisation in alkenes may take place by either by free radical or ionic mechanism. The free radical polymerisation takes place in the presence of an organic peroxide. Cationic polymerisation takes place in the presence of an acid while the anionic polymerisation proceeds in the presence of a base acting as a catalyst.

5.8 1, 2- AND 1, 4 – ADDITION TO CONJUGATED DIENES

In the case of isolated dienes (compounds containing two double bonds), addition of electrophiles takes place in a normal manner. But if we have a conjugated diene (alternate single and double bond), Then, apart from 1, 2-addition, 1, 4-addition also takes place. This can be illustrated as under:

$$CH_2=CH-CH=CH_2 + Br_2 \longrightarrow \underset{\text{1, 2-Dibromo product}}{CH_2(Br)-CH(Br)-CH=CH_2}$$

$$+ \underset{\text{1, 4-Dibromo product}}{CH_2(Br)-CH=CH-CH_2(Br)}$$

The expected result is 1, 2-addition product, but we also obtain a 1, 4-product where, the double bond shifts to a different position and the two bromine groups are linked at terminal positions. The percentage of the two products is temperature dependent. At 40°C, 85% of the product corresponds to 1, 4-addition while at 0°C, only about 30% is the 1, 4 product. Its mechanism can be explained as under:

$$Br_2 \longrightarrow Br^+ + Br^-$$

$$\overset{+}{Br} \curvearrowleft H_2\overset{1}{C}=\overset{2}{CH}-\overset{3}{CH}=\overset{4}{CH_2} \longrightarrow [Br\,H_2C-\overset{+}{CH}-CH=CH_2 \longleftrightarrow Br\,H_2C-CH=CH-\overset{+}{CH_2}]$$
$$\qquad\qquad\qquad\qquad\qquad\qquad\qquad\quad \text{I} \qquad\qquad\qquad\qquad\qquad \text{II}$$

The carbonium ion formed here gets stabilized by resonance.

If the carbocation I reacts with Br⁻ ion, we get 1, 2-dibromo product.

If the carbocation II reacts with Br⁻ in the final step, we get 1,4-dibromo product.

Diels-Alder Reaction

In this reaction, a conjugated diene bonds with an alkene to produce a cyclohexene molecule

1,3-Butadiene + Ethene →Δ Cyclohexene

The alkene that reacts with the diene is commonly referred to as dieneophile.

Diels Alder reactions are concerted, stereospecific and follow the endo rule. When we say that Diels Alder reaction is concerted, we mean it occurs in only one step and all of the atoms that are participating in the reaction form bonds simultaneously.

When we say the reaction is stereospecific, it means the substituents attached to the diene and the dieneophile retain their stereochemistry throughout the reaction. For example, if the functional groups on the dienophile are trans to each other, they will remain so in the product.

5.9 INDUSTRIAL APPLICATIONS OF ETHYLENE AND PROPENE

Ethylene

It is used in the manufacture of many industrial polymers viz. PVC, EPR, Sarans and polyethylene. These polymers are used in the manufacture of floor tiles, shoe soles, synthetic fibres, raincoats, purses, saran wraps and pipes

Propene

Propene is used in the manufacture of ethylene-propylene rubbers (EPR), acetone, isopropyl alcohol, n-propyl alcohol and allyl bromide

Solved Examples—Miscellaneous Examples

Example 11: How is it possible to know whether the given bexene is hexene-1, hexene-2 or hexene-3?

Solution: It is possible to distinguish between the three compounds by carrying out ozonolysis of the substance. An alkene forms an ozonide with ozone. The ozonide is subjected to hydrolysis when we obtain a mixture of carbonyl compounds. By analysing the carbonyl compounds, we can tell about the position of double bond in the molecule. The products obtained on ozonolysis of hexene-1 are as given below:

$$CH_3CH_2CH_2CH_2CH=CH_2 \xrightarrow{O_3} CH_3CH_2CH_2CH_2CH(O)(O-O)CH_2 \xrightarrow{Zn/\text{Acetic acid}}$$

HCHO + $CH_3CH_2CH_2CH_2CHO$
Methanal   Pentanal

Products obtained on ozonolysis of hexene-2 are as follows:

$$CH_3CH_2CH_2CH=CH_2 \xrightarrow{O_3} CH_3CH_2CH_2\underset{\underset{O-O}{|}}{\overset{\overset{O}{\diagup \diagdown}}{CH}}CH_2$$

$$\downarrow Zn / Acetic\ acid$$

$$CH_3CH_2CH_2CHO + CH_3CHO$$
$$\text{Butanal} \qquad \text{Ethanal}$$

Products obtained on ozonolysis of hexene-3 are as follows:

$$CH_3CH_2CH=CHCH_2CH_3 \xrightarrow{O_3} CH_3CH_2\underset{\underset{O-O}{|}}{\overset{\overset{O}{\diagup \diagdown}}{CH}}CHCH_2CH_3$$

$$\downarrow Zn / Acetic\ acid$$

$$CH_3CH_2CHO + CH_3CH_2CHO$$
$$\text{Propanal} \qquad \text{Propanal}$$

Thus by analysing the carbonyl compounds produced it is possible to tell whether the given hexene is hexene-1, hexene-2 or hexene-3.

Example 12: What happens when:

(*i*) 2, 2, 2-trimethyl-1-bromoethane reacts with alcoholic potassium hydroxide.

(*ii*) A mixture of ethylene (1 mole) and acetylene (1 mole) is treated with bromine (1 mole) in carbon tetrachloride solvent.

(*iii*) But-1-ene is treated with $CBrCl_3$ in presence of catalytic amount of benzoyl peroxide.

(*iv*) Propene-1 is heated with bromine followed by the treatment of product thus formed with sodium and methyl bromide.

Solution: (*i*) When 2, 2, 2-trimethyl-1-bromoethane reacts with alc. potassium hydroxide, 1, 2, 2-trimethyl ethene is obtained.

$$CH_3-\underset{\underset{CH_3}{|}}{\overset{\overset{CH_3}{|}}{C}}-CH_2-Br \xrightarrow[-Br]{alc.\ KOH} CH_3-\underset{\underset{CH_3}{|}}{\overset{\overset{CH_3}{|}}{C}}-\overset{\oplus}{C}H_2 \longrightarrow CH_3-\underset{\underset{CH_3}{|}}{\overset{\overset{CH_3}{|}}{C}}-\underset{\underset{CH_3}{|}}{\overset{\overset{CH_3}{|}}{C}}-CH_3$$
2, 2, 2-trimethyl-1-bromoethane

$$\downarrow$$

$$CH_3-\underset{\underset{CH_3}{|}}{\overset{\overset{CH_3}{|}}{C}}=\underset{\underset{CH_3}{|}}{\overset{\overset{H}{|}}{C}}-CH_3$$
1, 1, 2-trimethyl ethene

(*ii*) When a mixture of ethylene (1 mole) and acetylene (1 mole) is treated with bromine (1 mole) in carbon tetrachloride solvent, 1, 2-dibromoethene is obtained.

$$CH_2=CH_2 + CH\equiv CH \xrightarrow{Br_2 (1\,mole)} CHBr=CHBr + CH_2=CH_2$$
ethylene (1 mole)　　acetylene (1 mole)　　　　1, 2-dibromoethene

(*iii*) When But-1-ene is treated with $CBrCl_3$ in presence of benzoyl peroxide, 3-bromo-1, 1, 1-trichloropentane is formed.

$$CH_3CH_2CH=CH_2 + CBrCl_3 \xrightarrow{benzoyl\ peroxide} CH_3CH_2CH\underset{Br}{-}CH_2-CCl_3$$
but-1-ene　　　　　　　　　　　　　　3-bromo-1, 1, 1-trichloropentane

(*iv*) When propene-1 is treated with bromine, it yields 3-bromopropene which on treatment with sodium and methyl bromide yields butene-1.

$$CH_3CH=CH_2 \xrightarrow{Br_2} BrCH_2CH=CH_2 \xrightarrow[CH_3Br]{2Na} CH_3CH_2CH_2=CH_2$$
　　　　　　　　　　　　　　　　　　　　　　　butene-1

Example 13: How will you differentiate between 1-pentene and 2-pentene?

Solution: The products obtained on ozonolysis of pentene-1 are as given below:

$$CH_3CH_2CH_2CH=CH_2 \xrightarrow{O_3} CH_3CH_2CH_2CH\underset{O-O}{\overset{O}{\diagdown\!\!\diagup}}CH_2$$

$$\downarrow Zn/Acetic\ acid$$

$$HCHO + CH_3CH_2CH_2CHO$$
　Butanal　　　　Ethanal

Products obtained on ozonalysis of pentene-2 are as follows:

$$CH_3CH_2CH=CHCH_3 \xrightarrow{O_3} CH_3CH_2CH\underset{O-O}{\overset{O}{\diagdown\!\!\diagup}}CHCH_3$$

$$\downarrow Zn/Acetic\ acid$$

$$CH_3CH_2CHO + CH_3CHO$$
　Propanol　　　Ethanal

By identifying the reaction products, 1-pentane and 2-pentane can be differentiated.

Example 14: Compare with proper justification the stabilities of isobutene, Z-2-butene and E-2-butene.

Solution: In isobutene, two methyl groups are attached to same carbon atom due to which it experiences steric hinderance. Hence, it is least stable. In Z-2-butene, two methyl groups are attached to two different carbon atoms but on the same side of the double bond due to which it experiences steric hinderance but less than in isobutene. Hence, it is more stable than isobutene.

Alkenes (Carbon-Carbon Pi Bonds) 239

In E-2-butene, the two methyl groups are attached to two different carbon atoms on the opposite side of the double bond. Hence, it experiences least steric hinderance. So, it is stable to the maximum extent.

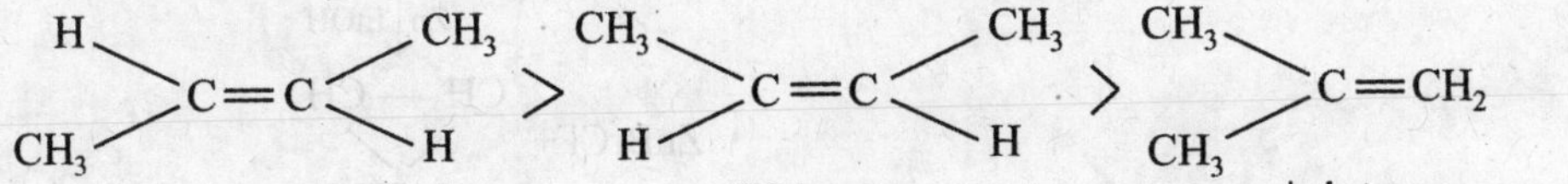

$$\underset{\text{Z-2-butene}}{\overset{H}{\underset{CH_3}{>}}C=C\overset{CH_3}{\underset{H}{<}}} > \underset{\text{E-2-butene}}{\overset{CH_3}{\underset{H}{>}}C=C\overset{CH_3}{\underset{H}{<}}} > \underset{\text{isobutene}}{\overset{CH_3}{\underset{CH_3}{>}}C=CH_2}$$

Example 15: Write the major product (products) in the following reaction:

$$CH_3-CH_2-\underset{\underset{CH_3}{|}}{\overset{\overset{CH}{|}}{C}}-CH_3 \xrightarrow{\text{alc. KOH}/\Delta}$$

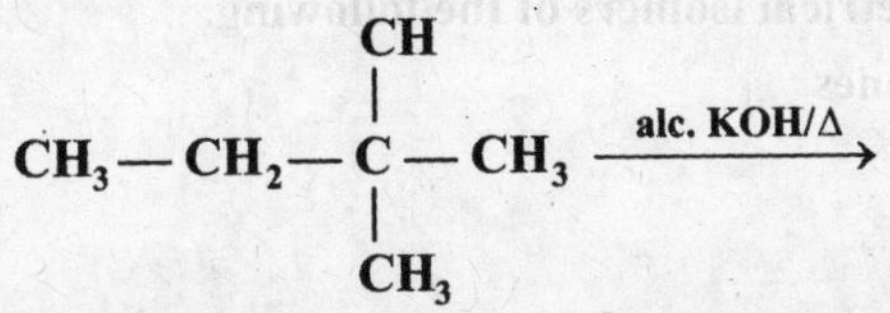

Solution: 2-methyl-butene-2 is the major product which is formed in the reaction as under:

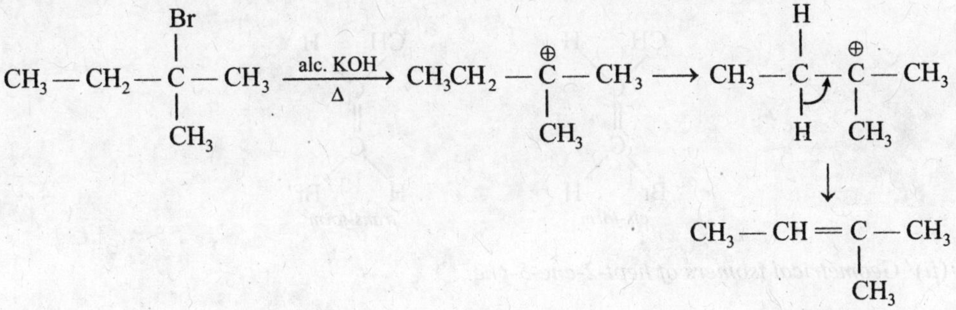

$$CH_3-CH_2-\underset{\underset{CH_3}{|}}{\overset{\overset{Br}{|}}{C}}-CH_3 \xrightarrow[\Delta]{\text{alc. KOH}} CH_3CH_2-\overset{\oplus}{\underset{\underset{CH_3}{|}}{C}}-CH_3 \longrightarrow CH_3-\underset{\underset{H}{|}}{\overset{\overset{H}{|}}{C}}-\overset{\oplus}{\underset{\underset{CH_3}{|}}{C}}-CH_3$$

$$\downarrow$$

$$CH_3-CH=\underset{\underset{CH_3}{|}}{C}-CH_3$$

2-methyl-butene-2
(major product)

Example 16: How will you bring about the following conversion?

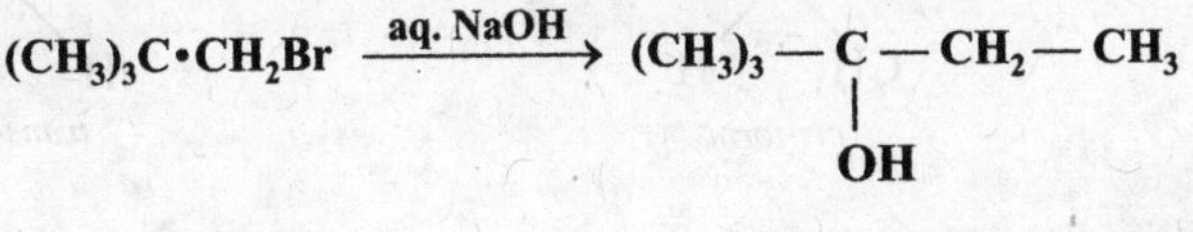

$$(CH_3)_3C\cdot CH_2Br \xrightarrow{\text{aq. NaOH}} (CH_3)_3-\underset{\underset{OH}{|}}{C}-CH_2-CH_3$$

Solution:

$$CH_3-\underset{\underset{CH_3}{|}}{\overset{\overset{CH_3}{|}}{C}}\underset{\underset{H}{|}}{\overset{\overset{H}{|}}{-C}}-Br \xrightarrow{\text{alc. NaOH}} CH_3-\underset{\underset{CH_3}{|}}{\overset{\overset{CH_3}{|}}{C}}-\overset{\oplus}{C}H_2 \longrightarrow CH_3-\underset{\oplus}{\overset{\overset{CH_3}{|}}{C}}-C-CH_3$$

$$\downarrow OH^-$$

$$(CH_3)_2C-CH_2-CH_3$$
$$\underset{OH}{|}$$

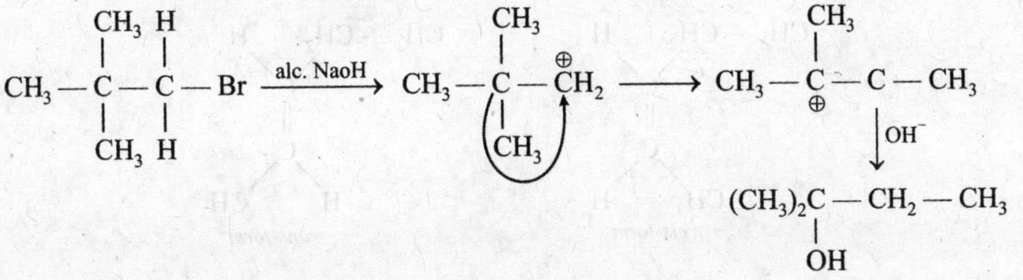

Example 17: Identify A and B in the following reaction:

$$CH_2=CH-CH_2-Cl + HBr \xrightarrow{\text{Peroxide}} A \xrightarrow{\frac{Zn}{Et}} B + ZnBrCl$$

Solution: The complete reaction is as under:

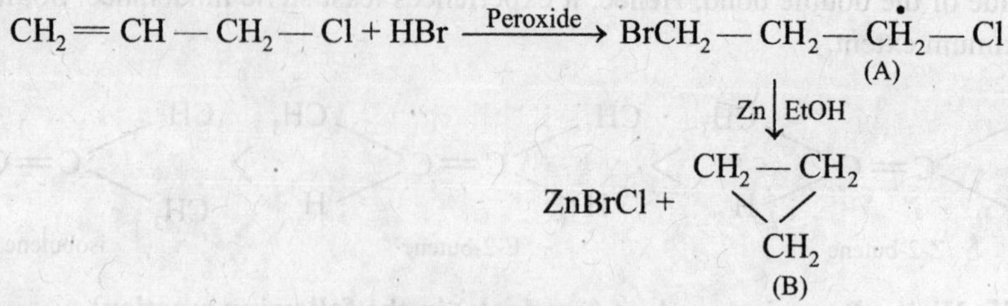

$$CH_2=CH-CH_2-Cl + HBr \xrightarrow{Peroxide} BrCH_2-CH_2-\overset{\bullet}{CH_2}-Cl$$
$$(A)$$

$$\downarrow Zn/EtOH$$

$$ZnBrCl + \underset{(B)}{\underset{CH_2}{\underset{\diagup\diagdown}{CH_2-CH_2}}}$$

Example 18: Draw geometrical isomers of the following:
  (*i*) Monobromopropenes
  (*ii*) Hept-2-en-5 yne
  (*iii*) Pentene-2

Solution:
  (*i*) Geometrical isomers of monobromopropene:

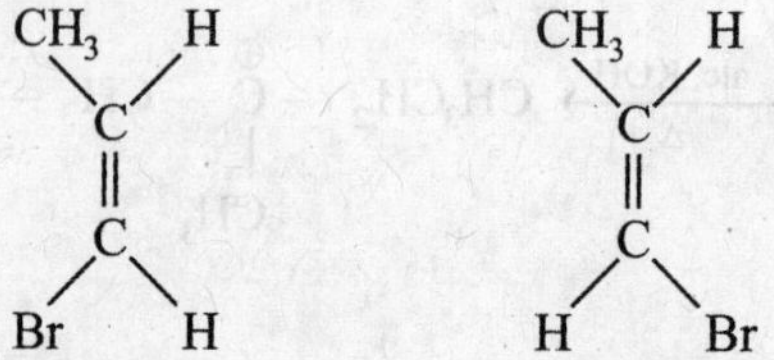

  cis-form   trans-form

  (*ii*) Geometrical isomers of hept-2-ene-5-yne:

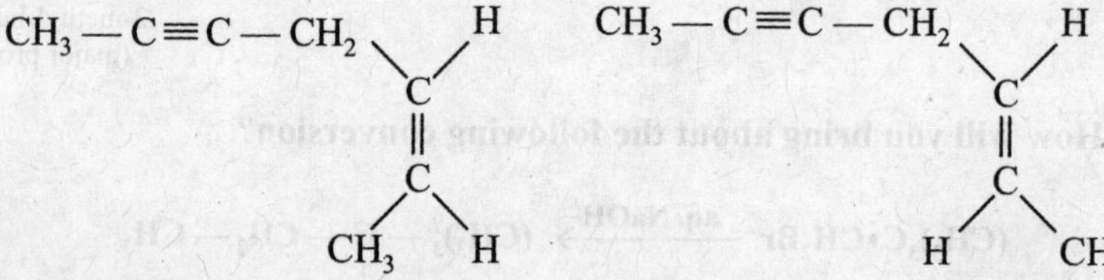

  cis-form   trans-form

  (*iii*) Geometrical isomers of pentene-2:

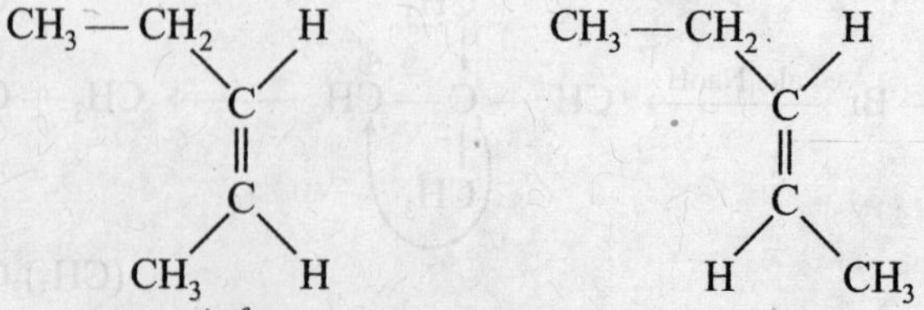

  cis-form   trans-form

Example 19: How will you distinguish between 1-hexene and *n*-hexane?

Solution: This can be done with the help of following tests.
  1. **Baeyer's test.** (reaction with cold dil. $KMnO_4$)

n-hexane does not react. 1-hexene reacts with Baeyer's reagent. Purple colour of $KMnO_4$ is discharged.

$$CH_2CH_2CH_2CH_2CH=CH_2 \xrightarrow[H_2O+O]{dil\ KMnO_4} CH_3CH_2CH_2CH_2\underset{OH}{CH}-\underset{OH}{CH_2}$$

2. **Reaction with Br_2 in CCl_4.**

   1-hexene will discharge the yellow colour of bromine

$$CH_3CH_2CH_2CH_2CH=CH_2 \xrightarrow[CCl_4]{Br_2\ in} CH_3CH_2CH_2CH_2CH\ BrCH_2Br$$
(Colourless)

   n-hexene will react only on heating to a high temperature.

Example 20: Dehydration of alcohols to alkenes is carried out with an acid but not with a base and the acid used is only H_2SO_4 and not HCl or HNO_3. Explain.

Solution: –OH group of the alcohol is a strong base and has low tendency to leave the molecule. To make it a better leaving group, a strong acid like H_2SO_4 is required. A base cannot help in the removal of –OH group. Therefore a base is not used for this purpose. Consider the reaction with H_2SO_4, HNO_3 and HCl. After the H^+ has been attached to the –OH of alcohol, the hydrogensulphate ion, nitrate ion or chloride ions are left in the solution as the case may be.

$H-O-\underset{\underset{O}{\|}}{\overset{\overset{O}{\|}}{S}}-O^-$ $O=\overset{+}{N}\underset{O^-}{\overset{O^-}{\diagup}}$ Cl^-

Hydrogen sulphate ion   Nitrate ion   Chloride

Out of the three anions, Cl^- is the best nucleophile. It will react with say $CH_3C^+H_2$ (in the dehydration of ethyl alcohol) to form CH_3CH_2Cl. We shall thus obtain the substitution product in place of elimination product. Therefore HCl is not used for dehydration of alcohols. NO_3^- and HSO_4^- are stabilised by resonance and they would not take part in substitution reaction. They would give the elimination product only. Out of HSO_4^- and NO_3^-, the former is better stabilized by resonance and hence H_2SO_4 is used for dehydration of alcohols to give alkene.

Example 21: How will you convert n-hexylamine into 1-hexene?

Solution: This conversion can be carried out using Hofmann elimination reaction.

$$\underset{n-\text{hexylamine}}{CH_3(CH_2)_3CH_2CH_2NH_2} \xrightarrow{CH_3I\ (excess)} CH_3(CH_2)_3\ CH_2CH_2-\overset{+}{N}(CH_3)_3I^- \xrightarrow[-AgI]{Ag_2O/H_2O}$$

$$CH_3(CH_2)_3CH_2CH_2-\overset{+}{N}(CH_3)_3\ \overline{OH} \xrightarrow[\substack{\text{Hofmann}\\ \text{elimination}}]{\Delta} \underset{1-\text{Hexene}}{CH_3(CH_2)_3\ CH=CH_2} + (CH_3)_3N + H_2O$$

Key Terms

- Oxymercuration-deoxymercuration
- Diels-Alder reaction
- Hydroboration-oxidation
- Saytzelf elimination

Evaluate Yourself

Multiple Choice Questions

1. Markovnikov's addition of HBr is not applicable to
   (a) Propene
   (b) 1-butene
   (c) I-pentene
   (d) 2-butene

2. Which of following compounds will react most readily with bromine in CCl_4?
   (a) $CH_3CH_2CH_3$
   (b) $(CH_3)_3CH$
   (c) $CH_3CH = CH_2$
   (d) $(CH_3)_4C$

3. Ethylene reacts with HI to give
   (a) Iodoethane
   (b) 2,2-Diiodoethane
   (c) 2,2-Diiodoethane
   (d) None of these

4. Which of the following compounds will have zero dipole moment?
   (a) *cis*-1,2-dibromoethylene
   (b) 1,1-dibromoethylene
   (c) *trans*-1,2-dibromoethylene
   (d) all of these

5. 2-Butene reacts with HBr to give
   (a) 1-Bromobutane
   (b) 2,3 Dibromobutane
   (c) 2-Bromobutane
   (d) 2,3-Dibromobutene

6. 2-Methylpropene reacts with HBr to give
   (a) *tert*-Butyl bromide
   (b) Isobutane
   (c) *n*-Butyl bromide
   (d) None of these

7. Which of the following alkenes reacts with HBr in the presence of a peroxide to give anti Markovnikov's product?
   (a) 1-Butene
   (b) 2,3-Dimethyl-2-butene
   (c) 2-Butene
   (d) 3-Hexene
8. In the addition of HX to a double bond, the hydrogen goes to the carbon that already has more hydrogens, is a statement of
   (a) Hund's rule
   (b) Markovnikov's rule
   (c) Huckel rule
   (d) Saytzeff rule
9. In the reaction of $CH_3CH_2CH = CH_2$ with HCl, the H of the HCl will become attached to which carbon?
   (a) C-1
   (b) C-2
   (c) C-3
   (d) C-4
10. The disappearance of the purple colour of $KMnO_4$ in its reaction with alkene is known as
    (a) Markovnikov test
    (b) Grignard test
    (c) Baeyer test
    (d) Wurtz test

Short Answer Questions

1. What is oxymercuration-demercuration? Does it give products corresponding to Markownikov addition or anti-Markownikov addition?
2. How does hot alkaline $KMnO_4$ react with following?
   (a) Ethene
   (b) 2-Methylpropene
3. Give the mechanism of addition of sulphuric acid to alkene.
4. What is Kharasch effect? Explain the mechanism of addition of HBr to propene.
5. Mention three physical properties of alkenes.

Long Answer Questions

1. Describe very briefly the following giving examples:
   (i) Markownikoff's rule
   (ii) Carbenes.
2. (a) Explain why chlorine forms a substituted product and not addition product upon heating with propene as 773 K.
   (b) Give the mechanism of free radical polymerisation of alkenes.

(c) Arrange the following in increasing order of their stability: 1-butene, *trans*-2-butene, *cis*-2-butene and isobutylene.

3. What is allylic substitution? Explain.
4. More highly alkylated alkenes are more stable. Explain why?
5. Explain the mechanism of hydroboration oxidation of propylene.
6. Discuss the mechanism of the following reactions:

   (i) $CH_3 - CH_2 - CH = CH_2 + HCl \longrightarrow$

   (ii) $\begin{array}{c} R \\ H \end{array} C = C \begin{array}{c} R \\ H \end{array} + CH_2 \longrightarrow$

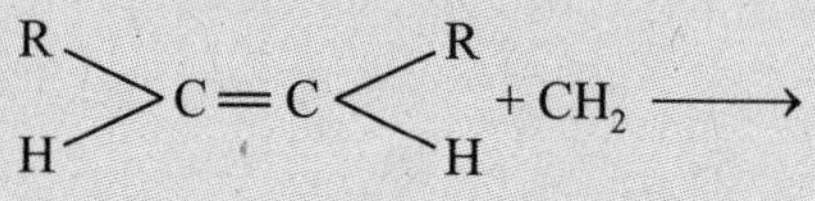

7. (a) Write short notes on Ozonolysis and hydroxylation of olefins.
   (b) Predict the main product and discuss the reaction mechanism.

   (i) $CH_3 - \underset{\underset{CH_3}{|}}{\overset{\overset{CH_3}{|}}{C}} = CH_2 - \underset{\underset{CH_3}{|}}{CH_2} \xrightarrow[\text{Heat}]{H_2SO_4}$

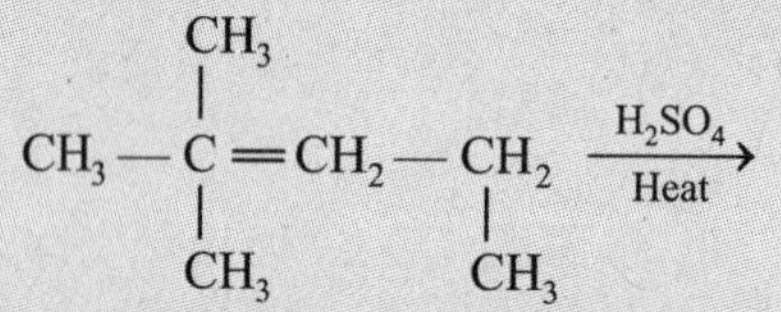

   (ii) $CH_3 - CH = CH_2 \xrightarrow{\text{Peroxide}}$

   (iii) Discuss the mechanism of addition of carbenes to alkenes.

8. (a) Give evidence in support of ionic mechanism of addition of halogens to alkenes. Illustrate by the addition of bromine to ethylene.
   (b) What is peroxide effect? Give its mechanism. Why is it only applicable to the addition of HBr and not to HCl and HI to unsymmetrical double bond?
   (c) Discuss the mechanism of free radical polymerisation of alkenes.
   (d) Explain why allyl free radical is more stable than *n*-propyl free radical.

9. Discuss the E_1 mechanism of dehydrohalogenation of alkyl halides.
10. Give a brief account of:
    (i) Markownikoff's rule
    (ii) Polymerisation
11. Discuss the mechanism of the following reactions:

    (i) $CH_3CH_2OH \xrightarrow[443 K]{H_2SO_4}$

    (ii) $CH_2 = CH_2 \xrightarrow{Br_2 + H_2O}$

    (iii) $CH_3 - \underset{\underset{CH_3}{|}}{\overset{\overset{CH_3}{|}}{C}} = CH_2 + CH_3 - \underset{\underset{CH_3}{|}}{\overset{\overset{CH_3}{|}}{C}} - H \xrightarrow{H_2SO_4}$

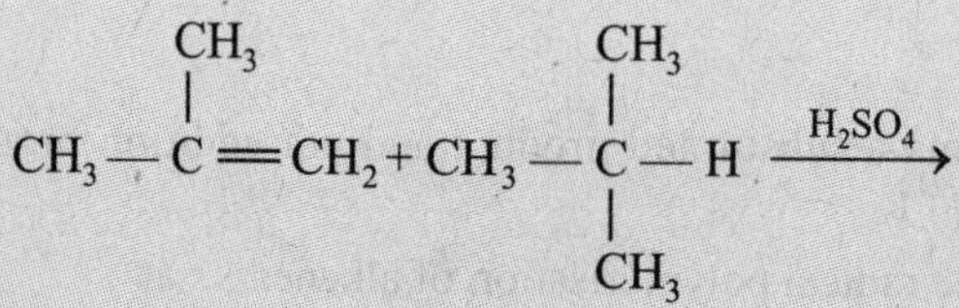

12. (a) How would you detect the presence of unsaturation in an organic compound.
    (b) Give the mechanism of the following reactions.
        (i) Hydroboration oxidation of alkenes
        (ii) Peroxide addition to alkenes
13. (a) Give the mechanism of dehydration of alcohols.
    (b) Explain that the addition of bromine to ethene is a two step process.
    (c) What is Markownikoff's rule? Discuss its mechanism?
14. Discuss the mechanism of hydroboration oxidation of alkenes.
15. (a) Discuss the mechanism of the following reactions:
        (i) Hydration of propene in acidic medium
        (ii) Electrophilic addition to alkenes
        (iii) Formation of halohydrins

Answers

Multiple Choice Questions

	1.	2.	3.	4.	5.	6.	7.	8.	9.	10.
(a)			■	■		■		■	■	
(b)								■		■
(c)		■		■	■					■
(d)	■									

Suggested Readings

1. Finar, I.L. Organic Chemistry (Vol. 1), Dorling Kindersley (India) Pvt. Ltd.
2. Morrison, R.N. & Boyd, R.N. Chemistry, Dorling Kindersley (India) Pvt. Ltd.

Chapter 6

Alkynes

LEARNING OBJECTIVES

After reading this chapter, you should be able to:
- learn the methods of preparation of alkynes
- understand the acidic nature of acetylenic hydrogen
- learn the electrophilic and nucleophilic addition reactions of alkynes
- understand hydration of alkynes to form carbonyl compounds
- learn the alkylation of terminal alkynes

6.1 NOMENCLATURE OF ALKYNES

Alkynes are unsaturated hydrocarbons having the general formula C_nH_{2n-2}. Such compounds are more unsaturated than alkenes. This is because alkynes contain a triple bond in a molecule. One triple bond is equivalent to two double bonds. First member of this series is acetylene C_2H_2 with two carbon atoms.

Nomenclature of some alkynes, according to common system and IUPAC system is given below:

Compound	Common Name	IUPAC Name
$CH \equiv CH$	Acetylene	Ethyne
$CH_3C \equiv CH$	Methyl acetylene	Propyne
$CH_3CH_2C \equiv CH$	Ethyl acetylene	But-1-yne
$CH_3C \equiv CCH_3$	Dimethyl acetylene	But-2-yne
$CH_3CH_2C \equiv CCH_3$	Ethyl methyl acetylene	Pent-2-yne

The following points may be borne in mind while naming alkynes:

1. The longest carbon chain containing the triple bond forms the parent chain and the alkyne is named accordingly.
2. The positions of the triple bond and other substituents (including double bond) are indicated by Arabic numerals.
3. Numbering of the parent chain is done such that the triple bond gets the lowest number. If the triple bond gets the same number from either side of the parent chain, then, the numbering is done in a manner that the substituents get the lowest numbers.
4. If both the double and triple bonds are present, then the compound is named as derivative of alkyne rather than alkene. The terminal 'e' from the suffix 'ene' is dropped while naming.
5. Numbering of the parent chain containing both double and triple bonds is done in such a manner that gives a lower number to the double or the triple bond i.e **lowest sum rule** for multiple bonds is followed. However, preference is given to the *double* bond over the *triple* bond. This is illustrated with the help of the following examples:

$$\overset{5}{H}C \equiv \overset{4}{C} - \overset{3}{C}H_2 - \overset{2}{C}H = \overset{1}{C}H_2$$
Pent-1-en-4-yne

$$\overset{6}{H}C \equiv \overset{5}{C} - \overset{4}{C}H = \overset{3}{C}H - \overset{2}{C}H = \overset{1}{C}H_2$$
Hexa-1, 3-dien-5-yne

Isomerism in Alkynes

Alkynes exhibit the following types of isomerism:

1. **Chain isomerism.** Alkynes having the same molecular formula but with different structures of the carbon chain are called chain isomers. For example, an alkyne with molecular formula C_5H_8 can exist in following two forms

$$\overset{5}{C}H_3\overset{4}{C}H_2\overset{3}{C}H_2\overset{2}{C} \equiv \overset{1}{C}H$$
Pent-1-yne

$$\overset{4}{C}H_3 - \overset{3}{\underset{\underset{CH_3}{|}}{C}H} - \overset{2}{C} \equiv \overset{1}{C}H$$
3-Methylbut-1-yne

2. **Position isomerism.** Alkynes with same carbon chain but having the triple bond in different positions are called position isomers. For example

$\overset{5}{C}H_3\overset{4}{C}H_2\overset{3}{C}H_2\overset{2}{C}\equiv\overset{1}{C}H$
Pent-1-yne

$\overset{1}{C}H_3\overset{2}{C}\equiv\overset{3}{C}\overset{4}{C}H_2\overset{5}{C}H_3$
Pent-2-yne

3. **Functional isomerism.** One triple bond is equivalent to two double bonds. The following compounds represent a case of functional isomerism.

$\overset{4}{C}H_3\overset{3}{C}H_2\overset{2}{C}\equiv\overset{1}{C}H$
But-1-yne

$\overset{1}{C}H_2=\overset{2}{C}H-\overset{3}{C}H=\overset{4}{C}H_2$
Buta-1, 3-diene

$\overset{1}{H_2}C=\overset{2}{C}=\overset{3}{C}H\cdot\overset{4}{C}H_3$
Buta-1, 2-diene

4. **Ring chain isomerism.** Alkynes and cycloalkenes could be isomers of each other.

$CH_3-C\equiv CH$ and △ $CH_3CH_2C\equiv CH$ and ☐
Propyne   Cyclopropene   But-1-yne   Cyclobutene

Solved Examples

Example 1: Draw the structures of all isomeric alkynes having the formula C_6H_{10} and give their IUPAC names.

Solution:

$CH_3-CH_2-CH_2-CH_2-C\equiv CH$
Hex-1-yne

$CH_3-CH_2-CH_2-C\equiv C-CH_3$
Hex-2-yne

$CH_3-CH_2-C\equiv C-CH_2-CH_3$
Hex-3-yne

$CH_3-CH-C\equiv C-CH_3$
     |
     CH_3
4-Methylpent-2-yne

$CH_3-CH_2-\underset{\underset{CH_3}{|}}{CH}-C\equiv CH$
3-methylpent-1-yne

$CH_3-\underset{\underset{CH_3}{|}}{CH}-CH_2-C\equiv CH$
4-methylpent-1-yne

Terminal and Non-terminal Alkynes

Alkynes can be classified into two types:

Terminal Alkynes

These alkynes contain the triple bond in the terminal position.

Non-terminal Alkynes

These alkynes contain the triple bond in the middle of the chain. It may be mentioned that the two types of alkynes show distinctly different behaviour towards a number of reagents which will be discussed in following sections.

Examples of terminal alkynes

$CH_3-C\equiv CH$ $CH_3-CH_2-C\equiv CH$ $CH_3CH_2CH_2C\equiv CH$
Propyne   But-1-yne   Pent-1-yne

Examples of non-terminal alkynes

$\overset{4}{C}H_3-\overset{3}{C}\equiv\overset{2}{C}-\overset{1}{C}H_3$
But-2-yne

$\overset{5}{C}H_3\overset{4}{C}H_2\overset{3}{C}\equiv\overset{2}{C}-\overset{1}{C}H_3$
Pent-2-yne

Structure of alkynes

Carbon-carbon triple bond is constituted of one sigma bond and two pi-bonds. The carbon atom forming the triple bond is *sp*-hybridised whch are linearly oriented in opposite directions making an angle of 180°. Therefore parts of the molecule that are linked to the carbon-carbon triple bond are linearly oriented.

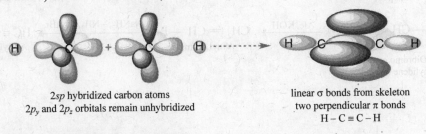

Sideways overlapping of unhybridised p-orbitals takes place forming two pi bonds. Due to small size of *sp* orbitals, the carbon-carbon bond distance in acetylene is 120 pm only as against the carbon-carbon single bond distance of 154 pm.

$C \equiv C$ has a bond strength of 823 kJ mol^{-1} in acetylene. It is stronger than $C = C$ of ethylene (610 kJ mol^{-1}) and $C - C$ of ethane (370 kJ mol^{-1})

2*sp* hybridized carbon atoms
$2p_y$ and $2p_z$ orbitals remain unhybridized

linear σ bonds from skeleton
two perpendicular π bonds
$H - C \equiv C - H$

6.2 PREPARATION OF ALKYNES

Methods of Preparation of Alkynes are described as under

(i) Dehydrohalogenation of vicinal dihalides

When 1, 2-dihaloalkane is heated with alcoholic potash, it undergoes dehydrohalogenation, yielding an alkyne. The reaction takes place in two stages involving the intermediate formation of vinyl halide (haloalkene).

The second stage of reaction generally requires a stronger base (sodium amide).

$$R-\underset{\underset{X}{|}}{\overset{\overset{H}{|}}{C}}-\underset{\underset{X}{|}}{CH}-R' \xrightarrow{KOH \text{ (alcoholic)}} R-CH=\underset{\underset{\text{Halo alkene}}{\text{(Vinyl halide)}}}{\overset{\overset{X}{|}}{C}}-R'$$

$$R-CH=\overset{\overset{X}{|}}{C}-R' \xrightarrow{NaNH_2} \underset{\text{Alkyne}}{RC \equiv CR'}$$

X = Cl, Br or I. R, R' may be H or alkyl group.

For example,

$$\overset{3}{CH_3}-\underset{\underset{Br}{|}}{\overset{\overset{H}{|}}{\underset{2}{C}}}-\underset{\underset{Br}{|}}{\overset{1}{CH}}-H \xrightarrow[-HBr]{KOH \text{ (Alc.)}} \underset{\text{1-Bromoprop-1-ene}}{\overset{3}{CH}-\overset{2}{CH}=\overset{1}{CHBr}} \xrightarrow[-HBr]{NaNH_2} \underset{\text{Methyl acetylene}}{CH_3-C \equiv CH}$$

1, 2-dibromopropane
(Propylene bromide)

From Gem dihalide. When 1, 1-dihaloalkane is heated with alcoholic potash or sodium amide, it undergoes dehydrohalogenation yielding an alkyne. The reaction proceeds in two stages involving the intermediate formation of haloalkene.

The sequence of the reactions is similar to that in dehydrohalogenation of *vic.*, dihalides.

$$R-CH_2-\underset{\underset{X}{|}}{\overset{\overset{R'}{|}}{C}}-X \xrightarrow{KOH\ (Alc.)} \underset{\text{1-Haloalkene (vinyl halide)}}{RCH=\overset{\overset{X}{|}}{C}-R'} \xrightarrow{NaNH_2,\ -NH_3,\ -NaX} \underset{\text{Alkyne}}{RC\equiv C-R'}$$

1, 1 Dihaloalkane
(gem dihalide)

For example,

$$\underset{\substack{\text{1, 1 Dibromo ethane}\\\text{or Ethylidene bromide}}}{CH_3-CH\diagup^{Br}_{\diagdown Br}} \xrightarrow[-HBr]{Alc.\ KOH} \underset{\text{Vinyl bromide}}{CH_2=CH-Br} \xrightarrow{NaNH_2,\ -NH_3,\ -NaBr} \underset{\text{Acetylene}}{HC\equiv CH}$$

(b)

$$CH_3-CH_2-\underset{\underset{Br}{|}}{\overset{\overset{H}{|}}{C}}-Br \xrightarrow[-HBr]{Alc.\ KOH} \underset{\text{1-Bromo prop-1-ene}}{CH_3-CH=CH-Br} \xrightarrow{NaNH_2,\ -NH_3,\ -NaBr} \underset{\text{Methyl acetylene}}{CH_3-C\equiv CH}$$

1, 1-Dibromopropane or
propylidene bromide

(ii) Dehalogenation of Tetrahalides

Dehalogenation of tetrahalides (1, 1, 2, 2-tetra haloalkanes) is carried out by passing their vapours over heated zinc and it results in the formation of alkyne.

$$R-\underset{\underset{X}{|}}{\overset{\overset{X}{|}}{C}}-\underset{\underset{X}{|}}{\overset{\overset{X}{|}}{C}}-R' \xrightarrow{Zn,\ -ZnX_2} \underset{X}{\overset{R}{\diagdown}}C=C\underset{R'}{\overset{X}{\diagup}} \xrightarrow{Zn,\ -ZnX_2} RC\equiv C-R'$$

$$H-\underset{\underset{Br}{|}}{\overset{\overset{Br}{|}}{C}}-\underset{\underset{Br}{|}}{\overset{\overset{Br}{|}}{C}}-H + 2\ Zn \longrightarrow 2\ ZnBr_2 + HC\equiv CH$$

1, 1, 2, 2-Tetrabromoethane

$$CH_3-\underset{\underset{Br}{|}}{\overset{\overset{Br}{|}}{C}}-\underset{\underset{Br}{|}}{\overset{\overset{Br}{|}}{C}}-H + 2\ Zn \longrightarrow 2\ ZnBr_2 + \underset{\substack{\text{Prop-1-yne}\\\text{Methyl acetylene}}}{CH_3-C\equiv CH}$$

1, 1, 2, 2-Tetrabromoethane

(iii) Action of Acetylides on Alkyl Halides (Alkylation of terminal alkynes)

Metallic acetylides yield higher alkynes by reacting with alkyl halides by S_N2 mechanism. It is a very good method of converting lower alkynes into higher alkynes.

Examples:

$$HC\equiv C^-Na^+ + CH_3-Br \xrightarrow{S_N2 \text{ reaction}} HC\equiv C-CH_3 + NaBr$$
Sod. acetylide / Bromoethane / Propyne

$$H-C\equiv C^-Na^+ + CH_3CH_2-I \xrightarrow{S_N2 \text{ reaction}} HC\equiv C-CH_2CH_3 + NaI$$
Sod. acetylide / Iodoethane / 1-Butyne

$$HC\equiv C^-Na^+ + CH_3CH_2CH_2-Br \xrightarrow{S_N2 \text{ reaction}} HC\equiv C-CH_2CH_2CH_3 + NaBr$$
Sod. acetylide / 1-Bromopropane / 1-Butyne

Acetylides used in the reaction are obtained from alkynes with terminal triple bond ($-C\equiv C-H$) by the action of sodium or sodium amide.

$$-C\equiv C-H + Na \longrightarrow -C\equiv C^-Na^+ + 1/2\,H_2$$

(iv) Hydrolysis of Calcium Carbides

Acetylene is prepared by the hydrolysis of calcium carbide. The latter is obtained by heating lime stone with coke at 2000°C in an electric furnace.

$$CaCO_3 + 3C \xrightarrow{2000°C} CaC_2 + CO_2 + CO$$

$$CaC_2 + 2H_2O \longrightarrow HC\equiv CH + Ca(OH)_2$$

(v) By dehalogenation of haloform

Chloroform and iodoform on heating with silver power undergo dehalogenation to form ethyne.

$$CH\,\underline{Cl_3 + 6Ag + Cl_3}\,CH \xrightarrow{\Delta} HC\equiv CH + 6AgCl$$
Chloroform / Ethyne

$$CH\,\underline{I_3 + 6Ag + I_3}\,CH \xrightarrow{\Delta} HC\equiv CH + 6AgI$$
Iodoform / Ethyne

(vi) Kolbe's electrolytic reaction

Acetylene can be prepared by electrolysis of a concentrated solution of sodium or potassium salt of maleic acid or fumaric acid. Thus,

$$\begin{matrix} H\diagdown\quad\diagup COOK \\ C \\ \| \\ C \\ H\diagup\quad\diagdown COOK \end{matrix} \text{ or } \begin{matrix} H\diagdown\quad\diagup COOK \\ C \\ \| \\ C \\ KOOC\diagup\quad\diagdown H \end{matrix} + 2H_2O \xrightarrow{\text{Electrolysis}} \begin{matrix} CH \\ \||| \\ CH \end{matrix} + 2CO_2 + H_2 + 2KOH$$

Pot. maleate (*cis*-isomer) / Pot. fumarate (*trans*-isomer) / Acetylene

Kolebe's electrolytic reaction is believed to occur by the following steps:

$$\underset{\text{Pot. maleate}}{\underset{\|}{\overset{\text{CHCOOK}}{\text{CHCOOK}}}} \xrightarrow{\text{Ionization}} \underset{\|}{\overset{\text{CHCOO}^-}{\text{CHCOO}^-}} + 2\,K^+$$

$$2H_2O \xrightleftharpoons{\text{Ionization}} 2\,OH^- + 2\,H^+$$

At anode:

$$\underset{\|}{\overset{\text{CHCOO}^-}{\text{CHCOO}^-}} \xrightarrow{-2e^-} \left[\underset{\|}{\overset{\text{CHCOO}}{\text{CHCOO}}}\right] \longrightarrow \underset{\text{Acetylene}}{\underset{\text{CH}}{\overset{\text{CH}}{\||\|}}} + 2\,CO_2$$

(unstable)

At cathode:

$$2H^+ + 2\,e^- \longrightarrow [2\,H] \longrightarrow H_2$$

(vii) Synthesis from carbon and hydrogen — Berthelot's synthesis

Acetylene can be prepared by passing a steam of hydrogen through an electric arc between carbon electrodes.

$$2C + H_2 \xrightarrow[3273\,K]{\text{Electric arc}} \underset{\text{Acetylene}}{HC \equiv CH}$$

Alkynes do not exhibit geometrical isomerism

The carbon atoms involved in the formation of triple covalent bond in alkynes are sp hybridised. As a result, alkynes have linear structure. Each triply bonded carbon atom is further linked to only one atom or group of atoms. Hence inspite of hindered rotations, the possibility of different relative spatial arrangement of the groups does not arise. Thus, alkynes do not exhibit geometrical isomerism. This type of isomerism is possible when there are at least two groups attached to one carbon.

$$\underset{Y}{\overset{X}{>}}C = C\underset{Y}{\overset{X}{<}} \qquad\qquad X-\!\!-C \equiv C-\!\!-X$$

Geometrical isomerism possible $\qquad\qquad$ (or Y) Geometrical isomerism not possible

Orbital structure of acetylene

In acetylene molecule both the carbon atoms are sp hybridised. One of the sp hybridised orbitals of carbon overlaps along internuclear axis with similar orbital of the othe carbon atom to form C — C, σ-bond. The second sp orbital of each carbon atom overlaps along internuclear axis with half filled $1s$ orbital of hydrogen atom to form C — H, σ-bond each. Since sp orbitals of each carbon atom lie along a straight line and overlapping of these orbitals takes place along their internuclear axis, all the four atoms of acetylene lie along the single straight line. Thus, acetylene is a linear molecule.

Each carbon atom is still left with two unhybridised p-orbitals which are perpendicular to each other as well as to the plane of carbon and hydrogen atoms. Each of these unhybridised orbitals of one carbon atom overlaps sideways with similar orbitals of the other carbon atom to form two π-bonds. There is overlapping between these two π electron clouds. As a result, the four lobes of two π-bonds merge to form a single electron cloud which is cylindrically symmetrical about the internuclear axis.

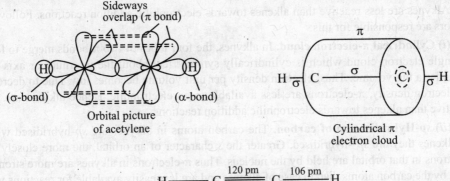

The carbon-carbon triple bond is thus made up of one strong σ-bond and two weak π-bonds.

Bond parameters. The various bond lengths and energies for the molecule of acetylene are given below:

(i) C C   bond length = 120 pm
(ii) C — H   bond length = 106 pm
(iii) H — C — C bond angle = 180°
(iv) C ≡ C   bond dissociation energy = 830 kJ mol^{-1}
(v) C — H   bond dissociation energy = 522 kJ mol^{-1}

6.3 PHYSICAL PROPERTIES OF ALKYNES

Physical properties of alkynes are given as under:

(*i*) **Physical state.** The first three members of this family (C_2 to C_4) are colourless gases, the next eight are liquids while the higer ones are solids.

(*ii*) **Smell.** All the alkynes are odourless. However, acetylene has garlic smell due to the presence of phosphine which is present as impurity.

(*iii*) **Melting and boiling points.** The melting points of alkynes are slightly higher than those of the corresponding alkenes and alkanes. Alkynes have linear structure and hence their molecules can be more closely packed in the crystal lattice as compared to those of corresponding alkenes and alkanes. That is why alkynes have higher melting points compared to alkenes and alkanes. Alkynes also show higher boiling points compared to corresponding alkenes and alkynes. This can be explained in terms of greater polarity in the molecules of alkynes compared to alkenes and alkynes.

(*iv*) **Density.** As we move to higher alkynes, the density increases with the increase in molecular mass. However, they are all lighter than water, with density much lower than one.

(*v*) **Solubility.** Alkynes, by and large, are non-polar compounds. Hence, they are insoluble in water but soluble in organic solvents.

6.4 CHEMICAL REACTIONS OF ALKYNES

Addition Reactions

Alkynes are less reactive than alkenes towards electrophilic addition reactions. Following factors are responsible for this:

(i) Cylindrical π-electron cloud. In alkynes, the four lobes of two π-bonds merge to form a single electron cloud which is cylindrically symmetrical about the internuclear axis and occupies a big volume. Thus, electron density per unit volume becomes low. Due to decrease in electron density, π-electrons are less available to an electrophile. Hence alkynes are less reactive than alkenes towards electrophilic addition reactions.

(ii) sp-Hybridisation of carbon. The carbon atoms in alkynes are sp-hybridised while in alkenes they are sp^2-hybridised. Greater the s character of an orbital, the more closely the electrons in that orbital are held by the nucleus. Thus π-electrons in alkynes are more strongly held by the carbon atoms than in case of alkenes and are less easily available for reactions with electrophiles. Thus makes alkynes less reactive than alkenes in electrophilic reactions

Bromine readily adds to propyne first forming *trans*-1, 2-dibromopropane and then 1, 1, 2, 2-tetrabromopropane.

$$CH_3—C\equiv C—H \xrightarrow{Br_2/CCl_4} \underset{\text{trans-1, 2-Dibromopropane}}{\begin{array}{c}Br\\CH_2\end{array}>C=C<\begin{array}{c}H\\Br\end{array}} \xrightarrow{Br_2/CCl_4} \underset{\text{1, 1, 2, 2-Tetrabromopropane}}{CH_2—\underset{Br}{\overset{Br}{C}}—\underset{Br}{\overset{Br}{C}}—H}$$

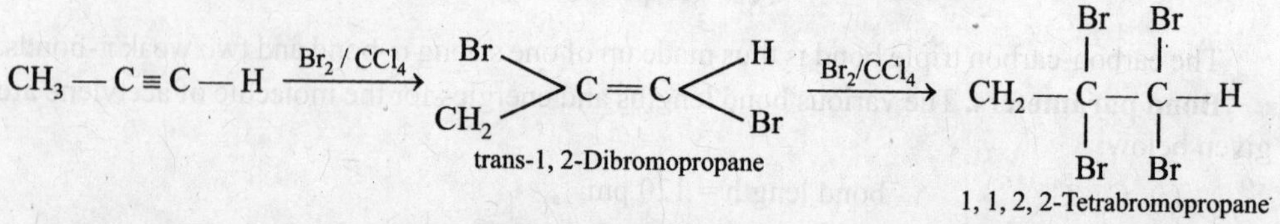

Mechanism. The mechanism of the reaction involves electrophilic addition. It takes place in two steps. This is known as halonium ion mechanism of addition. Bromine (or any halogen) gets polarised under the influence of π-electrons. Bromonium ion (Br$^+$) adds first forming a bridge bond, followed by the attachment of bromide ion.

First step

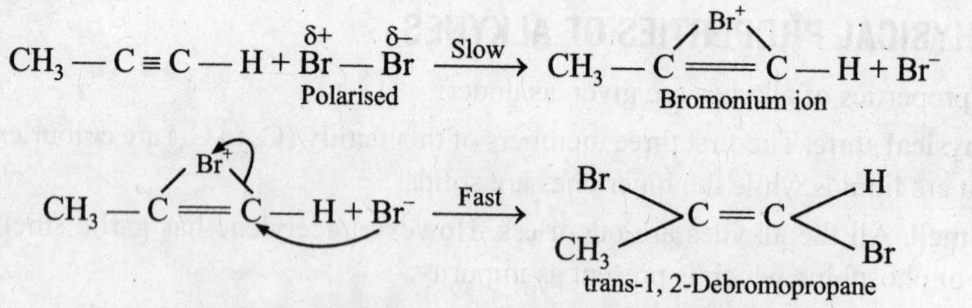

$$CH_3—C\equiv C—H + \overset{\delta+}{Br}—\overset{\delta-}{Br} \xrightarrow{\text{Slow}} \underset{\text{Bromonium ion}}{CH_3—C\overset{\overset{Br^+}{\diagup\diagdown}}{=\!=\!=}C—H} + Br^-$$

$$CH_3—\overset{\overset{Br^+}{\diagup\diagdown}}{C=\!=\!=C}—H + Br^- \xrightarrow{\text{Fast}} \underset{\text{trans-1, 2-Debromopropane}}{\begin{array}{c}Br\\CH_3\end{array}>C=C<\begin{array}{c}H\\Br\end{array}}$$

This sequence is repeated in the addition of another bromine molecule.

Second step

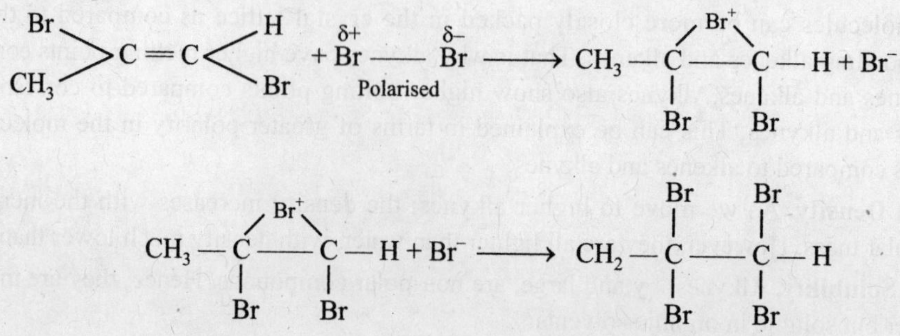

$$\begin{array}{c}Br\\CH_3\end{array}>C=C<\begin{array}{c}H\\Br\end{array} + \overset{\delta+}{Br}\text{———}\overset{\delta-}{Br} \longrightarrow CH_3—\underset{Br}{\overset{\overset{Br^+}{\diagup\diagdown}}{C\text{———}C}}—H + Br^-$$

$$CH_3—\underset{Br}{\overset{\overset{Br^+}{\diagup\diagdown}}{C\text{———}C}}—H + Br^- \longrightarrow \underset{\text{1, 1, 2, 2-Tetrabromopropane}}{CH_2—\underset{Br}{\overset{Br}{C}}—\underset{Br}{\overset{Br}{C}}—H}$$

Acidic nature of acetylenic protons

In acetylene and terminal alkynes, the hydrogen atom is attached to a *sp* hybridised carbon which is more electronegative because of increase in *s*-character. Due to greater electronegativity of the *sp* hybridised carbon, the electron pair of C — H bond gets displaced more towards carbon and this helps in the release of proton by strong bases. Consequently acetylene and terminal alkynes behave as acids.

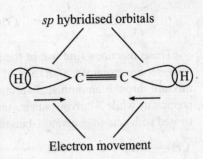

Following reactions illustrate the acidic nature of alkynes:

(i) Formation of alkali metal acetylides. Terminal alkynes react with sodium metal or sodamide in liquid ammonia to form sodium acetylides with the evolution of H_2 gas.

$$2HC \equiv C-H + 2Na \xrightarrow{475\ K} 2HC \equiv \overset{-}{C}\overset{+}{Na} + H_2$$
$$\text{Acetylene} \qquad\qquad\qquad \text{Sodium acetylide}$$

$$CH_3-C \equiv C-H + NaNH_2 \xrightarrow{196\ K} CH_3-C \equiv \overset{-}{C}\overset{+}{Na} + NH_3$$
$$\text{Propyne} \qquad\qquad\qquad\qquad \text{Sodium methyl acetylide}$$

Sodium acetylides is decomposed by water to regenerate acetylene. This reaction proves that H_2O is a stronger acid than acetylene because water displaces acetylene from sodium acetylide.

$$HC \equiv C^-Na^+ + H_2O \longrightarrow HC \equiv CH + NaOH$$
$$\text{Sod. acetylide} \qquad\qquad \text{Acetylene}$$

(ii) Formation of heavy metal acetylides. Due to acidic nature acetylenic hydrogens of alkynes can also be replaced by heavy metal ions such as Ag^+ and Cu^+ ions. For example, when treated with ammoniacal silver nitrate solution (*Tollen's reagent*), alkynes form a white precipitate of silver acetylides.

$$HC \equiv CH + 2[Ag(NH_3)_2]^+OH^- \longrightarrow AgC \equiv CAg + 2H_2O + 4NH_3$$
$$\text{Ethyne} \qquad \text{Tollen's reagent} \qquad\qquad \text{Silver acetylide}$$
$$\qquad\qquad\qquad\qquad\qquad\qquad\qquad (White\ ppt.)$$

$$R-C \equiv CH + [Ag(NH_3)_2]^+OH^- \longrightarrow R-C \equiv C-Ag + H_2O + 2NH_3$$
$$(Terminal\ alkyne) \quad \text{Tollen's reagent} \qquad\qquad \text{Silver alkynide}$$
$$\qquad\qquad\qquad\qquad\qquad\qquad\qquad (White\ ppt.)$$

Similarly, terminal alkynes react wtih ammoniacal cuprous chloride solution (Fehling solution) to form red ppt. of copper acetylide.

$$HC \equiv CH + 2[Cu(NH_3)_2]^+OH^- \longrightarrow CuC \equiv CCu + 2H_2O + 4NH_3$$
$$\text{Ethyne} \qquad\qquad\qquad\qquad\qquad\qquad \text{Copper acetylide}$$
$$\qquad\qquad\qquad\qquad\qquad\qquad\qquad (Red\ ppt.)$$

$$R-C \equiv CH + [Cu(NH_3)_2]^+OH^- \longrightarrow R-C \equiv C-Cu + H_2O + 2NH_3$$
$$(Terminal\ alkyne) \qquad\qquad\qquad\qquad \text{Copper acetylide}$$
$$\qquad\qquad\qquad\qquad\qquad\qquad\qquad (Red\ ppt.)$$

Silver and copper acetylides are not decomposed by water. They can, however, be decomposed with dilute mineral acids to regenerate the original alkynes as shown below:

$$AgC \equiv CAg + 2HNO_3 \longrightarrow HC \equiv CH + 2AgNO_3$$
$$\text{Silver acetylide} \qquad\qquad\qquad \text{Acetylene}$$

$$CuC \equiv CCu + 2\,HCl \longrightarrow HC \equiv CH + 2\,CuCl$$
Copper acetylide → Acetylene

These reactions find use in the separation of terminal and non-terminal alkynes from a mixture. For example, a mixture of 1-butyne and 2-butyne can be separated by passing the mixture through ammoniacal solution of silver nitrate. 1-Butyne forms an insoluble white precipitate while 2-butyne remains unaffected. The silver butynide thus formed is separated and treated with acid to generate 1-butyne.

$$CH_3-CH_2-C\equiv C-H \xrightarrow{Ag^+} CH_3CH_2-C\equiv C-Ag \xrightarrow{H^+} CH_3CH_2-C\equiv CH$$
1-Butyne — Silver butynide — 1-Butyne

$$CH_3-C\equiv C-CH_3 \xrightarrow{Ag^+} \text{No reaction}$$

Similarly, ammoniacal solution of cuprous chloride (or Fehling solution) can be used for the above separation.

(iii) Formation of alkynyl Grignard reagents. Acetylene and other terminal alkynes react with Grignard reagents to form the corresponding alkynyl Grignard reagents alongwith the formation of alkanes. For example,

$$HC \equiv CH + RMgX \xrightarrow[\text{ether}]{\text{Dry}} HC \equiv CMgX + RH$$
Acetylene — Grignard reagent — Acetylenic Grignard reagent — Alkane

$$R'-C\equiv CH + RMgX \xrightarrow[\text{ether}]{\text{Dry}} R'-C\equiv CMgX + RH$$
Terminal alkyne — Alkynyl Grignard reagent

Alkynl Grignard reagents are extensively used to prepare a variety of organic compounds. Metal acetylides can be used:

(i) For separation and purification of terminal alkynes from alkanes and alkenes.

(ii) To distinguish between terminal alkynes and non-terminal alkynes or alkenes.

Comparison of acidic strength of acetylene, ethylene and ethane

Strength of an acid depends upon the ease with which it can lose a proton. This will depend upon the type of bonding between carbon and hydrogen.

sp^3-s overlapping (Ethane: $H-CH_2-CH_2-H$)

sp^2-s overlapping (Ethylene: $H-CH=CH-H$)

$sp-s$ overlapping ($H-C\equiv C-H$)

(i) In the case of ethane, C—H bond involves overlapping of sp^3 hybrid orbital of carbon and s-orbital of H.

(ii) In case of ethylene, overlapping is between sp^2 hybrid orbital of carbon and s-orbital of H.

(iii) In case of acetylene, overlapping in C—H bond is between sp hybrid orbital of carbon and s-orbital of hydrogen, s-orbital has greater electron density compared to a p-orbital, which can be explained in terms of their shape. Out of sp^3, sp^2 and sp hybrid orbitals, the sp orbital has the greatest s-character followed by sp^2 and sp^3. Thus carbon atom possessing sp hybrid orbital will have maximum electronegativity followed by carbon having sp^2 hybrid orbitals and then followed by carbon having sp^3 hybrid orbital.

In view of this, the hydrogen atom in acetylene will be replaceable with maximum ease followed by that in ethylene and ethane. Hence acidic strength decreases in the order.

$$\text{acetylene} > \text{ethylene} > \text{ethane}$$

Addition of halogen acids on alkynes

Treatment of alkynes with halogen acid initially yields vinyl halides and finally alkylidene halides. Combination with second molecule takes place in accordance with Markownikoff's rule.

$$H-C\equiv C-H + HX \longrightarrow \underset{\text{Vinyl halide}}{H_2C=CHX} \xrightarrow[\text{(2nd molecule)}]{HX} \underset{\substack{\text{Ethylidene halide}\\ \text{(1, 1-Dihalo ethane)}}}{H_3C-CHX_2}$$

X = Cl, Br or I

In case of propyne (unsymmetrical alkyne), the first molecule of HX adds according to Markownikoff's rule.

For example,

$$CH_3-C\equiv CH + H-X \longrightarrow \underset{\text{2-Halo propene-1}}{CH_3-CX=CH_2} \xrightarrow{HX} \underset{\text{(2, 2-Dihalo propane)}}{CH_2-CX_2-CH_3}$$

Order of reactivity of halogen acids is HI > HBr > HCl.

Mechanism of the reaction (Electrophilic addition)

The addition of halogen acids to alkynes occurs through simple carbocations and not through cyclic cations.

Step 1. $\underset{\text{Ethyne}}{HC\equiv CH} + H^+ \xrightarrow[\text{Step}]{\text{Slow}} \underset{\text{Vinyl carbocation}}{CH_2=\overset{+}{CH}} \xrightarrow[\text{Fast Step}]{Cl^-} \underset{\text{1-Chloroethene}}{CH_2=CH-Cl}$

Step 2. $CH_2=CHCl + H^+ \xrightarrow[\text{Step}]{\text{Slow}} \begin{cases} \underset{\substack{\text{2° Carbocation I}\\ \text{(more stable)}}}{CH_3-\overset{+}{CH}-Cl} \\ \underset{\substack{\text{1° Carbocation II}\\ \text{(less stable)}}}{\overset{+}{CH_2}-CH_2Cl} \end{cases}$

Carbocation (I) being 2° is more stable than primary carbocation (II). Therefore, reaction occurs through carbocation (I) forming 1, 1-dichloroethane.

$$CH_3-\overset{+}{CH}-Cl + Cl^- \xrightarrow[\text{Step}]{\text{Fast}} \underset{\text{1, 1-Dichloroethane}}{CH_3-CHCl_2}$$

Free radical addition of hydrogen bromide

Addition of hydrogen bromide only in the presence of peroxides takes place contrary to Markownikoff's rule. This is called **peroxide effect.** For example,

$$CH_3C \equiv CH + HBr \xrightarrow{\text{Peroxide}} CH_3CH=CHBr \xrightarrow[\text{Peroxide}]{HBr} CH_3CH_2CHBr_2$$

Propyne   1-Bromopropene   1,1-Dibromopropane

Addition of hypohalous acids

Alynes add two molecules of hypohalous acids such as HOCl and HOBr to form dihalocarbonyl compounds. For example,

$$HC \equiv CH + Cl-OH \longrightarrow [Cl-CH=CH-OH] \xrightarrow{Cl-OH}$$

Ethyne   Hypochlorous acid   2-Chloroeth-1-en-1-ol

$$\begin{bmatrix} Cl-CH-CH-OH \\ | \quad\quad | \\ Cl \quad\quad Cl \end{bmatrix} \xrightarrow{-H_2O} \underset{Cl}{\overset{Cl}{>}}CH-CH=O$$

(Unstable)   2,2-Dichloroethanal

The addition takes place according to the Markownikoff's rule. It may be mentioned that oxygen being more electronegative than Cl, Br or I, the OH part of Cl–OH is partially negative.

$$CH_3-C \equiv CH \xrightarrow{\overset{\delta+}{Cl}-\overset{\delta-}{OH}} \begin{bmatrix} CH_3-C=CH \\ | \quad\quad | \\ OH \quad Cl \end{bmatrix} \xrightarrow{\overset{\delta+}{Cl}-\overset{\delta-}{OH}}$$

Propyne

$$CH_3-\underset{OH}{\overset{OH}{\underset{|}{\overset{|}{C}}}}-CHCl_2 \xrightarrow{-H_2O} CH_3-\overset{O}{\overset{||}{C}}-CHCl_2$$

(Unstable)   1,1-Dichloropropanone

To obtain hypohalous acid (HOX), we can use the mixture (X_2/H_2O)

Mechanism

Addition of hypohalous acids to alkynes is an **electrophilic addition** reaction and occurs in the following two steps.

Step 1. $HC \equiv CH + \overset{\delta+}{Cl} - \overset{\delta-}{OH} \xrightarrow{\text{Slow}} H\overset{+}{C}=CH-Cl + OH^-$

$HO^- + H\overset{+}{C}=CH-Cl \xrightarrow{\text{Fast}} [HO-\overset{1}{C}H=\overset{2}{C}H-Cl]$
                                                                  I
                                         (2-Chloroeth-en-1-ol)

Step 2. The addition of Cl^+ to (I) can occur to give either carbocation (I or III) both of which are stabilized by resonance. However, the reaction proceeds via carbocation II as explained below.

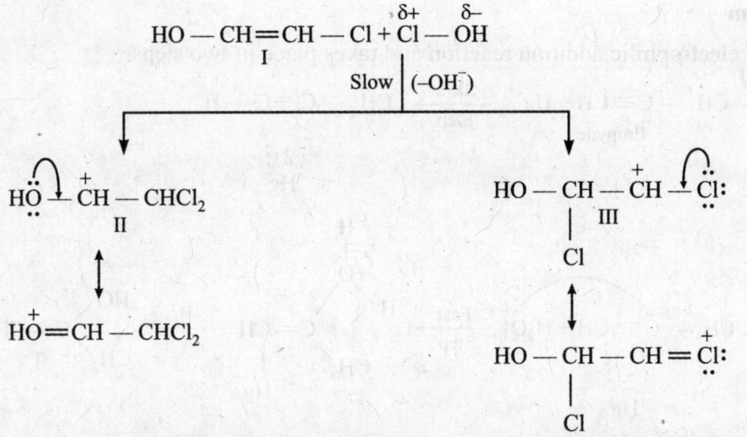

$$\text{HO}-\text{CH}=\text{CH}-\overset{\delta+}{\text{Cl}} + \overset{\delta-}{\text{Cl}}-\text{OH}$$
$$\text{I}$$
$$\text{Slow} \Big| (-\text{OH}^-)$$

$$\overset{+}{\underset{\text{II}}{\text{H}\overset{..}{\overset{..}{\text{O}}}-\overset{+}{\text{CH}}-\text{CHCl}_2}} \qquad \qquad \text{HO}-\text{CH}-\overset{+}{\text{CH}}-\overset{..}{\overset{..}{\text{Cl}}}:$$
$$\qquad\qquad\qquad\qquad\qquad\qquad\qquad\qquad \underset{\text{III}}{|}$$
$$\updownarrow \qquad \qquad\qquad\qquad\qquad\qquad\qquad\qquad \text{Cl}$$
$$\overset{+}{\text{HO}}=\text{CH}-\text{CHCl}_2 \qquad\qquad\qquad \updownarrow$$
$$\qquad\qquad\qquad\qquad\qquad\qquad \text{HO}-\text{CH}-\text{CH}=\overset{+}{\overset{..}{\text{Cl}}}:$$
$$\qquad\qquad\qquad\qquad\qquad\qquad\qquad\quad |$$
$$\qquad\qquad\qquad\qquad\qquad\qquad\qquad\quad \text{Cl}$$

In carbocation (II), the lone pairs of electrons on O are present in a 2p-orbital and the carbon atom with +ve charge has an empty 2p-orbital. Therefore, carbocation (II) is stabilized by 2p-2p overlapping. However, in carbocation (III), the lone pairs of electrons on Cl are in a 3p-orbital while the carbon atom having +ve charge has 2p-empty orbital. Since 2p-orbital of O and C are nearly of equal size and energy **therefore**, carbocation (II) is more resonance stabilized than carbocation (III). As a result, reaction occurs through carbocation (II) and the final product is 2, 2-dichloroethanal as shown below.

$$\underset{\text{II}}{\text{HO}-\overset{+}{\text{CH}}-\text{CHCl}_2} + \overset{-}{\text{OH}} \longrightarrow \left[\begin{array}{c}\text{HO}-\text{CH}-\text{CHCl}_2\\|\\\text{OH}\\\text{unstable}\end{array}\right] \xrightarrow{-\text{H}_2\text{O}} \underset{\substack{\text{1, 1-Dichloro}\\\text{ethanal}}}{\text{O}=\text{CH}-\text{CHCl}_2}$$

Addition of water

In the presence of $HgSO_4$ as catalyst, hydration of water in the presence of H_2SO_4 takes place at 333 K to give aldehydes and ketones.

$$\underset{\text{Ethyne}}{\text{HC}\equiv\text{CH}} + \text{H}-\text{OH} \xrightarrow[\substack{\text{H}_2\text{SO}_4\\333\text{ K}}]{\text{HgSO}_4} \underset{\text{Vinyl alcohol (unstable)}}{[\text{H}-\text{CH}=\text{CH}-\text{OH}]} \xrightarrow{\text{Tautomerises}} \underset{\text{Ethanal}}{\text{CH}_3-\overset{\overset{\text{O}}{\|}}{\text{C}}-\text{H}}$$

With *unsymmetrical terminal* alkynes, addition takes place as per Markownikoff's rule.

$$\text{CH}_3-\text{C}\equiv\text{CH} + \overset{\delta+}{\text{H}}\overset{\delta+}{\text{OH}} \xrightarrow[333\text{ K}]{\text{HgSO}_4,\ \text{H}_2\text{SO}_4} \text{CH}_3-\underset{|}{\overset{\overset{\text{OH}}{|}}{\text{C}}}=\text{CH}_2 \xrightarrow{\text{Tautomerises}} \underset{\text{Propanone}}{\text{CH}_3-\overset{\overset{\text{O}}{\|}}{\text{C}}-\text{CH}_3}$$

With *unsymmetrical but non-terminal* alkynes, a mixture of two ketones is possible.

$$\text{CH}_3\text{CH}_2-\text{C}\equiv\text{C}-\text{CH}_3 \xrightarrow[\text{H}_2\text{SO}_4,\ 33\text{K}]{\text{HgSO}_4} \underset{\text{Pentan-2-one}}{\text{CH}_3\text{CH}_2\text{CH}_2-\overset{\overset{\text{O}}{\|}}{\text{C}}-\text{CH}_3} + \underset{\text{Pentan-3-one}}{\text{CH}_3\text{CH}_2-\overset{\overset{\text{O}}{\|}}{\text{C}}-\text{CH}_2\text{CH}_3}$$

Mechanism

It is an electrophilic addition reaction and takes place in two steps.

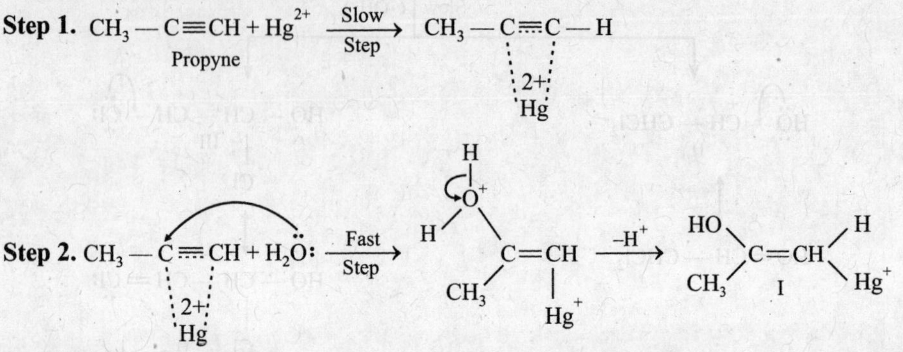

Step 1. $CH_3—C≡CH + Hg^{2+} \xrightarrow[\text{Step}]{\text{Slow}} CH_3—C≡C—H$ with Hg^{2+} bridging

Propyne

Step 2. $CH_3—C≡CH + H_2\ddot{O}:$ (with Hg^{2+}) $\xrightarrow[\text{Step}]{\text{Fast}}$ intermediate $\xrightarrow{-H^+}$ enol (I)

Protonation of intermediate (I) followed by loss Hg^{2+} gives the enol (II) which readily tautomerises to give carbonyl compound

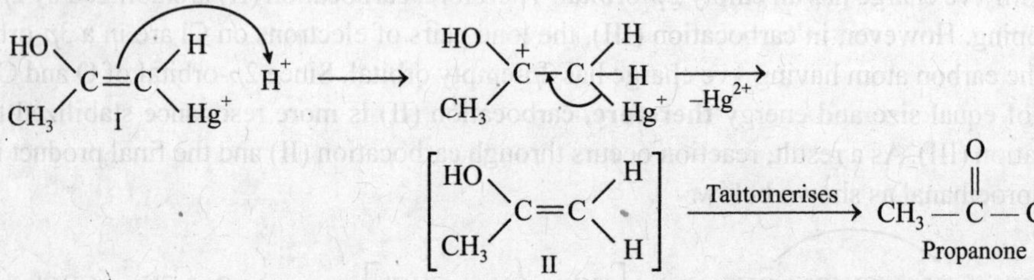

$$\text{(I)} \xrightarrow{H^+} \text{intermediate} \xrightarrow{-Hg^{2+}} \text{(II)} \xrightarrow{\text{Tautomerises}} CH_3—\overset{O}{\underset{\parallel}{C}}—CH_3$$

Propanone

Hydrogenation of alkynes

Catalytic hydrogenation of alkynes yields alkanes through the intermediate formation of an *alkene*.

The reaction can be stopped at the alkene stage with the proper choice of reagent. Predominantly, *trans* alkene is obtained if the reduction of alkyne is carried out in the presence of sodium or lithium in liquid ammonia. Almost entirely *cis* alkene is obtained if the hydrogenation of alkyne is carried out with Lindlar's catalyst (Palladium supoorted over $BaSO_4$ partially poisoned with quinoline) or Nickel Boride catalyst. Thus, the reduction of alkynes is a stereoselective reaction.

For example,

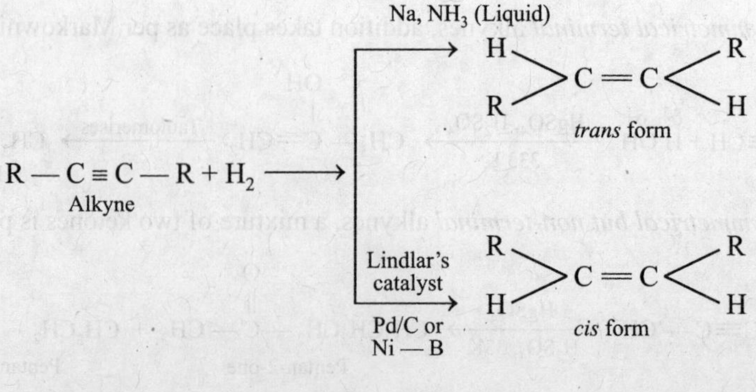

$$R—C≡C—R + H_2 \xrightarrow{\text{Na, NH}_3 \text{ (Liquid)}} \underset{\text{trans form}}{\overset{H}{\underset{R}{>}}C=C\overset{R}{\underset{H}{<}}}$$

$$\xrightarrow[\text{Pd/C or Ni—B}]{\text{Lindlar's catalyst}} \underset{\text{cis form}}{\overset{R}{\underset{H}{>}}C=C\overset{R}{\underset{H}{<}}}$$

Alkyne

Mechanism of Chemical Reduction

It is believed to occur through two electron transfers and two proton transfers involving formation of radicals.

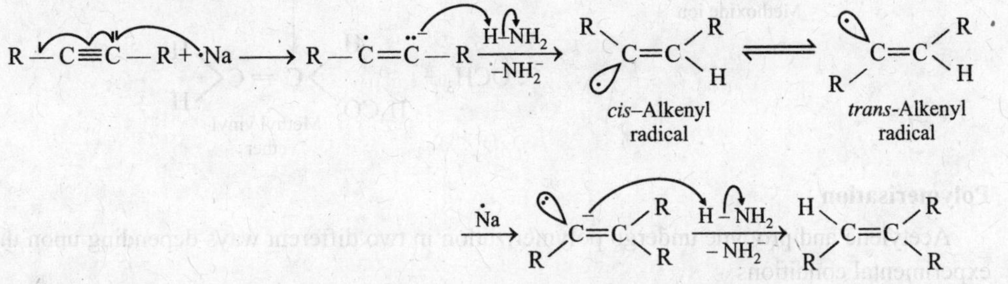

$$R-C\equiv C-R + \cdot Na \longrightarrow R-\dot{C}=\ddot{C}-R \xrightarrow[-NH_2^-]{H-NH_2} \underset{\substack{cis-\text{Alkenyl}\\\text{radical}}}{\overset{R}{\underset{R'}{>}}C=C\overset{R}{\underset{H}{<}}} \rightleftharpoons \underset{\substack{trans-\text{Alkenyl}\\\text{radical}}}{\overset{R}{\underset{R'}{>}}C=C\overset{R}{\underset{H}{<}}}$$

$$\xrightarrow{\dot{N}a} \overset{R}{\underset{R'}{>}}C=C\overset{R}{\underset{R'}{<}} \xrightarrow[-NH_2^-]{H-NH_2} \overset{H}{\underset{R'}{>}}C=C\overset{R}{\underset{R'}{<}}$$

The formation of *trans*-product is due to the rapid equilibrium between the *cis*- and *trans*-alkenyl radicals, in which the *trans*-radical being more stable is formed preferably.

Nucleophilic Addition Reactions

(i) Addition of HCN

Acetylene reacts with hydrogen cyanide in the presence of barium cyanide yielding vinyl cyanide (acrylonitrile).

$$H-C\equiv C-H + HCN \xrightarrow{Ba^{++}} \underset{\substack{\text{Vinyl cyanide}\\\text{(acrylonitrile)}}}{H_2C=CH-CN}$$

Acrylonitrile is used in the manufacture of Buna N.

The probable mechanism of the reaction is depicted below:

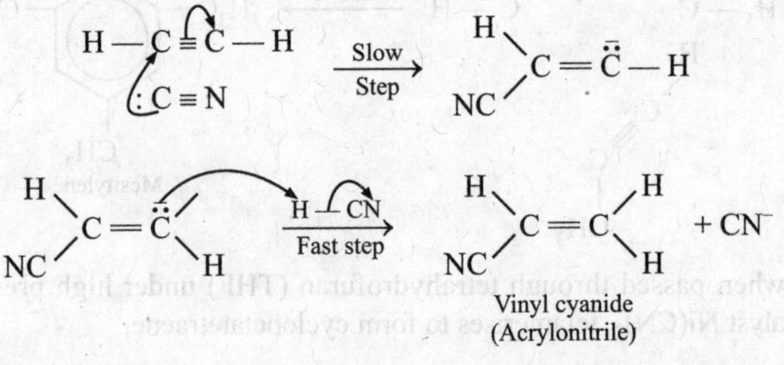

$$\underset{:C\equiv N}{H-C\equiv C-H} \xrightarrow[\text{Step}]{\text{Slow}} \overset{H}{\underset{NC}{>}}C=\ddot{C}-H$$

$$\overset{H}{\underset{NC}{>}}C=\ddot{C}\overset{}{\underset{H}{<}} \xrightarrow[\text{Fast step}]{H-CN} \underset{\substack{\text{Vinyl cyanide}\\\text{(Acrylonitrile)}}}{\overset{H}{\underset{NC}{>}}C=C\overset{H}{\underset{H}{<}}} + CN^-$$

Acrylonitrile is used in the manufacture of *orlon fibres* and synthetic rubber such as Buna-N.

(ii) Addition of Methanol

Acetylene adds methanol in the presence of sodium methoxide yielding vinyl ether.

$$H-C\equiv C-H + CH_3OH \xrightarrow{CH_3ONa} \underset{\text{Methyl vinyl ether}}{H_2C=CH.OCH_3}$$

Methyl vinyl ether is used in the manufacture of polyvinyl ether plastics.

The probable mechanism of the reaction is shown below:

Polymerisation

Acetylene and propyne undergo polymerization in two different ways depending upon the experimental conditions

Cyclic polymerisation

Acetylene and methyl acetylene when passed through red hot iron tube, polymerize to form benzene and mesitylene respectively.

Acetylene when passed through tetrahydrofuran (THF) under high pressure and in the presence of catalyst $Ni(CN)_2$, tetamerises to form cyclooctatetraene.

Linear polymerization. Acetylene when passed through a solution of cuprous chloride and ammonium chloride at 343 K, dimerises to form *vinyl acetylene* and finally trimerises to form *divinyl acetylene*.

$$HC \equiv CH + HC \equiv CH \xrightarrow{Cu_2Cl_2,\ NH_4Cl} \underset{\text{Vinyl acetylene}}{H_2C = CH - C \equiv CH}$$

$$HC \equiv CH + HC \equiv CH + HC \equiv CH \xrightarrow{Cu_2Cl_2,\ NH_4Cl} \underset{\text{Divinyl acetylene}}{H_2C = CH - C \equiv CH - CH = CH_2}$$

Oxidation Reactions

(i) Oxidation with cold dilute potassium permanganate. Alkynes are readily oxidised by cold dilute alkaline $KMnO_4$ solution to give α-dicarbonyl compounds.

In case of terminal alkynes, $\equiv CH$ part is oxidised to $-COOH$ group while in case of non-terminal alkynes, $\equiv CR$ part is oxidised to $R-\underset{\underset{O}{\|}}{C}=O$ group.

$$\underset{\text{Propyne}}{CH_3 - C \equiv CH} + 3[O] \xrightarrow[298-303\ K]{KMnO_4,\ H_2O} \left[\underset{\text{2-Oxopropanoic acid}}{CH_3 - \underset{\underset{O}{\|}}{C} - COOH} \right] \xrightarrow{[O]} \underset{\text{Acetic acid}}{CH_3COOH + CO_2}$$

$$\underset{\text{But-2-yne}}{CH_3 - C \equiv C - CH_3} + 2[O] \xrightarrow[298-303\ K]{KMnO_4,\ H_2O} \underset{\substack{\text{Butane-2, 3-dione} \\ (Biacetyl)}}{CH_3 - \underset{\underset{O}{\|}}{C} - \underset{\underset{O}{\|}}{C} - CH_3}$$

Acetylene, however, gives oxalic acid due to further oxidation of the initially formed glyoxal.

$$\underset{\text{Acetylene}}{\overset{CH}{\underset{CH}{\|}}} + 2[O] \xrightarrow[298-303\ K]{KMnO_4,\ H_2O} \left[\underset{\substack{\text{Ethane-1, 2-dial} \\ (Glyoxal)}}{\begin{array}{c} CH = O \\ | \\ CH = O \end{array}} \right] \xrightarrow[(Oxidation)]{+2[O]} \underset{\substack{\text{Ethane-1, 2-dioic acid} \\ (Oxalic acid)}}{\begin{array}{c} COOH \\ | \\ COOH \end{array}}$$

Pink colour of the $KMnO_4$ solution is discharged and a brown precipitate of manganese dioxide is obtained. This reaction is, therefore, used as a test for unsaturation under the name **Baeyer's test**.

(ii) Oxidation with hot $KMnO_4$ Solution. Terminal alkynes when treated with hot alkaline $KMnO_4$ solution, undergo cleavage of $C \equiv C$ bond leading to the formation of carboxylic acids and carbon dioxide depending upon the nature of the alkyne.

For example, acetylene gives CO_2 and H_2O while propyne gives acetic acid and CO_2.

$$\underset{\text{Acetylene}}{\overset{CH}{\underset{CH}{\|}}} + 4[O] \xrightarrow[373-383\ K]{KMnO_4,\ KOH} \left[\underset{\text{Oxalic acid}}{\begin{array}{c} COOH \\ | \\ COOH \end{array}} \right] \xrightarrow{[O]} 2CO_2 + H_2O$$

$$\underset{\text{Propyne}}{CH_3 - C \equiv CH} + 4[O] \xrightarrow[373-383\ K]{KMnO_4,\ KOH} \underset{\text{Ethanoic acid}}{CH_3COOH + CO_2}$$

However, non-terminal alkynes on oxidation with hot $KMnO_4$ solution give only carboxylic acids. For example,

$$CH_3-C\equiv C-CH_3 + 4[O] \xrightarrow[373-383\,K]{KMnO_4,\,KOH} CH_3COOH + HOOCCH_3$$
But-2-yne   Acetic acid
(2 molecules)

$$CH_3CH_2-C\equiv C-CH_3 + 4[O] \xrightarrow[373-383\,K]{KMnO_4,\,KOH} CH_3CH_2COOH + HOOCCH_3$$
Pent-2-yne   Propanoic acid   Ethanoic acid

It is possible to determine the position of the triple bond in an alkyne after identifying the products of oxidation.

(*iii*) **Combustion.** On combustion in free supply to oxygen, alkynes are oxidised to CO_2 and H_2O.

$$2CH\equiv CH + 5O_2 \longrightarrow 4CO_2 + 2H_2O$$

$$CH_3-C\equiv CH + 4O_2 \longrightarrow 3CO_2 + 2H_2O$$

With oxygen at high pressure, acetylene burns with a blue flame producing a temperature of 3000 K which is used for cutting and welding of metals.

Ozonolysis

Alkynes add on ozone to form ozonides. The ozonides are hydrolysed by zinc dust and water to form dicarbonyl compounds (1, 2-diketones). This is called reductive cleavage. Dicarbonyl compound undergoes oxidative cleavage to form acids. The identification of the acids formed helps to locate the position to triple bond in the original alkyne. Thus, ozonolysis followed by oxidative cleavage can be used as an unambiguous method for locating the position of triple bond in the original alkyne.

When acetylene is burnt in oxygen, very high temperatures are produced. The oxy-acetylene torch burns a mixture of acetylene and oxygen which produces a flame hot enough to cut or weld steel.

$$-C\equiv C- + O_3 \longrightarrow \underset{\text{Oxonide}}{-\overset{\displaystyle O}{\underset{\displaystyle O\!-\!O}{C\!-\!C}}-}$$

$$\downarrow \text{Zn dust/}H_2O\;(\text{Reductive cleavage})$$

$$\underset{\substack{+\\ -C-O-H\\ \|\\ O}}{\overset{\displaystyle -C-O-H}{\overset{\|}{O}}} \xleftarrow{\text{Oxidative cleavage}} \underset{\text{Diketone}}{-\overset{\|}{C}-\overset{\|}{C}-} + H_2O_2$$

For example,

$$HC\equiv CH \xrightarrow{O_3} \underset{\text{Ethyne ozonide}}{HC\!-\!\!\overset{O}{\overset{\diagup\,\diagdown}{}}\!\!-\!CH \atop | \quad\quad\quad | \atop O\!-\!-\!-\!O} \xrightarrow[\text{(reductive cleavage)}]{Zn/H_2O,\; -ZnO} \underset{\text{1, 2-Ethanedial (}Glyoxal\text{)}}{H-\overset{\|}{\underset{O}{C}}-\overset{\|}{\underset{O}{C}}-H}$$

Ethyne

$$CH_3-C\equiv CH \xrightarrow{O_3} \underset{\text{Propyne ozonide}}{CH_3-\underset{\underset{O\text{———}O}{|}}{\overset{\overset{O}{\diagup \diagdown}}{C}}-CH} \xrightarrow[-ZnO]{Zn/H_2O} \underset{\text{2-Oxopropanal}}{CH_3-\underset{\underset{O}{||}}{C}-\underset{\underset{O}{||}}{C}-H}$$
Propyne   *(reductive cleavage)*

$$CH_3-C\equiv C-CH_3 \xrightarrow{O_3} \underset{\text{But-2-yne ozonide}}{CH_3-\underset{\underset{O\text{———}O}{|}}{\overset{\overset{O}{\diagup \diagdown}}{C}}-C-CH_3} \xrightarrow[-ZnO]{Zn/H_2O} \underset{\text{Butane-2, 3-dione}}{CH_3-\underset{\underset{O}{||}}{C}-\underset{\underset{O}{||}}{C}-CH_3}$$
But-2-yne   *(reductive cleavage)*

Hydroboration-Oxidation of Alkynes

When treated with diborane, *terminal alkynes* (*i.e.*, alkynes which contain a —C≡C—H group) undergo addition in two stages, forming firstly a *vinylborane* and finally a *gem-dibora derivative*. It is not possible to stop the reaction at the stage of vinyl borane.

$$2R-C\equiv C-H + (BH_3)_2 \longrightarrow \left[\underset{\text{Vinyl borane}}{2\ RCH=CH \atop | \atop BH_2} \right] \xrightarrow{(BH_3)_2} \underset{\text{gem-dibora derivative}}{2RCH_2-CH-BH_2 \atop | \atop BH_2}$$
Terminal alkyne

In case of *internal alkynes* (*i.e.*, alkynes in which triply bonded carbons do not carry any hydrogen), the reaction can be stopped at the vinylic stage by using suitable amounts of the reagent.

If in place of diborane, a bulky sterically hindered borane such as bis (1, 2-dimethylpropyl) borane $\left[\begin{array}{c} CH_3 \quad CH_3 \\ | \quad\quad | \\ CH_3-CH-CH- \end{array} \right]_2$ BH commonly known as *disiamylborane* (abbreviated as Sia$_2$ BH) is used, the reaction can be stopped at the vinyl borane stage for all alkynes whether terminal or internal.

That is:

$$\underset{\substack{\text{Terminal}\\\text{alkyne}}}{R-C\equiv C-H} + Sia_2\ BH \longrightarrow R-CH=CH-BSia_2$$

$$\underset{\substack{\text{Internal}\\\text{alkyne}}}{R-C\equiv C-R} + Sia_2\ BH \longrightarrow R-CH=\underset{\underset{BSia_2}{|}}{C}-R$$

The vinylic boranes can then be oxidised with H_2O_2 in basic solution to form aldehydes or ketones via enolic intermediates.

$$-CH=\underset{|}{C}-B\ SiA_2 \xrightarrow{H_2O_2,\ OH^-} \left[-CH=\underset{\underset{\text{Enol}}{|}}{C}-OH \right] \longrightarrow \underset{\text{Aldehyde or ketone}}{-CH_2-\underset{|}{C}=O}$$

For example, hydroboration of 1-hexyne with Sia_2BH followed by oxidation with H_2O_2/OH^- to yield hexanal is shown below.

$$CH_3(CH_2)_3 C \equiv CH \xrightarrow{Sia_2BH} CH_3(CH_2)_3 CH = CHBSia_2 \xrightarrow{H_2O_2, OH^-}$$
1-Hexyne Vinylic borane

$$[CH_3(CH_2)_3 CH = CH - OH] \rightleftharpoons CH_3(CH_2)_3 CH_2 - \overset{O}{\underset{\|}{C}} - H$$
Enol Hexanal

Similarly hydroboration-oxidation of 4-octyne gives rise to 4-octanone.

$$C_3H_7 - C \equiv C - C_3H_7 \xrightarrow{Sia_2BH} C_3H_7 - \underset{\underset{BSia_2}{|}}{CH} = C - C_3H_7 \xrightarrow{H_2O_2, OH^-}$$
4-Octyne Vinylic borane

$$\left[C_3H_7 - \underset{\underset{OH}{|}}{CH} = C - C_3H_7 \right] \rightleftharpoons C_3H_7 - CH_2 - \underset{\underset{O}{\|}}{C} - C_3H_7$$
Enol 4-Octanone

It may be noted that hydroboration oxidation leads to **anti-Markownikoff** addition of H— and —OH. It may also be observed that hydroboration oxidation of **terminal alkynes** furnishes aldehydes while that of internal alkynes gives rise to ketones.

It may also be noted that hydroboration-oxidation is complementary to direct hydration of alkynes since different products are obtained. Direct hydration of terminal alkynes leads to a methyl ketone while hydroboration-oxidation of the same alkyne gives an aldehyde.

$$R - \overset{O}{\underset{\|}{C}} - CH_3 \xleftarrow[HgSO_4]{Dil. H_2SO_4} \boxed{R - C \equiv CH_3} \xrightarrow[(ii) H_2O_2/OH^-]{(i) B_2H_6} R - CH_2 - CHO$$
Ketone Alkyne Aldehyde

Addition of arsenic trichloride. Acetylene reacts with arsenic trichloride in the presence of anhyd. $AlCl_3$ to form **Lewisite** which is a very poisonous gas and was used in World War II.

$$HC \equiv CH + Cl - AsCl_2 \xrightarrow{Anhyd. AlCl_3} Cl - CH = CH - AsCl_2$$
Acetylene Lesisite
(β-Chlorovinyldichloroarsine)

Solved Examples

Example 2: What happens when:

(I) 2-butyne is treated with Pd/C (Lindlar calalyst and hydrogen)

(II) Propyne is treated with chlorine water

(III) Acetylene is treated with Br_2 dissolved in CCl_4

(IV) Acetylene is treated with ammoniacal cuprous chloride solution

(V) Acetylene is treated with ammoniacal $AgNO_3$

(VI) 2-Butyne is treated with sodium dissolved in liquid ammonia

(VII) Acetylene is subjected to ozonolysis.

Solution:

(I) $H_3C-C \equiv C-CH_3 + H_2 \xrightarrow{Pd/C}$ $\underset{cis\text{ 2-butene}}{\overset{CH_3 \quad\quad CH_3}{\underset{H \quad\quad\quad H}{>C=C<}}}$
   2-Butyne

(II) $Cl_2 + H_2O \longrightarrow HCl + HOCl$

$CH_3-C \equiv CH + \overset{\delta^-}{H}\overset{}{O}-\overset{\delta^+}{Cl} \longrightarrow CH_3-\underset{OH}{\overset{}{C}}=\underset{Cl}{\overset{}{CH}}$

$\xrightarrow{\overset{\delta^-}{H}\overset{}{O}-\overset{\delta^+}{Cl}}$

$\underset{\text{Dichloroacetone}}{CH_3-\overset{O}{\overset{||}{C}}-CHCl_2} \xleftarrow{-H_2O} \left[\underset{\text{Unstable}}{CH_3-\underset{OH}{\overset{OH}{\overset{|}{\underset{|}{C}}}}-CHCl_2} \right]$

(III) $CH \equiv CH \xrightarrow{Br_2/CCl_4} \underset{\substack{\text{1, 2-Dibromoethene} \\ \text{(Acetylene dibromide)}}}{\underset{Br \quad Br}{\overset{}{CH=CH}}} \xrightarrow{Br_2/CCl_4} \underset{\substack{\text{1, 1, 2, 2-Tetrabromoethane} \\ \text{(Acetylene tetrabromide)}}}{\underset{Br \quad Br}{\overset{Br \quad Br}{CH-CH}}}$

(IV) $CH \equiv CH + Cu_2Cl_2 + 2NH_4OH \longrightarrow \underset{\substack{\text{Copper acetylide} \\ \text{(Red ppt.)}}}{CuC \equiv CCu} + 2NH_4Cl + 2H_2O$

(V) $CH \equiv CH + 2NH_4OH + 2AgNO_3 \longrightarrow \underset{\text{Silver acetylide}}{AgC \equiv CAg} + 2NH_4NO_3 + 2H_2O$

(VI) $CH_3-C \equiv C-CH_3 + H_2 \xrightarrow{Na/liq.\ NH_3}$ $\underset{trans\text{-2-Butene}}{\overset{CH_3 \quad\quad H}{\underset{H \quad\quad\quad CH_3}{>C=C<}}}$

(VII) $CH \equiv CH + O_3 \longrightarrow \underset{\substack{\text{Acetylene} \\ \text{ozonide}}}{\underset{O-O}{\overset{O}{\overset{/\ \ \backslash}{CH-CH}}}} \xrightarrow[-H_2O_2]{H_2O/Zn} \underset{\text{Glyoxal}}{\overset{CHO}{\underset{CHO}{|}}}$

Example 3: How will you bring about the following conversions?

(I) Acetylene into oxalic acid
(II) Propyne into 2, 2-dibromopropane
(III) 2-Butyne into acetic acid
(IV) Propyne into propanone

(V) Acetylene into chloroprene

(VI) Acetylene into vinyl acetate.

Solution:

(I) Acetylene into oxalic acid

$$\underset{\text{Acetylene}}{\begin{matrix} CH \\ ||| \\ CH \end{matrix}} + 4[O] \xrightarrow{\text{Alk. } KMnO_4} \underset{\text{Oxalic acid}}{\begin{matrix} COOH \\ | \\ COOH \end{matrix}}$$

(II) Propyne into 2, 2-dibromopropane

$$\underset{\text{Propyne}}{CH_3-C\equiv CH} \xrightarrow{HBr} \underset{\text{2-Bromopropene}}{CH_3-\overset{\overset{\large Br}{|}}{C}=CH_2} \xrightarrow{HBr} \underset{\text{2, 2-dibromopropane}}{CH_3-\overset{\overset{\large Br}{|}}{\underset{\underset{\large Br}{|}}{C}}-CH_3}$$

(III) 2-butyne into acetic acid

$$\underset{\text{2-Butyne}}{CH_3-C\equiv C-CH_3} + H_2O + 3[O] \xrightarrow{\text{Alk. } KMnO_4} \underset{\text{Acetic acid}}{CH_3COOH} + \underset{\text{Acetic acid}}{HOOCCH_3}$$

(IV) Propyne into propanone

$$CH_3-C\equiv CH + HOH \xrightarrow[333\text{ K}]{HgSO_4/H_2SO_4} \underset{\text{Unstable}}{\left[CH_3-\overset{\overset{\large OH}{|}}{C}=CH_2\right]} \rightleftharpoons \underset{\text{Propanone}}{CH_3-\overset{\overset{\large O}{||}}{C}-CH_3}$$

(V) Acetylene into chloroprene

$$CH\equiv CH + CH\equiv CH \xrightarrow[NH_4Cl]{CuCl} \underset{\text{Vinyl acetylene}}{CH_2=CH-C\equiv CH}$$

$$CH_2=CH-C\equiv CH + HCl \longrightarrow \underset{\text{Chloroprene}}{CH_2=CH-\underset{\underset{\large Cl}{|}}{C}=CH_2}$$

(VI) $$CH\equiv CH + CH_3COOH \xrightarrow{Hg^{2+}} \underset{\text{Vinyl acetate}}{CH_2=CHOCOCH_3}$$

Example 4: Complete the following reactions

(*i*) $HC\equiv CH + H_2O \xrightarrow[H_2SO_4]{HgSO_4}$

(*ii*) $HC\equiv CH + HBr \longrightarrow$

Solution:

(*i*) $$\underset{\text{Acetylene}}{HC\equiv CH} + H_2O \xrightarrow{HgSO_4} \underset{\text{Vinyl alcohol}}{CH_2=CHOH} \xrightarrow{\text{Rearranges}} \underset{\text{Acetaldehyde}}{CH_3CHO}$$

(*ii*) $$HC\equiv CH + HBr \xrightarrow{HgSO_4} CH_2=CHBr \xrightarrow{HBr} \underset{\text{Ethylidene bromide}}{CH_3CHBr_2}$$

Example 5: Justify with example the following statement: Acetylene undergoes electrophilic as well as nucleophilic addition reactions.

Solution: Acetylene undergoes nucleophilic reaction like addition of water in the presence of dil. H_2SO_4 and $HgSO_4$ as follows:

$$CH \equiv CH + H_2O \xrightarrow[HgSO_4]{HgSO_4} CH_2CHO$$

Nucleophilic reaction with acetylene takes place because Hg^{2+} forms a complex with the π electrons of carbon-carbon triple bond thereby reducing election density. This consequently encourages attack by nucleophiles. Moreover the intermediate allylic carbanion in the nucleophilic attack gets stabilized.

Acetylene undergoes electrophilic reaction as given by alkenes because of π electrons of the carbon-carbon triple bond.

$$CH \equiv CH + Br_2 \longrightarrow CHBr = CHBr \xrightarrow{Br_2} CHBr_2 - CHBr_2$$

Example 6: How will you bring about the following conversion?

$$CH_3C \equiv CH \longrightarrow CH_3C \equiv CCH_3$$

Solution: $CH_3C \equiv CH + Na \longrightarrow CH_3C \equiv CNa \xrightarrow{CH_3Br} CH_3C \equiv CCH_3$

Example 7: How will you differentiate propyne-1 from butyne-2?

Solution: The compounds will be treated with ammoniacal cuprous chloride and ammoniacal silver nitrate separately. Propyne-1 gives a red precipitate with ammoniacal cuprous chloride and gives a white precipitate with ammoniacal silver nitrate. While butyne-2 does not give any precipitate with these reagents.

$$2CH_3C \equiv CH + Cu_2Cl_2 + 2NH_4OH \longrightarrow 2CH_3C \equiv C.Cu + 2NH_4Cl + 2H_2O$$
(red ppt.)

$$CH_3C \equiv CH + 2NH_4OH + 2AgNO_3 \longrightarrow 2CH_3C \equiv CAg + 2NH_4Cl + 2H_2O$$
(white ppt.)

$$CH_3.C \equiv C.CH_3 + Cu_2Cl_2 + 2NH_4OH \longrightarrow \text{No reaction}$$

$$CH_3.C \equiv C.CH_3 + 2NH_4OH + 2AgNO_3 \longrightarrow \text{No reaction}$$

Example 8: How will you synthesise acetone from propene?

Solution: Following steps are involved

$$\underset{\text{Propene}}{CH_3CH=CH_2} \xrightarrow{Br_2} \underset{}{CH_3\overset{Br}{\underset{|}{C}}H - \overset{Br}{\underset{|}{C}}H_2} \xrightarrow{NaNH_2} \underset{\text{Propyne}}{CH_3CH \equiv CH} \xrightarrow[H_2SO_4/HgSO_4]{H_2O} \underset{\text{Acetone}}{CH_3\overset{Br}{\underset{\|}{C}} - CH_3}$$

Example 9: How will you differentiate between but-1-yne and but-2-yne?

Solution: The compounds will be treated with ammoniacal cuprous chloride and ammoniacal silver nitrate. Butyne-1-gives red ppt and silver mirror respectively. Butyne-2 reacts with neither. For reactions, see text.

Example 10: Compare the relative acidities of water, alcohols, acetylene, ammonia and alkanes.

Solution: Comparision of acidities of the compounds can be made in terms of the following reactions:

1. $NaNH_2$ reacts with acetylene to form sodium acetylide and NH_3. A stronger acid displaces a weaker acid from its salts. Therefore, acetylene is a stronger and than NH_3.

$$HC \equiv CH + NaNH_2 \longrightarrow HC \equiv C^- Na^+ + NH_3$$
$$\text{(stronger acid)} \hspace{4cm} \text{(weaker acid)}$$

2. Lithium alkyls (RLi) react wtih ammonia to form alkanes. Therefore, ammonia is a stronger acid than alkanes.

$$NH_3 + RLi \longrightarrow R-H + LiNH_2$$
$$\text{(stronger acid)} \hspace{2cm} \text{(weaker acid)}$$

3. H_2O and alcohols decompose sodium acetylide to give back acetylene. Therefore water and alcohols are stronger acids than acetylene

$$HC \equiv C^- Na^+ + H_2O \longrightarrow HC \equiv CH + NaOH$$
$$\text{(stronger acid)} \hspace{2cm} \text{(weaker acid)}$$

$$HC \equiv C^- Na^+ + H_2O \longrightarrow HC \equiv CH + NaOH$$
$$\text{(stronger acid)} \hspace{2cm} \text{(weaker acid)}$$

Example 11: Why are conjugated dienes more reactive than alkenes or alkynes towards electrophillic addition reactions?

Solution: The relative reactivity of alkenes, alkynes and conjugated dienes depends upon the stability of the carbocation they generate on addition of an electrophile, which is compared as under:

$$CH_2 = CH_2 + Br^+ \longrightarrow \overset{+}{C}H_2 - CH_2 - Br$$
$$\text{Ethene} \hspace{4cm} \text{(I)}$$

$$CH \equiv CH + Br^+ \longrightarrow \overset{+}{C}H = CH - Br$$
$$\text{Ethene} \hspace{4cm} \text{(II)}$$

$$CH_2 = CH - CH = CH_2 + Br^+ \longrightarrow Br - CH_2 - \overset{+}{C}H - CH = CH_2$$
$$\text{Buta-1, 3-diene} \hspace{5cm} \text{(III)}$$

$$\updownarrow$$

$$Br - CH_2 - CH = CH - \overset{+}{C}H_2$$

Among the carbonations (I, II and III), the carbocation (III) resulting from buta-1, 3-diene is the most stable because it is stabilised by resonance. Out of carbocations (I and II), carbocation (II) is less stable because the +ve charge is located on a more electronegative sp^2-hybridised carbon. Thus, *the stability of carbocations follows the order: III > I > II*. Accordingly, the reactivity decreases in the same order, i.e., conjugated diene > alkene > alkyne.

Example 12: How will you distinguish between buta-1, 3-diene and but-1-yne?

Solution: But-1-yne is a terminal alkyne and hence can be distinguished from buta-1, 3-diene by Tollen's reagent which forms a white ppt. with but-1-yne but not with buta-1, 3-diene.

$$CH_3CH_2-C\equiv CH + [Ag(NH_3)_2]^+ OH^- \longrightarrow CH_3CH_2-C\equiv C\,Ag + 2NH_3 + H_2O$$
But-1-yne   Tollen's reagent   Silver but-1-ynide (white ppt.)

$$CH_2=CH-CH=CH_2 + [Ag(NH_3)_2]^+ OH^- \longrightarrow \text{No reaction}$$
Buta-1, 3-diene

Example 13: An organic compound (A), C_4H_6 absorbs two molecules of hydrogen on hydrogenation. It reacts with ammoniacal $AgNO_3$ solution to form a white precipitate. On treatment with dil. H_2SO_4 in presence of $HgSO_4$, it forms a ketone (B), C_4H_8O. Propose the structures of (A) and (B) and write the reactions involved.

Solution: (i) Compound (A), C_4H_6 on hydrogenation absorbs two molecules of H_2. Therefore, (A) can be either a cumulene, conjugated diene or an alkyne, because all of them take up two moles of H_2 for addition.

$$CH_2=C=CHCH_3 \qquad CH_2=CH-CH=CH_2$$
1, 2-Butadiene   1, 3-Butadiene

$$CH_3CH_2-C\equiv CH \quad \text{or} \quad CH_3C\equiv CCH_3$$
1-Butyne   2-Butyne

(ii) Compound (A) forms white precipitate with $AgNO_3$, solution. Therefore (A) must be a terminal alkyne *i.e.* 1-butyne. The reaction is given as under:

$$CH_3CH_2CH_2CH_3 \xleftarrow{2H_2, Ni} CH_3CH_2-C\equiv CH \xrightarrow{AgNO_3/NH_4OH} CH_3CH_2CH\equiv CAg\downarrow$$
n-Butane   1-Butyne (A)   White ppt.

(iii) Compound (A) on treatment with dil. H_2SO_4 in presence of $HgSO_4$, gives ketone (B). C_4H_8O. Therefore, (B) must be 2-butanone. The reaction is given as under:

$$CH_3CH_2-C\equiv CH + H_2O \xrightarrow{HgSO_4/H_2SO_4} CH_3CH_2-\overset{O}{\underset{\|}{C}}-CH_3$$
1-Butyne   2-Butanone

Example 14: How will you convert acetylene into (i) 2-pentanone and (ii) 1, 3-butadiene.

Solution: Conversion (i) and (ii) are given as under:

(i)
$$HC\equiv CH \xrightarrow{NaNH_2 \text{ in liq. } NH_3} HC\equiv C^- Na^+ \xrightarrow[-NaI]{CH_3CH_2I} CH_3CH_2C\equiv CH \xrightarrow{NaNH_2 \text{ in liq. } NH_3}$$
           Sod. acetylide   1-Butyne

$$CH_3CH_2\equiv C^- Na^+ \xrightarrow[-NaI]{CH_3-I} CH_3-C\equiv C-CH_2CH_3 \xrightarrow[333\,K]{Dil. H_2SO_4, HgSO_4} CH_3-\overset{O}{\underset{\|}{C}}-CH_2CH_2CH_3$$
Sod. butynide   2-Pentyne   2-Pentanone

(ii) $$HC\equiv CH \xrightarrow[-CH_3OH]{CH_3O^-Na^+} HC\equiv C^- Na^+ \xrightarrow{HCHO} HC\equiv C-CH_2OH \xrightarrow{HCHO/CH_3ONa}$$
                                                                                                                     Prop-2-yn-1-ol

$$HOCH_2-C\equiv C-CH_2OH \xrightarrow{H_2/Ni} HOCH_2-CH_2-CH_2-CH_2OH$$
But-2-yne-1, 4-diol

$$\xrightarrow[-2H_2O]{Al_2O_3, \Delta} CH_2=CH-CH=CH_2$$
                         1, 3-Butadiene

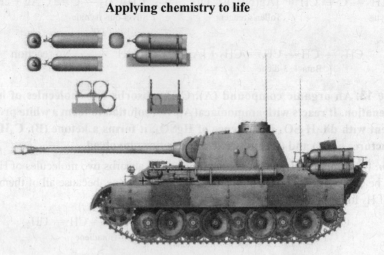

Applying chemistry to life

Acetylene powered war tank

Acetylene is a highly inflammable gas which can be generated relatively conveniently by dropping water on calcium carbide. It has been tried to run internal combustion engines fitted in defence tanks with some success

Key Terms

- Acidity of alkynes
- Nucleophilic addition
- Electrophilic addition
- Terminal alkynes

Evaluate Yourself

Multiple Choice Questions

1. Which of the following compounds on hydrolysis gives acetylene?
   (a) CaC_2
   (b) Mg_2C_3
   (c) Al_4C_3
   (d) Cu_2Cl_2

2. Which alkyne yields propanoic acid as the only product upon treatment with ozone followed by hydrolysis?
   (a) 1–Butyne
   (b) 2–Hexyne
   (c) 1–Pentyne
   (d) 3–Hexyne

3. Lindlar's Catalyst is
   (a) Li Al H$_4$
   (b) Pd/BaSO$_4$ in quinoline
   (c) NH$_2$ NH$_2$
   (d) HCl / Zn Cl$_2$
4. A triple bond consists of
   (a) 2 sigma bonds and 1 pi bond
   (b) 2 sigma bonds
   (c) 1 sigma bond and 2 pi bonds
   (d) 3 pi bonds
5. When acetylene is passed through hot iron tube at 400°C, it gives
   (a) benzene
   (b) toluene
   (c) o–xylene
   (d) mesitylene
6. Which of the following compounds does not react with ammoniacal AgNO$_3$ solution?
   (a) Acetylene
   (b) 1–Butyne
   (c) Propyne
   (d) 2-Butyne
7. Propyne reacts with aqueous H$_2$SO$_4$ in the presence of HgSO$_4$ to form
   (a) acetone
   (b) proponol-1
   (c) acetaldehyde
   (d) propanol-2
8. 1–Butyne can be distinguished from 2–butyne by using
   (a) potassium permanganate
   (b) bromine in CCl$_4$
   (c) Tollen's reagent
   (d) chlorine in CCl$_4$
9. Which of the following will give a negative test when treated with bromine in carbon tetrachloride?
   (a) Butane
   (b) 2-Butene
   (c) 1, 3-Butadiene
   (d) 3-Butyne
10. Propyne on passing through red hot copper tube forms
    (a) mesitylene
    (b) xylene
    (c) benzene
    (d) toluene

Short Answer Questions

1. Why $C \equiv C$ bond length shorter than the $C = C$ bond length?
2. What is the product of ozonolysis of 2-butyne?
3. Discuss the mechanism of nucleophilic addition reactions in ethyne.
4. Why are the hydrogen atoms in acetylene acidic in nature?
5. Give a brief account of nucleophilic addition reaction in alkynes.

Long Answer Questions

1. What happens when acetylene vapours are passed through dilute sulphuric acid in the presence of mercuric sulphate? Give the mechanism of the reaction.
2. Discuss keto-enol tautomerism with reference to the products of reaction of acetylene with water in the presence of $HgSO_4$ and H_2SO_4.
3. How will you convert:
    (i) Acetylene into acetic acid.
    (ii) Acetylene into chloroprene
    (iii) Acetylene into acetaldehyde
    (iv) Acetylene into Vinyl chloride
4. Write short notes on:
    (i) Acidity of treminal alkynes
    (ii) Nucleophilic additions of alkynes.
5. Justify with examples, the following statement:
   Acetylene undergoes electrophilic as well as nucleophilic addition.
6. (a) How will you account for acidic character of acetylene H-atoms?
   (b) An organic compound (A) C_4H_6 takes up two molecules of hydrogen. It reacts with ammoniacal $AgNO_3$ to form a white precipitate. On treatment with dilute H_2SO_4 in the presence of $HgSO_4$, (A) forms a ketone (B) C_4H_8O. Write the reactions involved and propose the structures of (A) and (B).
7. What are alkynes? Why are these unsaturated in nature? Illustrate with the help of orbital structure.
8. Discuss the mechanism of the addition of the following reactions in case of acetylene:
    (i) Hydrochloric acid
    (ii) Hypobromous acid.
9. How will you account for the fact that the alkynes take part in the nucleophilic addition reactions while alkenes do not?
10. What happens when acetylene is passed through dilute sulphuric acid containing mercuric sulphate at 333 K? Write the mechanism of this reaction.
11. Complete the following reactions:
    (i) $CH \equiv CH \xrightarrow{\text{Iron tubes}}$

    (ii) $CH \equiv CH \xrightarrow[NH_4Cl]{CuCl}$

(iii) $CH_3C \equiv CH \xrightarrow[HgSO_4]{H_2SO_4}$

(iv) $CH \equiv CH + O_3 \xrightarrow{Zn/CH_3COOH}$

12. Predict the products of the following reactions:

    (i) $CH \equiv CH + Cl_2 \longrightarrow$

    (ii) $CH \equiv CH + CH_3OH \longrightarrow$

    (iii) $CH \equiv C - CH = CH_2 + HCl \longrightarrow$

13. Alkynes participate in both electrophilic and nucleophilic addition reactions. Discuss.
14. Explain why alkynes are less reactive towards electrophilic addition than alkenes.
15. Write notes on:

    (i) Acidity of terminal alkynes.

    (ii) Nucleophilic addition reactions in alkynes.

Answers
Multiple Choice Questions

	1.	2.	3.	4.	5.	6.	7.	8.	9.	10.
(a)	■				■		■			
(b)			■							
(c)								■		
(d)		■				■				

Suggested Readings

1. Finar, I.L. Organic Chemistry (Vol. 1), Dorling Kindersley (India) Pvt. Ltd.
2. Morrison, R.N. & Boyd, R.N. Chemistry, Dorling Kindersley (India) Pvt. Ltd.

Chapter 7

Cycloalkanes

LEARNING OBJECTIVES

After reading this chapter, you should be able to:
- learn the types of cycloalkanes and their stability
- understand Baeyer Strain theory
- conformational analysis of alkanes
- draw the energy diagrams of cyclohaxane
- learn about chair, boat and twist boat diagrams
- understand relative stability from energy diagrams

7.1 CYCLOALKANES

Cycloalkanes are cyclic compounds containing closed hydrocarbon chains. They are also called *alicyclic* compounds and resemble alkanes in many respects. Examples of cycloalkanes are cyclopropane, cyclobutane etc. They can expressed by a general formula $(CH_2)_n$ where n is 3, 4, 5...

Nomenclature. Common and IUPAC systems of nomenclature of cycloalkanes are given below:

Compound	Common name	IUPAC name
CH_2 / CH_2—CH_2 (triangle)	Trimethylene	Cyclopropane
CH_2—CH_2 / CH_2—CH_2 (square)	Tetramethylene	Cyclobutane
(pentagon of CH_2 groups)	Pentamethylene	Cyclopentane
(hexagon of CH_2 groups)	Hexamethylene	Cyclohexane

For the sake of simplicity, cyclopropane, cyclobutane, cyclopentane and cyclohexane are represented by the following geometrical figures.

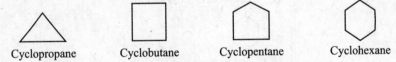

Cyclopropane   Cyclobutane   Cyclopentane   Cyclohexane

7.2 NOMENCLATURE OF SUBSTITUTED CYCLOALKANES

The following point should be kept in mind in naming cycloalkanes:

1. Nomenclature of the carbon atoms of the ring is done in such a way that the carbon atom holding the substituent gets the lowest possible number. In order to write the name of such compounds, the name of the alkyl group is prefixed to the name of the parent cycloalkane molecule. The following examples will illustrate

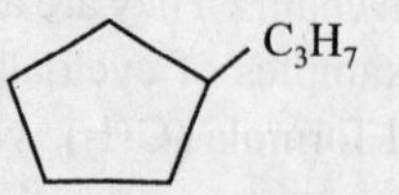

278 Chemistry for Degree Students—B.Sc. (Hons.) Semester II

Propyl cyclopentane

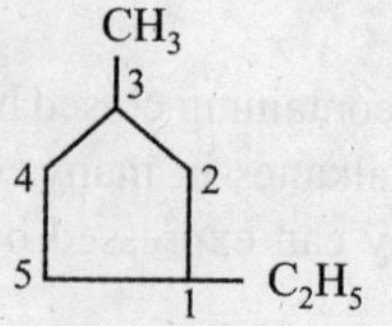

1-Ethyl-3-Methyl cyclopentane

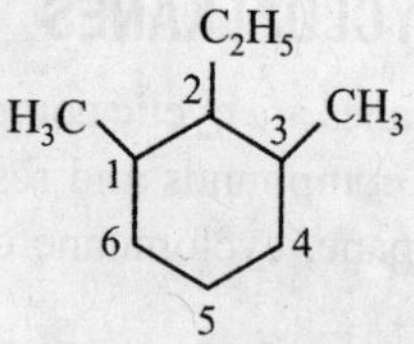

2-Ethyl-1, 3-Dimethyl cyclohexane

2. If the number of the carbon atoms in the ring is less the number of carbon atoms in the alkyl chain, then the ring structure is taken as the substituent and is named as cycloalkyl group. For example,

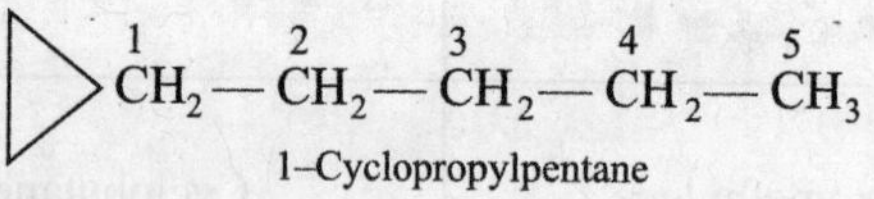

1–Cyclopropylpentane

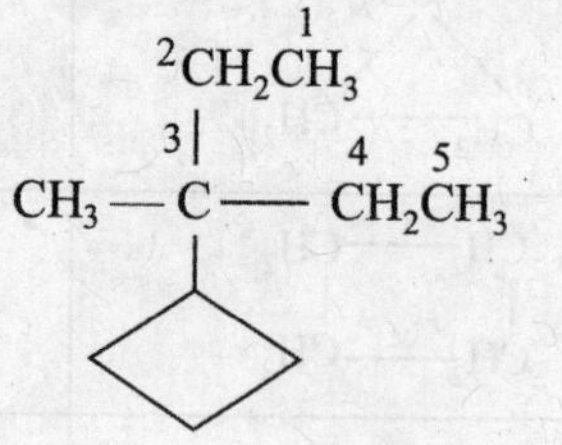

3–Methyl 3–cyclobutylpentane

Parenthesis are used whenever necessary to avoid confusion. Consider the following example:

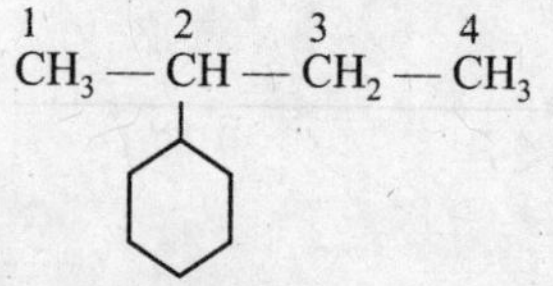

$$\overset{1}{CH_3} - \overset{2}{CH} - \overset{3}{CH_2} - \overset{4}{CH_3}$$

(2-Butyl) cyclohexane

2-Butyl in the parenthesis shows that the locant 2 refers to the substituent (side chain) attached to the ring and not to the ring itself.

3. If the side chain contains a multiple bond or a functional group, the alicyclic ring is treated as the substituent irrespective of the size of the ring. For example,

$$\overset{4}{CH} = \overset{3}{CH} - \overset{2}{CO} - \overset{1}{CH_3}$$

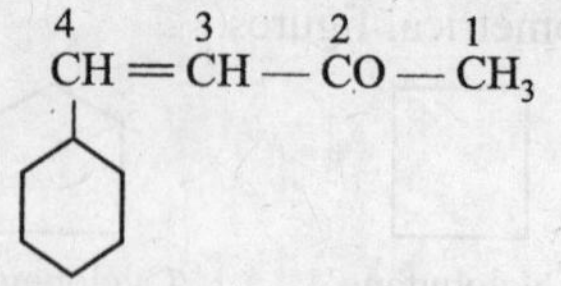

4-cyclohexylbut-3-en-2-one

4. If more than one alicyclic ring is attached to a single chain, the compound is named as a derivative of alkane irrespective of the size of the ring. For example,

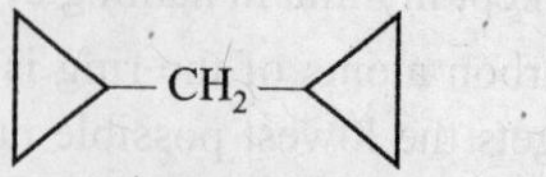

Dicyclopylmethane

5. If a multiple (double or triple) bond and some other substituents are present in the ring, the numbering is done in such a way that the carbon atoms of the multiple bond gets lowest locants 1 and 2 and the substituent groups get the lower locants at the first point of difference.

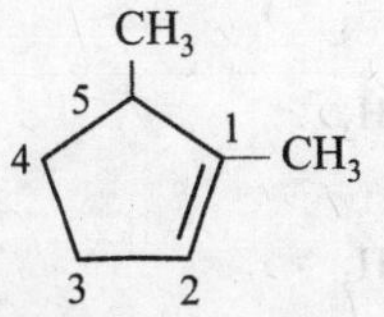

1,5-Dimethylcyclopent-1-ene

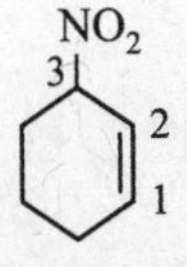

3-Nitrocyclohex-1-ene

6. If the ring contains a multiple bond and the side chain contains a functional group, then the ring is treated as a substituent and the compound is named as a derivative of the side chain.

$$\overset{3}{C}H_3 - \overset{2}{C}H - \overset{1}{C}H_2OH$$

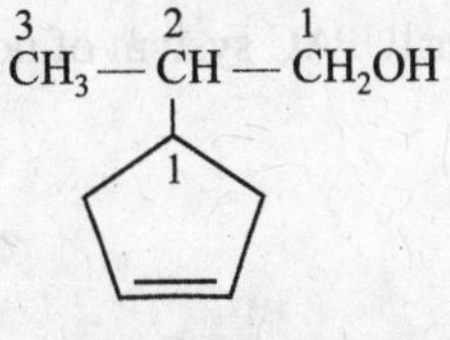

2-(cyclopen-3-en-1-yl) propan-1-ol

7. If the ring as well as the side chain contain functional groups, the compound is named as a derivative of the side chain or the alicyclic ring according as the side chain or the ring contains the principal functional group. For example,

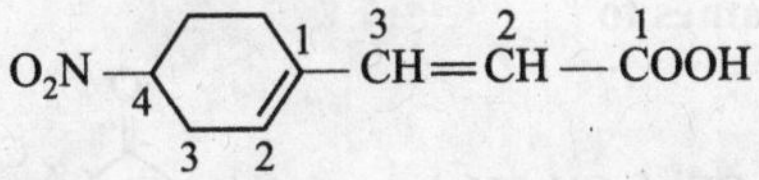

3-(4-nitrocyclohex-1-en-1-yl) prop-2-en-1-oic acid

However, if the alicyclic ring and the side chain contain the same functional group, the compound is named as a derivative of the side chain or the ring according as the side chain or the ring contains higher number of carbon atoms. For example,

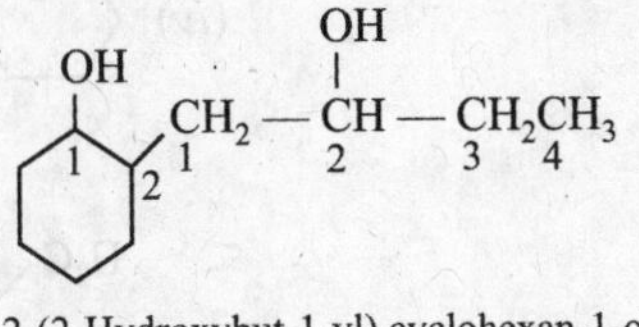

2-(2-Hydroxybut-1-yl) cyclohexan-1-ol

8. If the compound contains an alicyclic ring attached to directly to an aromatic ring, it is named as a derivative of benzene (or a compound having lowest rate of hydrogenation). For example,

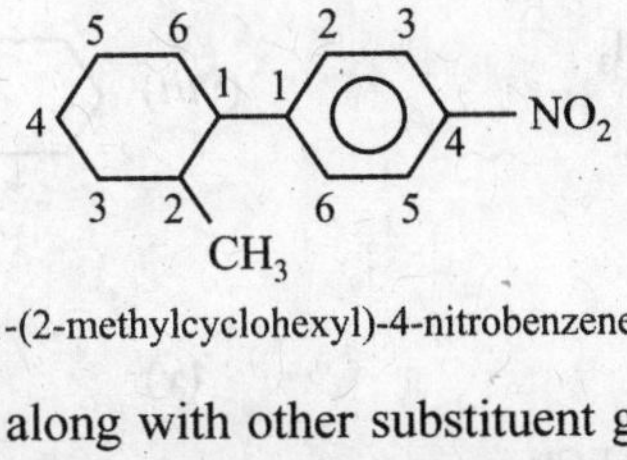

1-(2-methylcyclohexyl)-4-nitrobenzene

9. If some functional group along with other substituent groups are present in the ring, it is indicated by some appropriate prefix or suffix and its position is indicated by numbering the carbon atoms of the ring in such a way that the functional group gets the lowest number. For example,

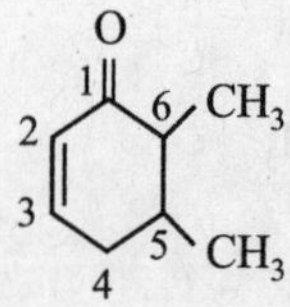

5,6-Dimethylcyclohex-2-en-1-one

10. If the alicyclic compound contains carbon-containing functional groups, the carbon atoms of the functional group are not included in the parent name of alicyclic system. For such system prefixes and suffixes as per IUPAC system of nomenclature are used

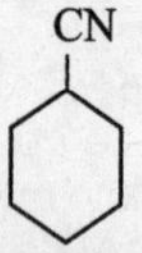

Cyclohexanecarbonitrile 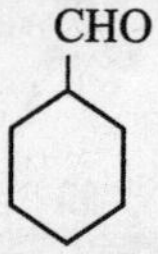   Cyclohexanecarbaldehyde

Solved Examples

Example 1: Assign IUPAC names to

(i) cyclopropyl—CH=CH—CH(CH$_3$)—CH$_2$CH$_3$

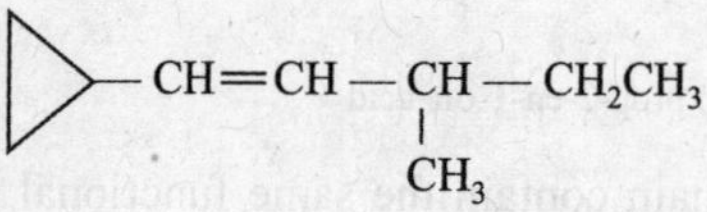

(ii) cyclopentanone with COOCH$_3$ substituent

(iii) cyclohex-2-en-1-ol

(iv) bicyclohexyl

(v) HOCH$_2$—C(CH$_3$)—CH=CHCH$_2$COOH

(vi) cyclobutene with H$_3$C and CH$_2$CHO substituents

(vii) cyclohexanone with —CH$_2$CH$_2$COCH$_3$ substituent

(viii) cyclohexyl—CONHC$_6$H$_5$

(ix) cyclohexane with [CH$_2$]$_9$CH$_3$, CH$_3$[CH$_2$]$_9$, [CH$_2$]$_9$CH$_3$ substituents

(x) cyclohexanone with COOH and OHC substituents

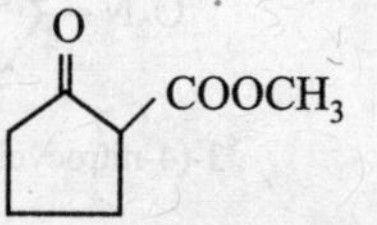

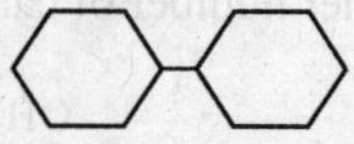

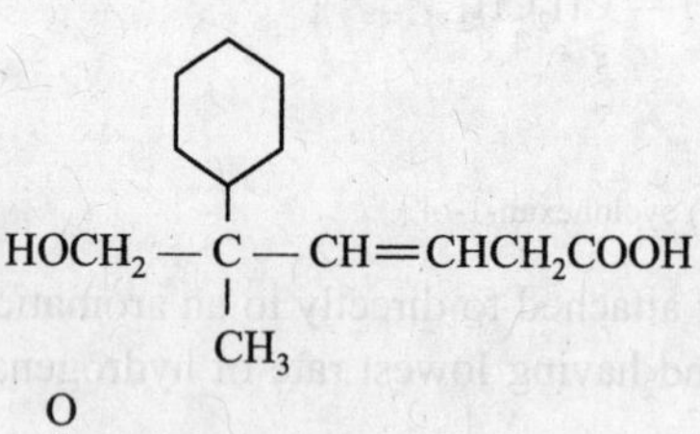

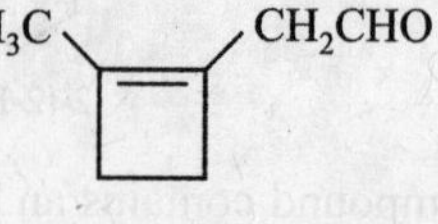

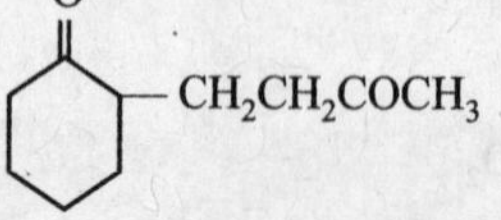

(xi) cyclohexyl=C=O

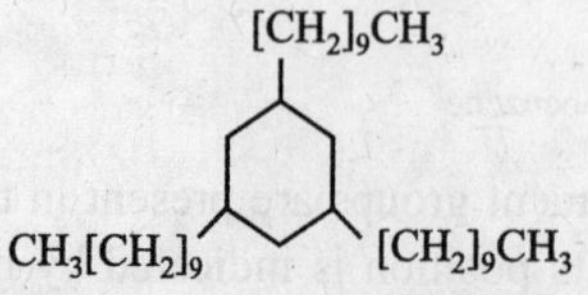

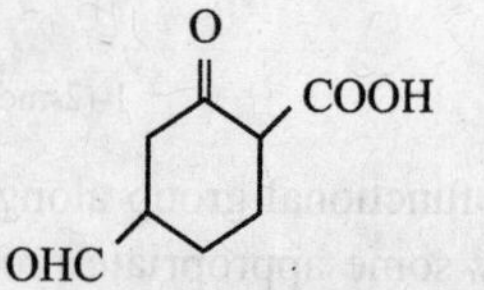

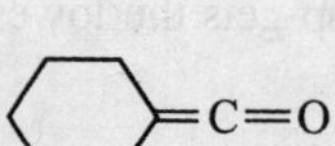

Answers

(i) 1-Cyclopropyl-3-methylpent-1-ene
(ii) Methyl (2-oxocyclopentane)-1-carboxylate
(iii) Cyclohex-2-en-1-ol
(iv) Cyclohexylcyclohexane
(v) 5-Cyclohexyl-6 hydroxy-5-methylhex-3-en-1-oic acid
(vi) 2-(2-Methylcyclobut-1-en-1-yl) ethanal
(vii) 2-(3-Oxobutyl)cyclohexan-1-one
(viii) Cyclohexanecarboxanilide
(ix) 1, 3, 5-Tris (decyl) cyclohexane
(x) 4-Formyl-2-oxocyclohexane-1-carboxylic acid
(xi) Cyclohexylidenemethanone

7.3 METHODS OF PREPARATION OF CYCLOALKANES AND ITS DERIVATIVES

(1) By the addition of Carbenes to Olefins:

Carbenes are the natural divalent carbon intermediates in which carbon is covalently bonded to two atoms and has two non bonded orbitals containing two electrons between them. Thus, a carbene has a sextet of electrons without any charge on it. Carbenes are of two types, singlet carbenes ($\uparrow \downarrow$) and triplet carbenes ($\uparrow \uparrow$). Both of them react with alkenes to form cycloalkane derivatives and the reaction is called **cycloaddition reaction**.

(a) Singlet carbenes. They react with alkanes in a stereospecific manner in which stereochemistry of alkane is retained in the cycloaddition product formed. This means that a *cis* alkene forms *cis* addition product while *trans* alkene gives rise to *trans* addition product explained below:

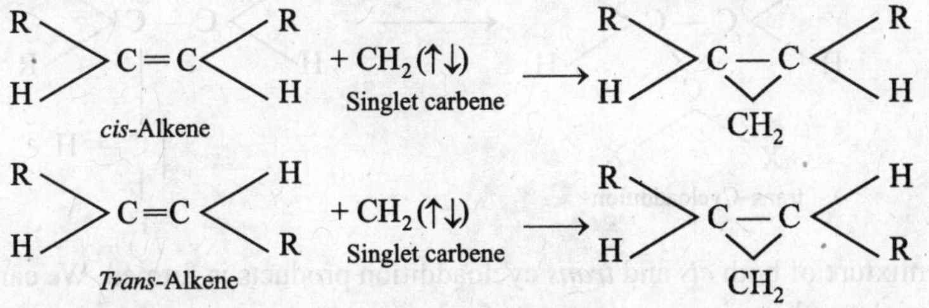

Mechanism. Addition of singlet carbene to an alkene is a **concerted** process and there is no spin restriction on the simultaneous formation of two new σ-bonds of the cyclopropane. Therefore, it is stereospecific in nature.

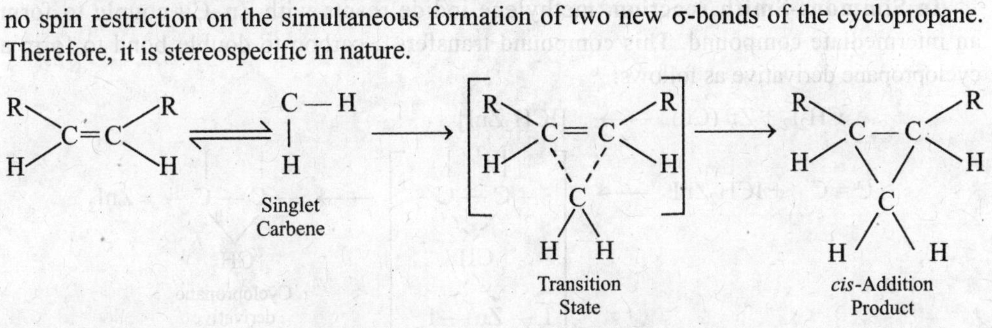

(b) Triplet carbenes. The addition of triplet carbene to alkane proceeds in a non-stereospecific manner, therefore, both *cis* and *trans* alkenes give mixture of cycloaddition products.

$$\underset{\text{Cis-Alkene}}{\overset{R}{\underset{H}{>}}C=C\overset{R}{\underset{H}{<}}} \quad \underset{\text{Triplet carbene}}{CH_2(\uparrow\uparrow)} \longrightarrow \underset{H}{\overset{R}{>}}C\underset{CH_2}{-}C\overset{R}{\underset{H}{<}} + \underset{H}{\overset{R}{>}}C\underset{CH_2}{-}C\overset{H}{\underset{R}{<}}$$

Mixture of products

Mechanism. The addition of triplet carbene to an alkene follows a different mechanism. The first step in the addition gives a biradical. It may undergo *spin inversion* before it cyclises into cyclopropane derivative. But at the same time, there is also possibility of free rotation about C — C bond. In case, this rotation is faster than the spin inversion then another cycloaddition product may also be formed as explained below:

$$\underset{H}{\overset{R}{>}}C=C\overset{R}{\underset{H}{<}} + \overset{C-H(\uparrow\uparrow)}{\underset{H}{|}} \longrightarrow \underset{H}{\overset{R}{>}}\overset{\uparrow}{C}-C\overset{R}{\underset{H}{<}} \xrightarrow{\text{Spin Inversion}} \underset{H}{\overset{R}{>}}\overset{\downarrow}{C}-C\overset{R}{\underset{H}{<}}$$

Biradical

$\downarrow$ (i) Rotation about C — C bond
$\downarrow$ (ii) Spin Inversion

$$\underset{H}{\overset{R}{>}}C-C\overset{R}{\underset{H}{<}} \qquad \underset{H}{\overset{R}{>}}C-C\overset{H}{\underset{R}{<}}$$

cis–Cycloaddition

$$\underset{H}{\overset{R}{>}}C-C\overset{H}{\underset{R}{<}} \longleftarrow \underset{H}{\overset{R}{>}}C-C\overset{H}{\underset{R}{<}}$$

trans–Cycloaddition

Thus, a mixture of both *cis* and *trans* cycloaddition products is formed. We can say that the addition of triplet carbene is non-stereospecific in nature.

In Simmons-Smith reaction, methylene iodide reacts with Zn–Cu couple to form an intermediate compound. This compound transfers a carbon to double bond to form a cyclopropane derivative as follows:

$$CH_2I_2 + Zn(Cu) \longrightarrow [ICH_2ZnI]$$

$$>C=C< + ICH_2ZnI \longrightarrow \begin{bmatrix} -C-C- \\ CH_2 \\ I-Zn-I \end{bmatrix} \longrightarrow -C-C- + ZnI_2$$

Cyclopropane derivative

2. By dehalogenation of α, ω-dihalogen derivatives of alkanes (Freund's method). A dihalogen compound containing halogens on terminal position on treatment with Zn, in the presence of NaI catalyst gets dehalogenated. The terminal halogen atoms are removed thereby closing the ring. The starting material to be taken depends upon the desired cycloalkane, as explained below:

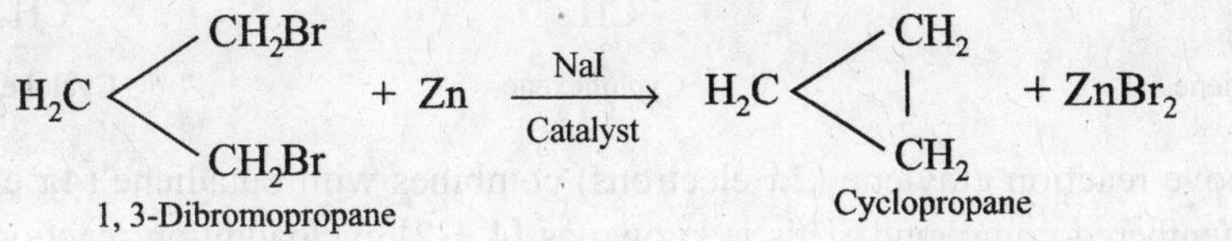

$$H_2C\begin{smallmatrix}CH_2Br\\ \\CH_2Br\end{smallmatrix} + Zn \xrightarrow{\text{NaI Catalyst}} H_2C\begin{smallmatrix}CH_2\\ | \\CH_2\end{smallmatrix} + ZnBr_2$$

1, 3-Dibromopropane   Cyclopropane

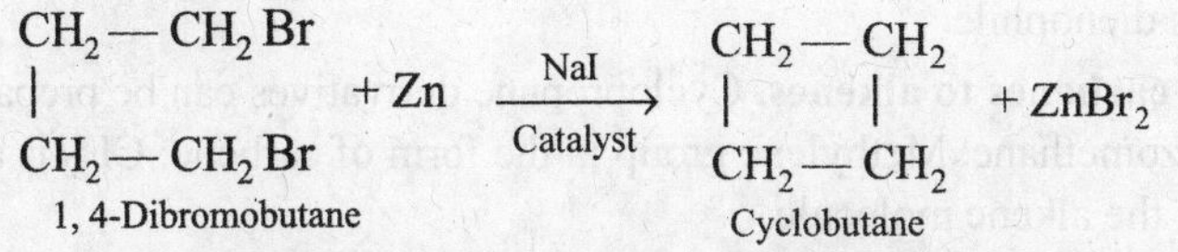

$$\begin{smallmatrix}CH_2-CH_2Br\\|\\CH_2-CH_2Br\end{smallmatrix} + Zn \xrightarrow{\text{NaI Catalyst}} \begin{smallmatrix}CH_2-CH_2\\|\quad\ \ |\\CH_2-CH_2\end{smallmatrix} + ZnBr_2$$

1, 4-Dibromobutane   Cyclobutane

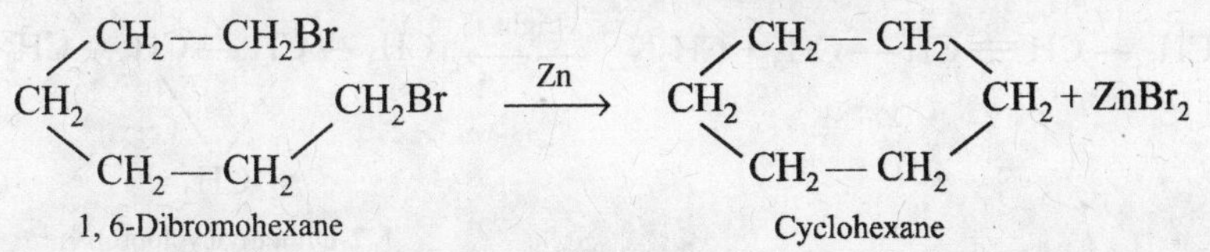

1, 6-Dibromohexane   Cyclohexane

3. Distillation of calcium or barium salts of dicarboxylic acids followed by reduction (Wislicenus method). A cyclic ketone results when a calcium or barium salt of a dicarboxylic acid is distilled. Ketone obtained is subjected to Clemmenson reduction, giving the cycloalkane as illustrated by the following reaction.

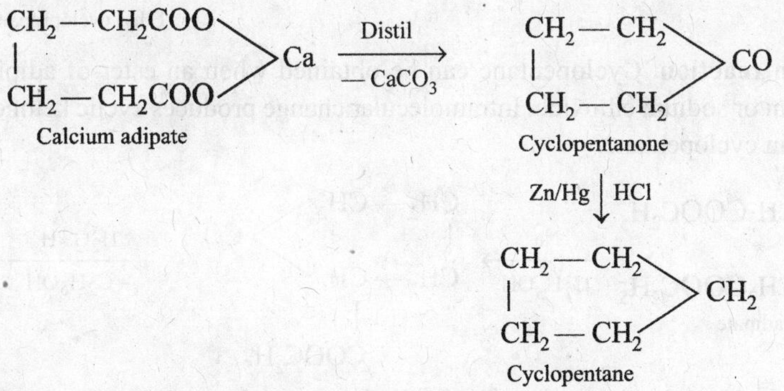

Calcium adipate $\xrightarrow[-CaCO_3]{\text{Distil}}$ Cyclopentanone $\xrightarrow[HCl]{Zn/Hg}$ Cyclopentane

4. Reduction of aromatic compounds. Cyclohexane and its derivatives can be prepared by passing hydrogen gas through benzene or its derivatives, in the presence of Ni at 423–473 K temperature.

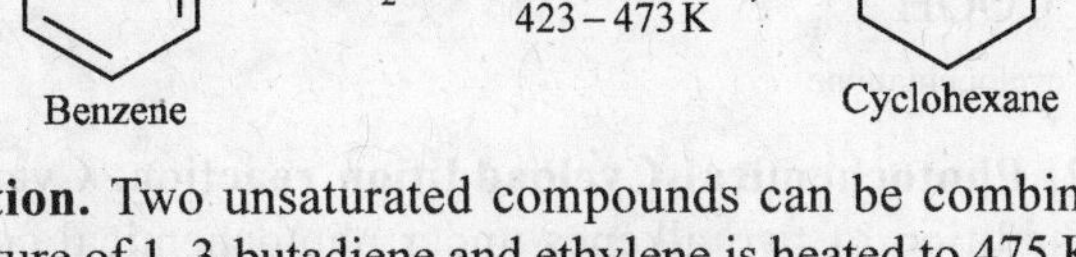

Benzene $+ 3H_2 \xrightarrow[423-473\,K]{Ni}$ Cyclohexane

5. Diels-Alder addition. Two unsaturated compounds can be combined to produce a cycloalkane. When a mixture of 1, 3-butadiene and ethylene is heated to 475 K, cyclohexene is obtained. It can be reduced with hydrogen to yield cyclohexane.

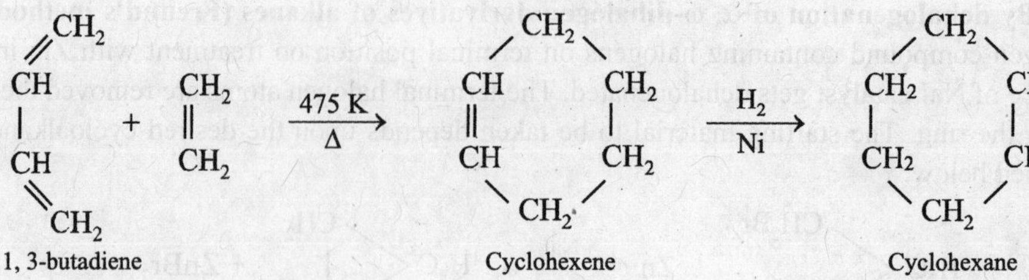

$$\underset{\text{1, 3-butadiene}}{\begin{array}{c}CH_2\\ \| \\ CH \\ | \\ CH \\ \| \\ CH_2\end{array}} + \underset{}{\begin{array}{c}CH_2\\ \| \\ CH_2\end{array}} \xrightarrow[\Delta]{475\text{ K}} \underset{\text{Cyclohexene}}{\begin{array}{c}CH_2 \\ / \quad \backslash \\ CH \quad CH_2 \\ \| \qquad | \\ CH \quad CH_2 \\ \backslash \quad / \\ CH_2\end{array}} \xrightarrow[Ni]{H_2} \underset{\text{Cyclohexane}}{\begin{array}{c}CH_2 \\ / \quad \backslash \\ CH_2 \quad CH_2 \\ | \qquad | \\ CH_2 \quad CH_2 \\ \backslash \quad / \\ CH_2\end{array}}$$

In the above reaction ethylene (2π electrons) combines with butadiene (4π electrons) to form a six membered compound. This is known as [4 + 2] cycloaddition reaction where the alkene is known as dienophile.

6. Addition of carbenes to alkenes. Cyclopropane derivatives can be prepared by treating an alkene with diazomethane. Methylene group in the form of carbene: CH_2 is attached across the double bond in the alkene molecule.

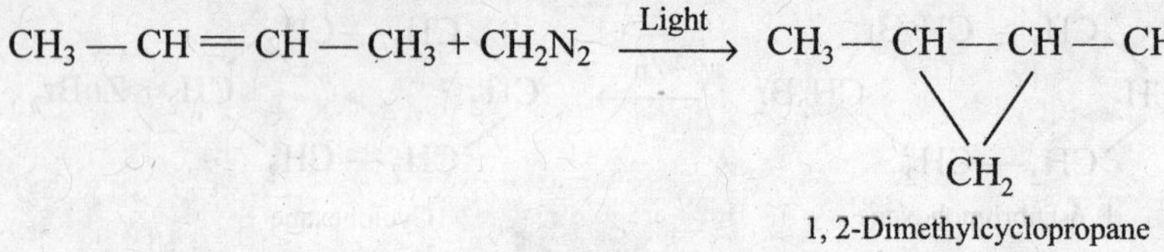

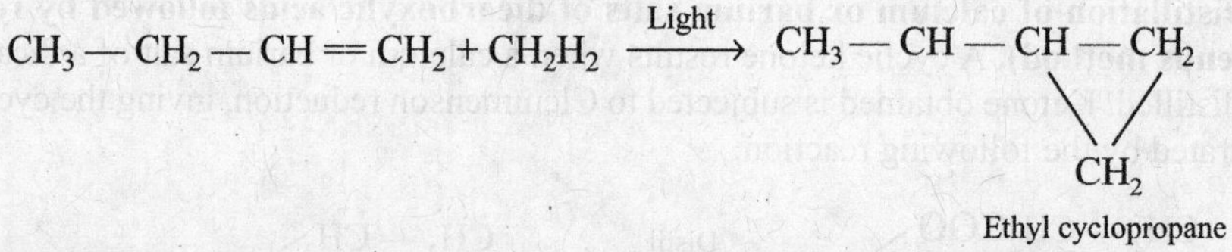

$$CH_3-CH=CH-CH_3 + CH_2N_2 \xrightarrow{\text{Light}} \underset{\text{1, 2-Dimethylcyclopropane}}{CH_3-CH-CH-CH_3 \atop \backslash \quad / \atop CH_2}$$

$$CH_3-CH_2-CH=CH_2 + CH_2H_2 \xrightarrow{\text{Light}} \underset{\text{Ethyl cyclopropane}}{CH_3-CH-CH-CH_2 \atop \backslash \quad / \atop CH_2}$$

7. Dieckmann reaction. Cyclopentane can be obtained when an ester of adipic acid is treated with sodium or sodium ethoxide. Intramolecular change produces cyclic ketone. Ketone is reduced to obtain cyclopentane.

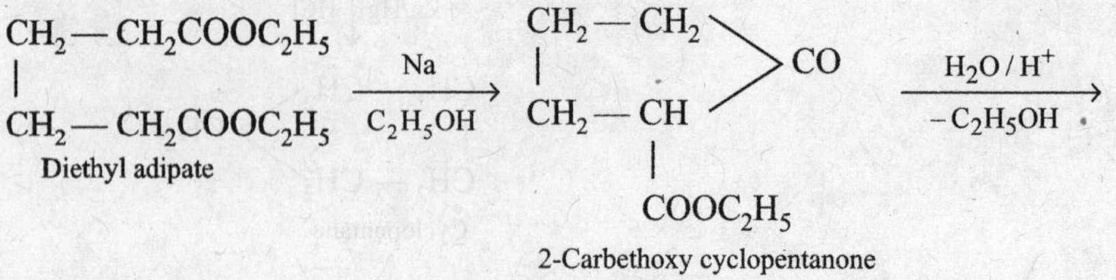

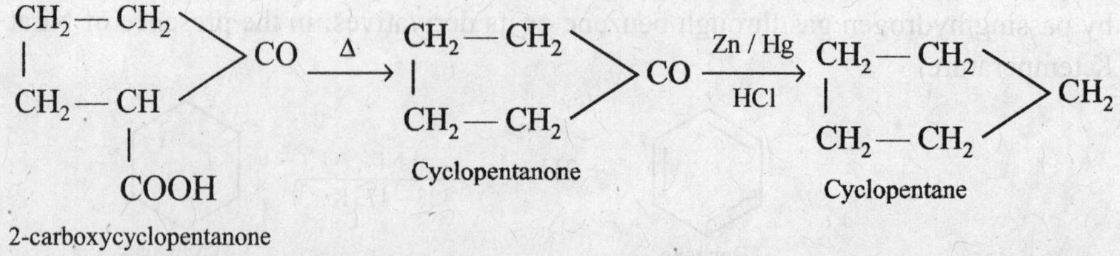

$$\underset{\text{Diethyl adipate}}{\begin{array}{c}CH_2-CH_2COOC_2H_5 \\ | \\ CH_2-CH_2COOC_2H_5\end{array}} \xrightarrow[C_2H_5OH]{Na} \underset{\text{2-Carbethoxy cyclopentanone}}{\begin{array}{c}CH_2-CH_2 \\ | \qquad \backslash \\ CH_2-CH \quad >CO \\ | \\ COOC_2H_5\end{array}} \xrightarrow[-C_2H_5OH]{H_2O/H^+}$$

$$\underset{\text{2-carboxycyclopentanone}}{\begin{array}{c}CH_2-CH_2 \\ | \qquad \backslash \\ CH_2-CH \quad >CO \\ | \\ COOH\end{array}} \xrightarrow{\Delta} \underset{\text{Cyclopentanone}}{\begin{array}{c}CH_2-CH_2 \\ | \qquad \backslash \\ | \qquad \quad >CO \\ CH_2-CH_2\end{array}} \xrightarrow[HCl]{Zn/Hg} \underset{\text{Cyclopentane}}{\begin{array}{c}CH_2-CH_2 \\ | \qquad \backslash \\ | \qquad \quad >CH_2 \\ CH_2-CH_2\end{array}}$$

8. By [2 + 2] Photochemical Cycloaddition reaction. Cyclobutane derivatives can be obtained by addition of two alkenes under photochemical conditions. [2 + 2] means combination of two molecules using two π electrons each.

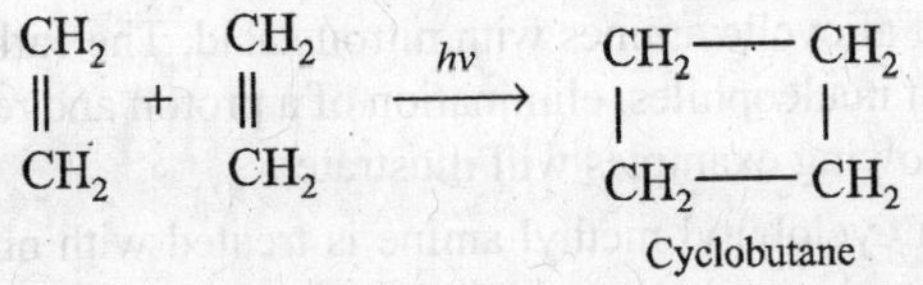

CH2 CH2 hv CH2 CH2
CH2 CH2 CH2 CH2
Cyclobutane

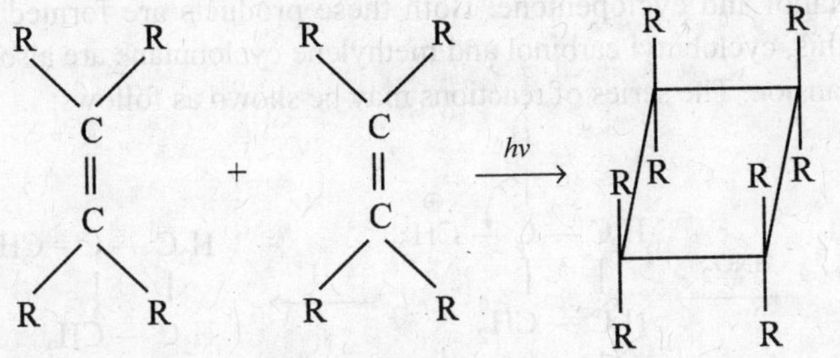

9. Tharpe-Ziegler reaction. A dinitrile is treated with LiN $(C_2H_5)_2$ in a large volume of benzene or toluene as solvent. A cyclic imino compound is obtained which on hydrolysis yields a cycloketone.

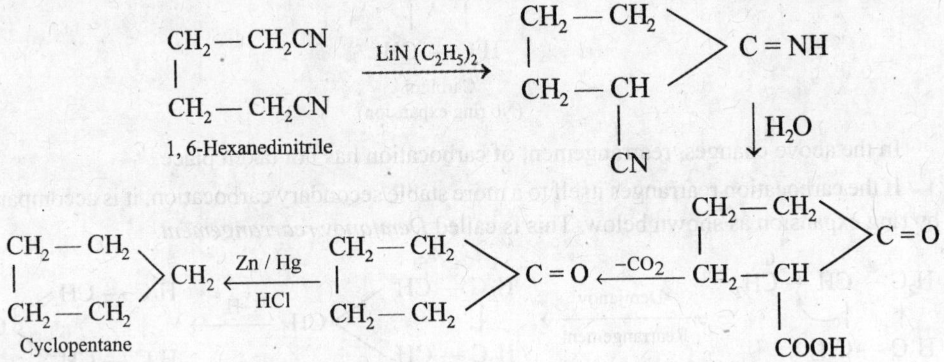

1, 6-Hexanedinitrile
LiN (C2H5)2
H2O
Zn / Hg
HCl
–CO2
Cyclopentane

10. From Malonic ester and Acetoacetic ester

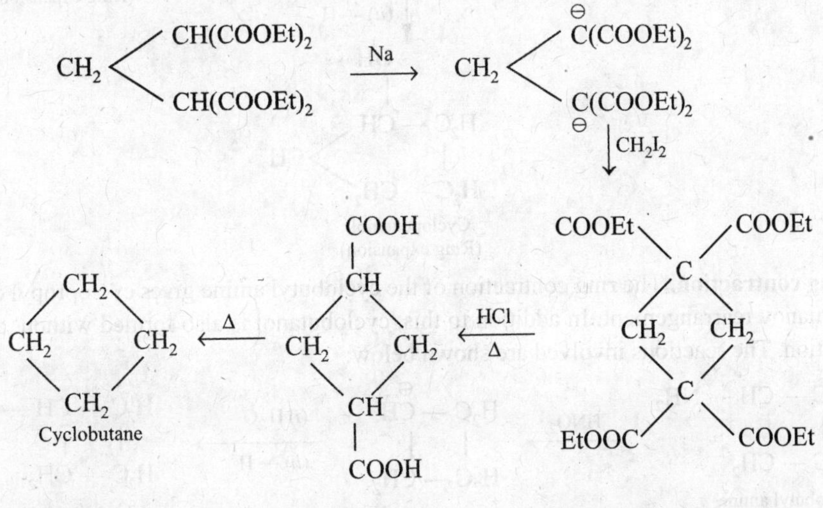

Na
CH2I2
HCl
Δ
Cyclobutane

11. By Demjanov rearrangement. This is a convenient method of expanding and contracting alicylic ring systems through the formation of carbocation intermediates. This is achieved by deamination of alicyclic amines with nitrous acid. The carbocations may undergo reactions such as addition of nucleophiles, elimination of a proton and rearrangement to a more stable carbocation. The following examples will illustrate.

Ring expansion. When cyclobutyl methyl amine is treated with nitrous acid, the product is a mixture of cyclopentanol and cyclopentene. Both these products are formed by ring expansion. In addition to this, cyclobutyl carbinol and methylene cyclobutane are also formed where there is no ring expansion. The series of reactions may be shown as follows:

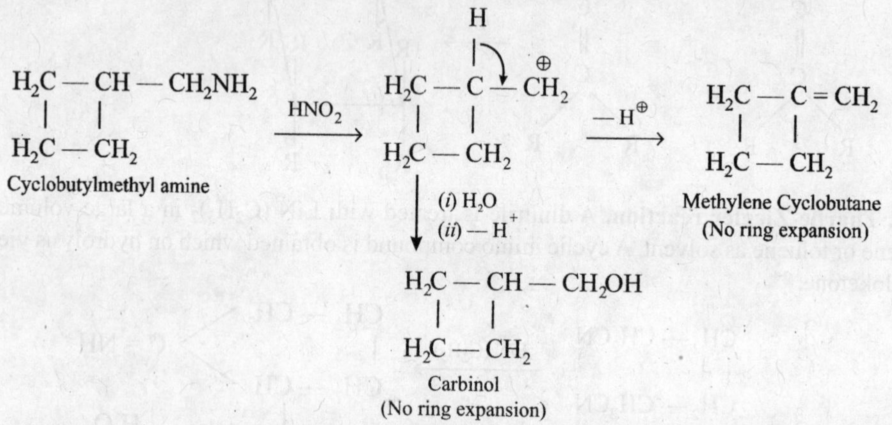

$$\underset{\text{Cyclobutylmethyl amine}}{\begin{array}{c} H_2C-CH-CH_2NH_2 \\ | \quad | \\ H_2C-CH_2 \end{array}} \xrightarrow{HNO_2} \begin{array}{c} H_2C-\overset{H}{\overset{|}{C}}\overset{\curvearrowleft}{-}\overset{\oplus}{C}H_2 \\ | \quad | \\ H_2C-CH_2 \end{array} \xrightarrow{-H^{\oplus}} \underset{\substack{\text{Methylene Cyclobutane} \\ \text{(No ring expansion)}}}{\begin{array}{c} H_2C-C=CH_2 \\ | \quad | \\ H_2C-CH_2 \end{array}}$$

$$\Big\downarrow \begin{array}{l}(i)\ H_2O \\ (ii)\ -H^+\end{array}$$

$$\underset{\substack{\text{Carbinol} \\ \text{(No ring expansion)}}}{\begin{array}{c} H_2C-CH-CH_2OH \\ | \quad | \\ H_2C-CH_2 \end{array}}$$

In the above changes, rearrangement of carbocation has not taken place.

If the carbocation rearranges itself to a more stable secondary carbocation, it is accompanied by ring expansion as shown below. This is called *Demjanov rearrangement*.

$$\begin{array}{c} H_2C-\overset{\curvearrowright}{C}H-\overset{\oplus}{C}H_2 \\ | \quad | \\ H_2C-CH_2 \end{array} \xrightarrow[\text{Rearrangement}]{\text{Demjanov}} \begin{array}{c} H_2C-\overset{\oplus}{C}H \\ | \qquad \diagdown \\ \qquad \qquad CH_2 \\ | \qquad \diagup \\ H_2C-CH_2 \end{array} \xrightarrow{-H^+} \underset{\substack{\text{Cyclopentane} \\ \text{(Ring expansion)}}}{\begin{array}{c} H_2C-CH \\ | \qquad \diagdown \\ \qquad \qquad CH \\ | \qquad \diagup \\ H_2C-CH_2 \end{array}}$$

$$\Big\downarrow \begin{array}{l}(i)\ H_2O \\ (ii)\ -H^+\end{array}$$

$$\underset{\substack{\text{Cyclopentanol} \\ \text{(Ring expansion)}}}{\begin{array}{c} \qquad OH \\ \qquad | \\ H_2C-CH \\ | \qquad \diagdown \\ \qquad \qquad CH_2 \\ | \qquad \diagup \\ H_2C-CH_2 \end{array}}$$

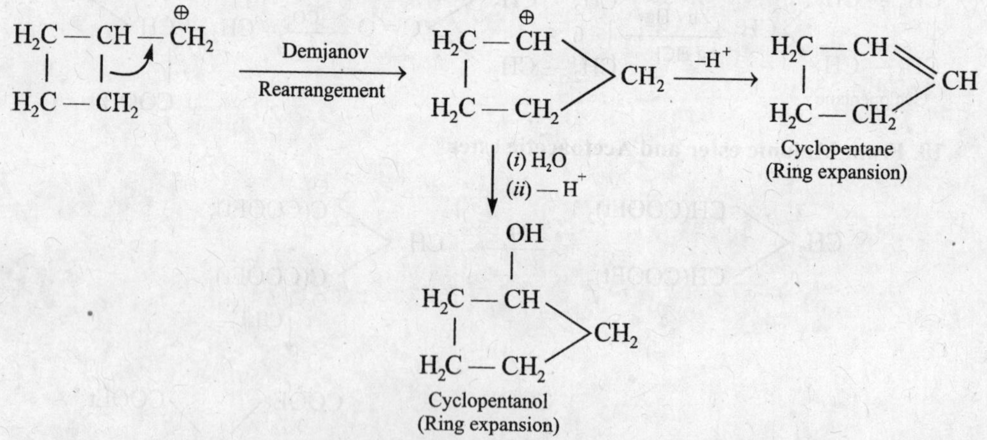

Ring contraction. The ring contraction of the cyclobutyl amine gives cyclopropyl carbinol by Demjanov rearrangement. In addition to this, cyclobutanol is also formed without any ring contraction. The reactions involved are shown below:

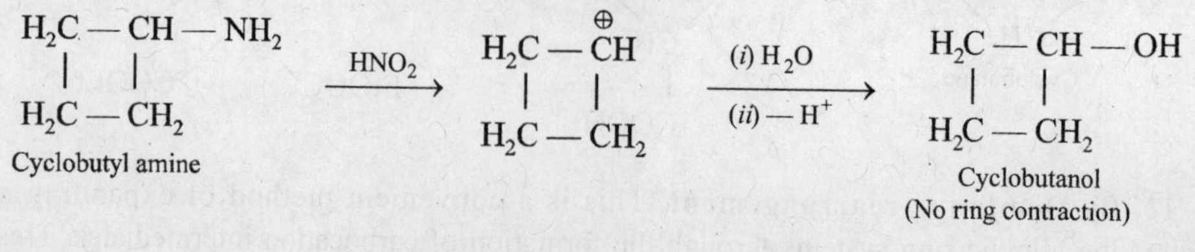

$$\underset{\text{Cyclobutyl amine}}{\begin{array}{c} H_2C-CH-NH_2 \\ | \quad | \\ H_2C-CH_2 \end{array}} \xrightarrow{HNO_2} \begin{array}{c} H_2C-\overset{\oplus}{C}H \\ | \quad | \\ H_2C-CH_2 \end{array} \xrightarrow[(ii)\ -H^+]{(i)\ H_2O} \underset{\substack{\text{Cyclobutanol} \\ \text{(No ring contraction)}}}{\begin{array}{c} H_2C-CH-OH \\ | \quad | \\ H_2C-CH_2 \end{array}}$$

$$H_2C\overset{\oplus}{-}CH \atop H_2C-CH_2 \xrightarrow{\text{Demjanov Rearrangement}} {H_2C-CH-\overset{+}{CH_2} \atop H_2C} \xrightarrow[(ii)-H^+]{(i) H_2O} {H_2C-CH-CH_2-OH \atop H_2C}$$

Cyclopropyl carbinol

It can thus be generalised that in the Demjanov rearrangement, ring contraction takes place when the carbocation intermediate has a positive charge on the alicyclic carbon atom. Similarly, ring expansion takes place when the carbocation has a positive charge on the carbon atom α- to the alicyclic ring.

7.4 BLANC'S RULE

Blanc's rule states that:

Pyrolysis of 1, 4- and 1, 5-linear carboxylic acids yields cyclic anhydrides while 1, 6- and 1, 7-dicarboxylic acids give cyclic ketones. 1, 8- and higher dicarboxylic acids remain unaffected.

Thus,

$$\begin{array}{c} CH_2COOH \\ | \\ CH_2COOH \end{array} \xrightarrow[-H_2O]{\Delta} \begin{array}{c} CH_2CO \\ | \diagdown \\   O \\ | \diagup \\ CH_2CO \end{array}$$

Succinic acid (1, 4-Dicarboxylic acid)   Succinic anhydride

$$\begin{array}{c} CH_2CH_2COOH \\ | \\ CH_2CH_2COOH \end{array} \xrightarrow[-H_2O]{\Delta \atop -CO_2} \begin{array}{c} CH_2CH_2 \\ | \diagdown \\   O \\ | \diagup \\ CH_2CH_2 \end{array}$$

Adipic acid (1, 6-Dicarboxylic acid)   Cyclopentanone

$$\begin{array}{c} CH_2CH_2CH_2COOH \\ | \\ CH_2CH_2CH_2COOH \end{array} \xrightarrow{\Delta} \text{No Reaction}$$

1, 8-dicarboxylic acid

Utility of the Blanc's rule

This rule finds use in the determination of size of the ring and chain length of the carboxylic acid. The number of carbons in the ring is reduced by one when the cyclization of the dicarboxylic compound takes place (Refer to cyclization of adipic acid into cyclopentanone).

In an unknown ring compound, whose size is desired to be determined, a double bond is created by a known method. The compound is then subjected to oxidation with $KMnO_4$ when the fission of double bond takes place and a dicarboxylic compound is obtained. This dicarboxylic compound is pyrolysed to obtain a cyclic compound and this can provide information about the size of the original ring.

Solved Examples

Example 2: How can you obtain
  (i) Ethyl cyclopentane
  (ii) Methyl cyclopentane?

Solution:

(i) Ethyl cyclopentane

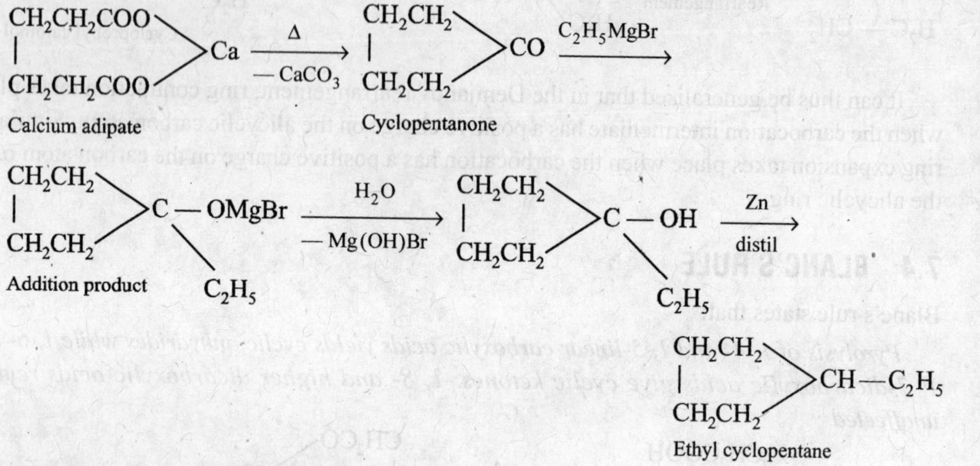

$$\text{Calcium adipate} \begin{array}{c} CH_2CH_2COO \\ | \\ CH_2CH_2COO \end{array}\!\!\!\!\Big> Ca \xrightarrow[-CaCO_3]{\Delta} \begin{array}{c} CH_2CH_2 \\ | \\ CH_2CH_2 \end{array}\!\!\!\!\Big> CO \xrightarrow{C_2H_5MgBr}$$

Cyclopentanone

$$\begin{array}{c} CH_2CH_2 \\ | \\ CH_2CH_2 \end{array}\!\!\!\!\Big> C \!\!-\!\! \begin{array}{c} OMgBr \\ \\ C_2H_5 \end{array} \xrightarrow[-Mg(OH)Br]{H_2O} \begin{array}{c} CH_2CH_2 \\ | \\ CH_2CH_2 \end{array}\!\!\!\!\Big> C\!\!-\!\!OH \xrightarrow[\text{distil}]{Zn}$$

Addition product

$$\begin{array}{c} CH_2CH_2 \\ | \\ CH_2CH_2 \end{array}\!\!\!\!\Big> CH\!\!-\!\!C_2H_5$$

Ethyl cyclopentane

It can alternatively, be prepared as under

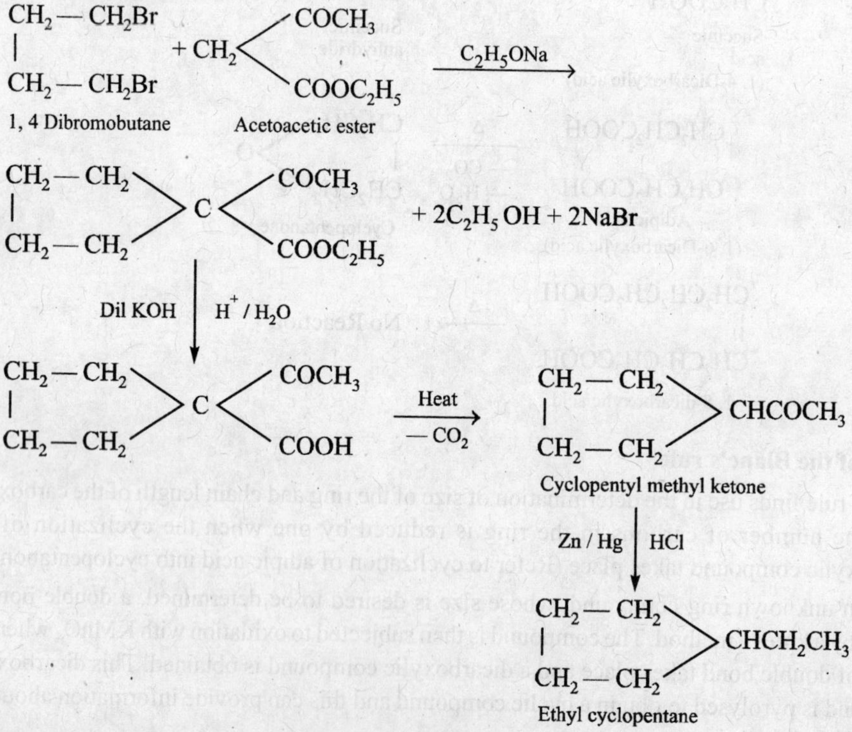

$$\begin{array}{c} CH_2\!\!-\!\!CH_2Br \\ | \\ CH_2\!\!-\!\!CH_2Br \end{array} + CH_2\!\!\!\Big< \begin{array}{c} COCH_3 \\ \\ COOC_2H_5 \end{array} \xrightarrow{C_2H_5ONa}$$

1, 4 Dibromobutane   Acetoacetic ester

$$\begin{array}{c} CH_2\!\!-\!\!CH_2 \\ | \\ CH_2\!\!-\!\!CH_2 \end{array}\!\!\!\!\Big> C\!\!\!\Big< \begin{array}{c} COCH_3 \\ \\ COOC_2H_5 \end{array} + 2C_2H_5OH + 2NaBr$$

$$\Bigg\downarrow \begin{array}{c} \text{Dil KOH} \\ H^+/H_2O \end{array}$$

$$\begin{array}{c} CH_2\!\!-\!\!CH_2 \\ | \\ CH_2\!\!-\!\!CH_2 \end{array}\!\!\!\!\Big> C\!\!\!\Big< \begin{array}{c} COCH_3 \\ \\ COOH \end{array} \xrightarrow[-CO_2]{\text{Heat}} \begin{array}{c} CH_2\!\!-\!\!CH_2 \\ | \\ CH_2\!\!-\!\!CH_2 \end{array}\!\!\!\!\Big> CHCOCH_3$$

Cyclopentyl methyl ketone

$$\Bigg\downarrow \begin{array}{c} Zn/Hg \quad HCl \end{array}$$

$$\begin{array}{c} CH_2\!\!-\!\!CH_2 \\ | \\ CH_2\!\!-\!\!CH_2 \end{array}\!\!\!\!\Big> CHCH_2CH_3$$

Ethyl cyclopentane

(ii) Methyl cyclopentane

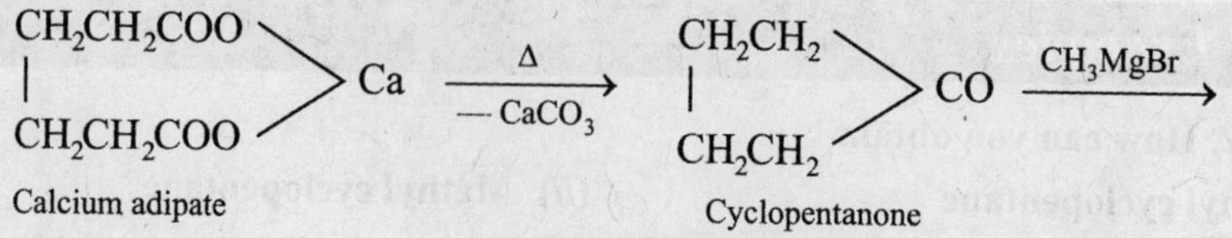

$$\text{Calcium adipate} \begin{array}{c} CH_2CH_2COO \\ | \\ CH_2CH_2COO \end{array}\!\!\!\!\Big> Ca \xrightarrow[-CaCO_3]{\Delta} \begin{array}{c} CH_2CH_2 \\ | \\ CH_2CH_2 \end{array}\!\!\!\!\Big> CO \xrightarrow{CH_3MgBr}$$

Cyclopentanone

$$\underset{\text{Addition product}}{\underset{CH_2CH_2}{\overset{CH_2CH_2}{\diagup}}\!\!\!\!\overset{|}{C}\!\!-\!\!\underset{CH_3}{\overset{OMgBr}{|}}} \xrightarrow[-Mg(OH)Br]{H_2O} \underset{CH_2CH_2}{\underset{|}{\overset{CH_2CH_2}{\diagup}}\!\!\!\!\overset{|}{C}\!\!-\!\!\underset{CH_3}{\overset{OH}{|}}} \xrightarrow[\text{distil}]{Zn}$$

$$\underset{\text{Methyl cyclopentane}}{\underset{CH_2CH_2}{\underset{|}{\overset{CH_2CH_2}{\diagup}}}\!\!\!\!CH\!-\!CH_3}$$

7.5 CHEMICAL PROPERTIES OF CYCLOALKANES

Cycloalkanes are saturated compounds like alkanes and therefore, exhibit substitution reactions. But cycloalkanes having ring of 3- or 4-carbon atoms are unstable and tend to form open chain aliphatic compounds by the addition of the reagent. Thus, they exhibit the chemical properties of both alkanes and alkenes.

(1) Reaction with halogens. Generally cycloalkanes undergo free radical substitution with halogens at high temperature or in the presence of light, for example,

Cyclohexane + Br ⟶ Bromocyclohexane + HBr

In case of cyclopropane, bromine and chlorine in the presence of $FeCl_3$ break the ring system and give open chain addition products. For example,

Cyclopropane + Br_2 ⟶ $CH_2Br-CH_2-CH_2Br$
1-3-Dibromopropane

(2) Reaction with hydrogen. When heated with hydrogen in the presence of nickel, cyclopropane and cyclobutane give addition products whereas higher members do not give this reaction.

Cyclopropane + H_2 $\xrightarrow[373K]{Ni}$ $CH_3-CH_2-CH_3$
n-Propane

Cyclopentane + H_2 ⟶ No reaction

(3) Reaction with halogen acids. Generally cycloalkanes do not react with halogen acids. However, cyclopropane and cyclobutane add on halogen acids to give open chain alkyl halides.

$$\underset{H_2C \,\text{---}\, CH_2}{\overset{CH_2}{\triangle}} + HI \longrightarrow CH_3\text{---}CH_2\text{---}CHI$$
$$\qquad\qquad\qquad\qquad\quad n\text{-Propyl iodide}$$

Cyclohexane + H_2 $\longrightarrow$ No reaction

Hence it may be pointed that cyclopropane is the most reactive cycloalkane because it exhibits many addition reactions accompanied by ring cleavage. Then comes cyclobutane which shows some addition reactions. The rest of the cycloalkanes do not form addition products.

7.6 BAEYER STRAIN THEORY

According to Le Bel and vant Hoff, the four valencies of a carbon atom are directed towards the corners of a regular tetrahedron and hence the angle between any two valencies (any two bonds) is 109° 28'. Baeyer held that any departure from this position leads to a strain in the molecule resulting in the decrease of the stability of the molecule.

In light of this concept, Baeyer gave a theory which is known as Baeyer's Strain Theory. **The main points of the theory are as follwing**:

(*i*) *The carbon atoms constituting the rings lie in the same plane. Hence bond angles between the adjacent carbon atoms of the ring no longer remain 109°28'. Different rings have different values of this angle.* For example, the cyclopropane ring is a triangle having C—C—C angle of 60° only.

(*ii*) *Any deviation, positive or negative from the normal tetrahedral bond angle of 109°28' during the formation of a ring creates a strain in the molecule which makes the molecule unstable.*

(*iii*) *The larger (more) the deviation from the normal angle, the greater is the strain and thus lesser is the stability. However, it should be noted that the sign of deviation does not make any difference.*

Adolf Von Baeyer (1835–1917)
Adolf Von Baeyer was a German chemist and is best known for his strain theory in ring compounds. He synthesised indigo and was awarded 1905 Nobel Prize in Chemistry.

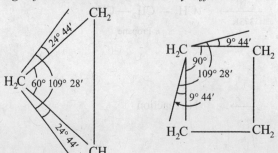

Baeyer worked out the deviation from normal angle or what can be called *angle strain* involved in the formation of cycloalkanes of various ring sizes. For instance the C—C—C angle of a cyclopropane ring is 60° because each carbon atom occupies a corner of an equilateral triangle. This shows that the angle strain in cyclopropane would be 1/2 (109° 28'-60°) = 24°-44'. In the same way C—C—C angle in cyclobutane would be 90° because each carbon atom is situated at the corner of the square. The angle strain in cyclobutane is 1/2 (109° 28'-90°) = 9°-44'. The angle strain for cycloalkanes calculated for various ring sizes are given in Table 7.1.

TABLE 7.1 Angle Strains in Various Cycloalkanes.

Compound	C—C—C Angle	Angle strain
Cyclopropane, C_3H_6	60°	1/2 (109°28'-60°) = + 24°44'
Cyclobutane, C_4H_8	90°	1/2 (109°28'-90°) = + 9°44'
Cyclopentane, C_5H_{10}	108°	1/2 (109°28'-108°) = + 0°44'
Cyclohexane, C_6H_{12}	120°	1/2 (109°28'-120°) = − 5°16'
Cycloheptane, C_7H_{14}	128°35'	1/2 (109°28'-128°34') = − 9°33'
Cyclooctane, C_8H_{16}	134°	1/2 (109°28'-135°) = − 14°46'

It is clear from the above data that angle strain decreases as we move from cyclopropane to cyclopentane. This predicts ease of formation and stability of the compound from cyclopropane to cyclopentane. This is actually found to be so.

Limitations

However the stability of cyclohexane and higher members cannot be explained by the angle strain values. According to angle strains, cyclohexane and higher members should be quite unstable as there is a lot of strain in the molecule. But this is not true. Actually cyclohexane and higher members are not planar molecules. Different carbon atoms in such molecules lie in different planes, thereby relieving the strain.

Solved Examples

Example 3: Giving examples of first three cycloalkanes, explain the relationship of relative stabilities with the heat of combustion.

Solution:

Relationship of relative stability with heat of combustion.			
Size of the ring	Heat of combustion per —CH_2 group kJ mol^{-1}	Size of the ring	Heat of combustion per —CH_2 group kJ mol^{-1}
3	697	6	658.5
4	686	7	662.3
5	664	8	663.6

Heat of combustion per CH_2 group of open chain alkanes = 658.0 kJ mol^{-1}.

Heat of combustion may be defined as the quantity of heat evolved when one mole of the compound is burnt in excess of air.

Heat of combustion per CH_2 group gives a fair idea of the relative stabilities of cycloalkanes.

If the heat of combustion is large, the compound contains more energy and is therefore less stable.

With this principle, let us examine the stabilities of cyclopropane, cyclobutane and cyclopentane. Heat of combustion of cyclopropane is 697 kJ mol^{-1}. It is 39 kJ more than that of open chain alkane. Thus it is less stable than the open chain alkane by this amount.

In the case of cyclobutane, the heat of combustion is more by an amount 28 kJ as compared to open-chain alkane. Thus cyclobutane is less stable than open chain alkanes but it is more stable than cyclopropane. Cyclopentane having heat of combustion 664 kJ is more stable than cyclobutane. The relative stability of first three members of cycloalkanes in decreasing order is given as under.

$$\text{Cyclopentane} > \text{Cyclobutane} > \text{Cyclopropane}$$

On having a look at the heat of combustion of cyclohexane and higher members, we observe that the values do not differ much. Thus higher members are expected to display almost uniform stability which is confirmed by their reactions.

If we draw a graph between ring strain and ring size, we get a curve of the shape as given alongside. From the graph, we observe that six-member rings do not suffer from any strain.

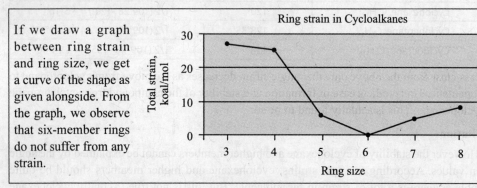

7.7 SACHSE MOHR THEORY OF STRAINLESS RINGS AND STABILITY OF CYCLOHEXANE

To explain the greater stability of cyclohexane as compared with cyclopentane and stability of large rings, Sachse in 1890 advanced the Theory of Strainless Rings. *According to this theory large rings due to their bigger size get twisted in different planes with the result that the carbon atoms do not lie in the same plane as is assumed by Baeyer.* Thus, because of carbon atoms being in the different planes, their valency bonds are not distorted from their original angle and consequently no strain is set up in the molecules. In other words, the rings with six or more carbon atoms get *puckered* and thus become strainless.

Thus according to Sachse cyclohexane exists in the following strain-free forms called *boat* and *chair* forms.

Mohr however suggested that the two forms cyclohexane are continuously undergoing transformation from one form to another.

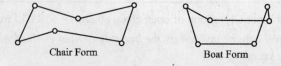

Chair Form   Boat Form

7.8 DIFFICULTIES IN THE SYNTHESIS OF LARGE-MEMBERED RINGS

It has been observed that heat of combustion per $-CH_2$ group in cycloalkanes from cyclopentane onwards remains nearly constant and is comparable to that of alkanes. Thus these cycloalkanes (5-membered and higher) should be as stable as corresponding alkanes. But still it is not easy to prepare such higher cyclohexane. To obtain a cycloalkane from a linear compound, it is necessary that the terminal groups of the linear compound should come sufficiently close to each other so that ring closure can take place.

In normal circumstances, the probability of end-groups coming close to provide cyclization are not very bright. This, then, results in intermolecular reaction leading to polymerisation. Baeyer advanced the views that because of tremendous strain, higher cycloalkanes are not formed easily. But this is not true. As postulated by Sachse-Mohr, such higher cycloalkanes are strain-free because the constituent atoms are in different planes and they retain the normal valency angle of 109°-28′. It is rather improbability of the end groups of the linear molecules coming closer that prevents the formation of ring compounds.

How is the difficulty overcome?

High dilution technique is resorted to in the preparation of higher cycloalkanes. The substance is taken in low concentration which minimises the collision between two molecules preventing polymerisation. Due to flopping, the two ends of the molecule come closer enough to react, giving rise to cyclization.

It may be remarked that during the synthesis of cyclic compounds, two processes are competing with each other. Process I is the ring closure by the coming together of the end groups and Process II is chain lengthening (polymerization) by the interaction of end group of one molecule with that of the other molecules as illustrated below. By reducing concentration of the reactant, the chances of Process II are minimised

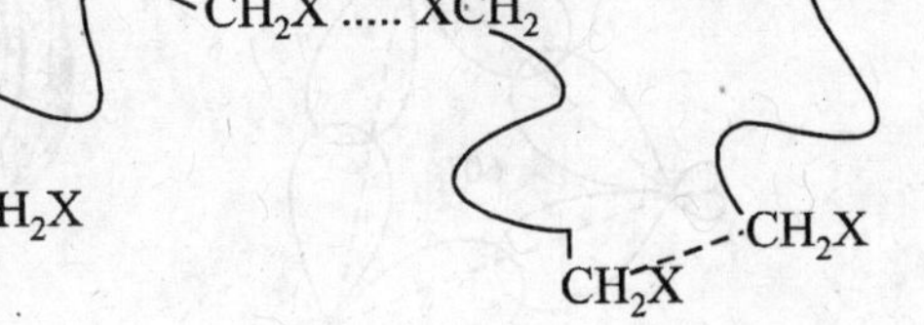

Process I (Ring closure)

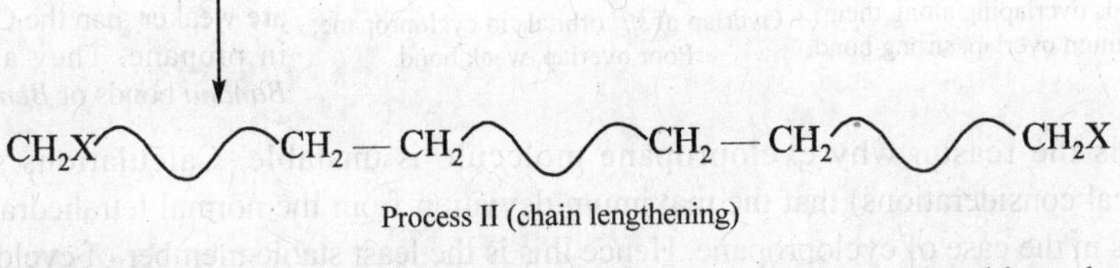

Process II (chain lengthening)

It is amazing that, once formed, these higher cyclic compounds are as stable as their straight chain counterparts.

Solved Examples

Example 4: Do the cycloalkanes exhibit isomerism with any class of open chain compounds? If so give examples.

Solution: Cycloalkanes, which can be represented by the general formula $(CH_2)_n$ are isomeric with alkenes (open-chain compounds) having the general formula $C_n H_{2n}$.

Thus cyclopropane is isomeric with propene.

Cyclopropane $CH_2 — CH = CH_2$ Propene

And cyclobutane is isomeric with butene-1 and butene-2

$CH_3 — CH = CH — CH_3$ $CH_3 — CH_2 — CH = CH_2$
Butene-2   Butene-1

Cyclobutane

Example 5: Cyclopropane is the least stable member of cycloalkanes. How do you justify this is terms of orbital picture of 3-membered rings?

OR

There are three banana bonds in cyclopropane. Explain.

Solution: In cyclopropane, the carbon atoms are sp^3 hybridised. The sp^3 hybrid orbitals are inclined at an angle of 109°28′ to one another. In order to form a single bond between two carbon atoms, the orbitals of the two carbon atoms must overlap to a maximum extent. Therefore when a carbon atom forms single bonds with two adjacent carbon atoms, the angle C—C—C should be 109°28′ as shown in the diagram below. This is not actually so in cyclopropane molecule. The angle C—C—C in this case is only 60°. This orientation will not allow perfect overlapping of the orbitals and therfore the bonds formed will be weaker.

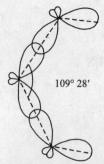

sp^3 orbitals overlaping along their axis; Maximum overlap, strong bond.

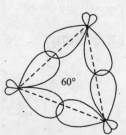

Overlap of sp^3 orbitals in cyclopropane; Poor overlap, weak bond.

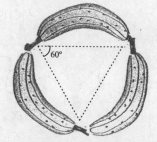

The C-C bonds in cyclopropane are weaker than the C-C bonds in propane. They are called *Banana* bonds or *Bent* bonds.

This is the reason why cyclopropane molecule is unstable. Calculations show (on geometrical considerations) that the maximum deviation from the normal tetrahedral angle of 109°28′ is in the case of cyclopropane. Hence this is the least stable member of cycloalkanes.

Example 6: Even though cycloheptane and higher members are free of angle strain, they cannot be synthesised easily. Explain.

Solution: It is necessary for the formation of a cyclic compound that its terminal carbon atoms should come closer. In the case of a large hydrocarbon chain, the possibility that end carbons will come sufficiently closer and form the closed ring is very small. That is why cycloheptane and higher members are difficult to synthesis.

Example 7: Give the methods of preparation of cyclopropene, cyclobutene cyclopentene and cyclohexene.

Solution:

1. **Cyclopropane,**

Cyclopropane may be prepared by the following sequence of reactions

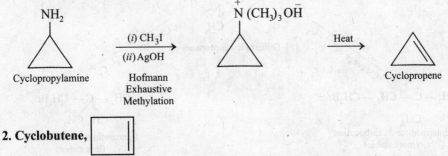

2. **Cyclobutene,**

Cyclobutene may be prepared by Hofmann exhaustive methylation of cyclobutylamine in a similar process as in cyclopropene

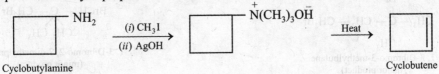

3. **Cyclopentene,**

(i) It may be prepared by dehydrating cyclopentanol

Cyclopentanol $\xrightarrow[-H_2O]{Conc.H_2SO_4}$ Cyclopentene

(ii) It may also be prepared by heating cyclopentyl bromide with ethanolic potassium hydroxide

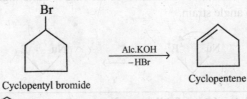

4. **Cyclohexene,**

It may be prepared by dehydrating cyclohexanol with sulphuric acid

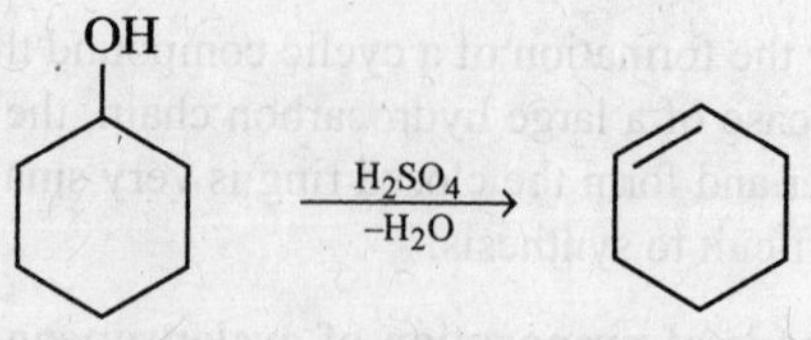

Cyclohexanol →[H$_2$SO$_4$][−H$_2$O] Cyclohexene

Example 8: Give two possible products of the reaction of 1, 1-dimethylcyclopropane with bromine at room temperature and predict the major product with explanation.

Solution: The two possible reaction paths are as under:

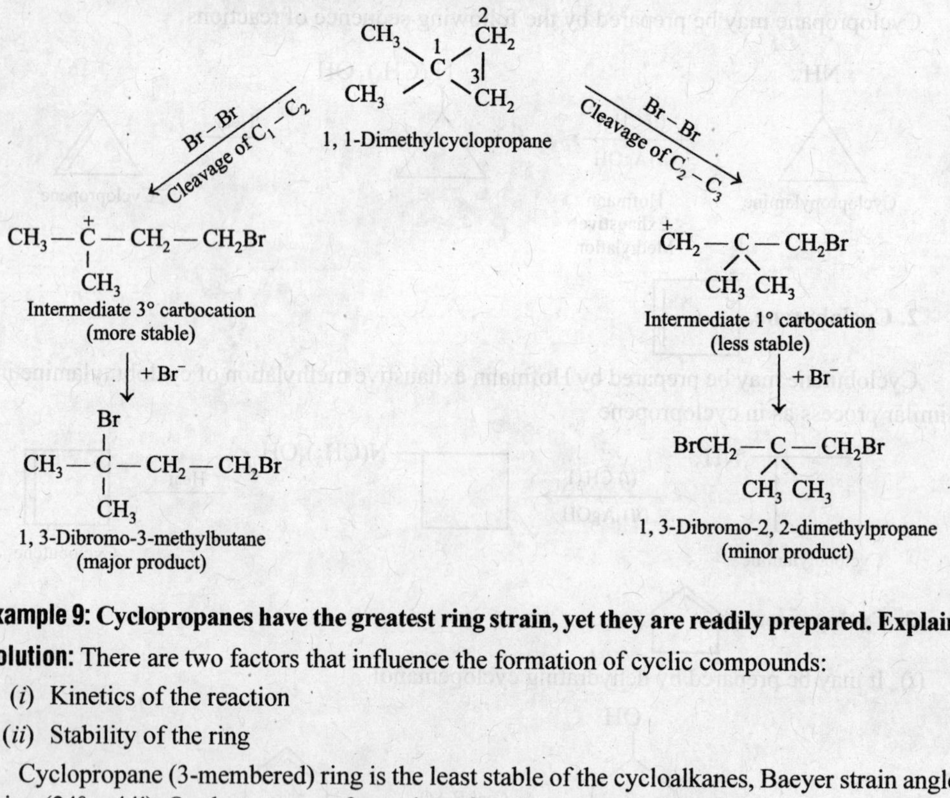

1, 1-Dimethylcyclopropane

Br–Br, Cleavage of C$_1$–C$_2$ →

$CH_3 — \overset{+}{C} — CH_2 — CH_2Br$
         |
         CH_3

Intermediate 3° carbocation
(more stable)

↓ + Br$^-$

         Br
         |
$CH_3 — C — CH_2 — CH_2Br$
         |
         CH_3

1, 3-Dibromo-3-methylbutane
(major product)

Br–Br, Cleavage of C$_2$–C$_3$ →

$\overset{+}{CH_2} — C — CH_2Br$
         / \
       CH_3 CH_3

Intermediate 1° carbocation
(less stable)

↓ + Br$^-$

$BrCH_2 — C — CH_2Br$
         / \
       CH_3 CH_3

1, 3-Dibromo-2, 2-dimethylpropane
(minor product)

Example 9: Cyclopropanes have the greatest ring strain, yet they are readily prepared. Explain

Solution: There are two factors that influence the formation of cyclic compounds:

(*i*) Kinetics of the reaction

(*ii*) Stability of the ring

Cyclopropane (3-membered) ring is the least stable of the cycloalkanes, Baeyer strain angle being (24° − 44′). On that account, formation of cyclopropane is difficult.

But kinetics factor dominates in the formation of cyclopropanes.

The two ends of the reagent forming the ring are close to each other. The entropy factor is favourable. Therefore the rate of formation of the ring is quite high, enough to oppose the instability factor due to angle strain.

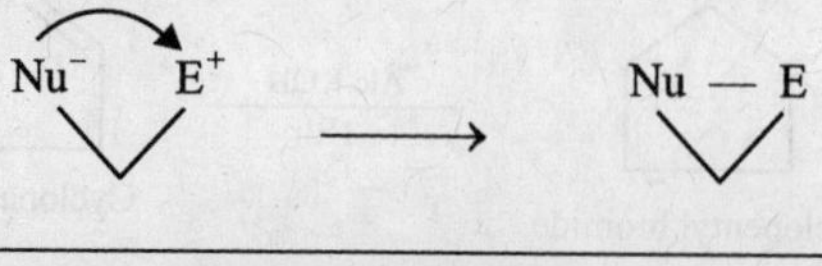

Nu$^-$ ⌒ E$^+$ ⟶ Nu — E

For Conformational Analysis of Alkanes, Energy Diagrams of Cyclohexane, Chair, Boat and Twist forms, Relative stability with Energy Diagrams, Refer to Chapter 3, Stereochemistry.

Key Terms

- Baeyer's strain theory
- Wislicenus method
- Chair, boat and twist forms
- Confirmational analysis

Evaluate Yourself

Multiple Choice Questions

1. Cycloalkanes have the same molecular formula as
    (a) alkanes
    (b) alkenes
    (c) alkynes
    (d) cycloalkenes
2. Which of the cycloalkanes is not expected to have ring strain?
    (a) Cyclopropane
    (b) Cyclobutane
    (c) Cycloheptane
    (d) None of these
3. Most stable conformation of cyclohexane is the
    (a) Haworth form
    (b) Boat form
    (c) Newman form
    (d) Chair form
4. Which of the following compounds will give cyclopropane on treatment with sodium in dry ether?
    (a) 1, 3–Dibromopropane
    (b) 1, 1–Dibromopropane
    (c) 1, 2–Dibromopropane
    (d) 2, 2–Dibromopropane
5. Which of the following compounds will react most readily with concentrated sulphuric acid?
    (a) Ethane
    (b) Cyclohexane
    (c) Propane
    (d) Cyclohexane
6. Cyclohexanol can be converted into cyclohexane by heating with
    (a) Zn (Hg) and HCl
    (b) Conc.H_2SO_4
    (c) $SOCl_2$
    (d) H_2 and Ni
7. What percentage of cyclohexane molecule is estimated to be in the boat form at any given time?
    (a) Over 99%
    (b) Between 90% and 99%
    (c) Approximately 50%
    (d) Less than 1%
8. Which of the following cycloalkanes is most reactive?
    (a) Cyclopropane
    (b) cyclohexane
    (c) Cyclobutane
    (d) Cycloheptane
9. Which of the following statements is wrong about cyclohexane?
    (a) It is a saturated cyclic hydrocarbon
    (b) All C – C – C bond angle are 109° – 28′
    (c) It is very unstable and strained compound
    (d) It can exist in two conformations designated as boat and chair forms

10. A compound of formula C_6H_{12} does not react with concentrated sulphuric acid. The compound could be
    (a) alkane
    (b) cycloalkane
    (c) alkene
    (d) cycloalkene

Short Answer Questions

1. What happens when
    (i) 1, 3-Dibromopropane is heated with sodium
    (ii) Diethyl adipate is treated with sodium ethoxide in toluene.
2. Large rings with more than six carbon atoms are stable but difficult to prepare. Explain.
3. Discuss briefly Simmons-Smith reaction for the preparation of cyclopropane derivatives.
4. Give two methods of preparation of cycloalkanes.
5. What is the cause of angle strain in cyclic compounds? Discuss in terms of oriental concept.

Long Answer Questions

1. What is Baeyer's strain theory? What are its limitations?
2. Give a brief account of Dieckmann cyclisation reaction. Discuss its mechanism.
3. (a) Explain the following :
    (i) Thorpe-Ziegler reaction
    (ii) Demjanov Rearrangement
   (b) Explain the validity of Sachse Mohr's theory using heat of combustion value of various cycloalkanes. Give orbital picture of angle strain.
4. (a) Explain Demjanov rearrangement pointing clearly the nature of the products formed.
   (b) Give the structure and names of main organic products in the following reactions :
      (i) Cyclopropane + Br_2 $\xrightarrow{(CCl_4)}$
      (ii) Cyclopentane + Br_2 $\xrightarrow{(575\ K)}$
      (iii) Cyclohexane + Br_2 $\longrightarrow$
5. Addition of singlet carbenes to alkenes is stereospecific in nature while that of triplet carbenes is non-stereospecific in nature. Explain.
6. What are cycloalkanes? How will you account for their reactive nature?
7. What type of compounds undergo ring expansion and what type of compounds undergo ring contraction in Demjanov rearrangement? Give details of the reactions.
8. (a) Write the following reactions with suitable examples.
    (i) Dieckmann Cyclisation
    (ii) Thorpe Ziegler reaction.
   (b) Explain Baeyer's strain theory. Why is it not applicable to cyclohexane?

9. (a) How would you prepare cyclopentane from diethyl adipate?
   (b) Cyclobutyl amine on reaction with HNO_2 yields two products. Explain with mechanism.
   (c) Choosing one suitable example differentiate between the terms "conformation" and "configuration'.
10. (a) Discuss Demjanov rearrangement in detail with suitable examples.
    (b) Write notes on the following:
        (i) Schase Mohr's theory of strainless rings.
        (ii) Orbital pircture of angle strain.
11. Give a brief account of Thorpe-Ziegler method and Dieckmann cyclisation reaction for the preparation of cycloalkanes.
12. How will you prepare cycloalkane derivatives by the following methods?
    (i) Simmons-Smith reaction
    (ii) Diels-Alder reaction
    (ii) Dehalogenation of dihalides.
13. What are α, ω dihalogen derivatives? How will you prepare cycloalkanes from them?
14. Give the main features of Baeyer's strain theory. Why does this theory fail to account for the stability of cyclohexane?
15. How does heat of combustion value of methylene group account for the stability of cycloalkanes?

Answers
Multiple Choice Questions

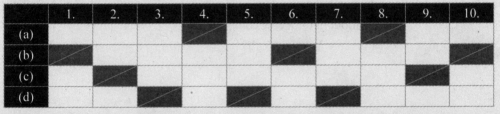

Suggested Readings

1. Finar, I.L. Organic Chemistry (Vol. 1), Dorling Kindersley (India) Pvt. Ltd.
2. Morrison, R.N. & Boyd, R.N. Chemistry, Dorling Kindersley (India) Pvt. Ltd.

Chapter 8

Aromatic Hydrocarbons

LEARNING
OBJECTIVES

After reading this chapter, you should be able to:
- learn aromatic nature of arenes
- learn Huckel's rule
- discuss cyclic carbocations/carbanions
- learn mechanism of aromatic electrophilic substitution, nitration, halogenation and sulphonation
- understand directing influence of groups
- learn about heterocyclic compounds

Aromatic Hydrocarbons

8.1 INTRODUCTION

Some sweet smelling compounds were obtained from natural sources during the study of organic chemistry in earlier times. These compounds had contrasting properties compared to aliphatic compounds. The sweet smelling compounds were called **aromatic** (Greek **aroma** = pleasant smell). Further studies on these compounds revealed that they contained benzene rings involving six carbon atoms in a ring. Later a large number of aromatic compounds were discovered which lacked pleasant smell. Some of them had even foul smell. Therefore, the word 'aroma' lost its significance in relation to aromatic compounds.

Aromatic compounds are now regarded as a class of compounds which contain at least one benzene ring. They are also called **benzenoids**. It may be mentioned that there are some compounds which do not contain benzene ring but they still behave as aromatic compounds. Such compounds are called **non-benzenoids**.

8.2 ARENES

Arenes are mixed aromatic aliphatic compounds. An alkyl, alkenyl or alkynyl group attached to a benzene ring constitutes an arene. Thus, arenes can be classified into the following types:

Alkyl benzenes

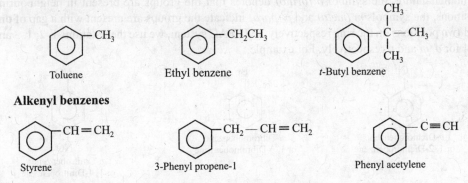

Alkenyl benzenes

8.3 NOMENCLATURE OF BENZENE DERIVATIVES

(a) Monosubstituted benzene derivatives

The following rules are observed in the naming of compounds:

(i) The monosubstitued benzene derivative is named by prefixing the name of the substituent to the word 'benzene'. For example,

However, there are cases when the name of the substituent is added after the root word 'benzene'. For example,

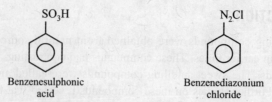

(*ii*) Some groups when present on the benzene ring provide a special name to the compound. For example,

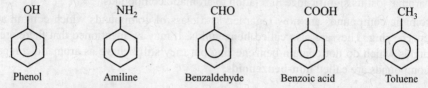

(b). Disubstituted benzene derivatives

The following rules are applied in the nomenclature of the compounds:

(*i*) If the two groups attached are similar, the prefix *di* is added before the name of the substituent. The relative positions of the groups needs to be indicated. In *common* system of nomenclature, the symbol *o (ortho)* denotes that the groups are present in neighbouring positions, the symbols *m (meta)* and *p (para)* indicate the groups are present with a gap of **one** and **two** positions on the ring respectively. In IUPAC system, we use the notations 1, 2; 1, 3 and 1, 4 for *o*, *m* and *p* respectively. For example,

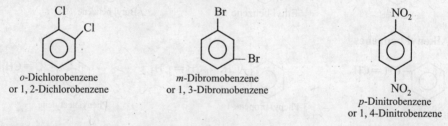

(*ii*) If the two substituents (functional groups) attached to the benzene ring are different, the compound is named as a derivative of the compound with the principle functional group at position 1. For example,

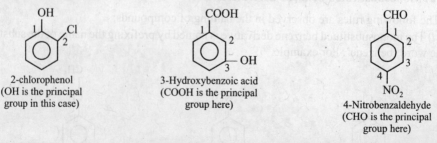

(*iii*) If both the groups attached to the ring are substituents (not functional groups), they are arranged alphabetically, the group that comes first in the alphabet is given the position 1. For example,

Aromatic Hydrocarbons **303**

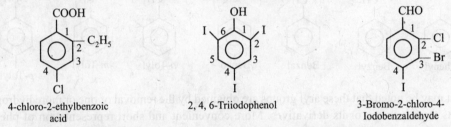

1-Chloro-2-nitrobenzene

1-Methyl-4-nitrobenzene

(c) Polysubstituted benzene derivatives.

(i) If one of the substituents (functional groups) gives a special name to the compound, that functional group is given the location 1 and the compound is named keeping in mind the **lowest sum rule** and naming the substituents in alphabetical order.

4-chloro-2-ethylbenzoic acid

2, 4, 6-Triiodophenol

3-Bromo-2-chloro-4-Iodobenzaldehyde

(ii) If a trisubstituted benzene contains all different functional groups, the principal functional group is given the locant (position) 1 and the compound is named keeping in view the alphabetical order of the groups and lowest locant rule. For example,

2-Hydroxy-4-nitrobenzoic acid

3-Ethoxy-4-hydroxylbenzonitrile

(iii) If all the functional groups in the compound are such that they are normally considered as substituent groups, the group that comes first in the alphabet is given the locant 1, keeping in mind that it does not violate lowest sum rule (sum of locants of all the substituents). Otherwise, that group may not be given the position 1

2-Bromo-1-chloro-3-iodibenzene
(And **not** 1-Bromo-2-chloro-6-iodobenzene)

2-Bromo-1-chloro-5-fluoro-4-iodo-3-nitrobenzene
(And **not** 1-Bromo-6-chloro-4-fluoro-3-iodo-2-nitrobenzene)

Aromatic Nucleus and Side Chain

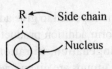

R — Side chain
Nucleus

The ring of six carbon atoms in benzene and its derivatives is called **nucleus**. R is the side chain provided at least one carbon from R is directly linked to the benzene nucleus. Thus, groups like – OH, – OCH$_3$, – NH$_2$, – X (Cl, Br, I) and – SO$_3$H do not come under the category of side chains.

Compounds in which the functional group is linked to the benzene ring are called *nuclear derivatives* while those in which it is present in the side chain are called *side chain derivatives*.

Aryl Groups

Some of the aryl groups that we come across in the study of organic reactions are given below:

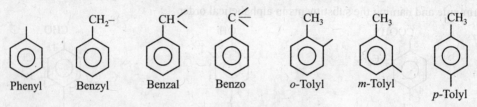

It may be noted that these aryl groups are obtained by the removal of one or more hydrogen atoms from benzene or its derivatives. More convenient and short representation of phenyl group are:

$$C_6H_5-, \phi- \text{ or } Ph-$$

8.4 STRUCTURE OF BENZENE

The structure of benzene was arrived at as under:

1. Molecular formula of benzene was found to be C_6H_6 based on elemental analysis and molecular mass determination. A saturated hydrocarbon with six carbon atom has the molecular formula C_6H_{14}. Benzene is thus short of eight hydrogen atoms per molecule. It appears that benzene is a highly unsaturated compound.

2. Addition of hydrogen to benzene takes place giving cyclohexane

$$\underset{\text{Benzene}}{C_6H_6} + 3H_2 \xrightarrow[473\text{-}523K]{\text{Raney Ni}} \underset{\text{Cyclohexane}}{C_6H_{12}}$$

3. Benzene forms an ozonide upon treatment with ozone

$$C_6H_6 + O_3 \longrightarrow C_6H_6O_9$$

4. Benzene adds three molecules of chlorine in presence of sunlight and in absence of a halogen carrier to form benzene hexachloride (BHC)

$$C_6H_6 + 3Cl_2 \xrightarrow{\text{Sun light}} C_6H_6Cl_6$$

The above reactions (2), (3) and (4) indicate the presence of three double bonds in the molecule.

5. An unsaturated hydrocarbon is expected to give addition reactions readily. But this is not so with benzene. Benzene do not form addition product easily. It rather gives the substitution products much more easily. Halogenation, nitration and sulphonation, which involve substitution of hydrogen with other groups are given readily. For example,

$$C_6H_6 \xrightarrow[Fe]{Br_2} \underset{\text{Bromobenzene}}{C_6H_5Br} + HBr$$

$$C_6H_6 \xrightarrow{H_2SO_4} \underset{\text{Benzenesulphonic acid}}{C_6H_5SO_3H} + H_2O$$

6. Benzene resists oxidation even with strong oxidising agents like chromic acid and potassium permanganate. This behaviour is typical of saturated compounds.

Kekule Structure of Benzene

According to Kekule, six carbon atoms of benzene are linked to each other by alternate single and double covalent bonds to form a hexagonal ring as shown in the following figure:

August Kekule (1829-1896)
August Kekule was a German organic chemist. From 1850s until his death, Kekule was one of the most prominent chemists in Europe especially in theoretical chemistry.

Kekule structure of benzene

Each carbon atom is linked to one hydrogen atom thus conforming to its molecular formula.

Drawbacks of Kekule structure. Kekule structure does not explain the following observations:

(*i*) **Chemical reactions.** Benzene does not give addition reactions and fails to decolorise Baeyer's reagent. It readily undergoes electrophilic substitution reactions in which the benzene ring is retained.

(*ii*) **Heat of combustion.** On the basis of Kekule structure, the heat of combustion of benzene is expected to be 3449.0 kJ mol^{-1}. But the experimental value is 3298.5 kJ mol^{-1}. Thus, benzene has 150.5 kJ mol^{-1} of energy less than Kekule structure. Benzene is, therefore, more stable than the structure proposed by Kekule.

(*iii*) **Heat of hydrogenation.** On the basis of Kekule structure, the heat of hydrogenation of benzene is expected to be 358.0 kJ mol^{-1}. But the experimental value is 208.5 kJ mol^{-1}. This again shows that benzene is more stable than expected from Kekule structure.

(*iv*) ***o*-Disubstituted product.** Benzene forms only one ortho disubstituted product whereas Kekule structure predicts two *o*-disubstituted products as shown below:

Presence of double bond between the substituents

Presence of single bond between the substituents

(*v*) **Carbon-carbon bonds lengths.** X-ray diffraction studies show that all the carbon-carbon bond lengths are identical, *viz.*, 139.7 pm and lie in between that of single and double bonds. This is not in accordance with Kekule structure which contains two kinds of carbon-carbon bonds.

Molecular orbital structure of benzene

Structure of benzene can be best described by using the orbital concept. Each carbon atom in benzene is sp^2 hybridised and thus forms three σ bonds, two with adjacent carbon atoms and one with hydrogen. Thus, all the six carbons and six hydrogen atoms lie in the same plane and the angle between two adjacent σ bonds is 120° [Fig. (a)].

Each carbon is still left with an unhybridised π-orbital lying above and below the plane of benzene ring. Each one of these π-orbitals overlaps sidewise on either sides to form two sets of π-electron clouds [Fig. (b) and (c)]. π electrons are delocalised as these can move over all the six carbon atoms [Fig. (d)]. As a result of this delocalisation two continuous ring like electron clouds one above and the other below the plane of carbon atoms [Fig. (e)] are formed. Bond angles and bond lengths in the molecule of benzene are given in [Fig. (f)].

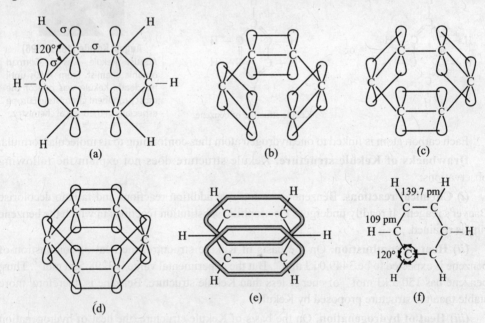

Benzene, a potential threat

United States Environmental Protection Agency classifies benzene as class A carcinogen and has concluded that Oregon's cancer risk is dominated by emissions of benzene. Studies show that breathing air contaminated with benzene inflicts genetic damage linked to childhood leukemia.

Dangers of benzene

Evidence in support of orbital structure of benzene

(i) Unusual stability. Benzene molecule exhibits unusual stability and resists the formation of addition products. This can easily be understood in terms of delocalisation of π-electrons which is responsible for aromaticity.

(ii) Isomer number. According to orbital concept all the six carbons in benzene are completely equivalent. Similarly, all the six hydrogen atoms also occupy identical positions. Thus, benzene should form only one monosubstituted and three disubstituted products. This has been found to be so in actual practice.

(iii) Electrophilic substitution reactions. There are two continuous ring like π-electron clouds one above and the other below the plane of carbon atoms. The π-electrons are easily attacked by electrophiles. Hence benzene undergoes electrophilic substitution reactions.

8.5 AROMATICITY

Organic compounds which resemble benzene in their chemical behaviour are called aromatic compounds. They exhibit certain characteristic properties which are quite different from those of aliphatic and alicyclic compounds. These characteristic properties are collectively referred to as **aromaticity or aromatic character.**

Characteristics

(i) Although their molecular formulae suggest a high degree of unsaturation, yet they do not respond to tests characteristic of unsaturated compounds. Thus, aromatic compounds fail to decolorise an aqueous solution of potassium permanganate (Baeyer's test).

(ii) They undergo readily certain electrophilic substitution reactions such as nitration, halogenation, sulphonation, Friedel-Crafts alkylation and acylation, etc.

(iii) Their molecules are flat or nearly flat as shown by physical methods such as X-ray and electron diffraction methods.

(iv) They are associated with high thermodynamic stability as is indicated by their low heats of combustion and hydrogenation.

8.6 HUCKEL'S RULE AND AROMATICITY

This rule is based upon molecular orbital treatment and is employed for predicting aromaticity in organic compounds. The main theoretical requirements for a substance to possess aromaticity are:

(i) The molecule or ion must be flat or nearly flat.

(ii) It must have cyclic clouds of delocalised π-electrons above and below the plane of the molecule. The π-electron clouds should encompass all the carbon atoms of the cyclic system.

(iii) The π-clouds in the molecule or ion must contain a total of $(4n + 2)$ π electrons where $n = 0, 1, 2, 3, ...,$ etc.

The above requirements are collectively known as *Huckel rule* or $(4n + 2)$ *rule*. This rule can be applied successfully to cyclic polyenes, polycyclic compounds and non-benzenoid compounds to predict aromaticity in them. Thus, molecules or ions having 2, 6, 10, 14 ... π electrons will be aromatic and others will be non-aromatic.

Theoretical Justification of Huckel Rule. On quantum mechanical grounds an aromatic system possesses a closed shell of π-electrons corresponding to insert gas configuration. Huckel pointed out that energies of molecular orbitals (MOs) of aromatic systems have a pattern where there is always one orbital of lowest energy followed by a degenerate (having same energy) pair of orbitals in order of increasing energy. Finally, there is one orbital of highest energy. Filling of the orbitals takes place as per **Hund's rule**.

For a conjugated planar monocyclic system having 2, 6, 10, 14... etc π-electrons, the lowest energy MO and all other degenerate pairs of MOs would be occupied by two electrons each. *Such systems are stable because of this closed shell filling of orbitals*. However, for a monocyclic planar conjugated system with 4, 8, 12 ... etc π-electrons, there will always be two singly occupied degenerate orbitals. Such systems are highly unstable and are named **antiaromatic**. Thus cyclobutadiene and cyclo-octatetraene with 4π and 8π electrons respectively possess two singly occupied degenerate orbitals (Fig 8.1) and are both antiaromatic.

It is thus not necessary that the molecule will contain a benzene ring in order to exhibit aromatic character. *Any planar ring compound containing cyclic cloud of delocalised (4n + 2) p electrons above and below the plane of the ring will show aromatic character*. We find that ring compounds with 2, 6, 10, ... etc π bonds behave aromatically.

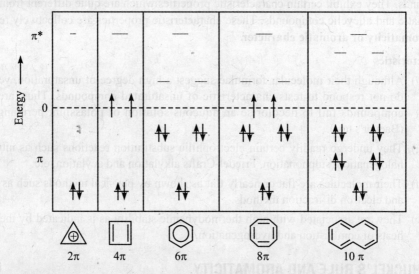

Fig. 8.1: Huckel Molecular Orbital (HMO) diagram for planar monocyclic conjugated systems

Two π-electron rings. If such rings can delocalise the electrons around the ring, they will show aromatic character, otherwise not.

Aromatic molecules containing 6π, 10π and 14π electrons

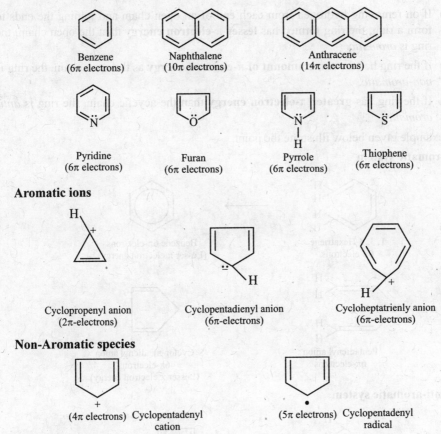

8.7 ANNULENES

Annulenes are monocyclic compounds having a system of single and double bonds in alternate positions. They may be called cyclic polyenes and can be represented by a general formula $(-CH = CH -)_n$ where n is a positive integer. They are named by writing the ring size followed by the word "annulene". Some examples of annulenes are given below:

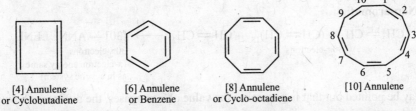

According to Huckel's rule, Annulenes containing $(4n + 2)$ π-electrons and having a coplaner ring should be aromatic in nature. The synthesis of a number of annulenes has confirmed this. Thus [6] annulene or benzene is aromatic. [4] annulene and [8] annulene are not aromatic because they do not conform to Huckel's rule. They are rather anti-aromatic. [10] annulene although expected to be aromatic as per Huckel's rule is not so because the ring is not coplaner.

Aromatic, Anti-aromatic and Non-aromatic compounds

(i) If on removing hydrogen from each end of an open chain and joining the ends to form a ring, the ring formed has **lesser** π-electron energy than the open chain, the ring is *aromatic*.

(ii) If the ring has the **same amount of** π-electron energy as the open chain, the ring is *non-aromatic*.

(iii) If the ring has **greater** π-electron energy than the acyclic chain, the ring is *anti-aromatic*.

The example given below illustrate the point.

(1) Aromatic system

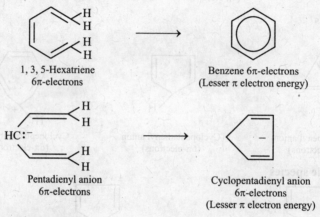

(2) Anti-aromatic system

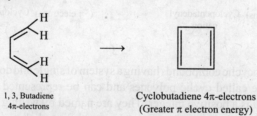

Similarly other systems having $4n\pi$ electrons are anti-aromatic.

(3) Non-aromatic

$$CH_2=CH-(CH=CH)_{15}-CH=CH_2 \longrightarrow [30]-\text{ANNULENE}$$

30π-electrons   30π-electrons
(π-electron energy same as in acyclic system)

It may be pointed out that in general as the value of n increases, the difference between the cyclic and corresponding acyclic system decreases both in system having $(4n)$ π-electrons and $(4n + 2)$ π-electrons. In other words, such cyclic systems show non-aromatic behaviour.

Solved Examples

Example 1: Which of the following species will exhibit aromatic character? Write their structural formulae.

Cyclooctatetraene, tropylium ion, cycloheptatrienyl anion, furan, cyclopropenyl anion, cyclopentadienyl cation, cyclobutadiene, anthracene, phenanthrene, pyridine.

Solution: (*a*) The compounds mentioned below are aromatic in nature in terms of Huckel's rule. No. of π-electrons in these compounds is $(4n + 2)$ where $n = 0, 1, 2$ etc.

I Tropylium ion
(cycloheptatrienyl cation)

(6π electrons)   II Furan (6π electrons)   III Anthracene (14π electrons)

IV Phenanthrene (14π electrons)   V Pyridine (10π electrons)

(*b*) Following compounds are **not aromatic** in nature as they do not comply with the $(4n + 2)$ rule.

Cyclooctatetraene (8π electrons)   Cycloheptatrienyl anion (8π electrons)

Cyclopropenyl anion (4π electrons)   Cyclopentadienyl cation (4π electrons)

Cyclobutadiene (4π electrons)

Example 2: Which of the following have aromatic character and why?

Solution: Apply $(4n + 2)$ Huckel's Rule

I

Cyclopentene

No. of π electrons = 2, Aromatic

II

Cyclopentadienyl cation

No. of π electrons = 4, Non-aromatic

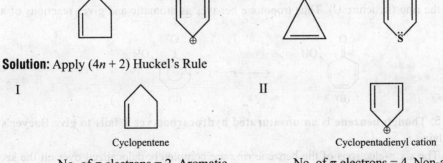

III

Cyclopropenyl cation
No. of π electrons = 2, Aromatic

IV.

Thiophene
No. of π electrons = 6, Aromatic

Example 3: **Cyclooctatetraene reacts with two equivalents of potassium to yield an unusually stable compound with the molecular formula, $2K^+ \ C_8H_8^{2-}$. Write the structure and account for its unusual stability.**

Solution: Cyclooctatetraene possesses 8 π-electrons. On treatment with two equivalents of potassium, two electrons are added to the system, thereby raising the number of π-electrons to 10.

Cyclooctatetraenly dianion thus formed has 10 electrons, therefore, it is aromatic and hence quite stable. The reaction is represented as under:

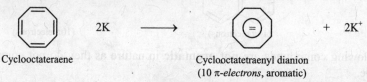

Cyclooctateraene   Cyclooctatetraenyl dianion
(10 π-electrons, aromatic)

Example 4: **Cycloheptatriene (A), though a cyclic nearly planar triene containing six π-electorns but does not exhibit usual stability of aromatic compounds whereas troplone (B), a derivative of (A), behaves like a typical phenol. Explain.**

A   B

Solution: Although cycloheptatriene (A) has six π-electrons, these are not completely delocalized over all the seven carbon atoms of the ring because it contains one sp^3-hybridized carbon atom. Thus, it is not aromatic and hence does not have the usual stability of aromatic compounds.

On the other hand, in tropolone (B), the oxygen atom of the keto group draws the π-electrons of the C = O towards itself thereby creating positive charge on the ring carbon atom. As a result, the six π-electrons of the propolone ring are completely delocalized over all the seven carbon atoms of the ring (structure C). Thus tropolone behaves as aromatic and gives reactions of a phenol.

(B)   ⟷   (C)

Example 5: **Though benzene is an unsaturated hydrocarbon yet it fails to give Baeyer's test. Explain.**

Solution: The six π-electrons of the benzene ring are completely delocalized over all the six carbon atoms of the ring. As a result of this delocalization, benzene gains extra stability and hence does not undergo addition reactions. Such reactions would destroy the delocalized π-electron cloud thereby decreasing the stability of the benzene ring.

Example 6: An aqueous solution of tropylium bromide (C_7H_7Br) on treatment with silver nitrate gives percipitate of silver bromide. How do you account for this observation?

Solution: In aqueous solution, tropylium bromide readily undergoes ionization to form tropylium cation and bromide ion which reacts with $AgNO_3$ solution to form a precipitate of AgBr. The ionisation of tropylium bromide is facilitated due to the formation of highly stable tropylium ion on account of high resonance energy.

Tropylium bromide $\xrightarrow{\text{Ionization}}$ Tropylium cation (*aromatic*) + Br^- $\xrightarrow{AgNO_3}$ ArBr↓

Example 7: 3-chlorocyclopropene reacts with $SbCl_5$ to form a salt but chlorocyclopropane does not. Explain.

Solution: Cyclopropenyl cation gets stabilised due to its attaining aromatic character. The cation than reacts forms a salt with $SbCl_5$

Cyclopropyl ion is not aromatic, hence it is not formed. Therefore salt formation does not take place.

Example 8: Comment on the aromatic character of 1, 3-cyclopentadiene and 1, 3-cyclopentadienyl anion.

Solution:

1, 3-cyclopentadiene (4π electrons)

1, 3-cyclopentadienyl anion (6π electrons)

1, 3-cyclopentadiene does not conform to (4n + 2) Huckel's rule.
Thus, it is not aromatic.

1, 3-cyclopentadienyl anion conforms to (4n + 2) Huckel's rule with n = 1. Thus, it is an aromatic species.

8.8 GENERAL MECHANISM OF ELECTROPHILIC SUBSTITUTION IN BENZENE

Electrophilic substitution reactions are initiated by substances which are either electrophilic themselves or which generate some electrophilic species. General mechanism involves the following sequence of steps:

(I) Generation of electrophile. To start with, there is preliminary reaction which generates an electrophile.

$$A-B \longrightarrow \underset{\text{Electrophile}}{A^{\oplus}} + B^{-}$$

(II) Formation of intermediate carbocation (carbonium ion). The electrophile attacks the π electron cloud of the benzene ring and thus brings about an electronic displacement. This results in the formation of **intermediate carbocation** which is resonance stabilized as shown below:

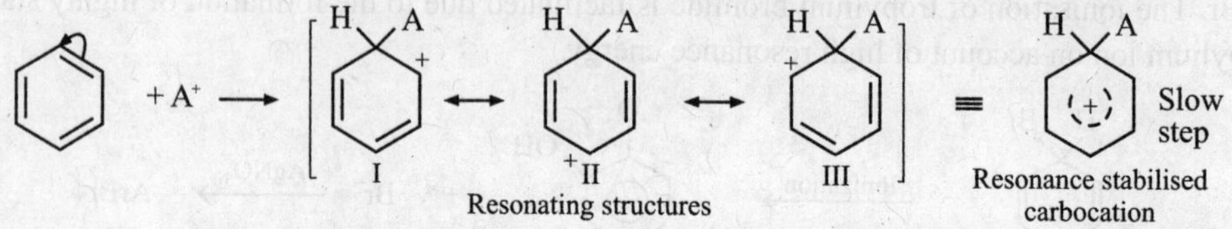

[Benzene + A⁺ → Resonating structures I, II, III ≡ Resonance stabilised carbocation — Slow step]

During the formation of the carbocation, the aromaticity of the benzene ring is destroyed. Consequently, the formation of carbocation is slow and hence is the **rate determining step**.

(III) Abstraction of a proton from the carbocation. The base present in the reaction mixture abstracts the proton from the carbocation to form the final product. Since the aromatic character of the benzene ring is restored, this step is **fast** and hence is not the rate determining step.

[Carbocation + :B⁻ → Product + BH, Fast step]

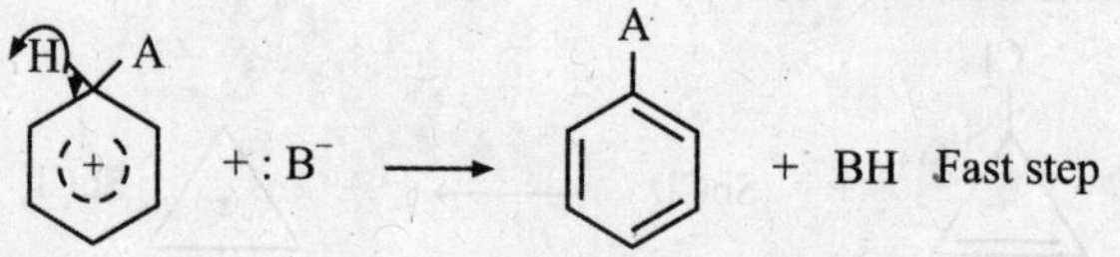

Evidence in support of the mechanism. Electrophilic aromatic substitution involves two steps viz. formation of carbocation and abstraction of proton from carbocation, after the generation of electrophile. The first step involving the formation of intermediate carbocation is slow and hence is the rate determining step of the reaction. The second step which involves the abstraction of a proton from the carbocation is fast. The above mechanism is supported by **isotope tracer technique and isolation of intermediates** as described below.

1. Isotope Tracer Technique

A carbon-deuterium bond is broken more slowly than a bond between carbon and hydrogen. Thus, if a hydgrogen is lost in the rate determining step of a reaction, it will show an **isotope effect** on replacing hydrogen by deuterium. In other words, a deuterated compound should undergo substitution more slowly than a non-deuterated compound. But this isotope effect has not been observed in aromatic electrophilic substitution reactions. For example, the rate of nitration of deutero-benzene is the same as the rate for benzene. Thus, proton elimination is not the rate determining step of the reaction. The step leading to the formation of carbocation does not involve the cleavage of carbon-hydrogen bond and hence it is the rate determining step of the reaction.

2. Isolation of intermediates

The intermediate formed in the reaction of benzotrifluoride with nitryl fluoride has been isolated as a yellow crystalline complex. On warming to 223 K, this complex decomposes and m-nitrobenzotrifluoride is formed.

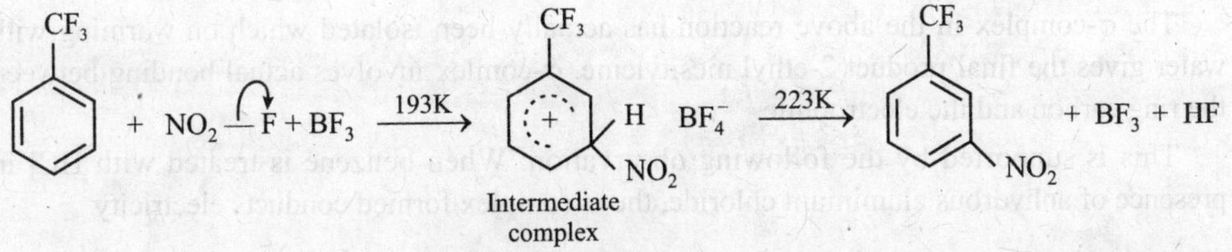

Aromatic Hydrocarbons

[Reaction scheme: Benzene + $NO_2-F + BF_3 \xrightarrow{193K}$ Intermediate complex (with H, NO_2, BF_4^-) $\xrightarrow{223K}$ nitrobenzene + BF_3 + HF]

Intermediate complex

The complex in this case attains stability with the formation of a σ-bond between the carbon atom of benzene ring and the electrophile and it becomes possible to isolate the intermediate. Such an intermediate complex is called **σ-complex**.

Substitution versus addition reactions

We could consider the possibility of the intermediate formed in the substitution reactions of benzene, undergoing addition reaction by the attack of nucleophile giving the addition product, instead of two substitution product.

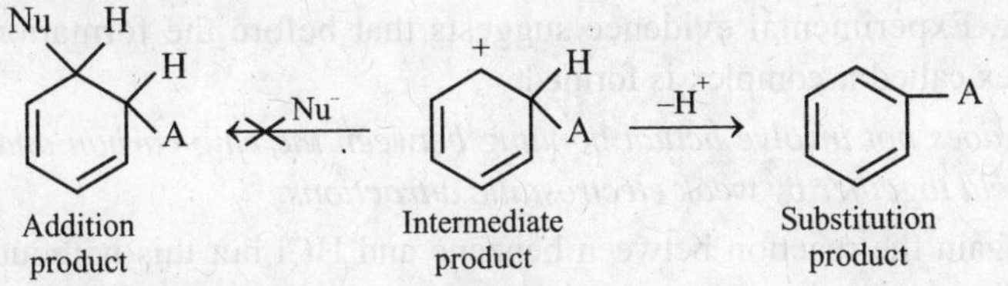

[Scheme: Addition product ⇍ Nu⁻ Intermediate product −H⁺→ Substitution product]

We observe that the addition product is not obtained. In such a change the aromatic character of benzene ring is lost. Losing the aromatic character involves an energy of around 150 kJ mol⁻¹ *i.e.* equal to resonance energy of benzene. The second step of the reaction is endothermic by around 150 kJ mol⁻¹ of energy. In constrast, *alkenes undergo addition reactions because no such resonance energy is involved.*

It may also be noted that the first step in the electrophilic attack on an alkene and benzene is the same, yet the rate of reaction of benzene is **slower** than that of alkene. This is because energy of activation (E_a) for the formation of carbocation intermediate derived from benzene is much higher than that from alkenes. An additional energy equivalent to resonance energy of benzene is involved in the case of benzene because the aromatic character is going to be lost.

8.9 σ- AND π-COMPLEXES

σ-complex. The intermediate formed in the electrophilic substitution reactions of benzene are also called σ-complexes because a σ-*bond is actually formed* between the carbon of the aromatic ring and the electrophile (NO_2, N^+O_2, Cl^+, SO_3 etc.).

Aromatic character of benzene is destroyed in the formation of σ-complex. This complex possesses higher energy than benzene at least to the extent of resonance energy (150 kJ mol⁻¹). *It is therefore reasonable to assume that the formation of σ-complex is a slow and rate-determining step.* Consider the following reaction.

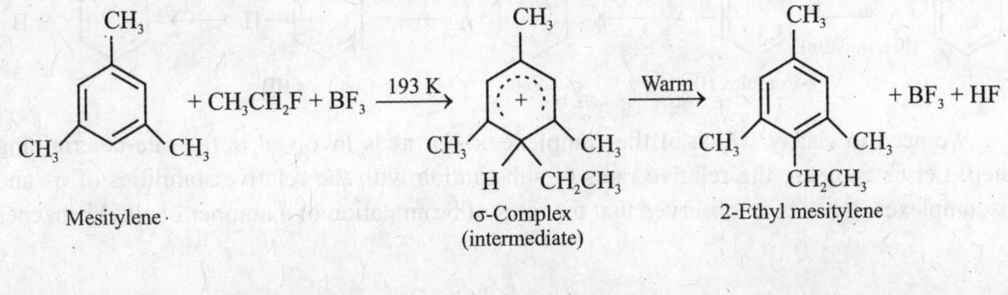

[Reaction: Mesitylene + $CH_3CH_2F + BF_3 \xrightarrow{193 K}$ σ-Complex (intermediate) $\xrightarrow{Warm}$ 2-Ethyl mesitylene + BF_3 + HF]

The σ-complex in the above reaction has actually been isolated which on warming with water gives the final product 2-ethyl mesityleme. σ-comlex involves actual bonding between the ring carbon and the electrophile.

This is supported by the following observation: When benzene is treated with HCl in presence of anhydrous aluminium chloride, the σ-complex formed conducts electricity.

$$\text{C}_6\text{H}_6 + \text{HCl} + \text{Anhy. AlCl}_3 \longrightarrow [\text{C}_6\text{H}_6\text{H}]^+ \text{AlCl}_4^-$$

σ-Complex
(Intermediate carbocation)

This is possible only when the electrophile (H^+ in this case) links to the benzene ring through a σ-bond.

π-complex. Experimental evidence suggests that before the formation of σ-complex, another complex called π-complex is formed.

π-complex does not involve actual bonding between the ring carbon and the electrophile but these are held together by weak electrostatic attractions.

Consider again the reaction between benzene and HCl but this without the presence of anhydrous $AlCl_3$.

$$\text{C}_6\text{H}_6 + \text{HCl} \longrightarrow \text{C}_6\text{H}_6 \cdots \text{H-Cl}$$

π-Complex

A 1:1 complex is still formed which does not conduct electricity. This is because σ-bond is not formed here. Electron displacement which is necessary for the formation of σ-bond does not occur. No positive and negative charges are created and hence the π-complex does not conduct electricity. The final picture that emerges during substitution in benzene rings is something like this:

(i) As the electrophile approaches the benzene ring, it is attracted by π-electron cloud to form a π-complex (I).

(ii) Later, the π-complex removes a pair of electrons from the benzene ring and forms a covalent bond with one of the ring carbon atoms to form a σ-complex.

(iii) This σ-complex loses a proton and this proton again forms a π-complex (II) with the ring.

(iv) The π-complex (II) finally loses the proton to form the substitution product.

The sequence of events can be represented as under:

$$\text{C}_6\text{H}_6 + A^+ \longrightarrow [\text{C}_6\text{H}_6 \cdots A^+] \longrightarrow [\text{C}_6\text{H}_6\text{-H-A}]^+ \longrightarrow [\text{C}_6\text{H}_5\text{A} \cdots H^+] \longrightarrow \text{C}_6\text{H}_5\text{A} + H^+$$

(Electrophile)   π-Complex (I)   σ-Complex   π-Complex (II)

We need to clarify which of the complexes σ- or π- is involved in the rate-determining step. Let us compare the relative rates of substitution with the relative stabilities of σ- and π-complexes. It has been observed that the rates of bromination of a number of alkyl benzenes move parallel to the stabilities of corresponding σ-complexes rather than those of π-complexes. We therefore conclude that *formation of σ-complex is the slow and rate-determining step in the substitution process.*

$$\text{C}_6\text{H}_6 + \text{Br–Br} \xrightarrow{\text{Fast}} \underset{\pi\text{-Complex}}{[\text{C}_6\text{H}_6\cdots\text{Br–Br}]} \xrightarrow{\text{Slow}} \underset{\sigma\text{-Complex}}{\left[\begin{array}{c}\text{H}\\ \text{Br}\end{array}\right]^+ \text{Br}^-}$$

To substantiate our results so far, we can make use of energy profile diagrams. If the formation of π-bond is the rate-determining step, we should get the diagram as shown in Fig 8.2. If formation of σ-bond is the rate determining step, we should get the diagram as shown in Fig 8.3. We indeed obtain a diagram as per Fig 8.3. *This proves beyond doubt that the formation in σ-bond is the rate-determining step.*

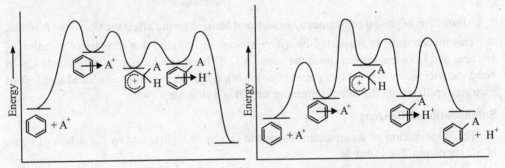

Fig. 8.2: *Energy profile diagram when formation of π-complex is the rate of determining step.*

Fig. 8.3: *Energy profile diagram when formation of σ-complex is the rate determining step.*

Electrophilic substitution
Nitration of Benzene

Benzene has a tendency to give electrophilic substitution reactions because of the π electron cloud above and below benzene ring. In these reactions, a hydrogen atom of benzene is substituted by $-\text{NO}_2$, $-\text{X}$ (halogen), $-\text{SO}_3\text{H}$, $-\text{R}$ (alkyl) or $-\text{COR}$ group, under different conditions and by taking a suitable electrophile. Such reactions are called electrophilic substitution reactions.

Nitration. This reaction involves the treatment of an aromatic compound with a mixture of nitric acid and sulphuric acid. For example:

$$\text{C}_6\text{H}_6 + \text{HONO}_2 \text{ (conc.)} \xrightarrow{\text{H}_2\text{SO}_4} \text{C}_6\text{H}_5\text{NO}_2 \text{ (Nitrobenzene)} + \text{H}_2\text{O}$$

Mechanism. The various steps in the nitration of benzene can be outlined as below:

(I) Generation of electrophile

$$\text{HONO}_2 + 2\text{H}_2\text{SO}_4 \rightleftharpoons \underset{\text{Nitronium ion}}{\text{NO}_2^+} + \text{H}_3\text{O}^+ + 2\text{HSO}_4^-$$

(II) Formation of carbocation

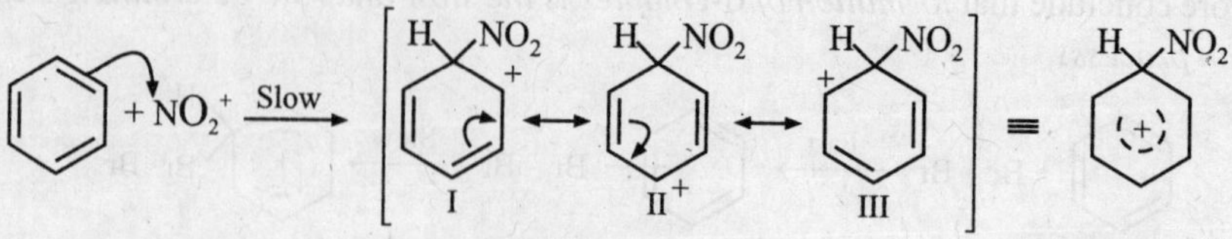

$$C_6H_6 + NO_2^+ \xrightarrow{Slow} [\text{I} \leftrightarrow \text{II} \leftrightarrow \text{III}] \equiv \text{Resonance stabilised carbocation}$$

Resonating structures

(III) Abstraction of proton from the carbocation

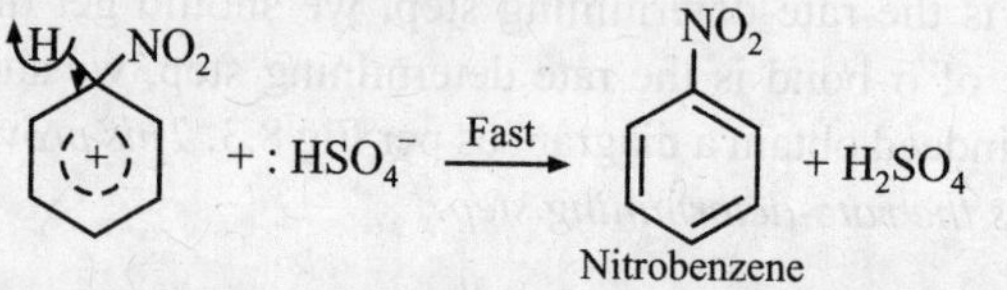

$$[C_6H_5(H)(NO_2)]^+ + :HSO_4^- \xrightarrow{Fast} C_6H_5NO_2 + H_2SO_4$$

Nitrobenzene

Abstraction of proton constitutes a *fast* step and hence does not affect the rate of the reaction.

This mechanism is supported by *tracer studies*. It is found that the rate of nitration of benzene and hexadeuterobenzene is the same. A C–D bond breaks more slowly than a C – H bond (heavier isotopes react comparatively slowly), therefore it can be concluded that loss of proton (step III) is not the rate determining step. It is a slow step.

Sulphonation of Benzene

The sulphonation of an aromatic compound can be brought about by the action of conc. H_2SO_4 or oleum. For example:

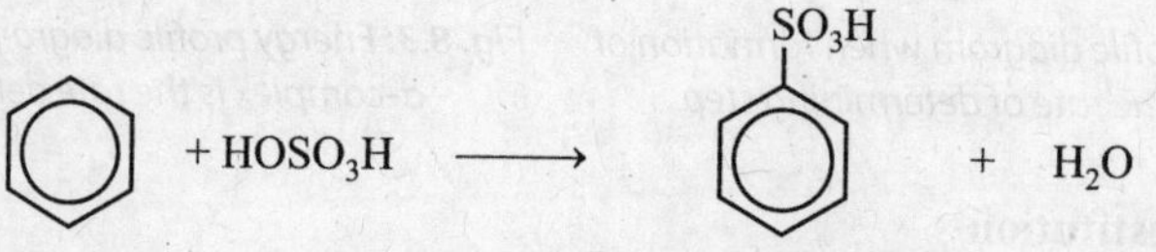

$$C_6H_6 + HOSO_3H \longrightarrow C_6H_5SO_3H + H_2O$$

Benzene sulphonic acid

Mechanism. The attacking electrophilic reagent in this reaction is believed to be sulphur trioxide which is present as such in oleum or may be formed by the dissociation of sulphuric acid. The electrophilic nature of SO_3 molecule is due to the presence of electron deficient sulphur atom.

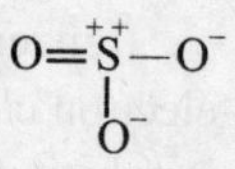

$$O=\overset{++}{\underset{\underset{O}{|}}{S}}-O^-$$

As the positive charge is concentrated on sulphur and negative charges are scattered on two oxygens, SO_3 molecule acts as electrophile.

The complete mechanism for this reaction may be outlined as follows:

(I) Generation of electrophile

$$2H_2SO_4 \rightleftharpoons SO_3 + HSO_4^- + H_3O^+$$

Electrophile

(II) Formation of carbocation

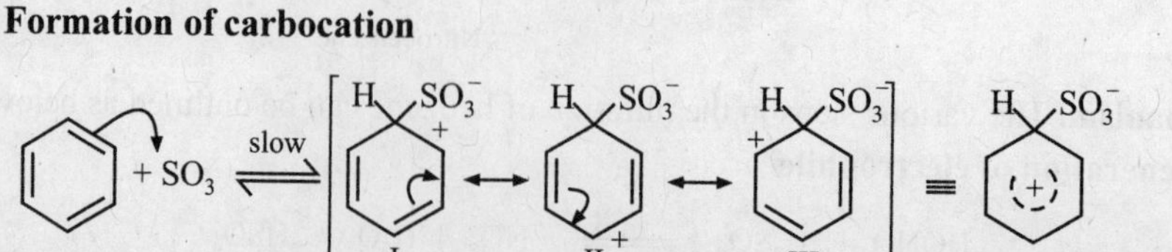

$$C_6H_6 + SO_3 \underset{slow}{\rightleftharpoons} [\text{I} \leftrightarrow \text{II} \leftrightarrow \text{III}] \equiv \text{Resonance stabilised carbocation}$$

Resonating structures

(III) Abstraction of proton from the carbocation

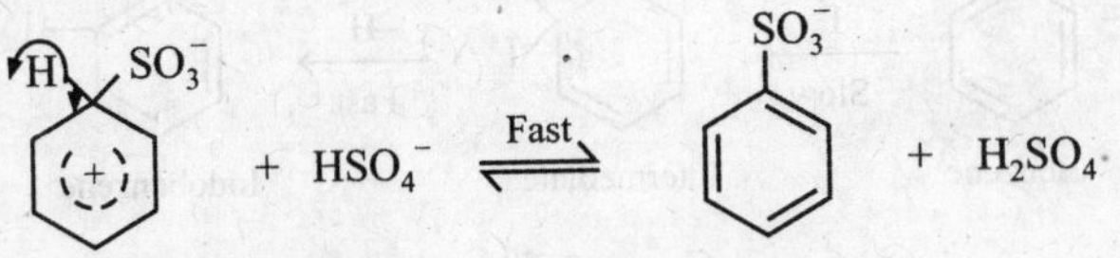

[Ar-H-SO₃⁻ carbocation] + HSO₄⁻ ⇌ (Fast) [Ar-SO₃⁻] + H₂SO₄

(IV) Formation of the final product

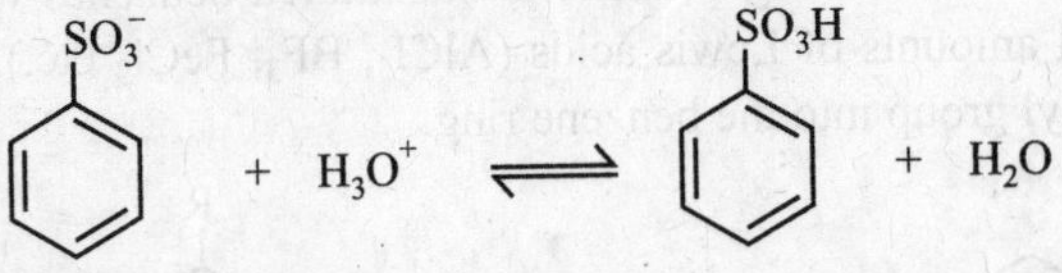

[Ar-SO₃⁻] + H₃O⁺ ⇌ [Ar-SO₃H] + H₂O

Halogenation of Benzene

Halogenation of benzene and other aromatic hydrocarbons can be brought about by treating with halogens in the presence of Lewis acids such as ferric halides, anhydrous aluminium chloride. Chlorination of benzene is considered below:

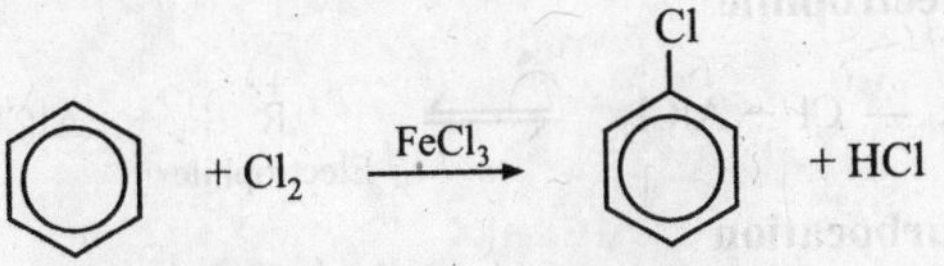

C₆H₆ + Cl₂ $\xrightarrow{FeCl_3}$ C₆H₅Cl (Chlorobenzene) + HCl

Mechanism. Various steps involved in the reaction are as follows:

(I) Generation of electrophile

$$Cl-Cl + FeCl_3 \rightleftharpoons \overset{+}{Cl} + FeCl_4^-$$

(II) Formation of intermediate complex

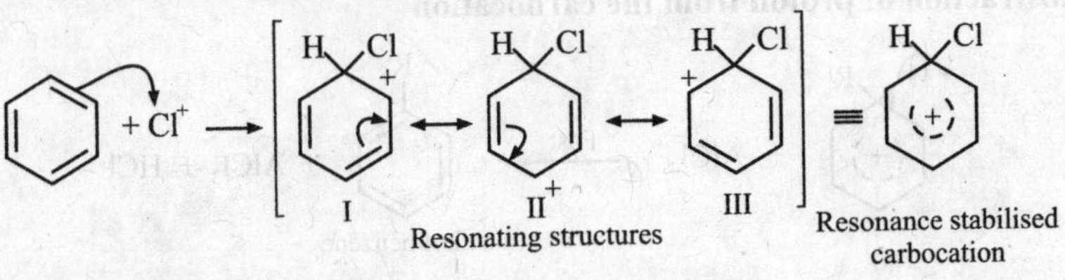

[Benzene] + Cl⁺ → [Resonating structures I ↔ II ↔ III] ≡ [Resonance stabilised carbocation]

(III) Loss of proton

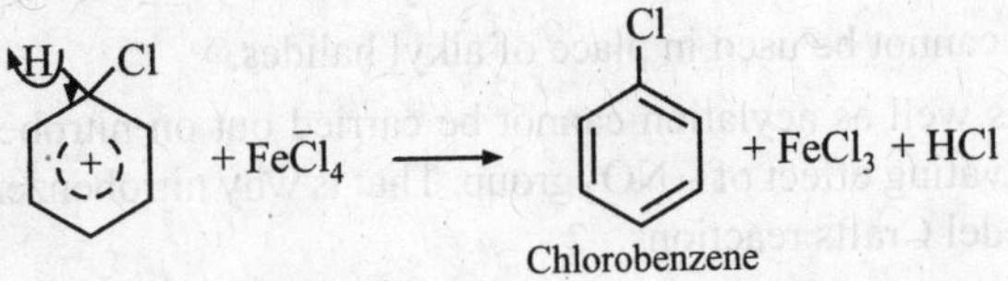

[H-Cl carbocation] + FeCl₄⁻ ⟶ [Chlorobenzene] + FeCl₃ + HCl

Some interhalogen compounds can be used for carrying out halogenation of aromatic compounds. The more positive of the halogens enters the ring.

$$I-Cl + AlCl_3 \rightarrow I^+ + AlCl_4^- \text{ (Cl}^+ \text{ is not formed)}$$

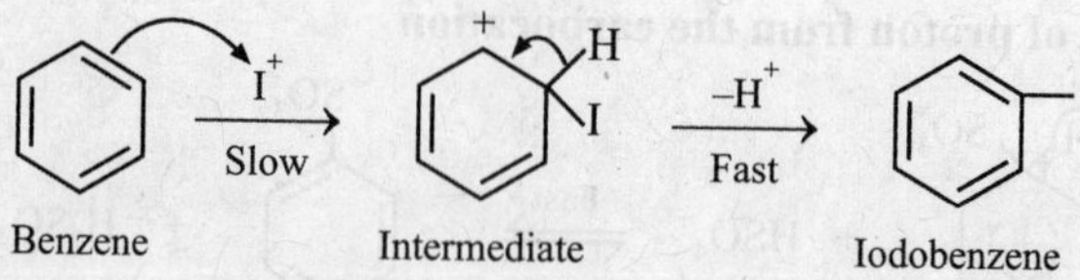

Friedel-Crafts Alkylation of Benzene

The reaction consists in treating benzene or substituted benzenes with an alkyl halide in the presence of small amounts of Lewis acids ($AlCl_3$, BF_3, $FeCl_3$ etc.). It leads to the direct introduction of an alkyl group into the benzene ring.

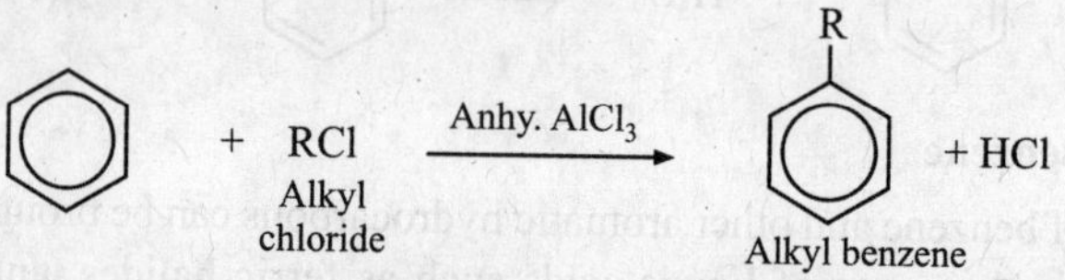

Mechanism. The reaction is believed to take place through the following steps:

(I) Generation of electrophile

$$R-Cl + AlCl_3 \rightleftharpoons \overset{+}{R} + AlCl_4^-$$
Electrophile

(II) Formation of carbocation

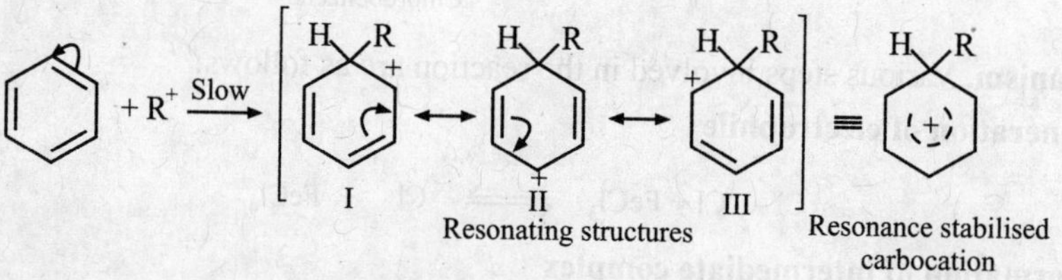

(III) Abstraction of proton from the carbocation

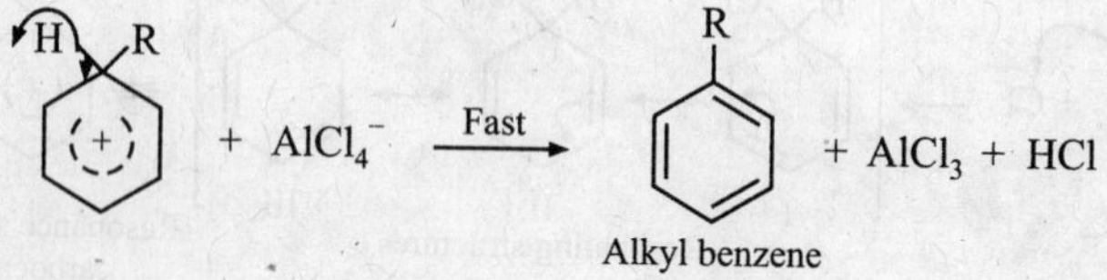

There are some limitations of Friedel Crafts alkylation.

1. Aryl halides cannot be used in place of alkyl halides.
2. Alkylation as well as acylation cannot be carried out on nitrobenzene because of the strong deactivating effect of $-NO_2$ group. That is why nitrobenzene is used as a solvent in many Friedel Crafts reactions.
3. Aniline does not undergo Fridel Crafts reaction due to the foundation of a complex between $-NH_2$ group and $AlCl_3$. Presence of positive charge on nitrogen deactivates the ring

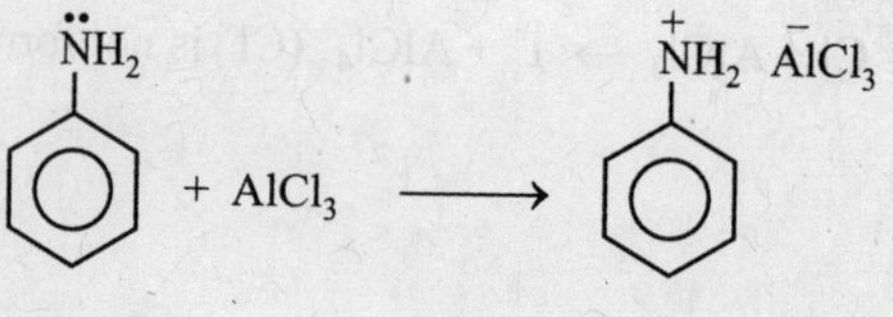

Charles Friedel (1832-1899)
Charles Friedel was a French chemist and mineralogist. He was a student of Louis Pasteur at Sorbonne.

James Mason Crafts (1839-1917)
James Mason Crafts was an American chemist, best known for developing the Friedal-Crafts alkylation and acylation reactions with Charles Friedel in 1876.

Friedel-Crafts Acylation of Benzene

When benzene is treated with acetyl chloride or acetic anhydride in the presence of anhydrous aluminium chloride, the formation of acetophenone takes place. This reaction is known as *Friedel-Crafts acylation*.

$$C_6H_6 + CH_3-CO-Cl \xrightarrow{\text{Anhydrous AlCl}_3} C_6H_5-CO-CH_3 + HCl$$
Acetyl chloride → Acetophenone

Benzene reacts with benzoyl chloride (C_6H_5COCl) in the presence of anhy. $AlCl_3$ to give benzophenone ($C_6H_5COC_6C_5$)

Mechanism. The most probable mechanism of the above reaction is discussed as under:

(I) Generation of electrophile

$$CH_3-CO-Cl + AlCl_3 \rightleftharpoons CH_3-\overset{\oplus}{C}=O + AlCl_4^-$$
Acetyl cation

$$(CH_3CO)_2O + AlCl_3 \longrightarrow CH_3-\overset{+}{C}=O + CH_3-CO-O\bar{A}lCl_3$$

(II) Formation of carbocation

[Resonating structures I, II, III] ≡ Resonance stabilised carbocation

(III) Abstraction of proton from the carbocation

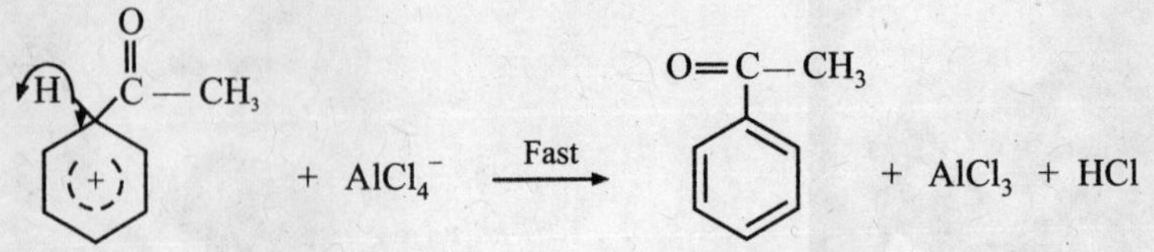

[Reaction: protonated intermediate with –C(=O)–CH₃ group and H on ring carbon + AlCl₄⁻ → Acetophenone (C₆H₅–CO–CH₃) + AlCl₃ + HCl, Fast]

Acetophenone

8.10 EFFECT OF SUBSTITUENTS ON ORIENTATION AND REACTIVITY OF BENZENE RING

It has been observed that substituents already attached to the benzene ring not only govern the orientation of further substitution but also affect the reactivity of the benzene ring. It is discussed in brief as below:

(I) Effect of substituents on orientation. As stated above, the nature of the group already attached to benzene ring determines the position of the incoming group. In general, groups have been classified into two categories:

(a) **Ortho-para directing groups.** The groups which direct the incoming group towards ortho and para positions are called ortho-para directing groups. Groups such as – R, – C₆H₅, – OH, – SH, – OR, – NH₂, – NHR, – NR₂, – Cl, – Br, – I, etc. are all ortho-para directing groups.

(b) **Meta directing groups.** The groups which direct the incoming groups towards meta position are called meta directing groups. Groups such as – COOH, – CHO, – CN, – NO₂, – COR, – SO₃H, etc., are all meta directing groups.

It may be mentioned that groups which contain double or triple bond are usually meta directing while those which do not contain multiple bonds are ortho-para directing. However, there are certain exceptions to this rule.

(II) Effect of substituents on reactivity. Reactivity of the benzene ring in electrophilic substitution reactions depends upon the tendency of the substituent group already present in the benzene ring to release or withdraw electrons. A group that releases electrons activates benzene ring while the one which draws electrons deactivates the benzene ring. It is found that except halogens all ortho-para directing groups activate the ring and all meta directing groups deactivate the ring towards further electrophilic substitution. Thus, nitration of toluene can be carried out at room temperature, while that of nitrobenzene requires more drastic conditions. This is illustrated as below:

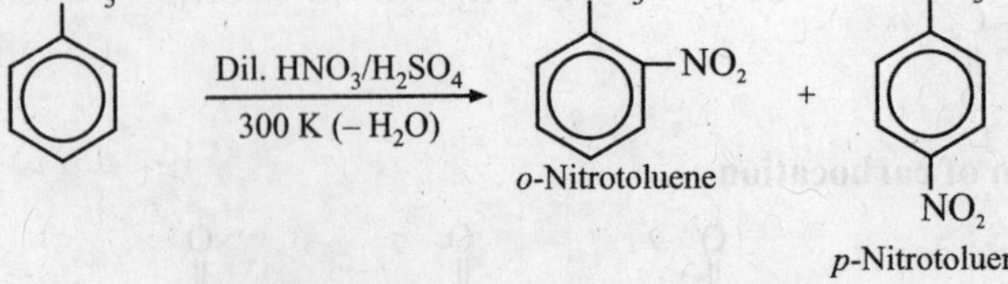

Toluene $\xrightarrow{\text{Dil. HNO}_3/\text{H}_2\text{SO}_4}_{300 \text{ K } (-\text{H}_2\text{O})}$ o-Nitrotoluene + p-Nitrotoluene

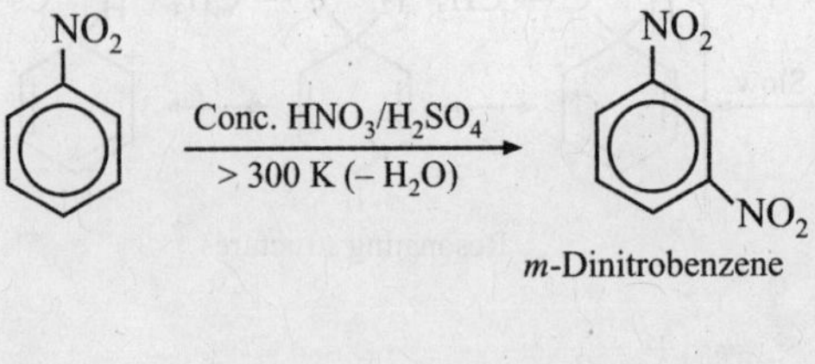

Nitrobenzene $\xrightarrow{\text{Conc. HNO}_3/\text{H}_2\text{SO}_4}_{> 300 \text{ K } (-\text{H}_2\text{O})}$ m-Dinitrobenzene

8.11 THEORY OF REACTIVITY IN AROMATIC COMPOUNDS ON THE BASIS OF INDUCTIVE AND RESONANCE EFFECTS

It has been observed that the rate determining step in electrophilic aromatic substitution is the formation of intermediate resonance stabilised carbocation.

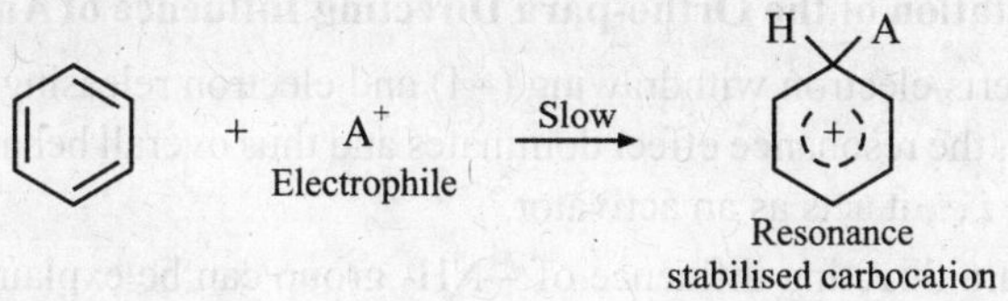

It is thus clear that *any factor which stabilises the intermediate carbocation will also stabilise the transition state leading to its formation.* Consequently, the carbocation will be formed more quickly and the rate of the overall reaction increases. On the other hand, *factors which destabilise the carbocation will decrease the ease of its formation and hence decrease the rate of the reaction.*

It is a familiar fact that *dispersal of charge leads to the stability of the system.* Now electron releasing groups tend to decrease the positive charge on the carbocation and thus stabilise the ion. As a result, the rate of the reaction increases. On the other hand, electron withdrawing groups destabilise the carbocation by intensifying the positive charge on it. Consequently, the rate of further electrophilic substitution reaction decreases. It may be mentioned that *release or withdrawal of electrons may occur due to inductive effect alone or through the net result of inductive and resonance effects.*

The effect of substituents on reactivity can be illustrated by considering electrophilic substitution in benzene, toluene and nitrobenzene. The relative rates in these three substitution reactions will depend upon the relative stabilities of their corresponding intermediate carbocations formed as shown below:

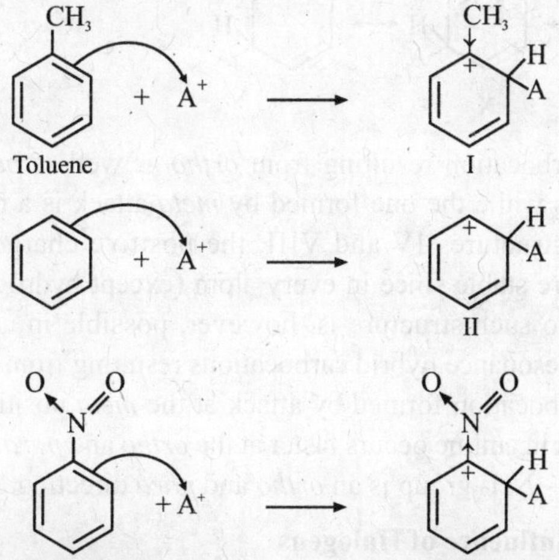

Due to its electron releasing inductive effect, methyl group tends to disperse the positive charge and thus stabilise the carbocation (I). As a aresult, the carbocation (I) is formed more quickly than the carbocation (II) from benzene. Consequently toluene undergoes electrophilic substitution at a faster rate than does benzene.

On the other hand, the nitro group tends to intensify the positive charge due to its electron withdrawing inductive and resonance effects. This destabilises the carbocation (III) which is, therefore, formed slowly than the carbocation (II) from benzene. Consequently nitrobenzene undergoes electrophilic substitution at a slower rate as compared to benzene.

Electronic Interpretation of the Ortho-para Directing Influence of Amino Group

Amino group exerts electron withdrawing (– I) and electron releasing (+M) effects. Of the two opposing effects, the resonance effect dominates and thus overall behaviour of $-NH_2$ group is electron releasing, *i.e.*, it acts as an activator.

The ortho and para directing influence of $-NH_2$ group can be explained by assuming that nitrogen can share more than a pair of electrons with the benzene ring and can accommodate a positive charge. Thus, consider the case of further electrophilic substitution in aniline. The various resonating structures of the carbocations formed by ortho, para and meta attack are given below:

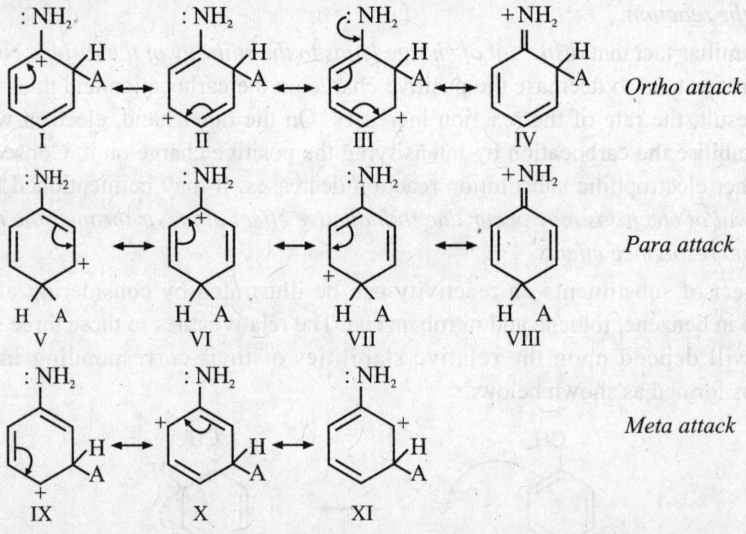

The intermediate carbocation resulting from *ortho* as well as *para* attack is a resonance hybrid of four structures while the one formed by *meta* attack is a resonance hybrid of three structures. Further, in structures IV and VIII, the positive charge is carried by nitrogen. These structures are more stable since in every atom (except hydrogen) has a complete octet of electrons in them. No such structure is, however, possible in case of *meta* attack. It is, therefore, clear that the resonance hybrid carbocations resulting from *ortho* and *para* attack are more stable than the carbocation formed by attack at the *meta* position. Consequently further electrophilic substitution in aniline occurs faster at the *ortho* and *para* positions than at the *meta* position. In other words $-NH_2$ group is an *ortho* and *para* directing group.

Ortho-Para Directing Influence of Halogens

Halogens exert electron withdrawing (–I) and electron releasing (+M) effects. Due to high electronegativities of halogens, the inductive effect predominates over the mesomeric effect and thus the overall behaviour of halogens is electron withdrawing. In other words, halogens act as deactivators for further substitution.

In order to account for their ortho and para directing nature, it has been assumed that halogens can share more than one pair of electrons with the benzene ring and can accommodate positive charge. Thus, consider the case of further electrophilic substitution in chlorobenzene. The various resonating structures of the carbocation formed by ortho, para and meta attack are given below:

Ortho attack (structures I, II, III, IV)

Para attack (structures V, VI, VII, VIII)

Meta attack (structures IX, X, XI)

It is clear that the intermediate carbocation resulting from ortho as well as para attack is a resonance hybrid of four structures while the one formed by meta attack is a resonance hybrid of three structures. Structures (I) and (VI) are highly unstable since in these the positive charge is carried by that carbon which is linked to electron withdrawing chlorine atom. However in structures (IV) and (VIII), the positive charge is carried by chlorine. These structures are extra stable since in every atom (except hydrogen) has a complete octet of electrons in them. No such structure is, however, possible in case of meta attack. Thus, the resonance hybrid carbocations resulting from ortho and para attack are more stable than that formed by attack at the meta position. As a result, chlorine is ortho and para directing. The same is true for other halogens. It may thus be concluded that in case of halogens, the reactivity is controlled by the stronger inductive effect and the orientation is determined by mesomeric effect.

Ortho and Para Directing Influence of Alkyl Groups

Alkyl groups exert electron releasing inductive effect (+I effect). Consider the case of further electrophilic substitution in toluene which contains an electron releasing methyl group. The various resonating structures of the carbocations formed by ortho, para and meta attack are given below:

Ortho attack (structures I, II, III)

[Structures IV, V, VI — Para attack]

[Structures VII, VIII, IX — Meta attack]

We find that in each case, the intermediate carbocation is a resonance hybrid of three structures. In structures (I) and (V), the positive charge is located on the carbon atom to which electrons releasing methyl group is attached. Therefore, the positive charge on such a carbon is highly dispersed and thus the corresponding structures (I) and (V) are more stable than all other structures. No such structure is, however, possible in case of meta attack. Hence the resonance hybrid carbocations resulting from ortho and para attack are more stable than the one formed by attack at the meta position. Therefore further electrophilic substitution in toluene occurs faster at the ortho and para position than at the meta position. In other words, methyl group is an ortho and para directing group. Other alkyl groups behave similarly.

Deactivating and Meta Directing Nature of Nitro Group Towards Electrophilic Aromatic Substitution.

Nitro group is electron withdrawing in nature. In this case, the electron withdrawal occurs through electron withdrawing effect ($-I$ effect) as well as electron withdrawing resonance effect ($-M$ effect). Due to its electron withdrawing character, nitro group deactivates the benzene ring towards further electrophilic substitution.

Let us examine the directing influence of $-NO_2$ group by considering electrophilic substitution in nitrobenzene. The various resonating structures of the carbocations formed by ortho, para and meta attack are given below:

[Structures I, II, III — Ortho attack]
I — Specially unstable

[Structures IV, V, VI — Para attack]
V — Specially unstable

[Structures VII, VIII, IX — Meta attack]

In the contributing structures (I) and (V), the positive charge is located on that carbon atom which is directly linked to electron withdrawing nitro group. Although $-NO_2$ group withdraws electrons from all positions, it does so most from the carbon directly attached to it. Hence this carbon atom, already made positive by nitro group has little tendency to accommodate the positive charge of the carbocation. Consequently structures (I) and (V) are unstable and their contribution towards stabilization of the carbocation is almost negligible. Thus carbocations formed by ortho and para attack are virtually resonance hybrids of only two structures while the one formed by meta attack is a resonance hybrid of three structures. Therefore the resonance hybrid carbocation resulting from meta attack is more stable than the carbocations resulting from ortho and para attack. Consequently, further electrophilic substitution takes place at the meta position. Thus, nitro group is meta directing.

Ortho-Para Directing Influence of Alkoxy Group

Alkoxy group is *o-p* directing because the lone-pair of electrons on oxygen takes part in resonance as illustrated below:

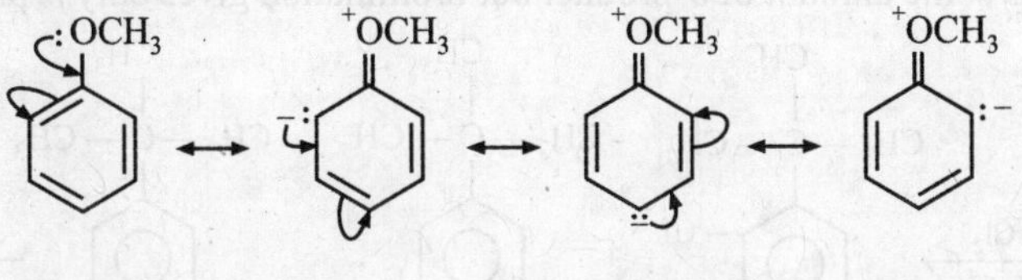

Ortho and para positions relative to methoxy group are negative. Hence in further substitution the electrophile will preferentially attach at these negative points.

8.12 ORTHO-PARA RATIO

There are two vacant ortho positions and one vacant para position in monosubstituted benzene. When we carry out further substitution, we expect the ortho/para ratio of the products as 2 : 1. But this does not happen actually. Ortho product is formed in much smaller amount than expected. Moreover, ortho/para ratio varies from substituent to substituent already present. A number of factors are responsible for this variation. Significant among them are steric factors, electronic factors and the reaction of the substituent already present with the attacking electrophile.

Steric Factors

Nitration of different alkylbenzenes has been carried out which reveals that the ortho/para ratio decreases from 1.5 to 0.25. For toluene, ehtylbenzene, isopropyl benzene and *tert* butylbenzene, the values are 1.5, 0.95, 0.5 and 0.25 respectively. We explain it on the basis of steric effects. A larger substituent will not easily allow the attachment of the electrophile to the ortho position and thus will lower the amount of ortho product. In moving from methyl group (in toluene) to tert butyl group, the size of the substituent increases progressively and therefore ortho/para ratio decreases.

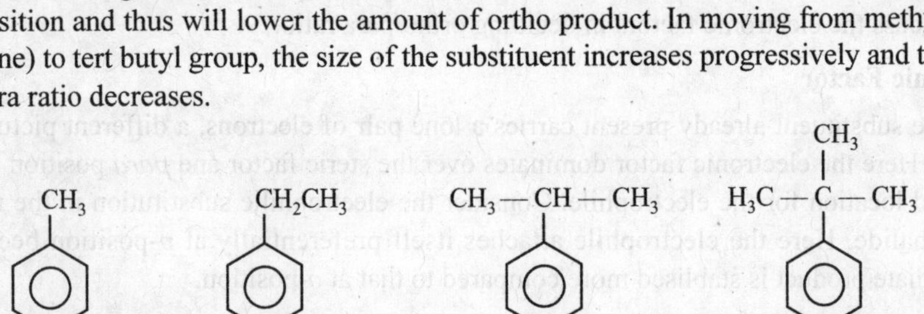

o/p = 1.5   o/p = 0.45   o/p = 1.5   o/p = 0.25

This occurs because the energy of activation of the intermediate product increases when we move from the substituent methyl to tert butyl, making it less stable. This is substantiated by the fact that nitration of acetanilide at 273 K almost exclusively yields p-product and the amount of o-product increases progressively with rise of temperature.

$$\text{NHCOCH}_3 \xrightarrow[\text{H}_2\text{SO}_4]{\text{HNO}_3, 273\ K} \text{NHCOCH}_3\text{-NO}_2\ (p\text{-product}) \xrightarrow{\text{Rise of Temp.}} \text{Mixture of } o\text{- and } p\text{- products}$$

Just as the size of the substituent affects the ortho/para ratio of the products, the size of the electrophile also matters. Keeping the size of the substituent same, the ortho/para ratio will decrease with increase in the size of electrophile. Thus, in the halogenation of tert butylbenzene, chlorination gives some amount of o-product but bromination gives only p-product.

[Reaction: tert-butylbenzene + Cl$_2$/FeCl$_3$ → Mix of o- and p- products; tert-butylbenzene + Br$_2$/Fe → Only p- product]

When two substituents of different sizes are already present in *para* positions, the new group will enter at *ortho* position to the smaller group because the electrophile will face greater hindrance *ortho* to the bigger group. For example

[Reaction: p-methyl-isopropylbenzene + A$^+$ (Electrophile) → substituted product + H$^+$]

So far we have discussed steric factors in deciding ortho/para ratio of the products. Now we shall discuss the electronic factors in deciding ortho/para ratio.

Electronic Factor

If the substituent already present carries a lone pair of electrons, a different picture might emerge. Here the electronic factor dominates over the steric factor and *para* position might be preferred location for the electrophile. Consider the electrophilic substitution in the molecule of aryl halide. Here the electrophile attaches itself preferentially at *p*-position because the intermediate product is stablised more compared to that at *o*-position.

Para Attack

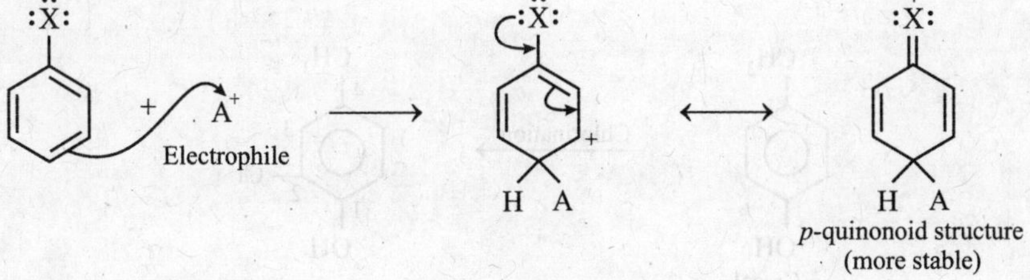

[Diagram: Para attack mechanism showing electrophile A⁺ attacking benzene with :X: substituent, forming intermediate with H and A at para position, resonance to p-quinonoid structure (more stable).]

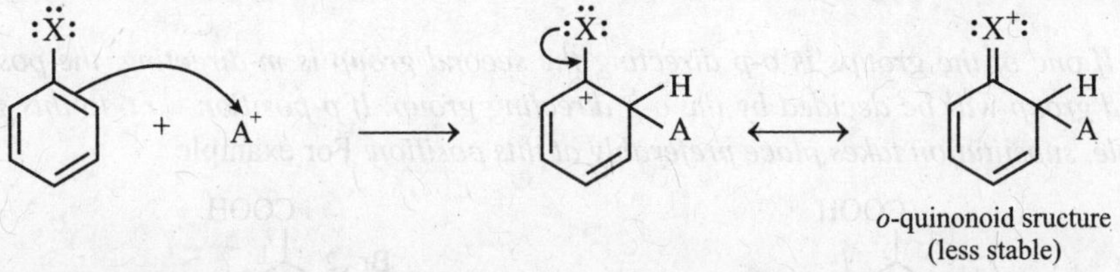

[Diagram: Alternative para attack pathway leading to o-quinonoid structure (less stable).]

Chemical Interaction

We observe that nitration of anisole with HNO_3/H_2SO_4 gives a mixture of o- and p-nitroanisole in the ratio of 31 : 67 but with N_2O_5 ($HNO_3 - Ac_2O$), the ratio of ortho-para isomers is 71 : 28. This is explained as under.

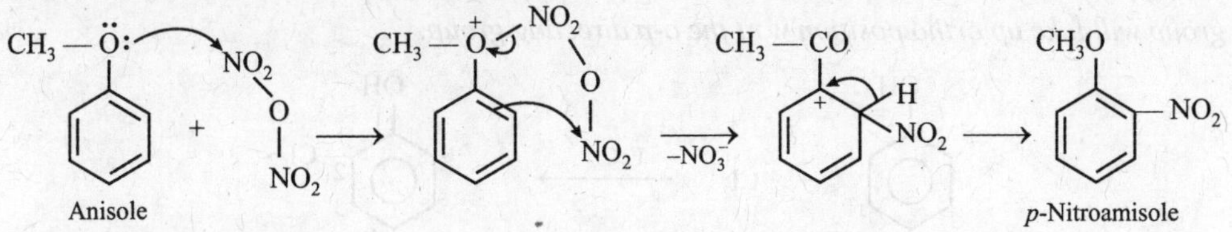

[Reaction scheme: Anisole + NO_2-O-NO_2 → intermediate with CH_3-O⁺-NO_2 attacking ring → CH_3-O with NO_2 displacement, $-NO_3^-$ → cyclohexadienyl cation with CH_3O and NO_2, H → CH_3O-C₆H₄-NO_2 p-Nitroanisole]

8.13 SUBSTITUTION OF THIRD GROUP IN THE BENZENE GROUP

The position taken by the third group in electrophilic substitution in benzene ring when two groups are already present on the ring depends upon the nature of the two groups already present. The following guidelines generally prove helpful in locating the position taken by the third incoming group.

(i) If the two groups already present are o-p directing, the position of the third group is decided by more powerful activating group. The activating effects of the ortho/para directing groups in the decreasing order are

$O^- > NH_2 > NR_2 > OH > NHCOCH_3, OCH_3, OCOCH_3 > CH_3 > C_6H_5 > Cl > Br > I$

If *para* position w.r.t. more powerful activating group is free, the new group preferably occupies *para* position compared to *ortho* position. For example

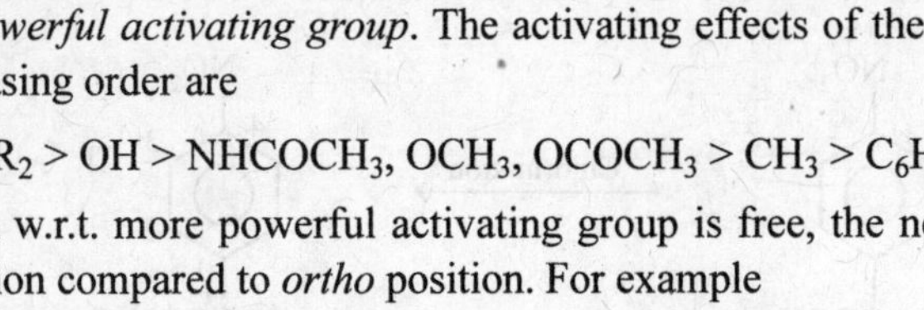

[Reaction: o-cresol (with CH₃ and OH) → Chlorination → 4-Chloro-2-methylphenol (positions labeled 1,2,3,4 with CH₃ at 2, OH at 1, Cl at 4)]

However, if the *para* position w.r.t. the more activating group is not vacant, the third group will take up *ortho* position w.r.t the more powerful activating group.

p-Cresol →(Chlorination)→ 2-Chloro-4-methylphenol

(ii) *If one of the groups is o-p directing the second group is m-directing, the position of the third group will be decided by the o-p directing group. If p-position w.r.t. to this group is available, substitution takes place preferably at this position.* For example

m-Hydroxybenzoic acid →(Br_2 / Fe)→ 2-Bromo-5-hydroxybenzoic acid

(iii) *If out of the two o-p and m-directing groups are located para to each other, the third group will take up ortho position w.r.t the o-p directing group.*

OH (with NO_2 para) + Cl_2 →($FeCl_3$)→ 2-Chloro-4-nitrophenol

(iv) *If both the groups already present are meta directing further substitution becomes very difficult. The position taken by the third group will be decided by the less powerful meta directing group.* The decreasing order of the *meta* directing effect of the groups is given below.

$$-\overset{+}{N}R_3 > \overset{+}{N}H_3 > NO_2 > CN > SO_3H > COOH > CONH_2 > COOR > CHO$$

NO_2 / COOH (para) →(Chlorination)→ 3-Chloro-4-nitrobenzoic acid

(v) *If both the groups already present are ortho-para directing but are located meta to each other, third group will not occupy the position between the two groups.* That is, third group will not occupy *ortho* position w.r.t. either group. The position taken will be *para* but w.r.t. more powerful activating group. For example,

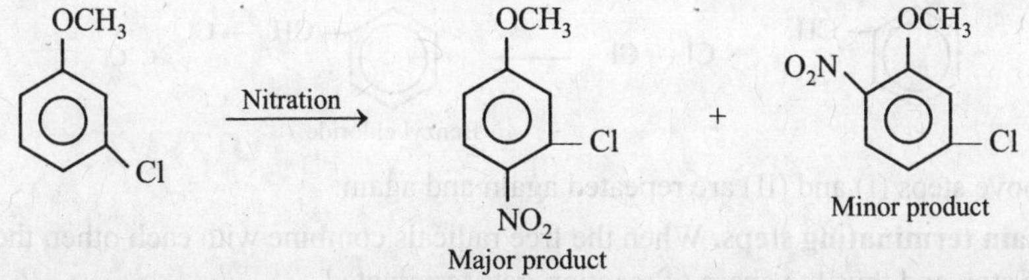

[Diagram: Nitration of 3-chloroanisole giving 4-chloro-2-nitroanisole (Major product) and 2-chloro-6-nitroanisole type (Minor product)]

Solved Example

Example 9: Benzyl chloride is more reactive towards nucleophilic substitution reactions than methyl chloride. Explain.

Or

Both benzyl and methyl carbocations are primary but benzyl cation is much more stable than methyl cation.

Solution: The greater reactivity of benzyl chloride is mainly due to the fact that it readily ionises to form a resonance stabilised benzyl carbocation. On the other hand, methyl chloride does not ionise to form methyl carbocation because there is no resonance.

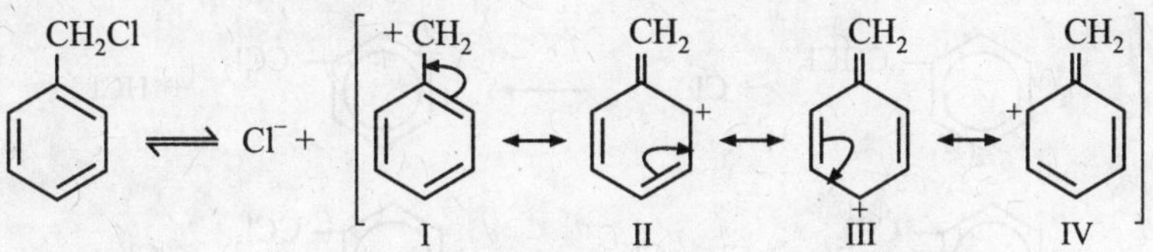

[Diagram: Ionization of benzyl chloride to benzyl carbocation with four resonance structures I, II, III, IV — Resonating structures of benzyl carbocation]

8.14 SIDE CHAIN HALOGENATION

The reaction in which one or more hydrogen atoms of the side chain are replaced by the halogen atoms is called *side chain halogenation*. It can be brought about by treating an arene with a suitable halogen in the presence of light or heat and the absence of halogen carriers.

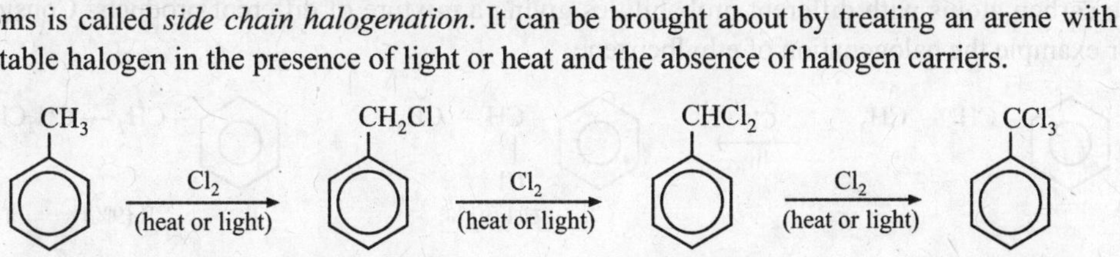

[Diagram: Toluene → Benzyl chloride → Benzyl dichloride → Benzotrichloride, each step with Cl_2 (heat or light)]

Mechanism of halogenation. Side chain halogenation of toluene occurs by a free radical mechanism as given below:

(a) Chain initiation step

$$Cl-Cl \xrightarrow[\text{(homolytic fission)}]{\text{Heat or light}} 2\,Cl\cdot$$

(b) Chain propagating steps

(I)  [Diagram: $C_6H_5-CH_3 + Cl\cdot \longrightarrow C_6H_5-\dot{C}H_2$ (Benzyl free radical) $+ HCl$]

(II)  $C_6H_5\text{-}\dot{C}H_2$ + Cl—Cl $\longrightarrow$ $C_6H_5\text{-}CH_2\text{-}Cl$ + Cl•
                                                                          Benzyl chloride

The above steps (I) and (II) are repeated again and again.

(c) **Chain terminating steps.** When the free radicals combine with each other, the reaction comes to a stop and thus the chain of reaction gets terminated.

$$Cl• + Cl• \longrightarrow Cl\text{—}Cl$$

When excess of chlorine is used, other hydrogen atoms are also replaced in the same manner. Thus,

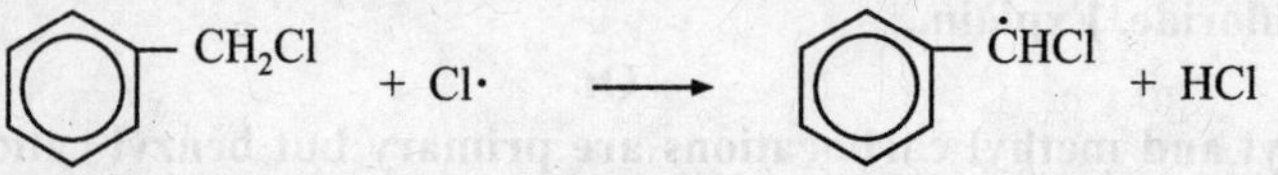

$C_6H_5\text{-}CH_2Cl$ + Cl• $\longrightarrow$ $C_6H_5\text{-}\dot{C}HCl$ + HCl
Benzyl chloride

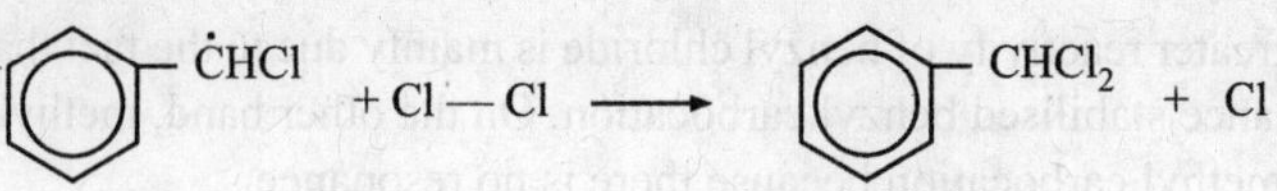

$C_6H_5\text{-}\dot{C}HCl$ + Cl—Cl $\longrightarrow$ $C_6H_5\text{-}CHCl_2$ + Cl•
                                                    Benzyl chloride

$C_6H_5\text{-}CHCl_2$ + Cl• $\longrightarrow$ $C_6H_5\text{-}\dot{C}Cl_2$ + HCl

$C_6H_5\text{-}\dot{C}Cl_2$ + Cl—Cl $\longrightarrow$ $C_6H_5\text{-}CCl_3$ + Cl•
                                                     Benzotrichloride

When the side chain contains more than one carbon, the halogen free radical can attack all the carbon atoms with different probabilities giving a mixture of different products. Consider for example the halogenation of ethylbenzene.

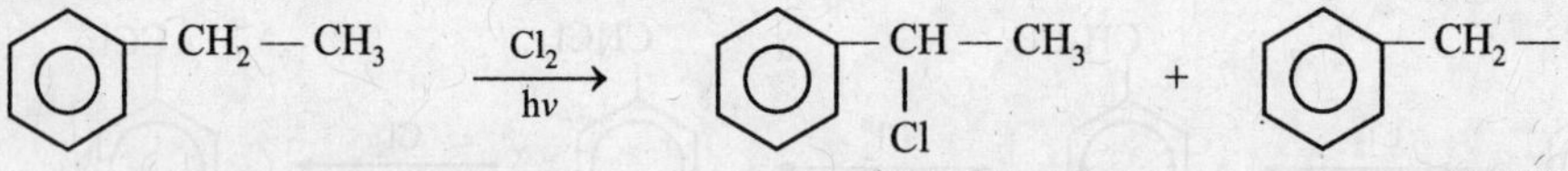

$C_6H_5\text{-}CH_2\text{-}CH_3$ $\xrightarrow{Cl_2, h\nu}$ $C_6H_5\text{-}CH(Cl)\text{-}CH_3$ + $C_6H_5\text{-}CH_2\text{-}CH_2Cl$
                                                              I (91%)   II (9%)

It may be noted that the probability of hydrogen abstraction from methyl and methylene carbon atoms is 3 : 2, but still hydrogen abstraction from benzylic carbon (linked to benzene ring directly) takes place preferentially giving I as the major product.

This is because the intermediate benzylic free radicals are more stable and hence easily formed. For example, benzylic free radical (C_6H_5 — $\dot{C}HCH_3$) obtained from ethylbenzene is a resonance hybrid of the following structures:

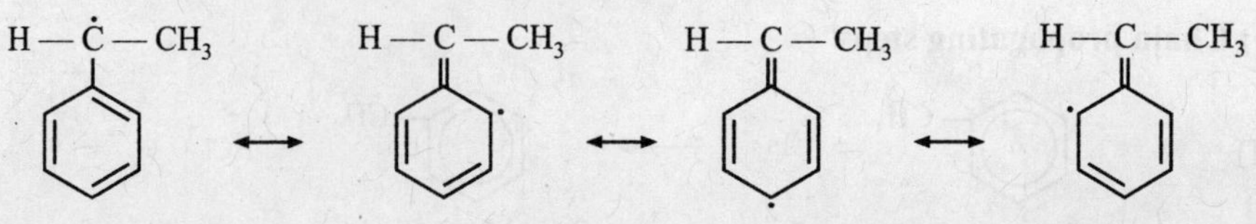

$H\text{-}\dot{C}\text{-}CH_3\text{-}C_6H_5$ $\longleftrightarrow$ $H\text{-}C\text{-}CH_3$ (radical on ring) $\longleftrightarrow$ $H\text{-}C\text{-}CH_3$ (radical on ring) $\longleftrightarrow$ $H\text{-}C\text{-}CH_3$ (radical on ring)

On the other hand, the second free radical ($C_6H_5-CH_2-\overset{\bullet}{C}H_2$) that can be obtained from ethylbenzene is a resonance hybrid of only two structures.

Due to its greater number of contributing structures, benzylic free radical is more resonance stabilised than non-benzylic free radical. Thus, bromination of ethylbenzene occurs exclusively at the **benzylic position.**

Bromine is more selective and less reactive than chlorine. Therefore bromination of ethylbenzene gives only benzylic product. The second product is not obtained at all.

$$\underset{\substack{\text{2-Bromo-1-phenylethane}\\(\text{zero}\%)}}{C_6H_5-CH_2CH_2Br} \longleftarrow \underset{\text{Ethylbenzene}}{C_6H_5-CH_2CH_3} \xrightarrow[h\nu]{Br_2} \underset{\substack{\text{1-Bromo-1-phenylethane}\\(100\%)}}{C_6H_5-CHCH_3 \atop |\;\;\;\;\;\;\;\; Br}$$

N-Bromosuccinimide (NBS)

This is a reagent to carry out benzylic bromination of alkylbenzenes.

$$C_6H_5-CH_2CH_3 \xrightarrow[CCl_4,\, 333\,K]{NBS} \underset{\text{1-Bromo-1-phenylethane}}{C_6H_5-\underset{Br}{\underset{|}{CH}}-CH_3}$$

8.15 SIDE CHAIN OXIDATION

Alkylbenzenes on heating with alkaline $KMnO_4$, acidified $K_2Cr_2O_7$ or dilute HNO_3 are oxidised to benzoic acid irrespective of the length of the side chain. For example,

$$\underset{\text{Toluene}}{C_6H_5-CH_3} + 3[O] \xrightarrow[\text{or dil } HNO_3,\, \Delta]{\text{Alk. } KMnO_4 \text{ or Acidified } K_2Cr_2O_7} \underset{\substack{\text{Benzoic}\\\text{acid}}}{C_6H_5-COOH} + H_2O$$

$$\underset{n\text{-Propylbenzene}}{C_6H_5-CH_2CH_2CH_3} + 9[O] \xrightarrow[\text{or dil } HNO_3,\, \Delta]{\text{Alk } KMNO_4,\, \text{Acidified } K_2Cr_2O_7} C_6H_5-COOH + 2CO_2 + 3H_2O$$

If two or more alkyl groups are attached, each of them is oxidised to – COOH group. For example,

$$\underset{p\text{-Xylene}}{\underset{\underset{CH_3}{|}}{\overset{\overset{CH_3}{|}}{C_6H_4}}} + 6[O] \xrightarrow[\Delta]{\text{Alk. KMnO}_4} \underset{\text{Terephthalic acid}}{\underset{\underset{COOH}{|}}{\overset{\overset{COOH}{|}}{C_6H_4}}} + 2H_2O$$

Mechanism. Exact mechanism of the reaction is a matter of speculation. It probably involves attack on benzylic hydrogen atom to form benzylic radical as intermediate. We can say this on the grounds that alkyl benzenes which do not possess benzylic hydrogen do not give this test.

$$\underset{\substack{tert\text{-Butylbenzene} \\ \text{(No benzylic hydrogen)}}}{C_6H_5-C(CH_3)_3} \xrightarrow{\text{KMnO}_4/\text{KOH}} \text{No reaction}$$

Utility. The oxidation of alkylbenzenes can be used to determine the number and relative positions of the alkyl group.

o-dialkylbenzene $\xrightarrow{\text{oxid}}$ Phthalic acid

m-dialkylbenzene $\xrightarrow{\text{oxid}}$ Isophthalic acid

p-dialkylbenzene $\xrightarrow{\text{oxid}}$ Terephthalic acid

Thus by identifying the dicarboxylic acid, we can estimate the number and positions of alkyl groups in the original compound.

Oxidation with Chromyl Chloride—Etard Reaction

Oxidation of toluene can be stopped at the aldehydic stage using the reagent chromyl chloride. With a strong oxidising agent, we would obtain benzoic acid.

$$C_6H_5CH_3 \xrightarrow[CS_2]{2CrO_2Cl_2} \underset{\text{Addition product}}{C_6H_5CH_3 \cdot 2CrO_2Cl_2} \xrightarrow{H_2O} \underset{\text{Benzaldehyde}}{C_6H_5CHO}$$

Benzyl chloride can be oxidised to benzoic acid or benzaldehyde using an appropriate oxidising agent.

Benzaldehyde ←[O]/Pb(NO₃)₂— Benzyl chloride —[O]/HNO₃→ Benzoic acid

Arthur Birch (1915-1995)
Arthur Birch was an organic chemist from Australia. He developed Birch reduction of aromatic rings which is widely used in synthetic organic chemistry

8.16 BIRCH REDUCTION

Partial reduction of benzene or its derivatives by electron transfer from metals like Li, Na or K in liquid ammonia in the presence of an alcohol is called Birch reduction.

In such reactions, 1, 4-addition of hydrogen to benzene takes place to form 3, 6-dihydrocyclohexa -1,. 4-diene.

Mechanism

benzene + e^- (From metal) → radical anion → ROH / –RO⁻ → cyclohexadienyl radical → + e^- → carbanion → ROH / –RO⁻ → 3, 4-Dihydro cyclohexa-1, 4-diene

Radical ions are formed at 1, 4-positions rather than 1, 2-positions due to electronic repulsions.

The rate of Birch reduction and its stereochemistry depends upon the nature of the substituent group present in the ring.

(a) *Electron donating groups such as alkyl, alkoxy etc. retard the rate of Birch reduction and addition of electrons occurs at 3, 6-positions w.r.t. the substituent* due to repulsion between the electrons on the electron-donating group and the incoming electrons.

CH₃-benzene (positions 1,3,6 labelled) —(i) e^- (ii) ROH→ intermediate —(i) e^- (ii) ROH→ 1-Methylcyclohexa-1, 4-diene

The electron will first attach itself at position 3. The shifting of π bonds of benzene will occur to give the radical at position 6 as shown in the reaction.

(b) Electron-withdrawing groups such as NO₂, – COOH, etc increase the rate of Birch reduction and the addition of electrons occurs at 1,4-positions w.r.t. the electron-withdrawing group, due to attraction between the electron withdrawing group and the incoming electron.

HOOC-benzene (positions 1,2,3,4 labelled) —(i) + e^- (ii) ROH→ intermediate —(i) + e^- (ii) ROH→ product

Thus, the electron will attach itself to position 1 in the beginning. Shifting of the π-bonds in benzene will take place to produce a radical at position 4 as shown in the reaction.

8.17 BIPHENYL OR DIPHENYL

Methods of Preparation

Biphenyl can be prepared by the following methods:

1. Ullmann synthesis. Iodobenzene on heating with copper powder in a sealed tube yields biphenyl.

$$\text{Ph–I} + \text{Cu} + \text{I–Ph} \xrightarrow[\Delta]{\text{Sealed tube}} \text{Ph–Ph} + 2\text{CuI}$$
(Biphenyl)

2. Fittig Reaction. Bromobenzene on treatment with sodium in dry ether gives biphenyl

$$\text{Ph–Br} + 2\text{Na} + \text{Br–Ph} \xrightarrow{\text{Dry ether}} \text{Ph–Ph} + 2\text{NaBr}$$
(Biphenyl)

3. Diazotisation. Benzidine on diazotisation followed by reduction with hypophosphorus acid produces biphenyl

$$H_2N\text{–Ph–Ph–}NH_2 \xrightarrow[273\text{-}278K]{\text{NaNO}_2 \text{ (HCl)}} ClN_2\text{–Ph–Ph–}N_2Cl \xrightarrow{H_3PO_2} \text{Ph–Ph}$$
(Benzidine) (Diazotisation) (Biphenyl)

4. Grignard Reaction. Phenylmagnesium on heating with bromobenzene in presence of $CoCl_2$ or $NiCl_2$ gives biphenyl.

$$\text{Ph–MgBr} + \text{Br–Ph} \xrightarrow[\text{or NiCl}_2, \Delta]{\text{CoCl}_2} \text{Ph–Ph} + \text{MgBr}_2$$
(Biphenyl)

Manufacture

Biphenyl is commercially prepared by passing vapours of benzene through red hot copper tube at 873-1073 K

$$2\,\text{Ph} \xrightarrow[873\text{-}1073\text{ K}]{\text{Red hot Cu tube}} \text{Ph–Ph} + H_2$$
(Biphenyl)

Properties of Biphenyl

(*i*) Biphenyl has a characteristic small with a melting point of 344 K.

(*ii*) Phenyl group is *o-p* directing group and is an activating group. *It activates the rings towards further substitution.* In the first instance, the substituent enters the less hindered 4-position and to a small extent 2-position.

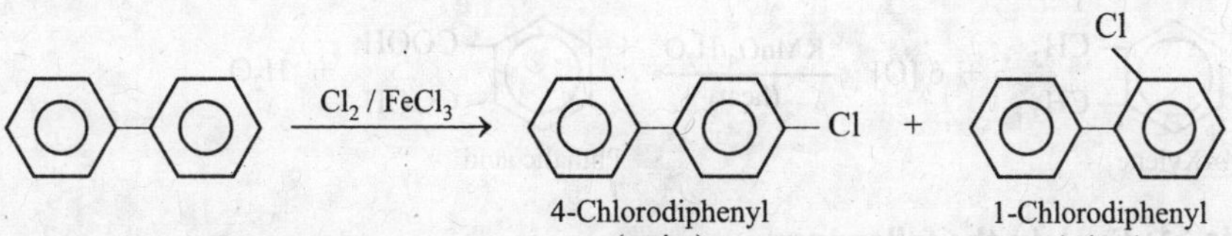

Aromatic Hydrocarbons   337

$$\text{Biphenyl} \xrightarrow{Cl_2/FeCl_3} \text{4-Chlorodiphenyl (major)} + \text{1-Chlorodiphenyl (minor)}$$

The second substituent always enters the second unsubstituted ring irrespective of whether the first substituent is electron donating or electron-withdrawing.

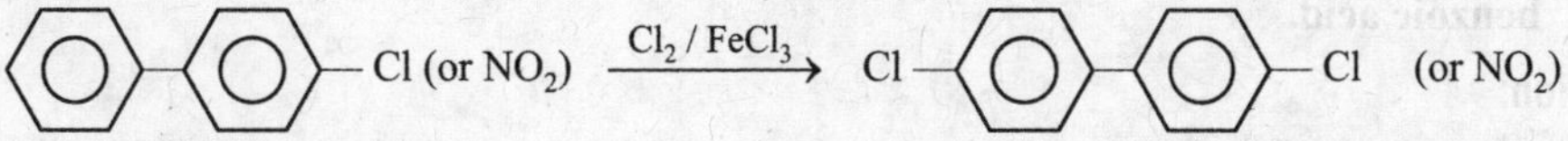

$$\text{Ph–Ph–Cl (or NO}_2\text{)} \xrightarrow{Cl_2/FeCl_3} \text{Cl–Ph–Ph–Cl (or NO}_2\text{)}$$

The positions taken by the first/second substituents can be explained in terms of the stability of the intermediate carbocations.

Uses
1. Chlorinated biphenyls are used as plasticizers and in transformer oils.
2. Benzidine (a derivative of diphenyl) is used as a dye intermediate.

Solved Examples

Example 10: What happens when:
   (i) Benzyl chloride is subjected to oxidation with lead nitrate.
   (ii) Toluene is treated with methyl chloride in the presence of anhydrous aluminium chloride.
   (iii) Benzyl chloride is treated with silver acetate.
   (iv) Chlorine is passed through boiling toluene.
   (v) o-Xylene is treated with hot aqueous $KMnO_4$ solution.

Solution:

(i) 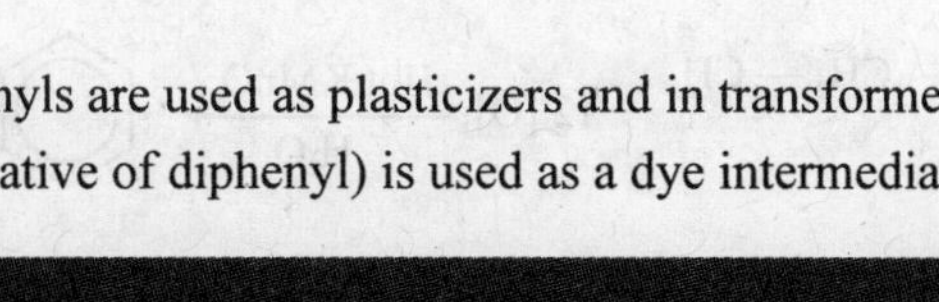
$$C_6H_5\text{–}CH_2Cl \xrightarrow[(O)]{Pb(NO_3)_2} C_6H_5\text{–}CHO \text{ (Benzaldehyde)}$$

(ii) Benzyl chloride

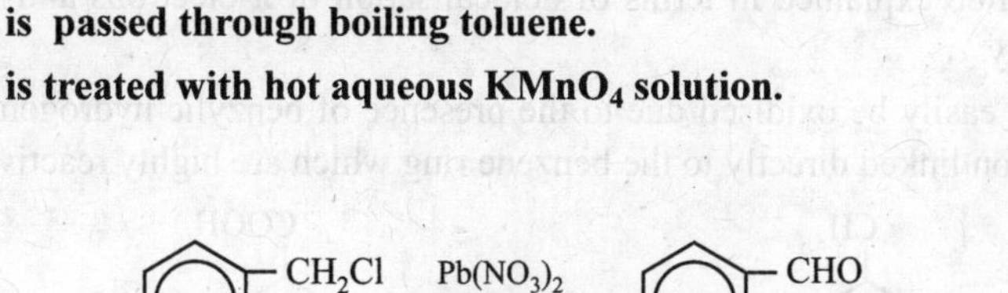

$$\text{Toluene} + CH_3Cl \xrightarrow[-HCl]{\text{Anhydrous } AlCl_3} o\text{-xylene} + p\text{-xylene}$$

(iii) $C_6H_5CH_2Cl + CH_3COOAg \longrightarrow AgCl + CH_3COOCH_2C_6H_5$
       Benzyl chloride   Silver acetate   Benzyl acetate

(iv) 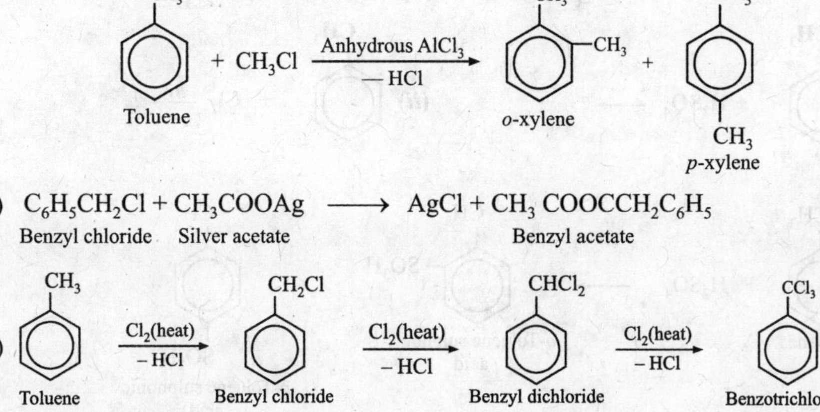
$$\text{Toluene} \xrightarrow[-HCl]{Cl_2(\text{heat})} \text{Benzyl chloride} \xrightarrow[-HCl]{Cl_2(\text{heat})} \text{Benzyl dichloride} \xrightarrow[-HCl]{Cl_2(\text{heat})} \text{Benzotrichloride}$$

(v) o-Xylene + 6 [O] $\xrightarrow{\text{KMnO}_4/\text{H}_2\text{O (heat)}}$ Phthalic acid (benzene with two COOH groups) + H_2O

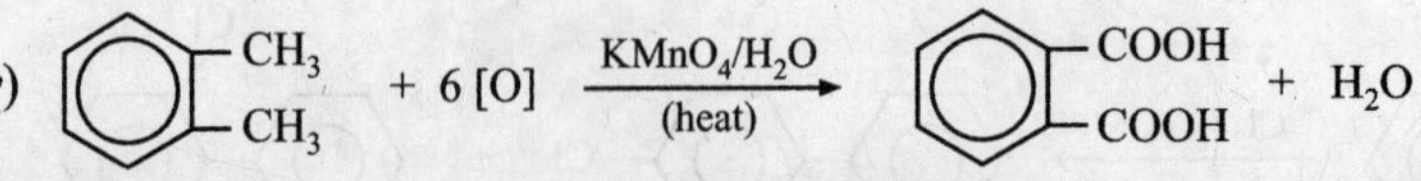

Example 11: Explain the following:

(i) n-Butylbenzene on oxidation gives benzoic acid but t-butylbenzene does not.

(ii) Benzene is stable towards oxidising agents but toluene can be oxidised easily to benzoic acid.

Solution:

(i) It is believed that oxidation involves the attack on benzylic hydrogen. t-Butylbenzene does not possess benzylic hydrogen and hence it can't undergo oxidation easily. On the other hand, n-butylbenzene is easily oxidised due to the presence of benzylic hydrogens. Thus,

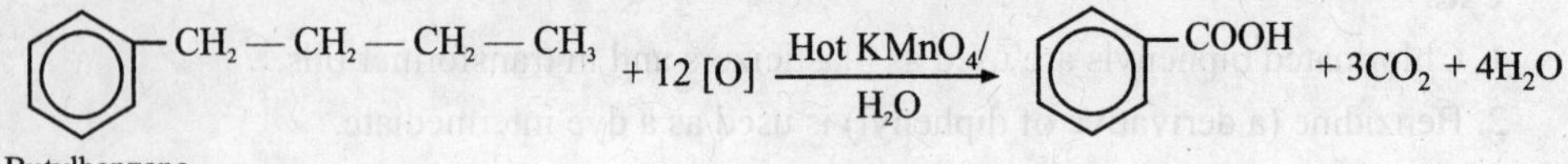

n-Butylbenzene (Ph—CH_2—CH_2—CH_2—CH_3) + 12 [O] $\xrightarrow{\text{Hot KMnO}_4/\text{H}_2\text{O}}$ Ph—COOH + $3CO_2$ + $4H_2O$

t-Butylbenzene (Ph—C(CH_3)$_3$) $\xrightarrow{\text{Hot KMnO}_4/\text{H}_2\text{O}}$ No reaction

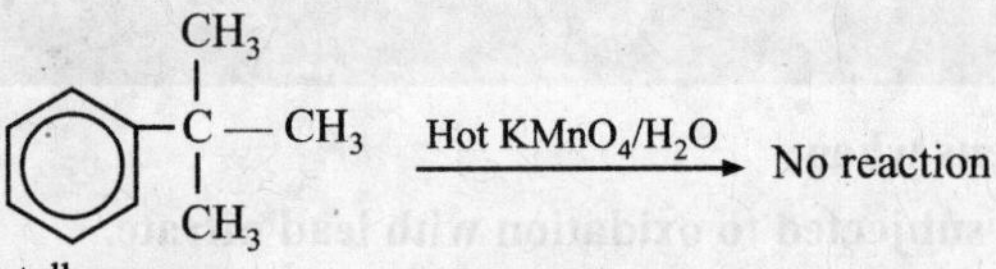

(ii) Benzene ring exhibits unusual stability and resists oxidation even by strong oxidising agents. This can be explained in terms of delocalisation of π electrons and its high value of resonance energy.

Toluene can easily be oxidised due to the presence of benzylic hydrogens i.e., hydrogens attached to carbon linked directly to the benzene ring which are highly reactive. Thus

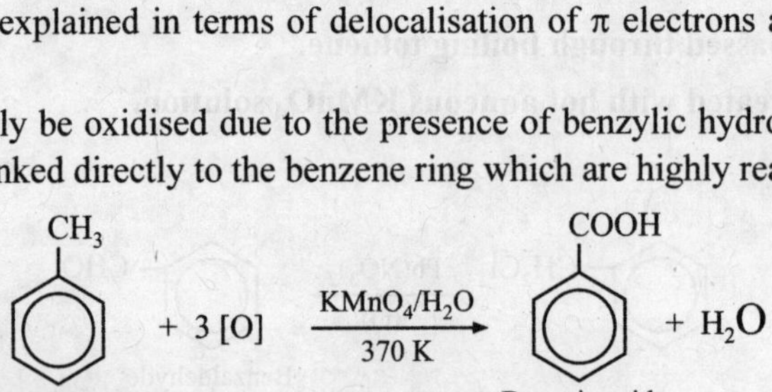

Toluene (Ph—CH_3) + 3 [O] $\xrightarrow[\text{370 K}]{\text{KMnO}_4/\text{H}_2\text{O}}$ Benzoic acid (Ph—COOH) + H_2O

Example 12: Fill in the blanks, giving names of products.

(i) Toluene (Ph—CH_3) + H_2SO_4 ⟶ ?

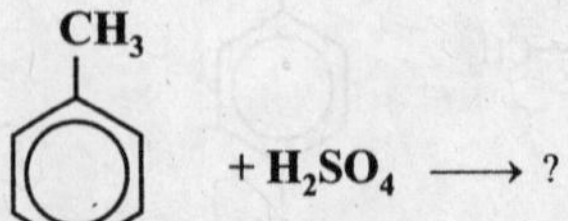

(ii) Toluene (Ph—CH_3) + Cl_2 $\xrightarrow{hv}$?

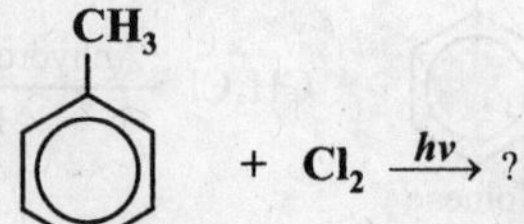

Solution:

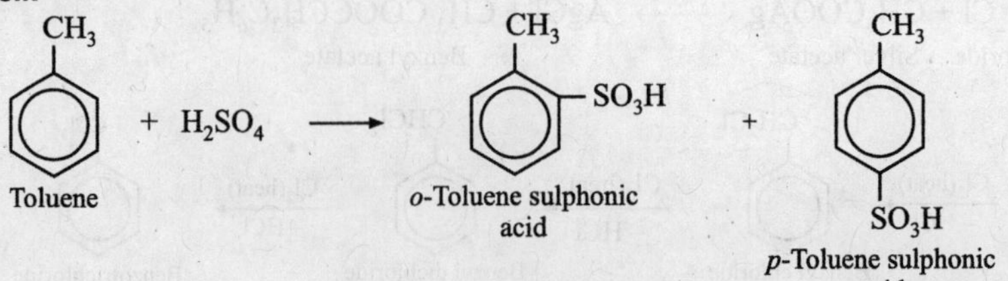

(i) Toluene + H_2SO_4 ⟶ o-Toluene sulphonic acid + p-Toluene sulphonic acid

(ii) Toluene + Cl$_2$ $\xrightarrow{h\nu}$ Benzyl chloride + HCl

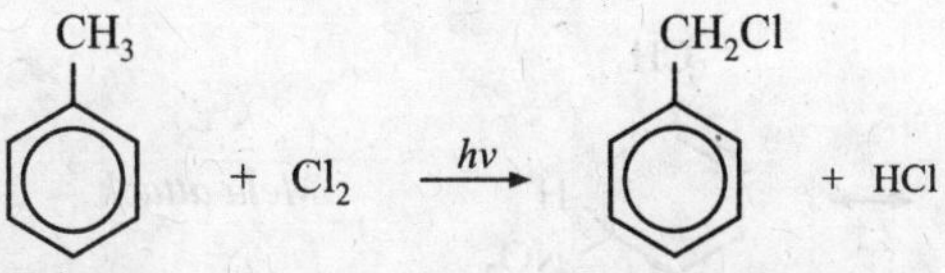

Example 13: Indentify the products A and B.

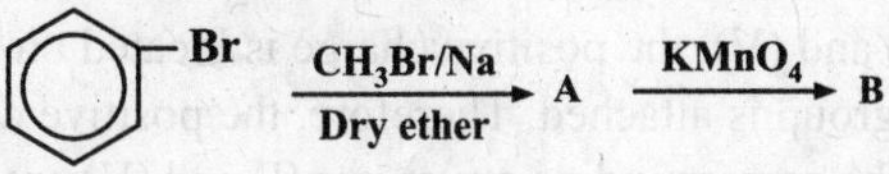

Bromo benzene $\xrightarrow[\text{Dry ether}]{\text{CH}_3\text{Br/Na}}$ A $\xrightarrow[]{\text{KMnO}_4}$ B

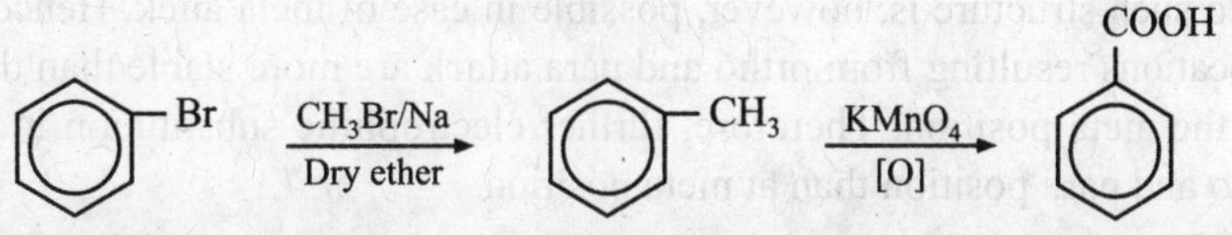

Solution:

Bromobenzene (C$_6$H$_5$Br) $\xrightarrow[\text{Dry ether}]{\text{CH}_3\text{Br/Na}}$ Toluene (C$_6$H$_5$CH$_3$) $\xrightarrow[\text{[O]}]{\text{KMnO}_4}$ Benzoic acid (C$_6$H$_5$COOH)

Example 14: Give the mechanism of nitration of toluene.

Solution:

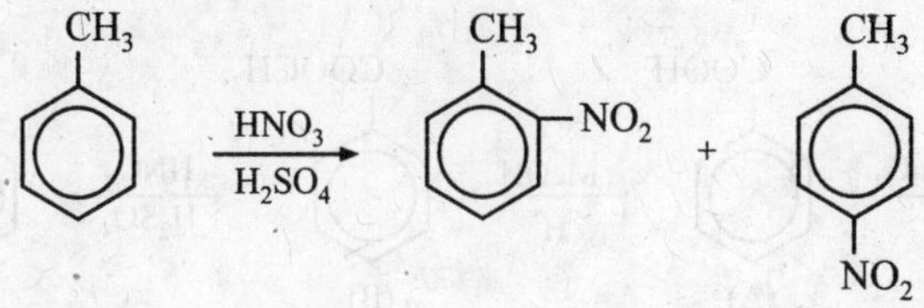

Toluene $\xrightarrow[\text{H}_2\text{SO}_4]{\text{HNO}_3}$ o-nitrotoluene + p-nitrotoluene

Mechanism.

(i) Generation of an electrophile

$$HONO_2 + 2H_2SO_4 \longrightarrow \overset{+}{N}O_2 + H_3\overset{+}{O} + 2HSO_4^-$$

Formation of carbocation

Alkyl groups exert electron releasing inductive effect (+I effect). Consider the case of further electrophilic substitution in toluene which contains an electron releasing methyl group. The various resonating structures of carbocations formed by ortho, para, meta attacks given below:

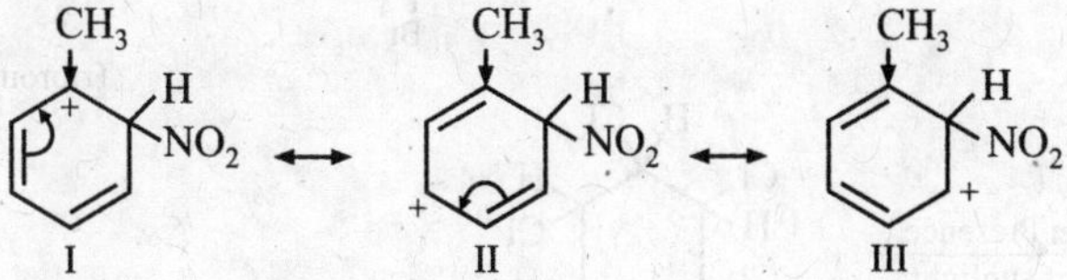

I ↔ II ↔ III   *Ortho attack*

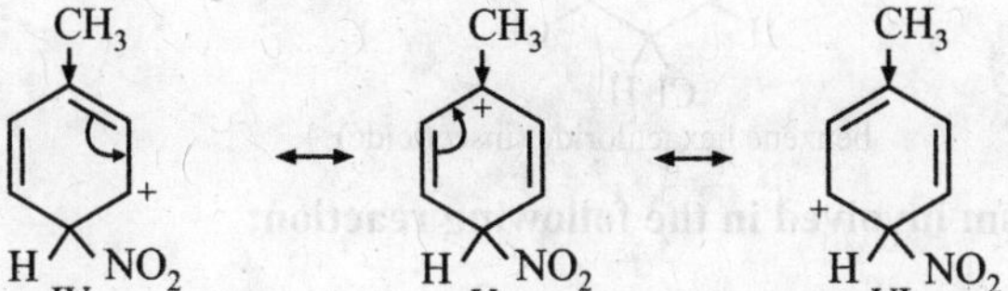

IV ↔ V ↔ VI   *Para attack*

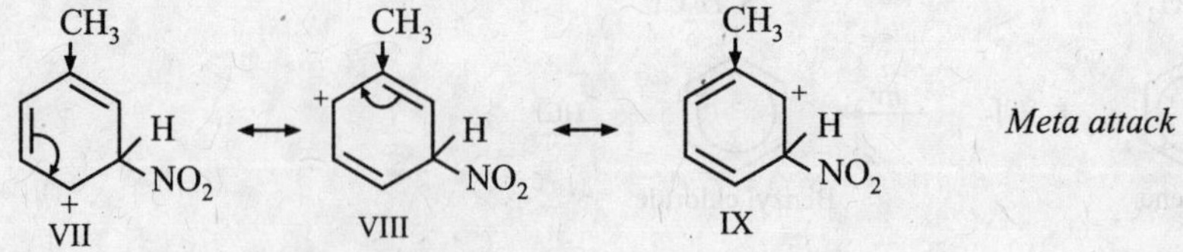

[Structures VII, VIII, IX showing meta attack resonance structures with CH₃, H, NO₂ groups]

Meta attack

We find that in each case, the intermediate carbocation is a resonance hybrid of three structures. In structures (I) and (V), the positive charge is located on the carbon atom to which electron releasing methyl group is attached. Therefore, the positive charge on such a carbon is highly dispersed and thus the corresponding structures (I) and (V) are more stable than all other structures. No such structure is, however, possible in case of meta attck. Hence the resonance hybrid carbocations resulting from ortho and para attack are more stable than the ones formed by attack at the meta position. Therefore, further electrophilic substitution in toluene occurs faster at ortho and para position than at meta position.

Example 15: Identify the compounds A, B and C.

$$\text{Toluene} \xrightarrow{KMnO_4} A \xrightarrow[H^+]{MeOH} B \xrightarrow{HNO_3/H_2SO_4} C$$

Solution:

$$\underset{\text{(A)}}{C_6H_5CH_3} \xrightarrow{KMnO_4} \underset{\text{(A)}}{C_6H_5COOH} \xrightarrow[H^+]{MeOH} \underset{\text{(B)}}{C_6H_5COOCH_3} \xrightarrow{HNO_3/H_2SO_4} \underset{\text{(C)}}{m\text{-}NO_2C_6H_4COOCH_3}$$

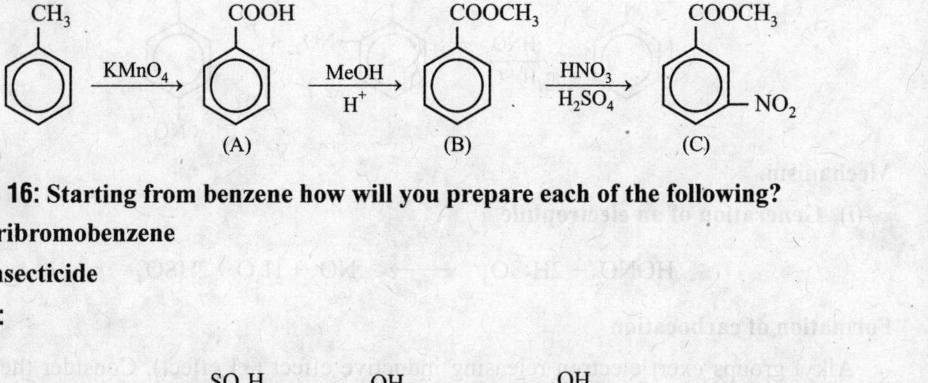

Example 16: Starting from benzene how will you prepare each of the following?

(i) Tribromobenzene

(ii) Insecticide

Solution:

(i)

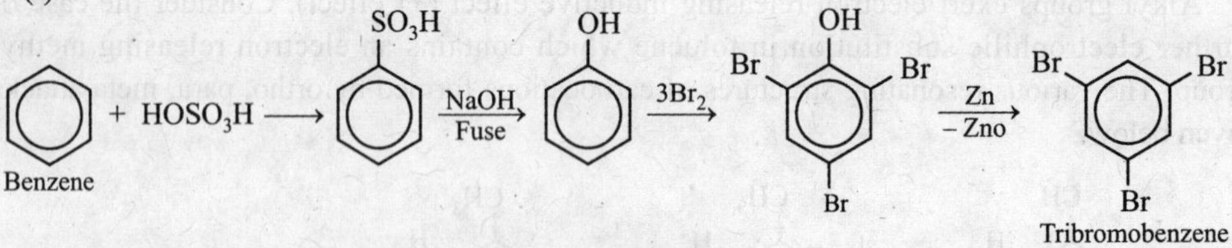

Benzene + HOSO₃H ⟶ C₆H₅SO₃H $\xrightarrow[\text{Fuse}]{NaOH}$ C₆H₅OH $\xrightarrow{3Br_2}$ 2,4,6-tribromophenol $\xrightarrow{Zn, -ZnO}$ Tribromobenzene

(ii) Benzene + 3 Cl₂ $\xrightarrow{\text{In Presence of sunlight}}$ benzene hexachloride (insecticide)

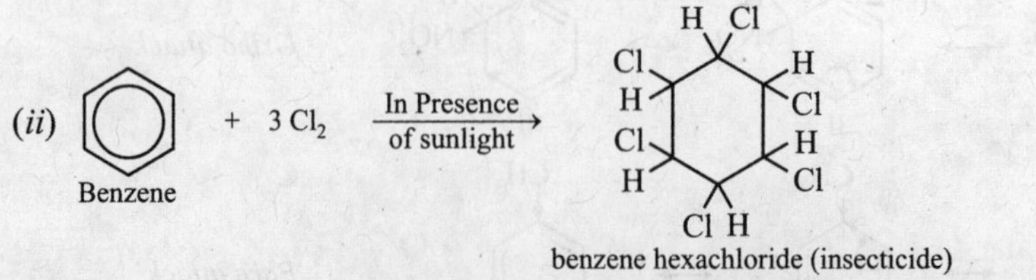

Example 17: Describe mechanism involved in the following reaction:

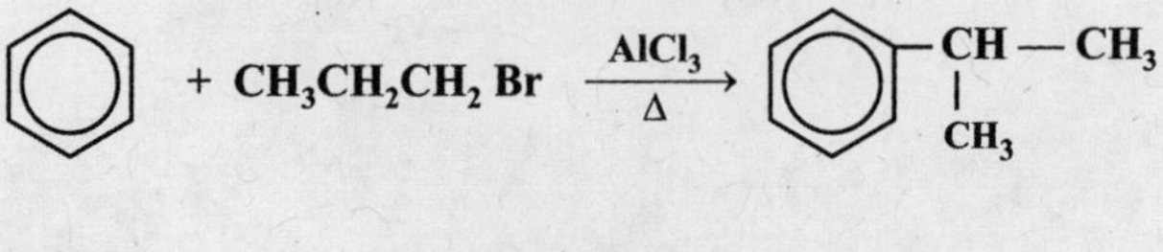

$$C_6H_6 + CH_3CH_2CH_2Br \xrightarrow[\Delta]{AlCl_3} C_6H_5\text{-}CH(CH_3)_2$$

Solution:

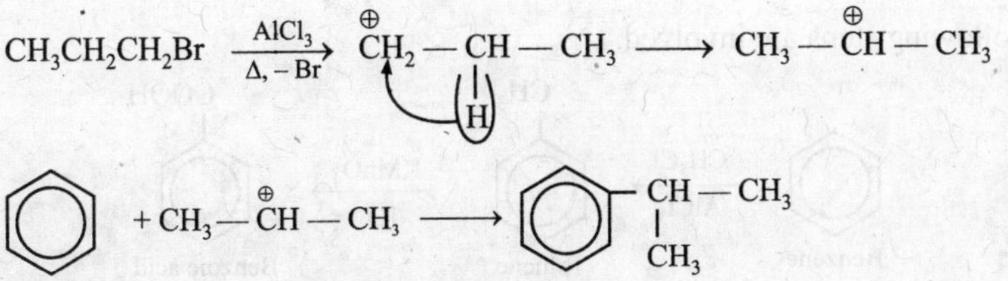

$$CH_3CH_2CH_2Br \xrightarrow[\Delta, -Br]{AlCl_3} \overset{\oplus}{C}H_2-CH-CH_3 \longrightarrow CH_3-\overset{\oplus}{C}H-CH_3$$

$$\text{(benzene)} + CH_3-\overset{\oplus}{C}H-CH_3 \longrightarrow \text{C}_6\text{H}_5-CH(CH_3)-CH_3$$

Example 18: When chlorine is passed into boiling toluene three different chlorinated products are obtained. What products will be obtained if each of these chlorinated products is separately treated with aqueous sodium hydroxide solution?

Solution: When chlorine is passed into boiling toluene chlorine displaces the hydrogen atoms of the side chain *i.e.*, methyl group, one by one.

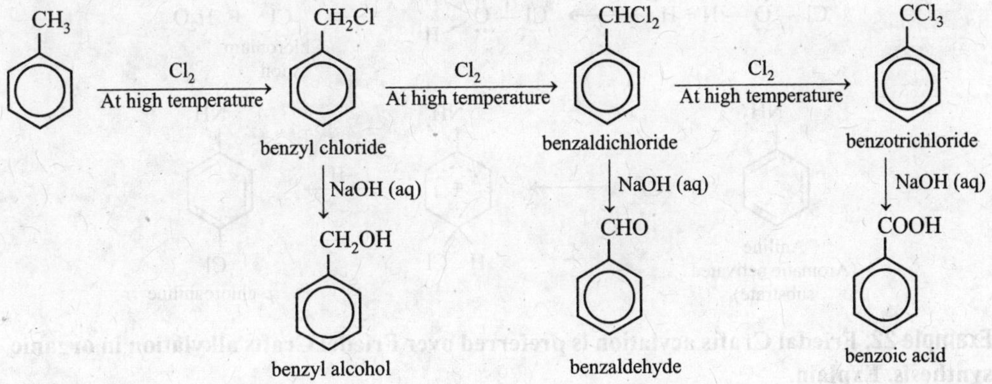

Toluene $\xrightarrow[\text{At high temperature}]{Cl_2}$ benzyl chloride (CH$_2$Cl) $\xrightarrow[\text{At high temperature}]{Cl_2}$ benzaldichloride (CHCl$_2$) $\xrightarrow[\text{At high temperature}]{Cl_2}$ benzotrichloride (CCl$_3$)

Each treated with NaOH (aq):
- benzyl chloride → benzyl alcohol (CH$_2$OH)
- benzaldichloride → benzaldehyde (CHO)
- benzotrichloride → benzoic acid (COOH)

Example 19: Give the preparation of toluene from (*i*) benzene (*ii*) chlorobenzene. How does toluene react with (*i*) Cl$_2$ in presence of FeCl$_3$ (*ii*) Cl$_2$ in sunlight?

Solution:

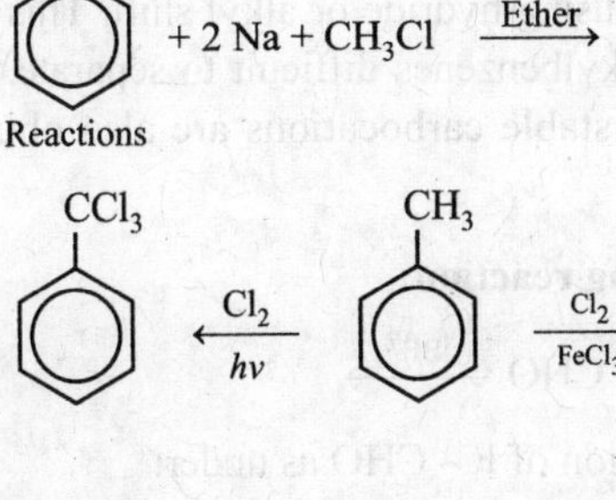

Benzene $+ CH_3Cl \xrightarrow{\text{Anhy AlCl}_3}$ Toluene

Chlorobenzene $+ 2Na + CH_3Cl \xrightarrow{\text{Ether}}$ Toluene

Reactions:

Toluene $\xrightarrow{Cl_2, h\nu}$ benzyl trichloride (CCl$_3$)

Toluene $\xrightarrow{Cl_2, FeCl_3}$ o-chlorotoluene + p-chlorotoluene

Example 20: How will you synthesise benzoic acid from benzene?

Solution: Following steps are involved.

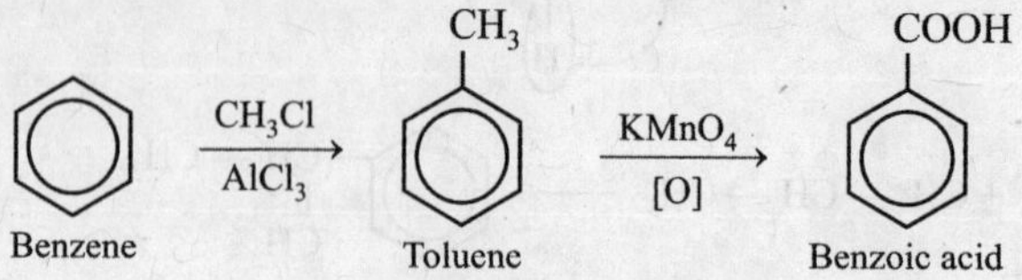

Benzene $\xrightarrow{\text{CH}_3\text{Cl}, \text{AlCl}_3}$ Toluene $\xrightarrow{\text{KMnO}_4, [O]}$ Benzoic acid

Example 21: It has been shown that chlorination of activated substrates can be achieved with hypochlorous acid in the presence of a proton catalyst. Suggest a suitable mechanism for the reaction.

Solution: The rate of the proton is to generate chloronium ion which is used in chlorination.

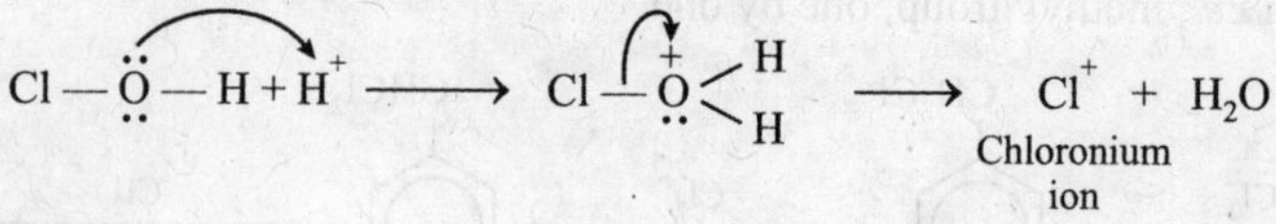

$$\text{Cl}-\ddot{\text{O}}-\text{H} + \text{H}^+ \longrightarrow \text{Cl}-\overset{+}{\text{O}}\!\!<^{\text{H}}_{\text{H}} \longrightarrow \text{Cl}^+ + \text{H}_2\text{O}$$
(Chloronium ion)

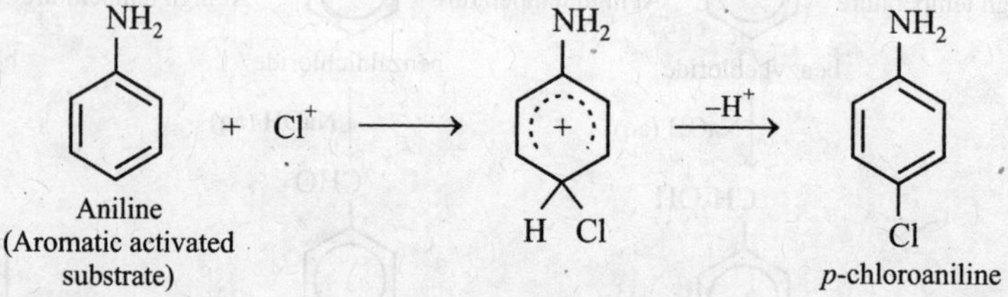

Aniline (Aromatic activated substrate) + Cl^+ ⟶ [intermediate] $\xrightarrow{-\text{H}^+}$ p-chloroaniline

Example 22: Friedal Crafts acylation is preferred over Friedel Crafts alkylation in organic synthesis. Explain.

Solution:

1. In Friedal Crafts alkylation, alkyl benzenes are obtained. Alkyl benzenes are more reactive than benzene due to electron-donating nature of the alkyl groups. This will result in further substitution in the benzene ring giving a mixture of mono, di and trialkylbenzenes which will pose the problem of separation of these components. On the other hand, if we carry out Friedal Crafts acylation, we shall obtain an acyl derivative. An acyl group is electron withdrawing and hence will deactivate the ring towards further substitution and a single pure compound will be formed.

2. Alkylation of benzene involves the formation of carbocations. These carbocations rearrange to more stable carbocations using hydride or alkyl shift. This will lead to the formation of a mixture of different alkylbenzenes difficult to separate. Some amounts of compounds corresponding to less stable carbocations are also obtained. No such problem is faced in acylation.

Example 23: Predict the product of following reaction:

$$\text{C}_6\text{H}_6 + \text{F}-\text{CHO} \xrightarrow{\text{BF}_3}$$

Solution: Lewis acid BF_3 causes the polarisation of F – CHO as under:

$$\text{BF}_3 + \text{F}-\text{CHO} \longrightarrow \text{BF}_3 \cdots \overset{\delta-}{\text{F}} \cdots \overset{\delta+}{\text{CHO}}$$

The electrophile C⁺HO attacks the benzene ring forming benzaldehyde

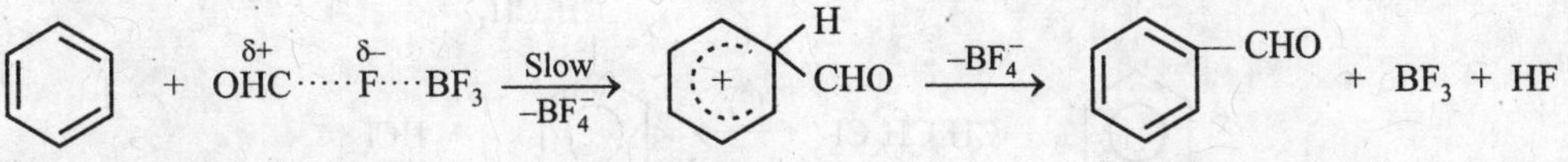

Example 24: Compared to aniline ($C_6H_5NH_2$), acetanilide ($C_6H_5NHCOCH_3$) is somewhat de-activated towards electrophilic aromatic substitution. What explanation do you offer for this difference?

Solution: In the case of acetanilide; the electron lone pair on N is involved in resonance with the acetyl group as shown below

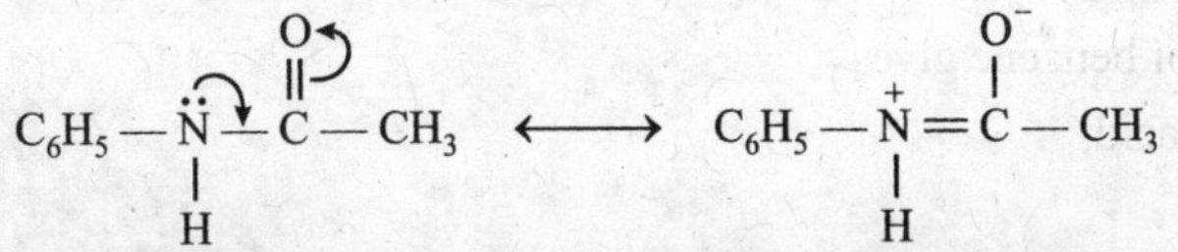

This resonance phenomenon creates a positive charge on N. Electrons are drawn towards positive N to neutralise the charge. Electrons density on benzene ring is thus reduced. That means acetanilide is deactivated towards electrophilic substitution. Amino group on the other hand is an activating group because electron pair on N takes part in resonance towards benzene ring and thus electron density on the benzene ring is elevated and hence amiline remains active towards electrophilic substitution.

Key Terms

- Huckel's rule
- Heterocyclic compounds
- Directing effect
- Friedel-crafts alkylation

Evaluate Yourself

Multiple Choice Questions

1. Benzene undergoes Friedel-Crafts reaction with isopropyl alcohol in the presence of H_2SO_4 catalyst to give
   (a) n-Propylbenezene
   (b) benzophenone
   (c) isopropylbenzene
   (d) nothing happens

2. Benzene reacts with chlorine in the presence of $FeCl_3$ catalyst to form
   (a) hexachlorobenzene
   (b) chlorobenzene
   (c) hexachlorocyclohexane
   (d) benzyl chloride

3. Consider the following reaction:

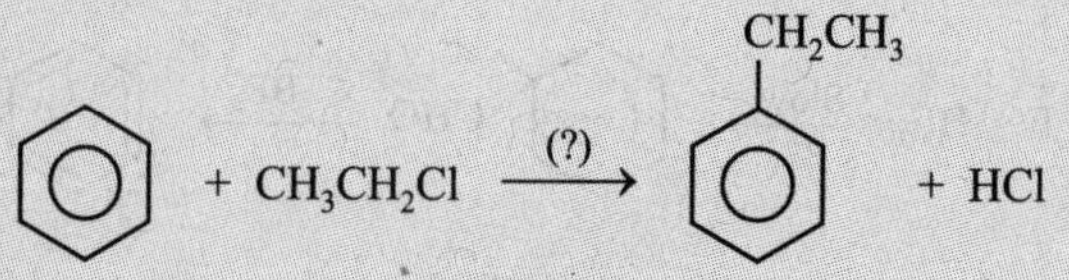

   benzene + CH_3CH_2Cl $\xrightarrow{(?)}$ ethylbenzene + HCl

   The catalyst used to complete the above reaction is
   (a) $LiAlH_4$
   (b) $AlCl_3$
   (c) Na
   (d) KOH

4. Ozonolysis of benzene gives
   (a) formic acid
   (b) glyoxal
   (c) formaldehyde
   (d) glycine

5. Benzene reacts with acetic anhydride in the presence of $AlCl_3$ to form
   (a) acetophenone
   (b) benzophenone
   (c) phenylacetic acid
   (d) phenyl acetate

6. Which of the following compounds uses only sp^2 hybridised carbon atoms for bond formation?
   (a) cyclohexane
   (b) benzene
   (c) cyclohexene
   (d) toluene

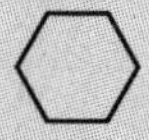

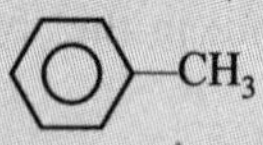

7. Which of the following compounds is aromatic?
   (a) cyclobutadiene
   (b) pyridine
   (c) pyridine (alternate)
   (d) cyclopentadiene

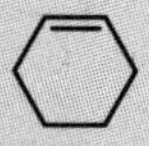

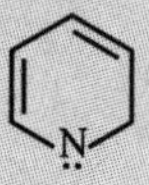

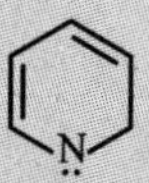

8. Benzene undergoes substitution reaction more readily then addition reaction because
   (a) it has a cyclic structure
   (b) it has three double bonds
   (c) it has six hydrogen atoms
   (d) there is delocalisation of electrons
9. Which of the following is used in order to make benzene react with concentrated nitric acid to give nitrobenzene?
   (a) Concentrated H_2SO_4
   (b) $FeCl_3$, Catalyst
   (c) Lindlar's catalyst
   (d) Ultraviolet ligh
10. Oxidation of toluene with chromyl chloride gives benzaldehyde. This reaction is known as
    (a) Perkin's reaction
    (b) Benzoin condensation
    (c) Etard's reaction
    (d) Ozonolysis

Short Answer Questions

1. Complete the following reactions

   (a) $C_6H_6 + CH_3COCl \xrightarrow{\text{Anhy. } AlCl_3}$

   (b) $C_6H_5NO_2 + HNO_3 \xrightarrow{H_2SO_4}$

   (c) $C_6H_6 + ICl \xrightarrow{FeX_3}$

   (d) $C_6H_6 + (CH_3)_3 COH \xrightarrow{H_2SO_4}$

2. Name the electrophile in each of the following reactions showing how the electrophile is formed.
   (a) Sulphonation
   (b) Nitration
   (c) Bromination
   (d) Friedal Crafts alkylation
3. Compare the relative stability of benzene, toluene and nitrobenzene towards electrophilic substitution.
4. Select the most reactive and least reactive substrate towards ring nitration in each of the following sets.
   (a) Benzene, mesitylene, m-xylene and toluene
   (b) m-Dinitrobenzene, m-nitrotoluene and toluene

5. Justify the following:
   (a) It is easier to nitrate toluene than benzene
   (b) Nitration of bromobenzene gives *ortho* and *para* bromonitrobenzene while bromination of nitrobenzene gives *meta* bromonitrobenzene.
6. Why does the side chain halogenation in alkylbenzenes take place preferentially at position alpha to the aromatic ring?
7. Sketch the following transformation:
   (a) Toluene $\longrightarrow$ $C_6H_5CH_2OCH_3$
   (b) Propylbenzene $\longrightarrow$ 1-Phenylpropane
8. Write short notes on the following:
   (a) Fittig reaction
   (b) Ullmann biaryl synthesis
9. Write the structures of major monosubstituted product formed when Br^+ attacks the molecule

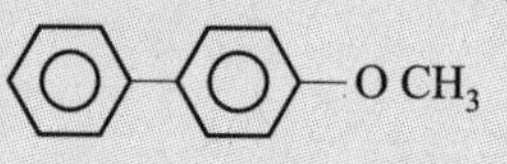

   $C_6H_5-C_6H_4-OCH_3$

10. What methods would you adopt to differentiate the three isomeric xylenes?

Long Answer Questions

1. (a) Complete the following reactions:

   (i) $C_6H_5-CH_2CH_2CH_3 + Cl_2 \xrightarrow{h\nu}$

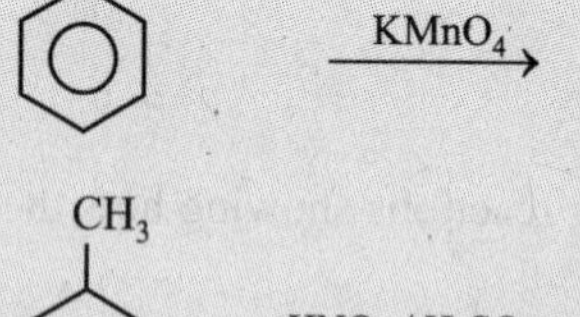

   (ii) $C_6H_5-CH_2CH_2CH_3 \xrightarrow{KMnO_4}$

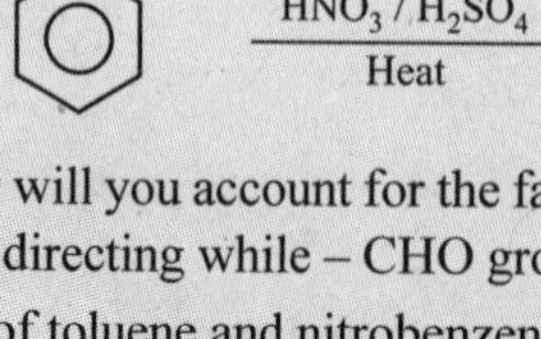

   (iii) $C_6H_5-CH_3 \xrightarrow[\text{Heat}]{HNO_3/H_2SO_4}$

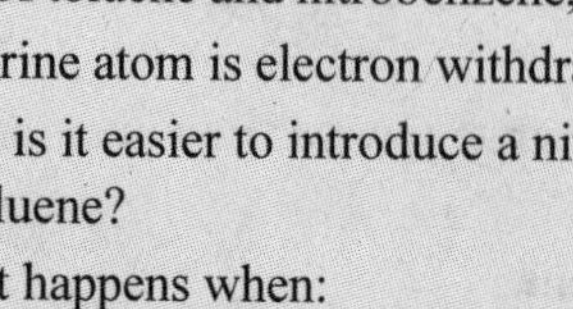

2. (a) How will you account for the fact that an –OH group in the benzene ring is ortho and para directing while – CHO group is meta directing in nature?
   (b) Out of toluene and nitrobenzene, which will be nitrated more easily?
3. (a) Chlorine atom is electron withdrawing but still ortho and para directing. Why?
   (b) Why is it easier to introduce a nitro group at ortho and para positions in phenol than in toluene?
   (c) What happens when:
      (i) o-Xylene is heated with hot $KMnO_4$ solution?

(ii) Chlorine is passed in boiling toluene?

(iii) Toluene is subjected to oxidation by chromyl chloride?

4. (a) Explain the stability of benzene on the basis of orbital structure.
   (b) Predict the products of the following reactions:
   (i) Nitration of chlorobenzene
   (ii) Methylation of toluene
   (iii) Chlorination of toluene
   (iv) Chlorination of nitrobenzene.
   (c) Halogens are electron withdrawing and yet they are ortho and para directing. Explain

5. Assign suitable reasons for the following:
   (i) Halogens are ortho and para directing though they are deactivating in nature.
   (ii) Methyl group has no lone pair of electrons, yet it is ortho and para directing in nature.
   (iii) Nitration of toluence can take place more easily as compared to benzene.
   (iv) In case of ortho and para disubstitution, the para isomer generally dominates.

6. What do you understand by the term aromaticity? How is it related with Huckel's rule? Give examples.

7. Comment on the statement
   "In case of halogens, the reactivity is determined by stronger inductive effect while the orientation is largely controlled by resonance effect".

8. (a) How will you prepare the following compounds from benzene?
   (i) Acetophenone
   (ii) Toluene
   (iii) Chlorobenzene
   (iv) Benzene hexachloride.
   (b) Explain the mechanism of nitration of benzene.
   (c) Explain why nitration of chlorobenzene gives ortho and para chloronitrobenzene but the chlorination of nitrobenzene gives meta chloronitrobenzene.

9. Predict the major products in the following reactions:

   (i) 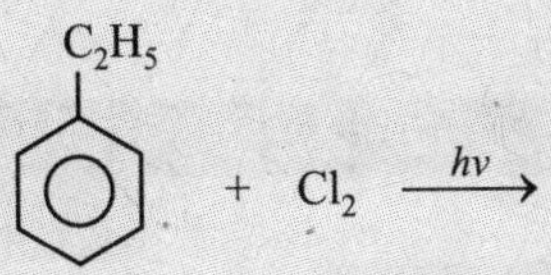 C_6H_5-C_2H_5 + Cl_2 $\xrightarrow{h\nu}$

   (ii) 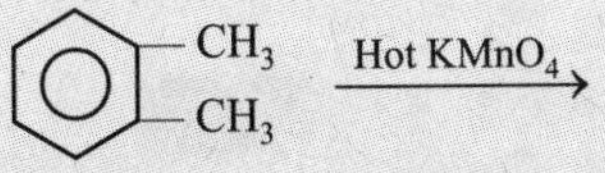 $C_6H_4(CH_3)_2$ $\xrightarrow{\text{Hot KMnO}_4}$

   (iii) 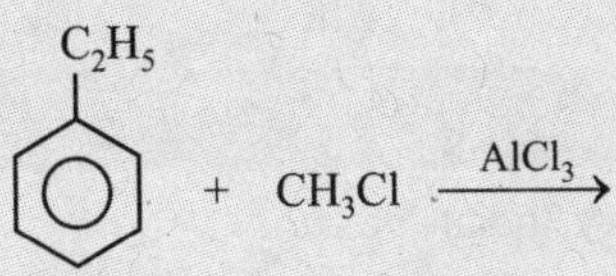 C_6H_5-C_2H_5 + CH_3Cl $\xrightarrow{AlCl_3}$

10. Predict the products of the following reactions:

   (i) 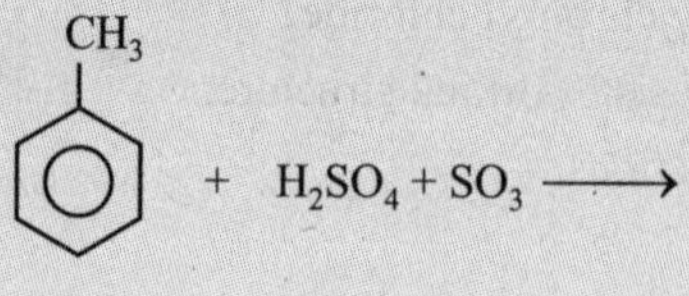 Toluene + H_2SO_4 + $SO_3 \longrightarrow$

   (ii) 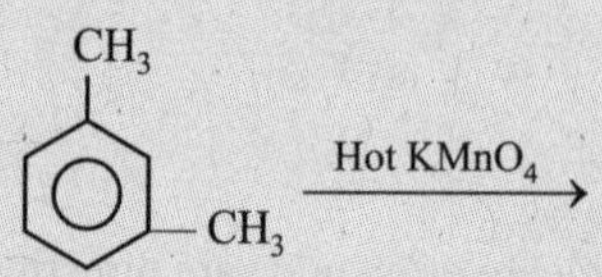 o-Xylene $\xrightarrow{\text{Hot KMnO}_4}$

   (iii) 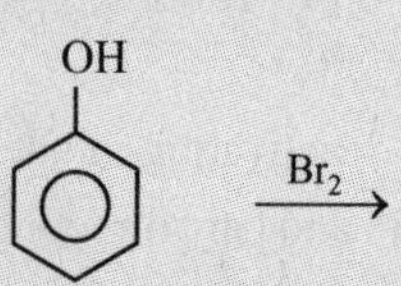 Phenol $\xrightarrow{Br_2}$

11. Discuss the orientation of electrophilic substitution in nitrobenzene.
12. What is Huckel's rule of aromaticity? Which of the following exhibit aromatic character?

   (i) Cyclooctatetraene (ii) Cyclopentadiene (iii) Tropyllium cation.

13. Discuss the mechanism of the electrophilic substitution for the following reactions in benzene:

   (i) Halogenation   (ii) Sulphonation

14. Give the mechanism for ortho and para directing influence of groups in benzenoids.
15. (a) Define aromaticity and state Huckel's rule. Will Cyclooctatetraene show aromatic character?

   (b) Explain why nitration of toluene is easier than that of benzene.

   (c) What happens when:

   (i) o-xylene is treated with hot $KMnO_4$ solution.

   (ii) Ethyl benzene is treated with chlorine in the presence of U.V. light.

   (iii) Toluene is treated with methyl chloride in the presence of anhydrous $AlCl_3$.

Answers

Multiple Choice Questions

	1.	2.	3.	4.	5.	6.	7.	8.	9.	10.
(a)										
(b)										
(c)										
(d)										

Suggested Readings

1. Finar, I.L. Organic Chemistry (Vol. 1), Dorling Kindersley (India) Pvt. Ltd.
2. Morrison, R.N. & Boyd, R.N. Chemistry, Dorling Kindersley (India) Pvt. Ltd.

Section II

Physical Chemistry

- Chemical Thermodynamics – I
  (Zeroth and First Laws of Thermodynamics and Thermochemistry)
- Chemical Thermodynamics – II
  (Second and Third Laws of Thermodynamics, Partial Molar Quantities)
- Chemical Equilibrium
- Solutions and Colligative Properties

Chapter 1

Chemical Thermodynamics – I
(Zeroth and First Laws of Thermodynamics and Thermochemistry)

LEARNING OBJECTIVES

After reading this chapter, you should be able to:

- Define terms like intensive and extensive properties, isolated, closed and open systems
- Get a view of zeroth law of thermodynamics
- Get introduced to terms like internal energy and heat capacity
- Differentiate between reversible, irreversible and free expansion of gases (ideal and van der Walls)
- Calculate bond energy, bond dissociation energy and resonance energy
- Derive Kirchoff's equation
- Define adiabatic flame temperature and explosion temperature

Chemical Thermodynamics – I (Zeroth and First Laws of Thermodynamics and Thermochemistry)

1.1 THERMODYNAMICS

Thermo means heat and *dynamics* means motion or mechanical work. Hence thermodynamics means that branch of science which deals with the conversion of heat into mechanical work and vice versa. In a wider sense *thermodynamics is a branch of science that deals with quantitative relationships between heat and other forms of energy*. Study of thermodynamics is based on three laws *viz.* first, second and third laws of thermodynamics. There is no theoretical proof of the laws, but nothing contrary to the laws has been reported so far. Zeroth law of thermodynamics which helps us to understand the concept of temperature was put forward later.

Objectives of Thermodynamic

(*i*) **To predict the feasibility of a process:** Whether or not, a process will occur under the given set of conditions, can be predicted by applying the principles of thermodynamics.

(*ii*) **To estimate the yield of the products:** Thermodynamic relations help to predict the yield of the products obtainable in a process or a reaction. It is possible to know the extent upto which the reaction takes place before attaining equilibrium.

(*iii*) **To deduce some important relationships:** It has been possible to deduce some important and useful results such as Raoult's law of lowering of vapour pressure and expressions for depression in freezing point, elevation in boiling point, distribution law and phase rule from the study of thermodynamics.

Limitations of Thermodynamics

Thermodynamics suffers from certain limitations. These are:

(*i*) It deals with the properties like temperature, pressure *etc.* of the matter in bulk or macroscopic quantities and not in microscopic quantities. Thus, it deals with large groups of atoms, molecules and ions rather than individual atoms, molecules or ions.

(*ii*) It helps to predict the feasibility of a process but gives no idea about the time taken for the process to complete or the rate at which the process would proceed.

(*iii*) Its treatment is limited to the initial and final states of a system and is silent about the path by which the change is brought about *i.e.* it does not reveal the mechanism of a process.

1.2 DEFINITION OF CERTAIN THERMODYNAMICS TERMS

1. System and surroundings: *The part of the universe considered for thermodynamic studies (i.e. to study the effect of temperature, pressure etc.) is called a system.*

Rest of the part of the universe, excluding the system, is called *surroundings*.

2. Open, closed and isolated systems

(*a*) **Open system:** *A system which can exchange both matter and energy with the surroundings is called open system.* If some water is kept in an open vessel (Fig. 1.1), exchange of both matter and energy takes place between the system and the surroundings. This will serve as an example of an opern system.

(*b*) **Closed system:** *A system which can exchange only energy with the surroundings but not matter, is called a closed system.* For example, if some water is placed in a closed metallic vessel (Fig. 1.2), then as the vessel is closed, no exchange of matter between the system and the surroundings can take place. But, with the conducting walls of the vessel, exchange of

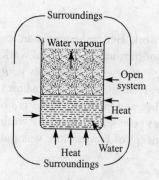 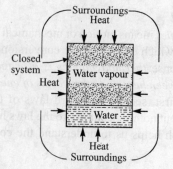 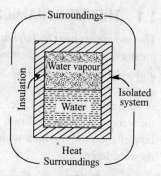

Fig. 1.1: Example of open system: Water kept in an open vessel

Fig. 1.2: Water placed in an closed vessel

Fig. 1.3: An example of an isolated system

energy can take place between the system and the surroundings. Thus, this provides us an example of closed system.

(c) **Isolated system:** *A system which can exchange neither matter not energy with the surroundings, is called an isolated system.* For example, water placed in a vessel which is closed as well as insulated, does not allow exchange of matter or energy between the system and the surroundings. (Fig. 1.3.) This provides us an example of isolated system.

3. Homogeneous system: *A system is called as homogeneous if it is uniform throughout.* A gas or a mixture of gases or a pure liquid or a pure solid or a solution of a solid in a liquid all provide us examples of homogeneous systems. Such a system has one phase only.

4. Heterogeneous system: *A system which is not uniform throughout, is said to be heterogeneous.* It consists of two or more phases which are separated from each other by definite boundaries. Common examples of the heterogeneous systems are (*a*) a system consisting of a liquid and its vapour, (*b*) two or more immiscible liquids, (*c*) a mixture of two or more solids.

5. Macroscopic system: *Macroscopic system means a system containing a large amount of the substance.* In other words, a macroscopic system is one containing large number of particles *i.e.* atoms, ions or molecules.

6. Macroscopic property: *A property related to the behaviour of particles in a macroscopic system is called a macroscopic property.* Pressure, volume, temperature, surface tension, viscosity, density, refractive index etc., are all examples of macroscopic property.

7. State of the system and state variables

State of system: The state of a system means a condition of the system when the various macroscopic properties like pressure, volume, temperature etc. of the system have been assigned definite values. The first and the latest states of the system are called *initial state* and *final state* respectively.

State variables: Macroscopic properties are known as state variables. The four most common macroscopic properties which are used to define the state of a system are *composition, pressure, volume* and *temperature*. If these properties are fixed, all other physical properties of the system are automatically assigned.

8. Thermodynamic equilibrium: A system is said to be in thermodynamic equilibrium if the macroscopic properties of the system in various phases do not show any variation with time.

Thermodynamic equilibrium implies that the following three different equilibria must exist simultaneously:

(i) **Thermal equilibrium:** This implies that there should be no flow of *heat* from one portion of the system to another. This is possible when the temperature of the system remains constant.

(ii) **Mechanical equilibrium:** This implies that no work should be done by one part of the system over another or no macroscopic movement of matter should take place within the system or with respect to its surroundings. This can be achieved when the pressure of the system remains constant.

(iii) **Chemical equilibrium:** This equilibrium implies that no change in composition should occur in any part of the system with passage of time.

9. Extensive and intensive properties

(i) **Extensive properties:** Properties which are dependent upon the quantity of the matter contained in the system are called extensive properties. Examples of extensive properties are *mass, volume, energy, heat capacity* etc. The net value of an extensive property is equal to the sum total of the values for the separate parts into which the system may be divided conveniently.

(ii) **Intensive properties:** Properties which are dependent only upon the nature of the substance and are independent of the amount of the substance present in the system are called intensive properties. The common examples of these properties are *temperature, pressure, refractive index, viscosity, density, surface tension, specific heat* etc. It is only because pressure and temperature are intensive properties (independent of the quantity of the matter present) that they are frequently used as variables to define the state of a system.

Interestingly an extensive property may become intensive property by specifying unit amount of the substance. Thus mass and volume are extensive properties but density is intensive property, heat capacity is an extensive property but specific heat is intensive.

10. Thermodynamic processes

(i) **Isothermal process:** *If a process is carried out in such a manner that the temperature remains constant throughout the process, it is called an isothermal process.* It can be understood that when such a process occurs, heat can flow from the system to the surroundings and vice versa in order to keep the temperature of the system constant.

(ii) **Adiabatic process:** *If a process is carried out in such a manner that no heat flows from the system to the surroundings* or *vice versa, or the system is completely insulated from the surroundings, it is called an adiabatic process.*

(iii) **Isochoric process:** *A process which involves no change in the volume of the system is called isochoric process.*

(iv) **Isobaric process:** *A process which involves no change in the pressure of the system is called isobaric process.*

(v) **A reversible process.** *It is defined as a process which is carried out infinitesimally slowly such that all changes occurring in the direct process can be exactly reversed and the system appears to be in a state of equilibrium at each moment. A reversible process may also be defined as a process which is carried out in such a manner that at every stage, driving force is only infinitesimally greater than the opposing force and which can be reversed by increasing the opposing force by an infinitesimal amount.*

Strictly speaking, a reversible process is almost impossible as it would require an infinite time for completion. A process can be made very nearly reversible in some cases. For example, if an opposing E.M.F. is applied to a cell, then no current flows if the opposing E.M.F. is exactly equal to that of the cell. If the opposing E.M.F. is very slightly (infinitesimally) less than that of the cell, a very small current is given out by the cell. If the opposing E.M.F. is slightly greater than that of the cell, a very small current flows in the opposite direction.

A reversible process can be demonstrated with the help of the following experiment as shown in Fig. 1.4.

Take a cylinder and fit a frictionless piston into it as shown in Fig. 1.4. The cylinder contains a gas at a moderate pressure, due to which the piston will start rising upwards. Balance it by placing a petri dish containing fine sand over the piston. In the balanced position, the piston will be static.

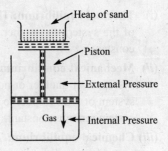

Fig. 1.4: Demonstration of a reversible process

Now if we add a very small sand grain to the petri dish, the piston will start moving downwards very slowly. On the other hand, if we take out a very small sand grain from the petri dish, the piston will start moving upwards very slowly. In both cases, it demonstrates reversible process.

(*vi*) **Irreversible process.** A process not meeting the above requirements is called an **irreversible process.** An irreversible process is defined *as that process which is not carried out infinitesimally slowly so that the successive steps of the direct process cannot be retraced and any change in the external conditions disturbs the equilibrium.* An irreversible process occurs when carried out rapidly or fast.

It is interesting to note that all naturally occurring processes like the flow of water down a hill or the flow of heat from a hot end to a cold end of an iron bar etc. are thermodynamically irreversible.

1.3 INTERNAL ENERGY AND CHANGE IN INTERNAL ENERGY

Internal energy: The evolution or absorption of energy in different processes indicates that every substance is associated with some definite amount of energy, whose actual value depends upon the nature of the substances like arrangement of atoms and electrons within the molecules and the conditions of temperature, pressure, volume and composition. The energy associated with a substance is called its *internal energy* and is normally denoted by the symbol E or U.

Change of internal energy: It is not possible to find the absolute value of internal energy of a substance because it involves certain quantities like translational, vibrational and rotational kinetic energies which are not easy to measure.

However, for practical purposes it is not required to know the absolute value of internal energy possessed by any substance (or a system). What is required in different processes is simply the *change of internal energy* when the reactants change into products or when system changes from initial state to the final state. This is easily measurable and is represented by ΔE (or ΔU).

Thus if the internal energy of a system in the initial state is E_1 (or U_1) and in the final state, it is E_2 (or U_2) then the change of internal energy (ΔE or ΔU) may be given by

$$\Delta E \text{ (or } \Delta U) = E_2 - E_1$$

Similarly, in a chemical reaction if E_R is the internal energy of the reactants and E_P is the internal energy of the products, then energy change accompanying the process would be

$$\Delta E = E_P - E_R \quad \text{or} \quad \Delta U = U_P - U_R$$

Sign conventions in energy: Obviously, if $E_1 > E_2$ (or $E_R > E_P$), the extra energy possessed by the system in the initial state (or the reactants) would be given out, and ΔE will be negative according to the above equations. Similarly if $E_1 < E_2$ (or $E_R < E_P$), energy will be absorbed in the process and ΔE will be positive.

Hence ΔE (or ΔU) *is negative* if energy *is evolved*

and ΔE (or ΔU) *is positive* if energy *is absorbed*

Units of U. The unit of energy is erg or joule, 1 Joule = 10^7 ergs

1.4 EXPRESSION FOR WORK OF EXPANSION AGAINST CONSTANT PRESSURE

The expression for such a work may be derived as follows:

Consider a gas enclosed in a cylinder fitted with a frictionless piston. (Fig. 1.5).

Area of cross-section of the cylinder = a sq cm

Pressure on the piston (which is slightly less than internal pressure of the gas so that the gas can expand) = P.

Distance through which gas expands = dl cm

As pressure is force per unit area, force (f) acting on the piston will be

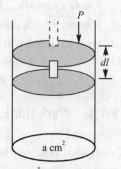

Fig. 1.5: Expansion of a gas

$$f = P \times a$$

∴ Work done by the gas (*i.e.* the system)

$$= \text{Force} \times \text{Distance} = f \times dl$$
$$= P \times a \times dl$$

But $a \times dl = dV$ is a small increase in the volume of the gas. Hence the small amount of work (δW) done by the gas can be written as

$$\delta W = P \, dV$$

Let us suppose the gas expands from initial volume V_1 to the final volume V_2, then the total work done (w) is given by

$$w = \int_{V_2}^{V_1} P \, dV$$

If the external pressure P against which the gas expands remains almost constant throughout the process, the above result may be written as

$$w = P \int_{V_2}^{V_1} dV = (V_2 - V_1)$$
$$= P \cdot \Delta V$$

where $\Delta V = (V_2 - V_1)$ is the total change in the volume of the gas.

If the external pressure (P) is slightly greater more than the pressure of the gas, the gas will contract *i.e.* the work will be done by the surroundings on the system. However, work done will be obtained by a similar method.

Since P is the *external* pressure, it is sometimes written as P_{ext} so that we have
$$w = P_{ext} \times \Delta V.$$
If the expansion of gas is against vacuum, $P_{ext} = 0$
$$\text{Work done} = 0 \times \Delta V = 0$$
Thus, no work is done when the gas is expanding against vacuum.

Note: According to the latest SI conventions of sign, w is taken as positive if work is done on the system (*i.e.* $w_{contraction}$ is positive) and it is taken as negative if work is done by the system (*i.e.* $w_{expansion}$ is negative). Thus $w_{exp} = -P\Delta V$ and $w_{contraction} = +P\Delta V$.

Electrical work. The electrical work done is expressed as under:

Electrical work done = Electromotive force (emf) × Quantity of electricity

Gravitational work. Gravitational work done when an object of mass (m) moves against the force of gravitation (g) and the displacement of the object is h, is given by:
$$\text{Work done} = m \times g \times h$$

1.4.1 Work Done in a Reversible Process is the Maximum Work Obtainable

The work done by a system is the maximum work if the process is carried out under reversible conditions. This may be proved as follows:

We have learnt earlier that work done by the system = $P_{ext} \times \Delta V$. Thus for a given change in volume (ΔV), the work done can have maximum value if P_{ext} is maximum. For expansion to take place, the external pressure should always be less than the internal pressure of the gas. Thus the maximum value that P_{ext} can have is that it should be *infinitesimally smaller* than the internal pressure of the gas P. But these conditions refer to a reversible process. Hence the maximum work is done in a process only when it is carried out under reversible conditions. Hence in reversible processes, w is sometimes replaced by w_{max}.

1.5 HEAT

Heat is a form of energy exchanged between the system and surroundings in expansion and contraction of gas because of difference in internal and external pressure. Heat may be defined as energy exchanged between the system and the surroundings as a result of the difference of temperature between them. It is usually represented by the letter q.

Sign of 'q'. When heat is given by the system to the surroundings, it is **given a negative** sign.

When heat is **absorbed by the system** from the surroundings, it is given a **positive** sign.

Units of 'q'. Heat is usually measured in terms of calories. *A calorie is defined as the quantity of heat required to raise the temperature of one gram of water through 1°C.*

In the S.I. units, heat is expressed in terms of *joules*. The two units are related to each other as under:
$$1 \text{ calorie} = 4.184 \text{ joules}$$

1.5.1 State and Path Functions and their Differentials

State function. A state function is that thermodynamic quantity whose change in value depends upon its values in the initial and final states and is independent of the path by which the change has been carried out. Internal value is a state function because change in internal energy (ΔU) will depend only upon its values in the initial state ($U_{initial}$) and the final state (U_{final}).

Path function. A path function is that thermodynamic quantity whose change in value during a process depends upon the path that is followed. Heat and work are path functions and not state functions.

We can obtain total change in a thermodynamic quantity, when the process occurs slowly by integration. But this is applicable only when the thermodynamic quantity (like internal energy) is a state function. Similarly, we can perform differentiation of only state functions and not path function. Thus differentiation and integration of only state functions and not path functions can be performed.

Let us remember that d stands for differentiation in calculus and therefore a small change in internal energy can be represented by dU. But we cannot represent small change in heat and work by dq and dw respectively. We should represent them appropriately by δq and δw or $đq$ and $đw$ respectively.

In view of the above we can say that internal energy is a **definite property** and is an **exact differential**. Heat and work do not qualify for these terms.

1.5.2 Equivalence Between Heat and Work

Joule carried out experiments to correlate the heat produced and work done in various processes. A scooter or car engine gets heated after working for some time. It shows heat must be evolved when work is done by a system. Joule observed that *whenever some definite amount of work was done, the heat produced was in the same amount.* According to Joule, work done (W) is proportional to heat produced (Q).

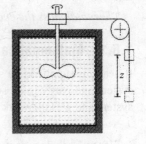

Joule's apparatus

$$W \propto Q \quad \text{or} \quad W = JQ$$

where J is proportionality constant called *Joule's mechanical equivalent of the heat.* If Q = 1 calorie, W = J

Thus, we can define Joule's mechanical equivalent, J as the amount of work required to produce 1 calorie of heat. Its value comes out to be 4.184×10^7 ergs or 4.184 joules. Thus, if 4.184 joules of work is done, the heat produced is equal to 1 calorie.

or   1 calorie = 4.184 joules

This relation gives us the equivalence between heat and work.

1.5.3 Zeroth law of Thermodynamics

The zeroth law of thermodynamics states: *When two bodies have equality of temperature with a third body, they in turn have equality of temperature with each other.*

This fact is so well known that it does not need any further experiments for its support. However, since this fact is not derivable from other laws and since it logically precedes the first and the second law of thermodynamics, it has been called as the zeroth law of thermodynamics.

Heat flows in the direction of decreasing temperature....

1.6 FIRST LAW OF THERMODYNAMICS

First law of thermodynamics can be stated in a number of ways. Some statements of the law are:

(*i*) Energy can neither be created nor destroyed, although it may be converted from one form to another.

(*ii*) The total energy of an isolated system remains constant, although it may undergo transformation from one form to the other.

(*iii*) Whenever certain quantity of some form of energy disappears, an exactly equivalent amount of some other form of energy must be produced.

(*iv*) It is impossible to construct a perpetual motion machine *i.e.* a machine which would produce work continuously without consuming energy.

Einstein in 1905 showed that the energy could be created by the destruction of mass. The energy produced (E) by the destruction of mass m is given by the equation

$$E = mc^2 \text{ (called Einstein equation)}$$

where c is the velocity of light. In the light of this idea, the law of conservation of energy has been modified and now it is called *law of conservation of mass and energy*. It may be stated as follows:

The total mass the energy of an isolated system remains constant.

1.6.1 Mathematical Formulation of the First Law of Thermodynamics

If heat q is supplied to a system, it may be used up partly to increase the internal energy of the system and partly to do some mechanical work. If ΔU is the increase in internal energy of the system and w is the work done by the system, then we have

$$\Delta U = q + w \qquad \ldots(i)$$

If the work done is work of expansion,

$$w = -P\Delta V$$

or $$q = \Delta U + P\Delta V$$

If the changes are very small, it may be written as

$$\delta q = dU + PdV \qquad \ldots(ii)$$

1.6.2 Internal Energy is a State Function but Work and Heat are Not

Internal energy: The internal energy is a state function, that is, when a process occurs, the change in its value depends only upon its value in the initial state and that in the final state and is independent of the path by which the change is brought about. For example, consider a system having a particular pressure, volume and temperature represented by the point A in Fig. 1.6. Suppose the pressure, volume and temperature are changed in such a way that the system is brought to the point B by path I. If the system is returned to the point A by path II, then the energy change involved in path I must be equal in magnitude to that involved in path II. If the energy changes involved in paths I and II are not the same, then suppose that the increase in internal energy by path I is greater than the decrease in internal energy

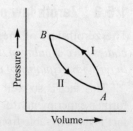

Fig. 1.6: Energy changes in direct and reverse paths

along path II. Thus by carrying out the process $A \to B$ by path I and the reverse process $B \to A$ by path II, though the original state has been restored, some energy would have been created. This is contrary to the first law of thermodynamics. Thus it may be inferred that the energy change involved in the process $A \to B$ must be equal in magnitude to that involved in the reverse process $B \to A$ by path II. In other words, *the energy change accompanying a process is independent of the path followed and depends only upon the initial and the final states.* Thus if U_A is the internal energy of the system in the initial state and U_B that in the final state, then the change in internal energy accompanying the process would be given by

$$\Delta U = U_B - U_A$$

Work and heat: Work done during a process does not depend merely upon the initial and the final states of the system but it depends upon the path followed. Hence work is **not** a state function but is a path function. This is explained as under :

Suppose we want to change the system from state A to state B. This may be achieved by one of the following different paths as shown in the different figures below (Fig. 1.7). If we restrict ourselves to the work of expansion only, we know that the work done for a small change in volume dV is PdV where P is the external pressure. Thus the total work done is the sum of PdV terms and this is equivalent to the area under the curve in the P–V diagrams. From the figures given below it is obvious that the areas under the curves are different and hence the work done is different when different paths are followed.

That heat is also not a state function follows directly from the mathematical formulation of the first law of thermodynamics viz.,

$$q = \Delta U - w$$

We thus find that heat q is the sum of two quantities *i.e.* the internal energy change, ΔU and the work done, w. Now ΔU is a state function whereas w is not. Now since w depends upon the path followed, therefore, q will also depend upon the path followed. In other words, q is also not a state function but is a path function.

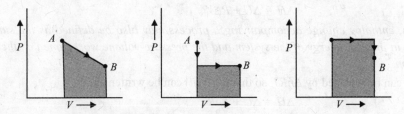

Fig. 1.7: *Different paths followed in going from state A to state B*

1.7 ENTHALPY OR HEAT CONTENT

For a process which is carried out at constant pressure, the work of expansion is given by

$$w = P\Delta V \qquad \ldots (i)$$

where ΔV is the increase in volume and P is the constant pressure.

According to first law of thermodynamics, we know that

$$q = \Delta U - w \qquad \ldots (ii)$$

where q is the heat absorbed by the system, ΔU is the increase in internal energy of the system and w is the work done by the system.

Under the condition of constant pressure, $w = -P\Delta V$. If we represent the heat absorbed by q_p, we get

$$q_p = \Delta U + P\Delta V \qquad \ldots (iii)$$

Let us suppose when the system absorbs q_p calories of heat, its internal energy increases from E_1 to E_2 and the volume increases from V_1 to V_2. Then we have

$$\Delta U = U_2 - U_1 \qquad \ldots (iv)$$

and

$$\Delta V = V_2 - V_1 \qquad \ldots (v)$$

Putting these values in equation (*iii*) above, we get

$$q_p = (U_2 - U_1) + P(V_2 - V_1)$$
$$q_p = (U_2 + PV_2) - (U_1 + PV_1) \qquad \ldots (vi)$$

As U, P and V are the functions of state, therefore, the quantity $U + PV$ must also be a state function. The thermodynamic quantity $U + PV$ is called the **heat content or enthalpy** of the system and is represented by the symbol H i.e. the enthalpy may be defined mathematically by the equation

$$H = U + PV$$

Thus if H_2 is the enthalpy of the system in the final state and H_1 is the value in the initial state, then

$$H_2 = U_2 + PV_2$$

and

$$H_1 = U_1 + PV_1$$

Putting these values in equation (*vi*), we get

$$q_p = H_2 - H_1$$

or

$$q_p = \Delta H \qquad \ldots (vii)$$

where $\Delta H = H_2 - H_1$ is the enthalpy change of the system. Hence **enthalpy change** *of a system is equal to the heat absorbed by the system at constant pressure.*

Putting the value of q_p from equation (*vii*) in equation (*iii*) we get

$$\Delta H = \Delta U + P\Delta V \qquad \ldots (viii)$$

Thus, *enthalpy change accompanying a process may also be defined as the sum of the increase in internal energy of the system and the pressure-volume work done i.e., the work of expansion.*

$P\Delta V$ can be replaced by ΔnRT, so that eq. (*viii*) can be written as

$$\Delta H = \Delta U + \Delta nRT$$

Physical concept of enthalpy or heat content. Total energy stored in a substance is called internal energy (U) while the energy store in a substance (or system) that is available for conversion into heat is called enthalpy or heat content (H). The teo quantities are related as under

$$H = U + PV$$

As U, P and V are extensive properties, H is also an extensive property. Just as we cannot find the absolute value of U, we cannot find the absolute value of H, too. However, our purpose is served by knowing ΔH *i.e.,* change in enthalpy, which we can determine experimentally.

Solved Example—Heat of Formation

Example 1: The heat of formation of methane at 27°C is –19.3 kcal when the measurements are made at constant pressure. What will be the heat of formation at constant volume?

Solution: Apply the relationship

$$\Delta H = \Delta U + \Delta n \cdot RT \qquad \ldots(i)$$

ΔH is the heat of reaction at constant pressure whereas ΔU is the heat of reaction at constant volume.

$$\Delta H = -19.3 \text{ kcal}$$

For the reaction

$$C(s) + 2H_2(g) \longrightarrow CH_4(g)$$

No. of moles of the gaseous reactants = 2
No. of moles of the gaseous products = 1
Thus $\Delta n = 1 - 2 = -1$

Substituting the values in eq. (i) above

$$-19.3 \times 1000 \text{ cal} = \Delta U + (-1) \times (1.987 \text{ cal}) \times 300$$

or $\qquad \Delta U = -19300 - 5961 \text{ cal}$

or $\qquad \Delta U = -13339 \text{ cal or } -13.339 \text{ kcal}.$

1.8 HEAT CAPACITY

We introduce here another term, heat capacity.

Heat capacity of a system is defined as the amount of heat required to raise the temperature of the system through 1°C.

If q is the amount of heat supplied to a system to raise the temperature from T_1 to T_2, then the heat capacity of the system is given by

$$C = \frac{q}{T_2 - T_1} = \frac{q}{\Delta T} \qquad \ldots(i)$$

The value of C has to be considered in a narrow range because the heat capacity varies with the temperature.

If δq is the small amount of heat absorbed by a system which raises the temperature of the system by a small amount dT, then the heat capacity of the system is given by

$$C = \frac{\delta q}{\delta T} \qquad \ldots(ii)$$

There are two types of heat capacities, viz.,
(i) Heat capacity at constant volume (C_v)
(ii) Heat capcity at constant pressure (C_p)
These are discussed separately as under

(i) Heat capacity at constant volume (C_v)

According to first law of Thermodynamics

$$\delta q = dU + PdV \qquad \ldots(iii)$$

Substituting for δq from eq. (iii) into eq. (ii)

$$C = \frac{dU + PdV}{dT} \qquad \ldots(iv)$$

When the volume is kept $dV = 0$ and the eq. (iv) becomes

$$C_v = \left(\frac{\partial U}{\partial T}\right)_v \qquad \ldots(v)$$

or *for an ideal gas,* this equation may be simplified as

$$C_v = \frac{dU}{dT} \qquad \ldots(vi)$$

C_v is the heat capacity at constant volume.

Thus *the heat capacity at constant volume may be defined as the rate of change of internal energy with temperature at constant volume.*

(iii) Heat capacity at constant pressure, C_p

When the pressure is kept constant during the absorption of heat, equation (iv) becomes

$$C_p = \left(\frac{\partial U}{\delta T}\right)_p + \left(\frac{\partial V}{\delta T}\right)_p \qquad \ldots(vii)$$

We know that the heat content or enthalpy of a system is given by

$$H = E + PV$$

Differentiating w.r.t. T at constant P, we get

$$\left(\frac{\partial H}{\partial T}\right)_p = \left(\frac{\partial U}{\partial T}\right)_p + P\left(\frac{\partial V}{\partial T}\right)_p \qquad \ldots(viii)$$

Combining equations (vii) and (viii), we get

$$C_p = \left(\frac{\partial H}{\partial T}\right)_p \qquad \ldots(ix)$$

or for any ideal gas, this equation may be written as

$$C_p = \frac{dH}{dT} \qquad \ldots(x)$$

C_p is the heat capacity at constant pressure.

Thus *the heat capacity at constant pressure may be defined as the rate of change of enthalpy with temperature at constant pressure.*

1.8.1 Relation Between C_p and C_v

We observe that if the volume of the system is kept constant when the heat is supplied to a system, then no work is done by the system. Thus the heat absorbed by the system is used up completely to increase the internal energy of the system. And if the pressure of the system is kept constant when the heat is supplied to the system, then some work of expansion is also done by the system in addition to the increase in internal energy. Thus if at constant pressure, the temperature of the system is to be raised through the same value as at constant volume, then some extra heat is required for doing the work of expansion. Thus $C_p > C_v$.

We have the following equations for C_v and C_p.

$$C_v = \frac{dU}{dT} \qquad ...(i)$$

$$C_v = \frac{dH}{dT} \qquad ...(ii)$$

To obtain the difference between the heat capacities of an ideal gas subtract equation (*i*) from equation (*ii*). So we have

$$C_p - C_v = \frac{dH}{dT} - \frac{dU}{dT} \qquad ...(iii)$$

But $\qquad H = U + PV \qquad$ (by definition)

and also $\qquad PV = RT \qquad$ (for 1 mole of an ideal gas)

∴ $\qquad H = U + RT.$

Differentiating this equation w.r.t *T*, we get

$$\frac{dH}{dT} = \frac{dU}{dT} + R \qquad ...(iv)$$

or $$\frac{dH}{dT} - \frac{dU}{dT} = R \qquad ...(v)$$

Combining equation (*iii*) and (*v*),

$$\boldsymbol{C_p - C_v = R} \qquad \text{(for 1 mole of an ideal gas)}$$

Thus C_p is greater then C_v by the gas constant *R*, *i.e.*, approximately 2 calories or 8.314 joules.

The ratio of specific heats (C_p/C_v) is represented by γ

Thus $\qquad \gamma = \dfrac{C_p}{C_v}$

It is found that the value of γ comes out to be about 1.67 for monoatomic gases, 1.40 for diatomic gases and 1.30 for triatomic gases.

1.8.2 Joule's Law

The law can be stated as: *The change of internal energy of an ideal gas with volume at constant temperature is zero.*

Mathematically, $\left(\dfrac{\partial U}{\partial V}\right)_T = 0$

This result has been obtained from the Joule Thomson experiment as explained hereunder.

A combination of bulbs A and B connected through a stop cock S is immersed in water taken in a trough (Fig. 1.8 *a*). Bulb A contains as ideal gas at certain pressure with the stop cock closed. There is complete vaccum in the flask B. The bulbs reach a thermal equilibrium after some time. The temperature of water is noted.

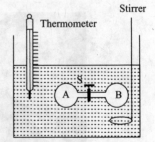

Fig. 1.8a: Joule Thomson effect

The stop cock S is now opened and the gas is allowed to move to bulb B. The bulbs reach a thermal equlibrium after some time. The temperture of water is again noted.

We find that there is no change in temperature. On opening the stop cock, the gas from the bulb A expands into the evacuated bulb B *i.e.* the opposing force $P_{opposing} = 0$. Thus work of expansion

$$dW = -P_{opposing}\, dV = 0$$

From the first law of thermodynamics, $dv = dq + dw$

As $dW = 0$, we have $dV = dq$

As the temperature of water has not changed, $dq = 0$. Hence $dU = 0$. The mathematical form of Joule's law can be arrived at as follows:

Taking U as a function of volume and temperature

$$U = f(V, T)$$

Differentiating

$$dU = \left(\frac{\partial U}{\partial T}\right)_V dT \left(\frac{\partial U}{\partial V}\right)_T dV \qquad ...(i)$$

We have found earlier that in such a situation $dU = 0$ and $dT = 0$.

Substituting for dU and dT in eq. (*i*) above, we have

$$\left(\frac{\partial U}{\partial V}\right)_T dV = 0 \quad \text{but} \quad dV \neq 0$$

$$\therefore \quad \left(\frac{\partial U}{\partial V}\right)_T = 0$$

However, we can apply the law in a strict sense to ideal and not real gases.

1.8.3 Joule-Thomson Effect

It was observed by Joule and Thomson that *when a real gas at a certain pressure expands adiabatically through a porous plug or a fine hole into a region of low pressure, it is accompanied by cooling (except for hydrogen and helium which get warmed up)*. This phenomenon is known as **Joule-Thomson effect**.

James P. Joule (1818-1889)
James Prescot Joule was an English physicist who studied the nature of heat and established a relationship to mechanical work. He also formulated the Joule's law which deals with transfer of energy

Sir William Thomson (1824-1907)
Scottish mathematician and physicist who contributed to many branches of science. He was one of the most famous scientists of his age

Fig. 1.8b: *Shows the experimental set-up of Joule and Thomson experiment*

A tube thoroughly insulated is fitted with a porous plug in the middle and pistons on either side as shown in the figure. This ensures adiabatic conditions for the experiment. The pressures of the gas are kept different on the two sides of the plug, being P_1 on the left side and P_2 on the right side where $P_1 > P_2$. By applying pressure on the piston on the left slowly enough (so as not to change the pressure P_1), a volume of the gas equal to V_1 is forced slowly through the porous plug and then allowed to expand to the pressure P_2 and volume V_2 by moving the piston on the right outward. When the process is being carried out as above, the temperature on the two sides of the porous plug is recorded accurately. The experiment is repeated with a different gases. It is observed that a fall in temperature of the gas takes place in case of all gases except hydrogen and helium when the experiments are carried out at ordinary temperature *i.e.* at room temperature.

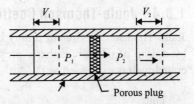

Fig. 1.8b: *Joule-Thomson experiment (Expansion of real gas through a porous plug)*

As the process is carried out adiabatically,
$$q = 0$$
∴ By first law of thermodynamics *viz.* $\Delta U = q + w$
$$\Delta U = w$$
or $\qquad w = \Delta U \qquad \qquad ...(i)$

It can be said that the work done, during the expansion of the gas under adiabatic conditions, is at the cost of the internal energy. In other words, when the work of expansion is done adiabatically, the internal energy of the gas decreases and hence the temperature decreases. In fact, a part of the internal energy is used up to overcome the forces of attraction existing among the molecules (van der Waals' forces).

In Joule-Thomson experiment, which is an example of adiabatic process, on the left side of the plug, the work is done *on* the system whereas on the right side of the plug, the work is done *by* the system.

Work done *on* the system on the left side $= P_1 V_1$
Work done *by* the system on the right side $= - P_2 V_2$
∴ Net work done by the system is
$$w = P_1 V_1 - P_2 V_2 \qquad \qquad ...(ii)$$
Putting the value in equation (*i*), we get
$$\Delta U = P_1 V_1 - P_2 V_2$$
Further putting $\qquad \Delta U = U_2 - U_1$
we get $\qquad U_2 - U_1 = P_1 V_1 - P_2 V_2$
or $\qquad U_2 + P_2 V_2 = U_1 + P_1 V_1$
or $\qquad H_2 = H_1$
i.e. $\qquad \Delta H = H_2 - H_1 = 0$

We therefore conclude that *when the expansion of a gas takes place adiabatically through a porous plug of a fine hole, the enthalpy of the system remains constant.*

1.8.4 Joule-Thomson Coefficient

Joule-Thomson coefficient μ may be defined as the temperature change in degrees produced by a drop of one atmospheric pressure when the gas expands under conditions of constant enthalpy. It is expressed as :

Thus $$\mu = \left(\frac{\partial T}{\partial P}\right)_H$$

(i) For cooling, μ will be positive (because dT and dP both will be negative).
(ii) For heating μ will be negative (because dT is positive while dP is negative).
(iii) If μ = 0, the gas gets neither heated up not cooled on adiabatic expansion (because μ = 0, only if $dT = 0$ for any value *of dP*).

1.8.5 Inversion Temperature

Every gas has a definite temperature (at a particular pressure) at which μ = 0. Below this temperature μ is positive and above this temperature μ is negative. *This temperature (at a particular pressure) at which μ = 0 i.e. the gas neither cooled down nor heated up on adiabatic expansion is called the* **inversion temperature.**

1.8.6 Helium and Hydrogen Show Heating on Adiabatic Expansio in Joule-Thomson Experiment

The inversion temperatures of most of the gases are very high (much higher than the room temperature). That is why these gases undergo cooling on adiabatic expansion under ordinary conditions of temperature. On the other hand, the inversion temperatures of hydrogen and helium are very low (being much below 0°C). Hence these gases undergo heating on adiabatic expansion under ordinary conditions of temperature. But, if these gases are first cooled to temperatures below their inversion temperatures and then allowed to expand adiabatically, these gases also show cooling effect.

1.9 RELATIONSHIP OF μ WITH OTHER THERMODYNAMIC QUANTITIES

Enthalpy H is a state function. Taking temperature T and pressure P as the independent variables.

$$H = f(T, P) \qquad ...(i)$$

∴ $$dH = \left(\frac{\partial T}{\partial P}\right)_T dP + \left(\frac{\partial H}{\partial T}\right) dT \qquad ...(ii)$$

In Joule-Thomson effect, the enthalpy of the system remains constant *i.e.,*

$$dH = 0$$

Substituting this value in the above equation, we get

$$\left(\frac{\partial T}{\partial P}\right)_T dP + \left(\frac{\partial H}{\partial T}\right) dT = 0 \qquad ...(iii)$$

Dividing this equation throughout by dP and imposing the condition of constant enthalpy, we get

$$\left(\frac{\partial H}{\partial P}\right)_T + \left(\frac{\partial H}{\partial T}\right)_P \left(\frac{\partial T}{\partial P}\right)_H = 0 \qquad \ldots(iv)$$

But $\left(\frac{\partial H}{\partial T}\right)_P = C_p$, heat capacity at constant pressure,

and $\left(\frac{\partial H}{\partial T}\right)_H = \mu$, Joule-Thomson coefficient $\qquad \ldots(v)$

Substituting these values in equation (iv), we get

$$\left(\frac{\partial H}{\partial P}\right)_T + \mu C_p = 0$$

or $\qquad \mu = -\frac{1}{C_P}\left(\frac{\partial H}{\partial P}\right)_T \qquad \ldots(vi)$

Thus μ can be calculated from $\left(\frac{\partial H}{\partial P}\right)_T$ and *vice versa*.

Solved Example—Joule-Thomson Coefficient

Example 2: Explain why an ideal gas shows neither heating nor cooling in Joule-Thomson experiment?

OR

Why is Joule-Thomson coefficient of an ideal gas zero?

Solution: An ideal gas, undergoes neither cooling nor heating on adiabatic expansion in Joule-Thomson experiment. This is expected as explained below:

The intermolecular forces of attraction (*i.e.* van der Waals forces) in an ideal gas are negligible. Hence no energy is used up in overcoming these forces of attraction when the gas expands adiabatically. Thus the internal energy of the gas does not fall and therefore the temperature also does not fall.

1.10 ISOTHERMAL REVERSIBLE EXPANSION OF A GAS

A general expression for the work of expression in an isothermal reversible process may be derived as follows :

Let P be the pressure of the gas within a system undergoing an isothermal expansion by a reversible process. Then in order that the expansion may take place very slowly (*i.e.* reversibly) the external pressure should be $P - dP$, where dP is very small quantity. The small amount of work (δw) done by the system when its volume increases by an infinitesimal amount, dV, is equal to the product of the external pressure and the volume change *i.e.*,

$$\delta w = -(P - dP)\, dV \qquad \ldots(i)$$

Neglecting the term $dP \times dV$ which is very small (being the product of two very small quantities), equation (i) becomes

$$\delta w = -P.\, dV \qquad \ldots(ii)$$

Hence the total work (w) done by the system when its volume increases from V_1 to V_2 will be given by

$$w = -\int_{V_1}^{V_2} P\,dV \qquad \ldots(iii)$$

Now, for n moles of an ideal gas, the equation of state is

$$PV = nRT \text{ or } P = \frac{nRT}{V}. \qquad \ldots(iv)$$

Substituting this value is equation (*iii*), we get

$$w = -\int_{V_1}^{V_2} \frac{nRT}{V} \qquad \ldots(v)$$

For an isothermal process, T is constant. Also n and R are constant quantities. Hence equation (*v*) may be written as

$$w = \int_{V_1}^{V_2} \frac{dV}{V} = -nRT\,\ln\frac{V_2}{V_1} \qquad \ldots(vi)$$

or

$$w = -2 \cdot 303\, nRT \log \frac{V_2}{V_1} \qquad \ldots(vii)$$

A relation can also be obtained in terms of pressure.

As the process is isothermal *i.e.* the temperature remains constant, therefore by Boyle's law

$$P_1 V_1 = P_2 V_2 \qquad \ldots(viii)$$

where P_1, V_1 are the initial pressure and volume and P_2, P_2 are the final pressure and volume of the system.

Equation (*viii*) can be written as

$$\frac{V_2}{V_1} = \frac{V_2}{V_1}$$

Substituting this value in equation (*vii*), we get

$$w = -2 \cdot 303\, nRT \log \frac{P_1}{P_2} \qquad \ldots(ix)$$

Work done in a *reversible process* is the *maximum* work obtainable from a system. Hence w in equations (*vii*) and (*ix*) may be replaced by w_{max}. Expressions (*vii*) and (*ix*) may thus be written as

$$w_{max} = -2 \cdot 303\, nRT \log \frac{V_2}{V_1} = -2 \cdot 303\, nRT \log \frac{P_1}{P_2}$$

Internal energy of a system is a function of temperature. As the temperature remains constant in an isothermal process, internal energy does not change or $\Delta U = 0$.

Expression for ΔU : Internal energy of an *ideal gas* is function of temperature only. Therefore, in an isothermal process as the temperature remains constant and the internal energy of the system (containing an ideal gas) also remains constant

i.e. $\qquad \Delta U = 0$

Expression for ΔH

$$\Delta H = \Delta(U + PV) = \Delta U + \Delta(PV)$$
$$= \Delta U + \Delta(nRT) = \Delta U + \Delta nR\,\Delta T$$
$$= 0 + 0 \begin{cases} \Delta U = 0 \text{ as explained above and} \\ \Delta T = 0 \text{ as the process is isothermal} \end{cases}$$

i.e. $\Delta H = 0$

Expression for q (heat absorbed): According to first law of thermodynamics

$$q = \Delta U - w$$

But $\Delta U = 0$ (as explained above)

$\therefore \quad q = -w$

Thus all the heat absorbed is used up completely in doing the work of expression only. Hence the expressions for the heat absorbed in an isothermal reversible expansion of an *ideal gas* are same as for the work done *i.e.*

$$q = -w = 2 \cdot 303\, nRT \log \frac{V_2}{V_1} = 2 \cdot 303\, nRT \log \frac{P_1}{P_2}$$

1.10.1 Values of q, w, ΔU and ΔH in Case of Isothermal Irreversible Expansion When

(i) opposing pressure is zero *(ii)* opposing pressure is constant and less than P_2 but greater than zero.

The expressions for w, q, ΔU and ΔH for the above expansion may be obtained as follows:

(i) **For free expansion:** $P_{opp} = 0$

$\therefore \qquad w = -\int P_{opp}\, dV = 0$

Hence $q = w = 0$

$\Delta U = \Delta H = 0$ (as the gas is *ideal* and expansion is *isothermal*)

(ii) **For intermediate expansion:** P_{opp} = constant and lies in the range $0 < P_{opp} < P_2$. Hence

$$w = -\int_{V_1}^{V_2} P_{opp}\, dV = -P_{opp}(V_2 = V_1)$$

$\therefore \qquad q = -w = P_{opp}(V_2 - V_1)$

$\Delta U = \Delta H = 0$ (as the gas is *ideal* and expansion is *isothermal*)

It may be noted that the work done in this process will be less than the work done in the reversible process.

Solved Examples—Isothermal Reversible Expansion of Gas

Example 3: Ten moles of an ideal gas at the initial pressure of one atmosphere at 0°C were expanded reversibly under isothermal conditions to a final pressure of 0·1 atmosphere. Calculate the work done by the gas, the change in internal energy and the heat absorbed by the system (R = 8·314 JK^{-1} deg^{-1} mol^{-1}).

Solution: Here we are given that

$$n = 10 \text{ moles} \quad P_1 = 1 \text{ atm} \quad P_2 = 0 \cdot 1 \text{ atm}$$
$$T = 0°C = 273 \text{ K} \quad R = 8.314 \text{ JK}^{-1} \text{mol}^{-1}$$

$$\therefore w_{max} = -2.303 \, nRT \log \frac{P_1}{P_2}$$

$$= -2.303 \times 10 \times 8 \cdot 314 \times 273 \times \log \frac{1}{0.1}$$

$$= -52271.7 \text{ J} \qquad (\because \log 10 = 1.0)$$

For an ideal gas during isothermal process

$$\Delta U = 0$$

$\therefore$ By first law of thermodynamics $(q = \Delta U - w)$

$$q = -w_{max} = 52271.7 \text{ J}$$

Example 4: One mole of an ideal gas is heated at constant pressure from 0°C to 100°C (*i*) Calculate the work involved (*ii*) If the gas were expanded isothermally and reversibly at 0°C from 1 atmosphere to some other pressure *P*, what must be the final pressure if the isothermal work is equal to the work involved in (*i*)?

Solution: Work involved when the gas is heated at constant pressure from 0°C to 100°C is obtained as follows:

(*i*)
$$w = -P\Delta V = -P(V_2 - V_1) = -(PV_2 - PV_1)$$
$$= -(nRT_2 - nRT_1) = -nR(T_2 - T_1)$$
$$= 1 \times 8.314 (373 - 273) = \mathbf{831 \cdot 4 \text{ J}}$$

(*ii*) Final pressure if the process is carried out isothermally involving the work as in step (*i*) is obtained as under

$$w = -2.303 \, nRT \log \frac{P_1}{P_2}$$

Put $P_1 = 1$ atm, $T = 273$ K, $R = 8.314$ JK^{-1} and $n = 1$
This on solving gives

$$P = \mathbf{0.6933 \text{ atm}}$$

Example 5: Calculate the pressure-volume work performed by the system during reversible isothermal expansion of two moles of an ideal gas from 2 litres to 10 litres at 20°C.

Solution: Here we are given that

$$n = 2 \text{ moles} \quad V_1 = 2 \text{ litres} \quad V_2 = 10 \text{ litres}$$
$$T = (20 + 273) \text{ K} = 293 \text{ K} \quad R = 8.314 \text{ Joules degree}^{-1} \text{mol}^{-1}$$

$$w_{max} = -2 \cdot 303 \, nRT \log \frac{V_2}{V_1}$$

$$= -2.303 \times 2 \times 8.314 \times 293 \times \log \frac{10}{2}$$

$$= -11220.2 \times \log 5 = -112202 \times 0.6990 = \mathbf{7842.9 \text{ joules}}$$

Chemical Thermodynamics – I (Zeroth and First Laws of Thermodynamics and Thermochemistry)

Problems for Practice

1. Find the work required to compress reversibly 1 mole of a perfect gas from 1 to 100 atmospheres at 27°C. **[Ans. 11488 J]**
2. Calculate the work done during isothermal expansion of one mole of an ideal gas from 10 atmoshpheres to 1 atmosphere at 27°C. **[Ans. – 5.74 kJ]**
3. Ten moles of an ideal gas are compressed isothermally and reversibly at 100°C from pressure of 1 atmosphere to a pressure of 10 atmosphere. Calculate the following quantities:
    (i) W and (ii) ΔH **[Ans. W = 71419 J, ΔH = 0]**
4. Calculate the work done by 10 litres of an ideal gas at 0°C and 10 atm in an isothermal reversible expansion to a final pressure of 1 atm (R = 8.314 joules).
    [**Hint.** (i) Reduce the given volume to volume at S.T.P.]
    $$\frac{P_1V_1}{T_1} = \frac{P_2V_2}{T_2} \quad \text{or} \quad \frac{10\times 10}{273} = \frac{1\times V_2}{273}$$
    or $V_2 = 100$ litres at S.T.P.
    (ii) Calculate the number of moles as follows:
    22.4 litres at S.T.P. = 1 mole
    100 litres at S.T.P = 4.46 moles
    $$w = -2.303 \, nRT \log \frac{P_1}{P_2}$$
    or $w = -23.3$ kJ]
5. Calculate the maximum work per mole when 5 moles of an ideal gas expand isothermally and reversibly from 20 litres to 100 litres at 27°C. [$R = 8.314 \times 10^7$ ergs mol^{-1} deg^{-1}].
    [Ans. – 200.76 $\times 10^5$ ergs]
6. Estimate the work done in joules when the volume of 16.0 g of oxygen at 300 K changes isothermally from 5.0 to 25.0 litres [$R = 8.3143$ Jk^{-1} mol^{-1}] **[Ans. – 2007.6 J]**
7. One mole of an ideal gas is heated at constant pressure from 0°C to 100°C.
    (i) Calculate the work involved (ii) If the gas were expanded isothermally and reversibly at 0°C from 1 atmosphere to some other pressure P, what must be the final pressure if the isothermal work is equal to the work involved in (i)
    [**Hint.** (i) $w = -P\Delta V = -P(V_2 - V_1) = (PV_2 - PV_1) = -(nRT_2 - nRT_1)$
    $= -nR(T_2 - T_1) = -1 \times 8.314 \, (373 - 273) = -831.4$ J]
    (ii) $w = 2.303 - nRT \log \frac{P_1}{P_2}$. Put P_1 = 1 atm, T = 273 K, R = 8.314 JK^{-1} mol^{-1}, n = 1
    On solving, we get $P_2 = 0.6933$ atm]
8. Calculate the value of w, ΔU and ΔH in the isothermal and reversible expansion of 64 grams of oxygen at 410 K from an initial volume of 20 litres to a final volume of 60 litres. Assume oxygen behaves ideally. **[Ans. w = –7490.8J, $\Delta H = \Delta U$ = zero]**
9. Estimate the maximum work which can be obtained by the isothermal expansion of 1 mole of an ideal gas from 2.24 litres to 22.4 litres at 273 K (R = 8.3143 JK^{-1} mol^{-1}). **[Ans. – 5227.4 J]**
10. Calculate the work done when 5 moles of an ideal gas at 27°C expand from an initial pressure of 5 atm to 1 atm reversibly.

1.11 VALUES OF w, q, ΔU AND ΔH IN ADIABATIC EXPANSION OF AN IDEAL GAS

As explained earlier an adiabatic process is the one in which no heat enters or leaves the system of any stage. For every infinitesimal change of the process,

$$\delta q = 0$$

Putting this value in the equation of the first law viz.

$$dU = \delta q + \delta w, \text{ we get}$$

or
$$dU = \delta w \qquad ...(i)$$

If it is the work of expansion only, then if dV is the small increase in volume and P is the pressure of the gas, then

$$w = -PdV \qquad ...(ii)$$

Substituting this value in equation (i), we get

$$dU = -PdV \qquad ...(iii)$$

For an ideal gas, $\quad C_v = \dfrac{dE}{dT}$

or $\quad dE = C_v dT$

Substituting this value in equation (iii) we get

$$C_v dT = -PdV \qquad ...(v)$$

Also
$$\Delta H = \Delta(U + PV) = \Delta U + \Delta(PV)$$
$$= \Delta U + \Delta(RT) = \Delta U + R\Delta T$$
$$= C_v \Delta T + R\Delta T = (C_v + R)\Delta T$$
$$= (C_v + R)(T_2 - T_1) = C_p(T_2 - T_1) \qquad ...(vi)$$

In a nutshell,

For any type of reversible or irreversible adiabatic process

$$q = 0$$
$$w = \Delta E$$
$$\Delta U = C_v \Delta T = C_v(T_2 - T_1)$$
$$\Delta H = C_p \Delta T = C_p(T_2 - T_1)$$

1.11.1 Reversible Adiabatic Expansion

I. Relationship between temperature and volume in reversible adiabatic expansion

For one mole of an *ideal gas*

$$PV = RT$$

or
$$P = \dfrac{RT}{V} \qquad ...(i)$$

Also $\quad C_v dT = -PdV \qquad ...(ii)$

Substituting for P from eq. (i) in (ii)

$$C_v dT = -\dfrac{RT}{V} dV$$

or
$$C_v \dfrac{dT}{T} = -R\dfrac{dV}{V} \qquad ...(iii)$$

Let us suppose the volume of the gas changes from V_1 to V_2 when the temperature changes from T_1 to T_2. Then assuming C_v to be independent of temperature, and integrating equation (iii) between the limits T_1, T_2 and V_1, V_2, we get

$$C_v \int_{T_1}^{T_2} \frac{dT}{T} = -R \int_{V_1}^{V_2} \frac{dV}{V}$$

$$C_v [\ln]_{T_1}^{T_2} = -R[\ln V]_{V_1}^{V_2} \quad \text{or} \quad C_v \ln \frac{T_2}{T_1} = \frac{V_2}{V_1}$$

or
$$C_v \ln \frac{T_2}{T_1} = R \ln \frac{V_1}{V_2}$$

For 1 mol of an ideal gas we know that
$$C_p - C_v = R$$

Substituting this value of R in equation (iv), we get
$$C_v \ln \frac{T_2}{T_1} = (C_p - Cv) \ln \frac{V_1}{V_2}$$

Dividing throughout by C_v we get
$$\ln \frac{T_2}{T_1} = \left(\frac{C_p}{C_v} - 1\right) \ln \frac{V_1}{V_2}$$

Putting $\frac{C_p}{C_v} = \gamma$, the ratio of the two heat capacities, we get

$$\ln \frac{T_2}{T_1} = (\gamma - 1) \ln \frac{V_1}{V_2} \quad \text{or} \quad \ln \frac{T_2}{T_1} = \ln \left(\frac{V_1}{V_2}\right)^{\gamma-1}$$

Taking antilogarithm of both sides, we get
$$\frac{T_2}{T_1} = \left(\frac{V_1}{V_2}\right)^{\gamma-1}$$

or $\quad T_1 V_1^{\gamma-1} = T_2 V_2^{\gamma-1}$

or $\quad \boldsymbol{TV^{\gamma-1} = \text{constant}}$

II. Relationship between temperature and pressure in adiabatic expansion of an ideal gas

For an ideal gas, $\quad P_1 V_1 = RT_1 \quad$ and $\quad P_2 V_2 = RT_2$

i.e. $\quad V_1 = \frac{RT_1}{P_1} \quad$ and $\quad V_2 = \frac{RT_2}{P_2}$

$\therefore \quad \dfrac{V_1}{V_2} = \dfrac{\frac{RT_1}{P_1}}{\frac{RT_2}{P_2}} = \dfrac{T_1 P_2}{T_2 P_1}$ ...(i)

Also $\quad \dfrac{T_2}{T_1} = \left(\dfrac{V_1}{V_2}\right)^{\gamma-1}$ ...(ii)

(The relation between T and V in case of adiabatic expansion)

From (i) and (ii)

$$\frac{T_1}{T_2} = \left(\frac{T_1 P_2}{T_2 P_1}\right)^{\gamma-1} \quad \text{or} \quad \frac{T_2}{T_1} \times \left(\frac{T_2}{T_1}\right)^{\gamma-1} = \left(\frac{P_2}{P_1}\right)^{\gamma-1}$$

or

$$\left(\frac{T_2}{T_1}\right)^{\gamma} = \left(\frac{P_2}{P_1}\right)^{\gamma-1} \quad \text{or} \quad \left(\frac{T_2}{T_1}\right)^{\gamma} = \left(\frac{P_1}{P_2}\right)^{1-\gamma}$$

Taking the γth root of both sides, we get

$$\frac{T_2}{T_1} = \left(\frac{P_1}{P_2}\right)^{(1-\gamma)/\gamma}$$

or $\quad T_1 P_1^{(1-\gamma)/\gamma} = T_2 P_2^{(1-\gamma)/\gamma}$ ...(iii)

or $\quad TP^{(1-\gamma)/\gamma} = \text{constant}$

III. Relationship between pressure and volume in adiabatic expansion of an ideal gas

Consider the relations between temperature and volume and temperature and pressure

$$\frac{T_2}{T_1} = \left[\frac{V_1}{V}\right]^{\gamma-1} \qquad \qquad \text{...(i)}$$

$$\frac{T_2}{T_1} = \left[\frac{P_1}{P_2}\right]^{(1-\gamma)/\gamma} \qquad \qquad \text{...(ii)}$$

From (i) and (ii)

$$\left(\frac{P_1}{P_2}\right)^{(1-\gamma)/\gamma} = \left(\frac{V_1}{V_2}\right)^{\gamma-1} \quad \text{or} \quad \left(\frac{P_1}{P_2}\right)^{(1-\gamma)/\gamma} = \left(\frac{V_2}{V_1}\right)^{1-\gamma}$$

Taking $(1 - \gamma)$th root of both sides, we have

$$\left(\frac{P_1}{P_2}\right)^{1/\gamma} = \frac{V_2}{V_1}$$

Raising both sides to the power γ, we have

$$\frac{P_1}{P_2} = \left(\frac{V_2}{V_1}\right)^{\gamma} \quad \text{or} \quad P_1 V_1^{\gamma} = P_2 V_2^{\gamma}$$

$$= \text{constant}$$

1.11.2 Work Done in an Adiabatic Reversible Expansion of an Ideal Gas

The expression for the work done in an adiabatic reversible expansion of an ideal gas may be obtained as under:

For adiabatic process, we have

$$PV^{\gamma} = \text{constant}$$

Differentiating this equation, we get

$$P(\gamma V^{\gamma-1}) dV + V^{\gamma} dP = 0$$

or $\gamma PV^{\gamma-1} dV + V^{\gamma} dP = 0$

Dividing throughout by $V^{\gamma-1}$, we get

$$\gamma P dV + V dP = 0$$

or $\qquad VdP = -\gamma PdV \qquad \ldots(i)$

For n moles of an ideal gas, $PV = nRT$

Differentiating this equation completely, we get

$$PdV + VdP = nRdT \qquad \ldots(ii)$$

Substituting for VdP from equation (i) in this equation we get

$$PdV - \gamma PdV = nRdT$$

or $\qquad (1-\gamma) PdV = nRdT$

or $\qquad PdV = \dfrac{nRdT}{1-\gamma} \qquad \ldots(iii)$

Work done in the adiabatic process is given by general equation

$$w = -\int_{V_1}^{V_2} PdV \qquad \ldots(iv)$$

Substituting for PdV from equation (iii) in this equation, and changing the limits from volumes to temperatures, we get

$$w = -\int_{T_1}^{T_2} \dfrac{nRdT}{1-\gamma} = -\dfrac{nR}{1-\gamma} \int_{T_1}^{T_2} dT$$

$$= -\dfrac{nR}{1-\gamma}(T_2 - T_1) \qquad \ldots(v)$$

Hence $\qquad w = -\dfrac{nR(T_2 - T_1)}{1-\gamma}$

Expression for work done may be obtained in a different form.

Putting $R = C_p - C_v$ and $\gamma = C_p/C_v$ equation (v) becomes

$$w = -\dfrac{n(C_p - C_v)(T_2 - T_1)}{1 - \dfrac{C_p}{C_v}} = -\dfrac{n(C_p - C_v)(T_2 - T_1)}{\dfrac{(C_p - C_v)}{C_p}}$$

$$= -\dfrac{n(C_p - C_v)(T_2 - T_1)}{\dfrac{-(C_p - C_v)}{C_p}}$$

$= nC_v(T_2 - T_1)$, (for n moles of the ideal gas)

$= C_v(T_2 - T_1)$, (for 1 mol of the ideal gas)

$\ldots(vi)$

For n moles of an ideal gas,

$\qquad P_1 V_1 = nRT_1 \qquad$ (for the initial state)

and $\qquad P_2 V_2 = nRT_2 \qquad$ (for the final state)

Substituting these values in equation (v), we get

$$w = -\frac{P_1V_2 - P_1V_1}{1-\gamma} \qquad \ldots(vii)$$

Alternatively, we can derive the expression for work done in adiabatic reversible expansion of an ideal gas in terms of C_v. We shall make use of the following two equations derived earlier.

$$w = -P\,dV \qquad \ldots(i)$$

and
$$C_v dT = -P\,dV \qquad \ldots(ii)$$

From (i) and (ii)
$$w = C_v\,dT \qquad \ldots(iii)$$

For the process when the volume changes from V_1 to V_2 and the temperature changes from T_1 and T_2, the work of expansion is given by

$$w = -\int_{V_1}^{V_2} P\,dV = \int_{T_1}^{T_2} C_v\,dT \qquad \ldots(iv)$$

Here, we make an assumption that C_v remains constant in the temperature range $T_1 \to T_2$. Equation (iv) reduces to

$$w = C_v\,(T_2 - T_1)$$

1.11.3 Irreversible Adiabatic Expansion

We have derived earlier expressions for irreversible isothermal expansion for an ideal gas Corresponding expressions for irreversible adiabatic expansion are derived as under

(a) Under the conditions of free expansion i.e., $P_{opp} = 0$

$$\delta w = -P_{opp}\,dV = 0$$

But
$$dU = \delta w$$

Refer to eq. (i) Sec 3.12.

$$\therefore \qquad dU = 0$$

For an ideal gas, U is a function of temperature. As $dU = 0$, that means dT is also equal to zero

$$dT = 0$$
$$dH = d(U + PV) = dU + d(PV)$$
$$= dU + d(RT) = dU + R\,dT = 0 + 0 = 0$$

Thus $\qquad dH = 0$

For a finite change, the values of w, ΔU and ΔH are thus given by

$$w = 0,\ \Delta U = 0,\ \Delta T = 0,\ \Delta H = 0$$

Thus, under the conditions of *free expansion*, the expressions for w, ΔU, ΔT and ΔH are the same as in the case of isothermal expansion.

(b) **Under the conditions of intermediate expansion**

P_{opp} = constant P_{opp} is greater than zero but less than V_2

$$w = -P_{opp}\,\Delta V = -P_{opp}\,(V_2 - V_1)$$

and
$$\Delta U = w = -P_{opp}\,(V_2 - V_1) \qquad \ldots(i)$$

Also
$$\Delta U = C_v\,\Delta T = C_v\,(T_2 - T_1) \qquad \ldots(ii)$$

and
$$\Delta H = C_p \Delta T = C_p(T_2 - T_1) \qquad ...(iii)$$

Dividing eq. (*iii*) by eq. (*ii*), we have

$$\frac{\Delta H}{\Delta U} = \frac{C_p}{C_v} = \gamma \quad \text{or} \quad \Delta H = \gamma \Delta U \qquad ...(iv)$$

Knowing the values of P_{opp}, V_2 and V_1; the values of w and ΔU can the calculated from eq. (*i*) above. Knowing the value of ΔU, ΔH can be calculated using eq. (*iv*).

1.11.4 Comparison of Isothermal and Adiabatic Expansion of an Ideal Gas

The following relation between pressure and volume exists for a *reversible adiabatic process,* as

$$PV^\gamma = \text{constant}$$

For a reversible isothermal process, the relationship is simply given by Boyle's law, *i.e.,*

$$PV = \text{constant}$$

We know C_p is always greater than C_v, therefore the ratio $\frac{C_p}{C_v} = \gamma$ is greater than unity.

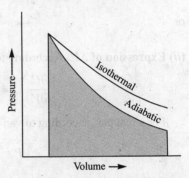

Fig. 1.9: *Isothermal and adiabatic curves*

That means the increase of volume for a given decrease of pressure will be less in an adiabatic than in an isothermal expansion. Consequently, plot of pressure against volume *i.e.,* pressure - volume curve will be steeper for an adiabatic than for an isothermal process, starting from the same point, as shown in Fig. 1.9.

Again, since the area under the pressure-volume curve gives the work of expansion, it may be observed that for the same increase of volume, the work done in an adiabatic process is less than in an isothermal process, as revealed from Fig. 1.9.

1.12 CALCULATION OF w, q, ΔU AND ΔH IN THE EXPANSION OF A REAL (VAN DER WALLS) GAS IN ISOTHERMAL REVERSIBLE EXPANSION

(*i*) Expression for w,

$$dw = -P_{ext}\, dV$$

For the reversible expansion

$$P_{ext} = P_{int} - dP = P_{int}$$

where P_{int} is the pressure (internal pressure) of the gas

Hence
$$dw = -P_{ext}\, dV \qquad ...(i)$$

As the gas is real, P_{int} can be obtained from van der Waal's equation *i.e.*

$$\left(P + \frac{an^2}{V^2}\right)(V - nb) = nRT$$

We get
$$P_{int} = \frac{nRT}{V - nb} - \frac{an^2}{V^2} \qquad ...(ii)$$

Putting this value in eqn. (*i*), we get

$$-dw = \left(\frac{nRT}{V-nb} - \frac{an^2}{V^2}\right)dV$$

$$-w = \int_{V_1}^{V_2}\left(\frac{nRT}{V-nb} - \frac{an^2}{V^2}\right)dV$$

$$= nRT\left[\ln(V_2-nb)\right]_{V_1}^{V_2} - an^2\left[-\frac{1}{V}\right]_{V_1}^{V_2}$$

or $$-w = nRT\ln\frac{V_2-nb}{V_1-nb} + an^2 + \left(\frac{1}{V_2} - \frac{1}{V_1}\right) \quad\ldots(iii)$$

(ii) Expression of ΔU: According to thermodynamic equatio of state

$$\left(\frac{\partial U}{\partial V}\right)_T = T\left(\frac{\partial U}{\partial V}\right)_V - P \quad\ldots(iv)$$

For a real gas, according to van der Waal's equation

$$P = \frac{nRT}{V-nb} - \frac{an^2}{V^2} \quad\ldots(v)$$

Differentiating w.r.t. T at constant volume, we get

$$\therefore \left(\frac{\partial P}{\partial T}\right)_V = \frac{nR}{V-nb} \quad\ldots(vi)$$

Substituting the values from eqns. (v) and (vi) in eqn. (iv) we get

$$\left(\frac{\partial U}{\partial V}\right)_T = \frac{nRT}{V-nb} - \left(\frac{nRT}{V-nb} - \frac{an^2}{V^2}\right) = \frac{an^2}{V^2} \quad\ldots(vii)$$

For dU, we can write

$$dU = \left(\frac{\partial U}{\partial T}\right)_V dT + \left(\frac{\partial U}{\partial V}\right)_T dV \quad\ldots(viii)$$

For the isothermal process, $dT = 0$

$$\therefore dU = \left(\frac{\partial U}{\partial T}\right)_T dV = \left(\frac{an^2}{V^2}\right)dV \quad \text{[Substituting value from eqn (vii)]}$$

Hence for a finite process

$$\Delta U = \int_{V_1}^{V_2}\frac{an^2}{V^2}dV$$

or $$\Delta U = -an^2\left(\frac{1}{V_2} - \frac{1}{V_1}\right) \quad\ldots(ix)$$

(iii) Expression of q: According to first law of thermodynamics

$$q = \Delta U - w$$

Substituting the value of w and ΔE and eqns. (iii) and (ix), we get

$$q = -an^2\left(\frac{1}{V_2} - \frac{1}{V_1}\right) + nRT\ln\frac{V_2-nb}{V_1-nb} + an^2\left(\frac{1}{V_2} - \frac{1}{V_1}\right)$$

Chemical Thermodynamics – I (Zeroth and First Laws of Thermodynamics and Thermochemistry)

or
$$q = nRT \ln \frac{V_2 - nb}{V_1 - nb} \qquad ...(x)$$

(iv) Expression for ΔH: We have
$$\Delta H = \Delta(U + PV)$$
$$= \Delta U + \Delta(PV)$$

Substituting the value of ΔU and P from eqns (ix) and (ii), we get

$$\Delta H = -an^2\left(\frac{1}{V_2} - \frac{1}{V_1}\right) + \Delta\left[\left(\frac{nRT}{V - nb} - \frac{an^2}{V^2}\right)V\right]$$

$$= -an^2\left(\frac{1}{V_2} - \frac{1}{V_1}\right) + nRT\left(\frac{V_2}{V_2 - nb} - \frac{V_1}{V_1 - nb}\right) - an^2\left(\frac{1}{V_2} - \frac{1}{V_1}\right)$$

$$= -2an^2\left(\frac{1}{V_2} - \frac{1}{V_1}\right) + nRT\left[\frac{V_2 - nb + nb}{V_2 - nb} - \frac{V_1 - nb + nb}{V_1 - nb}\right]$$

$$= -2an^2\left(\frac{1}{V_2} - \frac{1}{V_1}\right) + nRT\left[\left(1 + \frac{nb}{V_2 - nb}\right) - \left(1 + \frac{nb}{V_1 - nb}\right)\right]$$

$$\Delta H = -2an^2\left(\frac{1}{V_2} - \frac{1}{V_1}\right) + n^2 RT\left(\frac{1}{V_2 - nb} - \frac{1}{V_1 - nb}\right) \qquad ...(xi)$$

1.13 $q, w, \Delta E$ AND ΔH IN ADIABATIC REVERSIBLE EXPANSION OF A REAL (OR VAN DER WAALS) GAS

(i) Value for q: For the adiabatic process, as already explained,
$$q = 0$$

(ii) Expression for w and ΔU: By firist law of thermodynamics,
$$\Delta U = q + w$$
As $q = 0$, therefore $\quad \Delta U = w$

The epxression for ΔU and hence for w can be obtained as follows:
$$dU = \left(\frac{\partial U}{\partial T}\right)_V dT + \left(\frac{\partial U}{\partial V}\right)_T dV$$

Putting $(\partial U/\partial T)_v = nC_v$ (for n moles) and $\left(\frac{\partial U}{\partial V}\right)_T = \frac{an^2}{V^2}$, we get

$$dU = nC_v dT + \frac{an^2}{V^2} dv \qquad ...(i)$$

Hence for a finite process,
$$\Delta U = \int_{T_1}^{T_2} nC_v\, dT + \int_{V_1}^{V_2} \frac{an^2}{V^2} dV$$

Taking C_v to be independent of temperature, we have

$$\Delta U = nC_v(T_2 - T_1) - an^2\left(\frac{1}{V_2} - \frac{1}{V_1}\right) \qquad \ldots(ii)$$

This expression can be used for calculation of ΔU if T_1, T_2, V_1 and V_2 are known. However, if T_2 is not given, it is first calculated as follows:

$$dU = dw = -PdV$$

or
$$dU = -PdV \qquad \ldots(iii)$$

For a real gas
$$P = \frac{nRT}{V-nb} - \frac{an^2}{V^2}$$

and
$$dU = nC_v dT + \frac{an^2}{V^2}dV \qquad \text{Refer Eqn. }(i)\text{ above}$$

Substituting these values in eqn. (iii), we get

$$nC_v dT + \frac{an^2}{V^2}dV = -\frac{nRT}{V-nb}dV + \frac{an^2}{V^2}dV$$

$$nC_v dT = -\frac{nRT}{V-nb}dV$$

$$C_v dT = -\frac{RT}{V-nb}dV$$

Separating the variables, it can be rewritten as

$$C_v \frac{dT}{T} = -R\frac{dV}{V-nb}$$

Integrating between suitable limits, we get

$$\int_{T_1}^{T_2} C_v \frac{dT}{T} = -R\int_{V_1}^{V_2}\frac{dV}{V-nb}$$

or
$$C_v [\ln T]_{T_1}^{T_2} = -R[\ln(V-nb)]_{V_1}^{V_2}$$

or
$$C_v \ln \frac{T_2}{T_1} = -R\ln\frac{V_2-nb}{V_1-nb}$$

or
$$\ln\left(\frac{T_2}{T_1}\right)^{C_v} = \ln\left(\frac{V_1-nb}{V_2-nb}\right)^R$$

or
$$\left(\frac{T_2}{T_1}\right)^{C_v} = \left(\frac{V_1-nb}{V_2-nb}\right)^R$$

or
$$\frac{T_2}{T_1} = \left(\frac{V_1-nb}{V_2-nb}\right)^{R/C_v}$$

or
$$T_2 = T_1\left(\frac{V_1-nb}{V_2-nb}\right)^{R/C_v} \qquad \ldots(iv)$$

Knowing V_1, V_2 and T_1, T_2 (temperature after expansion) can be calculated. This may be substituted in eqn. (ii) to get ΔU and hence w.

(iii) Expression for ΔH

$$\Delta H = \Delta(U + PV) = \Delta U + \Delta(PV)$$

$$= \Delta U + \Delta\left[\left(\frac{nRT}{V-nb} - \frac{an^2}{V^2}\right)V\right]$$

$$= \Delta U + nR\left(\frac{V_2 T_2}{V_2 - nb} - \frac{V_1 T_1}{V_1 - nb}\right) - an^2\left(\frac{1}{V_2} - \frac{1}{V_1}\right)$$

Substituing the value of ΔU from eqn. (ii), we get

$$\Delta H = nC_v(T_2 - T_1) - 2an^2\left(\frac{1}{V_2} - \frac{1}{V_1}\right) + nR\left(\frac{V_2 T_2}{V_2 - nb} - \frac{V_1 T_1}{V_1 - nb}\right) \quad \ldots(v)$$

Solved Examples—Adiabatic Reversible Expansion of an Ideal Gas

Example 6: For a certain gas $C_p = 8.58$ J. Two moles of the gas are expanded adiabatically from an initial temperature of 20°C to a final temperature of –45.4°C. Calculate the work done in the above case.

Solution: $\qquad C_p = 8.58$ J, $C_v = 8.58 - 2 = 6.58$ J

Therefore $\qquad r = \dfrac{C_p}{C_v} = \dfrac{8.58}{6.58} = 1.3$

$n = 2$, $R = 8.314$ JK^{-1} mol^{-1}

Work done $\qquad w = -\dfrac{nR(T_2 - T_1)}{1-\gamma} = -\dfrac{2 \times 8.314 \times (-65.4)}{(1-1.3)}$

$$= -\dfrac{2 \times 8.314 \times 65.4}{0.3} \text{ J} = -3624.9 \text{ J}$$

Example 7: Three moles of a perfect gas with $C_v = 5.0$ cal/mol/degree at 1.0 atmosphere are to be compressed adiabatically and reversibly from a volume of 75 litres at 1.0 atmosphere to a pressure of 100 atmospheres. Predict (a) the final volume of the gas (b) the final temperature of the gas.

Solution: Here we are given that

$n = 3$ moles $\qquad C_v = 5.0$ cal/mol/degree
$P_1 = 1.0$ atm $\qquad P_2 = 100$ atm $\qquad V_1 = 75$ litres

For a perfect gas $\qquad C_p = C_v + R$

$= 5.0 + 2.0 = 7.0$ cal (R $\cong$ 2 cal)

$\therefore \qquad \gamma = \dfrac{C_p}{C_v} = \dfrac{7.0}{5.0} = 1.4$

(a) To calculate the final volume

$$P_1 V_1^\gamma = P_2 V_2^\gamma$$

or $\qquad \left(\dfrac{V_1}{V_2}\right)^\gamma = \dfrac{P_2}{P_1}$ or $\left(\dfrac{75}{V_2}\right)^{1.4} = \dfrac{100}{1.0} = 100$

$\therefore \qquad 1.4\,[\log 75 - \log V_2] = \log 100$

or $\quad 1.4 (1.8751 - \log V_2) = 2$
or $\quad\quad 1.4 \log V_2 = 0.62514$
or $\quad\quad \log V_2 = 0.4465$ or $V_2 =$ antilog $(0.4465) = \mathbf{2 \cdot 796}$ **litres**

(b) *To calculate the final temperature*
For a perfect gas,
$$PV = nRT$$
When
$\quad\quad P = 100$ atm
$\quad\quad V = 2.796$ litres (calculated above)
$\quad\quad n = 3$ moles (given)
$\quad\quad R = 0 \cdot 0821$ litre atm deg^{-1} mol^{-1}

$$\therefore \quad T = \frac{P \times V}{n \times R} = \frac{100 \times 2.796}{3 \times 0.0821} = 1135.2 \text{ K}$$

Example 8: Two moles of H_2 are compressed adiabatically from S.T.P. conditions to occupy a volume of 4.48 litres. Calculate the final temperature (γ for $H_2 = 1.41$).

Solution: We are given that $n = 2$

so that
$$\left. \begin{array}{l} V_1 = 2 \times 22.4 = 44.8 \text{ litres} \\ T_1 = 273 \text{ K} \end{array} \right\} \text{at S.T.P.}$$

$V_2 = 4.48$ litres, $T_2 = ?$ $\gamma = 1.41$

Substituting the values in the formula

$$T_1 V_1^{\gamma-1} = T_2 V_2^{\gamma-1} \text{ or } \frac{T_2}{T_1} = \left(\frac{V_1}{V_2}\right)^{\gamma-1}$$

We get
$$\frac{T_2}{273} = \left(\frac{44.8}{4.48}\right)^{1.41-1} = (10)^{0.41}$$

or
$\quad\quad T_2 = 273 \times (10)^{0.41}$

$\therefore \quad \log T_2 = \log 273 + 0.41 \log 10$
$\quad\quad\quad\quad = 2 \cdot 4362 + 0.41 = 2 \cdot 8462$

or
$\quad\quad T_2 =$ antilog $2 \cdot 8462 = \mathbf{701.8}$ **K**

Example 9: A dry gas at NTP is expanded adiabatically from 1 litre to 5 litre. Calculate the final temperature assuming ideal behaviour ($C_p/C_v = 1.4$)

Solution: Apply the relation $TV^{\gamma-1} =$ constant

$$T_1 V_1^{\gamma-1} = T_2 V_2^{\gamma-1} \text{ or } \frac{T_1}{T_2} = \left(\frac{V_2}{V_1}\right)^{\gamma-1}$$

$T_1 = 273$ K, $T_2 = ?$ $V_1 = 1$ litre, $V_2 = 5$ litre

Substituting the values in the equation above,

$$\frac{273}{T_2} = 5^{1.4-1} \quad \text{or} \quad \frac{273}{T_2} = 5^{0.4}$$

Taking logarithm of both sides

	$\log 273 - \log T_2 = 0.4 \log 5$
or	$\log T_2 = \log 273 - 0.4 \log 5$
or	$\log T_2 = 2.4362 - (0.4 \times 00.6990) = 2.1566$
or	$T_2 = 143.4$ K

Problems for Practice

1. Two moles of hydrogen at S.T.P are compressed adiabatically to a volume of 10 litres. Calculate the final pressure and temperature of the gas. Given that γ for hydrogen is 1.41.

   [**Ans.** Final Pressure = 8.30 atm, Final temperature = 505 K]

   [**Hint:** $V_1 = 2 \times 22.4 = 44.8$ litres, $V_2 = 10$ litres]

2. 14 grams of N_2 at 290 K are compressed adiabatically from 8 to 5 litres. Calculate the final temperature and the work done on the gas. Assume $C_p = \frac{7}{2}R$.

   [**Ans.** $T = 349.9$ K, $w = 622.5$ J]

3. A quantity of air ($\gamma = 1.4$) at 300 K is compressed adiabatically to one-third of its volume. Find the change in its temperature.

   [**Ans.** 465.5 K]

4. Calculate the change in energy for 2 moles of H_2 warmed at constant volume from 25°C to 50°C. Given that for the gas near room temperature, C_v is constant and is about 5 cal/deg.

   [**Ans.** 250 ergs.]

5. For a certain ideal gas $C_p = 8.58$ cal mol^{-1}. What will be the final volume and temperature when 2 moles of the gas at 20°C and 15 atm are allowed to expand adiabatically and reversibly to 5 atm pressure? Also calculate the work done in the above process.

   [**Ans.** $V = 7.44$ litres $t = -45.4°C$, $w = -3625$ J]

   [**Hint.** First calculate T_2, then find out w using the formula $w = -\dfrac{nR(T_2 - T_1)}{1 - \gamma}$

   $T_1 \gamma = 293$ K, $T_2 = 227.6$ K

   $\gamma = \dfrac{8.58}{6.58} = 1.30, n = 2, R = 8.314$ JK^{-1} mol^{-1}]

1.14 THERMOCHEMISTRY

The branch of chemistry which deals with heat changes accompanying chemical reactions is called **thermochemistry.**

A chemical reaction involves rearrangement of atoms, *i.e.*, some bonds within the reactant molecules are broken while some new bonds in molecules of the products are formed. Energy is consumed in the dissociation of bonds while energy is evolved when the bonds are formed. If the total energy evolved is greater than the total energy consumed, the net result is the *release of energy*. On the other hand, if the energy consumed is greater than the energy evolved, the net result is the *absorption of energy*.

All chemical reactions are usually accompanied by energy changes. These energy changes appear in the form of heat, light, work, electricity, etc. These energy changes are of great practical utility both for domestic and industrial purposes. A few reactions alongwith the form in which the accompanying energy changes appear and the use to which they are put are listed below:

(i) *Burning of coal in air* produces *heat*. That is why it is used as a fuel.
$$C(s) + O_2(g) \longrightarrow CO_2(g) + \text{heat}$$

(ii) *Digestion and metabolism of carbohydrates* in biological systems produce *heat* which maintains the body temperature.

(iii) *Burning of candle (wax) in air* produces heat and light. It is, therefore, used as a source of light. Wax is a hydrocarbon.

(iv) *Reaction that takes place in a galvanic cell* produces electrical energy which may be used to run an electrical motor or to ring a bell, etc.
$$Zn(s) + Cu^{2+}(aq) \longrightarrow Zn^{2+}(aq) + Cu(s)$$

Units of Heat and Work. The heat changes are measured in calories, kilocalories, joules or kilojoules.

$$1 \text{ calorie} = 4.184 \text{ J} \qquad\qquad 1 \text{ kcal} = 4.184 \text{ kJ}$$

The S.I. unit of heat is joule or kilojoule.

Work is measured in ergs or joules. The S.I. unit of work is joule. $\qquad$ 1 Joule = 10^7 ergs

1.14.1 Enthalpy and Enthalpy Change

When a process is carried out at constant pressure and temperature, there is a change in volume.

If the volume increases, some mechanical work is done by the system so that the internal energy decreases. Similarly, if the volume decreases, some mechanical work is done on the system so that the internal energy increases. In all such processes occurring at constant temperature when pressure remains constant, the total energy of a system is given by a thermodynamic quantity called **heat content** or **enthalpy** of the system and not by internal energy. It is denoted by H and is defined mathematically by the equation

$$H = E + PV \qquad\qquad ...(i)$$

where E is the internal energy, P and V are the pressure and volume of the system respectively.

The **enthalpy** *of a system may be defined as the sum of the internal energy and the product of pressure and volume of the system.*

The *physical significance of enthalpy* may be put as under:

The **enthalpy** *of a substance is the amount of energy stored within the substance, i.e., available for conversion into heat.*

Like internal energy, *the heat content or enthalpy of a substance is also a state function, i.e.,* depends only on the state of the system and is independent of the manner by which the state has been attained.

Enthalpy change. Like internal energy, the absolute value of enthalpy of a substance cannot be determined. However, the change in the enthalpy when a system undergoes a change can be determined experimentally accurately and that serves our purpose.

A relation between enthalpy change and internal energy change exists as follows :

$$\Delta H = \Delta E + P \Delta V \qquad\qquad ...(ii)$$

where $\qquad\qquad \Delta E = E_2 - E_1$ represents the change in internal energy
and $\qquad\qquad \Delta V = V_2 - V_1$ represents the change in volume of the system.

Thus, **enthalpy change** *of a system may be defined as the sum of internal energy change and the pressure-volume work done.*

If q_p is the heat absorbed at constant pressure, then $q_p = \Delta H$ ...(iii)

Thus, enthalpy change *is the heat absorbed or evolved at constant pressure.*

Standard enthalpy change. The enthalpy change of a reaction when all the reactants and the products are in their standard states, *i.e.*, at 298 K and under a pressure of one atmosphere is known as the standard enthalpy change. It is usually represented by $\Delta H°$.

As a chemical reaction involves breaking of bonds of the reactants and the formation of new bonds in the products, some net energy change (called the enthalpy of reaction) takes place.

$$\text{Enthalpy change of a reaction} = \begin{bmatrix} \text{Energy consumed to} \\ \text{break the bonds of} \\ \text{the reactants} \end{bmatrix} - \begin{bmatrix} \text{Energy released in the} \\ \text{formation of the bonds in} \\ \text{the products} \end{bmatrix}$$

If energy consumed > energy released, the net result is the absorption of energy and the reaction is *endothermic*.

If energy released > energy consumed, the net result is the evolution of energy and the reaction is *exothermic*.

Latest Sign Conventions for q or ΔH, ΔU and w.

(*i*) When heat is absorbed by the system, q is positive.

(*ii*) When heat is given out by the system, q is negative.

(*iii*) When energy is absorbed by the system, ΔU is positive.

(*iv*) When energy is given out by the system, ΔU is negative.

(*v*) When work is done on the system, w is positive.

(*vi*) Work done by the system, w is negative.

1.14.2 Exothermic and Endothermic Reactions

Exothermic Reactions *are those reactions which are accompanied by the evolution of heat.*

The quantity of heat produced is shown alongwith the products with a 'plus' sign. A few examples of exothermic reactions are given below:

$$C(s) + O_2(g) \longrightarrow CO_2(g) + 393.5 \text{ kJ}$$

$$H_2(g) + \frac{1}{2}O_2(g) \longrightarrow H_2O + 248.8 \text{ kJ}$$

$$CH_4(g) + 2O_2(g) \longrightarrow CO_2(g) + 2H_2O(g) + 890.4 \text{ kJ}$$

Endothermic Reactions *are those reactions which are accompanied by absorption of heat.*

Heat absorbed can be written with the products with a 'minus' sign. A few examples of endothermic reactions are given below:

$$N_2(g) + O_2(g) \longrightarrow N_2(g) + O_2(g)$$
$$C(s) + H_2O(g) \longrightarrow CO(g) + H_2(g) - 131.4 \text{ kJ}$$
$$C(s) + 2S(s) \longrightarrow CS_2(l) - 92.0 \text{ kJ}$$

The enthalpy change (ΔH) accompanying a reaction is given by

ΔH = Heat content of products − Heat content of reactants = $H_P - H_R$

A reaction is exothermic, if the total heat content of the reactants is more than that of the products, i.e., $H_R > H_P$. Thus, ΔH will be negative for an exothermic process.

Thus, the exothermic reactions given above may be written in terms of ΔH as:

$$C\,(s) + O_2\,(g) \longrightarrow CO_2\,(g),\ \Delta H = -393.5\ kJ$$

$$H_2\,(g) + \frac{1}{2}O_2\,(g) \longrightarrow H_2O\,(l),\ \Delta H = 285.\ kJ$$

$$CH_4\,(g) + 2O_2\,(g) \longrightarrow CO_2\,(g) + 2H_2O\,(l),\ \Delta H = -890.4\ kJ$$

A reaction is endothermic if the total heat content of reactants is less than that of the products, i.e.,

$$H_R < H_P$$

Thus, ΔH will be positive for an endothermic reaction.

Endothermic reactions given above may be written in terms of ΔH as:

$$N_2\,(g) + O_2\,(g) \longrightarrow 2NO\,(g),\qquad \Delta H = +180.7\ kJ$$
$$C\,(s) + H_2O\,(g) \longrightarrow CO\,(g) + H_2\,(g),\qquad \Delta H = +131.4\ kJ$$
$$C\,(s) + 2S\,(s) \longrightarrow CS_2\,(l),\qquad \Delta H = +92.0\ kJ$$

Exothermic and endothermic reactions may be represented graphically as shown in Fig. 1.10 (a) and 1.10 (b).

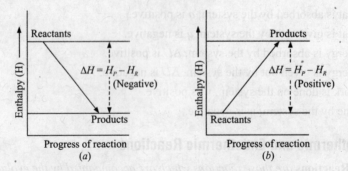

Fig. 1.10 (a): Exothermic reaction (b) Endothermic reaction

1.14.3 Thermochemical Equation

A balanced chemical equation not only indicating the quantities of the different reactants and products but also indicating the amount of heat evolved or absorbed is called a **thermochemical equation.**

Fractional coefficients are allowed in writing a thermochemical equation. For example, the formation of water is written as

$$H_2\,(g) + \frac{1}{2}O_2\,(g) \longrightarrow H_2O\,(l) \qquad \Delta H = -285.8\ kJ$$

Thus, 285.8 kJ of heat is produced when 1 mole of hydrogen reacts with 0.5 mole of oxygen. If the quantities of reactants are doubled, the heat produced will also be doubled. For example, in the above case, we may write

$$2H_2\,(g) + O_2\,(g) \longrightarrow 2H_2O\,(l), \qquad \Delta H = -571.6\ kJ$$

Some Points to Remember about Thermochemical Equations

(i) Unless otherwise mentioned, ΔH values are for the standard state of the substances (i.e., 298 K and 1 atm pressure).

(ii) The physical state (s, l, g, etc.) of the different substances must be mentioned as the heat evolved or absorbed depends upon the physical state of the substances.

(iii) If the coefficients of the substances are multiplied or divided by some number, the value of ΔH is also multiplied or divided by the same number.

(iv) If the reaction is reversed, the sign of ΔH changes but the magnitude remains the same.

1.15 HEAT OF REACTION (OR ENTHALPY OF REACTION)

The amount of heat evolved or absorbed in a chemical reaction when the number of moles of the reactants as represented by the chemical equation have completely reacted, is called the **heat of reaction.**

Let us consider the following two examples:

$$C\ (s) + O_2\ (g) \longrightarrow CO_2\ (g) \qquad \Delta H = -393.5 \text{ kJ}$$
$$C\ (s) + 2S\ (s) \longrightarrow CS_2\ (l), \qquad \Delta H = +92.0 \text{ kJ}$$

The first equation indicates that when 1 mole of solid carbon (i.e., 12 g) combines completely with 1 mole of oxygen gas (i.e., 32 g), 393.5 kJ of heat are produced. Similarly, the second equation tells that when 1 mole of solid carbon (i.e., 12 g) reacts completely with 2 moles of sulphur (64 g), 92 kJ of heat are absorbed.

Heat of reaction, ΔH = Heat content of products – Heat content of reactants

If the reaction is reversed, the sign of ΔH is also reversed.

1.15.1 Factors Influencing Heat of Reaction

The heat of a reaction depends upon a number of factors as explained below :

(i) *Quantities of the reactants involved.* If the quantities of the reactants are changed, the heat of reaction changes proportionally.

(ii) *Physical state of the reactants and products.* Since latent heat is involved in the change of state, therefore, the physical state of the reactants and products affects the heat of reaction. For example, when hydrogen and oxygen gases combine to form liquid water, the heat of reaction is different from the value when they combine to form water vapours.

$$H_2\ (g) + \frac{1}{2} O_2\ (g) \longrightarrow H_2O\ (l), \qquad \Delta H = -285.8 \text{ kJ}$$
$$H_2\ (g) + \frac{1}{2} O_2\ (g) \longrightarrow H_2O\ (g), \qquad \Delta H = -248.8 \text{ kJ}$$

It is, therefore, essential to write the symbols s, l, g or aq to indicate whether a particular substance is solid, liquid, gaseous or an aqueous solution.

(iii) *Allotropic modifications.* Heat of reaction changes if a different allotropic form is involved in the reaction. For example,

$$C\ (\text{diamond}) + O_2\ (g) \longrightarrow CO_2\ (g) \qquad \Delta H = -395.4 \text{ kJ}$$
$$C\ (\text{graphite}) + O_2\ (g) \longrightarrow CO_2\ (g) \qquad \Delta H = -393.5 \text{ kJ}$$

(*iv*) *Concentration of solutions.* If the solutions are involved in a reaction, their concentrations affect the heat of reaction.

(*v*) *Temperature.* The heat of reaction depends upon the temperature at which the reactants and products are taken. Generally, the heat of reaction increases with temperature.

(*vi*) *Conditions of constant pressure or constant volume, i.e.,* whether the reaction takes place at constant pressure or at constant volume.

If q_p and q_v represent the heats of reaction at constant pressure and volume respectively, then the following equation is applicable.

$$q_p = q_v + \Delta n_g RT$$

where
$$\Delta n_g = n_p - n_r$$

n_p and n_r represent the number of moles of gaseous reactants and products respectively. R and T have the usual meanings.

1.15.2 Heat (or Enthalpy) of Combustion

The heat of combustion of a substance is defined as the heat change (usually the heat evolved) when 1 mole of substance is completely burnt or oxidised in oxygen. For example,

$$C\,(s) + O_2\,(g) \longrightarrow CO_2\,(g), \qquad\qquad \Delta H = -393.5 \text{ kJ}$$

When one mole of carbon (12 g) burns completely in oxygen, 393.5 kJ of heat are evolved. Thus, heat of combustion of carbon is 393.5 kJ.

The significance of the words *burns completely* must be clearly understood. For example, carbon may be oxidised to carbon monoxide (CO) or carbon dioxide (CO_2). The heats evolved in the two cases are different, *viz.*,

$$C\,(s) + \frac{1}{2}O_2\,(g) \longrightarrow CO\,(g), \qquad\qquad \Delta H = -110.5 \text{ kJ}$$

$$C\,(s) + O_2\,(g) \longrightarrow CO_2\,(g), \qquad\qquad \Delta H = -110.5 \text{ kJ}$$

Complete oxidation means oxidation to CO_2 and not to CO. Hence, the heat of combustion of carbon is 393.5 kJ mol^{-1}.

1.15.3 Calorific Value of Foods and Fuels

For the working of the human machine, we eat carbohydrates, fats, etc., in the form of food. The carbohydrates are first decomposed in our body by the enzymes to form glucose which is then oxidised by the oxygen that we inhale to produce energy as given by the following equation.

$$C_6H_{12}O_6\,(s) + 6O_2\,(g) \longrightarrow 6CO_2\,(g) + 6H_2O\,(g) + 2900 \text{ kJ}$$

This oxidation reaction is usually called '*combustion of food*'.

Different foods have different heats of combustion. These are usually expressed in terms of their calorific values.

The calorific value of a fuel or food is the amount of heat in calories or joules produced from the complete combustion of one gram of the fuel or the food.

Thus, according to the above reaction, 1 mole of glucose, *i.e.,* 180 g of glucose produce 2900 kJ of heat.

Hence, calorific value of glucose $= \dfrac{2900}{180} = 16.1$ kJ g^{-1}

Calorific values of some foods and fuels are given in Tables 1.1 and 1.2.

Chemical Thermodynamics – I (Zeroth and First Laws of Thermodynamics and Thermochemistry)

TABLE 1.1 — Calorific values of foods per gram

Food	Calorific value kJ/gram	Food	Calorific value kJ/gram
Ghee	37.65	Fish	5.2
Butter	30.50	Meat	12.0
Cheese	14.56	Egg (yolk)	5.4
Milk (buffalo's)	4.90	Egg (white)	14.0
Milk (cow's)	2.20	Sugar	16.0
Curd	2.51	Rice	16.7
Soyabean	18.07	Potatoes	4.0

TABLE 1.2 — Calorific values of some common fuels in kJ per kg

Fuel	Calorific value
Wood	16,000
Charcoal	29,000
Natural gas	46,000
Kerosene	38,000

A normal youngman needs 10,000 kJ of energy per day for normal functioning.

Applying Chemistry to Life

With the development of civilization and general rise in standard of living we tend to eat more and more. We consume more than what is needed by our body. This leads to obesity which is the major cause of a number of diseases like blood pressure, diabetes, arthritis, etc. Add to this our sluggish and sedentary lifestyle. This is going to cause havoc to our bodies. The antidote for such problems is extensive and intensive exercise. We require at least five hours of brisk jogging, swimming, yoga or cycling per weak to keep fit.

1.15.4 Adiabatic Flame Temperature

This temperature is defined as the temperature that is achieved by allowing a fuel to react with an oxidiser and approach equilibrium by means of an isobaric and adiabatic process.

Adiabatic flame temperature is the highest temperature a system can achieve at contant pressure (but not a constant volume). This temperature is achieved by a combustion system at constant pressure when there are no heat losses from the system. In other words, all the heat from the combustion reaction is used to heat up the products of combustion.

Adiabatic flame temperature for a given fuel-oxidiser combination is determined by finding the final state temperature (*i.e.* the adibatic flame temperature) for which the sum of the enthalpies of the reactants is equal the sum of the enthalpies of the products.

For methane combustion in air at 1 atmosphere, the adiabtic flame temperature is 2328 K or 2055°C. For hydrogen burning in air at 1 atmosphere, the adiabatic flame temperature is 2400 K or 2127°C.

Explosion Temperature

There are certain reactions which produce inflammable products. Such reactions are carried out under extremely stringent conditions to prevent any untoward incident. But accidents do take place when the heating limits are crossed.

The minimum temperature at which the reactants or products of a reaction in progress, catch fire resulting in explosion, is called the explosion temperature.

Solved Example—Calorific Value

Example 10: A cylinder of indane gas contains 11.2 kg of butane. A normal family needs 25000 kJ energy per day for cooking. How long will the cylinder of gas last assumping 100% combustion?

If the efficiency of combustion is 85%, how long will the cylinder last? Heat of combustion of butane is 2658 kJ mol^{-1}.

Solution: Mol. mass of butane is 58 g mol^{-1}.

Thus, 58 g butane on complete combustion produces = 2658 kJ

∴ 11200 g butane on complete combustion produces = $\dfrac{2658 \times 11200}{58}$ kJ = 513268.9 kJ.

The cylinder will last = $\dfrac{513268.96}{25000}$ = 20.5 days

If the efficiency of combustion is only 85%, the cylinder will last = 0.85 × 20.5 ≈ 17.5 days.

1.16 STANDARD STATE

The standard molar enthalpy of every element at 1 atm pressure and 298.15 K in the most stable state is taken as zero. These conditions of 1 atm pressure and 298.15 K are called standard state conditions. It is not possible to determine the absolute value of molar enthalpy, but we can use the above convention to find out the standard molar enthalpy of a substance. The most stable form of hydrogen is gaseous H_2, that of bromine is liquid Br_2 while that of iodine is solid I_2. Their molar standard enthalpies are taken as zero at 1 atmospheric pressure and 298.15 K temperature. In case of elements exhibiting allotrophy, we consider the most stable form under the conditions of standard state.

1.16.1 Enthalpy of Formation and Standard Enthalpy of Formation

Heat of Formation (or Enthalpy of Formation) : *The enthalpy of formation of a substance is defined as the heat change that takes place when 1 mole of the substance is formed from its elements under given conditions of temperature and pressure. It is usually represented by ΔH_f.*

Standard enthalpy of formation *of a substance is defined as the heat change accompanying the formation of 1 mole of the substance in the standard state from its elements, also taken in the standard state (i.e., 298 K, and 1 atmospheric pressure). It is usually represented by* ΔH°_f.

For example, in the reaction

$$C\,(s) + O_2\,(g) \longrightarrow CO_2\,(g), \qquad \Delta H^\circ_f = -393.5 \text{ kJ}$$

When 1 mole of CO_2 (g) is formed from its elements, viz., C (s) and O_2 (g) (all substances being taken in the standard state), 393.5 kJ of heat is produced. Hence, the standard heat of formation of gaseous CO_2 is 393.5 kJ mol^{-1}.

Importance of standard enthalpies of formation. Knowing the standard enthalpies of formation of the different compounds involved in a chemical reaction, the standard enthalpy change of the given reaction can be obtained using the formula.

$$\Delta H^\circ_{reaction} = \begin{bmatrix} \text{Sum of the standard heats} \\ \text{of formation of products} \end{bmatrix} - \begin{bmatrix} \text{Sum of the standard heats} \\ \text{of formation of reactants} \end{bmatrix}$$

$$\Delta H^\circ_{reaction} = \Sigma \Delta H^\circ_f \text{ (Products)} - \Delta H^\circ_f \text{ (Reactants)}$$

Thus, for a general reaction

$$aA + bB \longrightarrow cC + dD$$

$$\Delta H^\circ_{reaction} = [c\,\Delta H^\circ_f\,(C) + d\,\Delta H^\circ_f\,(D)] - [a\,\Delta H^\circ_f\,(A) + b\,\Delta H^\circ_f\,(B)]$$

Remember ΔH°_f is taken as zero for the elements in a reaction. Suppose we are interested in finding the heat of combustion of methane. The reaction is represented as:

$$CH_4\,(g) + 2O_2\,(g) \longrightarrow CO_2\,(g) + 2H_2O\,(l)$$

$$\Delta H^\circ_{reaction} = [c\,\Delta H^\circ_f \text{ for } CO_2 + (2 \times \Delta H^\circ_f \text{ for } H_2O)] - [\Delta H^\circ_f \text{ for } CH_4 + (2 \times \Delta H^\circ_f \text{ for } O_2)]$$

Putting the values of heats of formation in the standard state

$$\Delta H^\circ_{reaction} = [-393.5 + 2 \times (-285.8)] - [-74.9 + 0] = -965.1 + 74.9 = -890.2 \text{ kJ}$$

Standard Enthalpy of Combustion

It is defined as the amount of heat evolved or absorbed when one mole of the substance under standard conditions of 298.15 K temperature and 1 atmosphere pressure is burnt completely to form the products also under the standard conditions. It is represented by $\Delta_c H^\circ$, ΔH°_c or $\Delta H^\circ_{reaction}$.

Consider, for example, the combustion of methane to give carbon dioxide and water as represented hereunder.

$$CH_4\,(g) + 2O_2\,(g) \longrightarrow CO_2\,(g) + 2H_2O\,(g), \Delta H^\circ = -890.4 \text{ kJ mol}^{-1}$$

Thus, standard heat of combustion of methane *i.e.,* $\Delta H^\circ_{combustion} = -890.4$ kJ mol^{-1}. The above statement does not mean that the combustion is taking place at 298.15 K. In fact at this temperature the combustion of methane cannot happen. This only implies that it is the enthalpy change that occurs when the reactants and products are at 298.15 K. During combustion, the temperature could be much higher than that.

Solved Example—Enthalpies of Formation

Example 11: Enthalpies of formation of CO_2 (g) and H_2O (l) under standard conditions are 394.65 kJ and 285.84 kJ per mole. If the standard enthalpy of combustion of acetaldehyde (CH_3CHO) be 1167.62 kJ per mole, find its enthalpy of formation.

Solution: From the given data

(a) $\quad\quad C + O_2 \longrightarrow CO_2 \text{ (g)};\quad\quad\quad\quad\quad\quad\quad \Delta H = -495.65 \text{ kJ}$

(b) $\quad\quad H_2 + \dfrac{1}{2} O_2 \longrightarrow H_2O \text{ (l)};\quad\quad\quad\quad\quad \Delta H = -285.84 \text{ kJ}$

(c) $\quad\quad CH_3CHO + \dfrac{1}{2} O_2 \longrightarrow 2CO_2 + 2H_2O \text{ (g)};\quad \Delta H = -1167.62 \text{ kJ}$

Our aim is to find out the ΔH value for the reaction

$$2C + 2H_2 + \dfrac{1}{2} O_2 \longrightarrow CH_3CHO$$

Using standard enthalpy values

ΔH_f^o for $CO_2 = -394.65$ kJ, ΔH_f^o for $H_2O = -285.84$ kJ and ΔH_f^o for $O_2 = -0$
($\therefore$ Enthalpy of formation of elements is taken as zero).

Let ΔH_f^o for CH_3CHO be x kJ.

ΔH value for any reaction = Enthalpies of the products – Enthalpies of the reactants.

From the equation

$$CH_3CHO + \dfrac{5}{2} O_2 \longrightarrow 2CO_2 \text{ (g)} + 2H_2O; \Delta H = -1167.62 \text{ kJ}$$

$\therefore -1167.62 = [2 \times \Delta H_f^o(CO_2) + 2 \times \Delta H_f^o(H_2O)] - [\Delta H_f^o(CH_3CHO) + \dfrac{5}{2} \times \Delta H_f^o(O_2)]$

$\quad\quad\quad\quad = 2 \times (-394.65) + 2 \times (-285.84) - x + 0 = -789.30 - 571.62 - x$

$\quad\quad\quad\quad -1176.62 = -1360.89 - x \quad \text{or} \quad x = -193.36 \text{ kJ}.$

Hence, enthalpy of formation of CH_3CHO is **–193.36 kJ.**

Example 12: Calculate ΔH^o for the reaction : $CO_2(g) + H_2 (g) \longrightarrow CO \text{ (g)} + H_2O \text{ (g)}$ given that ΔH^o_f for $CO_2 \text{ (g)}$, $CO \text{ (g)}$ and $H_2O \text{ (g)}$ are $-393.5, -111.3$ and -241.8 kJ mol^{-1} respectively.

Solution: ΔH^o for the reaction $CO_2 \text{ (g)} + H_2 \text{ (g)} \longrightarrow CO \text{ (g)} + H_2O \text{ (g)}$ is

$\Delta H^o = [\Delta H^o_{f(\text{products})} - \Delta H^o_{f(\text{reactants})}]$

$\quad\quad = [(-111.3) + (-241.8)] - [(-393.5) + 0] = (-111.3 - 241.8) - (-393.5)$

$\quad\quad = -353.1 + 393.5$

$\Delta H^o = +40.4$ kJ. [Enthalpy of every element in standard state is assumed as zero].

Example 13: Enthalpies of formation of C_2H_5OH (l), CO_2 (g) and H_2O (l) are -277.0, -393.5 and -285.8 kJ mole^{-1} respectively. Calculate enthalpy change for reaction.

$$C_2H_5OH \text{ (l)} + 3O_2 \text{ (g)} \longrightarrow 2CO_2 \text{ (g)} + 3H_2O \text{ (l)}$$

Solution: ΔH^o for the reaction C_2H_5OH (l) + $3O_2$ (g) $\longrightarrow$ $2CO_2$ (g) + $3H_2O$ (l) is

$\Delta H^o = [\Delta H^o_{f(\text{products})} - \Delta H^o_{f(\text{reactants})}]$

$\quad\quad = [2 \times \Delta H^o_f(CO_2) + 3 \times \Delta H^o_f(H_2O)] - [\Delta H^o_f(C_2H_5OH) + 3 \times \Delta H^o_f(O_2)]$

$\quad\quad = [2 \times (-393.5) + 3 \times (-285.8)] - [-277 + 0] = -787.0 - 857.4 + 277$

$\quad\quad = -1644.4 + 277 =$ **–1367.4 kJ.**

1.16.2 Heat (or Enthalpy) of Neutralization

The heat of neutralization of an acid by a base is defined as the heat change when one gram equivalent of the acid is neutralized by a base, the reaction being carried out in dilute aqueous solution.

The heat of neutralization of a base by an acid is defined in a similar manner.

When one gram equivalent of HCl is neutralized by NaOH or one gram equivalent of NaOH is neutralized by HCl, in dilute and aqueous solution, 57.1 kJ of heat is produced.

$$NaOH + HCl \longrightarrow NaCl + H_2O, \Delta H = -57.1 \text{ kJ}$$

Hence, the heat of neutralization of HCl with NaOH or NaOH with HCl is 57.1 kJ.

The heat of neutralization of any strong acid (HCl, HNO_3, H_2SO_4) with a strong base (NaOH, KOH) or vice versa, is always the same, *i.e.*, 57.1 kJ. This is because the strong acids, strong bases and the salts formed by them are all completely ionized in dilute aqueous solution. Thus, the reaction between a strong acid and a strong base may be written as

$$Na^+ + OH^- + H^+ + Cl^- \longrightarrow Na^+ + Cl^- + H_2O, \qquad \Delta H = -57.1 \text{ kJ}$$

Cancelling out common ions on the two sides, we have

$$H^+ \text{ (aq)} + OH^- \text{ (aq)} \longrightarrow H_2O \text{ (l)}, \qquad \Delta H = -57.1 \text{ kJ}$$

Thus, neutralization may be regarded as a reaction between the H^+ ions given by the acid with the OH^- ions given by the base. Strong acids and strong bases ionize completely in dilute aqueous solution, therefore the number of H^+ ions and OH^- ions produced by one gram equivalent of the strong acid and the strong base is always the same. Hence, the heat of neutralization between a strong acid and a strong base is always the same. *i.e.*, –57.1 kJ.

If the acid or the base is weak, the heat of neutralization is usually less than 57.1 kJ. Consider the neutralization of a weak acid like acetic acid with a strong base like NaOH. Acetic acid ionizes to a small extent whereas NaOH ionizes completely as

(i) $\qquad CH_3COOH \rightleftharpoons CH_3COO^- + H^+$
(ii) $\qquad NaOH \longrightarrow Na^+ + OH^-$

When H^+ ions given by the acid combine with the OH^- ions given by the base, the equilibrium (i) shifts to the right, *i.e.*, more of acetic acid dissociates. A part of the heat produced during the combination of H^+ ions and OH^- ions is used up for the complete dissociation of acetic acid. Hence, the net heat evolved in the above reaction is less than 57.1 kJ.

Compared to a strong acid, a weak acid produces a smaller number of H^+, hence, heat of neutralization is smaller.

Example 14: Determine the amount of heat released when:

(*a*) 0.75 mole of nitric acid is neutralised by 0.75 mole of sodium hydroxide.

(*b*) 0.75 mole of hydrochloric acid is neutralised by 0.5 mole of potassium hydroxide.

(*c*) 900 ml of 0.5 M hydrochloric acid solution is mixed with 500 ml of 0.4 M sodium hydroxide solution. Assuming the specific heat of water as 4.18 J K^{-1} g^{-1}, also calculate the rise in temperature.

(*d*) 200 ml of 0.2 M sulphuric acid is mixed with 300 ml of 0.1 M sodium hydroxide solution. Calculate the rise in temperature.

Solution: (*a*) The reaction is:

$$(0.75 \text{ moles}) HNO_3 + (0.75 \text{ moles}) NaOH \longrightarrow NaNO_3 + H_2O.$$

When one mole of HNO_3 is neutralised by 1 mole of NaOH, heat released is 57.1 kJ. Therefore, when 0.75 mole of HNO_3 is neutralised by 0.75 mole of NaOH, heat released will be

$$57.1 \times 0.75 = \mathbf{42.8 \text{ kJ}}.$$

(b) Out of 0.75 mole of HCl, 0.5 mole will react with 0.5 mole of KOH and 0.25 mole of HCl will be left unreacted.

$$\text{Heat released} = 57.1 \times 0.5 = \mathbf{28.5 \text{ kJ}}.$$

(c) 1000 ml of 1 M HCl = 1 mole

$$\therefore \quad 900 \text{ ml of } 0.5 \text{ M HCl} = \frac{900 \times 0.5}{1000} = 0.45 \text{ mole}$$

Similarly,

$$500 \text{ ml of } 0.4 \text{ M NaOH} = \frac{500 \times 0.4}{1000} = 0.2 \text{ mole}$$

Out of 0.45 mole of HCl, 0.2 mole of HCl reacts with 0.2 mole of NaOH

$$\therefore \quad \text{Heat evolved} = 57.1 \times 0.2 = 11.4 \text{ kJ}.$$

To calculate the rise in temperature, use the relation

Mass × Sp. heat × Rise in temp. = Heat evolved

Heat evolved is 11.4 kJ or 11400 J, mass of the mixture is 900 g + 500 g (assuming the densities of the solution as unity) and sp. heat is 4.18 J.

$$\therefore \quad \text{Rise in temp.} = \frac{\text{Heat evolved}}{\text{Mass} \times \text{Sp. heat}} = \frac{11.4 \times 1000}{1400 \times 4.18} = 1.95°$$

(d) 1 molecule of H_2SO_4 releases two H^+ ions

$$200 \text{ ml. of } 0.2 \text{ M } H_2SO_4 = \frac{2 \times 200 \times 0.2}{1000} \text{ moles of } H^+ = 0.08 \text{ moles of } H^+$$

300 ml of 0.1 M NaOH = 0.03 mole OH^-

Out of 0.08 mole of H^+, 0.03 mole H^+ will react with 0.03 mole OH^-.

$$\therefore \quad \text{Heat evolved} = 57.1 \times 0.03 = 1.7 \text{ kJ}.$$

$$\therefore \quad \text{Rise in temperature} = \frac{1.7 \times 1000}{500 \times 4.18} = 0.8°.$$

Example 15: Calculate the amount of heat evolved when:

(i) 600 cm^3 of 0.1 M hydrochloric acid is mixed with 200 cm^3 of 0.2 M sodium hydroxide solution.

(ii) 250 cm^3 of 0.2 M sulphuric acid is mixed with 400 cm^3 of 0.5 M potassium hydroxide solution.

Assuming that the specific heat of water is 4.18 J K^{-1} g^{-1} and ignoring the heat absorbed by the container, thermometer, stirrer, etc., what would be the rise in temperature in each of the above cases?

Solution: (i)

$$600 \text{ cm}^3 \text{ of } 0.1\text{M HCl} = \frac{0.1}{1000} \times 600 \text{ mole of HCl}$$

$$= 0.06 \text{ mole of HCl} = 0.06 \text{ mole of } H^+ \text{ ions}$$

$$200 \text{ cm}^3 \text{ of } 0.2\text{M NaOH} = \frac{0.2}{1000} \times 200 \text{ mole of NaOH}$$

$$= 0.04 \text{ mole of NaOH} = 0.04 \text{ mole of OH}^- \text{ ions}$$

Thus, 0.04 mole of H^+ ions will combine with 0.04 mole of OH^- ions to form 0.04 mole of H_2O and 0.02 mole of H^+ ions will remain unreacted.

Heat evolved when 1 mole of H^+ ions combines with 1 mole of OH^- ions = 57.1 kJ.

∴ Heat evolved when 0.04 mole of H^+ ions combines with 0.04 mole of OH^- ions.

$$= 57.1 \times 0.04 = \mathbf{2.284 \text{ kJ}}.$$

(ii) $$250 \text{ cm}^3 \text{ of } 0.2\text{M H}_2\text{SO}_4 = \frac{0.2}{1000} \times 250 \text{ mole of H}_2\text{SO}_4$$

$$= 0.05 \text{ mole of H}_2\text{SO}_4 = 0.10 \text{ mole of H}^+ \text{ ions}$$

$$400 \text{ cm}^3 \text{ of } 0.5\text{M KOH} = \frac{0.5}{1000} \times 400 \text{ mole of KOH}$$

$$= 0.2 \text{ mole of KOH} = 0.2 \text{ mole of OH}^- \text{ ions}$$

Thus, 0.10 mole of H^+ ions will neutralize 0.10 mole of OH^- ions, (out of 0.2 mole of OH^- ions) to form 0.10 mole of H_2O.

Hence, heat evolved = $57.1 \times 0.1 = \mathbf{5.71 \text{ kJ}}$.

In case (i), heat produced = 2.284 kJ = 2284 J

Total mass of the solution = 600 + 200 = 800 g

Specific heat = 4.18 $JK^{-1}g^{-1}$

$$Q = m \times s \times \Delta t$$

∴ $$\Delta t = \frac{Q}{m \times s} = \frac{2284}{800 \times 4.18} = \mathbf{0.68°}.$$

In case (ii), heat produced = 5710 J

Total mass of the solution = 250 + 400 = 650 g

∴ $$\Delta t = \frac{Q}{m \times s} = \frac{4568}{650 \times 4.18} = \mathbf{1.68°}.$$

1.17 DIFFERENT KINDS OF HEAT (OR ENTHALPY) OF REACTION

1. Heat of Solution (or Enthalpy of solution): *The heat of solution of a substance in a particular solvent is defined as the heat change (i.e., the amount of heat evolved or absorbed) when 1 mole of the substance is dissolved in such a large volume of the solvent that further addition of the solvent does not produce any more heat change.*

Integral and Differential Enthalpy of Solution

There are two kinds of heat of solution as discussed below:

Integral heat of solution: *It is the heat charge that occurs when a specific quantity of solute is dissolved in a specified quantity of a specified solvent under conditions of constant temperature and pressure.*

The state of solution of a substance is described in terms of moles of solvent used per mole of solute dissolved at atmospheric pressure and room temperature.

If the amount of solvent used is so large that a further addition of solvent no longer causes a change in the integral heat of solution per mole of the solute, the integral heat of solution is said to be heat of solution at infinite dilution.

Differential heat of solution: *Differential heat of solution of a solute at a specified concentration of a solution for a specified temperature and pressure condition is the heat change observed per further mole of solute dissolved in such a large quantity of solution that no significant change in concentration occurs.*

Mathematically, differential heat of solution of a solute for a given solution, ΔH_1 is given by

$$\left[\frac{\partial(\Delta H)}{\partial n_{solute}}\right]$$

It is given by the partial molal enthalpy of solute in the solution less the molal enthalpy of solute used.

$$\Delta H_1 = \left[\frac{\partial(\Delta H)}{\partial n_1}\right]_{n_2}$$

Where n_1 and n_2 stand for number of moles of solute and solvent respectively. Differential heat of solution at a given concentration of solute is slope of the tangent to the ΔH vs n_{solute} curve for a specified amount of the solvent, $n_{solvent}$.

The thermochemical equations for the dissolution of KCl and $CuSO_4$ in water may represented as

$$KCl\ (s) + aq \longrightarrow KCL\ (aq), \Delta H = +18.6\ kJ$$
$$CuSO_4\ (s) + aq \longrightarrow CuSO_4\ (aq), \Delta H = -66.5\ kJ$$

Thus,

the heat of solution of $KCl = +18.6$ kJ mol^{-1}

and heat of solution of $CuSO_4 = -66.5$ kJ mol^{-1}

It is interesting to note that the salts like copper sulphate, calcium chloride, etc., when present in the hydrated state (*i.e.*, $CuSO_4.5H_2O$, $CaCl_2.6H_2O$, etc.) dissolve with the absorption of heat. For example,

$$CuSO_4.5H_2O + aq \longrightarrow CuSO_4\ (aq), \Delta H = +11.7\ kJ$$

Thus, it can be generalized that the process of dissolution is usually endothermic for :

(*i*) salts which do not form hydrates like NaCl, KCl, KNO_3, etc.

(*ii*) hydrated salts like $CuSO_4.5H_2O$, $CaCl_2.6H_2O$, etc.

2. Heat of Hydration. *The amount of heat change (i.e., the heat evolved or absorbed) when one mole of the anhydrous salt combines with the required number of moles of water so as to change into the hydrated salt, is called the heat of hydration.*

For example, the heat of hydration of copper sulphate is -78.2 kJ mol^{-1}. This is represented as follows:

$$CuSO_4\ (s) + 5H_2O \longrightarrow CuSO_4.5H_2O\ (s), \Delta H = -78.2\ kJ$$

3. Heat of Fusion. *Heat of fusion is the heat change that takes place when one mole of a solid substance changes into its liquid state at its melting point.*

Heat of fusion (ΔH_{fus}) of ice (m.p. = 273 K) is 6.0 kJ mol^{-1}. It may be represented as follows:

$$\underset{\text{Ice}}{H_2O(s)} \longrightarrow \underset{\text{Water}}{H_2O(l)}, \quad \Delta H = +6.0 \text{ kJ}$$

4. Heat of Vaporisation. *It is the heat change that takes place when one mole of a liquid changes into its gaseous state at its boiling point.*

For example, the heat of vaporisation ($\Delta H_{vap.}$) of water into its gaseous state at the boiling point (373 K) is 40.7 kJ. It may be represented as follows:

$$\underset{\text{Water}}{H_2O(l)} \longrightarrow \underset{\text{Steam}}{H_2O(g)}, \quad \Delta H = +40.7 \text{ kJ}$$

5. Heat of Sublimation. *Heat of sublimation of a substance is the heat change that takes place when 1 mole of a solid changes directly into vapour phase at a given temperature below its melting point.*

Sublimation is a process in which a solid on heating changes directly into gaseous state below its melting point. For example, the heat of sublimation of iodine is 62.39 kJ mol^{-1}. It is represented as follows:

$$I_2(s) \longrightarrow I_2(g), \Delta H = +62.39 \text{ kJ}$$

Most solids, that sublime readily are molecular solids, *e.g.*, iodine and naphthalene, etc.

1.18 HESS'S LAW OF CONSTANT HEAT SUMMATION

G.H. Hess, in 1840, enunciated a law about the heats of reactions which states as follows:

The total amount of heat evolved or absorbed in a reaction is the same whether the reaction takes place in one step or in a number of steps. In other words, the total amount of heat change in a reaction depends only upon the nature of the initial reactants and the nature of the final products and not upon the path or the manner by which this change is brought about.

This is known as Hess's law of constant heat summation.

G.H. Hess (1802–1850)
G.H. Hess was a Swiss born Russian chemist. He is noted today for two fundamental laws of thermochemistry — the law of constant heat summation and the law of thermoneutrality

Hess's law may be illustrated with the help of following examples:

(1) Sulphur (Rhombic) burns to form SO_3 to evolve 395.4 kJ of heat.

$$S_R + \frac{3}{2}O_2(g) \longrightarrow SO_3(g), \Delta H = -395. \text{ kJ}$$

Sulphur may change to SO_3 in two steps as given below:

$$S_R + O_2(g) \longrightarrow SO_2(g), \Delta H = -297.5 \text{ kJ}$$

$$SO_2(g) + \frac{1}{2}O_2(g) \longrightarrow SO_3(g), \Delta H = -97.9 \text{ kJ}$$

(See accompanying figure)

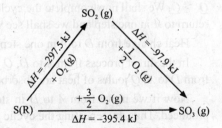

Total heat evolved in the two steps is $-297.5 + (-97.9) = -395.4$ kJ.

This is the same as for the direct reaction in one direct step.

(2) When carbon (graphite) burns to form carbon dioxide directly in one step, 393.5 kJ mol^{-1} of heat is evolved.

$$C(s) + O_2(g) \longrightarrow CO_2(g), \Delta H = -393.5 \text{ kJ}$$

If carbon burns to form carbon monoxide first which then burns to form carbon dioxide, the heats evolved in the two steps are as follows (see accompanying figure).

$$C(s) + \frac{1}{2}O_2(g) \longrightarrow CO(g), \Delta H = -111.5 \text{ kJ}$$

$$CO(g) + \frac{1}{2}O_2(g) \longrightarrow CO_2(g), \Delta H = -282.0 \text{ kJ}$$

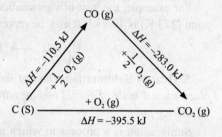

Thus, the total heat evolved in the two steps comes out to be $(-111.5) + (-282.0) = -393.5$ kJ, which is the same when the reaction takes place directly in one step.

Proof of Hess's Law. Consider the general reaction: A ⟶ D

Suppose the heat evolved in this reaction directly is Q joules.

Now suppose the same reaction takes place in three steps as follows:

$$A \xrightarrow{\text{Step 1}} B \xrightarrow{\text{Step 2}} C \xrightarrow{\text{Step 3}} D$$

Suppose the heats evolved in these three steps are q_1, q_2 and q_3 joules respectively (Fig. 1.11).

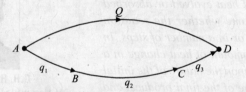

Fig. 1.11: Theoretical proof of Hess's law

Thus, the total heat evolved from A to D = $q_1 + q_2 + q_3 = Q'$ joules (say)

According to Hess's law, we must have $Q = Q'$

Suppose Hess's law is not correct. In that case either $Q' < Q$ or $Q' > Q$. Let us suppose $Q' > Q$. We shall now complete the cycle, *i.e.*, we shall go from A to D through three steps and return to A in one step and we shall see what heat changes take place.

Heat change from D to A in one step is reverse of heat change from A to D in one step.

In the direct process from A to D, Q joules of heat are evolved. Therefore, in direct process from D to A, Q joules of heat are absorbed.

Now if we move from A to D in steps and return to A directly. $Q' - Q$ joules of heat is produced. Thus, by repeating the cyclic process a number of times, a large amount of heat can be created. This is, however, against the law of conservation of energy. Hence, Q' cannot be greater than Q. Similarly, Q' cannot be less than Q. It has to be equal to Q. Thus, Hess's law is proved.

Thermodynamically also, Hess's law is justified. This is because enthalpy change is a state function. It depends upon the enthalpy difference in the initial and final states only and not upon the manner how the final state was reached. Thus whether S is converted to SO_3 directly or through the steps (*i*) $S \longrightarrow SO_2$ and (*ii*) $SO_2 \longrightarrow SO_3$, net enthalpy change or heat of reaction is the same.

1.18.1 Applications of Hess's Law

Hess's law is very useful in determining the heats (enthalpies) of many reactions, which cannot otherwise be measured by direct methods. The thermochemical equations are added, subtracted, multiplied or divided just like other algebraic equations. Some important applications of Hess's law are:

(*i*) **Determination of enthalpy of formation.** The heat of formation of some compounds can not be determined directly by experiment. This can be calculated by the application of Hess's law.

Suppose we want to determine the heat of formation of CO

$$C + \frac{1}{2}O_2 \longrightarrow CO; \Delta H = ?$$

By direct experiment, it is not possible because in the reaction between carbon and oxygen, we may get a mixture of CO and CO_2. But we require only CO. Therefore, from the experiment it is not possible to determine the heat of formation.

We know experimentally, the heat of combustion of pure carbon and carbon monoxide. It is represented by the following equations:

(*a*) $\quad\quad\quad\quad C + O_2 \longrightarrow CO_2; \Delta H = -393.5 \text{ kJ}$

(*b*) $\quad\quad\quad\quad CO + \frac{1}{2}O_2 \longrightarrow CO_2; \Delta H = -283.0 \text{ kJ}$

Now, applying Hess's law on (*a*) and (*b*), we get the required equation for the heat of formation of CO. This can be achieved by subtracting (*b*) from (*a*), *i.e.*, by carrying out (*a* – *b*).

$$C + \frac{1}{2}O_2 \longrightarrow CO; \Delta H = -110.5 \text{ kJ}$$

Therefore, enthalpy of formation of CO is – 110.5 kJ.

(*ii*) **Determination of heat (enthalpy) of transition.** Transition of an element from one allotropic form to another is a slow process and requires heating of one allotropic form of the element for a long time at constant temperature. Hence, enthalpies of transition of elements cannot be determined experimentally. They are, however, determined indirectly by means of Hess's law. Thus, Hess's law is useful in determining the enthalpies of transition of the allotropic forms such as rhombic sulphur to monoclinic sulphur, yellow phosphorus to red phosphorus, graphite to diamond, etc.

Suppose we want to determine the enthalpy of transition of sulphur (rhombic) into sulphur (monoclinic). Each variety is combustioned and heats of combustion are found out. This is represented by the following equations:

(*a*) $\quad\quad\quad\quad S(R) + O_2 \longrightarrow SO_2; \Delta H = -295.1 \text{ kJ}$

(*b*) $\quad\quad\quad\quad S(M) + O_2 \longrightarrow SO_2; \Delta H = -296.4 \text{ kJ}$

Subtracting (*b*) from (*a*), we get

$$S(R) \longrightarrow S(M); \Delta H = -295.1 - (-296.4) = +1.3 \text{ kJ}$$

Thus, transition of one gram atom of rhombic sulphur to monoclinic sulphur absorbs 1.3 kJ of heat.

(iii) **Calculation of enthalpies from given data.** Thermochemical equations can be added, subtracted, multiplied or divided by any numerical factor like algebraic equations. Hess's law is, thus, used for the calculation of enthalpy changes for the reactions for which direct experimental measurements are not available.

Solved Examples

Example 16: Calculate the heat of formation of acetic acid if its heat of combustion is 867 kJ mol^{-1}. The heats of formation of CO_2 (g) and H_2O (l) are –393.5 kJ mol^{-1} and –285.9 kJ mol^{-1} respectively.

Solution: The equations given are:

(a) $CH_3COOH + 2O_2 \longrightarrow 2CO_2 + 2H_2O$; $\quad\quad \Delta H = -867$ kJ

(b) $C + O_2 \longrightarrow CO_2$; $\quad\quad \Delta H = -393.5$ kJ

(c) $H_2 + \frac{1}{2}O_2 \longrightarrow H_2O$; $\quad\quad \Delta H = -205.9$ kJ

Our aim is to find out ΔH value for the reaction

$$2C + 2H_2 + O_2 \longrightarrow CH_3COOH\ ;\ \Delta H = ?$$

To obtain this equation, multiply equations (b) and (c) by 2 and add. We get equation (d)

$$2C + 2O_2 + 2H_2 + O_2 \longrightarrow 2CO_2 + 2H_2O$$

or

(d) $\quad 2C + 2H_2 + 3O_2 \longrightarrow 2CO_2 + 2H_2O$; $\quad\quad \Delta H = -1358.8$ kJ

Subtracting (a) from (d), we get

$2C + 2H_2 + 3O_2 \longrightarrow 2CO_2 + 2H_2O$; $\quad\quad \Delta H = -1358.8$ kJ
$CH_3COOH + 2O_2 \longrightarrow 2CO_2 + 2H_2O$; $\quad\quad \Delta H = -867$ kJ
― ― ― ― ―
$2C + 2H_2 + O_2 \longrightarrow CH_3COOH$; $\quad\quad \Delta H = -491.8$ kJ

∴ Heat of formation of acetic acid = **– 491.8 kJ**.

Example 17: Find the heat of formation of SO_2 from the following thermochemical equations

(a) $\quad S\ (s) + \frac{3}{2}O_2\ (g) \longrightarrow SO_3\ (g) + X$ kJ

(b) $\quad SO_2\ (s) + O_2\ (g) \longrightarrow SO_3\ (g) + Y$ kJ

Solution: Our aim is to find out ΔH value for the reaction

$$S + O_2 \longrightarrow SO_2 = \Delta H = ?$$

This can be obtained by subtracting (b) from (a), *i.e.,* by carrying out (a – b)

(a) $\quad S\ (s) + \frac{3}{2}O_2\ (g) \longrightarrow SO_3\ (g)$; $\quad\quad \Delta H = -X$ kJ

(b) $\quad SO_2\ (g) + O_2\ (g) \longrightarrow SO_3\ (g)$; $\quad\quad \Delta H = -Y$ kJ
― ― ―
$S + O_2 \longrightarrow SO_2\ (g)$ $\quad \Delta H = -X - (-Y) = -X + Y = -(X - Y)$ kJ

Example 18: Calculate enthalpy of formation of anhydrous Al_2Cl_6 from the following data:

(a) $2Al (s) + 6 HCl (aq) \longrightarrow Al_2Cl_6 (aq) + 3H_2$; $\Delta H = 1004.4$ kJ
(b) $H_2 (g) + Cl_2 (g) \longrightarrow 2HCl (g)$; $\Delta H = 754.74$ kJ
(c) $HCl (g) + aq \longrightarrow HCl (aq)$; $\Delta H = -73.24$ kJ
(d) $Al_2Cl_6 (s) + aq \longrightarrow Al_2Cl_6 (aq)$; $\Delta H = -642.69$ kJ

Solution: Thermochemical equation for the formation of anhydrous Al_2Cl_6 is

$$2Al + 3Cl_2 \longrightarrow Al_2Cl_6 (s); \Delta H = ?$$

To obtain ΔH value for this equation from the given data we perform as

$$[a + 3 \times b + 6 \times c] - d$$

Rearranging the given equations, we get

$2Al (s) + 6HCl (aq) + 3H_2 + 3Cl_2 \longrightarrow Al_2Cl_6 (aq) + 3H_2 (g) + 6HCl (g)$
$+ 6 HCl (g) + aq - Al_2Cl_6 (s) - aq \quad\quad + 6HCl (aq) - Al_2Cl_6 (aq)$

$2Al(s) + 3Cl_2 (g) \longrightarrow Al_2Cl_6 (s), \Delta H = 1004.4 + 3 \times 754.74 + 6$
$\times (73.24) - (-642.69) = 3471.87$ kJ

Example 19: Calculate enthalpy of formation of methane. Given:

(a) $C (graphite) + O_2 (g) \longrightarrow CO_2 (g)$; $\Delta H = -393.5$ kJ
(b) $2H_2 (g) + O_2 (g) \longrightarrow 2H_2O (l)$; $\Delta H = -571.8$ kJ
(c) $CH_4 (g) + 2O_2 (g) \longrightarrow CO_2 (g) + 2H_2O (l)$; $\Delta H = -890.3$ kJ

Solution: We want to find out the value of ΔH for the reaction

$$C (graphite) + 2H_2 (g) \longrightarrow CH_4; \Delta H = ?$$

This can be obtained by the operation $(a + b) - c$, we get

$C (graphite) + O_2 + 2H_2 + O_2 \longrightarrow CO_2 + 2H_2O$ ...$(a+b)$

or $\quad C (graphite) + 2O_2 + 2H_2 \longrightarrow CO_2 + 2H_2O$; $\Delta H = -965.3$ kJ
$\quad\quad\quad CH_4 + 2O_2 \longrightarrow CO_2 + 2H_2O$; $\Delta H = -890.3$ kJ
$\quad\quad\quad\quad\quad - \quad\quad\quad - \quad\quad\quad -$
_____

$\quad\quad\quad C (graphite) + 2H_2 \longrightarrow CH_4 \quad\quad \Delta H = -75.0$ kJ ...$(a+b) - c$

Example 20: From the thermochemical equations:

(a) $N_2 (g) + 3H_2 (g) \longrightarrow 2NH_3 (g)$; $\Delta H = -92.40$ kJ
(b) $2H_2 (g) + O_2 (g) \longrightarrow 2H_2O$; $\Delta H = -483.79$ kJ

Find ΔH_f for NH_3 (g) and ΔH_f for H_2O (g).

Solution: Heat of formation (ΔH_f) is for one mole

$\therefore \quad\quad \Delta H_f$ for $NH_3 = \dfrac{-92.4}{2} = -46.2$ kJ

$\therefore \quad\quad \Delta H_f$ for $H_2O = \dfrac{-483.79}{2} = -241.895$ kJ.

Example 21: Calculate the heat of formation of carbon disulphide given that the heats of combustion of carbon disulphide, carbon and sulphur are –1109 kJ, –394.6 kJ and –298.7 kJ respectively.

Solution: The thermochemical equations for the combustion of CS_2, C and S are

(a) $CS_2 + 3O_2 \longrightarrow CO_2 + 2SO_2$; $\Delta H = -1109$ kJ
(b) $C + O_2 \longrightarrow CO_2$; $\Delta H = -394.6$ kJ
(c) $S + O_2 \longrightarrow SO_2$; $\Delta H = -298.7$ kJ

Our aim is to find out the value of ΔH for the following equation,

$$C + 2S \longrightarrow CS_2; \quad \Delta H = ?$$

Multiplying (c) by 2 and adding to (b), we get

$$2S + 2O_2 + C + O_2 \longrightarrow 2SO_2 + CO_2$$

or

(d) $2S + C + 3O_2 \longrightarrow 2SO_2 + CO_2$; $\Delta H = (-298.7) \times 2 + (-394.6) = -992.0$ kJ

Subtracting (a) from (d), we get

$2S + C + 3O_2 \longrightarrow 2SO_2 + CO_2$; $\Delta H = -992.0$ kJ
$CS_2 + 3O_2 \longrightarrow CO_2 + 2SO_2$; $\Delta H = -1109$ kJ

$2S + C \longrightarrow CS_2$; $\Delta H = -992 - (-1109) = +117$ kJ

∴ Heat of formation of CS_2 = **+ 117 kJ mol⁻¹**

Example 22: The enthalpy of combustion of ethylene (C_2H_4) is – 1410 kJ. What will be calorific value of ethylene per gram?

Solution: The thermochemical equation for the combustion of C_2H_4 is

$$\underset{\text{28 grams}}{C_2H_4} + O_2 \longrightarrow 2CO_2 + 2H_2O \; ; \; \Delta H = -1410 \text{ kJ}$$

It is evident from the above equation that

28 grams of C_2H_4 give heat = –1410 kJ

∴ 1 gram of C_2H_4 will give heat = $\dfrac{-1410}{28}$ = –50.36 kJ

∴ Calorific value of ethylene = **– 50.36 kJ.**

Example 23: One gram of an organic liquid A (mol. wt. = 128) liberates 150 joules of heat on solidification. Calculate the enthalpy of fusion of A.

Solution: Molecular weight of liquid A = 128 g

Heat evolved when 1 gram of liquid A solidifies = 150 J

∴ Heat evolved when 128 grams (1 mole) of liquid A solidify = 150 × 128 = 19200 J

$$\text{Liquid (A)} \longrightarrow \text{Solid (A)}; \Delta H = -19200 \text{ J}$$

Rearranging this equation, we get

$$\text{Solid (A)} \longrightarrow \text{Liquid (A)}; \Delta H = +19200$$

∴ Enthalpy of fusion of A = **+ 19200 joules.**

Example 24: Calculate the heat of formation of KCl from the following data :

(i) $KOH (aq) + HCl (aq) \longrightarrow KCl (aq) + H_2O (l)$; $\Delta H = -57.3$ kJ

(ii) $H_2 (g) + \dfrac{1}{2} O_2 (g) \longrightarrow H_2O (l)$; $\Delta H = -286.2$ kJ

(iii) $\dfrac{1}{2} H_2 (g) + Cl_2 (g) + aq \longrightarrow HCl (aq)$; $\Delta H = -164.4$ kJ

(iv) $K(s) + \frac{1}{2}O_2(g) + \frac{1}{2}H_2(g) + aq \longrightarrow KOH(aq);$ $\Delta H = -487.4$ kJ

(v) $KCl(s) + aq \longrightarrow KCl(aq);$ $\Delta H = +18.4$ kJ

Solution: Our aim is to find out ΔH for the following reaction.

(vi) $K(s) + \frac{1}{2}Cl_2(g) \longrightarrow KCl(s), \Delta H = ?$

In order to get this thermochemical equation, we follow the following two steps:

Step 1. Adding Eqns. (*iii*) and (*iv*) and subtracting Eqn. (*v*), we have

(vii) $K(s) + \frac{1}{2}Cl_2(g) + H_2(g) + \frac{1}{2}O_2(s) \longrightarrow KCl(s) + HCl(aq) - KCl(aq)$

$$\Delta H = -487.4 + (-164.4) - (18.4) = -670.2 \text{ kJ}.$$

Step 2. To cancel out the terms of this equation which do not appear in the required equation (*vi*), add Eqn. (*i*) to Eqn. (*vii*) and subtract Eqn. (*ii*) from their sum. This gives

$K(s) + \frac{1}{2}Cl_2(g) \longrightarrow KCl(s); \Delta H = -670.2 + 57.3 - (-286.2) = \mathbf{-441.3}$ **kJ.**

Example 25: Calculate the heat of hydration of anhydrous copper sulphate ($CuSO_4$) into hydrated copper sulphate ($CuSO_4.5H_2O$). Given that the heats of solutions of anhydrous copper sulphate and hydrated copper sulphate are -66.5 and $+11.7$ kJ mol^{-1} respectively.

Solution: Data provided is:

(i) $CuSO_4(s) + aq \longrightarrow CuSO_4(aq);$ $\Delta H = -66.5$ kJ
(ii) $CuSO_4.5H_2O(s) + aq \longrightarrow CuSO_4(aq);$ $\Delta H = +11.7$ kJ

Our aim is to find out heat of the following reaction.

$CuSO_4(s) + 5H_2O(l) \longrightarrow CuSO_4.5H_2O(s);$ $\Delta H = ?$

Equation (*i*) can be written in two steps as :

(iii) $CuSO_4(s) + 5H_2O(l) \longrightarrow CuSO_4.5H_2O(s);$ $\Delta H = q_1$ kJ
(iv) $CuSO_4.5H_2O(s) + aq \longrightarrow CuSO_4(aq);$ $\Delta H = q_2$ kJ

According to Hess's law $q_1 + q_2 = -66.5$ kJ

Further, equations (*ii*) and (*iv*) are same $\longrightarrow q_2 = +11.7$ kJ.

Putting this value in eq. (*viii*) above, we get

$q_1 + 11.7 = -66.5$ or $q_1 = -66.5 - 11.7$ kJ $= -78.2$ kJ

Thus, equation (*iii*) may be written as

$CuSO_4(s) + 5H_2O(l) \longrightarrow CuSO_4.5H_2O(s);$ $\Delta H = -78.2$ kJ

This is what we aimed at. Hence, the required value of the heat of hydration is $\Delta H = -78.2$ kJ.

Example 26: Calculate the heat of formation of carbon monoxide (CO) from the following data :

(i) $C(s) + O_2(g) \longrightarrow CO_2(g);$ $\Delta H = -393$ kJ
(ii) $CO(g) + \frac{1}{2}O_2(g) \longrightarrow CO_2(g);$ $\Delta H = -282$ kJ

Solution: Our aim is to find out ΔH for the following reaction:

(i) $\quad C(s) + \frac{1}{2} O_2(g) \longrightarrow CO(g); \Delta H = ?$

Subtracting equation (ii) from (i), we get

$C(s) + \frac{1}{2} O_2(g) - CO(g) \longrightarrow 0; \qquad \Delta H = -393.3 - (-282.8) = -110.5$ kJ

or $\quad C(s) + \frac{1}{2} O_2(g) \longrightarrow CO(g); \qquad \Delta H = -110.5$ kJ

∴ Heat of formation of CO, $\Delta H_f =$ **110.5 kJ**.

Example 27: Calculate the heat change accompanying the transformation of C (graphite) to C (diamond). Given that the heats of combustion of graphite and diamond are 393.5 and 395.4 kJ mol^{-1} respectively.

Solution: Data provided is:
(i) $\quad C(graphite) + O_2(g) \longrightarrow CO_2(g); \qquad \Delta H = -393.5$ kJ
(ii) $\quad C(diamond) + O_2(g) \longrightarrow CO_2(g); \qquad \Delta H = -395.4$ kJ

Our aim is to find out ΔH for the following reaction

$\qquad C(graphite) \longrightarrow C(diamond); \qquad \Delta H = ?$

Subtracting equation (ii) from (i), we get

$C(graphite) - C(diamond) \longrightarrow 0; \qquad \Delta H = -393.5 - (-395.4) = +1.9$ kJ

or $\qquad C(graphite) \longrightarrow C(diamond); \qquad \Delta H =$ **+ 1.9 kJ.**

Example 28: The heat of combustion of ethyl alcohol (C_2H_5OH) is 1380.7 kJ mol^{-1}. If the heats of formation of CO_2 and H_2O are 394.5 and 286.6 kJ mol^{-1} respectively, calculate the heat of formation of ethyl alcohol.

Solution: Data provided is :
(i) $\quad C_2H_5OH + 3O_2 \longrightarrow 2CO_2 + 3H_2O; \qquad \Delta H = -1380.7$ kJ
(ii) $\quad C + O_2 \longrightarrow CO_2; \qquad \Delta H = -394.5$ kJ
(iii) $\quad H_2 + \frac{1}{2} O_2 \longrightarrow H_2O; \qquad \Delta H = -286.6$ kJ

Our aim is to determine ΔH for the following reaction

$\qquad 2C + 3H_2 + \frac{1}{2} O_2 \longrightarrow C_2H_5OH$

In order to get this thermochemical equation multiply Eqn. (ii) by 2 and Eqn. (iii) by 3 and substract Eqn. (i) from their sum, i.e., operating : 2 × Eqn. (ii) + 3 × Eqn. (iii) – Eqn. (i), we get

$2C + 3H_2 + \frac{1}{2} O_2 \longrightarrow C_2H_5OH; \Delta H = 2(-394.5) + 3(-286.6) - (-1380.7) = -268.1$ kJ

Thus, the heat of formation of ethyl alcohol = $\Delta H_f =$ **–268.1 kJ mol^{-1}**.

Example 29: Calculate the standard enthalpy of formation of n-butane given that the standard enthalpies of combustion of *n*-butane (g), C (graphite) and $H_2(g)$ are – 2878.5 kJ mol^{-1}, 393.5 kJ mol^{-1} and – 285.3 kJ mol^{-1} respectively.

Solution: We aim at the equation

$\qquad 4C + 5H_2 \longrightarrow C_4H_{10}; \qquad \Delta H = ?$

The given data can be expressed in the form of the following equations.

(i) $\quad C_4H_{10} + 6\frac{1}{2}O_2 \longrightarrow 4CO_2 + 5H_2O;$ $\quad\quad \Delta H = -2878.5$ kJ

(ii) $\quad 4C + 4O_2 \longrightarrow 4CO_2;$ $\quad\quad \Delta H = 4 \times (-393.5)$ kJ

(iii) $\quad 5H_2 + \frac{5}{2}O_2 \longrightarrow 5H_2O;$ $\quad\quad \Delta H = 5 \times (-285.3)$ kJ

Add (ii) and (iii) and subtract (i)

$$(-1574.0) + (-1426.5) - (-2878.5) = -122.0 \text{ kJ}$$

Example 30: Heats of solution (ΔH) for $BaCl_2.2H_2O$ and $BaCl_2$ are 8.8 and -20.6 kJ respectively. Calculate the heat of hydration of $BaCl_2$ to $BaCl_2.2H_2O$.

Solution: We are given

$BaCl_2.2H_2O$ (s) + aq $\longrightarrow BaCl_2$ (aq); $\quad\quad \Delta H = 8.8$ kJ ...(i)

$BaCl_2$ (s) + aq $\longrightarrow BaCl_2$ (aq); $\quad\quad \Delta H = -20.6$ kJ ...(ii)

We aim at:

$BaCl_2$ (s) + $2H_2O \longrightarrow BaCl_2.2H_2O$ (s); $\quad\quad \Delta H = ?$ ...(iii)

Equation (ii) may be written in two steps as

$BaCl_2$ (s) + $2H_2O \longrightarrow BaCl_2.2H_2O$ (s); $\quad\quad \Delta H = \Delta H_1$ (say) ...(iv)

$BaCl_2.2H_2O$ (s) + aq $\longrightarrow BaCl_2$ (aq); $\quad\quad \Delta H = \Delta H_2$ (say) ...(v)

Then according to Hess's law

$\quad\quad \Delta H_1 + \Delta H_2 = -20.6$ kJ

But $\quad\quad \Delta H_2 = 8.8$ kJ $\quad\quad$ [$\because$ Equation (i) = Equation (v)]

$\therefore \quad\quad \Delta H_1 = -20.6 - 8.8 = -29.4$ kJ

But Equation (iii) = Equation (iv)

Hence, the heat of hydration of $BaCl_2 = -29.4$ **kJ**.

Example 31: Given the following thermochemical equations:

(i) $\quad$ S (rhombic) + O_2 (g) $\longrightarrow SO_2$ (g); $\quad\quad \Delta H = -297.5$ kJ

(ii) $\quad$ S (monoclinic) + O_2 (g) $\longrightarrow SO_2$ (g); $\quad\quad \Delta H = -300.0$ kJ

Calculate ΔH for the transformation of one gram atom of rhombic sulphur into monoclinic sulphur.

Solution: We aim at:

$\quad\quad$ S (rhombic) $\longrightarrow$ S (monoclinic); $\quad\quad \Delta H = ?$

Equation (i) – Equation (ii) gives

S (rhombic) – S (monoclinic) $\longrightarrow 0;$ $\quad\quad \Delta H = 297.5 - (-300.0) = 2.5$ kJ

or $\quad\quad$ S (rhombic) $\longrightarrow$ S (monoclinic); $\quad\quad \Delta H = +2.5$ **kJ.**

Thus, for the transformation of one gram atom of rhombic sulphur into monoclinic sulphur, 2.5 kJ of heat is absorbed.

Example 32: Calculate the enthalpy of formation of N_2O_5(g):

$$N_2 \text{ (g)} + \frac{5}{2}O_2 \text{ (g)} \longrightarrow N_2O_5 \text{ (g)}$$

The following information is given:

(i) $N_2(g) + O_2(g) \longrightarrow 2NO(g)$; $\Delta H° = +180$ kJ
(ii) $4NO_2(g) + O_2(g) \longrightarrow 2N_2O_5(g)$; $\Delta H° = -120.6$ kJ
(iii) $2NO(g) + O_2(g) \longrightarrow 2NO_2(g)$; $\Delta H° = -114.0$ kJ

Solution: Perform the operation $(i) + \dfrac{1}{2}(iii)$ to get the desired equation as

$$N_2(g) + \dfrac{5}{2}O_2(g) \longrightarrow N_2O_5(g)$$

Applying Hess's law, perform the same operation on enthalpies to obtain the enthalpy of formation of $N_2O_5(g)$

$$\Delta H_f = [180 - \dfrac{1}{2}(120.6) - 114.0] \text{ kJ} = 5.7 \text{ kJ}$$

Problems for Practice

1. Calculate the enthalpy of formation of acetic acid from the following data:
   (i) $C(s) + O_2(g) \longrightarrow CO_2(g)$, $\Delta H = -393.7$ kJ
   (ii) $H_2(g) + \dfrac{1}{2}O_2(g) \longrightarrow H_2O(l)$, $\Delta H = -285.8$ kJ
   (iii) $CH_3COOH(l) + 2O_2(g) \longrightarrow 2CO_2(g) + 2H_2O(l)$, $\Delta H = -873.2$ kJ
   [Ans. -485.8 kJ mol^{-1}]

2. Calculate the standard enthalpy of formation of SO_3 at 298 K using the following reactions and enthalpies.
   $S_8(s) + 8O_2(g) \longrightarrow 8SO_2(g)$, $\Delta H° = -2775$ kJ mol^{-1}
   $2SO_2(g) + O_2(g) \longrightarrow 2SO_3(g)$, $\Delta H° = -198$ kJ mol^{-1}
   [Ans. -445.9 kJ mol^{-1}]

3. From the following thermochemical equations, calculate the standard enthalpy of formation of HCl (g).
   (A) $H_2(g) \longrightarrow 2H(g)$, $\Delta H = +436.0$ kJ mol^{-1}
   (B) $Cl_2(g) \longrightarrow 2Cl(g)$, $\Delta H = +242.7$ kJ mol^{-1}
   (C) $HCl(g) \longrightarrow H(g) + Cl(g)$, $\Delta H = +431.8$ kJ mol^{-1}
   [Ans. -92.45 kJ mol^{-1}]

4. Calculate the enthalpy of formation of acetic acid if the enthalpy of combustion to $CO_2(g)$ and $H_2O(l)$ is -867.0 kJ mol^{-1} and enthalpies of formation of $CO_2(g)$ and $H_2O(l)$ are respectively -393.5 and -285.9 kJ mol^{-1}
   [Ans. -491.8 kJ mol^{-1}]

5. Calculate the enthalpy of formation of sucrose ($C_{12}H_{22}O_{11}$) from the following data:
   (i) $C_{12}H_{22}O_{11} + 12O_2 \longrightarrow 12CO_2 + 11H_2O$; $\Delta H = -5200.7$ kJ mol^{-1}
   (ii) $C + O_2 \longrightarrow CO_2$, $\Delta H = -394.5$ kJ mole^{-1}
   (iii) $H_2 + \dfrac{1}{2}O_2 \longrightarrow H_2O$, $\Delta H = -285.8$ kJ mol^{-1}
   [Ans. -2677.1 kJ mol^{-1}]

6. Ethylene on combustion gives carbon dioxide and water. Its enthalpy of combustion is 1410.0 kJ mol^{-1}. If the enthalpy of formation of CO_2 and H_2O are 393.3 kJ and 286.2 kJ respectively. Calculate the enthalpy of formation of ethylene. **[Ans. + 51.0 kJ mol^{-1}]**

7. Calculate the enthalpy of formation of benzene, given that enthalpies of combustion of benzene, carbon and hydrogen are –3281.5 kJ, –394.9 kJ and –286.1 kJ/mol, respectively.
[Ans. +53.8 kJ mol^{-1}]

8. Calculate the enthalpy of formation of *n*-butane from the following data:
   (*i*) $2C_4H_{10}$ (g) + $13O_2$ (g) $\longrightarrow$ 8 CO_2 (g) + 10 H_2O (l), $\Delta H = -5757.2$ kJ mol^{-1}
   (*ii*) C (s) + O_2 (g) $\longrightarrow$ CO_2 (g), $\Delta H = -405.4$ kJ mol^{-1}
   (*iii*) $2H_2$ (g) + O_2 (g) $\longrightarrow$ $2H_2O$ (l), $\Delta H = -572.4$ kJ mol^{-1}
   On what law are your calculations based? **[Ans. –174 kJ mol^{-1}]**

9. Calculate the enthalpy of formation of anhydrous Al_2Cl_6 from the following data:
   (*i*) 2Al (s) + 6HCl (aq) $\longrightarrow$ Al_2Cl_6 (aq) + 3 H_2 (g) + 1004.2 kJ mol^{-1}
   (*ii*) H_2 (g) + Cl_2 (g) $\longrightarrow$ 2HCl (g) + 184.1 kJ mol^{-1}
   (*iii*) HCl (g) + aq $\longrightarrow$ HCl (aq) + 73.2 kJ mol^{-1}
   (*iv*) $2Al_2Cl_6$ (s) + aq $\longrightarrow$ Al_2Cl_6 (aq) + 643.1 kJ mol^{-1} **[Ans. –1352.6 kJ mol^{-1}]**

10. Calculate the enthalpy of formation of carbon disulphide given that the enthalpy of combustion of carbon disulphide is 110.2 kJ mol^{-1} and those of sulphur and carbon are 297.4 kJ and 394.5 kJ/g atom respectively. **[Ans. –879.1 kJ mol^{-1}]**

1.18.2 Relationship Between Heat of Reaction at Constant Volume and That at Constant Pressure

We have the relation of first law of thermodynamics
$$\Delta U = q + w$$
If the work involved is that of expansion of a gas the above equation reduces to
$$\Delta U = q - P dV \text{ or } q = \Delta U + P \Delta V$$
(*a*) If volume remains constant during the reaction, $\Delta V = 0$
Hence $\qquad q_v = \Delta U$ ...(*i*)

i.e., heat of reaction at constant volume is equal to internal energy change

(*b*) If the pressure remains constant during the reaction,
$\qquad q_p = \Delta U + P \Delta V$ ...(*ii*)
or $\qquad q_p = \Delta H \qquad (\because \Delta U + P \Delta V = \Delta H)$
or $\qquad \Delta H = \Delta U + P \Delta V$ ...(*iii*)

Substituting ΔV by $V_2 - V_1$ in eq. (*ii*), we have
$$q_p = \Delta U + P(V_2 - V_1) = \Delta U + (PV_2 - PV_1)$$
For an ideal gas $PV = nRT$, therefore, we have
$\qquad PV_1 = n_1 RT$ ...(*iv*)
$\qquad PV_2 = n_2 RT$ ...(*v*)

where n_1 and n_2 stand for number of moles of gaseous reactants and gaseous products respectively. Substituting for PV_1 and PV_2 in eq. (*iii*), we have
$$\Delta H = \Delta U + (n_2 RT - n_1 RT) = \Delta U + (n_2 - n_1) RT$$
or $\qquad \Delta H = \Delta U + \Delta n_g RT$ ... (*vi*)

Δn_g denotes the difference between the number of moles of gaseous products and gaseous reactants.

Substituting the value of ΔU from (i), eq. (vi) can be written as

$$q_p = q_v + \Delta n_g RT$$

Special Conditions

(a) When a reaction is carried out in a closed vessel so that there is no change in volume, $\Delta V = 0$.
(b) In case of reactions involving only solids or only liquids or only solutions but no gaseous reactants or products, $q_p = q_v$.
(c) In case of reaction involving same number of moles of gaseous reactants and products ($\Delta n_g = 0$), the values of q_p and q_v are the same.

Heat of combustion is determined using a bomb calorimeter. We have the relation

$$H = U + PV \text{ or } \Delta H = \Delta U + P\Delta V + VdP \quad \ldots \text{(vii)}$$

At constant volume $\Delta V = 0$, hence eq. (vii) reduces to $\Delta H = \Delta U + VdP$
At constant pressure $\Delta P = 0$, hence eq. (vii) reduces to $\Delta H = \Delta U + PdV$.

Solved Example

Example 33: When NH_4NO_2 (s) decomposes at 373 K, it forms N_2 (g) and H_2O (g). The ΔH for the reaction at one atmosphere pressure and 373 K is -223.6 kJ mol^{-1} of NH_4NO_2 (s) decomposed. What is the value of ΔU for the reaction under the same conditions? Given R = 8.31 JK^{-1}mol^{-1}.

Solution: The reaction involved is

$$NH_4NO_2 \text{ (s)} \longrightarrow N_2 \text{ (g)} + 2H_2O \text{ (g)}$$

In this reaction, there is no gaseous substance in the reactants and there are two gaseous substances in the products.

$$\Delta n_g = n_p - n_r = 1 + 2 - 0 = 3$$
$$\Delta H \text{ (or } q_p) = -223.6 \text{ kJ mol}^{-1}, T = 373 \text{ K}$$
$$R = 8.3 \text{ JK}^{-1}\text{mol}^{-1} = 8.3 \times 10^{-3} \text{ kJ K}^{-1}\text{mol}^{-1}$$
$$\Delta U = \Delta H - \Delta n_g RT \quad \text{Substituting the values, we have}$$
$$= -223.6 - 3 \times 8.3 \times 10^{-3} \times 373 = -223.6 - 9.28$$
$$= -232.9 \text{ kJ mol}^{-1}$$

Problems for Practice

1. The enthalpy change (ΔH) for the reaction N_2 (g) + $3H_2$ (g) $\longrightarrow$ $2NH_3$ (g) is -92.38 kJ at 298 K. What is ΔU for the reaction? **[Ans. -87.42 kJ]**
2. The heat of combustion of benzene in a bomb calorimeter (constant volume) was found to be 3263.9 kJ mol^{-1} at 25°C. Claculate the heat of combustion of benzene at constant pressure. **[Ans. -3267.6 kJ mol^{-1}]**
3. The internal energy change (ΔU) for the reaction CH_4 (g) + $2O_2$ (g) $\longrightarrow$ CO_2 (g) + $2H_2O$ (l) is -885 kJ mol^{-1} at 298 K. What is ΔH at 298 K? **[Ans. -889.96 kJ mol^{-1}]**
4. When 0.532 g of benzene, boiling point 353 K is burnt with excess of oxygen in a constant volume system, 22.3 kJ of heat is given out. Calculate ΔH for the combustion process. **[Ans. -3274.2 kJ mol^{-1}]**

1.19 BOND ENTHALPY OR BOND ENERGY

The bond energy of a particular bond is defined as the average amount of energy released when one mole of bonds are formed from isolated gaseous atoms or the amount of energy required when 1 mole of bonds are broken so as to get the separated gaseous atoms.

For diatomic molecules (like H_2, HCl, etc.), the bond energy is equal to the dissociation energy of the molecules but for a *polyatomic molecule* like CH_4, the bond dissociation energies of the four C — H bonds are different. Hence, an average value is taken. Bond energies of some common bonds are given in Table 1.3.

Heat of reaction can be calculated from bond energy data. The following formula can be used directly in these calculations.

$\Delta H_{reaction} = \Sigma$ Bond energies of reactants $- \Sigma$ Bond energies of products

TABLE 1.3 *Bond energies of some common bonds*

Bond	Bond energy (kJ mol^{-1})	Bond	Bond energy (kJ mol^{-1})
H — H	436	C — H	414
H — F	565	O — H	463
H — Cl	431	N — H	389
H — Br	364	C — C	347
H — I	297	C = C	619
F — F	155	C ≅ C	812
Cl — Cl	242	C — Cl	326
Br — Br	190	C — O	335
I — I	149	C = O	707
O = O	494	C — N	293
N ≅ N	941	C = N	616
		C ≅ N	879

Solved Examples—Calculation of Bond energy

Example 34: Calculate the enthalpy change for the reaction

$$H_2 \text{ (g)} + Br_2 \text{ (g)} \longrightarrow 2HBr \text{ (g)}$$

Given that the bond energies of H — H, Br — Br and H — Br are 435, 192 and 364 kJ mol^{-1} respectively.

Solution: First method. *By calculating the total energy absorbed and released.*

Energy absorbed in the dissociation of 1 mole of H – H bonds = 435 kJ
Energy absorbed in the dissociation of 1 mole of Br – Br bonds = 192 kJ
Total energy absorbed = 435 + 192 = 627 kJ
Energy released in the formation of 1 mole of H – Br bonds = 364 kJ
∴ Energy released in the formation of 2 moles of H – Br bonds = 2 × 364 = 728 kJ
Energy released is greater than energy absorbed.

Hence, net result is the release of energy.
$$\text{Energy released} = 728 \text{ kJ} - 627 \text{ kJ} = 101 \text{ kJ}$$
Thus, for the given reaction,
$$\Delta H = -101 \text{ kJ}.$$

Second method. *Using the formula directly.*

$\Delta H_{reaction}$ = Σ Bond energies of reactants − Σ Bond energies of products
= [Bond energy (H — H) + Bond energy (Br — Br)]
   − [2 × Bond energy (H— Br)]
= 435 + 192 − (2 × 364) = 627 −128 = **−101 kJ**.

Hess's law can also be applied to solve the problem.

Example 35: Calculate the bond energy of C — H bond, given the heat of combustion of methane (CH_4), graphite and hydrogen are 891 kJ, 394 kJ and 286 kJ respectively while the heat of sublimation of graphite is 717 kJ and the heat of dissociation of hydrogen molecule is 436 kJ.

Solution: The data given is:

$$CH_4 (g) + 2O_2 (g) \longrightarrow CO_2 (g) + 2H_2O (g); \qquad \Delta H = -891 \text{ kJ} \quad ...(i)$$

$$C (s) + O_2 (g) \longrightarrow CO_2 (g); \qquad \Delta H = -394 \text{ kJ} \quad ...(ii)$$

$$H_2 (g) + \frac{1}{2} O_2 (g) \longrightarrow H_2O (g); \qquad \Delta H = -286 \text{ kJ} \quad ...(iii)$$

$$C (s) \longrightarrow C (g); \qquad \Delta H = +717 \text{ kJ} \quad ...(iv)$$

$$H_2 (g) \longrightarrow 2H (g); \qquad \Delta H = +436 \text{ kJ} \quad ...(v)$$

We want ΔH for the following reaction

$$\frac{1}{4} [CH_4 (g)] \longrightarrow C (g) + 4H (g)]; \qquad \Delta H = ? \quad ...(vi)$$

The problem is solved in two steps.

First we calculate ΔH for $CH_4 (g) \longrightarrow C (g) + 4H (g)$ and then divide the result by 4.

Operating Eqn. (*i*) + Eqn. (*iv*) + 2 × Eqn. (*v*), we get

$$CH_4 (g) + 2O_2 (g) + C (s) + 2H_2 (g) \longrightarrow CO_2 (g) + 2H_2O (g) + C (g) + 2H (g) \quad ...(vii)$$

To cancel out the terms not needed now operate Eqn. (*vii*) − 2 × Eqn. (*iii*) − Eqn. (*ii*).

$$CH_4 (g) \longrightarrow C (g) + 4H (g);$$

$\Delta H = -891 + 717 + 2 \times 436 - 2 \times (-286) - (-394) = + 1664 \text{ kJ}.$

This is the energy required for the dissociation of 4 moles of C—H bonds.

Bond dissociation energy for C—H bond = $\frac{1664}{4}$ kJ = **416 kJ mol^{-1}**.

Resonance Energy from Thermochemical Data

Some molecules get stabilized because of resonance going on in the molecule. **The extent of stabilization of a substance per mole as a consequence of resonance is called resonance energy or resonance stabilization energy.** Take for example the case of benzene, the molecule of benzene has three double bonds in the ring at alternate positions. The molecule shows exceptional stability in the sense that it has no tendency for addition reactions, under ordinary conditions inspite of the fact that it has three double bonds. It prefers to undergo substitution reactions, in which case, it is able to preserve its aromatic character having three double bonds in alternate positions.

It is possible to calculate resonance energy from heat of hydrogenation as under:

Heat of hydrogenation of cyclohexene which contains one double bond = –29 k cal/mol

(Heat of hydrogenation is always negative)

Heat of hydrogenation of three double bonds in a six-membered ring compound = 3 × (–29) k cal/mol = –87 k cal/mol

Heat of hydrogenation of benzene (experimental value) = –50 k cal/mol

Therefore, we can say that

Resonance energy of benzene = –50 – (–87) = 37 k cal/mol

Problems for Practice

1. ΔH for the reaction

$$H-C\cong N\,(g) + 2H_2\,(g) \longrightarrow H-\underset{H}{\overset{H}{\underset{|}{\overset{|}{C}}}}-N-H\,(g)$$

is –150 kJ. Calculate the bond energy of $C \cong N$ bond.
[Given bond energies of $C-H = 414$ kJ mol^{-1}; $H-H = 435$ kJ mol^{-1}; $C-N = 293$ kJ mol^{-1}, $N-H = 396$ kJ mol^{-1}] **[Ans. 893 kJ mol^{-1}]**

2. Calculate the enthalpy of hydrogenation of $C_2H_2\,(g)$ to $C_2H_4\,(g)$.
Given bond energies. $C-H = 414.0$ kJ mol^{-1}, $C \cong C = 827.6$ kJ mol^{-1}, $C = C = 606.0$ kJ mol^{-1}; $H-H = 430.5$ kJ mol^{-1}) **[Ans. –175.9 kJ mol^{-1}]**

3. Calculate the $C-C$ bond energy from the following data:
   (i) $2C\,(graphite) + 3H_2\,(g) \longrightarrow C_2H_6\,(g)$, $\Delta H = -84.67$ kJ
   (ii) $C\,(graphite) \longrightarrow C\,(g)$, $\Delta H = 716.7$ kJ
   (iii) $H_2\,(g) \longrightarrow 2H\,(g)$, $\Delta H = 435.9$ kJ
   Assume the $C-H$ bond energy as 416 kJ. **[Ans. –964 kJ mol^{-1}]**

4. Propane has the structure $H_3C-CH_2-CH_3$. Calculate the change in enthalpy for the following reaction:
$$C_3H_8\,(g) + 5O_2\,(g) \longrightarrow 3CO_2\,(g) + 4H_2O\,(g)$$
Given that average bond enthalpies are :

$C-C$	$C-H$	$C=O$	$O=O$	$O-H$
347	414	741	498	464 kJ mol^{-1}

[Ans. –1662 kJ mol^{-1}]

5. Calculate $\Delta H°$ for the reaction
$$CH_2 = CH_2 + 3O_2 \longrightarrow 2CO_2 + H_2O$$

Given that the average bond energies of the different bonds are

Bond	C—H	O=O	C=O	O—H	C=C
Bond energy (kJ mol^{-1})	414	499	724	460	619

[**Ans.** – 964 kJ mol^{-1}]

6. Calculate the enthalpy change for the following reaction

$$H_2 (g) + Cl_2 (g) \longrightarrow 2HCl (g)$$

Given that the bond dissociation energies of H—H, Cl—Cl and H—Cl are 437 kJ, 244 kJ and 433 kJ mol–1 respectively.

[**Ans.** – 185 kJ]

Solved Examples—Miscellaneous Examples

Example 36: If a man takes a diet which gives him energy equal to 9500 kJ per day and he expends energy in all forms to a total of 12000 kJ per day, what is the change in internal energy per day? If the energy lost was stored as sucrose (1632 kJ per 100 g), how many days should it take to lose 1 kg? Ignore water loss.

Solution: Loss of energy per day = 12000 – 9500 = 2500 kJ

For a loss of 1632 kJ of energy, sucrose ($C_{12}H_{22}O_{11}$) lost = 100 g (given)

∴ For a loss of 2500 kJ of energy, sucrose lost = $\dfrac{100}{1632} \times 2500$ g = 153.2 g

Thus, loss of 153.2 g in weight takes place in 1 day.

∴ Time taken to lose 1 kg or 1000 g glucose = $\dfrac{1}{153.2} \times 1000 = $ **6.5 days.**

Example 37: From the thermochemical equation

$$C_6H_6 (l) + 7\tfrac{1}{2} O_2 (g) \longrightarrow 3H_2O (l) + 6CO_2 (g) ; \Delta H = -3264.64 \text{ kJ mol}^{-1}$$

Calculate the energy evolved when 39 g of C_6H_6 are burnt in an open container.

Solution: 1 mole of benzene = 78 g

When 78 g of C_6H_6 burn, heat evolved = 3264.64 kJ

∴ When 39 g of C_6H_6 burn, heat evolved = $\dfrac{3264.64}{78} \times 39 = $ **1632.32 kJ.**

Example 38: The heat evolved in the combustion of methane is given by the following equation:

$$CH_4 (g) + 2O_2 (g) \longrightarrow CO_2 (g) + 2H_2O (l); \Delta H = -890.3 \text{ kJ}$$

(*a*) How many grams of methane would be required to produce 445.15 kJ of heat on combustion?

(*b*) How many grams of carbon dioxide would be formed when 445.15 kJ of heat are evolved?

(*c*) What volume of oxygen at STP would be used in the combustion process (*a*) or (*b*)?

Solution: (*a*) From the given equation,

Heat produced from 1 mole of CH_4, *i.e.,* 16 g of CH_4 = 890.3 kJ

Heat produced from 8 g of CH_4 = **445.15 kJ.**

(b) From the given equation,
When 890.3 kJ of heat are evolved, CO_2 formed = 44 g
∴ When 445.15 kJ of heat are evolved, CO_2, formed = 22 g.
(c) O_2 used in the production of 890.3 kJ of heat = 44.8 litres at STP.
Hence, O_2 used in the production of 445.15 kJ of heat = 22.4 litres at STP.

Example 39: The heat evolved in the combustion of glucose is shown in the following equation:

$$C_6H_{12}O_6 + 6O_2 (g) \longrightarrow 6CO_2 (g) + 6H_2O (g); \Delta H = -2840 \text{ kJ}$$

What is the energy requirement for production of 0.36 g of glucose by the reverse reaction?

Solution: Data provided is

$$C_6H_{12}O_6 + 6O_2 (g) \longrightarrow 6CO_2 (g) + 6H_2O (g); \qquad \Delta H = -2840 \text{ kJ}$$

Writing the reaction in reverse direction

$$6CO_2 (g) + 6H_2O (g) \longrightarrow C_6H_{12}O_6 (s) + 6O_2 (g); \qquad H = +2840 \text{ kJ}$$

For production of 1 mole of $C_6H_{12}O_6$ (180 g) heat required (absorbed) = 2840 kJ

∴ For production of 0.36 g of glucose, heat absorbed = $\dfrac{2840}{180} \times 0.36$ = **5.68 kJ.**

Example 40: The thermochemical equation for solid and liquid rocket fuel are given below :

$$2Al (s) + 1\tfrac{1}{2} O_2 (g) \longrightarrow Al_2O_3 (s); \qquad \Delta H = -1667.8 \text{ kJ}$$

$$H_2 (g) + \tfrac{1}{2} O_2 (g) \longrightarrow H_2O (l); \qquad \Delta H = -285.9 \text{ kJ}$$

(a) If equal masses of aluminium and hydrogen are used, which is better rocket fuel?

(b) Determine ΔH for the reaction $Al_2O_3 (s) \longrightarrow 2Al (s) + 1\tfrac{1}{2} O_2 (g)$

Solution: 2 moles of Al = 54 g
54 g of Al on combustion gives heat = 1667.8 kJ
∴ 1 g of Al on combustion gives heat = $\dfrac{1667.8}{54}$ = 30.9 kJ
1 mole of H_2 = 2 g
2 g of H_2 on combustion give = 285.9 kJ of heat
∴ 1 g of H_2 on combustion will give = $\dfrac{285.9}{2}$ = 142.95 kJ of heat

Thus, H_2 is a better rocket fuel.
(b) Writing the reverse of the first reaction, we have

$$Al_2O_3 (s) \longrightarrow 2Al (s) + 1\tfrac{1}{2} O_2 (g); \quad H = +1667.8 \text{ kJ}$$

Thus, for the reaction given in part (b) of the problem,
$$\Delta H = +1667.8 \text{ kJ.}$$

1.20 VARIATION OF HEAT OF REACTION WITH TEMPERATURE KIRCHOFF'S EQUATION

Heat evolved or absorbed in a reaction depends upon the temperature at which it is carried out. The relationship showing the variation of the heat of reaction with temperature was given by Kirchoff in 1856 and is stated as

The change in the heat of reaction at constant pressure for every degree change of temperature is equal to the change in the heat capacity at constant pressure, accompanying the reaction.

It can be derived as under.

Consider the general reaction

$$A \longrightarrow B$$

Gustav Kirchoffs (1824–1887)
Gustav Kirchoff's was a German physicist who contributed to many branches of physics

Case I. When the reaction is carried out at constant pressure

Suppose the heat of reaction (heat evolved) at temperature $T_1 = \Delta H_1$ and the heat of reaction (heat evolved) at temperature $T_2 = \Delta H_2$. (Fig. 1.11)

Starting from A at temperature T_1, we can reach B at temperature T_2 by two different routes:

Route I : We perform the reaction at temperature T_1 to obtain B at the same temperature and then heat this up to the temperature T_2. Then

Heat **evolved** in the first stage = ΔH_1

Heat **absorbed** in the second stage = $(C_p)_B (T_2 - T_1)$

where $(C_p)_B$ is the average molar heat capacity of the products.

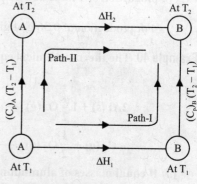

Fig 1.11: Changing A to T_1 to B at T_1 by two different paths

∴ Net heat evolved = $\Delta H_1 - (C_p)_B (T_2 - T_1)$

Route II: We first heat up the reactants to the temperature T_2 and then perform the reaction to give the products at the same temperature. Then if $(C_p)_A$ is the average molar heat capacity of the reactant A,

Heat **absorbed** in the first stage = $(C_p)_A (T_2 - T_1)$

Heat **evolved** in the second stage = ΔH_2

∴ Net heat evolved = $\Delta H_2 - (C_p)_A (T_2 - T_1)$.

Initial and the final states are the same in both the cases. Therefore, by law of conservation of energy, we must have

Net heat evolved by route I = Net heat evolved by route II

i.e. $\quad \Delta H_1 - (C_p)_B (T_2 - T_1) = \Delta H_2 - (C_p)_A (T_2 - T_1)$

or $\quad \dfrac{\Delta H - \Delta H_1}{T_2 - T_1} = \Delta C_p$

Chemical Thermodynamics – I (Zeroth and First Laws of Thermodynamics and Thermochemistry)

When the reaction is carried out at constant volume: We have the thermodynamic relation

$$\Delta H = \Delta U + P\Delta V$$

Under conditions of constant volume, $\Delta V = 0$

$$\therefore \quad \Delta H = \Delta U$$

Hence equation derived above takes the form

$$\frac{\Delta U_2 - \Delta U_1}{T_2 - T_1} = \Delta C_v$$

where $\Delta C_v = (C_v)_B - (C_v)_A$ is the change in the heat capacity at constant volume accompanying the reaction, $(C_v)_B$ and $(C_v)_B$ are the mean molar heat capacities at constant volume of the reactants and the products respectively, ΔU_1 and ΔU_2 are the internal energy changes (*i.e.*, heats of reaction at constant volume) when the reaction is carried out at temperatures T_1 and T_2 respectively. Thus Kirchoff's equation may also be expressed as follows :

The change in the heat of reaction at constant volume for every degree change of temperature is equal to the change in the heat capacity at constant volume, accompanying the reaction.

Solved Examples

Example 41: Calculate the heat of formation of H_2O (l) at 383 K if ΔH for the reaction

$$H_2 (g) + \frac{1}{2} O_2 (g) \longrightarrow H_2O (l)$$

at 298 K is –298.06 kJ. The average values of heat capacities between the two temperatures for H_2 (g), O_2 (g) and H_2O (l) are 27.61, 29.50 and 75.31 $JK^{-1}mol^{-1}$ respectively.

Solution:
$$H_2 (g) + \frac{1}{2} O_2 (g) \longrightarrow H_2O (l)$$

$$\Delta C_p = C_p(H_2O) - C_p(H_2) - \frac{1}{2} C_p (O_2)$$

$$= 75.31 - 27.61 - \frac{1}{2} \times 29.50 = 75.31 - 27.61 - 14.75$$

$$= 32.95 \; JK^{-1}mol^{-1} = 32.95 \times 10^{-3} \; kJK^{-1}mol^{-1}$$

Applying Kirchoff's equation

$$\frac{\Delta H_2 - \Delta H_1}{T_2 - T_1} = \Delta C_p \quad \text{Substituting the values, we have}$$

$$\frac{\Delta H_2 - (-298.06)}{383 - 298} = 32.95 \times 10^{-3} \quad \text{or} \quad \frac{\Delta H_2 + 298.06}{85} = 32.95 \times 10^{-3}$$

or $\quad \Delta H_2 + 298.06 = 85 \times 32.95 \times 10^{-3}$

or $\quad \Delta H_2 = (85 \times 32.95 \times 10^{-3}) - 298.06$

$$= 2.80 - 298.06 = -295.26 \; kJ$$

Example 42: Calculate the heat of formation of HCl at 348 K from the following data:

$$\frac{1}{2} H_2 (g) + \frac{1}{2} Cl_2 (g) \longrightarrow HCl (g); \; \Delta H^0_{298} = -92300 \; J$$

The mean heat capacities over this temperature range are

$$H_2 \text{ (g)}; C_p = 28.53 \text{ JK}^{-1} \text{ mol}^{-1}$$
$$Cl_2 \text{ (g)}; C_p = 32.26 \text{ JK}^{-1} \text{ mol}^{-1}$$
$$HCl \text{ (g)}; C_p = 28.49 \text{ JK}^{-1} \text{ mol}^{-1}$$

Solution: The reaction under consideration is

$$\frac{1}{2} H_2 \text{ (g)} + \frac{1}{2} Cl_2 \text{ (g)} \longrightarrow HCl \text{ (g)}$$

For this reaction,

$$\Delta C_p = (C_p)_{HCl} - \frac{1}{2}(C_p)_{H_2} - \frac{1}{2}(C_p)_{Cl_2}$$

$$= 28.49 - \frac{1}{2}(28.53) - \frac{1}{2}(32.26)$$

$$= 1.91 \text{ JK}^{-1} \text{ mol}^{-1}$$

We are given that

$$\Delta H_1 = -92300 \text{ J}, T_1 = 298 \text{ K}, \Delta H_2 = ?$$

According to Kirchoff's equation

$$\frac{\Delta H_2 - \Delta H_1}{T_2 - T_1} = \Delta C_p$$

or

$$\Delta H_2 = \Delta H_1 + \Delta C_p (T_2 - T_1)$$
$$= -92300 + (-1.91)(348 - 298) = -923300 - 95.5$$
$$= -92395.5 \text{ J}$$

Problems for Practice

1. If the standard heat of formation of HCl gas at 25°C is -22.66 kcal mol^{-1}, calculate the value at 60°C. C_p for H_2 (g) = 6.82 cal mol^{-1} deg^{-1}.

2. The heat of solution of ammonia at 25°C was found to be -11.04 k cal. Calculate the heat of solution at 348 K. Given that the mean heat capacities of N_2, H_2 and NH_3 are 6.80, 6.77 and 8.86 cal/degree/mol respectively.

   [Hint: $\frac{1}{2} N_2 + \frac{3}{2} H_2 \longrightarrow NH_3$

   $$\Delta C_p = (C_p) NH_3 - \frac{1}{2}(C_p) N_2 - \frac{3}{2}(C_p) H_2$$

   $T_1 = 25 + 273 = 298$ K, $T_2 = 348$ K]

3. The molar heat capacities at constant pressure of H_2 (g), Cl_2 (g) and HCl (g) are respectively 29.3, 34.7 and 28.9 JK^{-1}. If the heat of formation of HCl (g) at constant pressure at 293 K is -91.2 kJ, what will be its heat of formation at 313 K? [**Ans.** -91.27 kJ]

4. Consider the reaction

$$H_2O \text{ (g)} \longrightarrow H_2 + \frac{1}{2} O_2 \text{ ; } \Delta H° = +57780 \text{ cal at } 18°C.$$

What would be $\Delta H°$ value at 25°C? The C_p values are

$C_p (H_2O) = 8.02$, $C_p (H_2) = 6.89$ and $C_p (O_2) = 7.96$ cal deg^{-1} mol^{-1} [**Ans.** 57796.45 kcal]

Some Solved Conceptual Problems

Problem 1: Under what conditions, an extensive property may become intensive property? Give an example.

Solution: An extensive property a_s might become intensive when unit amount of the substance is specified. For example, mass is extensive, but mass per unit volume (*i.e.* density) is intensive. Similarly, heat capacity is extensive but specific heat (which is heat capacity per unit mass) is intensive.

Problem 2: State the variables which are kept constant in the following processes

    (*a*) Isothermal   (*b*) Isobaric   (*c*) Isocaloric

Solution: (*i*) Temperature (*ii*) Pressure (*iii*) Volume

Problem 3: Which of the following expressions represent work done and under what conditions?

    (*a*) $P\Delta V$ (*b*) $\Delta(PV)$ (*c*) $P_{ext} \int_{V_1}^{V_2} dV$

Solution:

(*i*) $P\Delta V$ denotes work done against constant external pressure, P when the volume changes from V_1 to V_2 such that $\Delta V = V_2 - V_1$

(*ii*) It does not stand for work done

(*iii*) It represents the work done when there is reversible expansion of the gas from the volume V_1 to V_2 against extend pressure P_{ext}

Problem 4: What are the conditions under which *q* and *w* become state functions. Discuss briefly.

Solution:

(*i*) At constant volume, there is no work of expansion, hence $w = 0$. By first law of thermodynamics $\Delta U = q_v$. As ΔU is a state function, q_v is also a state function.

(*ii*) At constant external pressure $w = -P\Delta V$. Hence *w* becomes a state function.

(*iii*) $\Delta U = q_p + w$. As ΔU is a state function, *w* under constant pressure is also a state function [see (*ii*) above]. Therefore q_p is also a state function.

Problem 5: What are standard state conditions? Which form of carbon is assigned a value of zero for molar enthalpy in the standard state and why?

Solution: A substance is considered to be in standard state if it is at a temperature of 298.15 K and a pressure of one atmosphere. Graphite is assigned a value of zero for enthalpy in the standard state because it is the stable form.

Problem 6: In the case of polyatomic molecules, why is the bond energy of a particular type of bond, the average value?

Solution: Bond energy for the dissociation of a particular bond is not the same in different molecules. It is not the same even in the same molecule like CH_4. Bond energies of four C — H are not equal because after the dissociation of every C — H bond, the electronic environments change. Hence the average value is taken.

Problem 7: What is the Thermodynamic basis of Hess's law?

Solution: Enthalpy is a state function. That means enthalpy change during a reaction does not depend upon the reaction path in which the change is brought about. It depends only upon the nature of the reactants and the final products.

Key Terms

- Kirchoff's temperature
- Heat capacity
- Internal energy
- Isothermal and adiabic

Evaluate Yourself

Multiple Choice Questions

1. First law of thermodynamics is the law of
   - (a) entropy
   - (b) free energy
   - (c) conservation of energy
   - (d) none of these

2. The mathematical form of first law of thermodynamics is
   - (a) $dq = dU + dw$
   - (b) $dq = dU - dw$
   - (c) $dU = dq + dw$
   - (d) none of these

3. Standard state refers to
   - (a) One atmosphere pressure and 25°C temperature
   - (b) One atmosphere pressure and 100°C temperature
   - (c) One atmosphere pressure and 0°C temperature
   - (d) None of these

4. Endothermic reaction is one in which
   - (a) heat is converted into electricity
   - (b) heat is absorbed
   - (c) heat is given out
   - (d) heat is converted into mechanical work

5. Thermochemistry is the study of
   - (a) heat changes accompanying chemical reaction
   - (b) net entropy change in a reaction
   - (c) net free energy change in a chemical reaction
   - (d) none of these

6. A process during which there is no heat change is called
   - (a) an adiabatic process
   - (b) reversible process
   - (c) irreversible process
   - (d) none of these

7. A process in which a system comes to its initial position after a series of operations is called
   (a) reversible
   (b) cyclic
   (c) adiabatic
   (d) none of these

8. A reaction whose heat of reaction shows the bond energy of HCl is
   (a) $HCl\,(g) \longrightarrow H\,(g) + Cl\,(g)$
   (b) $2\,HCl\,(g) \longrightarrow H_2\,(g) + Cl_2\,(g)$
   (c) $HCl\,(g) \longrightarrow \frac{1}{2}H_2\,(g) + \frac{1}{2}Cl_2\,(g)$
   (d) $HCl\,(g) \longrightarrow H^+\,(g) + Cl^-\,(g)$

9. A process which proceeds infinitesimally slowly is called
   (a) irreversible
   (b) reversible
   (c) isothermal
   (d) adiabatic

10. The equations showing the effect of temperature change on the heat of reaction is known as
    (a) Arrhenius equation
    (b) Kirchoff's equation
    (c) Ostwald's equation
    (d) none of these

Short Answer Questions

1. What are extensive and intensive properties? Give three examples of each of them.
2. Give in brief the objectives and limitations of thermodynamics.
3. What is the relationship between Heat and Mechanical work? Define Joule's mechanical equivalent of heat. What is its value?
4. State First law of thermodynamics in two different ways. Derive its mathematical formulation.
5. Briefly explain the terms Enthalpy and Enthalpy change. Is it intensive or extensive property?

Long Answer Questions

1. Explain the terms (i) 'Enthalpy' and 'Enthalpy change' (ii) Heat capacity. Derive expression for heat capacity at constant volume and that at constant pressure. Derive the relationship between them.

2. Prove thermodynamically that Joule-Thomson coefficient for an ideal gas is zero. Also deduce expression for Joule-Thomson coefficient for real gases in terms of van der Waals constants 'a' and 'b'

3. What do you understand by Heat of reaction at constant volume and that at constant pressure. Derive the relationship between them. Under what conditions, the two are equal?

4. Prove that for a monatomic gas (having transtational kinetic energy only), $C_v = \dfrac{3}{2} R$.

5. Derive expressions for ΔU and ΔH for adiabatic reversible expansion of real gas.

6. Briefly explain the terms enthalpy of neutralisation. Comment on their values if both are strong or one of them is weak or both are weak?

7. Derive thermodynamically Kirchoff's equation giving the variation of heat of reaction with temperature.

8. Derive thermodynamically Kirchoff's equation.

   Or

   Show that the temperature dependence of heat of reaction is given by the relation

   $$\left(\dfrac{\partial \Delta H}{\partial T}\right)_P = \Delta C_p.$$

   Name the equation.

9. (a) Distinguish between reversible isothermal expansion and reversible adiabatic expansion.

   (b) Show that for adiabatic expansion of an ideal gas,
   $$PV^\gamma = \text{constant}$$

10. Derive the following expression for the reversible adiabatic expansion of a real gas.
    $$T_2 = T_1 \left(\dfrac{V_1 - nb}{V_2 - nb}\right)^{R/C_p}$$

11. Define 'Bond energy' for a diatomic molecule and for a polyatomic molecule. How bond energy data helps to calculate the enthalpy change of a reaction?

12. How can the reversible isothermal expansion of an ideal gas be brought about? Derive an expression for the work of expansion of such a process.

13. What are the values/expressions for q, w, ΔU and ΔH for
    (i) irreversible isothermal free expansion of an ideal gas?
    (ii) irreversible isothermal intermediate expansion of an ideal gas?

    Out of reversible and irreversible expansion which one will absorb less heat and why?

14. Derive the expression for the work done in an adiabatic expansion. Will the expression be different for adiabatic expansion along a reversible and an irreversible path?

15. The work done in a reversible isothermal expansion of an ideal gas is greater than the work done in reversible adiabatic expansion. Explain.

Answers

Multiple Choice Questions

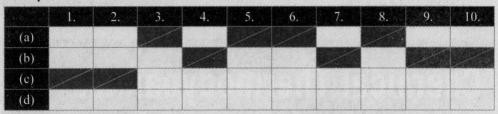

Suggested Readings

1. Peter, A. & Paula, J. de. Physical Chemistry, Oxford University Press.
2. Metz, C.R. 2000 Solved Problems in Chemistry, Schaum Series.

Chapter 2

Chemical Thermodynamics – II
(Second and Third Laws of Thermodynamics and Partial Molar Properties)

LEARNING OBJECTIVES

After reading this chapter, you should be able to:

- learn the concept of entropy
- understand thermodynamic scale of temperature
- state second law of thermodynamics in different ways
- calculate entropy change for reversible and irreversible process
- understand the concept of residual entropy
- calculate absolute entropy of molecules
- correlate free energy change and sponteity
- derive a relation between Joule-Thomson coefficient and thermodynamic parameters
- derive Maxwell relations
- introduced to partial molar quantities
- derive Gibbs' Duhem equation
- calculate chemical potential of ideal mixtures

2.1 INTRODUCTION – NEED FOR THE SECOND LAW

A major **limitation of the first law of thermodynamics** is that it provides no information about the feasibility of a process. For example, it does not tell us whether water can run uphill itself, whether a gas diffuse from a region of low pressure to a region of high pressure, whether heat can flow from a cold body to a hot body etc. It does tell that there is an exact equivalence between various forms of energy and that heat gained is equal to heat lost. But that does not serve the whole purpose. Second law of thermodynamics answers the questions that remain unanswered by first-law.

The first law states that when heat is converted into work, the work performed is equivalent to heat absorbed. However, this is not precisely correct. It has been seen from our day-to-day exprience that heat absorbed cannot be completely converted into work without leaving some change in the system or surroundings.

In order to sort out these issues, second law of thermodynamics was proposed. It has a number of statements. Different statements of second law of thermodynamics supply answers to various questions.

2.1.1 Different Statements of Second Law of Thermodynamics

1. *All spontaneous processes like the flow of heat from hot end to cold end, diffusion of gas from high pressure to low pressure or the flow of water down a hill etc. are thermodynamically irreversible.*

Spontaneous process means a process which can take place without the help of any external agency. All *natural processes,* some of which are mentioned in the definition above, are spontaneous processes.

Further, the first law states that when heat is converted into work, the work obtained is equivalent to the heat absorbed. However it has been seen from experience that the heat absorbed cannot be completely converted into work without leaving some change in the system or the surroundings. Hence the second law is also stated as follows:

2. *The complete conversion of heat into work is impossible without leaving some effects elsewhere.*

3. *It is impossible to construct a machine, functioning in cycles, which can convert heat completely into the equivalent amount of work without producing changes elsewhere.*

4. *Without the use of an external agency, heat cannot by itself pass from a colder to a hotter body.*

2.1.2 Efficiency of a Machine

The fraction of the heat absorbed by a machine that it can transform into work is called the **efficiency** *of the machine.* Thus if Q is the heat absorbed and W is the work done, then the efficiency of the machine is given by

$$\eta = \frac{W}{Q}$$

The machine used for the conversion of heat into work is called **Heat Engine**. In order to bring about this conversion, the engine absorbs heat from a heat reservoir at a higher temperature, called the **source**, converts a part of it into work and returns the remainder to the heat reservoir at a lower temperature, called the **sink**.

2.2 CARNOT CYCLE

A *Carnot cycle is a process in which a system returns to its original state after a number of successive changes. A process conducted in this manner is called a cyclic process.* The Carnot cycle consists of four different operations which can be shown on pressure-volume diagram as shown in Fig. 2.1.

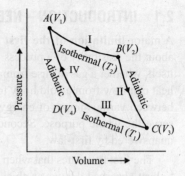

Fig. 2.1: *Carnot cycle*

A Carnot engine consists of a cylinder containing 1 mole of an ideal gas as the working substance and fitted with weightless, frictionless piston so shat all the operations in the cycle are carried out reversibly. For this reason, it is also called *reversible heat engine*. The cylinder is supposed to be insulated on all sides except at the bottom so that heat can flow to or from the system only through the bottom. Further, it is supposed that there are two heat reservoirs, one at a higher temperature T_2 (called the *source*) and the other at a lower temperature T_1 (called the *sink*). If some operation is carried out by placing the cylinder in the source or the sink, it can exchange heat with it and hence the temperature remains constant so that the process is *isothermal*. On the other hand, if the cylinder is placed on an insulating material, no heat exchange can take place between the system and the surroundings and hence the process is *adiabatic*.

Four different operations are carried out as shown in Fig. 2.2 and are described below:

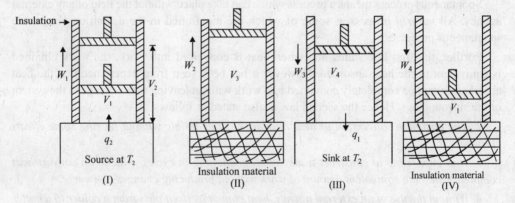

Fig. 2.2: *Different operations of the Carnot cycle*

I. Isothermal expansion: The cylinder containing one mole of the ideal gas, occupying a volume V_1, is placed in contact with the source (*i.e.* heat reservoir at temperature T_2). The gas absorbs heat q_2 from the source and expands *isothermally* and *reversibly* till its volume has increased to V_2. Since the gas is ideal, the work done w_1 by one mole of the ideal gas will be given by

$$-w_1 = RT_2 \ln \frac{V_2}{V_1} \qquad \qquad ...(i)$$

Further for an ideal gas, the work done is equal to the heat absorbed, q_2 so that

$$-w_1 = q_2 = RT_2 \ln \frac{V_2}{V_1} \qquad \qquad ...(ii)$$

Here, w_1 work done by the system has been taken as negative and q_2 as positive as per latest SI conventions.

II. Adiabatic expansion: The cylinder is now removed from the source and placed in perfectly insulating material so that the gas now expands adiabatically and reversibly. Work is done in the expansion but since no heat enters or leaves the system, the temperature must fall. The reversible adiabatic expansion is continued till the temperature has fallen to T_1 which is the temperature of the sink. Suppose the corresponding volume becomes V_3. The path is shown by the adiabatic curve BC in Fig. 1.1. The work done w_2 is be given by

$$-w_2 = C_v(T_1 - T_2)$$

or $$w_2 = -C_v(T_2 - T_1) \qquad \ldots(iii)$$

where C_v is the heat capacity of the ideal gas. Here, again w_2 *is negative* by sign conventions.

III. Isothermal compression: The cylinder is now removed from the insulating material and placed in contact with the sink (*i.e.* the heat reservoir at the lower temperature T_1). The gas is compressed isothermally and reversibly till the volume decreases from V_3 to V_4. The process is represented by the path CD in Fig. 2.1. The work done on the system is taken as *positive* by new conventions and will be given by

$$w_3 = RT_1 \ln \frac{V_4}{V_3} \qquad \ldots(iv)$$

During the operation, an amount of heat q_1, exactly equal to w_3, will be returned to the sink at T_1. According to sign conventions, q_1 will be negative and w_3 will be positive.

$$-q_1 = w_3 = RT_1 \ln \frac{V_4}{V_3} \qquad \ldots(v)$$

IV. Adiabatic compression: The cylinder is now removed from the sink and placed again on the insulating material. The gas is then compressed adiabatically and reversibly along DA till the initial state A is regained. The temperature of the gas rises from T_1 to T_2. The work done on the system is *positive* and will be given by

$$w = C_v(T_2 - T_1) \qquad \ldots(vi)$$

As a result of these four operations, the system has returned to its original state so that a reversible cycle has been completed. The *net work* (w) done by the system will be given by

$$w = (-w_1) + (-w_2) + (w_3) + (w_4) \qquad \ldots(vii)$$

$$= RT_2 \ln \frac{V_2}{V_1} - C_v(T_2 - T_1) + RT_1 \ln \frac{V_4}{V_3} + C_v(T_2 - T_1)$$

$$= RT_2 \ln \frac{V_2}{V_1} + RT_1 \ln \frac{V_4}{V_3} \qquad \ldots(viii)$$

V_1 and V_4 lie on one adiabatic curve and V_3 and V_2 lie on another, applying volume-temperature relationships, we have

$$\left(\frac{V_4}{V_1}\right)^{\gamma-1} = \frac{T_2}{T_1} \qquad \ldots(ix)$$

and $$\left(\frac{V_3}{V_2}\right)^{\gamma-1} = \frac{T_2}{T_1} \qquad \ldots(x)$$

Comparing equations (*ix*) and (*x*), we get

$$\left(\frac{V_4}{V_1}\right)^{\gamma-1} = \left(\frac{V_3}{V_2}\right)^{\gamma-1} \quad \text{or} \quad \frac{V_4}{V_1} = \frac{V_3}{V_2} \quad \text{or} \quad \frac{V_4}{V_3} = \frac{V_1}{V_2} \qquad ...(xi)$$

Substituting this value in equation (viii), we get

$$w = RT_2 \ln\frac{V_2}{V_1} + RT_1 \ln\frac{V_1}{V_2} = RT \ln\frac{V_2}{V_1} - RT_1 \ln\frac{V_2}{V_1}$$

$$= R(T_2 - T_1)\ln\frac{V_2}{V_1} \qquad ...(xii)$$

Dividing equation (xii) by equation (ii), we have

$$\frac{W}{q_2} = \frac{T_2 - T_1}{T_2} \qquad ...(xiii)$$

This gives the *efficiency* of the Carnot cycle or engine.

From equation (xiii), it is clear that the efficiency of the reversible heat engine depends only upon the temperatures of the source and the sink and is independent of the nature of the working substance.

Since the quantity $\frac{T_2 - T_1}{T_2}$, which represents efficiency is always less than unity, the efficiency of the heat engine is thus always less than unity.

2.2.1 Carnot Theorem

The result that follows from Carnot cycle is called Carnot theorem.

Carnot theorem is stated as follows:

The efficiency of a reversible heat engine depends only upon the temperatures of the source and the sink and is independent of the nature of the working substance. In other words, all reversible heat engines working between the same two temperatures have the same efficiency.

Combining equations (ii), (v) and (viii) in the previous section, we get

$$w = q_2 - q_1$$

i.e. net work done by the system is equal to net heat absorbed by the system. Putting the value of w in equation (xiii) above, we have

$$\frac{q_2 - q_1}{q_2} = \frac{T_2 - T_1}{T_2}$$

Hence the efficiency of a heat engine may be given by any one of the following expressions:

$$\eta = \frac{w}{q_2} = \frac{q_2 - q_1}{q_2} = \frac{T_2 - T_1}{T_2}$$

Obviously, the efficiency of the engine can be increased by widening the difference in the temperatures of the source and sink.

Solved Examples—Carnot Cycle

Example 1: Calculate the amount of heat supplied to Carnot's cycle working between 368 K and 288 K if the maximum work obtained is 895 joules.

Solution: It is given that
$$T_1 = 368 \text{ K}, \quad T_2 = 288 \text{ K}, \quad w = 895 \text{ Joules}$$

Substituting these values in the formula,
$$\eta = \frac{W}{q_2} = \frac{T_2 - T_1}{T_2}$$

we have
$$\frac{895}{q_2} = \frac{368 - 288}{368} \quad \text{or} \quad q_2 = \frac{895 \times 368}{80} = \mathbf{4117 \text{ Joules}}$$

Example 2: Calculate the maximum efficiency of a steam engine operating between 110° and 25°C. What would be the efficiency of the engine if the boiler temperature is raised to 140°C, the temperature of the sink remaining the same.

Solution: First case. Given that
$$T_2 = 110°C = 110 + 273 \text{ K} = 383 \text{ K}$$
$$T_1 = 25°C = 25 + 273 \text{ K} = 298 \text{ K}$$

$$\therefore \quad \eta = \frac{T_2 - T_1}{T_2} = \frac{383 - 298}{383} = 0.222 = \mathbf{22.2\%}$$

Second case. Given that
$$T_2 = 140°C = 140 + 273 \text{ K} = 413 \text{ K}$$
$$T_1 = 25°C = 25 + 273 \text{ K} = 298 \text{ K}$$

$$\therefore \quad \eta = \frac{T_2 - T_1}{T_2} = \frac{413 - 298}{413} = 0.278 = \mathbf{27.8\%}$$

Example 3: Compare the thermodynamic efficiencies to be expected:

(*a*) when an engine is allowed to operate between 1000 K and 300 K;

(*b*) when an engine is allowed to operate between 1000 K and 600 K and then the waste heat is passed on to another engine which operates between 600 K and 300 K.

Solution: Case (*a*)
$$\eta = \frac{1000 - 300}{10000} = 0.70 = \mathbf{70\%}$$

Case (*b*) (*i*)
$$\eta = \frac{1000 - 600}{10000} = 0.40 = \mathbf{40\%}$$

(*ii*)
$$\eta = \frac{600 - 300}{600} = 0.50 = 50\% \text{ of the rejected } 60\% \text{ of } (i) = 30\%$$

∴ Total $\eta = 40 + 30 = 70$

Problems for Practice

1. What percentage T_1 is of T_2 for a heat engine whose efficiency is 10%. **[Ans. 90%]**
2. What will be the minimum amount of work required to operate a refrigerator machine which removes 4.2 kJ of heat at 273 K and rejects it at 323 K? **[Ans. 650 Joules]**
3. The boiling point of water at 50 atmospheres is 265°C. Compare the theoretical efficiencies of a steam engine operating between the boiling point of water at

    (*i*) 1 atmosphere (*ii*) 50 atmosphere, assuming the temperature of sink in each case as 35°C.

    [Ans. (*i*) 17.43% (*ii*) 42.35%]

2.3 THERMODYNAMIC SCALE OF TEMPERATURE

An ideal gas or a perfect gas ceases to exist at a temperature of –273.15°C. This temperature is called zero degree absolute. The scale developed on this basis is called perfect gas temperature scale. It is found to be independent of the ideal gas taken.

Lord Kelvin was the first to develop a scale of temperature that was independent of the nature of the gas. This was based on the efficiency of a reversible heat engine. This scale of temperature is called thermodynamic scale of temperature or Kelvin scale of temperature. It is found to be similar to the ideal gas temperature scale. Thermodynamic scale of temperature is developed as under.

Lord Kelvin (1824-1907)
He was a British mathematical physicist and engineer. He is known among other things for formulation of first and second laws of thermodynamics. For his work on transatlantic telegraph, he was knighted (given the title 'Lord') by Queen Victoria. Absolute temperatures are stated in units of kelvin in his honour.

Assume two heat reservoirs, one acting as a source and the other acting as a sink. Imagine a reversible heat engine operating between them. We further assume that the temperature of source on the new scale is proportional to the quantity of heat absorbed from it and the temperature of the sink is proportion to the heat lost to it.

Let the
Heat absorbed from the source $= Q_2$
Heat lost to the sink $= Q_1$
Temperature of source on the thermodynamic scale $= \theta_2$
Temperature of sink on the thermodynamic scale $= \theta_1$
According to the above discussion,

$$\frac{Q_2}{Q_1} = \frac{\theta_2}{\theta_1}$$

Taking reciprocal of both sides and subtracting the result from 1, we get

$$1 - \frac{Q_1}{Q_2} = 1 - \frac{\theta_1}{\theta_2} \quad \text{or} \quad \frac{Q_2 - Q_1}{Q_2} = \frac{\theta_2 - \theta_1}{\theta_2}$$

Putting $\theta_1 = 0$ i.e. zero of the new scale,

$$\frac{Q_2 - Q_1}{Q_2} = 1$$

This signifies that zero on the Kelvin scale is the temperature of the sink for a reversible heat engine whose efficiency is 1 i.e. complete conversion of heat into work. But this is possible only at absolute zero on the perfect gas scale of temperature. We conclude from here that gas scale and Kelvin scale are similar, provided the gas taken is ideal (perfect).

Thermodynamic scale of temperature derived from second law of thermodynamics is more fundamental than that derived from ideal gas because we can use any fluid in the former.

2.4 CONCEPT OF ENTROPY

For a reversible Carnot cycle working between temperatures T_2 and T_1,

$$\frac{q_2 - q_1}{q_2} = \frac{T_2 - T_1}{T_2} \qquad \ldots(i)$$

where q_2 is the heat absorbed isothermally and reversibly at temperature T_2 and q_1 is the heat lost isothermally and reversibly at temperature T_1.

The above equation may be written as

$$1 - \frac{q_1}{q_2} = 1 - \frac{T_1}{T_2}$$

or
$$\frac{q_2}{T_2} = \frac{q_1}{T_1} \qquad \ldots(ii)$$

or
$$\frac{q}{T} = \text{constant}$$

Thus heat absorbed or lost isothermally and reversibly divided by the temperature at which the heat is absorbed or lost is a constant quantity for a particular system.

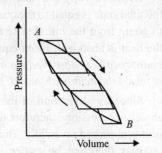

In equation (ii), q_2 is the heat absorbed at temperature T_2 and q_1 is the heat lost at temperature T_1. If q_1 were the heat absorbed at temperature T_1, equation (ii) would be rewritten as

$$\frac{q_2}{T_2} = -\frac{q_1}{T_1} \quad \text{or} \quad \frac{q_2}{T_2} + \frac{q_1}{T_1} = 0$$

i.e.
$$\sum \frac{q}{T} = 0 \qquad \ldots(iii)$$

Consider any reversible cycle ABA. It may be regarded as being made up of a number of Carnot cycles drawn across the diagram as shown in Fig. 2.3. Starting from A and going through all the cycles successively from A to B it can be shown that all paths inside the closed curve ABA cancel each other leaving only the zig-zag outer path. This zig-zag path is almost the same as the path ABA of the reversible cycle. By making each Carnot cycle smaller and increasing their number, it can be made to match more closely to ABA. The reversible cycle can be regarded as being constituted of an infinite number of Carnot cycles. For each of these cycles, we have

$$\sum \frac{q}{T} = 0.$$

Fig. 2.3: *Carnot cycles drawn across the reversible cycle.*

Hence for an infinite number of Carnot cycles,

$$\sum \frac{\delta q}{T} = 0. \qquad \ldots(iv)$$

where δq is a very small quantity of the heat absorbed at temperature T during the small isothermal change of the reversible process.

The above term may be divided into two parts, one for the path A to B and the other for the back path B to A. Thus

$$\sum \frac{\delta q}{T} = \sum_{A \to B} \frac{\delta q}{T} + \sum_{B \to A} \frac{\delta q}{T} = 0$$

or we may write it as follows:

$$\sum_{A \to B} \frac{\delta q}{T} = -\sum_{B \to A} \frac{\delta q}{T}$$

or
$$\left(\frac{q}{T}\right)_{A \to B} = -\left(\frac{q}{T}\right)_{B \to A} \qquad ...(v)$$

This means that the total value of the quantity $\frac{q}{T}$ for the path A to B is equal to the total value of the quantity $\frac{q}{T}$ for the reverse path B to A.

Thus *quantity* $\frac{q}{T}$ *is a state function, i.e. its value depends only upon the initial state* (A) *and the final state* (B) *and is independent of the path*. Obviously $\frac{q}{T}$ is the change in the value of a quantity or a function having some definite values at A and B, depending upon the values of pressure, volume and temperature. This quantity or the function is called **entropy** and is usually represented by the symbol S. Thus if S_A is the value of the entropy at A and S_B is the value at B, then we must have

$$\frac{q}{T} = S_B - S_A = \Delta S \qquad ...(vi)$$

where ΔS represents the total change in entropy in going from the initial state A to the final state B.

Hence **entropy** is a state function, the change in the value of which from the initial state to the final state is equal to the quantity q/T i.e. the total heat absorbed reversibly and isothermally in going from the initial state to the final state divided by the absolute temperature at which the heat is absorbed. From equation (*vi*), the **entropy change** *may be defined directly as the quantity of heat absorbed isothermally and reversibly divided by the absolute temperature* (T) *at which the heat is absorbed.*

Since the derivation of the above result is based upon Carnot cycle in which the heat is absorbed reversibly, therefore the expression (*vi*) for the entropy change is valid only when the heat is absorbed reversibly. Thus it is better to use the symbol q_{rev} in place of q. The expression (*vi*) may, therefore, be written as

$$\Delta S = \frac{q_{rev}}{T} \qquad ...(vii)$$

Further, as entropy is a state function, a *small change* in its value can be represented by dS. Hence we may write

$$dS = \frac{q_{rev}}{T}$$

2.4.1 Entropy Change in a Reversible Process (or at equilibrium)

If q_{rev} is the heat absorbed by the system reversibly, then the heat lost by the surroundings will also be q_{rev}. If the process takes place isothermally at the absolute temperature T, then

(*i*) Entropy change *of the system* is given by

$$\Delta S_{system} = -\frac{q_{rev}}{T} \qquad ...(i)$$

(*ii*) Entropy change *of the surroundings* is given by

$$\Delta S_{surroundings} = -\frac{q_{rev}}{T} \qquad ...(ii)$$

Thus the *total entropy change* for the combined system and the surroundings will be

$$\Delta S_{system} + \Delta S_{surroundings} = \frac{q_{rev}}{T} - \frac{q_{rev}}{T} = 0 \qquad ...(iii)$$

We may conclude that *in a reversible process, the net entropy change for the combined system and the surroundings is zero.*

2.4.2 Entropy Change in an Irreversible Process

Suppose that the total heat lost by the surroundings is q_{irrev}. This heat is absorbed by the system but the *entropy change* of the system *does not depend upon the heat actually absorbed but it depends upon the heat absorbed reversibly i.e.,* on the quantity q_{rev}. Thus if the heat is absorbed isothermally by the system at the absolute temperature T, the *entropy change of the system* is given by

$$\Delta S_{system} = \frac{q_{rev}}{T} \qquad ...(i)$$

Further suppose that the loss of heat (q_{irrev}) by the surroundings takes place infinitesimally slowly because the surroundings are much bigger in size and magnitude compared to the system (*i.e.* reversibly) and isothermally at the temperature T. Then the *entropy change of the surroundings* is given by

$$\Delta S_{surroundings} = -\frac{q_{irrev}}{T} \qquad ...(ii)$$

The *total entropy change for the combined system and the surroundings will, become*

$$\Delta S_{system} + \Delta S_{surroundings} = \frac{q_{irrev}}{T} - \frac{q_{irrev}}{T} \qquad ...(iii)$$

We know that the work done in a reversible process is the maximum work *i.e.*

$$w_{rev} > w_{irrev} \qquad ...(iv)$$

Further, as the internal energy (U) is state function, the value of ΔU is same whether the process is carried out reversibly of irreversibly. Therefore

$$\Delta U = q_{rev} - w_{rev} = q_{irrev} - w_{irrev} \qquad ...(v)$$

Combining results (*iv*) and (*v*), we come to the conclusion

$$q_{rev} > q_{irrev} \qquad ...(vi)$$

$$\therefore \quad \frac{q_{rev}}{T} > \frac{q_{irrev}}{T} \qquad ...(vii)$$

or $\qquad \frac{q_{rev}}{T} - \frac{q_{irrev}}{T} > 0 \qquad ...(viii)$

From (*iii*) and (*viii*), we have

$$\Delta S_{system} + \Delta S_{surroundings} > 0 \qquad ...(ix)$$

Thus it may be concluded that *in an irreversible process, the entropy change for the combined system and the surroundings is greater than zero i.e., an irreversible process is accompanied by a net increase of entropy.*

Since all spontaneous processes are thermodynamically irreversible it may be stated that *All spontaneous processes are accompanied by a net increase of entropy.*

From Art 2.4.1 and 2.4.2, we can say that sum of entropy changes for the system and surroundings for a process in thermodynamic equilibrium (reversible process) is zero whereas for a process *not* in equilibrium (irreversible process), the sum is greater than zero.

2.4.3 Clausius Inequality

For an irreversible process, we can write

$$\Delta S_{system} + \Delta S_{surroundings} > 0 \qquad \ldots(i)$$

or

$$\Delta S_{system} > - \Delta S_{surroundings} \qquad \ldots(ii)$$

If q_{irrev} is the heat change in surroundings at the temperature T, then

$$\Delta S_{surroundings} = - \frac{q_{irrev}}{T} \qquad \ldots(iii)$$

Substituting this value in (ii) above, we get

$$\Delta S_{system} > \frac{q_{irrev}}{T} \qquad \ldots(iv)$$

If the change in entropy is small, eq. (iv) can be written as

$$dS_{system} > \frac{q_{irrev}}{T} \qquad \ldots(v)$$

Expressions (iv) and (v) are known as Clausius inequality. Clausius assumed that the universe could be taken as isolated system, in which all naturally occurring processes are irreversible.

2.4.4 Entropy of the Universe is Increasing

All processes occurring in nature are thermodynamically irreversible and these are accompanied by increase of entropy. Hence it may be concluded that

The entropy of the universe is continuously increasing.

Thus the main ideas of the *first and the second law of thermodynamics* may be summed up as follows:

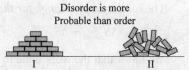

If you unload bricks from a truck onto the ground, the heap of bricks will more likely appear as shown in II than in I. This proves that disorder is more probable than order.

The energy of the universe is constant whereas the entropy of the universe is continuously increasing and tends to a maximum value.

2.5 ENTROPY CHANGE FOR AN IDEAL GAS UNDER DIFFERENT CONDITIONS

Let us consider one mole of an ideal gas enclosed in a cylinder fitted with a frictionless piston. If a small amount of heat δq_{rev} is supplied to the system reversibly and isothermally at the temperature T, then the entropy change accompanying the process is given by

$$dS = \frac{\delta q_{rev}}{T} \qquad \ldots(i)$$

According to first law of thermodynamics, for an infinitesimal process,

$$\delta q = dE + \delta w \qquad \ldots(ii)$$

If the process is carried out *reversibly*, equation (ii) may be written as

$$\delta q_{rev} = dE + \delta w \qquad \ldots(iii)$$

Also, if the *work is restricted to the work of expansion,*

$$\delta w = PdV \qquad \ldots(iv)$$

where dV is the small increase in volume and P is the pressure of the system.

Putting the value of dw from equation (iv) in equation (iii), we get

$$\delta q_{rev} = dE + PdV \qquad \ldots(v)$$

Substituting this value in equation (i), we have

$$dS = \frac{dE + PdV}{T}$$

or it may be written as

$$TdS = dE + PdV \qquad \ldots(vi)$$

For 1 mole of an ideal gas, we know that

$$C_v = \frac{dE}{dT}$$

i.e,

$$dE = C_v dT \qquad \ldots(vii)$$

where C_v is the molar heat capacity at constant volume and

$$PV = RT$$

i.e.

$$P = \frac{RT}{V} \qquad \ldots(viii)$$

where V is the volume of the system at temperature T and pressure P and R is the gas constant.

Substituting the values of dE and P from equation (vii) and (viii) in equation (vi), we get

$$TdS = C_v dT + \frac{RT}{V} dV$$

or

$$dS = C_v \frac{dT}{T} + R \frac{dV}{V} \qquad \ldots(ix)$$

If the volume changes from V_1 to V_2 when the temperature changes from T_1 to T_2, then the entropy change accompanying the complete process is given by the equation

$$\int_{S_1}^{S_2} dS = \int_{T_1}^{T_2} C_v \frac{dT}{T} + \int_{V_1}^{V_2} R \frac{dV}{V} \qquad \ldots(x)$$

Assuming that C_v remains constant in the temperature range T_1 to T_2, equation (x) may be put as

$$\int_{S_1}^{S_2} dS - = C_v \int_{T_1}^{T_2} \frac{dT}{T} + R \int_{V_1}^{V_2} R \frac{dV}{V}$$

$$\Delta S = C_v \ln \frac{T_2}{T_1} + R \ln \frac{V_2}{V_1} \qquad \ldots(xi)$$

This is an expression for the calculation of entropy change of 1 mole of an ideal gas accompanying a process when temperature changes from T_1 to T_2 and the volume changes from V_1 to V_2.

Further for 1 mole of an ideal gas, we may write

$$P_1 V_1 = RT_1 \text{ for the initial state} \qquad \ldots(xii)$$

and

$$P_2 V_2 = RT_2 \text{ for the final state} \qquad \ldots(xiii)$$

Dividing equation (xiii) by equation (xii), we have

$$\frac{P_2 V_2}{P_1 V_1} = \frac{T_2}{T_1}$$

or
$$\frac{V_2}{V_1} = \frac{T_2 P_1}{T_1 P_2} \qquad \text{...(xiv)}$$

Substituting this value in equation (xi), we get

$$\Delta S = C_v \ln \frac{T_2}{T_1} + R \ln \frac{T_2 P_1}{T_1 P_2} \qquad \text{...(xv)}$$

Also, we know that

$$C_p - C_v = R \text{ or } C_v = C_p - R$$

Putting this value in equation (xv), we get

$$\Delta S = (C_p - R) \ln \frac{T_2}{T_1} + R \ln \frac{T_2 P_1}{T_1 P_2}$$

$$= C_p \ln \frac{T_2}{T_1} - R \ln \frac{T_2}{T_1} + R \ln \frac{T_2}{T_1} + R \ln \frac{P_1}{P_2}$$

or
$$\Delta S = C_p \ln \frac{T_2}{T_1} + R \ln \frac{P_1}{P_2} \qquad \text{...(xvi)}$$

This equation gives the entropy change of 1 mole of an ideal gas accompanying a process when temperature changes from T_1 to T_2 and the pressure changes from P_1 to P_2.

(i) *If temperature is kept constant* (**isothermal process**)

$$T_1 = T_2$$

Equations (xi) and (xvi) are reduced to

$$\Delta S = R \ln \frac{V_2}{V_1} = R \ln \frac{P_1}{P_2} \qquad \text{...(xvii)}$$

(ii) *If pressure is kept constant* (**isobaric process**)

$$P_1 = P_2$$

Equation (xvi) is reduced to

$$\Delta S = C_p \ln \frac{T_2}{T_1} \qquad \text{...(xviii)}$$

(iii) *If volume is kept constant* (isochoric process)

$$V_1 = V_2$$

Equation (xi) converts to

$$\Delta S = C_p \ln \frac{T_2}{T_1} \qquad \text{...(xix)}$$

2.5.1 Entropy Change During Phase Change

A phase change involves the following:

(i) Melting of solid into liquid and freezing of liquid into solid.

(ii) Boiling of liquid into vapour and condensation of vapour into liquid

Entropy of melting

$$\Delta S_m = \frac{\Delta H_m}{T_m}$$

where ΔH_m is the latent heat of melting and T_m is the melting temperature.

Entropy of vaporisation

$$\Delta S_v = \frac{\Delta H_v}{T_b}$$

where ΔH_v is the latent heat of vaporisation and T_b is the boiling temperature.

Processes like melting and evaporation involve *absorption of heat* which is taken as *positive* by conventions of sign, therefore ΔS for these processes is positive *i.e.* these processes are accompanied by *increase of entropy*. On the other hand, the reverse processes like freezing and condensation involve *evolution of heat which* is taken as *negative*, therefore for these processes ΔS is negative *i.e.* these processes are accompanied by *decrease of entropy*.

Solved Examples—Entropy Change

Example 4: Calculate the increase in entropy in the evaporation of one mole of water at 3.73 K. Latent heat of vaporisation of water is 2.26 kJ/g (molecular mass of water = 18)

Solution: Given that

$$T_1 = 373 \text{ K}$$
$$\Delta H_v = 2260 \text{ Jg}^{-1} = 2260 \times 18 \text{ Jmol}^{-1} = 40680 \text{ J mol}^{-1}$$

$$\therefore \quad \Delta S = \frac{\Delta H_v}{T_b} = \frac{40680}{373} = 109 \text{ JK}^{-1} \text{ mol}^{-1}$$

Example 5: Calculate the entropy change for the fusion of 1 mole of a solid which melts at 300 K. The latent heat of fusion is 2.51 kJ mol⁻¹.

Solution: Given that

$$\Delta H_m = 2510 \text{ J mol}^{-1}$$
$$T_m = 300 \text{ K}$$

$$\therefore \quad \Delta S = \frac{\Delta H_m}{T_m}$$

$$= \frac{2510}{300} = 8.37 \text{ JK}^{-1} \text{ mol}^{-1}$$

Example 6: Calculate the temperature at which the reaction Ag_2O (s) $\rightarrow$ 2Ag (s) + $\frac{1}{2}O_2$ (g) at 1 atmospheric pressure will be in equilibrium. The values of ΔH and ΔS for the reaction are 30.58 kJ and 66.11 JK⁻¹ respectively and these values do not change much with temperature.

Solution: For a reaction in equilibrium

$$\Delta S = \frac{\Delta H}{T}$$

Given that $\Delta H = 30.58 \text{ kJ} = 30580 \text{ J}$ and $\Delta S = 66.11 \text{ JK}^{-1}$

Substituting the values in the above equation, we have

$$66.11 = \frac{30580}{T} \quad \text{or} \quad T = 462.6 \text{ K}$$

Example 7: Calculate the entropy change involved in the isothermal reversible expansion of 5 moles of an ideal gas from a volume of 10 litres to a volume of 100 litres at 300 K.

Solution: We have the following expression for ΔS in terms of temperature and volume

$$\Delta S = C_v \ln \frac{T_2}{T_1} + R \ln \frac{V_2}{V_1} \quad \text{for 1 mole of an ideal gas}$$

For isothermal process $\quad T_1 = T_2$.

$\therefore \qquad\qquad \Delta S = R \ln \dfrac{V_2}{V_1} \qquad\qquad$ for 1 mole of an ideal gas

or $\qquad\qquad \Delta S = nR \ln \dfrac{V_2}{V_1} \qquad\qquad$ for n mole of an ideal gas

or $\qquad\qquad \Delta S = 2.303 \, nR \log \dfrac{V_2}{V_1}$

Given that

$$n = 5, \, V_1 = 10 \text{ litres}, \, V_2 = 100 \text{ litres}$$
$$R = 8.314 \text{ JK}^{-1} \text{mol}^{-1}$$

$\therefore \qquad\qquad \Delta S = 2.303 \times 5 \times 8.314 \times \log \dfrac{100}{10} = 95.75 \text{ JK}^{-1} \text{mol}^{-1}$

Example 8: 100 g of ice at 0°C was added to an insulated beaker containing 150 g of water at 100°C. Calculate

    (i) The final temperature of the system and

    (ii) The entropy change of the process.

Heat of fusion of ice is 80 cal/g and heat capacity of water is 1.0 cal/g/degree (Ignore the heat capacity of the beaker).

Solution: (i) Let the final temperature after mixing be $= t\,°C$

$\qquad$ Heat lost by water $= 150 \times (100 - t)$ cal

$\qquad$ Heat gained by ice $= 100 \times 80 + 100 \times (t - 0)$

$\qquad\qquad\qquad\qquad\quad = 8000 + 100\, t$ cal

$\qquad$ Heat gained $=$ Heat lost

$\qquad\qquad 8000 + 100t = 150 (100 - t)$

or $\qquad\qquad\qquad t = \mathbf{28°C}$

(ii) the process involves the following steps:

    (a) 100 g of ice melts at 0°C to give 100 g of liquid water.

    (b) 100 g of liquid water gets heated up from 0°C to 28°C.

    (c) 150 g of water gets cooled from 100°C to 28°C.

When temperature remains constant (*i.e.* for isothermal process), $\Delta S = \dfrac{q_{\text{rev}}}{T}$. When temperature varies (but volume or pressure remains constant), $\Delta S = C \ln \dfrac{T_2}{T_1}$, where C is C_v or C_p.

Entropy change in step (a) $= \dfrac{(100 \text{ g}) \times (80 \text{ cal/g})}{273} = \dfrac{8000}{273} = 29.30 \text{ cal deg}^{-1}$

Entropy change in step (b) $= (100 \text{ g})(1 \text{ cal/g}) \times 2.303 \times \log \dfrac{28 + 273}{0 + 273}$

$$= 230.3 \log \frac{301}{273} = 230.3 \times 0.0424$$

$$= 9.765 \text{ cal deg}^{-1}$$

Entropy change in step (c)

$$= (150 \text{ g}) \times (1 \text{ cal/g}) \times 2.303 \log \frac{28+273}{100+273}$$

$$= 345.45 \log \frac{301}{373} = 345.45 \times (-0.0931)$$

$$= -32.16 \text{ cal deg}^{-1}$$

∴ Net entropy change in the process is given by

$$= 29.30 + 9.765 - 32.16$$

$$= 6.905 \text{ cal deg}^{-1}$$

Example 9. Write expressions for entropy changes in the following cases:

(i) Temperature changes from T_1 to T_2 and volume changes from V_1 to V_2 for 1 mole of an ideal gas

(ii) Temperature changes from T_1 to T_2 and pressure changes from P_1 to P_2 for 1 mole of an ideal gas

(iii) 1 mole of a solid melts at its melting point.

(iv) 1 mole of a liquid vaporises at its boiling point.

Solution: (i) $\Delta S = C_v \ln \dfrac{T_2}{T_1} + R \ln \dfrac{V_2}{V_1}$

(ii) $\Delta S = C_p \ln \dfrac{T_2}{T_1} + R \ln \dfrac{P_1}{P_2}$

(iii) $(\Delta S)_{fus} = \dfrac{(\Delta H)_{fus}}{T_f}$

(iv) $(\Delta S)_{vap} = \dfrac{(\Delta H)_{vap}}{T_b}$

Problems for Practice

1. Calculate the change in entropy when 40 g of water at 50°C are mixed with 80 g of water at −20°C. **[Ans. 0.12 cal deg⁻¹]**

2. For the reaction

$$H_2O (l) \longrightarrow H_2O (g)$$

$\Delta H = 9710$ cal mol−1, $P = 1$ atm; $\Delta S = 26$ cal deg^{-1} mol^{-1}

Find out the temperature at which liquid water and water vapour at 1 atmospheric pressure are in equilibrium. **[Ans. 100.46°C]**

3. One mole of an ideal gas expands from a volume of 5 dm³ to a volume of 60 dm³ in an isothermal reversible process at 300 K. Calculate the entropy change during the process ($R = 8.314$ J K⁻¹ mol⁻¹). **[Ans. 20.67 JK⁻¹ mol⁻¹]**

4. Calculate the entropy change involved in the conversion of 1 mole of ice at 273 K to water at the same temperature.
   Latent heat of fusion of ice is 335 joules per gram. [**Ans.** 22.1 J K^{-1}]

5. A heated copper block at 130°C loses 340 J of heat to the surroundings which are at room temperature of 32°C. Calculate
   (i) entropy change of the copper block
   (ii) entropy change of the surroundings
   (iii) total entropy change of the universe due to this process.
   Assume that the temperature of the block and the surroundings remains constant.
   [**Ans.** (i) –0.84 J K^{-1} (ii) +1.11 J K^{-1} (iii) +0.27 J K^{-1}]

6. Calculate the entropy change when 1 mole of solid SO_2 melts at –76°C. The latent heat of fusion is 1769 cal mol^{-1}. [**Ans.** 8.98 cal K^{-1} mol^{-1}]

7. 1 g of ice at 0°C is added to 10 g of water at the boiling point. What will be the final temperature and what is the entropy change accompanying this process ? Assume that the heat of fusion of ice is 80 cal/g and the specific heat of water is 1 cal/degree.
   [**Ans.** t = 83.6°C, ΔS = 0.11 cal deg^{-1}]

8. Calculate the entropy change when 2 moles of an ideal gas are allowed to expand from a volume of 1.0 litre to a volume of 10.0 litres at 27°C. [**Ans.** 38.29 J K^{-1} mol^{-1}]

9. For a certain ideal gas, $C_p = \frac{5}{2}R$ joules mol^{-1} deg^{-1}. Calculate the change in entropy suffered by 3 moles of the gas on being heated from 300 K to 600 K at (a) constant pressure (b) constant volume. [**Ans.** (a) 43.2 J K^{-1} (b) 25.9 J K^{-1}]

2.5.2 Physical Significance of Entropy

Various processes which are accompanied by a net increase of entropy are associated with an increased randomness of distribution. For example :

(i) **The melting of a solid** is accompanied by a net increase of entropy. We know that the molecules, atoms or ions in a solid have fixed positions and they become free to move about in the molten state. This is expressed by saying that there is no *disorder* among the molecules, atoms or ions in the solid but the disorder sets in when the solid changes into liquid. In other words, we can say that the *randomness* has increased.

(ii) **The vaporization of a liquid** is accompanied by a net increase of entropy. The disorder or the randomness also increases because the molecules of vapour are more random than the molecules in the liquid.

Thus

Entropy is a measure of randomness or disorder of the system.

This concept may be further understood with the help of the following analogy.

In a college when all classes are being held, all the students are sitting in their respective class rooms seriously and the disorder is minimum. As soon as the bell goes, the students of different classes move out to go to other rooms and disorder sets in. In other words, the disorder or the randomness increases.

Solved Examples—Second Law of Thermodynamics

Example 10: Is it possible to construct a reversible heat engine of unit efficiency? Give detailed reason for your answer.

Solution: $\eta = \dfrac{T_2 - T_1}{T_2}$. For η to be equal to unity, T_1 should be 0 K, *i.e.*, temperature of the sink should be zero degree absolute which is impossible. Hence, **it is not** possible to construct a reversible heat engine of unit efficiency.

Example 11: Write the expression for the efficiency of Carnot cycle in terms of work done, heat absorbed and temperatures of the source and sink.

Solution:
$$\eta = \dfrac{w}{q_2} = \dfrac{q_2 - q_1}{q_2} = \dfrac{T_2 - T_1}{T_2}$$

where
- w = Net work done
- q_2 = Total heat absorbed
- $q_2 - q_1$ = Net heat absorbed $\qquad (\because w = q_2 - q_1)$
- T_2 = Temperature of source
- T_1 = Temperature of sinks

Example 12: How can the efficiency of Carnot engine be increased?

Solution: $\eta = \dfrac{T_2 - T_1}{T_2}$. Hence, to increase the efficiency of the heat engine, the temperature of the sink (T_1) should be minimum possible for a particular temperature of the source (T_2).

Example 13: Can the efficiency of a heat engine be greater than unity? If not why?

Solution: $\eta = \dfrac{T_2 - T_1}{T_2}$. Efficiency ($\eta$) cannot be greater than unity because temperature of sink (T_1) has to be less than the temperature of the source (T_2).

Example 14: What is the basis of thermodynamic scale of temperature? Why is it more basic than that based on perfect or ideal gas?

Solution: Thermodynamic scale of temperature as defined on the basis of second law of thermodynamics is more basic than that based on the ideal gas. This is because it is independent of the working substance.

Example 15: Which of the following have the higher value of S?

(*i*) CO_2 at 15°C, 1 atm or dry ice at 1 atm.

(*ii*) a coiled spring or a spring relaxed.

(*iii*) 1 g of liquid water at 25°C, or 1 g of water vapour at 25°C.

(*iv*) silica glass or quartz (crystalline silica).

Solution: (*i*) CO_2 at 15° because gas is more random than solid.

(*ii*) A spring relaxed is more random.

(*iii*) Water vapour have greater randomness than liquid water.

(*iv*) Silica is more random than quartz because in the former SiO_4 tetrahedra are random while in the latter, they are arranged in a perfect order.

Example 16: Tell which will have greater entropy in each of the following cases. Justify your answer.

(*i*) Magnetised iron piece or non-magnetised piece of iron.

(*ii*) A mole of gas contained in a vessel of one litre capacity or that contained in a vessel of two-litre capacity.

(*iii*) A mole of $CaCO_3$ (s) or a system containing one mole of each of CaO (s) and CO_2 (g).

Solution: (*i*) Non-magnetised piece of iron.

(*ii*) A mole of gas contained in a vessel of 2 litre capacity has greater randomness and hence greater entropy.

(*iii*) One of mole of CaO + CO_2 will have greater entropy because gas has greater randomness.

Example 17: Write expressions for the entropy change in the following cases:

(*i*) Temperature changes from T_1 to T_2 and volume changes from V_1 to V_2 for 1 mole of an ideal gas.

(*ii*) Temperature changes from T_1 to T_2 and pressure changes from V_1 to V_2 for 1 mole of an ideal gas.

(*iii*) 1 mole of a solid melts at the melting point.

(*iv*) 1 mole of a liquid vaporises at the boiling point.

Solution: (*i*) $\Delta S = C_v \ln \dfrac{T_2}{T_1} + R \ln \dfrac{V_2}{V_1}$ (*iii*) $\Delta_{fus} S = \dfrac{\Delta_{fus} H}{T_f}$

(*ii*) $\Delta S = C_v \ln \dfrac{T_2}{T_1} + R \ln \dfrac{P_1}{P_2}$ (*iv*) $\Delta_{vap} S = \dfrac{\Delta_{vap} H}{T_b}$

Example 18: Justify the 'Second Law of Thermodynamics' which states that "The net entropy of the universe tends to increase."

Solution: All spontaneous processes occurring in nature are thermodynamically irreversible and every irreversible process is accompanied by increase in entropy. Hence entropy of the universe is continuously increasing.

Example 19: In the following processes, state whether the entropy of the system increases, decreases or remains constant:

(*i*) Vaporization of a mole of water into steam at its boiling point.

(*ii*) Solidification of a mole of liquid at its freezing point.

(*iii*) Isothermal compression of a liquid.

(*iv*) Separation of a mixture into its two constituents.

Solution: (*i*) Entropy increases   (*ii*) Entropy decreases

(*iii*) Entropy remains almost constant,   (*iv*) Entropy decreases.

Example 20: Select the most random system of the following:

(*i*) sugar cubes in a box   (*ii*) sugar cubes scattered on a table

(*iii*) sugar dissolved in water   (*iv*) finely powdered sugar.

Solution: Sugar dissolved in water is the most random system because the molecules are free to move about in the solution.

2.6 ENTROPY CHANGE ON MIXING OF IDEAL GASES

Suppose at constant temperature, n_1 moles of an ideal gas 1 at the initial pressure $P_1°$ are mixed with n_2 moles of another ideal gas 2 at initial pressure $P_2°$. After mixing let their partial pressures in the mixture be P_1 and P_2 respectively. We know, at constant temperature, the entropy change of an ideal gas when its pressure changes from initial pressure P_i to final pressure P_f is given by

$$\Delta S = R \ln \frac{P_i}{P_f} \text{ mol}^{-1}$$

∴ Entropy change of the first gas when the pressure of n_1 moles of the gas changes from $P_1°$ to P_1 is given by

$$\Delta S_1 = n_1 R \ln \frac{P_1^0}{P_1} \qquad \qquad ...(i)$$

Similarly, entropy change of the second gas when pressure of n_2 moles of the gas changes from $P_2°$ to P_2 is given by

$$\Delta S_2 = n_2 R \ln \frac{P_2^0}{P_2} \qquad \qquad ...(ii)$$

Total entropy change on mixing the two gases will be the sum of the above two changes.

$$\Delta S_{mixing} = n_1 R \ln \frac{P_1^0}{P_1} + n_2 R \ln \frac{P_2^0}{P_2} \qquad \qquad ...(iii)$$

Let P be the total pressure of the mixture. Then, $P = P_1 + P_2$. Further let x_1 and x_2 be the mole fractions of gases 1 and 2 in the mixture. By Dalton's law of partial pressures,

$$P_1 = x_1 P \text{ and } P_2 = x_2 P$$

Substituting these values in eqn. (iii), we get

$$\Delta S_{mixing} = n_1 R \ln \frac{P_1^0}{x_1 P} + n_2 R \ln \frac{P_2^0}{x_2 P} \qquad \qquad ...(iv)$$

Taking the simplest case in which each gas is taken at the same initial pressure. Under these conditions, after mixing, volume of the mixture will be the sum of their initial volumes, i.e., $V = V_1 + V_2$. And final pressure of the mixture will be nearly the same as initial pressure of each gas, i.e., $P_1° = P_2° = P$ as shown in the diagram below. Equation (iv) then is simplified to

$$\Delta S_{mixing} = -n_1 R \ln x_1 - n_2 R \ln x_2$$

or

$$\Delta S_{mixing} = -R(n_1 \ln x_1 + n_2 \ln x_2) \qquad \qquad ...(v)$$

Gas 1	Gas 2		On mixing	
Pressure = P Temp = T Volume = V_1 No. of units = n_1	+	Pressure = P Temp = T Volume = V_2 No. of units = n_2	→	Pressure = P Temp = T Volume = $V_1 + V_2$ No. of units = $n_1 + n_2$

For a mixture of a number of gases, eqn. (v) can be written in the general form as under:

$$\Delta S_{mixing} = -R \Sigma n_i \ln x_i \qquad \qquad ...(vi)$$

Entropy change for *one mole* of the mixing is obtained by dividing eqn. (v) by $(n_1 + n_2)$.

$$\Delta S_{mixing} = -R\left(\frac{n_1}{n_1+n_2}\ln x_1 + \frac{n_2}{n_1+n_2}\ln x_2\right)$$

or
$$\Delta S_{mixing} = -R(x_1 \ln x_1 + x_2 \ln x_2) \qquad ...(vii)$$

which can be generalized to the form

$$\boxed{\Delta S_{mixing} = -R\sum x_i \ln x_i} \qquad ...(viii)$$

We make the following conclusions from the above equations.

1. ΔS_{mixing} is independent of temperature.

2. As $x_i < 1$, ΔS_{mixing} will always be positive *i.e.* mixing of gases is accompanied by increase in entropy.

Solved Examples—Calculation of Entropy of Mixing of Ideal Gases

Example 21: Calculate the molar entropy change of mixing 0.2 mole of oxygen and 0.6 mole of nitrogen at 25°C assuming that they are ideal gases.

Solution: $n_1 = 0.2$ mole, $n_2 = 0.6$ mole

$$\therefore x_1 = \frac{0.2}{0.8} = 0.25, \; x_2 = \frac{0.6}{0.8} = 0.75$$

$$\therefore \Delta S_{mixing} = -R(x_1 \ln x_1 + x_2 \ln x_2)$$
$$= -2.303\, R(x_1 \log x_1 + x_2 \log x_2)$$
$$= -2.303 \times 8.314\, (0.25 \log 0.25 + 0.75 \log 0.75)$$
$$= +2.303 \times 8.314 \times 0.2443 \text{ J K}^{-1} \text{ mol}^{-1}$$
$$= 4.678 \text{ J K}^{-1} \text{ mol}^{-1}$$

Example 22: Write expressions for entropy change when n_1 moles of an ideal gas 1 are mixed with n_2 moles of another ideal gas 2. How can you modify the equation in terms of mole fraction? Comment on the effect of temperature.

Solution: Entropy of mixing of two gases is given by

$$\Delta S_{mix} = -R\left[n_1 \ln \frac{n_1}{n_1+n_2} + n_2 \ln \frac{n_2}{n_1+n_2}\right] \qquad ...(i)$$

It may be written as

$$\Delta S_{mix} = -R(n_1 \ln x_1 + n_2 \ln x_2) \qquad ...(ii)$$

Where x_1 and x_2 stand for mole fractions of gas 1 and gas 2 respectively. The total number of moles of the two components in the above mixture is $n_1 + n_2$. So to calculate ΔS_{mix} for 1 mole of the mixture, we need to divide the R.H.S. of eq. (*ii*) by $n_1 + n_2$

$$\Delta S_{mix} \text{ (for 1 mole of mixture)} = -R\left[\frac{n_1}{n_1+n_2}\ln x_1 + \frac{n_2}{n_1+n_2}\ln x_2\right] \qquad ...(iii)$$

$$= -R(x_1 \ln x_1 + x_2 \ln x_2)$$

We find that temperature T is not involved in the equation

Hence ΔH_{mix} is independent of temperature.

2.7 HELMHOLTZ FUNCTION OR WORK FUNCTION

Another thermodynamic quantity which is used in the study of thermodynamics is called Helmholtz function or work function or Helmholtz free energy.

This is denoted by A and is defined by the equation

$$A = U - TS \qquad \ldots(i)$$

where U is the internal energy of the system, T is the temperature and S is the entropy.

Since, U, T and S are the functions of the state of the system only and do not depend upon its previous history, therefore A also must be a state function.

In order to understand the *physical significance* of the work function, consider an isothermal change taking place at temperature T. then

$$A_1 = U_1 - TS_1 \text{ for the initial state} \qquad \ldots(ii)$$

and
$$A_2 = U_2 - TS_2 \text{ for the final state} \qquad \ldots(iii)$$

where A_1, U_1 and S_1 are respectively the values of work function, internal energy and entropy of the system in the initial state and A_2, U_2 and S_2 are the corresponding values in the final state.

∴ Change in the function A accompanying the process is given by

$$A_2 - A_1 = (U_2 - TS_2) - (U_1 - TS_1)$$
$$= (U_2 - U_1) - T(S_2 - S_1)$$

or
$$\Delta A = \Delta U - T\Delta S \qquad \ldots(iv)$$

According to the definition of entropy,

$$\Delta S = \frac{q_{rev}}{T} \qquad \ldots(v)$$

According to the first law of thermodynamics as applied to an isothermal reversible process, we have

$$\Delta U = q_{rev} - w_{max} \qquad \ldots(vi)$$

Substituting the values of ΔS and ΔU from equation (v) and (vi) in equation (iv), we get

$$\Delta A = (q_{rev} - w_{max}) - T \cdot \frac{q_{rev}}{T}$$

or
$$-\Delta A = w_{max} \qquad \ldots(vii)$$

Thus *for a process occurring at constant temperature, the decrease in the Helmholtz function work function A is equal to the maximum work done by the system.* It is for this reason that this thermodynamic quantity has been termed as work function.

2.8 GIBB'S FUNCTION OR GIBB'S FREE ENERGY*

This is another thermodynamic quantity.

This is usually denoted by G and is defined the equation

$$G = H - TS \qquad \ldots(i)$$

where H, T and S are the heat content, temperature and entropy of the system respectively.

---

*As per the latest recommendation of IUPAC (International Union of Pure and Applied Chemists), we should use the terms Gibb's energy in place of Gibb's free energy).

Since H, T and S are the function of the state only, therefore G is also function of the state of the system only. Hence for the isothermal process occurring at temperature T, we can write

$$G_1 = H_1 - TS_1 \text{ for the initial state} \quad \ldots(ii)$$
$$G_2 = H_2 - TS_2 \text{ for the final state} \quad \ldots(iii)$$

or
$$G_2 - G_1 = (H_2 - TS_2) - (H_1 - TS_1)$$
$$= (H_2 - H_1) - T(S_2 - S_1)$$

or
$$\Delta G = \Delta H - T\Delta S \quad \ldots(iv)$$

where
ΔG = Change in Gibb's function free energy of the system
ΔH = Enthalpy change of the system

and
ΔS = Entropy change of the system

The physical significance of Gibb's function may be understood as follows:

At constant temperature (T),

$$\Delta S = \frac{q_{rev}}{T}$$

i.e.
$$T\Delta S = q_{rev} \quad \ldots(v)$$

At constant pressure (P), $\Delta H = \Delta U + P\Delta V$ $\quad \ldots(vi)$

Substituting the values of $T\Delta S$ and ΔH from equation (v) and (vi) in equation (iv), we get

$$\Delta G = (\Delta U + P\Delta V) - q_{rev}$$
$$= (\Delta U - q_{rev}) + P\Delta V \quad \ldots(vii)$$

Now, according to first law of thermodynamics

$$\Delta U = q_{rev} - w_{max} \quad \text{(work done by a system carries negative sign)}$$

or $\Delta U - q_{rev} = -w_{max}$

Substituting this value in equation (vii), we get

$$\Delta G = -w_{max} + P\Delta V$$
$$\Delta G = w_{max} - P\Delta V$$

But $P\Delta V$ is the *work of expansion* done by the system corresponding to the increase in volume ΔV. Hence ($w_{max} - P\Delta V$) gives the maximum work other than the work of expansion. This is called the *maximum useful work* available from the process. We conclude that *for a process occurring at constant temperature and constant pressure, the decrease in Gibb's function is equal to the* **maximum useful work** *obtainable from the process i.e. the total work minus the pressure volume work* (*or work of expansion*).

2.8.1 Variation of Helmholtz Function with Temperature and Volume

The work function is given by the equation

$$A = U - TS \quad \ldots(i)$$

Complete differentiation of this equation gives

$$dA = dU - TdS - SdT \quad \ldots(ii)$$

Entropy change is given by

$$dS = \frac{\delta q_{rev}}{T} \quad \ldots(iii)$$

From first law of thermodynamics,

$$\delta q_{rev} = dU - \delta w \qquad \ldots(iv)$$

and if the work is restricted to the work of expansion only,

$$-\delta w = PdV \qquad \ldots(v)$$

Putting this value in equation (iv), we have

$$\delta q_{rev} = dU + PdV \qquad \ldots(vi)$$

Putting this value in equation (iii), we get

$$dS = \frac{dU + PdV}{T}$$

or

$$TdS = dU + PdV \qquad \ldots(vii)$$

Putting this value in equation (ii), we obtain

$$dA = dU - dU - PdV - SdT$$
$$= -PdV - SdT \qquad \ldots(viii)$$

(a) *If temperature is kept constant*, $dT = 0$. Equation (viii) takes the form

$$(dA)_T = -(PdV)_T \qquad \ldots(ix)$$

or

$$\left(\frac{\partial A}{\partial A}\right)_T = -P \qquad \ldots(x)$$

(b) *If volume is kept constant*, $dV = 0$. Equation (viii) becomes

$$(dA)_V = -(SdT)_V$$

$$\left(\frac{\partial A}{\partial A}\right)_V = -S \qquad \ldots(xi)$$

Thus variation of Helmholtz function with temperature and volume is given by eqs. (x) and (xi).

2.8.2 Variation of Gibb's Function (free energy) with Temperature and Pressure

The free energy is given by the equation

$$G = H - TS \qquad \ldots(i)$$

But,

$$H = U + PV \qquad \ldots(ii)$$

Substituting this value in equation (i), we get

$$G = U + PV - TS \qquad \ldots(iii)$$

Complete differentiation of this equation gives

$$dG = dU + PdV + VdP - TdS - SdT \qquad \ldots(iv)$$

But

$$dS = \frac{\delta q_{rev}}{T} = \frac{dU + PdV}{T}$$

or

$$TdS = dU + PdV \qquad \ldots(v)$$

Substituting this value in equation (iv), we get

$$dG = dU + PdV + VdP - (dU + PdV) - SdT = VdP - SdT$$

i.e.

$$dG = VdP - SdT \qquad \ldots(vi)$$

This is expression gives the change in free energy with change in pressure and change in temperature in a reversible process. Equation (vi) is called **total differential equation.**

(a) *If temperature is kept constant,* $dT = 0$. Equation (vi) takes the form

$$(dG)_T = (VdP)_T \quad \ldots(vii)$$

or

$$\left(\frac{\partial G}{\partial P}\right)_T = V \quad \ldots(viii)$$

(b) *If pressure is kept constant,* $dP = 0$. Equation (vi) becomes

$$(dG)_p = -(SdT)_p$$

or

$$\left(\frac{\partial G}{\partial P}\right)_P = -S \quad \ldots(ix)$$

Thus variation of Gibb's energy with pressure and temperature is given equations (viii) and (ix) respectively.

2.8.3 Change in Gibb's Function or Free Energy for a Process Under Isothermal Condition

The change in Gibb's function for a complete process under isothermal conditions can be calculated as under:

For an infinitesimal change, under isothermal conditions, we have the equation

$$(dG)_T = VdP \quad \ldots(i)$$

For a complete process under isothermal conditions, when the pressure changes from P_1 to P_2, we have

$$(\Delta G)_T = \int_{P_1}^{P_2} VdP \quad \ldots(ii)$$

For n moles of an ideal gas $PV = nRT$, i.e., $V = \dfrac{nRT}{P}$

Substituting for V in eq. (ii), we have

$$(\Delta G)_T = \int_{P_1}^{P_2} \frac{nRT}{P} dP = nRT \int_{P_1}^{P_2} \frac{dP}{P}$$

$$= nRT \ln \frac{P_2}{P_1} \quad \ldots(iii)$$

Further for an isothermal process,

$$P_1V_1 = P_2V_2 \text{ or } \frac{P_2}{P_1} = \frac{V_1}{V_2}$$

Substituting this value in equation (iii), we get

$$\boxed{(\Delta G)_T = nRT \ln \frac{P_2}{P_1} = nRT \ln \frac{V_1}{V_2}} \quad \ldots(iv)$$

Solved Examples—Free Energy Change

Example 23: Calculate the change in Gibb's function accompanying the compression of 1 mole of CO_2 at 57°C from 5 atm to 50 atm. Assume that CO_2 behaves like an ideal gas.

Solution: Given that $n = 1$ mole

$T = 57 + 273$ K $= 330$ K

Taking
$P_1 = 5$ atm
$P_2 = 50$ atm
$R = 1.987$ cal deg^{-1} mol^{-1}, we have

$$(\Delta G)_T = 2.303\, nRT \log \frac{P_2}{P_1}$$

$$= 2.303 \times 1 \times 1.987 \times 303 \times \log \frac{5}{30} = \mathbf{1510.1\ cal}$$

Example 24: Calculate the change in Gibb's function which occurs when one mole of an ideal gas expands reversibly and isothermally at 300 K from the initial volume of 5 litres to 50 litres.

Solution: Given that
$n = 1$ mole
$T = 300$ K
$V_1 = 5$ litres
$V_2 = 50$ litres

Taking $R = 8.314$ JK^{-1} mol^{-1}

Substituting the values in the expression for $(\Delta G)_T$, we have

$$(\Delta G)_T = 2.303\, nRT \log \frac{V_1}{V_2}$$

$$= 2.303 \times 1 \times 8.314 \times 300 \times \log \frac{5}{30} = \mathbf{-5744.1\ joules}$$

Example 25: Write expressions for the following.

(i) Variation of Helmholtz function with volume at constant temperature.

(ii) Variation of Helmholtz function with temperature at constant volume.

(iii) Total change in Helmholtz function at constant temperature when the volume changes from V_1 to V_2.

Solution: (i) $\left(\dfrac{\partial A}{\partial V}\right)_T = -P$ (ii) $\left(\dfrac{\partial A}{\partial V}\right)_V = -S$ (iii) $(\Delta A)_T = -nRT \ln \dfrac{V_2}{V_1}$

Problems for Practice

1. During an isothermal reversible compression of one mole of an ideal gas, its volume decreases from 10 litres to 1 litre. The process was carried out at 27°C. Calculate the change in entropy and change in free energy (Gibb's function) of the gas. ($R = 2$ cal/degree/mol).
   [**Ans.** $\Delta S = -4.606$ cal deg^{-1}, $\Delta G = +1381.8$ cal]

2. 1 mole of an ideal gas is allowed to expand at 25°C till its pressure falls to one-fifth of its original pressure. Find out the free energy change accompanying the process.
   [**Ans.** -953.2 cal]

3. The pressure of one mole of an ideal gas at 25°C falls from 5.0 bar to 0.2 bar. Evaluate the change in free energy. ($R = 8.314$ J K^{-1} mol^{-1}) [**Ans.** 7976.2 J]

4. Four moles of an ideal gas expand isothermally from 1 litre to 10 litres at 300 K. Calculate the change in free energy of the gas. ($R = 8.314$ J K^{-1} mol^{-1}) [**Ans.** -22.98 kJ]

2.9 CRITERIA FOR FEASIBILITY OR SPONTANEITY OF A PROCESS

In the earlier sections in this chapter, the following generalisations were derived.

(i) If $\Delta S_{system} + \Delta S_{surroundings} > 0$; the process is irreversible, *i.e.*, it is spontaneous or feasible.

(ii) If $\Delta S_{system} + \Delta S_{surroundings} = 0$; the process is reversible *i.e.*, the system is in equilibrium.

The above two results may be combined and represented as

$$\Delta S_{system} + \Delta S_{surroundings} > 0 \qquad ...(i)$$

where the sign *greater than* refers to an irreversible process (*i.e.* spontaneous process) and the sign *equal to* for a reversible process.

Eq. (*i*) requires the knowledge of change in the entropy of the surroundings which is not so convenient to determine. Hence, we would search other criteria which involves change in entropy of the system only, in order to arrive at the criteria for feasibility of a process.

For an infinitesimal change, the above criterion may be written as

$$dS_{system} + dS_{surroundings} \geq 0 \qquad ...(ii)$$

where the sign, > stands for irreversible process and the sign = for the reversible process.

If the surroundings lose heat δq reversibly and isothermally at the temperature T (which may be absorbed by the system reversibly or irreversibly), then

$$dS_{surrounds} = -\frac{\delta q_{rev}}{T} \qquad ...(iii)$$

According to first law of thermodynamics,

$$\delta q_{rev} = dU + pdV \qquad ...(iv)$$

Substituting this value in equation (*iii*), we get

$$dS_{surrounding} = -\frac{dU + PdV}{T} \qquad ...(v)$$

Substituting this value in equation (*ii*), we get

$$dS_{system} - \frac{dU + PdV}{T} \geq 0$$

or $\qquad TdS - dU - PdV \geq 0$

or $\qquad \boldsymbol{TdS \geq dU + PdV} \qquad ...(vi)$

where dS stands for the entropy change of the system without the subscript *system*.

Equation (*vi*) is the basic equation which leads to a number of criteria for predicting the feasibility of a process. In this equation, the sign '=' stands for the reversible process and the sign '>' for the irreversible process. From equation (*vi*), the various criteria may be deduced as under.

(*i*) *In terms of entropy change of the system*. If the internal energy and the volume of the system are kept constant,

$$dU = 0 \text{ and } dV = 0$$

Then from equation (*vi*), we have

$$(TdS)_{U,V} \geq 0$$

or $\qquad \boldsymbol{(dS)_{U,V} \geq 0} \qquad ...(vii)$

where the sign '=' is for the reversible process and the sign '>' for the irreversible process. The subscripts U and V indicate that these properties remain constant.

(ii) *In terms of internal energy change of the system.* If the entropy and volume of the system are kept constant,
$$dS = 0, dV = 0$$
The equation (vi) becomes
$$0 \geq (dU)_{S,V}$$
or $\quad\quad (dU)_{S,V} \leq 0 \quad\quad$ ...(viii)

where the sign '=' stands for the reversible process and the sign '<' for the irreversible process. This implies that *if a process under constant entropy and volume is accompanied by a decrease of internal energy, the process is irreversible; if no change of internal energy takes place, the process is reversible.*

(iii) *In terms of enthalpy change of the system.*
$$H = U + PV$$
Complete differentiation of this equation gives
$$dH = dU + PdV + VdP$$
or $\quad\quad dU + PdV = dH - VdP$

Putting this value in equation (vi), we get
$$TdS \geq dH - VdP$$
If entropy and pressure of the system are kept constant,
$$dS = 0 \text{ and } dP = 0, \text{ so that}$$
$$0 \geq (dH)_{S,P}$$
or $\quad\quad (dH)_{S,P} \leq 0 \quad\quad$ ...(ix)

where the sign '=' refers to the reversible process and the sign '<' refers to the irreversible process.

(iv) *In terms of change in work function of the system.*
$$A = U - TS$$
Complete differentiation of this equation gives
$$dA = dU - TdS - SdT$$
or $\quad\quad TdS = dU - SdT - dA$

Putting this value in equation (vi), we get
$$dU - SdT - dA \geq dU + PdV$$
$$-SdT - dA \geq PdV$$
or $\quad\quad SdT + dA \leq -PdV$

If temperature and volume of the system are kept constant, $dT = 0, dV = 0$, so that we have
$$(dA)_{T,V} \leq 0 \quad\quad ...(x)$$
whereas usual sign '=' is for the reversible process and the sign '<' for the irreversible process.

(v) *In terms of free energy change of the system.*
$$G = H - TS$$
Further $\quad\quad H = U + PV$
$\therefore \quad\quad G = U + PV - TS$

Complete differentiation of this equation gives
$$dG = dU + PdV + VdP - TdS - SdT$$
or $\quad\quad dU + PdV = dG - VdP + TdS + SdT$

Substituting this value in equation (vi), we get
$$TdS \geq dG - VdP + TdS + SdT$$
or
$$0 \geq dG - VdP + SdT$$
or
$$dG - VdP + SdT \leq 0$$

If pressure and temperature of the system are kept constant, $dP = 0$, $dT = 0$, so that we obtain

$$(dG)_{P,T} \leq 0 \qquad ...(xi)$$

where, as before, the sign '=' refers to the reversile process and the sign '<' refers to the irreversible process.

The criterion in terms of free energy change is most important because most of the processes take place at constant temperature and pressure.

(i) If $(dG)_{P,T} < 0$, the process is irreversible, *i.e.*, it is feasible.
(ii) If $(dG)_{P,T} = 0$, the process is reversible, *i.e.*, the system is in equilibrium.
(iii) If $(dG)_{P,T} > 0$, the process does not occur, *i.e.*, it is not feasible.

2.10 MAXWELL RELATIONSHIPS

Certain variables in thermodynamics such as entropy are hard to measure experimentally Maxwell relations provide a way to exchange variables. Maxwell relationships are a set of equations which are derivable from the simple equations involving Helmboltz function (A), Gibb's function (G), internal energy (U), enthalpy (H), entropy (S) and pressure, volume, temperature etc.

The four basic equations that we have derived in previous sections of this chapter are

$$dU = TdS - PdV \qquad ...(i)$$
$$dH = TdS + VdP \qquad ...(ii)$$
$$dA = -SdT - PdV \qquad ...(iii)$$
$$dG = -SdT + VdP \qquad ...(iv)$$

Consider eq. (i). If V is constant, then $dV = 0$. Eq (i) gives the following result

$$\left(\frac{\partial U}{\partial S}\right)_V = T \qquad ...(1)$$

If S is constant, then eq. (i) gives the result

$$\left(\frac{\partial U}{\partial V}\right)_S = -P \qquad ...(2)$$

Differentiating equation (1) with respect to V keeping S constant, we get

$$\frac{\partial^2 U}{\partial S \partial V} = \left(\frac{\partial T}{\partial V}\right)_S \qquad ...(3)$$

Differentiating eq. (2) with respect to S keeping V constant, we get

$$\frac{\partial^2 U}{\partial V \partial S} = -\left(\frac{\partial P}{\partial S}\right)_V \qquad ...(4)$$

L.H.S. of equations (3) and (4) are equal, therefore we can equate their R.H.S.

Chemical Thermodynamics – II (Second and Third Laws of Thermodynamics and Partial Molar Properties)

$$\left(\frac{\partial T}{\partial V}\right)_S = -\left(\frac{\partial T}{\partial V}\right)_V \quad \ldots(5)$$

Following the same mathematical procedure with equations (*ii*), (*iii*) and (*iv*), we can obtain equations listed below

$$\left(\frac{\partial T}{\partial P}\right)_S = \left(\frac{\partial V}{\partial S}\right)_P \quad \ldots(6)$$

$$\left(\frac{\partial S}{\partial V}\right)_T = \left(\frac{\partial P}{\partial T}\right)_V \quad \ldots(7)$$

$$\left(\frac{\partial S}{\partial P}\right)_T = -\left(\frac{\partial V}{\partial T}\right)_P \quad \ldots(8)$$

Equations (5), (6), (7) and (8) are called **Maxwell relations**.

Using equations (*i*) and (*ii*), (*i*) and (*iii*), (*ii*) and (*iv*) & (*iii*) and (*iv*) respectively, we can obtain another set of equations as given below.

$$\left(\frac{\partial U}{\partial S}\right)_V = \left(\frac{\partial H}{\partial S}\right)_P \quad \ldots(9)$$

$$\left(\frac{\partial U}{\partial V}\right)_S = \left(\frac{\partial A}{\partial V}\right)_T \quad \ldots(10)$$

$$\left(\frac{\partial H}{\partial P}\right)_S = \left(\frac{\partial G}{\partial P}\right)_T \quad \ldots(11)$$

$$\left(\frac{\partial A}{\partial T}\right)_V = \left(\frac{\partial G}{\partial T}\right)_P \quad \ldots(12)$$

Solved Examples

Example 26: Show that $\left(\frac{\partial U}{\partial S}\right)_U = T - P\left(\frac{\partial T}{\partial P}\right)_S$.

Solution: We have the combined form of first and second laws of thermodynamics as

$$dU = TdS - PdV \quad \ldots(i)$$

Divide by dS, keeping P constant

$$\left(\frac{\partial U}{\partial S}\right)_P = T - P\left(\frac{\partial V}{\partial S}\right)_P \quad \ldots(ii)$$

But $\left(\frac{\partial V}{\partial S}\right)_P = \left(\frac{\partial T}{\partial P}\right)_S$ from Maxwell equation (6) above

$$\therefore \left(\frac{\partial U}{\partial S}\right)_P = T - P\left(\frac{\partial T}{\partial P}\right)_S$$

Example 27: Derive two thermodynamics equations of state from the combined form of first and second laws of thermodynamics and appropriate maxwell equations.

Solution: We have the relation from thermodynamics

$$dU = TdS - PdV$$

Dividing by dV and keeping T constant, we have

$$\left(\frac{\partial U}{\partial V}\right)_T = T\left(\frac{\partial S}{\partial V}\right)_T - P \qquad \ldots(i)$$

But
$$\left(\frac{\partial S}{\partial V}\right)_T = \left(\frac{\partial P}{\partial T}\right)_V \qquad \text{Maxwell eq. (7)}$$

Therefore eq. (i) reduces to
$$\left(\frac{\partial U}{\partial V}\right)_T = T\left(\frac{\partial P}{\partial T}\right)_V - P$$

This is **first** thermodynamic equation of state.

Again, we have the following thermodynamic relation
$$dH = TdS + VdP$$

Divide by dP, keeping T constant
$$\left(\frac{\partial H}{\partial P}\right)_T = T\left(\frac{\partial S}{\partial P}\right)_T + V \qquad \ldots(ii)$$

But
$$\left(\frac{\partial S}{\partial P}\right)_T = -\left(\frac{\partial V}{\partial T}\right)_P \qquad \text{Maxwell relation (8)}$$

Therefore eq (ii) reduces to
$$\left(\frac{\partial H}{\partial P}\right)_T = -T\left(\frac{\partial H}{\partial P}\right)_P + V$$

or
$$\left(\frac{\partial H}{\partial P}\right)_T = V - T\left(\frac{\partial V}{\partial T}\right)_P$$

This is **second** thermodynamic equation of state.

Example 28: Establish the following thermodynamic relations.

$$\left(\frac{\partial U}{\partial S}\right)_V = T \qquad \left(\frac{\partial U}{\partial V}\right)_S = -P$$

$$\left(\frac{\partial H}{\partial S}\right)_P = T \qquad \left(\frac{\partial H}{\partial P}\right)_S = V$$

$$\left(\frac{\partial A}{\partial T}\right)_V = -S \qquad \left(\frac{\partial A}{\partial V}\right)_T = -P$$

$$\left(\frac{\partial G}{\partial T}\right)_P = -S \qquad \left(\frac{\partial G}{\partial P}\right)_T = V$$

Solution: Writing the fundamental equations of thermodynmics
$$dU = TdS - PdV \qquad \ldots(i)$$
$$dH = TdS + VdP \qquad \ldots(ii)$$
$$dA = -SdT - PdV \qquad \ldots(iii)$$
$$dG = -SdT + VdP \qquad \ldots(iv)$$

If V is kept constant, $dV = 0$, eq. (i) becomes

$$\left(\frac{\partial U}{\partial S}\right)_V = T$$

If S is constant, $dS = 0$, eq. (*i*) takes the form

$$\left(\frac{\partial U}{\partial V}\right)_S = -P$$

If P is constant, $dP = 0$, eq. (*ii*) takes the form

$$\left(\frac{\partial H}{\partial S}\right)_P = T$$

If S is constant, $dS = 0$, eq. (*ii*) takes the form

$$\left(\frac{\partial H}{\partial P}\right)_S = V$$

If V is constant, $dV = 0$, eq. (*iii*) takes the form

$$\left(\frac{\partial A}{\partial T}\right)_V = -S$$

If T is constant, $dT = 0$, eq. (*iii*) takes the form

$$\left(\frac{\partial A}{\partial V}\right)_T = -P$$

If P is constant, $dP = 0$, eq. (*iv*) takes the form

$$\left(\frac{\partial G}{\partial T}\right)_P = -S$$

If T is constant, $dT = 0$, eq. (*iv*) takes the form

$$\left(\frac{\partial G}{\partial P}\right)_T = V$$

2.11 GIBB'S–HELMHOLTZ EQUATION

The Gibbs free energy (*G*) is given by the equation,

$G = H - TS$ ...(*i*)

Hence for an *isothermal* process, we can write

$G_1 = H_1 - TS_1$ for the initial state ...(*ii*)

and $G_2 = H_2 - TS_2$ for the final state ...(*iii*)

Subtracting equation (*ii*) from equation (*iii*), we get

$G_2 - G_1 = (H_2 - TS_2) - (H_1 - TS_1)$

$= (H_2 - H_1) - T(S_2 - S_1)$

or $\Delta G = \Delta H - T\Delta S$ ...(*iv*)

where ΔG is the change in free energy of the system

ΔH is the change in enthalpy of the system

and ΔS is the entropy change of the system.

H.V. Helmholtz (1821-1894)
Hermann Von Helmholtz was a German Scientist and Philosopher who made fundamental contributions to physiology, optics, electrodynamics and mathematics.

Putting $H = U + PV$, in eq. (i) we get
$$G = U + PV - TS \qquad \text{...(v)}$$
Differentiating this equation completely, we get
$$dG = dU + PdV + VdP - TdS - SdT \qquad \text{...(vi)}$$
But
$$dS = \frac{\delta q_{rev}}{T} = \frac{dU + PdV}{T} \quad \text{(From first law of thermodynamics)}$$
or
$$TdS = dU + PdV$$
Substituting this value in equation (vi), we get
$$dG = VdP - SdT$$
If pressure is kept constant, $dP = 0$, so that the above equation reduces to
$$(dG)_P = -(SdT)_P$$
or
$$\left(\frac{\partial G_1}{\partial T}\right)_P = -S \qquad \text{...(vii)}$$

$\therefore$ For the initial state, $\left(\dfrac{\partial G_1}{\partial T}\right)_P = -S_1$ ...(viii)

For the final state, $\left(\dfrac{\partial G_1}{\partial T}\right)_P = -S_2$ ...(ix)

Subtracting equation (viii) from (ix), we get
$$\left(\frac{\partial G_2}{\partial T}\right)_P - \left(\frac{\partial G_1}{\partial T}\right)_P = -S_2 - (-S_1)$$
$$= -S_2 + S_1 = -(S_2 - S_1)$$
$$= -\Delta S$$

or it may be written as
$$\left[\frac{\partial (G_2 - G_1)}{\partial T}\right]_P = -\Delta S$$
$$\left[\frac{\partial (\Delta G)}{\partial T}\right]_P = -\Delta S \qquad \text{...(x)}$$

Substituting this value of $-\Delta S$ in equation (iv), we get
$$\Delta G = \Delta H + T\left[\frac{\partial (\Delta G)}{\partial T}\right]_P \qquad \text{...(xi)}$$

This is the most common form of Gibbs-Helmholtz equation.

An analogous equation involving work function A can be obtained using the equation
$$A = U - TS$$
and proceeding in a similar manner as above :
$$\Delta A = \Delta U + T\left[\frac{\partial (\Delta A)}{\partial T}\right]_V \qquad \text{...(xii)}$$

Gibbs Helmholtz equation can be obtained in another form as under:

Writing the Gibb's-Helmholtz equation in its common form

$$\Delta G = \Delta H + T\left[\frac{\partial(\Delta G)}{\partial T}\right]_P$$

Dividing the above equation by T^2.

$$\frac{\Delta G}{T^2} = \frac{\Delta H}{T^2} + \frac{1}{T}\left[\frac{\partial(\Delta G)}{\partial T}\right]_P$$

or

$$\frac{1}{T}\left[\frac{\partial(\Delta G)}{\partial T}\right]_P = \frac{\Delta G}{T} - \frac{\Delta G}{T^2}$$

or

$$\left[\frac{\partial(\Delta G/T)}{\partial T}\right]_P = -\frac{\Delta H}{T^2}$$

2.11.1 Applications of Gibbs-Helmholtz Equation

Gibbs-Helmholtz equation has the following applications.

1. Knowing the free energy change at any two different temperatures, the enthalpy change accompanying the process can be calculated.

2. When a redox reaction (*e.g.* Zn + $CuSO_4$ ⟶ $ZnSO_4$ + Cu) is allowed to take place in a single beaker, heat is evolved. It was believed earlier that when the same reaction is allowed to take place in an electrochemical cell, the electrical energy produced was equivalent to the heat evolved. However, it was revealed later that in a number of cases, the electrical energy produced is not equivalent to the heat evolved. This was explained by Gibbs and Helmholtz by stating that the *electrical energy produced was equal to the decrease in the free energy* ($-\Delta G$) *accompanying the cell reaction*. They pointed out that if n is the number of electrons given out by one of the electrodes (*e.g.* for Zn Zn^{+2} + $2e$, $n = 2$), or taken up by the other electrode (*e.g.* for Cu^{+2}+ $2e$ Cu, $n = 2$), then the quantity of electricity produced = nF faradays, where $F = 1$ Faraday = 96,500 coloumbs. If E is the E.M.F. of the cell, then

Electrical energy produced by the cell = nFE

∴ Decrease in free energy, $-\Delta G = nFE$

Substituting this value in the Gibbs-Helmholtz equation

$$\Delta G = \Delta H + T\left[\frac{\partial(\Delta G)}{\delta T}\right]_P$$

we have

$$-nFE = \Delta H + T\left[\frac{\partial(-nFE)}{\delta T}\right]_P$$

or

$$nFE = -\Delta H + TnF\left(\frac{\partial F}{\partial T}\right)_P$$

Thus knowing the temperature coefficient of the E.M.F. of the cell *i.e.* $(\partial E/\partial T)_P$, ΔH can be calculated if E is known or *vice versa*. The electrical energy produced (nFE) is equal to the heat evolved ($-\Delta H$) only when $(\partial E/\partial T)_p = 0$.

Solved Examples—Gibb's Helmholtz Equation

Example 29: The free energy change involved in a process is -1235 J at 300 K and -1200 J at 310 K. Calculate the change in enthalpy of the process at 305 K.

Solution: Using Gibb's Helmholtz equation

$$\Delta G = \Delta H + T\left[\frac{\partial(\Delta G)}{\partial T}\right]_P$$

where $\left[\frac{\partial(\Delta G)}{\partial T}\right]_P = \frac{\Delta G_2 - \Delta G_1}{T_1 - T_1}$. Substituting values, we have

$$\left[\frac{\partial(\Delta G)}{\partial T}\right]_P = \frac{-1200 - (-1235)}{310 - 300} = \frac{35}{10} = 3.5 \text{ JK}^{-1}$$

The free energy change at 306 K can be taken as the average values of free energy change at 300 K and 310 K.

Thus, $\Delta G_{(at\ 305\ K)} = \frac{1}{2}[-1200 + (-1235)] = \mathbf{-1217.5\ J}$

Gibbs Helmholtz equation can be written as

$$\Delta H = \Delta G - T\left[\frac{\partial(\Delta G)}{\partial T}\right]_P$$

Substituting values, we get

$$\Delta H = -1217.5 - 305 \times 3.5 = \mathbf{-2285\ J.}$$

Example 30: The free energy change of a reaction at 298 K is found to be 18.0 kJ. Calculate the free energy at 350 K. Given that the enthalpy change for the reaction is +80 kJ. It is assumed that ΔH is independent of temperature.

Solution: Writing Gibbs Helmholtz equation in the form

$$\frac{\partial\left(\frac{\Delta G}{T}\right)}{\partial T} = -\frac{\Delta H}{T^2}$$

It can also be expressed as

$$d\left(\frac{\partial G}{T}\right) = -\frac{\Delta H}{T^2}dT$$

Integrating both sides of above equation, we get

$$\frac{\Delta G_2}{T_2} - \frac{\Delta G_1}{T_1} = \Delta H\left(\frac{1}{T_2} - \frac{1}{T_1}\right)$$

Given that

$$\Delta G_1 = 18.0 \text{ kJ}, \Delta H = 80.00 \text{ kJ}, T_1 = 298 \text{ K}, T_2 = 350 \text{ K}$$

Substituting the values, we get

$$\frac{\Delta G_2}{398} - \frac{18.0}{298} = 80\left(\frac{1}{350} - \frac{1}{298}\right)$$

$$\frac{\Delta G_2}{398} = -0.03988 + 0.060402$$

or $\Delta G_2 = 0.020522 \times 398$ or $\Delta G_2 = \mathbf{8.168\ kJ.}$

Problems for Practice

1. The free energy change accompanying a given process is – 85.8 kJ at 25°C and – 82.6 kJ at 35°C. Calculate the enthalpy change of the process at 25°C. **[Ans. –82.56 kJ]**
2. Assuming that, for a certain reaction ΔH remains constant in the temperature range 427° to 627°C. Calculate ΔG for this reaction at 627°C from the following data:
$\Delta H = -25$ kJ, $\Delta G (427°C) = -40$ kJ] **[Ans. –44.26 kJ]**

2.12 NERNST HEAT THEOREM

This theorem gives the variation of enthalpy change (ΔH) and free energy change (ΔG) of a system with decrease of temperature.

According to Gibbs-Helmholtz equation,

$$\Delta G = \Delta H + T\left(\frac{\partial(\Delta G)}{\partial T}\right)_P$$

From this equation, it is evident that at the absolute zero *i.e.* when $T = 0$, $\Delta G = \Delta H$. Nernst observed that as the temperature is lowered towards absolute zero, the value of $\partial(\Delta G)/\partial T$ decreases and then approaches zero asymptotically. This means that ΔG and ΔH *are not only equal at absolute zero but the values approach each other asymptotically near this temperature.* This result is known as Nernst heat theorem. Mathematically, it may be expressed as

$$\lim_{T \to 0} \frac{d(\Delta G)}{dT} = \lim_{T \to 0} \frac{d(\Delta H)}{dT} \qquad ...(i)$$

Graphically, the result may be represented by Fig. 2.4.

In Fig. 2.4. ΔG has been shown as greater than ΔH at temperature away from absolute zero. However the reverse is also possible because $\partial(\Delta G)/\partial T$ can be both positive or negative.

Further $\left[\frac{\partial(\Delta G)}{\partial T}\right]_P = -\Delta S \qquad ...(ii)$

and $\left[\frac{\partial(\Delta H)}{\partial T}\right]_P = \Delta C_p \qquad ...(iii)$
(Kirchoff's equation)

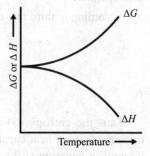

Fig. 2.4: *Nernst heat theorem*

From equations (*i*), (*ii*) and (*iii*), we can say, as T → 0

$$\Delta S = 0 \text{ and } \Delta C_p = 0$$

i.e. as the temperature is lowered to absolute zero, the enthalpy change of the reaction and difference in the heat capacities of products and reactants also tends to be zero.

Since gases do not exist at the absolute zero, this means that the heat theorem is not applicable to gases. Similarly, it has been found to be inapplicable to liquids also. Nernst heat theorem is applicable to solids only.

2.13 THIRD LAW OF THERMODYNAMICS

Third law of thermodynamics states:

The entropy of all perfectly crystalline solids may be taken as zero at the absolute zero temperature.

The law is also stated as follows :

Every substance has finite positive entropy but at the absolute zero of temperature, the entropy may become zero and in fact it does become zero in case of perfectly crystalline solids.

Importance of the Third Law

Calculation of absolute entropies: It helps in the calculation of the absolute entropies of chemical compounds at any desired temperature. This is explained as under:

The infinitesimal entropy change is given by

$$dS = \frac{dq}{T} \qquad ...(i)$$

But $C_p = \frac{dq}{T}$ so that $dq = C_p\, dT$

Substituting this value in equation (i), we have

$$dS = \frac{C_p\, dq}{T}$$

Entropy change of a substance when its temperature changes from absolute zero to the temperature (T) can be calculated using the equation

$$\int_{S=S_0}^{S=S} dS = \int_{T=T_0}^{T=T} C_p \frac{dT}{T} \quad \text{or} \quad S - S_0 = \int_0^T C_p \frac{dT}{T} \qquad ...(ii)$$

where S_0 is the entropy of the substance at absolute zero and S is the entropy of the substance at temperature T.

According to third law of thermodynamics, $S_0 = 0$.

$$\therefore \quad S_T = \int_0^T C_p \frac{dT}{T}$$

$$= \int_0^T C_p\, d\ln T \qquad ...(iii)$$

Thus the entropy (S) of the substance at temperature T can be calculated from the measurements of heat capacities (C_p), at a number of temperatures between 0 K to T K. The integral in equation (iii) can be evaluated by plotting C_p vs $\ln T$ i.e. $2.303 \log T$ and then measuring the area under the curve between $T = 0$ to $T = T$ as shown in Fig. 2.5.

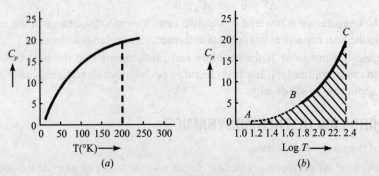

Fig. 2.5: Plot of heat capacities (Cp) vs T in the first case and vs log T in the second case.

If C_p is supposed to remain almost constant in the temperature range 0 to TK, equation (*iii*) can be simplified to

$$S = C_p \ln T = 2 \cdot 303 \, C_p \log T \qquad \ldots(iv)$$

This equation helps to calculate the value of S directly from the values of C_p and T.

However, heat capacities cannot be measured with accuracy below a temperature of 15 K. Thus part CB of the curve is experimental curve using values of C_p. The curve is extrapolated to get the value at 0K. Between the range 0–15 K, heat capacity is obtained by Debye T-cubed law

$$C_p = a\,T^3 \qquad \ldots(v)$$

where a is a constant. It is determined from the value of C_p at some low temperature.

Let T_1 be the temperature above which heat capacity can be measured. This corresponds to the point B in Fig 2.5 (*b*). C corresponds to the temperature T at which the entropy of the solid is to be determined. In the Fig. 2.5 point A corresponds to 0K. The integral in eq. (*iii*) can be split as under

$$S_T = \int_0^{T_1} C_p \frac{dT}{T} + \int_{T_1}^{T} C_p \frac{dT}{T} \qquad \ldots(vi)$$

The first integral in eq. (*vi*) can be calculated using eq. (*v*)

$$\int_0^{T_1} C_p \frac{dT}{T} = \int_0^{T_1} aT^3 \frac{dT}{T} = \int_0^{T_1} aT^2 \, dT = \frac{1}{3}aT_1^3$$

or
$$\Delta S_1 = \frac{1}{3} C_p \text{ (at } T_1) \qquad \ldots[\text{See eq. } (v)]$$

The second integral in eq. (*vi*) can be evaluated by graphical method. C_p is plotted versus ln T. The area under the curve between the limits T_1 and T gives the value of the integral. Therefore, entropy of the solid between 0K and TK, is given by

$$S_T = \frac{1}{3} C_p \text{ (at } T_1) + \int_{T_1}^{T} C_p \, d\ln T \qquad \ldots(vii)$$

If there is some allotropic change between the temperature range 0K – TK, then the entropy of transition $\Delta H_t / T_t$ is added to eq. (*vii*).

Thus, we have

$$S_T = \frac{1}{3} C_p \text{ (at } T_1) \int_{T_1}^{T} C_p d\ln T + \frac{\Delta H_t}{T_t}$$

For liquid and gaseous substances, the total absolute entropy of the substance at the given temperature will be the total of all the entropy changes which the substance undergoes in order to reach that particular state starting from the crystalline solid at absolute zero. Thus if a substance is gaseous at 1 atm pressure and 25°C, the entropy of the gas at 25°C will be the sum of the following entropies involved at different stages :

(*i*) the entropy of heating the crystalline solid from $T = 0$ to $T = T_m$ the melting point;

(*ii*) $\dfrac{\Delta H_m}{T_m}$ the entropy of melting, where ΔH_m is latent heat of melting;

(*iii*) the entropy of heating the liquid from T_m to T_b, the boiling point;

(*iv*) the entropy of vaporization, $\dfrac{\Delta H_v}{T_b}$ where ΔH_v is latent heat of vaporisation;

(*v*) the entropy of heating the gas from T_b to 25°C (*i.e.* 298 K).

The complete expression for the calculation of entropy at temperature T may be expressed as

$$S = \underbrace{\int_0^{T_m} C_{p(s)}\, d\ln T}_{\text{I}} + \underbrace{\frac{\Delta H_m}{T_m}}_{\text{II}} + \underbrace{\int_{T_m}^{T_b} C_{p(l)}\, d\ln T}_{\text{III}} + \underbrace{\frac{\Delta H_v}{T_b}}_{\text{IV}} + \underbrace{\int_{T_b}^{T} C_{p(g)}\, d\ln T}_{\text{V}}$$

2.13.1 Testing Validity of Third Law of Thermodynamics

Validity of the third law can be tested as under :

(1) By calculation of equilibrium constants: The equilibrium constant (K) is related to standard free energy change ($\Delta G°$) according to the equation

$$-\Delta G° = RT \ln K \qquad \ldots(i)$$

Also, the standard free energy change is related to the standard enthalpy change ($\Delta H°$) and standard entropy change ($\Delta S°$) according to the equation

$$\Delta G° = \Delta H° - T\Delta S° \qquad \ldots(ii)$$

Substituting the experimental value of $\Delta H°$ and calculated value of $\Delta S°$, using the absolute values of entropies obtained from the third law of thermodynamics, and applying the relation

$$\Delta S° = \Sigma S°_{\text{products}} - \Sigma S°_{\text{reactants}}$$

$\Delta G°$ can be calculated for a reaction. Putting this value of $\Delta G°$ in Eq. (i), the equilibrium constant K can be calculated. Value of equilibrium constant as calculated above comes out to be in close agreement with the experimental value. That proves the validity of the law.

(2) By comparing with entropies calculated from statistical mechanics: The absolute entropies calculated on the basis of the third law are found to be in complete agreement with those calculated using statistical mechanics. This again proves the validity of third law.

2.14 RESIDUAL ENTROPY

Entropy calculated by using third law of thermodynamics as given in the previous section is called thermal entropy. We can also calculate the entropy of a substance by using the following relation of statistical mechanics:

$$S = k \ln W$$

where S = statistical entropy

W = Number of possible arrangement of atoms, molecules or ions in a crystal (also called thermodynamic probability)

k = Boltzmann constant

We observe that, in some cases, the experimental values of thermal entropy (obtained from third law of thermodynamics) are not in perfect agreement with those calculated from statistical methods. Thermal entropies are found to be slightly lower than statistical entropies. The lower values of third law entropies are due to the fact that even at $T = 0$, there is some disorder in the solid and it possesses a finite value of entropy.

The entropy which the crystal of a substance possesses at $T = 0$ is called residual entropy.

Some of the crystals, which do not show zero entropy at 0K are CO, NO, H_2 crystals and crystalline ice.

Calculation of residual entropy

Residual entropy can be calculated using the relation
$$S = k \ln W \qquad ...(i)$$
If W at 0K is known, we can calculate residual entropy.

Residual entropy of CO

Suppose, a sample of CO contains N molecules. Each molecule of CO can have two orientations.
$$CO \text{ or } OC$$
Let there be N molecules in one mole of the substance.
Therefore N molecules can have 2^N orientations *i.e.* $W = 2^N$
Applying eq. (*i*) above, the entropy can be calculated as
$$S = k \ln 2^N \text{ or } S = kN \ln 2 \qquad ...(ii)$$
kN can be written as equal to R.
Thus, for one mole of the substance, the expression would be
$$S = R \ln 2 = 2.303 \times 8.314 \times 0.3010 \text{ Jk}^{-1} \text{ mol}^{-1}.$$
$$= 5.85 \text{ JK}^{-1} \text{ mol}^{-1}.$$

The number of orientations for a molecule of the type ABO_3 (A and B are two different elements, O is oxygen) is 4. Hence $W = 4$ in this case.

In general, if there are x possible orientations of a molecule with nearly the same energy, the residual molar entropy will be
$$S = R \ln x$$

Solved Example

Example 31: $ClBrO_3$ molecule can have four possible arrangements with nearly the same energy. Calculate the residual entropy of $ClBrO_3$ molecule

Solution: Residual entropy $S = R \ln W = 2.303 \, R \log W$

In this case $W = 4, R = 8.314$

Substituting the values in the above equation, we have
$$S = 2.303 \times 8.314 \times 0.6021$$
$$= 11.5 \text{ Jk}^{-1} \text{ mol}^{-1}.$$

Solved Examples—Third Law of Thermodynamics

Example 32: Write expressions for the following:

(*i*) Variation of Helmholtz function with volume at constant temperature.

(*ii*) Variation of Helmholtz function with temperature at constant volume.

(*iii*) Total change in Helmholtz function at constant temperature when volume changes from V_1 to V_2.

Solution: (*i*) $\left(\dfrac{\partial A}{\partial V}\right)_T = -P$ (*ii*) $\left(\dfrac{\partial A}{\partial T}\right)_V = -S$ (*iii*) $(\Delta A)_T = -nRT \ln \dfrac{V_2}{V_1}$.

Example 33: Write expression for total change in Gibbs function when at constant temperature, volume changes from V_1 to V_2 or pressure changes from P_1 to P_2.

Solution:
$$(\Delta G)_T = nRT \ln \frac{P_2}{P_1} = nRT \ln \frac{V_1}{V_2}.$$

Example 34: What is the difficulty in determining absolute entropy of a substance ? How has the problem been solved by Debye ?

Solution: In the determination of absolute entropy of a substance, heat capacity values are required at different temperatures, starting from 0K. But experimentally, heat capacities at low temperatures (< 15 K) cannot be determined. The problem was solved by Debye who gave third power law for heat capacities of solids according to which $C_p = a\, T^3$.

Example 35: What is the difference between Helmholtz function and Gibbs function? Under what condition, ΔG becomes equal to ΔA?

Solution: Helmholtz function $A = U - TS$
Gibbs function $G = H - TS$
$-\Delta A = w_{max}$ and $-\Delta G = w_{max} - P\,\Delta V$

Thus, if during the process there is no change in volume, $P\,\Delta V = 0$ so that $-\Delta G = -\Delta A$.

Example 36: What is total differential equation ? How does Gibbs function vary with pressure at constant temperature and with temperature at constant pressure?

Solution. Total differential equation is $dG = V\,dP - S\,dT$
$$\left(\frac{\partial G}{\partial P}\right)_T = V, \quad \left(\frac{\partial G}{\partial T}\right)_P = -S.$$

Example 37: Under what condition electrical energy produced in a cell reaction is equal to the heat evolved in the cell reaction?

Solution:
$$n\,FE = -\Delta H + TnF \left(\frac{\partial E}{\partial T}\right)_P$$

Thus, $n\,FE = -\Delta H$ when $\left(\frac{\partial E}{\partial T}\right)_P = 0$, *i.e.*, when the temperature coefficient of the E.M.F. of the cell = zero.

Example 38: Write criteria for spontaneity and equilibrium conditions in terms of S, U, H, A and G.

Solution: $(dS)_{U,V} \geq 0, \quad (dU)_{S,V} \leq 0, \quad (dH)_{S,P} \leq 0, \quad (dA)_{T,V} \leq 0, \quad (dG)_{P,T} \leq 0$

Sign '=' refers to reversible process (*i.e.*, equilibrium) and > or < refers to irreversible process (spontaneity).

2.15 RELATIONSHIP AMONG MOLARITY, MOLALITY AND MOLE FRACTION

Molarity. It is the number of moles of a solutes dissolved in one litre of the solution. It may also be expressed as

$$\text{Molarity} = \frac{\text{Strength of the solute in grams per litre of the solution}}{\text{Molecular mass of the solute}}$$

Molality. It is the number of moles of the solute dissolved in 1000 g of the solvent. It may also be written as

$$\text{Molality} = \frac{\text{Strength of the solute in grams per 1000 g of the solvent}}{\text{Molecular mass of the solute}}$$

Mole fraction. It is the fraction of the number of moles of a component to the total number of moles of all the components in a mixture. Thus

$$\text{Mole fraction of solute} = \frac{\text{Number of moles of the solute in a solvent}}{\text{Total number of moles of solute and solvent}}$$

Relationship between molarity and molality. Let us say we have a solution whose molarity is x i.e., it contains x moles of the solution dissolved in 1 litre (1000 ml) of the solution. Its molality will be calculated as under:

Let the molecular mass of the solute be m

Let the density of the solution be d g/ml

Mass of 1000 ml (1 litre) of the solution = Density × Volume

$$= d \times 1000 \text{ g}$$

Mass of the solute = No. of moles × molecular mass

$$= x \times m \text{ grams}$$

Mass of the solvent = Mass of the solution − Mass of the solute

$$= 1000\, d - (x \times m) \text{ g}$$

$1000\, d - (x \times m)$ g of the solvent contain = x moles of the solute

$$1000 \text{ g of the solvent contain} = \frac{x \times 1000}{(1000 - d) - (x \times m)} \text{ moles}$$

or   Molality of the solution $(y) = \dfrac{x \times 1000}{(1000 - d) - (x \times m)}$

This is the relation between molarity (x) and molality (y) of a solution. Interconversion of the molarity and molality is thus possible using this relation.

Relationship between molarity and mole fractions. If we know the molarity of a solution, it is possible to calculate mole fraction of the solute and the solvent as follows:

Let the molarity of the solution be x and the molecular mass of the solute be m and the density of the solution be d g/ml

That means number of moles of the solution in one litre of the solution = x

Mass of 1 litre (1000 ml) of the solution = $d \times 1000$ g

Mass of the solute in 1000 ml of the solution = $x \times m$

Mass of the solvent in 1000 ml of the solution = $(d \times 1000) - (x \times m)$ g

Let the molecular mass of the solvent be m_1

Mole fraction of the solute $n_1 = \dfrac{\text{No. of moles of the solute}}{\text{No. of moles of the solute + No. of moles of the solvent}}$

$$= \frac{x}{x + \left[\dfrac{(d \times 1000) - (x \times m)}{m_1}\right]}$$

By substituting the values of x, d, m and m_1, we can obtain the value of n_1.

Mole fraction of the solvent $n_2 = 1 - n_1$

Relationship between molality and mole fraction. We can calculate the mole fractions of solute and solvent from the molality of the solution.

Let the molality of the solution be y

Molecular mass of the solute $= m$

Molecular mass of the solvent $= m_1$

As the molality of the solution is y, the solution contains y moles of the solute in 1000 g of the solvent

No. of moles of the solute $= y$

No. of moles of the solvent $= \dfrac{1000}{m_1}$

Mole fraction of the solute, $n_1 = \dfrac{y}{y + \dfrac{1000}{m_1}}$

Mole fraction of the solvent $= 1 - n_1$

2.16 PARTIAL MOLAR QUANTITIES

Let an extensive property such as volume, free energy, entropy, energy content etc be represented by X. Suppose there are n constituents in the system having $n_1, n_2, n_3 \ldots$ moles of individual components. Then the property X is a function of temperature, pressure as well as of the amounts of different constituents.

Thus $\qquad X = f(T, P, n_1, n_2, n_3 \ldots) \qquad \ldots(i)$

If there is a small change in the temperature, pressure and the amounts of the constituents, then change in property X is given by

$$dX = \left(\dfrac{\partial X}{\partial T}\right)_{P, n_1, n_2 \ldots} dT + \left(\dfrac{\partial X}{\partial P}\right)_{T, n_1, n_2 \ldots} dP + \left(\dfrac{\partial X}{\partial n_1}\right)_{T, P, n_2, n_3, \ldots} dn_1$$

$$+ \left(\dfrac{\partial X}{\partial n_2}\right)_{T, P, n_1, n_2, \ldots} dn_2 + \ldots \quad \ldots(ii)$$

The first term on R.H.S. gives the change in the value of X with temperature when pressure and composition are kept constant. The second term on R.H.S. gives the change in the value of X with pressure when temperature and composition are kept constant. The remaining quantities give the change in the value of X with a change in the amount of a constituent, when temperature, pressure and the amounts of other constituents are kept constant.

If the temperature and pressure of the system are kept constant, then

$$dT = 0 \text{ and } dP = 0$$

so that equation (ii) becomes

$$(dX)_{T,P} = \left(\dfrac{\partial X}{\partial n_1}\right)_{T, P, n_2, n_3 \ldots} dn_1 + \left(\dfrac{\partial X}{\partial n_2}\right)_{T, P, n_1, n_3 \ldots} dn_2 + \ldots \qquad \ldots(iii)$$

First quantity on right hand side is partial molar quantity of component 1, the second quantity for the component 2 and so on.

These are represented by putting a bar over the symbol of that particular property i.e. $\overline{X}_1, \overline{X}_2$ for the 1st, 2nd component etc. respectively. Thus

$$\left(\frac{\partial X}{\partial n_1}\right)_{T,P,n_2,n_3,...} = \overline{X}_1, \left(\frac{\partial X}{\partial n_2}\right)_{T,P,n_1,n_3,...} = \overline{X}_2 \text{ etc.}$$

In general, for any component i

$$\left(\frac{\partial X}{\partial n_i}\right)_{T,P,n_1,n_2,...} = \overline{X}_i \qquad ...(iv)$$

2.17 PARTIAL MOLAR FREE ENERGY (CHEMICAL POTENTIAL)

Consider the *extensive* property, free energy. Let it be represented by G. Suppose that the system consists of n constituents (in a solution or in a heterogeneous system), the amounts of which present in the system are $n_1, n_2, n_3 ....$ moles. Then the property G is a function not only of temperature and pressure but of the amounts of the different constituents as well, so that we can write

$$G = f(T, P, n_1, n_2, n_3 .....) \qquad ...(i)$$

Now, if there is a small change in the temperature, pressure and the amounts of the constituents, then the change in the property F is given by

$$dG = \left(\frac{\partial G}{\partial t}\right)_{P,n_1,n_2,...} dT + \left(\frac{\partial G}{\partial P}\right)_{T,n_1,n_2,...} dP$$

$$+ \left(\frac{\partial G}{\partial n_1}\right)_{T,P,n_2,n_3,...} dn_1 + \left(\frac{\partial G}{\partial n_2}\right)_{T,P,n_1,n_3,...} dn_2 + ... \qquad ...(ii)$$

On the right hand side, the first quantity gives the change in the value of G with temperature when pressure and composition are kept constant; the second quantity gives the change in the value of G with pressure when temperature and composition are kept constant; the remaining quantities give the change in the value of G with a change in the amount of a constituent, when temperature, pressure and the amounts of other constituents are kept constant.

If the temperature and pressure of the system are kept constant, then

$$dT = 0 \text{ and } dP = 0$$

so that equation (*ii*) becomes

$$(dG)_{T,P} = \left(\frac{\partial G}{\partial n_1}\right)_{T,P,n_2,n_3,...} dn_1 + \left(\frac{\partial G}{\partial n_2}\right)_{T,P,n_1,n_3,...} dn_2 + ... \qquad ...(iii)$$

Each derivative on the right hand side is called partial molar property and is represented by putting a bar over **the** symbol of that particular property i.e. $\overline{G}_1, \overline{G}_2$ for the 1st, 2nd component etc. respectively. Thus,

$$\left(\frac{\partial G}{\partial n_1}\right)_{T,P,n_2,n_3,...} = \overline{G}_1; \quad \left(\frac{\partial G}{\partial n_2}\right)_{T,P,n_1,n_3,...} = \overline{G}_2 \text{ etc.}$$

In general, for any component i

$$\left(\frac{\partial G}{\partial n_i}\right)_{T,P,n_1,n_2,\ldots} = \overline{G}_1 \qquad \ldots(iv)$$

This quantity is called *partial molar free energy or chemical potential* and is usually represented by the symbol μ. Thus

$$\mu_i = \overline{G}_1 = \left(\frac{\partial G}{\partial n_i}\right)_{T,P,n_1,n_2,\ldots} \qquad \ldots(v)$$

If $dn_i = 1$ mole, $\mu_i = (dG)T, P, n_1, n_2, \ldots$

Hence chemical potential may be defined as follows:

The chemical potential of a constituent in a mixture is the increase in the free energy which takes place at constant temperature and pressure when 1 mol of that constituent is added to the system, keeping the amounts of all other constituents constant i.e. when 1 mol of the constituent is added to such a large quantity of the system that its composition remains almost unchanged.

Eq. (*iii*) can be written as:

$$(dG)_{T,P} = \mu_1 dn_1 + \mu_2 dn_2 + \ldots \qquad \ldots(vi)$$

For a system of definite composition represented by the number of moles n_1, n_2, n_3 etc., equation (*vi*) on integration gives

$$G_{T,P,N} = n_1\mu_1 + n_2\mu_2 + \ldots$$

where the subscript N stands for constant composition.

On the right hand side, the first term gives contribution of the first constituent to the total free energy of the system, the second term gives the contribution of the second constituent and so on. Obviously, μ_1, μ_2 etc. give the *contribution per mole* to the total free energy. Hence chemical potential may also be defined as under:

The chemical potentical of a constituent in a mixture is its contribution per mol to the total free energy of the system of a constant composition at constant temperature and pressure. It may be noted that whereas free energy is an **extensive property,** the chemical potential is an **intensive property** because it refers to one mole of the substance.

2.18 GIBBS-DUHEM EQUATION

Free energy (G), being an extensive property, depends not only upon the temperature and pressure of the system but also upon the composition of the system. If the system consists of a number of constituents, the amounts of which are $n_1\ n_2\ n_3\ \ldots$ moles respectively, then we can write

$$G = f(T, P, n_1, n_2, n_3, \ldots) \qquad \ldots(i)$$

For small changes in temperature, pressure and the quantity of different constituents, the small change in free energy can be obtained by the partial differentiation of equation (*i*). This gives

$$(dG) = \left(\frac{\partial G}{\partial T}\right)_{P,n_1,n_2,\ldots} dT + \left(\frac{\partial G}{\partial T}\right)_{T,n_1,n_2,\ldots} dP + \left(\frac{\partial G}{\partial n_1}\right)_{T,P,n_1,n_2,\ldots} dn_1 + \left(\frac{\partial G}{\partial n_2}\right)_{T,P,n_1,n_3,\ldots} dn_2 + \ldots \qquad \ldots(ii)$$

Chemical Thermodynamics – II (Second and Third Laws of Thermodynamics and Partial Molar Properties)

If temperature and pressure are kept constant
$$dT = 0, dP = 0$$
so that equation (ii) becomes

$$(dG)_{T,P} = \left(\frac{\partial G}{\partial n_1}\right)_{T,P,n_2,n_3,...} dn_1 + \left(\frac{\partial G}{\partial n_2}\right)_{T,P,n_1,n_3,...} dn_2 + ... \quad ...(iii)$$

Putting $\left(\frac{\partial G}{\partial n_1}\right)_{T,P,n_2,n_3,...} = \mu_1,$

$$\left(\frac{\partial G}{\partial n_2}\right)_{T,P,n_1,n_3,...} \text{ and so on}$$

we get $\quad (dG)_{T,P} = \mu_1 dn_1 + \mu_2 dn_2 + ... \quad ...(iv)$

For a system of definite composition, represented by the number of moles n_1, n_2, n_3 etc., equation (iv) on integration gives

$$G_{T,P,N} = n_1\mu_1 + n_2\mu_2 ...$$

Differentiating this equation under conditions of constant temperature and pressure but varying composition, we get

$$(dG)_{T,P} = (n_1 d\mu_1 + \mu_1 dn_1) + (n_2 d\mu_2 + \mu_2 dn_2) + ...$$
$$= (\mu_1 dn_1 + \mu_2 dn_2 + ...) + (n_1 d\mu_1 + n_2 d\mu_2 + ...) \quad ...(v)$$

Comparing equations (iv) and (v), we get
$$n_1 d\mu_1 + n_2 d\mu_2 + ... = 0$$

$$\boxed{\sum n_i d\mu_i = 0}$$

This equation which is applicable to a system under constant temperature and pressure is called **Gibbs-Duhem equation**.

2.18.1 Variation of Chemical Potential with Temperature and Pressure

In order to study the variation of chemical potential with temperature and pressure, let us first derive expressions for the *variation of* free energy *with temperature and pressure* to be used therein.

Free energy is a function of temperature, pressure and composition of the system i.e.
$$G = f(T, P, n_1, n_2, n_3 ...)$$

$$\therefore \quad (dG) = \left(\frac{\partial G}{\partial T}\right)_{P,N} dT + \left(\frac{\partial G}{\partial P}\right)_{T,N} dP + \left(\frac{\partial G}{\partial n_1}\right)_{T,P,n_2,n_3,...} dn_1$$
$$+ \left(\frac{\partial G}{\partial n_2}\right)_{T,P,n_1,n_3,...} dn_1 + ... \quad ...(i)$$

If the system is a **closed** one, there is no change in the number of moles of the various constituents present i.e. dn_1, dn_2, dn_3 etc. will be all zero, so that the above equation reduces to

$$(dG) = \left(\frac{\partial G}{\partial T}\right)_{P,N} dT + \left(\frac{\partial G}{\partial P}\right)_{T,N} dP \quad ...(ii)$$

For a closed system, we also know that the *total differential equation* is

$$dG = VdP - SdT = -SdT + VdP \qquad \ldots(iii)$$

where V is the volume and S is the entropy of the system.

Comparing equations (ii) and (iii), we get

$$\left(\frac{\partial G}{\partial T}\right)_{P,N} = -S \qquad \ldots(iv)$$

$$\left(\frac{\partial G}{\partial P}\right)_{T,N} = V \qquad \ldots(v)$$

These two equations give the variation of free energy with temperature and pressure, for a closed system.

Variation of chemical potential with temperature: By definition, chemical potential of any constituent of a system is given by

$$\mu_i = \left(\frac{\partial G}{\partial n_1}\right)_{T,P,n_1,n_2\ldots}$$

Differentiating this equation with respect to temperature, we get

$$\left(\frac{\partial \mu_i}{\partial T}\right)_{P,N} = \frac{\partial^2 G}{\partial n_i \partial T} \qquad \ldots(vi)$$

Differentiating equation (iv) with respect to n_i, we get

$$\frac{\partial^2 G}{\partial T \partial n_i} = -\left(\frac{\partial S}{\partial n_i}\right)_{T,P,n_1,n_2,\ldots} = \overline{S_i} \qquad \ldots(vii)$$

where $\left(\dfrac{\partial S}{\partial n_i}\right)_{T,P,n_1,n_2,\ldots} = \overline{S_i}$ is the partial molar entropy of the ith constituent.

Combining equations (vi) and (vii), we get

$$\boxed{\left(\frac{\partial \mu_i}{\partial T}\right)_{P,N} = \overline{S_i}} \qquad \ldots(viii)$$

Thus the rate of variation of chemical potential, of any constituent i of a system, with temperature at constant pressure and composition is equal to the partial molar entropy of that constituent.

Variation of chemical potential with pressure: Chemical potential of a component i is given by

$$\mu_i = \left(\frac{\partial G}{\partial n_i}\right)_{T,P,n_1,n_2,\ldots}$$

Differentiating this equation with respect to pressure, we get

$$\left(\frac{\partial \mu_i}{\partial P}\right)_{T,N} = \frac{\partial^2 G}{\partial n_i \partial P} \qquad \ldots(ix)$$

Differentiating equation (v) with respect to n_i we get

$$\frac{\partial^2 G}{\partial P \partial n_i} = \left(\frac{\partial V}{\partial n_i}\right)_{T,P,n_1,n_2,\ldots} \qquad \ldots(x)$$

where $\left(\dfrac{\partial V}{\partial n_i}\right)_{T,P,n_1,n_2,\ldots} = \overline{V_i}$, represents the partial molar volume of the ith constituent.

Combining equations (ix) and (x), we get

$$\boxed{\left(\frac{\partial \mu_i}{\partial P}\right)_{T,N} = \overline{V}_i} \qquad \ldots(xi)$$

Hence *the rate of variation of chemical potential of any constituent i of the system, with pressure is equal to the partial molar volume of that constituent.*

2.19 CHEMICAL POTENTIAL OF A COMPONENT IN A SYSTEM OF IDEAL GASES

For an ideal gas, we know that

$$PV = nRT \qquad \ldots(i)$$

As the system consists of n_1, n_2, n_3 etc. moles of the different constituents,

$$n = n_1 + n_2 + \ldots n_i + \ldots \qquad \ldots(ii)$$

Substituting this value in equation (i), we get

$$PV = (n_1 + n_2 + \ldots n_i + \ldots)RT \qquad \ldots(iii)$$

or

$$V = (n_1 + n_2 + \ldots n_i + \ldots)\frac{RT}{P} \qquad \ldots(iv)$$

Differentiating this equation with respect to n_i, keeping all other n's constant as well as temperature and pressure constant, we get

$$\left(\frac{\partial V}{\partial n_i}\right)_{T,P,n_1,n_2,\ldots} = \frac{RT}{P} \qquad \ldots(v)$$

or

$$\overline{V}_i = \frac{RT}{P} \qquad \ldots(vi)$$

We have another relation showing change of chemical potential with pressure

$$\left[\frac{\partial \mu_i}{\partial P}\right]_{T,N} = \overline{V}_i \qquad \ldots(vii)$$

From eq. (vi) and (vii)

$$\boxed{\left(\frac{\partial \mu_i}{\partial P}\right)_{T,N} = \frac{RT}{P}} \qquad \ldots(viii)$$

This equation can be rewritten as

$$d\mu_i = \frac{RT}{P} dP = RT\, d\ln P \qquad \ldots(ix)$$

In the mixture of gases, if p_i is the partial pressure of the ith constituent, we have

$$p_i V = n_i RT \qquad \ldots(x)$$

where n_i is the number of moles of the ith constituent.

For the complete mixture of gases containing n moles, for ideal behaviour, we know that

$$PV = nRT \qquad \ldots(xi)$$

Dividing equation (x) by equation (xi), we get

$$\frac{p_i}{P} = \frac{n_i}{n}$$

$$p_i = \frac{n_i}{n} P = K.P$$

where $K \left(= \dfrac{n_i}{n} \right)$, is a constant.

$$\therefore \quad \ln p_i = \ln K + \ln P$$

Hence $\quad d \ln p_i = d \ln P \qquad [\because d \ln K = 0]$

Substituting this value in equation (ix), we get

$$d\mu_i = RT d \ln p_i$$

Integrating this equation, we get

$$\boxed{\mu_i = \mu_i^0 + RT \ln p_i} \qquad ...(xii)$$

where μ_i^0 is the integration constant, the value of which depends upon the *nature of the gas* and also on the *temperature*.

From equation (xii), it is obvious that when $p_i = 1$ atm, $\mu_i = \mu_i^0$. Thus μ_i^0 is the chemical potential of the ith constituent when its partial pressure is unity in the mixture of ideal gases.

2.20 CRITERIA FOR PHASE EQUILIBRIUM FOR MULTICOMPONENT SYSTEM

Let us consider the simplest case consisting of only two phases A and B. Suppose a number of components ($C_1, C_2,...C_c$) are distributed between them (Fig. 2.6). Further suppose that at constant temperature and pressure, a small amount dn_1 of component 1 is transferred from phase A to phase B. If $(\mu_1)_A$ and $(\mu_1)_B$ represent the chemical potentials of component 1 in phases A and B respectively, we have

Decrease in the free energy of component 1 in phase $A = (\mu_1)_A \delta n_1$

Increase in the free energy of component 1 in phase $B = (\mu_1)_B \delta n_1$

Since the system is a closed one, $(\Delta, G)_{T,P} = 0$

$$\therefore \quad -(\mu_1)_A \delta n_1 + (\mu_1)_B \delta n_1 = 0$$

(minus sign has been used to represent the decrease)

or $\quad (\mu_1)_A = (\mu_1)_B$

It means *for a system consisting of two phases in equilibrium, the chemical potential of any given component is same in both the phases.*

Similarly, if a system consists of three phases, A, B and C in equilibrium, the above criterion will apply to the equilibrium between phases A and B as well as to the equilibrium between phases B and C so that we will have

$$(\mu_1)_A = (\mu_1)_B = (\mu_1)_C$$

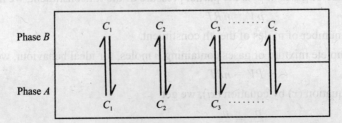

Fig. 2.6: Distribution of a number of components between two phases

In general for a system consisting of a number of phases P_1, P_2, P_2 etc in equilibrium (*i.e.*, a multiphase equilibria) for any component i,

$$(\mu_i)_{P1} = (\mu_i)_{P2} = (\mu_i)_{P3} = ...$$

This result may be expressed as:

For a multiphase equilibria containing a number of components distributed among them, the chemical potential of any component is the same in all the phases.

2.20.1 Criteria for Two-phase Equilibrium for One-component System

For a system consisting of one pure substance only which may exist in two different phases in equilibrium, *e.g.*, liquid water $\rightleftharpoons$ water vapour or ice $\rightleftharpoons$ liquid water or sulphur (monoclinic) $\rightleftharpoons$ sulphur (rhombic) etc., the criterion is obtained in a simple manner in terms of *free energy*. In such a case, if a certain amount of the substance is transferred from one phase to the other, the molar free energy of one phase decreases while that of the other phase increases by an equal amount. Hence the net result is that there is no **change** in free energy, *i.e.*

$$\Delta G = 0$$

This implies that if at equilibrium, G_1 is the molar free energy of phase I and G_2 that of the phase II, we will have

$$G_2 - G_1 = 0 \text{ or } G_1 = G_2$$

Thus, *whenever two phases of the same single substance (one component system) are in equilibrium, at a given temperature and pressure, the molar free energy is the same for each phase.*

2.21 CLAUSIUS-CLAPEYRON EQUATION

When a system is in equilibrium, the free energy change accompanying the process is zero *i.e.*

$$\Delta G = 0, \text{ for a system in equilibrium} \qquad ...(i)$$

Further, the variation of free energy with temperature and pressure is given by the total differential equation, *viz.*

$$dG = VdP - SdT \qquad ...(ii)$$

Consider the system consisting of two phase I and II in equilibrium at temperature T and pressure P. Under these conditions, suppose that free energy of phase I is G_1 and that of phase II is G_2. If the temperature of the system is changed from T to $T + dT$ and pressure from P to $P + dP$, suppose the corresponding change in free energy of phase I is dG_1 and that of phase II is dG_2 (Fig. 2.7). Then according to equation (*ii*)

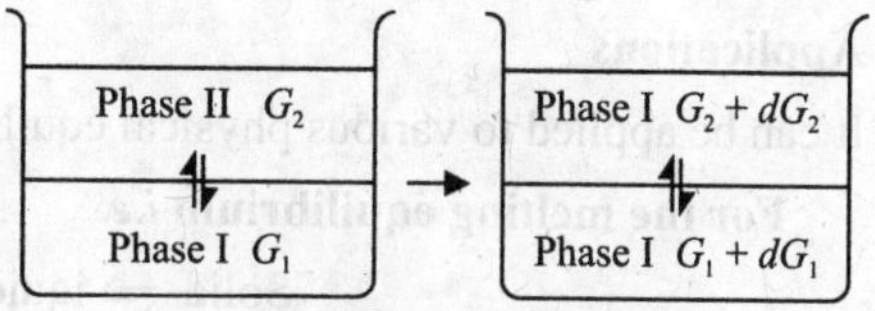

Fig. 2.7: *Phase equilibria at T and P and at T + dT and P + dP*

$$dG_1 = V_1 dP - S_1 dT \qquad ...(iii)$$
and
$$dG_2 = V_2 dP - S_2 dT \qquad ...(iv)$$

where V_1 and S_1 are the molar volume and entropy of phase I and V_2 and S_2 are the corresponding values for phase II.

But when phase I is in equilibrium with phase II,
$$\Delta G = 0$$
i.e. $\qquad G_2 - G_1 = 0$

where G_1 is the molar free energy of phase I and G_2 that of phase II at temperature T and pressure P.

Similarly, for the equilibrium at temperature $T + dT$ and pressure $P + dP$, we must have
$$(G_2 + dG_2) - (G_1 + dG_1) = 0$$
But $\qquad G_2 - G_1 = 0$ (as already stated above)
$\therefore \qquad dG_2 - dG_1 = 0$
or $\qquad dG_2 = dG_1 \qquad \qquad \ldots(v)$

Substituting the values of dG_1 and dG_2 from equations (iii) and (iv), we get
$$V_2 dP - S_2 dT = V_1 dP - S_1 dT \text{ or } (V_2 - V_1) dP = (S_2 - S_1) dT$$
or $\qquad \Delta V . dP = \Delta S . dT$
or $\qquad \dfrac{dP}{dT} = \dfrac{\Delta S}{\Delta V} \qquad \qquad \ldots(vi)$

where $\qquad \Delta V = V_2 - V_1$ is the molar volume change

and $\qquad \Delta S = S_2 - S_1$ is the molar entropy change

where phase I changes into phase II.

If this change occurs at the temperature T (i.e. melting point, boiling point etc.) and ΔH is latent heat of transformation, then
$$\Delta S = \dfrac{\Delta H}{T}$$
Putting this value in equation (vi), we get
$$\boxed{\dfrac{dP}{dT} = \dfrac{\Delta H}{T \Delta V}} \qquad \ldots(vii)$$

This equation was first derived by Clapeyron (1934) and is called **Clapeyron equation**.

Applications

It can be applied to various physical equilibria like melting, vaporisation and sublimation.

For the melting equilibrium i.e.
$$\text{Solid} \rightleftharpoons \text{liquid}$$
$\Delta H = \Delta H_m$, latent heat of melting and $T = T_m$, the melting point

Equation (vii) may then be written as
$$\dfrac{dP}{dT} = \dfrac{\Delta H_m}{T_m \Delta H} \qquad \ldots(viii)$$

Further $\qquad \Delta V = V_l - V_s$

where V_s = molar volume of the solid and V_l = molar volume of the liquid (melt)

Hence equation (viii) may be written as

$$\frac{dP}{dT} = \frac{\Delta H_m}{T_m(V_l - V_s)}$$

or $\quad\quad\quad\quad\dfrac{dT}{dP} = \dfrac{T_m(V_l - V_s)}{\Delta H_m}$ ...(ix)

This equation gives the variation of melting point with pressure.

For the vaporisation equilibrium *i.e.*

$$\text{Liquid} \rightleftharpoons \text{Vapour}$$

The Clapeyron equation, *i.e.* equation (*viii*) takes the form

$$\frac{dP}{dT} = \frac{\Delta H_v}{T_b(V_v - V_l)} \quad\quad\quad ...(x)$$

where ΔH_v = latent heat of vaporisation,

T_b = boiling point of the liquid

V_l = molar volume of the liquid, and

V_v = molar volume of the vapour.

Volume of the liquid (V_l) is very small as compared to the volume of the vapour (V_v), V_l can be neglected in comparison to V_v. Hence equation (*x*) may be written as (omitting the subscript in T_b).

$$\frac{dP}{dT} = \frac{\Delta H_v}{TV_v} \quad\quad\quad ...(xi)$$

Further assuming that the vapour behave like an ideal gas, we will have

$$PV_v = RT$$

or $\quad\quad\quad\quad V_v = \dfrac{RT}{P}$ ...(xii)

Substituting this value in equation (*xii*), we get

$$\frac{dP}{dT} = \frac{\Delta H_v}{RT^2}.P \quad \text{or} \quad \frac{1}{P}.\frac{dP}{dT} = \frac{\Delta H}{RT^2}$$

or $\quad\quad\quad\quad \dfrac{d \ln P}{dT} = \dfrac{\Delta H_v}{RT^2}$ ...(xiii)

This equation is called **Clausius-Clapeyron equation** as it was derived by Clausius during a detailed study of Clapeyron equation.

The equation (*xiii*) is very commonly written in another form called **integrated form of Clausius-Clapeyron equation**. This may be derived as follows:

Equation (*xiii*) may be rewritten as

$$d \ln P = \frac{\Delta H_v}{RT^2} dT \quad\quad\quad ...(xiv)$$

If the temperature changes from T_1 to T_2 when the vapour pressure changes from P_1 to P_2 then integrating equation (*xiv*) between the appropriate limits, we get

$$\int_{P_1}^{P_2} d\ln P = \int_{T_1}^{T_2} \frac{\Delta H_v}{RT^2} dT \quad \text{or} \quad \ln \frac{P_2}{P_1} = \frac{\Delta H_v}{R} \int_{T_1}^{T_2} \frac{1}{T^2} dT$$

$$= \frac{\Delta H_v}{R}\left(\frac{1}{T_1} - \frac{1}{T_2}\right)$$

Changing the natural logarithm to the logarithm to base 10, we get

$$2.303 \log \frac{P_2}{P_1} = \frac{\Delta H_v}{R}\left(\frac{T_2 - T_1}{T_1 T_2}\right)$$

or
$$\log \frac{P_2}{P_1} = \frac{\Delta H_v}{2.303\, R}\left(\frac{T_2 - T_1}{T_1 T_2}\right) \qquad \ldots(xv)$$

This is called the *integrated form of Clausius-Clapeyron equation*, and has a number of applications.

(*i*) If the vapour pressure of a liquid is known at one particular temperature, the value at any other temperature can be calculated, if the latent heat of vaporization of the liquid is known.

(*ii*) If the vapour pressures of a liquid at two different temperatures are known, the latent heat of vaporization of the liquid can be calculated.

(*iii*) If the boiling point of a liquid is known at some particular pressure, the boiling point at any other pressure can be calculated, provided the latent heat of vaporization of the liquid is known.

Applying Chemsitry to Life

The wonderful recreational sport of ice-skating involves the principles of thermodynamics. The thin blades of the skates exert pressure on ice causing a very thin layer of ice to melt. The thin layer of water produced acts as a lubricant for the smooth movement of skates.

Ice-skating

Solved Examples—Clausius-Clayperon Equation

Example 40: The normal boiling point of water is 100°C. Its vapour pressure at 80°C is 0.4672 atmosphere. Calculate the enthalpy of vaporisation per mole of water.

Solution: Here, we have

$T_1 = 100°C = 100 + 273 = 373$ K, $P_1 = 1$ atm, $T_2 = 80°C = 80 + 273 = 353$ K

$P_2 = 0.4672$ atm, $\Delta H_v = $ to be calculated

Substituting the values in the equation,

$$\log \frac{P_2}{P_1} = \log \frac{0.4672}{1}$$

we get
$$\log \frac{0.4672}{1} = \frac{\Delta H_v}{2.303 \times 8.314} \left(\frac{353 - 373}{353 \times 373} \right)$$

or
$$-0.3305 = \frac{\Delta H_v}{2.303 \times 8.314} \times \left[\frac{-20}{353 \times 373} \right] \text{ or } \Delta H_v = 41.66 \text{ kJ mol}^{-1}$$

Example 41: Water boils at 373 K at one atmospheric pressure. At what temperature will it boil when atmospheric pressure becomes 528 mm of Hg at some space station.

Latent heat of H_2O = 2.28 kJ g^{-1}, R = 8.314 JK^{-1} mol^{-1}

Solution: We are given that

$$T_1 = 373 \text{ K}, P_1 = 1 \text{ atm} = 760 \text{ mm}$$
$$P_2 = 528 \text{ mm}, T_2 = \text{To be calculated}$$
$$\Delta H_v = 2.28 \text{ kJ/g} = 2.28 \times 18 \text{ kJ mol}^{-1} = 41.04 \text{ kJ mol}^{-1} = 41040 \text{ J mol}^{-1}$$
$$R = 8.314 \text{ JK}^{-1} \text{ mol}^{-1}$$

Substituting the values in the equation,

$$\log \frac{P_2}{P_1} = \frac{\Delta H_v}{2.303 R} \left(\frac{T_2 - T_1}{T_2 T_1} \right)$$

we get
$$\log \frac{521}{760} = \frac{41040}{2.303 \times 8.314} \times \left(\frac{T_2 - 373}{373 T_2} \right)$$

or
$$-0.1582 - 2143 \times \left(\frac{T_2 - 373}{373 T_2} \right)$$

or $-59.0 \, T_2 = 2143 \, T_2 - 799488.2$ or $2202.4 \, T_2 = 799488.2$ or $T_2 = \mathbf{363 \text{ K}}$

Example 42: Water boils at 100°C at a pressure of 1 atm. Calculate the vapour pressure of water at 90°C. The heat of vaporisation of water is 9.80 kcal mol^{-1}.

Solution: Here, we have
$$T_1 = 100 + 273 = 373 \text{ K}, P_1 = 1 \text{ atm}$$
$$T_2 = 90 + 273 = 363 \text{ K}, P_2 = \text{To be calculated}$$
$$\Delta H_v = 9.80 \text{ kcalmol}^{-1} = 9800 \text{ cal mol}^{-1},$$
$$R = 1.987 \text{ cal deg}^{-1} \text{ mol}^{-1}$$

Substituting these values in the equation

$$\log \frac{P_2}{P_1} = \frac{\Delta H_v}{2.303 R} \left(\frac{T_2 - T_1}{T_1 T_2} \right)$$

we get
$$\log \frac{P_2}{P_1} = \frac{9800}{2.303 \times 1.987} \times \frac{363 - 373}{363 - 373}$$

or
$$\log P_2 = -0.1582 = \overline{1}.8418 \text{ or } P_2 = \mathbf{0 \cdot 6947 \text{ atm}}$$

Key Terms

- Entropy
- Gibbs'-Duhem equation
- Gibbs'-Helmholtz equation
- Partial molar quantities

Evaluate Yourself

Multiple Choice Questions

1. For an isothermal process the total change in work function is given by the expression:
   (a) $(\Delta A)_T = -nRT \ln \dfrac{V_1}{V_2}$

   (b) $(\Delta A)_T = -nRT \ln \dfrac{P_2}{P_1}$

   (c) $(\Delta A)_T = -nRT \ln \dfrac{V_2}{V_1}$

   (d) $(\Delta A)_T = -nRT \ln \dfrac{V_1}{V_2}$

2. Which of the following is not the criterion for spontaneity of a process:
   (a) $(dG)_{T,P} \geq 0$
   (b) $(dA)_{T,V} \leq 0$
   (c) $(dG)_{T,P} \leq 0$
   (d) $(dS)_{U,V} \geq 0$.

3. Nernst heat theorem can be mathematically stated as:
   (a) $\lim\limits_{T \to 0} \dfrac{d(\Delta G)}{dT} = \lim\limits_{T \to 0} \dfrac{d(\Delta H)}{dT} = 0$

   (b) $\lim\limits_{T \to 0} \dfrac{d(\Delta H)}{dT} = \lim\limits_{T \to 0} \dfrac{d(\Delta S)}{dT} = 0$

   (c) $\lim\limits_{T \to 0} \dfrac{d(\Delta U)}{dT} = \lim\limits_{T \to 0} \dfrac{d(\Delta A)}{dT} = 0$

   (d) $\lim\limits_{T \to 0} \dfrac{d(\Delta A)}{dT} = \lim\limits_{T \to 0} \dfrac{d(\Delta H)}{dT} = 0$

4. Under what conditions change in work function is equal to change in Gibbs free energy i.e. $\Delta A = \Delta G$:
   (a) When there is no change in volume, i.e. $\Delta V = 0$.
   (b) When there is no change in enthalpy, i.e. $\Delta H = 0$.
   (c) When there is no change in temperature.
   (d) When there is no change in pressure.

5. The relationship between decrease in Gibbs free energy and electrical energy is :
   (a) $-\Delta G = nF.E_{cell}$
   (b) $\Delta G = nRT \ln E_{cell}$
   (c) $-\Delta G = -nF E_{cell}$
   (d) $\Delta G = -nRT \ln E_{cell}$

6. Point out the correct relationship:
   (a) $\left(\dfrac{\partial A}{\partial V}\right)_P = -T$
   (b) $\left(\dfrac{\partial A}{\partial T}\right)_P = -P$
   (c) $\left(\dfrac{\partial G}{\partial P}\right)_T = -V$
   (d) $\left(\dfrac{\partial G}{\partial T}\right)_P = -H$

7. Gibbs Helmholtz equation is:
   (a) $\Delta G = \Delta H - T\Delta S$
   (b) $\Delta G = \Delta H - T\left[\dfrac{(\partial G)}{\partial T}\right]_P$
   (c) $\Delta G = \Delta U + T\left[\dfrac{\partial(\Delta G)}{\partial T}\right]_P$
   (d) $\Delta G = \Delta H + T\left[\dfrac{\partial(\Delta G)}{\partial T}\right]_P$

8. At very low temperatures near the range of 0 K, the C_P of a solid can be calculated using the equation:
   (a) $C_P \approx \alpha T^3$
   (b) $C_P = \alpha T^2$
   (c) $C_P = \dfrac{1}{3}\alpha T^3$
   (d) $C_P = \dfrac{1}{3}\alpha T$

9. Residual entropy is:
   (a) The entropy possessed by the in crystalline substance at $T = 0$.
   (b) The entropy arising out of defects in crystalline substance
   (c) The entropy which is in excess over the normal value
   (d) The remaining entropy of a substance.

10. Which of the following is not a state function:
    (a) H (b) S (c) U (d) W

Short Answer Questions

1. What is Carnot cycle ? How does it lead to the definition of entropy?
   Justify that $\Sigma \dfrac{q}{T} = 0$ for Carnot cycle.
2. Explain the term 'Entropy'. Show that it is a state function.
   Write down the units of 'Entropy'.
3. Explain that in a reversible process, there is no net entropy change.

4. Explain that the entropy of the system and the surroundings increases in an irreversible process but remains constant in a reversible process.
5. Why there was a need for introduction of second law of thermodynamics ? Illustrate with a suitable example.

Long Answer Questions

1. Explain the term 'free energy'. What information does the free energy of a reaction give about the spontaneity of the reaction to occur ?
2. What is Nernst heat theorem ? What result follows from it regarding entropy change and heat capacity change ? State 'third law of thermodynamics'. Briefly desribe its importance.
3. What is Residual entropy ? What is its origin and how can it be calculated ? Explain taking suitable examples.
4. Show that $\Delta G = \Delta H + T\left[\dfrac{\partial(\Delta G)}{\partial T}\right]_P$
5. First law of thermodynamics leads to the concept of energy content and second law leads to the concept of entropy. Does third law lead to any new concept ? If not, then why is it called a law ?
6. How can you test the validity of the third law of thermodynamics ?
7. What is Residual entropy ? How the concept of residual entropy originated ? How is it calculated ?
8. Explain the term 'Gibbs function'. Deduce that decrease in its value is equal to the maximum net work obtainable from the system. Also explain how does it vary with temperature and pressure ?
9. Bring out clearly the criteria for reversibility and irreversibility in terms of S, U, H, A and G. What do you mean by a reversible and an irreversible process ?
10. Derive an expression relating the changes in free energy and enthalpy of a closed thermodynamic system at constant pressure.
11. Derive 'total differential equation' and hence derive the following expression for an isothermal process when the volume changes from V_1 to V_2 or pressure changes from P_1 to P_2

$$(\Delta G)_T = nRT \ln\dfrac{P_2}{P_1} = nRT \ln\dfrac{V_1}{V_2}$$

12. What are the applications of Gibbs-Helmholtz equation ?
13. How the results of Nernst theorem led to the enunciation of third law of thermodynamics?
14. Prove that the entropy of any substance at very low temperature is one third of its molar heat capacity at that temperature.
15. Explain the term 'Helmholtz function'. How can you deduce that for a process occurring at constant temperature, the decrease in Helmholtz function 'A' is equal to the maximum work done by the system. Also explain how it varies with temperature and volume.

Chemical Thermodynamics – II (Second and Third Laws of Thermodynamics and Partial Molar Properties)

Answers

Multiple Choice Questions

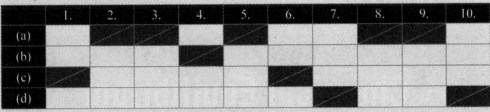

Suggested Readings

1. Castellan, G.W Physical Chemistry, Narosa.
2. Engel, T. & Reid, P. Physical Chemistry, Prentice-Hall.

Chapter 3
Chemical Equilibrium

LEARNING OBJECTIVES

After reading this chapter, you should be able to:
- Establish criteria for thermodynamic equilibrium
- Understand the concept of fugacity
- Derive the relation between Gibb's free energy and reaction quotient
- Couple exoergic and endoergic reactions
- Learn the dependence of equilibrium constant on temperature, pressure and concentration
- Understand the relation between free energy of mixing and spontaneity
- Learn Le Chatelier principle

3.1 INTRODUCTION

Most of the chemical reactions, when carried out in closed vessels do not go to completion. A process starts of its own or on initiation, continues for sometime with falling rate and ultimately appears to have stopped. Although, reactants are still there, but they don't appear to react anymore. The process or the reaction is said to be in a state of *equilibrium*. The composition of the reaction mixture at equilibrium at a given temperature is the same irrespective of whether we start from the reactant side or product side. This may be illustrated with the help of the following examples.

1. Decomposition of calcium carbonate: Solid calcium carbonate on heating in a closed vessel starts decomposing into solid calcium oxide and gaseous carbon dioxide. Pressure built up due to carbon dioxide can be recorded on a manometer attached to the vessel. The pressure goes on increasing and then becomes constant and remains so after that. Calcium carbonate is still present in the reaction mixture but does not appear to decompose further. A state of equilibrium, as represented by the following equation, is attained.

$$CaCO_3 \text{ (s)} \rightleftharpoons CaO \text{ (s)} + CO_2 \text{ (g)}$$

2. Evaporation of water in a closed vessel: If a small amount of water is placed in a closed vessel at room temperature, evaporation starts. A pressure due to water vapours is built up in the vessel, which can be recorded by the attached manometer. After some time, we find that the pressure of water vapours remains constant, if the temperature does not change. Sufficient water is still present, but it does not appear to evaporate further. We say that a state of equilibrium between evaporation of water and condensation of vapours has been established. This can be represented as

$$H_2O \text{ (l)} \rightleftharpoons H_2O \text{ (g)}$$

3.1.1 Characteristics of Chemical Equilibrium

1. Chemical equilibrium at a given temperature, is characterised by constant values of certain observable properties such as pressure, concentration or colour. It is well known that hydrogen gas and iodine vapours react at higher temperature to give hydrogen iodide gas as per the following equation

$$H_2 \text{ (g)} + I_2 \text{ (g)} \rightleftharpoons 2HI \text{ (g)}$$

It is also known that hydrogen iodide gas is not very stable and it dissociates into hydrogen and iodine as follows:

$$2HI \text{ (g)} \rightleftharpoons H_2 \text{(g)} + I_2 \text{(g)}$$

The composition of the reaction mixture at equilibrium, is *1.56 moles of* HI, *0.22 mole of* H_2 *and 0.22 mole of* I_2 whether we carry out the reaction by taking one mole each of H_2 and Cl_2 as the reactants or two moles of HI as the reactant at 448°C.

This shows that *when equilibrium is attained at a given temperature, each reactant and each product has a fixed concentration and this is independent of the fact whether we start the reaction with the reactants or with the products. Thus, equilibrium can be attained from either side.*

2. *A catalyst can hasten the approach of equilibrium but does not alter the state of equilibrium. In other words, the relative concentrations of the products and the reactants remain the same irrespective of the presence or absence of a catalyst.*

3. *Chemical equilibrium is* **dynamic** *in nature*. Equilibrium involves two reactions proceeding in opposite directions. One of these reactions proceeds *from the reactants towards the products* and is known as the *forward reaction*. The other proceeds *from the products towards the reactants* and is known as the *reverse reaction*. When equilibrium is attained, there is no further change in the concentrations of the products or the reactants. This gives the impression that the reaction has come to a stand-still. But this is *not* the case. Actually, the two opposing reactions, the forward reaction and the reverse reaction, are proceeding simultaneously at *equal rates*.

3.1.2 Experimental Proof of Dynamic Nature of Chemical Equilibrium

Consider the following reaction

$$H_2(g) + I_2(g) \rightleftharpoons 2HI(g)$$

When the equilibrium is attained at 448°C, a small amount of radioactive isotope of iodine is introduced into the reaction mixture. It is well known that radioactive isotopes have the same *chemical* properties as their non-radioactive counterparts. After sometime if the mixture is examined, it will be seen that HI *contains radioactive iodine*. There is, however, no change in the relative amounts of HI, H_2 and I_2. This shows that although there has been no change in the relative amounts of the reactants and the products, *chemical reaction has been taking place from left to right*. But since there has been no increase whatsoever in the amount of HI, it is evident that *the reaction has been taking place from the right to the left as well, at the same rate*. Thus, the two opposing reactions have been proceeding. This proves the dynamic nature of equilibrium.

Two analogies from the real life can be cited to explain the phenomenon of chemical equilibrium.

1. Take a cylindrical vessel fitted with a tap (Fig. 3.1 *a*). We have another water tap located above the vessel. If we regulate the flow of water into the vessel such that it is equal to the outflow of water from the vessel through its tap, we shall find that the level of water in the vessel remains the same. The same thing applies to a reaction at equilibrium. Both the forward reaction and backward reactions are taking place but the two rates are equal and opposite.

2. A child is running up an escalator with some speed. If the escalator moves downward with the same speed, then the location of the child will not change (Fig. 3.1 *b*). Although he is running upward, he will be found stuck at the same position. This is comparable to the situation when the forward and backward reactions are taking place at equal speeds in opposite directions and it appears that the reaction has stopped.

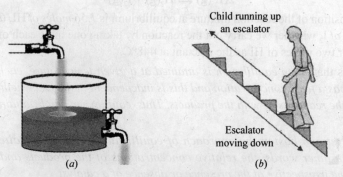

Fig. 3.1: Equilibrium

3.2 THE STATE OF CHEMICAL EQUILIBRIUM

In a general reversible reaction

$$A + B \rightleftharpoons C + D$$

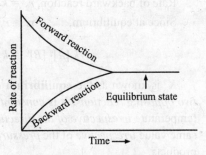

Fig. 3.2: Variation of rate of forward and backward reactions with time

the reactants A and B react to form products. With the passage of time the concentrations of reactants decrease. Therefore the rate of forward reaction decreases. On the other hand, the concentration of C and D increases. Therefore rate of backward reaction increases. A stage is reached when the rate of backward reaction becomes equal to rate of forward reaction and the system attains a state of equilibrium. The variation of rate of forward and backward reaction is shown in Fig. 3.2.

Thus, *the state of equilibrium is a state in which the measurable properties of the system do not undergo any noticeable change under a given set of conditions*. That, means the properties like concentration, pressure etc. of the system become constant. But this does not mean that the reaction has stopped. In fact, both the forward and backward reactions carry on even after attainment of equilibrium. However, the rates of forward and backward reactions become equal as shown in Fig. 3.2. Thus, *the equilibrium is dynamic in nature*.

3.3 LAW OF MASS ACTION

According to the law of mass action, *the rate at which a substance reacts is proportional to its 'active mass' and the rate of a chemical reaction is proportional to the product of the active masses of the reacting substances*.

The term *active mass* used in the above statement implies *activity* which, for the sake of simplicity, may be taken as equal to *molar concentration*.

Consider the following reversible reaction, taking place at a constant temperature:

$$A + B \rightleftharpoons C + D$$

According to the law of mass action, the rate (r_1) at which A and B react is given by the equation

$$r_1 = k_1 [A][B] \quad \ldots(i)$$

where k_1 is the proportionality constant, known as the **rate constant** of the reaction and the square brackets indicate molar concentrations. As the reaction proceeds further, the molar concentrations of A and B continue to decrease. Consequently, the velocity of the forward reaction goes on decreasing with time. The rate of the back reaction (r_2) is given by the equation

$$r_2 = k_2 [C][D] \quad \ldots(ii)$$

where k_2 is the rate constant of the reverse reaction.

Thus, ultimately, a dynamic equilibrium is attained when the rate of the forward reaction becomes equal to that of the reverse reaction, *i.e.*, $r_1 = r_2$.

3.3.1 Law of Chemical Equilibrium

Consider the reaction

$$a\text{A} + b\text{B} \rightleftharpoons c\text{C} + d\text{D}$$

Rate of forward reaction, $r_1 = k_1 [A]^a [B]^b$
Rate of backward reaction, $r_2 = k_2 [C]^c [D]^d$
Since at equilibrium, $r_1 = r_2$, it follows that

$$k [A]^a [B]^b = k_2 [C]^c [D]^d \quad \text{or} \quad \frac{k_1}{k_2} = K_c = \frac{[C]^c [D]^d}{[A]^a [B]^b} \qquad ...(i)$$

K_c is known as the **equilibrium constant.** It is simply the *ratio of the rate constants of two opposing reactions*. It is *constant at a given temperature*. It has the same value at a given temperature *irrespective of the direction* from which the equilibrium is reached. It has also the same value *irrespective* of the *pressure* on the system or the *concentrations* of the reactants and products.

In a general reaction represented by the equation

$$aA + bB + cC + \cdots \rightleftharpoons lL + mM + nN + \cdots,$$

the equilibrium constant is given by

$$K_c = \frac{[L]^l [M]^m [N]^n \cdots}{[A]^a [B]^b [C]^c \cdots} \qquad ...(ii)$$

For gaseous reactions, it is more convenient to use *partial pressures* instead of concentrations. The equilibrium constant in that case is represented by K_p. Thus, for the general gaseous reaction

$$aA + bB + \cdots \rightleftharpoons lL + mM + \cdots,$$

$$K_p = \frac{[p_L]^l \times [p_M]^m \cdots}{[p_A]^a \times [p_B]^b \cdots} \qquad ...(iii)$$

where p_L and p_M, etc stand for partial pressures of the products and p_A, p_B, etc. for the partial pressures of the reactants.

3.3.2 Types of Equilibrium Constants and their Units

In eq. (*i*) above, the concentrations of reactants and products are generally expressed in terms of moles/litre.

1. The equilibrium constant is then written as K_c where the subscript c refer to the equilibrium constant expressed in terms of molar concentrations.

2. When the molar concentrations are replaced by activities of reactants and products, then; the equilibrium constant is written in *terms of activities* and is called K_a.

i.e., $\qquad K_a = \dfrac{a_C^c \cdot a_D^d}{a_A^a \cdot a_B^b} \cdot K_c = \dfrac{[C]^c [D]^d}{[A]^a [B]^b}$

As activity is dimensionless, K_a has no units.

3. In gaseous reactions, the concentration terms in eq. (*i*) are replaced by the partial pressure. Then, the equilibrium constant may be expressed as

$$K_p = \frac{p_C^c \cdot p_D^d}{p_A^a \cdot p_B^b}$$

where p_A, p_B, p_C and p_D are the partial pressures of gases A, B, C and D respectively. K_p may not always be equal to K_c. Units of K_p are (Atmospheres)$^{\Delta n}$.

As concentrations are expressed in moles-litre, according to eq. (i), dimensions of K_c are given by

$$\left(\frac{\text{Moles}}{\text{llitre}}\right)^{\Delta n}$$

where Δn is the difference is the number of moles of products and the reactants [say $\Delta n = (c + d) - (a + b)$].

when $\Delta n = 0$, K_c and K_p both will become dimensionless i.e. will have no units, simply numbers.

4. When equilibrium constant is expressed in terms of mole fractions of reactants and products, it is expressed as K_x.

$$K_x = \frac{(x_C)^c (x_D)^d}{(x_A)^a (x_B)^b}$$

where x_A, x_B, x_C and x_D represent the mole fractions of A, B, C and D.

3.3.3 Relationship between K_p and K_c

For a general reaction,

$$aA + bB \rightleftharpoons cC + dD$$

The equilibrium constant in terms of concentrations (moles/litre) is

$$K_c = \frac{[C]^c [D]^d}{[A]^A [B]^B} \qquad \text{...(i)}$$

If the reactants and products are gaseous, then the concentration terms may be replaced by partial pressure. The equilibrium constant K_p is written as

$$K_p = \frac{p_C^c \, p_D^d}{p_A^a \, p_B^b} \qquad \text{...(ii)}$$

For an ideal gas, $pV = nRT$ or $p = \frac{n}{V}RT = CRT$,

where C is the molar concentration. For different gases A, B, C, D, we may write

$$p_A = C_A RT, \; p_B = C_B RT, \; p_C = C_C RT, \; p_D = C_D RT$$

Putting these values in equation (ii), we get

$$K_p = \frac{(C_C RT)^c (C_D RT)^d}{(C_A RT)^a (C_B RT)^b} \quad \text{or} \quad K_p = \frac{C_C^c C_D^d}{C_A^a C_B^b}(RT)^{(c+d)-(a+b)}$$

or

$$\boxed{K_p = K_c (RT)^{\Delta n}} \qquad \text{...(iii)}$$

where $\Delta n = (c + d) - (a + b)$

= [Number of moles of products] − [Number of moles of reactants]

3.3.4 Relationship between K_x, K_p and K_c

The partial pressure p of a gas in an ideal gaseous mixture is related to total pressure P as

$$p = xP$$

where x is the mole fraction of the gas in the mixture

Substituting for the partial pressures of A, B, C and D in equation (*ii*), we get

$$K_p = \frac{p_C^c \, p_D^d}{p_A^a \, p_B^b} = \left[\frac{x_C^c \, x_D^d}{x_A^a \, x_B^b}\right] p^{(c+d)-(a+b)}$$

or $\qquad K_p = K_x (P)^{\Delta n}$ ...(*iv*)

From eq. (*iii*) and (*iv*), we have

$$K_c (RT)^{\Delta n} = K_x P^{\Delta n} \quad \text{or} \quad K_c = K_x \left(\frac{P}{RT}\right)^{\Delta n}$$

or $\qquad \boxed{K_c = K_x V^{\Delta n}}$

Where V is the volume of system containing 1 mole of ideal gas.
When $\Delta n = 0$, K_p, K_c and K_x become equal.

or $\qquad K_c = K_x V^{\Delta n}$

3.3.5 Relationship between K_a and K_p or K_c

Activity and molar concentration are related as

$$a = \gamma C \qquad \qquad ...(i)$$

where γ is activity coefficient

$$K_a = \frac{a_C^c \, a_D^d}{a_A^a \, a_B^b} \qquad \qquad ...(ii)$$

Substituting for activities of A, B, C and D in eq. (*ii*), we have

$$K_a = \left(\frac{\gamma_C^c \, \gamma_D^d}{\gamma_A^a \, \gamma_B^b}\right) \left(\frac{C_C^c \, C_D^d}{C_A^a \, C_B^b}\right) = K_\gamma \cdot K_c$$

or $\qquad \boxed{K_a = K_\gamma K_c}$ ...(*iii*)

For very dilute solutions as $C \to 0$, $\gamma \to 1$
Then $\qquad K_a \to K_c$
The activity of any gaseous component is related to its partial pressure as

$$a = \gamma_p \qquad \qquad ...(iv)$$

Substituting (*iv*) in eq. (*ii*), we get

$$K_a = K_\gamma K_p$$

For gases at very low pressure $\gamma = 1$
$\therefore \qquad K_\gamma = 1$, hence $K_a = K_p$

3.4 PARTIAL MOLAR QUANTITIES

The thermodynamic properties such as free energy, enthalpy, entropy, free energy etc. are extensive properties. Their values change with change in mass of the system. Consider an open system containing n_1, n_2, n_3 ...... moles of various components. Let the property X be a function of temperature., pressure and the number of moles of various constituents *i.e.*

$$X = f(T, P, n_1, n_2, n_3, \ldots n_i) \qquad \qquad ...(i)$$

where $n_1 + n_2 + n_3 + \ldots n_i = N$

For a small change in temperature, pressure and the number of moles of constituents, the change dX in property X will be given by the following equation:

$$dX = \left(\frac{\partial X}{\partial T}\right)_{P,n_1,n_2,...} dT + \left(\frac{\partial X}{\partial P}\right)_{T,n_1,n_2,...} dP$$

$$+ \left(\frac{\partial X}{\partial n_1}\right)_{T,P,n_2,n_3,...} dn_1 + \left(\frac{\partial X}{\partial n_2}\right)_{T,P,n_1,n_3,...} dn_2 + ... \quad ...(ii)$$

At constant temp. and pressure of the system, $dT = 0$ and $dP = 0$.

Equation (ii), then becomes

$$(dX)_{T,P} = \left[\frac{\partial X}{\partial n_1}\right]_{T,P} dn_1 + \left[\frac{\partial X}{\partial n_2}\right]_{T,P} dn_2 + ... \quad ...(iii)$$

Each term on the right hand side of equation (iii) is called partial molar property of that particular component. It is represented by putting a bar over the concerned property i.e. $\overline{X}_1, \overline{X}_2$ for the first and second component respectively.

Thus, $\left[\frac{\partial X}{\partial n_1}\right]_{T,P,n_2,n_3,...} = \overline{X}_1$ and $\left[\frac{\partial X}{\partial n_2}\right]_{T,P,n_1,n_3,...} = \overline{X}_2$

In general, for an ith constituent, we can write

$$\left[\frac{\partial X}{\partial n_i}\right]_{T,P,n_1,n_2,...} = \overline{X}_i$$

3.4.1 Partial Molar Free Energy: Concept of Chemical Potential

The most important partial molar property is partial molar free energy. It is also called Chemical potential and represented by symbol μ. For an ith component, it is written as

$$\mu = \overline{G}_i = \left[\frac{\partial X}{\partial n_i}\right]_{T,P,n_1,n_2,...} \quad ...(i)$$

If dn_i is taken as 1 mole,

Then $\mu = (dG)_{T,P,n_1,n_2}$, ...(ii)

The chemical potential of a given substance may be defined as the change in free energy of the system which results on the addition of **one mole** of that particular substance at a constant temperature and pressure, to such a large quantity of the system that there is no appreciable change in the overall composition of the system.

Taking free energy as the extensive property, change in property i.e. dG may be written as

$$(dG)_{T,P} = \left[\frac{\partial G}{\partial n_1}\right]_{T,P,n_2,n_3,...} dn_1 + \left[\frac{\partial G}{\partial n_2}\right]_{T,P,n_1,n_3,...} dn_2 + ...$$

or $(dG)_{T,P} = \mu_1 dn_1 + \mu_2 dn_2 + ...$ ...(iii)

where $\mu_1, \mu_2, ...$ are chemical potentials of constituents 1, 2, ...

For a system of definite composition having $n_1, n_2, ...$ moles of components, equation (iii) on integration gives

$$G_{T,P,N} = n_1\mu_1 + n_2\mu_2 + ...$$

The subscript N stands for total number of moles.

Here μ_1, μ_2 etc. give the contribution per mole of components 1 and 2 to the total free energy of the system under conditions of constant temperature and pressure.

Thus, *the chemical potential of a constituent in a mixture is its contribution per mole to the total free energy of the system of a constant composition at constant temperature and pressure.*

It is observed that for one mole of pure component $G = \mu$ *i.e.* free energy is the same as chemical potential.

3.4.2 The Free Energy Change of a Reaction

For a general reaction
$$aA + bB \rightleftharpoons cC + dD$$
the free energy change is given by

ΔG = Free energy of products – Free energy of reactants

In terms of chemical potentials of reactants and products, the free energy change is given by the following equation
$$\Delta G = (c\,\mu_C + d\,\mu_D) - (a\,\mu_A + b\,\mu_B)$$
where μ_A, μ_B, μ_C and μ_D are the chemical potentials of A, B, C and D respectively.

Standard Free Energy Change of a Reaction

When all the reactants and products of a chemical reaction are taken in their standard state, the free energy change accompanying the reaction is called standard free energy change. It is represented by $\Delta G°$. Thus,
$$\Delta G° = G°_{Products} - G°_{Reactants}$$
In terms of chemical potentials, we can write it in the form
$$\Delta G° = (c\mu°_C + d\mu°_D) - (a\mu°_A + b\mu°_B)$$
where $\mu°_A$, $\mu°_B$, $\mu°_C$ and $\mu°_D$ are the standard chemical potentials of A, B, C and D respectively.

3.5 FREE ENERGY CHANGE AS A CRITERION OF SPONTANEITY

It has been discussed earlier that change in free energy can be used as a criterion of feasibility of a process. If

(*i*) $\Delta G < 0$, The process is spontaneous in forward direction.

(*ii*) $\Delta G > 0$, The process is non-spontaneous in forward direction. In fact it will proceed in backward direction *i.e.* products wil get converted into reactants.

(*iii*) $\Delta G = 0$, The reaction will be at equilibrium.

Consider a reversible reaction : Reactants $\rightleftharpoons$ Products

Free energy changes that take place when the reaction is in progress are depicted in Fig. 3.3.

In the beginning the concentration of reactants is maximum, therefore, the free energy associated, with them is also maximum while the free energy associated with products is minimum. As the reaction proceeds, the concentration of reactants decreases

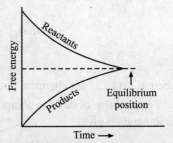

***Fig. 3.3:** Free energy change when reactants are converted into products*

and concentration of products increases. Hence the free energy associated with products increases.

Equilibrium is reached when the concentration of reactants and products becomes constant and the free energy of reactants is equal to the free energy of products. Thus, at equilibrium, $\Delta G = 0$.

In a reversible reaction, both the forward and backward reactions take place. The condition of spontaneity is that ΔG should be negative. Thus ΔG is negative for the forward as well as backward reaction. This is explained with the help of the following example.

Consider the following homogeneous gaseous reaction,

$$N_2O_4 (g) \rightleftharpoons 2NO_2 (g)$$

The variation of free energy of N_2O_4 and NO_2 versus the extent of reaction at constant temperature and pressure is shown in Fig. 3.4. At point A (where the extent of reaction is zero) only the reactant N_2O_4 is present. Point A represents the standard free energy of the reactant.

At point B, only products i.e. 2 moles of NO_2 are present. Point B therefore represents standard free energy of products. Point C is the equilibrium state where 16.6% of N_2O_4 and 83.40% NO_2 exist together. At this point the $\Delta G = 0$. The difference between the standard free energy of products (point B) and the standard free energy of reactants (point A) is $\Delta G° = +5.40$ kJ. The free energy of equilibrium mixture (point C) is 0.84 kJ lower than point A. That implies that when one mole of N_2O_4 (g) changes into equilibrium mixture, the value of $\Delta G° = -0.84$ kJ. Similarly when 2 moles of NO_2 change into equilibrium mixture, the value of $\Delta G°$ is given by

$$\Delta G° = -5.40 + (-0.84) = -6.24 \text{ kJ}^*$$

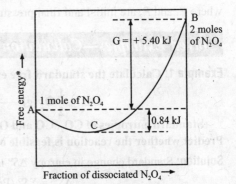

Fig. 3.4: *Graph of free energy of system showing spontaneity of forward as well as backward reaction*

Thus, we find that in both the cases the value of $\Delta G°$ is negative. Therefore both the processes are spontaneous.

3.5.1 Calculation of Standard Free Energy Change in a Reaction

The standard free energy change in a reaction may be calculated by one of the following methods

1. Using the relation: $\Delta G° = \Delta H° - T\Delta S°$

Heat of reaction $\Delta H°$ is determined experimentally and entropy change $\Delta S°$ is calculated from the absolute entropies of the reactants and products, using the relation

$$\Delta S° = \Sigma S° \text{ (Products)} - \Sigma S° \text{ (Reactants)}$$

2. Using the relation: $\Delta G° = -RT \ln K_p$ or $\Delta G° = -RT \ln K_c$.

Knowing the values of K_p or K_c, it is possible to calculate standard free energy change using the above relationships.

---

*No scale is indicated on the Y-axis as free energy absolute values are not known. It is the change in free energy values that can be determined. However, such a curve will ensure accurate representation of its shape.

3. Using the relation: $\Delta G° = \Sigma \Delta_f G°$ **(Products)** $- \Sigma \Delta_f G°$ **(Reactants)**

Standard free energy of formation is defined as the free energy change accompanying a reaction in which one mole of the substance in its standard state is formed from its elements in their standard states. Free energy of formation of an element in its standard state is taken as zero.

4. Free energy change ΔG of a gaseous reaction, at constant temperature can also be obtained, using the following relation

$$\Delta G = n\,RT \ln \frac{P_2}{P_1}$$

where P_1 and P_2 are initial and final pressures respectively.

Solved Examples—Calculation of Standard Free Energy Change

Example 1: Calculate the standard free energy change for the reaction

$$CO + \frac{1}{2}O_2(g) \longrightarrow CO_2(g), \quad \Delta H° = 270 \text{ kJ}$$

Standard entropies of CO_2, CO and O_2 are 205, 190 and 200 J/degree/mole respectively. Predict whether the reaction is feasible or not.

Solution: Standard change in entropy $\Delta S°$ is given by

$$\Delta S° = \Sigma S° \text{ (Products)} - \Sigma S° \text{ (Reactants)}$$

or
$$\Delta S° = S°(CO_2) - [S°(CO) + \frac{1}{2}S°(O_2)]$$

Substituting the values, we have

$$\Delta S° = 205 - [190 + \frac{1}{2} \times 200] = -85 \text{ J/degree/mole}$$

Apply the relation

$$\Delta G° = \Delta H° - T\Delta S°$$

Here $\Delta H° = -270$ kJ $= -270000$ J, $\quad T = 25 + 273 = 298$ K, $\quad \Delta S° = -85$ J/degree/mole

$\therefore \qquad \Delta G° = -270000 - 298\,(-85)$ J $= -244670$ J $= -244.67$ kJ

Since $\Delta G°$ is negative, hence the reaction is feasible.

Example 2: Calculate the standard free energy change of the reaction

$$NO(g) + \frac{1}{2}O_2(g) \longrightarrow NO_2(g)$$

Standard free energies of formation of NO (g) and NO_2 (g) are 87 kJ and 52 kJ mol^{-1}.

Solution:
$$\Delta_f G° = \Sigma \Delta_f G° \text{ (Products)} - \Sigma \Delta_f G° \text{ (Reactants)}$$

or
$$\Delta_f G° = \Delta_f G°(NO_2) - [\Delta_f G°(NO) + \frac{1}{2}\Delta_f G°(O_2)]$$

Substituting the values, we have

$$\Delta_f G° = 52 - [87 + \frac{1}{2} \times (0)] \quad \text{or} \quad \Delta_f G° = -35 \text{ kJ} \quad \therefore \Delta_f G°(O_2) = 0$$

Problems for Practice

1. Calculate the standard free energy change at 25°C for the reaction
$$N_2 (g) + 3H_2 (g) \longrightarrow 2NH_3 (g), \quad \Delta H° = -22.08 \text{ kcal}$$
Given that $\Delta S°$ for the reaction is -47.4 cal/degree
Predict the feasibility of the above reaction at the given state.
   [Ans. $\Delta G° = -7954.8$ kcal, Feasible]

2. Calculate the standard free energy change for the reaction
$$4NH_3 (g) + 5O_2 (g) \longrightarrow 4NO (g) + 6H_2O (l)$$
Given that the standard free energies of formation ($\Delta_f G°$) for NH_3 (g), NO (g) and H_2O (l) are -16.74, 86.61 and -237.23 kJ mol-1 respectively.
Predict the feasibility (or spontaneity) of the above reaction at the standard state.
   [Ans. $\Delta G° = -1009.98$ kJ, Feasible]

3. Calculate the standard free energy for the reaction
$$Fe_2O_3 (s) + 3H_2 (g) \longrightarrow 2Fe (s) + 3H_2O (l)$$
Given that the standard free energies of formation of Fe_2O_3 and H_2O are -741.0 kJ and -237.2 kJ per mole respectively. Can the above reaction occur at the standard state?
   [Ans. $\Delta G° = +29.4$ kJ, No]

3.6 THERMODYNAMIC DERIVATION OF THE LAW OF CHEMICAL EQUILIBRIUM

Consider a general reversible reaction,
$$aA + bB \rightleftharpoons cC + dD$$
where the reactants and products are taken to be ideal gases. If μ_A, μ_B, μ_C, μ_D are the chemical potentials of A, B, C and D respectively at constant temperature and pressure, then

Total free energy of reactants = $G_{\text{Reactants}} = a\mu_A + b\mu_B$
Total free energy of products = $G_{\text{Products}} = c\mu_C + d\mu_D$

At constant temperature and pressure, the free energy change of the reaction is given by
$$\Delta G = G_{\text{Products}} - G_{\text{Reactants}}$$
or
$$\Delta G = (c\mu_C + d\mu_D) - (a\mu_A + b\mu_B) \qquad ...(i)$$

When the system is in equilibrium state at constant temperature and pressure, $\Delta G = 0$, therefore equation (1) can be written as
$$(c\mu_C + d\mu_D) - (a\mu_A + b\mu_B) = 0$$
or
$$c\mu_C + d\mu_D = a\mu_A + b\mu_B \qquad ...(ii)$$

This represents a general thermodynamic condition for equilibrium applicable to all systems.

If the reactants and products are ideal gases, the chemical potential of any component in the mixture (Refer Art 3.4.6) is given by
$$\mu = \mu° + RT \ln p \qquad ...(iii)$$
where p is the partial pressure of the component and $\mu°$ is the chemical potential in the standard state (i.e. at 1 atm pressue). $\mu°$ depends only on temperature.

Substituting the value of μ from equation (iii) into equation (ii) we get

$$c(\mu_C^\circ + RT \ln p_C) + d(\mu_D^\circ + RT \ln p_D) = a(\mu_A^\circ + RT \ln p_A) + b(\mu_B^\circ + RT \ln p_B)$$

$$RT\, c(\ln p_C + d \ln p_D) + c(c\mu_C^\circ + d \ln p_D) = RT\,(a \ln p_A + b \ln p_B) + (a\mu_A^\circ + b\mu_B^\circ)$$

$$RT[(\ln P_C^c + \ln P_D^d) - (\ln P_A^a + \ln p_B^b)] = a(\mu_A^\circ + b\mu_B^\circ) + c(\mu_C^\circ + d\mu_D^\circ)$$

or
$$RT \ln \frac{p_C^c \cdot p_D^d}{p_A^a \cdot p_B^b} = G^\circ_{\text{Reactants}} - G^\circ_{\text{Products}} = (G^\circ_{\text{Products}} - G^\circ_{\text{Reactants}} -) \qquad \ldots(iv)$$

$$= -\Delta G^\circ \qquad \ldots(v)$$

or
$$\ln \frac{p_C^c \cdot p_D^d}{p_A^a \cdot p_B^b} = \frac{-\Delta G^\circ}{RT} \quad \text{or} \quad \frac{p_C^c \cdot p_D^d}{p_A^a \cdot p_B^b} = e^{-\Delta G^\circ / RT} \qquad \ldots(vi)$$

The standard free energy change of a reaction ΔG° depends only on temperature, hence at constant temperature, the term $e^{-\Delta G^\circ/RT}$ is constant. Thus equation (vi) can be written as

$$\frac{p_C^c \cdot p_D^d}{p_A^a \cdot p_B^b} = \text{Constant }(K_p) \qquad \ldots(vii)$$

This is thermodynamic derivation of law of chemical equilibrium in terms of partial pressures.

From equations (vi) and (vii), we arrive at the result

$$\boxed{\Delta G^\circ = -RT \ln K_p} \qquad \ldots(viii)$$

If molar concentrations are taken in place of partial pressures, the chemical potential of a component may be expressed as

$$\mu = \mu^\circ + RT \ln C$$

where C is the molar concentration of the given component in the reaction mixture and μ° is the chemical potential of the substance in the standard state (when the concentration is unity). Repeating the same steps as described above, we can show that

$$\frac{C_C^c \cdot C_D^d}{C_A^a \cdot C_B^b} = \text{Constant }(K_c)$$

or
$$\frac{[C]^c \cdot [D]^d}{[A]^a \cdot [B]^b} = K_c$$

Similarly, we can obtain a relation between ΔG° and K_c as under

$$\boxed{\Delta G^\circ = -RT \ln K_c} \qquad \ldots(ix)$$

To derive the law of equilibrium in terms of activities, we write the chemical potential of each component as

$$\mu = \mu_x^\circ + RT \ln a$$

where a is the activity of the component and μ_x° depends both on temperature and pressure.

Substituting the value of μ in eq. (ii), we get

$$\frac{a_C^c \cdot b_D^d}{a_A^a \cdot a_B^b} = \text{Constant }(K_a)$$

Also
$$\boxed{\Delta G^\circ = -RT \ln K_a} \qquad \ldots(x)$$

To derive the equation in terms of mole fractions, we can use the relation

$$\mu = \mu_x^\circ + RT \ln x$$

where x is the mole fraction and as before μ_x° depends on temperature and pressure. Substituting the values in eq. (ii) and repeating the steps, we obtain the equation in terms of mole fractions of the reactants and products as

$$\frac{x_C^c \cdot x_D^d}{x_A^a \cdot x_B^b} = K_x \text{ (constant)} \qquad ...(xi)$$

By knowing, ΔG° of the reaction we can calculate K_p, K_c, K_a or K_x and vice versa. $K_p = K_c$ only when the number of moles of reactants and products are equal. Therefore, the value of ΔG° obtained in equations (vii) and (ix) are different.

Solved Examples—Standard Free Energy Change and Equilibrium Constant

Example 3: At 2000 K, the standard state free energy change (ΔG°) for the reaction $N_2 + O_2 \rightleftharpoons 2NO$ is given by $\Delta G^\circ = 92048 - 10.46$ T J. Calculate K_p for the reaction at 2000 K (R = 8.314 J K^{-1} mol^{-1}).

Solution:
$$\Delta G^\circ = -RT \ln K_p = -2.303\, RT \log K_p$$

$$\log K_p = -\frac{\Delta G^\circ}{2.303\, RT} = \frac{(-92048) + 10.46 \times 2000 \text{ J mol}^{-1}}{(2.303)(8.314 \text{ J K}^{-1} \text{ mol}^{-1})(2000 \text{ K})}$$

$$= 1.857$$

$$K_p = 1.39 \times 10^{-2} \text{ atm}$$

Example 4: The value of K_p, for the water gas reaction

$$CO + H_2O \rightleftharpoons CO_2 + H_2$$

is 1.06×10^5 at 25°C. Calculate the standard state free energy change (ΔG°) of the reaction at 25°C. (R = 8.314 J K–1 mol^{-1})

Solution:
$$\Delta G^\circ = -RT \ln K_p = -2.303\, RT \log K_p$$
$$= -(2.303)(8.314 \text{ J K}^{-1})(298 \text{ K}) \log(1.06 \times 10^{-5})$$
$$= -28673.6 \text{ J} = -28.674 \text{ kJ}$$

Example 5: At 2000 K, the standard free energy change (ΔG°) for the reaction

$$N_2 + O_2 \rightleftharpoons 2NO$$

is given by $\Delta G^\circ = 22000 - 2.0$ T cal.

Calculate K_p for the reaction at 2000 K (R = 1.98 calorie)

Solution: Given
$$\Delta G^\circ = 22000 - 2.0 \text{ T and T} = 2000 \text{ K}$$
$$\therefore \quad \Delta G^\circ = 220000 - 2.0 \times 2000 = 18000 \text{ cal}$$

Using the relationship between ΔG° and K_p
$$\Delta G^\circ = -TR \ln K_p = -2.303\, RT \log K_p,$$

Substituting the value, we have
$$18000 = -2.303 \times 1.98 \times 2000 \times \log K_p$$

or $\quad K_p = -1.9737$ or $\log K_p = \overline{2}.0263$

or $\quad K_p = \text{antilog}\,(\overline{2}.0263) = \mathbf{1.063 \times 10^{-2}}$

Example 6: Calculate the standard state free energy change ($\Delta G°$) of the reaction
$$2H_2O\ (g) \rightleftharpoons 2H_2\ (g) + O_2\ (g)$$
given that at 1000 K, water vapour at 1 atmosphere pressure has a degree of dissociation $= 3 \times 10^{-5}$ per cent (R = 8.314 J K^{-1} mol^{-1}).

Solution: The partial pressures of the various reactants in the equilibrium mixture will be evidently as below:
$$pH_2 = 3 \times 10^{-7}\ \text{atm},\ pO_2 = \frac{3}{2} \times 10^{-7}$$

$$K_p = \frac{p_{H_2}^2}{p_{H_2O}} = \frac{9 \times 10^{14} \times \frac{3}{2} \times 10^{-7}}{1} = 13.5 \times 10^{-21}\ \text{atm}$$

$$\Delta G° = -RT \ln K_p = -2.303\ RT \log K_p$$
$$= -2.303\ (8.314\ \text{J K}^{-1}\ \text{mol}^{-1})\ (1000\ \text{K}) \log (13.5 \times 10^{-21})$$
$$= 380448\ \text{J} = 380.45\ \text{kJ}$$

Problems for Practice

1. At 27°C, the standard free energy change for the reaction
$$CO\ (g) + H_2O\ (g) \rightleftharpoons CO_2\ (g) + H_2\ (g)$$
is – 6909 cal. Calculate K_p for the equilibrium. **[Ans. 1.078 × 10^5]**

2. The standard state free energy of a reaction at 298.15 K is –16.65 kJ. Calculate the value of equilibrium constant of the reaction at 298.15 K. **[Ans. 8.252 × 10^2]**

3. For the reaction
$$CCl_4\ (l) + H_2\ (g) \longrightarrow HCl\ (g) + CHCl_3\ (l)$$
at 25°C, $\Delta H° = -21.83$ kcal/mol and $\Delta S° = 9.92$ kcal K^{-1} mol^{-1}. Determine whether the reaction is favourable at 25° under standard conditions.
Calculate the value of K_p at 25°C.
[Ans. $\Delta G° = -24.79$ kcal, Reaction is favourable, $K_p = 1.5 \times 10^{18}$]

4. Calculate K_p for the following reaction at 25°C
$$2SO_2\ (g) + O_2\ (g) \rightleftharpoons 2SO_3\ (g)$$
$$\Delta_f G°\ (SO_2) = -71.79\ \text{kcal/mol}$$
$$\Delta_f G°\ (SO_3) = -88.52\ \text{kcal/mol}$$
[Ans. 1.2 × 10^{22}]

5. The values of standard state free energy for a gaseous reaction
$$CO\ (g) + H_2O\ (g) \rightleftharpoons CO_2\ (g) + H_2\ (g)$$
at 298 K is – 28.53 kJ.
Calculate the value of the equilibrium constant K_p (R = 8.314 JK^{-1} mol^{-1}) **[Ans. 1.0 × 10^5]**

3.7 VAN'T HOFF REACTION ISOTHERM

Consider the following reaction
$$aA + bB + \cdots \rightleftharpoons lL + mM + \cdots$$
taking place *under any conditions* of temperature, pressure and composition. The free energy change of the reaction is given by the expression

$$\Delta G = (G)_{products} - (G)_{reactants} \qquad ...(i)$$
$$= (l\mu_L + m\mu_M + \cdots) - (a\mu_A + b\mu_B + \cdots)$$

where μ_A, μ_B, μ_L, μ_M, etc., represent chemical potentials of the species concerned.

Let us suppose that various reactants and products are in gaseous state. Chemical potential of a gaseous substance in *any state* is given by the equation

$$\mu_{i(p)} = \mu^°_{i(p)} + RT \ln p_i \qquad ...(ii)$$

where $\mu^°_{i(p)}$ is the chemical potential of a gaseous component i, in the *standard state* (partial pressure = 1 atmosphere) and $\mu_{i(p)}$ is the chemical potential of the gaseous component i in any state of partial pressure p_i.

Substituting the values of chemical potentials of various species at their partial pressures from eq. (*ii*) in (*i*)

$$\Delta G = \{l(\mu^°_{L(p)} + RT \ln p_L) + m(\mu^°_{M(p)} + RT \ln p_M) + \cdots\}$$
$$- \{a(\mu^°_{A(p)} + RT \ln p_A) + b(\mu^°_{B(p)} + RT \ln p_B) + \cdots\} \quad ...(iii)$$

Re-arranging, we have

$$\Delta G = \{l\mu^°_{L(p)} + m\mu^°_{M(p)} + \cdots\} - (a\mu^°_{A(p)} + b\mu^°_{B(p)} + \cdots)\}$$
$$+ RT \frac{(p_L)^l (p_M)^m \cdots}{(p_A)^a (p_B)^b \cdots} \qquad ...(iv)$$

The first expression on the right hand side, evidently, is the free energy change of the reaction when the products and the reactants are all in their respective standard states. This expression may be substituted by $\Delta G°$. Hence,

$$\Delta G = \Delta G° + RT \ln Q_p \qquad ...(v)$$

where Q_p stands for the **reaction quotient** of partial pressures of the products and the reactants, viz.

$$\frac{(p_L)^l (p_M)^m \cdots}{(p_A)^a (p_B)^b \cdots}$$

Eq. (*v*) is known as **van't Hoff reaction isotherm**. This gives the *free energy change of a reaction* in terms of standard free energy change and a factor Q depending upon the partial pressures of the components in the reaction mixture under the given state at constant temperature.

Q_p is different from K_p. Here the values p_L, p_M, p_A, p_B .... etc. are not the partial pressures at equilibrium, but for any state. If the partial pressure values correspond to equilibrium state, then Q_p will become equal to K_p. From eq. (*viii*) in section 3.6, we have

$$\Delta G° = -RT \ln K_p$$

Substituting for $\Delta G°$ in eq. (*v*), we have

$$\Delta G = -RT \ln K_p + RT \ln Q_p \qquad ...(vi)$$

Eq. (*v*) and (*vi*) are two forms of van't Hoff reaction isotherm.

Solved Example

Example 7: At 25°C, the $\Delta G°$ for the reaction
$$H_2(g) + I_2(g) \longrightarrow 2HI(g),$$
is –3.1 kcal. Calculate ΔG when hydrogen gas at 0.1 atm and iodine vapour at 0.1 atm react to form hydrogen iodide at 10 atm, and 25°C.

Solution: For the given reaction,

$$K = \frac{p^2(HI)}{p(H_2)\, p(I_2)}$$

where p terms represent partial pressures of the various species in the reaction mixture.

$$\Delta G = \Delta G° + RT \ln K = \Delta G° + 2.303\, RT \log \frac{p^2(HI)}{p(H_2)\, p(I_2)}$$

$$= -3100 \text{ cal} + 2.303\,(1.987 \text{ cal K}^{-1}\text{mol}^{-1})(298 \text{ K}) \log \frac{(10)^2}{(0.1)(0.1)}$$

$$= -3100 \text{ cal} + 1363.7 \log(10^4) \text{ cal} = -3100 \text{ cal} + 4\,(1363.7) \text{ cal}$$

$$= +2354.8 \text{ cal} = \mathbf{2.355 \text{ kcal}}$$

3.8 RELATION BETWEEN K_p AND K_c

The value of K_p for the reaction

$$aA + bB + cC + \cdots \rightleftharpoons lL + mM + \cdots$$

is given by the expression

$$K_p = \left[\frac{(p_L)^l (p_M)^m \cdots}{(p_A)^a (p_B)^b \cdots}\right]_{eq} \qquad \ldots(i)$$

The value of K_c for the above reaction would be given by the expression

$$K_c = \left(\frac{(c_L)^l (c_M)^m \cdots}{(C_A)^a (c_B)^b \cdots}\right)_{eq} \qquad \ldots(ii)$$

where p and c stand for partial pressure and concentration, respectively. Now, for an ideal gas,

$$pV = nRT, \quad \text{or} \quad p = (n/V)\, RT = cRT$$

or
$$c = p/RT \qquad \ldots(iii)$$

where c is the concentration. Thus, for any component i whose partial pressure is p_i,

$$c_i = p_i/RT \qquad \ldots(iv)$$

Substituting the concentrations of various species in terms of their partial pressures in Eq. (*ii*) we have

$$K_c = \left\{\frac{\left(\frac{p_L}{RT}\right)^l \left(\frac{p_M}{RT}\right)^m \cdots}{\left(\frac{p_A}{RT}\right)^a \left(\frac{p_B}{RT}\right)^b \cdots}\right\}_{eq} = \left(\frac{(p_L)^l (p_M)^m \cdots}{(p_A)^a (p_B)^b \cdots}\right)_{eq} \times \left(\frac{1}{RT}\right)^{\{(l+m+\cdots)-(a+b+\cdots)\}}$$

$$= K_p \left(\frac{1}{RT}\right)^{\Delta n} \qquad \ldots(v)$$

where Δn = Number of moles of products — Number of moles of reactants

or
$$\boxed{K_p = K_c (RT)^{\Delta n}} \qquad \ldots(vi)$$

Solved Example

Example 8: The value of K_p for the equilibrium

$$2H_2O\,(g) + 2Cl_2\,(g) \rightleftharpoons 4HCl\,(g) + O_2(g)$$

is 0.035 atm at 400°C when the partial pressures are expressed in atmospheres. Calculate the value of K_c for the same reaction.

Solution: $\Delta n = 4 + 1 - 2 - 2 = 1$

∴ We shall use the relation

$$K_p = K_c (RT)^{\Delta n} \quad \text{or} \quad K_c = \frac{K_p}{(RT)^{\Delta n}}$$

Substitute the values

$$K_c = \frac{0.035 \text{ atm}}{(0.0820 \text{ l atm K}^{-1} \text{ mol}^{-1} \times 673 \text{ K})} = 6.342 \times 10^{-4} \text{ mol l}^{-1}$$

3.9 VAN'T HOFF EQUATION FOR THE TEMPERATURE DEPENDENCE OF EQUILIBRIUM CONSTANT (VAN'T HOFF REACTION ISOCHORE)

The equation for reaction isotherm when the reactants as well as the products are *gaseous* and are also in their *standard states,* is represented as

$$\Delta G° = -RT \ln K_p$$

Differentiating with respect to temperature at constant pressure, we have

$$\left\{\frac{\partial(\Delta G°)}{\partial T}\right\}_p = -R \ln K_p - RT \frac{d(\ln K_p)}{dT} \qquad \ldots(i)$$

Multiplying throughout by T, we get

$$T\left\{\frac{\partial(\Delta G°)}{\partial T}\right\}_p = -RT \ln K_p - RT^2 \frac{d(\ln K_p)}{dT} \qquad \ldots(ii)$$

Substituting $\Delta G°$ for $-RT \ln K_p$, we have

$$T\left\{\frac{\partial(\Delta G°)}{\partial T}\right\}_p = \Delta G° - RT^2 \frac{d(\ln K_p)}{dT} \qquad \ldots(iii)$$

The well-known Gibbs-Helmholtz equation for substances in *standard states* may be written as

$$\Delta G° = \Delta H° + \left\{\frac{\partial(\Delta G°)}{\partial T}\right\}_p \qquad \ldots(iv)$$

Comparing Eqs. (*iii*) and (*iv*), we get

$$RT^2 \frac{d(\ln K_p)}{dT} = \Delta H° \qquad \ldots(v)$$

or

$$\frac{d(\ln K_p)}{dT} = \frac{\Delta H°}{RT^2} \qquad \ldots(vi)$$

Eq.(*vi*) is known as the **van't Hoff's equation**[†]. $\Delta H°$ is the enthalpy change for the reaction at constant pressure when the reactants as well as the products are in their standard states.

[†] The equation was earlier called *reaction isochore*. The word 'isochore' refers to constant volume. Initially van't Hoff derived the equation for changes at constant volume. But since the equation (*vi*) above involves $\Delta H°$ which is at constant pressure, the name isochore is misleading and is no longer being used.

It is known from experiments that the enthalpy change, ΔH, accompanying a chemical reaction does not vary appreciably with change in partial pressures of the reactants or products. Therefore, *we may take $\Delta H°$ as equal to ΔH, where ΔH is the enthalpy change of the reaction whatever may be the partial pressures of the reactants or products.* Hence, the van't Hoff equation may be written as

$$\frac{d(\ln K_p)}{dT} = \frac{\Delta H}{RT^2} \qquad ...(vii)$$

Integrating Eq. (*vii*) between temperatures T_1 and T_2 at which the equilibrium constants are K_p' and K_p'', respectively and assuming that ΔH remains constant over this range of temperature, we get

$$\int_{K_p'}^{K_p''} d(\ln K_p) = \frac{\Delta H}{R} \int_{T_1}^{T_2} \frac{dT}{T^2} \quad \text{or} \quad \ln K_p'' - \ln K_p' = \frac{-\Delta H}{R}\left[\frac{1}{T_1} - \frac{1}{T_2}\right]$$

$$= \frac{\Delta H}{R}\left[\frac{T_2 - T_1}{T_1 T_2}\right] \qquad ...(viii)$$

or $\qquad \log K_p'' - \log K_p' = \log \dfrac{K_p''}{K_p'} = \dfrac{\Delta H}{2.303}\left[\dfrac{T_2 - T_1}{T_1 T_2}\right] \qquad ...(ix)$

Knowing the equilibrium constant at a temperature it is possible to calculate the equilibrium constant at another temperature provided the heat of reaction (ΔH) is known.

Alternatively, knowing the equilibrium constants of a reaction at two temperatures, the heat of reaction (ΔH) can be calculated.

Van't Hoff's equation in terms of K_c. K_p and K_c are related as

$$K_p = K_c (RT)^{\Delta n} \qquad ...(x)$$

Taking logarithm of both sides of eq. (*x*), we have

$$\ln K_p = \ln K_c + \Delta n \ln R + \Delta n \ln T$$

Differentiating the above equation w.r.t. T, we have

$$\frac{d \ln K_p}{dT} = \frac{d \ln K_c}{dT} + \frac{\Delta n}{T} \qquad ...(xi)$$

Equation (*vi*) derived above is rewritten as

$$\frac{d \ln K_p}{dT} = \frac{\Delta H°}{RT^2} \qquad ...(xii)$$

From equation (*xi*) and (*xii*)

$$\frac{d \ln K_c}{dT} + \frac{\Delta n}{T} = \frac{\Delta H°}{RT^2} \quad \text{or} \quad \frac{d \ln K_c}{dT} = \frac{\Delta H°}{RT^2} - \frac{\Delta n}{T}$$

or $\qquad \dfrac{d \ln K_c}{dT} = \dfrac{\Delta H° - \Delta n(RT)}{RT^2} \qquad ...(xiii)$

We have the thermodynamic relation

$$\Delta H = \Delta U + P\Delta V$$

For standard state, we may write the equation as

$$\Delta H° = \Delta U° + P\Delta V = \Delta U° + \Delta nRT$$

or $\qquad \Delta H° - \Delta nRT = \Delta U° \qquad ...(xiv)$

Chemical Equilibrium **499**

From equation (xiii) and (xiv), we have

$$\frac{d \ln K_c}{dT} = \frac{\Delta U°}{RT^2} \qquad ...(xv)$$

$\Delta U°$ represents the internal energy change when the reactants and products are taken in their standard states.

Solved Examples—Van't Hoff Equation

Example 9: The equilibrium constant K_p, for the reaction $H_2(g) + S(g) \rightleftharpoons H_2S(g)$ is 20.2 atm at 945°C and 9.21 atm at 1065°C. Calculate the heat of reaction.

Solution: Applying van't Hoff equation

$$\log K_p^* - \log K_p' = \frac{\Delta H}{2.303 R}\left[\frac{T_2 - T_1}{T_1 T_2}\right]$$

$$T_1 = 945 + 273 = 1218 \text{ K}, \quad T_2 = 1065 + 273 = 1338 \text{ K}$$

$$K_p' = 20.2 \text{ atm}, \quad K_p'' = 9.21 \text{ atm}$$

Substituting the values, we have

$$\log 9.21 - \log 20.2 = \frac{\Delta H}{(2.303)(8.314 \, JK^{-1} \, mol^{-1})}\left[\frac{1338 - 1218}{(1218 \text{ K})(1338 \text{ K})}\right]$$

$$\Delta H = -88126.3 \text{ J} = -88.826 \text{ kJ}$$

Example 10: The equilibrium constant K_p for the reaction $N_2 + 3H_2 \rightleftharpoons 2NH_3$ is 1.64×10^{-4} atm at 400°C. What will be the equilibrium constant at 500°C if the heat of reaction in this temperature range is –105185.8 J ?

Solution: $\log K_p^* - \log K_p' = \frac{\Delta H}{2.303 R} \frac{T_2 - T_1}{T_1 T_2}$

$$T_1 = 400 + 273 = 673 \text{ K},$$
$$T_2 = 500 + 273 = 773 \text{ K}, \Delta H = -105185.8 \text{ J}$$

$$\log K_p'' = \log (1.64 \times 10^{-4}) + \frac{-105185.8 \text{ J mol}^{-1}}{(2.303)(8.314 \, JK^{-1} \, mol^{-1})}\left(\frac{773 \text{ K} - 673 \text{ K}}{673 \text{ K} \times 773 \text{ K}}\right)$$

$$\therefore \quad K_p' = 0.144 \times 10^{-4} \text{ atm}$$

Example 11: The equilibrium constant K_p for a reaction at 600 K and 620 K are 1×10^{-12} and 5×10^{-12} respectively. Considering ΔH to be constant in the above temperature range, calculate ΔH and ΔS for the reaction. ($R = 8.314$ J K^{-1} mol^{-1})

Solution: Given that

$$K'_p = 1 \times 10^{-12}, \quad T_1 = 600 \text{ K}, \quad K''_p = 5 \times 10^{-12}, \quad T_2 = 620 \text{ K}$$
$$R = 8.314 \text{ J K}^{-1} \text{ mol}^{-1}$$

Substituting the values in van't Hoff equation

$$\log \frac{K_p^*}{K_p'} = \frac{\Delta H}{2.303 R}\left[\frac{T_2 - T_1}{T_1 T_2}\right] \text{ we have}$$

$$\log \frac{5 \times 10^{-12}}{1 \times 10^{-12}} = \frac{\Delta H}{2.303 \times 8.314}\left[\frac{620 - 600}{600 \times 620}\right]$$

$$\log 5 = \frac{\Delta H}{2.303 \times 8.314} \times \frac{20}{600 \times 620}$$

or
$$\Delta H = \frac{2.303 \times 8.314 \times 600 \times 620 \times \log 5}{20} \text{ Joules} = 24894 \text{ J}$$

To calculate ΔS, use the relation, $\Delta G = \Delta H - T\Delta S$

At equilibrium $\Delta G = 0$, therefore $\Delta S = \dfrac{\Delta H}{T}$

$$\Delta S \text{ at } 600 \text{ K} = \frac{24890}{600} = 414.9 \text{ J K}^{-1} \text{ mol}^{-1}$$

$$\Delta S \text{ at } 620 \text{ K} = \frac{24890}{620} = 401.5 \text{ J K}^{-1} \text{ mol}^{-1}$$

Problems for Practice

1. Calculate the ethalpy change for the reaction
$$N_2 + O_2 \rightleftharpoons 2NO$$
Given that the equilibrium constant for this reaction is 4.08×10^{-4} at 2000 K and 3.60×10^{-3} at 2500 K. **[Ans. 181.1 kJ]**

2. For a certain reaction $K_p = 1.21 \times 10^{-4}$ at 1800 K and 2.31×10^{-4} at 1900 K. Calculate K_p at 2000 K. **[Ans. 4.13×10^{-4}]**

3. The equilibrium constant K_p for a reaction is 3.0 at 400°C and 4.0 at 500°C. Calculate the value of $\Delta H°$ for the reaction ($R = 8.314$ J K^{-1} mol^{-1}). **[Ans. 12.44 kJ]**

4. For the reaction $N_2O_4 \rightleftharpoons 2NO_2$, $\Delta H = 14.6$ kcal.
if $K_p = 0.141$ at 298 K, calculate K_p at 338 K. **[Ans. 2.607]**

Solved Example

Example 12: Explain the terms

(i) Homogeneous equilibrium (ii) Heterogeneous equilibrium.

Solution: (i) An equilibrium which has all the reactants and products in the same phase is called homogeneous equilibrium. For example
$$H_2(g) + I_2(g) \rightleftharpoons 2HI(g)$$
Here H_2, I_2 and HI all exist in the same gaseous phase.

similarly $\quad CH_3COOH\ (l) + C_2H_5OH\ (l) \rightleftharpoons CH_3COOC_2H_5\ (l) + H_2O\ (l)$

is also an example of homogeneous equilibria as all the reactants and products are in the liquid phase.

(ii) An equilibrium which has the reactants and products in two or more phases is called heterogeneous equilibrium. For example
$$CaCO_3(s) \rightleftharpoons CaO\ (s) + CO_2(g)$$

3.10 COUPLING OF EXOERGIC AND ENDOERGIC REACTIONS

Many chemical reactions are endoergic. They are not spontaneous becuase ΔG in such cases is positive. Remember endoergic reactions are also named as endoergonic by a section of scientists. Such reactions require energy to be supplied externally in order that the reaction takes place. However, we can do something for these reactions to take place. We can couple these reactions with other reactions which are exoergic (i.e. reverse of endoergic). Exoergic reactions have negative value of ΔG, which is a criterion for spontaneity of reactions. Thus, if we couple the first reaction with the second reaction and the net ΔG value of two reactions is negative, the first reaction (as well as the second reaction) can be made to occur. This happens via the formation of a **shared intermediate**.

An example of a reaction that can be made to take place through this procedure is the decomposition of calcium carbonate

$$CaCO_3 (s) \rightleftharpoons CaO (s) + CO_2 (g) \qquad \Delta G° = 130.4 \text{ kJ/mol}$$

As the ΔG is +ve, the reaction would not take place on mild heating.

Consider a different reaction that can be coupled with this reaction.

$$C (s) + O_2 (g) \rightleftharpoons CO_2 (g) \qquad \Delta G° = -394.36 \text{ kJ/mol}$$

This reaction has a large negative value of ΔG compared to positive ΔG value of the first reaction. So, if we combine the two reactions, we can expect decomposition of $CaCO_3$ easily.

$$CaCO_3 (s) + C (s) + O_2 \rightleftharpoons CaO (s) + 2CO_2 (g) \qquad \Delta G° = -263.96 \text{ kJ/mol}$$

As Gibb's energy is a state fuction, the ΔG of the combined reaction is the net of the ΔG values of the two reactions.

This concept of coupling two reactions has its applications in biology.

Coupled Reactions in Biology

In biological systems, some enzyme-catalysed reactions can be explained to be taking place as two coupled half-reactions one of which is spontaneous ($\Delta G < 0$) and the second one is non-spontaneous ($\Delta G > 0$). Hydrolysis of ATP (adenosine triphosphate) to generate ADP (adenosine diphosphate) is spontaneous coupling reaction.

$$ATP + H_2O \rightleftharpoons ADP + P_i$$

where P_i represents phosphate ion.

ATP is a major energy molecule produced by metabolism and it works as energy source in cell. ATP is dispatched to wherever it is required in the body where a non-spontaneous reaction needs to occur. Then the two reactions couple and the overall reaction becomes thermodynamically possible.

A reaction will proceed if the products have lower energy than the reactants. We call such reactions as exoergic or exergonic reactions. A reaction where the products have higher energy than the reactants (endoergic) can proceed if some energy is supplied. Exoergic reactions like burning of glucose drives the ATP synthesis. The ATP molecules are used to energise other endoergic reactions like protein synthesis. This is illustrated as under in Fig. 3.5.

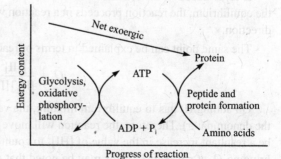

Fig. 3.5: Coupling reaction of glycolysis and protein formation

3.11 LE-CHATELIER'S PRINCIPLE

Le Chatelier, a noted French chemist, studied the effect of *concentration, temperature and pressure* on a large number of chemical equilibria. He summed up his observations in the form of a generalisation known as **Le-Chatelier's principle,** which states as follows:

Le-Chatelier (1850-1936)
He was an an influential Frech chemist of late 19th and early 20th century. He is famours for devising the principle known after his name, used by chemists to predict the effect of changing conditions on a system in chemical equilibrium.

If an equilibrium is subjected to a stress, the equilibrium shifts in such a way as to reduce the stress.

According to this principle, *if a system at equilibrium is subjected to a change of concentration, pressure or temperature, the equilibrium shifts in the direction that tends to* **undo** *the effect of the change.*

Effect of Change of Concentration

Consider the following equilibrium:

$$Fe^{3+} (aq) + (SCN)^- (aq) \rightleftharpoons [Fe(SCN)]^{2+} (aq)$$
Pale yellow   Colourless   Dark brown

Suppose, some ferric salt is added to this equilibrium. The colour of the solution will darken immediately showing that there is increase in the concentration of the coloured complex ion-$[Fe(SCN)]^{2+}$. This change is in accordance with Le-Chatelier's principle. Addition of more of Fe^{3+} ions has resulted in increasing the concentration of the complex ferrisulphocyanide.

Now, suppose a small amount of potassium ferrisulphocyanide capable of giving the complex ion, $[Fe(SCN)]^{2+}$ is added to the equilibrium. The solution will be seen to become less dark showing that the dark coloured $[(Fe(SCN)]^{2+}$ ion has been changed into Fe^{3+} and $(SCN)^-$ ions. Thus, increasing the concentration of the *product* shifts the equilibrium in favour of the *reactants*.

Consider the following reaction:

$$H_2 (g) + I_2 (g) \rightleftharpoons 2 HI (g), \quad K_c = \frac{[HI]^2}{[H_2][I_2]}$$

If H_2 is added to the reaction mixture at equilibrium, the reaction is disturbed. To restore the equilibrium, the reaction proceeds in a reaction wherein H_2 is consumed *i.e.*, in the forward direction.

The same point can be explained in terms of reaction quotient, Q_c

$$Q_c = \frac{[HI]^2}{[H_2][I_2]}$$

On adding H_2 gas to equilibrium mixture, the value of Q_c becomes less than K_c ($[H_2]$ is in the denominator). Therefore, the reaction will move in the forward direction because there will be a resultant increase in the value of $[HI]^2$ to counter the increase in the value of $[H_2]$ thereby bringing Q_c at par with K_c. This may be noted that Q_c always has a tendency to match K_c, the equilibrium constant. If we remove HI from the equilibrium mixture, the value of Q_c will decrease. Therefore, the reaction will again move in the forward direction to keep the value of Q_c equal to K_c.

Effect of Change of Temperature

Consider the equilibrium

$$\underset{\text{Colourless}}{N_2O_4 (g)} \rightleftharpoons \underset{\text{Brown}}{2NO_2(g)}; \quad \Delta H = +59.0 \text{ kJ}$$

In this equilibrium, the reaction favouring the product (NO_2) is seen to be endothermic. Therefore, the opposing reaction favouring the reactant (N_2O_4) must be exothermic. Now suppose the system is *heated* and its temperature is allowed to rise. According to Le Chatelier's principle, the equilibrium will shift in the direction which tends to undo the effect of heat, *i.e.*, which tends to produce *cooling*. Therefore, the equilibrium will shift in favour of NO_2, *i.e.*, the dissociation of N_2O_4 into NO_2 will increase. If the system is cooled, on the other hand, the equilibrium will shift in the direction which tends to produce *heat*. Therefore, the equilibrium will shift in the reverse direction, *i.e.*, in favour of N_2O_4. The dissociation of N_2O_4 will decrease (Fig. 3.6).

$2NO_2 (g) \rightleftharpoons N_2O_4 (g) + \text{heat}$

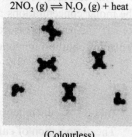

(Colourless)
Low temperature

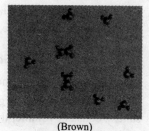

(Brown)
High temperature

Fig. 3.6: *Equilibrium reaction* $N_2O_4 \rightleftharpoons 2NO_2$

The combination of nitrogen and hydrogen to give ammonia is represented by the following thermochemical equation :

$$N_2(g) + 3H_2 (g) \rightleftharpoons 2NH_3(g); \Delta H = -92.38 \text{ kJ}$$

The forward reaction is *exothermic*. Therefore, the back reaction must be endothermic. By Le Chatelier's principle, the increase of temperature will favour the back reaction, *i.e.*, the dissociation of ammonia. Therefore, in order to get a better yield of ammonia, the temperature at equilibrium should be low. But since the reaction rate varies with the temperature, the time taken to reach the equilibrium state becomes very long if the temperature is kept low. Therefore, a temperature close to 500°C, which is neither too low nor too high, is maintained.

Effect of Change of Pressure

Consider another gaseous reaction involving *combination of nitrogen and hydrogen to give ammonia:*

$$\underset{\underset{\text{4 moles}}{\underbrace{\hspace{5cm}}}}{\underset{\text{1 mole}}{N_2(g)} + \underset{\text{3 moles}}{3H_2 (g)}} \rightleftharpoons \underset{\text{2 moles}}{2NH_3}$$

As is evident from the above equation, the forward reaction is accompanied by a decrease in the number of moles. If the pressure is *increased,* the volume will decrease and, therefore, *the number of moles per unit volume will increase.* According to Le Chatelier's principle, therefore, the equilibrium will shift in the direction in which there is *decrease* in the number of moles, *i.e.*, in favour of the formation of ammonia. Thus, *the higher the pressure, the greater would be the* yield of ammonia. It is on account of this consideration that a pressure of about 200 atmospheres is maintained in the Haber's process for the manufacture of ammonia.

The combination of sulphur dioxide and oxygen to give sulphur trioxide:
$$2SO_2(g) + O_2(g) \rightleftharpoons 2SO_3(g)$$
is also accompanied by a *decrease* in the number of moles. According to Le-Chatelier's principle, therefore, a high pressure would give a better yield of sulphur trioxide.

This can also be understood by using reaction quotient, Q. Let $[SO_2]$, $[O_2]$ and $[SO_3]$ be the molar concentrations at equilibrium. When the pressure is doubled the volume is reduced to half, the partial pressure and concentrations are doubled. We obtain the reaction quotient by replacing equilibrium concentration by double its value.

$$Q_c = \frac{[SO_3(g)]^2}{[SO_2(g)]^2 [O_2(g)]}$$

As $Q_c < K_c$, the reaction proceeds in the forward reaction.

The dissociation of calcium carbonate
$$CaCO_3(s) \rightleftharpoons CaO(s) + CO_2(g)$$
is accompanied by an *increase* in the number of gaseous moles. The dissociation of calcium carbonate would, therefore, be *suppressed* by an *increase* of pressure.

Effect of Addition of Inert Gas

The effect of addition of inert gas can be studied under two different conditions:

(*i*) *Addition of an inert gas at constant volume.* When an inert gas is added to the equilibrium state at constant volume, then the total pressure will increase. But the partial pressure of each component will remain unchanged. Under these conditions, there will be no effect on the equilibrium on addition of the inert gas.

(*ii*) *Addition of an inert gas at constant pressure.* When an inert gas is added to the system at constant pressure there will be an increase in the volume. As a result, the number of moles per unit volume of different components will decrease. The equilibrium will shift to the side where the number of moles are increased. For example consider the following equilibrium.

$$2SO_2(g) + O_2(g) \rightleftharpoons 2SO_3(g) \text{ at constant pressure}$$

The addition of inert gas at constant pressure will shift the equilibrium to backward direction. For the equilibrium

$$PCl_5(g) \rightleftharpoons PCl_3(g) + Cl_2(g) \text{ at constant pressure}$$

the addition of inert gas at constant pressure will shift the equilibrium to the forward direction.

However, addition of an inert gas to the following equilibrium

$$N_2(g) + O_2(g) \rightleftharpoons 2NO(g) \text{ at constant pressure}$$

will have no effect because the number of moles of reactants and products are same.

Effect of Catalyst

There is no effect of addition of a catalyst on the equilibrium state. This is because catalyst increases the rate of forward as well as backward reaction to the same extent. It simply helps to achieve the equilibrium quickly. It may be further noted that catalyst has no effect on the equilibrium concentration of a reaction mixture.

3.11.1 Applications of Le-Chatelier's Principle to Physical Equilibrium

The following examples illustrate the application of Le-Chatelier's principle to physical equilibrium.

1. Liquid-vapour equilibrium. Consider water-vapour equilibrium

$$H_2O \text{ (g)} \rightleftharpoons H_2O \text{ (vapour)} \quad \text{(Endothermic process)}$$
(Less volume) (More volume)

The conversion of water to water vapour is an endothermic process.

Also as liquid water changes to vapour phase, it is accompanied by increase in volume. According to Le-Chatelier's principle, the increase in temperature shifts the equilibrium towards the right i.e. more of liquid water wil evaporate. This wil absorb the heat supplied.

On increasing pressure, the equilibrium shifts in a direction so as to lower the pressure. Thus increase of pressure favours the condensation of water vapour into liquid water. On the other hand decreasing the pressure will favour the vaporisation of water into water vapour.

2. Effect of pressure on boiling point of a liquid. Consider the equilibrium

$$\text{Liquid} \rightleftharpoons \text{Vapour}$$

The vaporisation of liquid is accompanied by increase of pressure. Thus, if pressure is increased, the equilibrium shifts towards condensation of vapour into liquid state at a given temperature. The vapour pressure will decrease. Higher temperature is thus needed to make the liquid boil. This is the reason that the boiling point of a liquid increases with increase in external pressure.

3. Effect of pressure on the freezing point of a liquid or melting point of a solid. Consider the equilibrium

$$\text{Ice} \rightleftharpoons \text{Water} \quad \text{(Endothermic process)}$$
(More volume) (less volume)

When a solid melts, there is a change in volume. For example, when ice melts, there is decrease in volume.

Increase of pressure on ice ⇌ water equilibrium will cause the equilibrium to shift towards right. Thus if at constant temperature, the pressure is increased, more of ice will melt. In order to retain ice in equilibrium with water, it is necessary to lower the temperature. Hence, the increase of pressure will lower the melting point of ice. As the process is endothermic, increase of temperature will shift the equilibrium to RHS. Thus more ice will melt.

Most of the solids show increase in volume on melting, e.g., sulphur (solid) ⇌ sulphur (liquid) equilibrium. Therefore we can say that if the pressure on this equilibrium is increased, the melting point of sulphur is raised.

4. Effect of temperature on solubility. We can explain the effect of temperature on solubility of solids with the help of Le-Chatelier's principle. During the dissolution process if heat is evolved, the solubility decreases with increase in temperature. $CaCl_2$, NaOH, etc. are the examples of such solutes. The solubility of such substances decreases on increase of temperature. On the other hand dissolution of substances like NH_4Cl, $NaNO_3$ etc. is accompanied by absorption of heat. Their solubility increases with increase in temperature as per Le-Chatelier's principle.

5. Effect of pressure on solubility of gases in liquids. Consider the equilibrium involving dissolution of CO_2 in water.

$$CO_2 \text{ (g)} \rightleftharpoons CO_2 \text{ (aq)}$$

On increasing the pressure of CO_2 gas the equilibrium shifts in the direction which causes a decrease in the pressure of CO_2. The pressure of CO_2 will be lowered only if it dissolves more in water to form CO_2 (aq). Thus, according to Le-Chatelier's principle, the solubility of a gas in a liquid will increase with increase in pressure of the gas in equilibrium with its solution.

6. Manufacture of ammonia. During the manufacture of ammonia from the combination of nitrogen and hydrogen, ammonia is liquefied and removed from the reaction mixture, so that in accordance with the Le Chatelier's principle, the reaction keeps moving in the forward direction and gives the maximum yield of ammonia.

7. Manufacture of lime. Lime is used as an important building material. During the preparation of lime by heating calcium carbonate in the kiln, carbon dioxide is constantly removed from the reaction mixture so that the reaction keeps going in the forward reaction to yield maximum amount of calcium oxide (lime)

$$CaCO_3 \overset{\Delta}{\rightleftharpoons} CaO + CO_2$$

Example: State Le-Chatelier's principle and apply it for the following equilibrium :

$$PCl_5 \text{ (g)} \rightleftharpoons PCl_3 \text{ (g)} + Cl_2 \text{ (g)} - 17 \text{ k cal}$$

Solution: For the statement of Le-Chatelier's principle, see relevant section

$$PCl_5 \text{ (g)} \rightleftharpoons PCl_3 \text{ (g)} + Cl_2 \text{ (g)} - 17 \text{ k cal}$$

The above reaction is endothermic and takes place with increase of pressure (number of gaseous product molecules is greater than the number of gaseous reactant molecules) in the forward direction.

Effect of pressure. On increasing the pressure, the reaction will shift in a direction where the effect can be neutralized i.e. in a direction where the pressure is decreased. Thus, the reaction will move in the backward direction. On decreasing the pressure, the reverse effect will be observed.

Effect of temperature. On increasing the temperature, the reaction will shift in a direction where it is endothermic i.e. heat can be absorbed. Thus, the reaction will shift in the forward direction. On decreasing the temperature, however, the reverse action will take place.

3.11.2 Quantitative Aspect of Le Chatelier Principle

Consider the following equilibrium reaction

$$SO_2 \text{ (g)} + NO_2 \text{ (g)} \rightleftharpoons SO_3 \text{ (g)} + NO \text{ (g)}$$

Suppose, we increase the concentration of SO_3 in the above reaction. This will disturb the equilibrium and produce a stress in the reaction. The stress has to be removed by taking an appropriate step. When the concentration of SO_2 is increased, the denominator in the following equation increases.

$$K_{eq} = \frac{[SO_3][NO]}{[SO_2][NO_2]}$$

Now the experimental reaction quotient Q is less than K_{eq}. To re-establish the equilibrium, the numerator must increase. This can happen if the reaction shifts to the right and increases the yield of product.

Problem: At 200°C, analysis of the equilibrium mixture of the reaction:

$$SO_2\,(g) + NO_2\,(g) \rightleftharpoons SO_3\,(g) + NO\,(g)$$

Shows the following equilibrium concentrations:

$$[SO_2] = 4.0\ M,\quad [NO_2] = 0.50\ M,$$
$$[SO_3] = 3.0\ M,\quad [NO] = 2.0$$

What is the new equilibrium concentration of NO when 1.5 moles of NO_2 are added to the equilibrium mixture above?

Solution: First find the value of K_{eq}. This value will remain the same for equilibrium mixture and the stressed mixture. Substitute the values in the following equation:

$$K_{eq} = \frac{[SO_3][NO]}{[SO_2][NO_2]}$$

$$= \frac{(3.0\ M) \times (2.0\ M)}{(4.0\ M) \times (0.50\ M)}$$

$$= 3.0$$

For the stressed mixture

	SO_2 (g)	+	NO_2 (g)	$\rightleftharpoons$	SO_3 (g)	+	NO (g)
Initial	4.0 M		0.5 M		3.0 M		2.0 M
Final	(4.0 – x)		(0.5 + 1.5 – x)		(3.0 + x)		(2.0 + x)

where x is the number of moles of SO_2 that will react with x moles of NO_2 to give the product. This has been necessitated because, we added 1.5 moles of NO_2. So the reaction has to proceed to the right.

Substituting the new values in the equation, we get

$$K_{eq} = \frac{[SO_2][NO]}{[SO_3][NO_2]}$$

$$3.0 = \frac{[3.0 + x][2.0 + x]}{[4.0 - x][2.0 - x]}$$

We shall get a quadratic eqation on simplifying the above equation.

$$2.0x^2 - 23x + 18 = 0$$

On solving this quadratic equation, we get the value of $x = 0.75$

Hence, $[NO] = 2.0 + 0.75 = 2.75$ mol L^{-1}
$[SO_2] = 3.0 + 0.75 = 3.75$ mol L^{-1}
$[SO_3] = 4.0 - 0.75 = 3.25$ mol L^{-1}
$[NO_2] = 2.0 - 0.75 = 1.25$ mol L^{-1}

3.12 PHYSICAL EQUILIBRIA

We come across solid-liquid and liquid-vapour equilibrium processes in our daily life quite often. For example Ice $\rightleftharpoons$ Water and Water $\rightleftharpoons$ Steam equilibrium processes. These processes can be studied precisely using Clausius-Clapeyron equation which has been described in detail in the chapter Chemical Thermodynamics-II in this book.

Some Solved Conceptual Examples

Example 13: Under what condition K_p, K_c, K_a and K_x are all equal?

Solution: It is when $\Delta n = 0$, *i.e.*, $n_p = n_r$ (gaseous).

Example 14: How thermodynamic equilibrium constant in terms of activities (K_a) is related to K_p or K_c? Under what condition K_a becomes equal to K_p or K_c?

Solution: For the reaction, $aA + bB + \ldots \rightleftharpoons mM + nN + \ldots$,

$$K_a = \frac{a_M^m \, a_N^n \ldots}{a_A^a \, a_B^b \ldots}$$

Activity = activity coefficient × composition variable (conc. or pressure)

i.e. $a = \gamma X$

Substituting the value of a in the above equation.

$$K_a = \left(\frac{\gamma_M^m \, \gamma_N^n \ldots}{\gamma_A^a \, \gamma_B^b \ldots}\right)\left(\frac{X_M^m \, X_N^n \ldots}{X_A^a \, X_B^b \ldots}\right) = K_\gamma K_X$$

$\therefore$ $K_a = K_\gamma K_c$ or $K_a = K_\gamma K_p$

For ideal gases or real gases at very low pressures, $\gamma = 1$ for each gas.

Then $K_\gamma = 1$ and hence $K_a = K_p$

Similarly, for very dilute solutions, as $C \to 0$, $\gamma \to 1$ and $K_a \to K_c$.

Example 15: Can the equilibrium $CaCO_3 (s) \rightleftharpoons CaO (s) + CO_2 (g)$ be attained in an open vessel? Why or why not?

Solution: No, this equilibrium cannot be attained in open vessel because one of products, viz., CO_2 is gaseous which escapes out.

Example 16: How is that a reversible reaction is spontaneous in the forward as well as backward direction?

Solution: This is because the free energy of the equilibrium mixture is lower than that of reactants as well as that of products. Hence, ΔG is negative both for the forward direction and for the backward direction.

Example 17: How is the free energy of a gas determined (*i*) by studying its isothermal expansion (*ii*) from its equilibrium constant K_p (*iii*) from enthalpy change and entropy data?

Solution: (*i*) $\Delta G = nRT \ln \dfrac{P_2}{P_1}$ (*ii*) $\Delta G° = -RT \ln K_p$

(*iii*) $\Delta G° = \Delta H° - T\Delta S°$ and $\Delta S° = \Sigma S°$ (Products) $- \Sigma S°$ (Reactants).

Example 18: On the basis of Le-Chatelier's principle discuss the favourable reaction conditions for the following reaction

$$N_2 (g) + O_2 (g) \rightleftharpoons 2NO(g) \quad (\Delta H = 180.5 \text{ kJ})$$

Solution: The given reaction is endothermic and therefore, it will be favoured by high temperature. Δn for the reaction is zero and therefore increase of pressure has no effect on equilibrium.

Example 19: Why $\Delta G°$ obtained from K_p and K_c has different values?

Solution: $\Delta G° = -RT \ln K_p$, Also $\Delta G° = -RT \ln K_c$. As $K_p \neq K_c$ unless $\Delta n = 0$, therefore, $\Delta G°$ values are different in the two cases.

Example 20: Which of the properties remain constant when equilibrium is attained?

Solution: Temperature, pressure and chemical potential of a particular substance remain constant in all the phases.

Example 21: Write van't Hoff equation. Explain that it leads to the same effect of temperature on equilibrium constant as predicted by Le-Chatelier.

Solution: van't Hoff equation is $\dfrac{d \ln K_p}{dT} = \dfrac{\Delta H°}{RT^2}$

For endothermic reactions, $\Delta H°$ is +ve. Hence, $d \ln K_p/dT$ is +ve. This means that $\ln K_p$ and hence K_p increases with increase of temperature, *i.e.*, equilibrium shifts towards products.

For exothermic reactions, $\Delta H°$ is –ve. Hence, $d \ln K_p/dT$ is –ve. This means that $\ln K_p$ and hence K_p decreases with increase of temperature, *i.e.*, equilibrium shifts towards reactants. This is what is stated by Le-Chatelier's principle.

Example 22: How is free energy change of a reaction in a given state related to its reaction quotient (pressure quotient) Q_p in that state and the equilibrium constant K_p?

Solution: $\Delta G = -RT \ln K_p + RT \ln Q_p$.

Example 23: What is van't Hoff reaction isotherm? Why is it so called?

Solution: van't Hoff reaction isotherm is $\Delta G = \Delta G° + RT \ln Q_p$

It gives free energy change of the reaction in terms of standard free energy change and the value of the reaction quotient Q_p under the given state at constant temperature. That is why it is called reaction isotherm.

Example 24: Write van't Hoff equation giving the variation of equilibrium constant with temperature both in the differential form and in the integrated form?

Solution: Differential form is $\dfrac{d \ln RT}{dT} = \dfrac{\Delta H°}{RT^2}$

Integrated form is $\log \dfrac{(K_p)_2}{(K_p)_1} = \dfrac{\Delta H}{2.303 R} \left(\dfrac{1}{T_1} - \dfrac{1}{T_2} \right)$

Key Terms

- Le Chatelier principle
- Free energy of mixing
- Exoergic and endoergic reactions
- Fugacity

Evaluate Yourself

Multiple Choice Questions

1. According to law of Mass Action, for the reaction
   $$2A + B \longrightarrow \text{Products, which equation holds good?}$$
   (a) Rate = k $[A]^2$ [B]
   (b) Rate = k [A] $[B]^2$
   (c) Rate = k [A] [B]
   (d) Rate = k [A] $[B]^{1/2}$

2. In which of the following reactions, K_p is less than K_c?
   (a) N_2 (g) + $3H_2$ (g) $\rightleftharpoons$ $2NH_3$ (g)
   (b) PCl_5 (g) $\rightleftharpoons$ PCl_3 (g) + Cl_2 (g)
   (c) H_2 (g) + I_2 (g) $\rightleftharpoons$ 2HI (g)
   (d) $2SO_3$ (g) $\rightleftharpoons$ $2SO_2$ (g) + O_2 (g)

3. What will be the equilibrium constant at 717 K for the reaction :
   $$2HI (g) \rightleftharpoons \frac{1}{2} H_2 (g) + \frac{1}{2} I_2 (g)$$
   if its value for the reaction
   $$H_2 (g) + I_2 (g) \rightleftharpoons 2HI (g) \text{ at 717 K is 64?}$$

   (a) 8
   (b) 64
   (c) $\frac{1}{64}$
   (d) $\frac{1}{8}$.

4. With increase in temperature, equilibrium constant of a reaction:
   (a) always decreases
   (b) always increases
   (c) may increase or decrease depending upon whether $n_p < n_r$ or $n_p > n_r$.

5. The relation between K_p and K_c for the reaction
   $$2NO (g) + Cl_2 (g) \rightleftharpoons 2NOCl (g) \text{ is}$$
   (a) $K_p = K_c / RT$
   (b) $K_p = K_c (RT)$
   (c) $K_p = K_c / (RT)^2$
   (d) $K_p = K_c$

6. K_1 and K_2 are equilibrium constants for reactions (1) and (2)
   $$N_2 (g) + O_2 (g) \rightleftharpoons 2NO (g) \qquad \text{...(1)}$$

$$NO(g) \rightleftharpoons \frac{1}{2}N_2(g) + \frac{1}{2}O_2(g) \qquad ...(2)$$

Then,

(a) $K_1 = K_2^2$ \qquad (b) $K_1 = \dfrac{1}{K_2}$

(c) $K_1 = \left(\dfrac{1}{K_2}\right)^2$ \qquad (d) $K_1 = (K_2)^0$.

7. Which of the following equilibria will shift to right side on increasing the temperature?
   (a) $2SO_2(g) + O_2(g) \rightleftharpoons 2SO_3(g)$
   (b) $H_2O(g) \rightleftharpoons H_2(g) + \dfrac{1}{2}O_2(g)$
   (c) $4HCl(g) + O_2(g) \rightleftharpoons 2H_2O(g) + 2Cl_2(g)$
   (d) $CO(g) + H_2O(g) \rightleftharpoons CO_2(g) + H_2O(g)$

8. At 490° C, the equilibrium constant for the synthesis of HI is 50, the value of K for the dissociation of HI will be:
   (a) 2.0 \qquad (b) 20.0
   (c) 0.002 \qquad (d) 0.02.

9. Consider the following reaction equilibrium.
   $$2SO_2(g) + O_2(g) \rightleftharpoons 2SO_3(g); \quad \Delta H° = -198 \text{ kJ}.$$
   On the basis of Le-Chatelier's principle, the condition favourable for the forward reaction is:
   (a) decreasing the temperature and increasing the pressure.
   (b) increasing temperature as well as pressure.
   (c) lowering of temperature as well as pressure.
   (d) any value of temperature and pressure.

10. In which of the following cases, does the reaction go farthest to completion?
    (a) K = 10 \qquad (b) K = 1
    (c) K = 10^3 \qquad (d) K = 10^{-2}.

Short Answer Questions

1. Define the law of chemical equilibrium.
2. State the units in which K_p and K_c are expressed.
3. What do you understand by a reversible reaction? State with an example.
4. Explain the meaning of the term equilibrium constant.
5. What are irreversible reactions? Give an example.
6. State the law of mass action.
7. How is the standard free energy change of a chemical reaction related to its equilibrium constant?

Long Answer Questions

1. Derive van't Hoff equation in the integrated form.
2. Starting from basic principles, derive the relationship $\Delta G° = - RT \ln K_p$.
3. Define law of chemical equilibrium. How can it be derived thermodynamically?
4. Write expression for equilibrium constant
   (a) in terms of pressures (K_p)
   (b) in terms of concentrations (K_c)
   (c) in terms of activities (K_a)
   (d) in terms of mole fractions (K_x)
   Derive the relationship between
   (i) K_p and K_c
   (ii) K_a and K_p or K_c
   (iii) K_c and K_x.
5. Briefly describe at least three different ways by means of which the standard free energy change of a reaction can be calculated.
6. For a gaseous reaction, derive the following relationship:
$$\Delta G = - RT \ln K_p + RT \ln Q_p.$$
7. Derive van't Hoff equation in terms of equilibrium constant K_c and standard internal energy change, $\Delta U°$.
8. What do you understand by reversible and irreversible reactions? What do you mean by saying that equilibrium is dynamic in nature? State the law of chemical equilibrium. Name at least three different types of equilibrium constants. Give the relationship between them. What are their units?
9. A reversible reaction is spontaneous in the forward direction as well in the backward direction (*i.e.*, ΔG is negative for both). Explain with a suitable example.
10. Derive the following for the gaseous reaction $aA + bB + ... \rightleftharpoons mM + nN + ...$
$$\frac{p_M^m \, p_N^n \, ...}{p_A^a \, p_B^b \, ...} = e^{-\Delta G°/RT}$$
11. For a gaseous reaction (not in equilibrium) derive the relationship $\Delta G = \Delta G° + RT \ln Q_p$.
12. How is free energy change and standard free energy change of a reaction calculated from the chemical potentials of the reactants and products involved? Taking suitable example, explain how free energy change of forward and backward reaction of a reversible reaction takes place.
13. Apply Le-Chatelier principle to predict suitable conditions for getting maximum yield of the product in each of the following cases :
    (*i*) manufacture of ammonia by Haber's process.
    (*ii*) manufacture of nitric oxide in Birkland–Eyde process for manufacture of nitric acid.
    (*iii*) manufacture of hydrogen by Bosch process.

Answers
Multiple Choice Questions

	1.	2.	3.	4.	5.	6.	7.	8.	9.	10.
(a)	■	■							■	
(b)					■		■			
(c)						■		■		■
(d)			■	■						

Suggested Readings

1. Levine, I.N. Physical Chemistry, Tata McGraw Hill.
2. Metz, C.R. 2000 Solved Problems in Chemistry. Schaum Series.

Chapter 4

Solutions and Colligative Properties

LEARNING OBJECTIVES

After reading this chapter, you should be able to:

- Know the meaning of dilute solution
- Learn Raoult's law and Henry's law
- Understand excess thermodynamic functions
- Learn how to measure lowering of vapour pressure
- Derive thermodynamically relationship between four colligative properties and amount of solute
- Calculate molecular masses of normal, dissociated and associated solutes in solution

4.1 RAOULT'S LAW

Raoult's law states:

In a solution the vapour pressure of a component at a given temperature is equal to the mole fraction of that component in the solution multiplied by the vapour pressure of that component in the pure state. Mathematically,

$$p_i = x_i \times p^o$$

Thus in a *binary solution*,

Vapour pressure of the solvent in the solution

= Mole fraction of the solvent in solution × Vapour pressure of the pure solvent ...(i)

We can write a similar equation for the solute if it is volatile.

Now if the solute is **non-volatile**, it will not contribute to the total vapour pressure of the solution. Thus the vapour pressure of the solution will be the vapour pressure due to solvent only in the solution.

For such solutions,

Vapour pressure of the solution

= Mole fraction of the solvent in solution × Vapour pressure of the pure solvent

or in terms of symbols, we can write

$$p_s = x_1 \times p^o \qquad ...(ii)$$

This can be rewritten in the form

$$\frac{p_s}{p^o} = x_1 \qquad ...(iii)$$

If the solution contains n_2 moles of the solute dissolved in n_1 moles of the solvent, we have

Mole fraction of the solvent in solution (x_1)

$$= \frac{n_1}{n_1 + n_2}$$

Substituting this value in equation (iii), we get

$$\frac{p_s}{p_o} = \frac{n_1}{n_1 + n_2}$$

Subtracting each side from 1, we get

$$1 - \frac{p_s}{p^o} = 1 - \frac{n_1}{n_1 + n_2}$$

or

$$\frac{p^o - p_s}{p_o} = \frac{n_2}{n_1 + n_2} \qquad ...(iv)$$

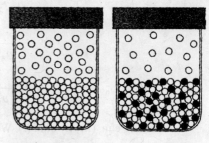

Representation of lowering of vapour pressure
○ *Solvent molecules*
● *Solute molecules*
Number of solvent molecules that leave the surface of a solution decreases with the increase in the concentration of the solution. Maximum solvent molecules leave the surface of pure solvent.

In this expression, $p^o - p_s$ is the lowering of vapour pressure, $\frac{p^o - p_s}{p^o}$ is called relative lowering of vapour pressure, $\frac{n_2}{n_1 + n_2}$ represents the mole fraction of the solute in the solution.

Hence the expression (v) may be expressed in words as follows:

The relative lowering of vapour pressure of a solution containing a non-volatile solute is equal to the mole fraction of the solute in the solution.

This is another definition of Raoult's law.

4.1.1 Relationship Between the Mole Fractions of the Components in the Liquid Phase and in the Vapour Phase

Consider a binary solution of two components A and B. Let us suppose x_A and x_B are their mole fractions in the liquid phase and $p°_A$ and $p°_B$ are their vapour pressures in the pure state. The partial pressures of A and B will be given $p_A = x_A p°_A$ and $p_B = x_B p°_B$. Let us further suppose that y_A and y_B are the mole fractions of A and B in the vapour phase. y_A and y_B can be calculated as under:

$$y_A = \frac{p_A}{p_A + p_B} = \frac{x_A p°_A}{x_A p°_A + x_B p°_B} \qquad \ldots (i)$$

$$= \frac{(1-x_B) p°_A}{(1-x_B) p°_A + x_B p°_B} = \frac{(1-x_B) p°_A}{(p°_B - p°_A) x_B + p°_A} \qquad \ldots (ii)$$

$$y_B = \frac{p_B}{p_A + p_B} = \frac{x_B p°_B}{x_A p°_A + x_B p°_B} \qquad \ldots (iii)$$

$$= \frac{x_B p°_B}{(p°_B - p°_A) x_B + p°_A} \qquad \ldots (iv)$$

On dividing eqn. (*iii*) by eqn. (*i*). We get

$$\frac{y_B}{y_A} = \frac{x_B p°_B}{x_A p°_A} = \frac{x_B}{x_A} \times \frac{p°_B}{p°_A} \qquad \ldots (v)$$

Thus if B is more volatile than A, i.e., $p°_B > p°_A$ or $p°_B/p°_A > 1$, then $y_B/y_A > x_B/x_A$. This means that vapour phase is richer in the more volatile component B than the liquid phase from which it vaporises. This result is known as **Konowaloff's rule**.

Solved Examples— Raoult's Law

Example 1: Liquid A (molecular mass 46) and liquid B (molecular mass 18) form an ideal solution. At 293 K the vapour pressures of pure A and B are 44.5 and 17.5 mm of Hg respectively. Calculate (*a*) the vapour pressure of a solution of A in B containing 0.2 mole fraction of A, and (*b*) the composition of the vapour phase.

Solution: (*a*) $p°_A = 44.5$ mm, $x_A = 0.2$, $p°_B = 17.5$ mm, $x_B = (1 - 0.2) = 0.8$

According to Raoult's law

$$p_A = p°_A \cdot x_A$$

Partial pressure of A, $\quad p_A = p°_A x_A = 44.5 \times 0.2 = 8.9$ mm.

Partial pressure of B, $\quad p_B = p°_B x_B = 17.5 \times 0.8 = 14.0$ mm.

Total pressure $\quad p = p_A + p_B = 8.9 + 14 = 22.9$ mm.

(*b*) In vapour phase

Mole fraction of $\quad A = \dfrac{\text{Partial pressure of } A}{\text{Total pressure}}$

or $\quad x_A = \dfrac{p_A}{p_A + p_B} \dfrac{8.9}{8.9+14} = \dfrac{8.9}{22.9} = 0.39$

$x_B = 1 - 0.39 = 0.61$.

Example 2: The vapour pressure of two pure liquids A and B are 15000 and 30000 Nm^{-2} at 298 K. Calculate the mole fraction of A and B in the vapour phase when an equimolar solution of the liquids is made.

Solution: $\quad p_A^o = 15000 \; Nm^{-2}, \; p_B^o = 30000 \; Nm^{-2}$

In equimolar solution, mole fractions of A and B, i.e., x_A and x_B are equal. Let

$$x_A = x_B = 0.5, \; x_A = 0.5, \; x_B = 1 - 0.5 = 0.5$$

Applying Raoult's law of ideal solution.

$$p_A = p_A^o x_A = 15000 \times 0.5 = 7500 \; Nm^{-2}$$

and $\quad p_B = p_B^o x_B = 30000 \times 0.5 = 15000 \; Nm^{-2}$

Total pressure $\quad p = p_A + p_B = 7500 + 15000 = 22500 \; Nm^{-2}$

In the vapour phase

Mole fraction of $\quad A = \dfrac{\text{Parital pressure } A}{\text{Total pressure}} = \dfrac{7500}{22500} = 0.3333$

Mole fraction of $\quad B = 1 - 0.3333 = 0.6667$

Example 3: Benzene C_6H_6 (b.p. 353.1 K) and toluene C_7H_8 (b.p. 383.6 K) are two hydrocarbons that form a very nearly ideal solution. At 313 K, the vapour pressure of pure liquids are 160 mm Hg and 60 mm Hg respectively. Assuming an ideal solution behaviour, calculate the partial pressures of benzene and toluene and the total pressure over the following solutions:

(*i*) One made by combining equal number of toluene and benzene molecules.

(*ii*) One made by combining 4 mol of toluene and 1 mol of benzene.

(*iii*) One made by combining equal masses of toluene and benzene.

Solution: (*i*) When the number of molecules of toluene and benzene are equal, that means the number of moles of the two liquids are also equal:

Thus mole fraction of benzene $\quad x_A = \dfrac{1}{1+1} = 0.5$

Mole fraction of toluene $\quad x_B = 1 - 0.5 = 0.5$

According to Raoult's law

$$p = p^o . x$$

Partial pressure of benzene $\quad p_A = p_A^o . x_A = 160 \times 0.5 = 80 \; mm$

Partial pressure of toluene $\quad p_B = p_B^o x_B = 60 \times 0.5 = 30 \; mm$

Total vapour pressure $\quad p_A + p_B = 80 + 30 = 110 \; mm$

(*ii*) Mole fraction of benzene $\quad x_A = \dfrac{1}{1+4} = \dfrac{1}{5} = 0.2$

Mole fraction of toluene $\quad x_B = 1 - 0.2 = 0.8$

Partial pressure of benzene $\quad p_A = p_A^o x_A = 160 \times 0.2 = 32 \; mm$

Partial pressure of toluene $\quad p_B = 60 \times 0.8 = 48 \; mm$

Total vapour pressure $\quad p_A + p_B = 32 + 48 = 80 \; mm$

(*iii*) Here the masses of the two liquids are the same. Let the amount of each be m g

Then mole fraction of benzene $\quad x_A = \dfrac{m/78}{m/78 + m/92} = 0.541$

Mole fraction of toluene $\quad x_B = 1 - 0.541 = 0.459$

Partial pressure of benzene $\quad p_A = p_A^o\, x_A = 160 \times 0.541$
$\qquad\qquad\qquad\qquad\qquad = 86.56$ mm of Hg

Partial pressure of toluene $\quad p_B = p_B^o\, x_B$
$\qquad\qquad\qquad\qquad\qquad = 60 \times 0.459 = 27.54$ mm of Hg

Total vapour pressure $\qquad\qquad = 86.56 + 27.54 = 114.1$ mm

Example 4: The vapour pressures of pure components A and B are 120 mm and 96 mm Hg. What will be the partial pressures of the components and the total pressure when the solution contains 1 mole of component A and 4 mole of component B and the solution is ideal? What will be the composition in the vapour phase?

Solution: $\qquad\qquad\qquad\qquad\qquad p_A^o = 120$ mm, $p_B^o = 96$ mm

Mole fraction of A, $\qquad x_A = \dfrac{\text{No. of moles of } A}{\text{No. of moles of } A + \text{No. of moles of } B}$

$\qquad\qquad\qquad\qquad\qquad = \dfrac{1}{1+4} = \dfrac{1}{5} = 0.2$

Mole fraction of B, $\qquad x_B = 1 - 0.2 = 0.8$

Partial pressure of component A, $\quad p_A = p_A^o \times x_A$
$\qquad\qquad\qquad\qquad\qquad\qquad = 120 \times 0.2 = 24.0$ mm Hg

Partial pressure of B, $\qquad p_B = p_B^o \times x_B$
$\qquad\qquad\qquad\qquad\qquad = 96 \times 0.8 = 76.8$ mm

Total pressure $\qquad\qquad p_A + p_B = 24 + 76.8$ mm $= 100.8$ mm

Mole fraction of A in the vapour phase $= \dfrac{p_A}{\text{Total pressure}} = \dfrac{24}{100.8} = 0.238$

Mole fraction of B in the vapour phase $= 1 - 0.238 = 0.762$

Example 5: A solution containing 6.0 gram of benzoic acid in 50 gram of ether ($C_2H_5.OC_2H_5$) has a vapour pressure equal to 5.466×10^4 Nm^{-2} at 300 K. Given that vapour pressure of ether at the same temperature is 5.893×10^4 NM^{-2}, calculate the molecular mass of benzoic acid.

Solution: Vapour pressure of ether (Solvent) $= p^o = 5.893 \times 10^4$ Nm^{-2}

Vapour pressure of ether solution $= p = 5.466 \times 10^4$ Nm^{-2}

Molecular mass of solvent ($C_2H_5 - O - C_2H_5$) (M) = 74

Mass of solute (benzoic acid) $w = 6$ grams

Mass of solvent (ether) $\quad W = 50$ grams

Let molecular mass of solute (benzoic acid) $= m$

Substituting the values in the relation, we have

$$\dfrac{p_A^o - p}{p_A^o} = \dfrac{w}{m} \times \dfrac{M}{W}$$

$$\frac{5.893\times 10^4 - 5.466\times 10^{-4}}{5.893\times 10^4} = \frac{6\times 74}{m\times 50} \quad \text{or} \quad \frac{0.427\times 10^4}{5.893\times 10^4} = \frac{6\times 74}{m\times 50}$$

$$\frac{0.427}{5.893} = \frac{6\times 74}{m\times 50} \quad \text{or} \quad m = \frac{6\times 74\times 5.983}{50\times 0.427} \quad \text{or} \quad m = 122.5$$

i.e., the molecular mass of solute (benzoic acid) = 122.55 amu.

Example 6: The vapour pressure of water at 293 K is 17.51 mm, lowering of vapour pressure of sugar solution is 0.0614 mm.

Calculate

(a) Relative lowering of vapour pressure.

(b) Vapour pressure of the solution.

(c) Mole fraction of water.

Solution: Vapour pressure of solvent (water) = 17.51

Let vapour pressure of the solution = p (to be calculated)

∴ Lowering of vapour pressure = $p° - p$ = 0.0614 mm

(a) ∴ Relative lowering of vapour pressure $= \dfrac{p_o - p}{p°} = \dfrac{0.0614}{17.51} = 0.00351$

(b) Vapour pressure of the solution $\quad p = p° - (p° - p)$
$$= 17.51 - (0.0614) = 17.4486 \text{ mm}$$

Now according to Raoult's Law

$$\frac{p_o - p}{p°} = \text{mole fraction of the solute}$$

$$\frac{p_o - p}{p°} = \frac{n_1}{n_1 + n_2} = x_2$$

∴ mole fraction of the solute $= \dfrac{p° - p}{p°} = \dfrac{0.0614}{17.51} = 0.00351$

(c) Hence, mole fraction of the solvent $= (1 - 0.00351) = 0.99649$

Mole fraction of water = 0.99649

Example 7: The vapour pressure of a 5% aqueous solution of non-volatile organic substances at 373 K is 745 mm. Calculate the molecular mass of the solute.

Solution: Weight of non-volatile organic solute, w = 5 g

Weight of solvent (water), W = 95 g

Molecular mass of solvent (water) M = 18

Molecular mass of non-volatile solute m = ?

$p°$, the vapour pressure of the pure solvent (water) at 373 K = 760 mm

Vapour pressure of the solution p = 745 mm

Substituting the values in the relation,

$$\frac{p° - p}{p°} = \frac{w}{m}\times \frac{M}{W} \quad \text{or} \quad \frac{760 - 745}{760} = \frac{5\times 18}{m\times 95}$$

or $\quad m = \dfrac{5\times 18\times 760}{15\times 95} = 48$

Example 8: At 298 K, the vapour pressure of water is 23.75 mm of Hg. Calculate the vapour pressure at the same temperature over 5% aqueous solution of urea ($NH_2 CONH_2$).

Solution: This solution may be considered as a dilute solution and the approximate relation given below may be used

$$\frac{p_A^o - p_A}{p_A^o} = \frac{wM}{mW}$$

In the present case $p_A^o = 23.75$, $w = 5$ g Therefore $W = 100 - 5 = 95$ g
$M = 18$, $m = 60$ (mol. wt of urea)
Substituting these values in the equation above

$$\frac{23.75 - p_A}{23.75} = \frac{5 \times 18}{60 \times 95} \quad \text{or} \quad p_A = 23.375 \text{ mm.}$$

Problems for Practice

1. The vapour pressure of an aqueous solution of cane sugar (mol. wt. 342) is 756 mm at 100° C. How many grams of sugar are present in 1000 g of water? **[Ans. 100.4 g]**

2. Dry air was passed through a solution containing 40 g of a solute in 90 g water and then through water. The loss in weight of water was 0.05 g. The wet air was then passed through a sulphuric acid, whose weight increased by 2.0 g. What is the molecular weight of the dissolved substance? **[Ans. 320]**

3. The vapour pressure of water is 92 mm at 50°C. 18.1 g of urea are dissolved in 100 g of water. The vapour pressure is reduced by 5 mm. Calculate the molecular weight of urea.

    [**Hint.** The solution is not dilute. Apply the relation $\dfrac{p^o - p_s}{p^o} = \dfrac{n_2}{n_1 + n_2}$]

4. The vapour pressure of 2.1% of an aqueous solution of a non-electrolyte at 100°C is 755 mm. Calculate the molecular weight of the solute. **[Ans. 58.68]**

5. The vapour pressure of water of 20°C is 17 mm. Calculate the vapour pressure of a solution containing 2 g of urea (mol. wt = 60) in 50 g of water. Assume that the solution is not dilute. **[Ans. 16.799 mm]**

6. A current of dry air was passed through a series of bulbs containing a solution of 3.458 g of a substance in 100 g of ethyl alcohol and then through pure ethyl alcohol. The loss in weight of former was 0.9675 g and in the later 0.055 g. Calculate the molecular weight of the solute. **[Ans. 29.6]**

7. Calculate the vapour pressure at 22°C of a 0.1 M solution of urea. The density of the solution may be taken as 1g/ml. The vapour pressure of pure water at 22° is 20 mm.
[Ans. 19.96 mm]

    [**Hint.** 0.1 M solution of urea means 0.1 mole *i.e.*, 6.0 g of urea dissolved per litre of solution *i.e.*, in 1000 g of solution ($\because$ density = 1 g/ml). Hence $w_2 = 60$ g, $w_1 = 1000 - 6.0 = 994.0$ g, $p^o = 20$ mm (Given), $p_s = $ to be calculated]

8. A solution containing 6 g of benzoic acid in 50 g of ether ($C_2H_5OC_2H_5$) has a vapour pressure of 410 mm of mercury at 20°C. Given that the vapour pressure of ether at the same temperature is 442 mm of mercury, calculate the molecular weight of benzoic acid.
[Ans. 122.56]

9. A current of dry air was passed through a bulb containing 26.66 g of an organic substance in 200 g of water, then through a bulb at the same temperature containing pure water and finally through a tube containing fused calcium chloride. The loss in weight of water bulb was 0.0870 g and the gain in the weight of $CaCl_2$ tube was 2.036 g. Calculate the molecular weight of the organic substance in the solution. [Ans. 53.8]

4.2 HENRY'S LAW

At a constant temperature, the solubility of a gas in a liquid is directly proportional to the partial pressure of the gas present above the surface of liquid or solution. This is the statement of Henry's law.

In terms of mole fraction, we can say that **mole fraction of a gas in the solution is proportional to the partial pressure of the gas over the solution.** Another form of Henry's law which is most commonly used states:

The partial pressure of the gas in vapour phase (p) is proportional to the mole fraction of the gas (x) in the solution. Mathematically, it can be written as

$$p = K_H x$$

K_H is called Henery's constant.

K_H has different values for different gases. Therefore, it is a function of nature of the gas. Thus, higher the value of K_H at a given pressure, lower is the solubility of the gas. K_H values for N_2 and O_2 increase with increase of temperature indicating that solubility of gases increases with decrease of temperature. Due to this reason, aquatic species are more comfortable in cold water rather than in warm water. Aquatic animals get more oxygen for breatheing.

Applications of Henry's Law

1. Soft drink bottles are sealed under high pressure of CO_2 to increase their solubility.
2. Scuba divers or deep sea divers have to deal with high concentration of dissolved gases while breathing air from the cylinder underwater. Increased pressure increases the solubility of atmospheric gases in blood. When the diver comes up to the surface, the pressure decreases. This releases the dissolved gases and causes the formation of bubbles of nitrogen in the blood. It is a painful process. To handle the situation, the air tankes are filled with 11.7% helium, 56.2% nitrogen and 32.1% oxygen.

Applying Chemsitry to Life

Scuba diving (or deep sea diving) is a recreational sport and is gaining greater participation worldwide

Deep sea diving

4.3 EXCESS THERMODYNAMIC FUNCTIONS

Excess Thermodynamic function is defined as the difference between the thermodynamic function of mixing for a real system and an ideal system at the same temperature and pressure.

If the thermodynamic function is Y, then

$$Y^E = \Delta Y_{mix} \text{ (real)} - \Delta Y_{mix} \text{ (ideal)}$$

where Y^E is the excess function for the quantity Y.

Excess Chemical Potential (μ^E)

$\mu_i^E = \mu_{real}^M - \mu_{ideal}^M$ (E represents excess and M represents mixing)

We know, $\mu_{real} = \mu° + RT \ln a_i$ (for real system)   ...(i)

and $\mu_{ideal} = \mu° + RT \ln x_i$ (for ideal system)   ...(ii)

Also $a_i = \gamma_i x_i$ ...(iii)

μ_{real} and μ_{ideal} represent chemical potential for real and ideal system. a_i and x_i represent activity and concentration. γ_i reprsent the activity coefficient.

Subtract eq. (ii) from (i)

$$\mu_i^E = (\mu_i° + RT \ln a_i) - (\mu_i° + RT \ln x_i) \quad ...(iv)$$

substituting for a_i from eq. (iii), eq. (iv) becomes

$$\mu_i^E = (\mu_i^E + RT \ln x_i + RT \ln \gamma_i) - (\mu_i° + RT \ln x_i)$$

or $\boldsymbol{\mu_i^E = RT \ln \gamma_i}$.

Excess Gibb's Free Energy, G^E

$$G^E = \Delta G^M_{real} - \Delta G^M_{ideal}$$

$$= RT \Sigma n_i \ln a_i - RT \Sigma n_i \ln x_i \quad ...(v)$$

Substituting $\gamma_i x_i$ for a_i, eq. (v) reduced to

$$G^E = RT [\Sigma n_i \ln x_i + \Sigma n_i \ln \gamma_i - \Sigma n_i \ln x_i]$$

or $G^E = RT \Sigma n_i \ln \gamma_i$ ...(vi)

For a binary system, $i = 2$

$\therefore \quad G^E = RT [n_1 \ln \gamma_1 + n_2 \ln \gamma_2]$ ...(vii)

Excess Entropy Function, S^E

We have the relation for entropy

$$S^E = -\left(\frac{\partial G^E}{\partial T}\right) \quad \text{(By definition)}$$

Substituting for G from eq. (vii)

$$S^E = -RT\left[n_1\left(\frac{\partial \ln \gamma_1}{\partial T}\right)_P + n_2\left(\frac{\partial \ln \gamma_2}{\partial T}\right)_P\right] \quad ...(viii)$$

or $S^E = -R[n_1 \ln \gamma_1 + n_2 \ln \gamma_2]$ ...(ix)

Excess Enthalpy of Mixing, H^E

$$H^E = \Delta H^M_{real} - \Delta H^M_{ideal}$$

But $\Delta H^M_{ideal} = 0$ (we know enthalpy of mixing for an ideal system is zero)

$\therefore \quad H^E = \Delta H^M_{real}$ ...(x)

Also $\quad H^E = G^E + TS^E$ ...(xi)

From (viii) and (xi)

$$H^E = -RT^2\left(n_1 \frac{\partial \ln \gamma_1}{\partial T} + n_2 \frac{\partial \ln \gamma_2}{\partial T}\right)$$

Excess volume, V^E

$$V^E = \left(\frac{\partial G}{\partial P}\right)_P \quad \text{(By definition)}$$

Using eq. (vii), it becomes

or $\quad V^E = RT\left[n_1 \frac{\ln \gamma_1}{\partial P} + n_1 \frac{\ln \gamma_2}{\partial P}\right]$

4.4 COLLIGATIVE PROPERTIES

It has been observed that certain properties of dilute solutions are not dependent on the nature of the solute (non-volatile) presence in the solution. These properties depend upon the number of particles (or concentration) of the solute in the solution. Such properties are known as colligative properties and are listed below:

1. Lowering of vapour pressure of the solvent
2. Osmotic pressure of the solution
3. Elevation in boiling point of the solvent
4. Depression in freezing point of the solvent.

It has been observed that the validity of the above behaviour is limited to dilute solutions when they behave nearly as ideal solutions. As these properties are observed with non-volatile solutes, it can be visualised that the escaping tendency of the solvent is reduced in the presence of such solutes and consequently the vapour pressure of the solvent is lowered. With the lowering of vapour pressure of the solvent, we can explain the elevation in boiling point and depression in freezing point. An important application of the study of colligative properties is to determine the molecular masses of unknown substances and in the study of their molecular state in solution. In the following reactions, we shall take up a detailed study of each of the colligative properties.

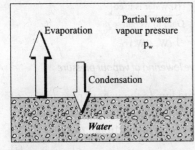

(a) Rates of evaporation and condensation become equal as the equilibrium is attained

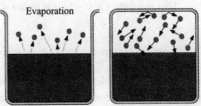

(b) Representation of evaporation in an open and closed vessel

4.5 DETERMINATION OF VAPOUR PRESSURE OF A LIQUID

Manometric Method

The vapour pressure of a liquid of solution can be measured with the help of a manometer (see Fig. 4.1). The bulb B is filled with the liquid or solution. The air in the connecting tube is then removed with a vacuum pump. When the stopcock is closed, the pressure inside is due only to the vapour evaporating from the solution or liquid. This method is generally used for aqueous solutions. The manometric liquid is usually mercury which has low volatility.

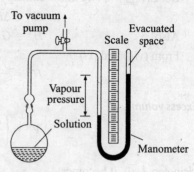

Fig. 4.1: Measurement of vapour pressure of aqueous solutions with a manometer

4.6 DETERMINATION OF LOWERING OF VAPOUR PRESSURE OF THE SOLVENT

Ostwald and Walker's Dynamic Method (Gas Saturation Method)

In this method the relative lowering of vapour pressure can be determined straightaway. The measurement of the individual vapour pressures of a solution and solvent is thus eliminated.

Procedure. The apparatus used by Ostwald and Walker is shown in Fig. 4.2. It consists of two sets of bulbs:

(a) Set A containing the solution   (b) Set B containing the solvent

Each set is weighed separately. A slow stream of dry air is then drawn by suction pump through the two sets of bulbs. At the end of the operation, these sets are reweighed. From the loss of weight in each of the two sets, the lowering of vapour pressure is calculated. The temperature of the air the solution and the solvent must be kept constant throughout.

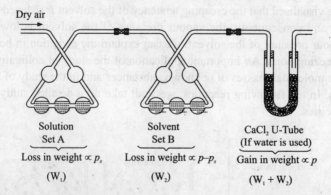

Fig. 4.2: Ostwald-Walker's method of measuring the relative lowering of vapour pressure

Calculations. As the air bubbles through set A it is saturated up to the vapour pressure p_s, of solution and then up to vapour pressure p of solvent in set B. Thus, the amount of solvent taken up in set A is proportional to p_s and the amount taken up in set B is proportional to $(p - p_s)$.

$$w_1 \propto p_s \qquad \qquad \ldots(i)$$
$$w_2 \propto p - p_s \qquad \qquad \ldots(ii)$$

Adding (i) and (ii), we have
$$w_1 + w_2 \propto p_s + p - p_s$$
$$\propto p \qquad \ldots(iii)$$

Dividing (ii) and (iii), we can write
$$\frac{p - p_s}{p} = \frac{w_2}{w_1 + w_2} \qquad \ldots(iv)$$

Knowing the loss of mass in set B (w_2) and the total loss of mass in the two sets ($w_1 + w_2$), we can find the relative lowering of vapour pressure from equation (iv).

If water is the solvent used, a set of calcium chloride tubes is attached to the end of the apparatus to catch the escaping water vapour. Thus, the gain in mass of the $CaCl_2$- tubes is equal to ($w_1 + w_2$), the total loss of mass in sets A and B.

Solved Example—Lowering of Vapour Pressure

Example 9: A stream of dry air was passed through a bulb containing a solution of 7.50 g of an aromatic compound in 75.0 g of water and through another globe containing pure water. The loss in mass in the first globe was 2.810 g and in the second globe it was 0.054 g. Calculate the molecular mass of the aromatic compound (Mol mass of water = 18).

Solution: According to the theory of Ostwald-Walker method,
$$\frac{p - p_s}{p} = \frac{w_2}{w_1 + w_2}$$

In the present case,

w_1, loss of mass of solution = 2.810 g, w_2, loss of mass of solvent (water) = 0.054 g

Substituting values in the above equation
$$\frac{p - p_s}{p} = \frac{0.054}{2.810 + 0.054} = \frac{0.054}{2.864} = 0.0188$$

According to Raoult's law, $\frac{p - p_s}{p} = \frac{w/m}{w/m + W/M}$

Substituting the values, we have $0.0188 = \frac{7.50/m}{7.50/m + 75.0/18}$ or $m = 93.6$

4.7 RELATION BETWEEN THE RELATIVE LOWERING OF VAPOUR PRESSURE AND THE MOLECULAR MASS OF THE SOLUTE (RAOULT'S LAW)

Suppose vapour pressure of the pure solvent $A = p^o$
Let the solute be B.
Let the vapour pressure of the solvent in the solution $= p_s$
Since $p^o > p_s$
∴ Lowering of vapour pressure $= p^o - p_s$

Relative lowering of vapour pressure $= \dfrac{p^o - p_s}{p^o}$

According to Raoult's Law, relative lowering of vapour pressure is equal to the mole fraction of the solute in the solution.

If the solution contains n_2 moles of the solute dissolved in n_1 moles of the solvent, we have

Mole fraction of the solute $= \dfrac{n_2}{n_1 + n_2}$ ...(i)

According to Raoult's law, relative lowering of vapour pressure is equal to the mole fraction of the solute in the solution.

∴ $$X_B = \dfrac{n_2}{n_1 + n_2}$$

∴ $$\dfrac{p^o - p_s}{p^o} = \dfrac{n_2}{n_1 + n_2}$$ ...(ii)

It is evident from (ii) that relative lowering of vapour pressure depends only upon mole fraction or molar concentration of the solute. Therefore, relative lowering of vapour pressure is a colligative property.

Molecular Mass of the Non-volatile Solute

Now, number of moles of the solute $n_2 = \dfrac{w}{m}$

where w = mass of the solute and m = molecular mass of the solute.

And number of moles of the solvent $n_1 = \dfrac{W}{M}$

where W = mass of the solvent and M = molecular mass of the solvent.

The expression (ii) becomes

$$\dfrac{p^o - p_s}{p^o} = \dfrac{\dfrac{w}{m}}{\dfrac{w}{m} + \dfrac{W}{M}}$$ ...(iii)

In a dilute solution, n_2 is negligible as compared to n_1. Therefore neglecting n_2 (or w/m) in the denominator, we get from expression (iii)

$$\dfrac{p^o - p_s}{p^o} = \dfrac{\dfrac{w}{m}}{\dfrac{W}{M}} = \dfrac{w}{m} \times \dfrac{M}{W} \text{ or } \dfrac{p^o - p_s}{p^o} = \dfrac{wM}{mW}$$ ...(iv)

Thus by measuring the lowering of vapour pressure of a solution, the molecular mass m of a solute in a given solution of a known concentration can be determined, if other quantities are known.

4.8 THERMODYNAMIC DERIVATION OF THE EXPRESSION FOR RELATIVE LOWERING OF VAPOUR PRESSURE

Relative lowering of vapour pressure of a solution can be obtained from thermodynamics as under:

Consider a dilute solution of a non-volatile solute in a suitable solvent at a particular temperature and pressure. The following equilibrium exists between the solution and vapour:

Solution ⇌ Vapour (At constant temperature and pressure)

Chemical potential of the solvent in the liquid phase, i.e., in the solution is equal to that of the solvent in the vapour phase under this condition of equilibrium. So we can write

$$\mu_1^l = \mu_1^v \qquad \ldots(i)$$

where the symbol 1 indicates solvent, l the liquid phase and v the vapour phase

or
$$d\mu_1^l = d\mu_1^v \qquad \ldots(ii)$$

Thus any small change in the chemical potential of the solvent in the liquid phase is equal to that of the solvent in the vapour phase.

We know that the chemical potential of the solvent in the solution is a function of T, P and concentration. This is expressed as under:

$$d\mu_1^l = f(T, P, x_1)$$

For small changes in T, P and x_1, the change in the value of chemical potential is given by

$$d\mu_1^l = \left(\frac{\partial \mu_1^l}{\partial T}\right)_{P, x_1} dT + \left(\frac{\partial \mu_1^l}{\partial P}\right)_{T, x_1} dP + \left(\frac{\partial \mu_1^l}{\partial x_1}\right)_{T, P} dx_1 \qquad \ldots(iii)$$

At constant T and P, dT and dP become zero, therefore, eqn. (iii) can be written as

$$d\mu_1^l = \left(\frac{\partial \mu_1^l}{\partial x_1}\right)_{T, P, x_1} dx_1 \qquad \ldots(iv)$$

From thermodynamic relation for an ideal solvent

$$\mu_1^l = (\mu_1^0)^l + RT \ln x_1$$

$\therefore \qquad d\mu_1^l = d(RT \ln x_1) = RTd = \ln x_1 = RT \dfrac{dx_1}{x_1} \qquad \ldots(v)$

or
$$\frac{d\mu_1^l}{dx_1} = \frac{RT}{x_1} \qquad \ldots(vi)$$

However, the chemical potential of the solvent in the vapour phase depends only on T and P because there is only one component. Therefore, we can write

$$\mu_1^v = f(T, P)$$

or
$$d\mu_1^v = \left(\frac{\partial \mu_1^v}{\partial T}\right)_P dT + \left(\frac{\partial \mu_1^v}{\partial P}\right)_T dP \qquad \ldots(vii)$$

At constant temperature, equation (vii) becomes

$$d\mu_1^v = \left(\frac{\partial \mu_1^v}{\partial P}\right)_T dP \qquad \ldots(viii)$$

From thermodynamics, we have the relation $\left(\dfrac{\partial \mu_1^v}{\partial P}\right)_T = \overline{V}_1$, called partial molar volume. Hence eqn. (viii) can be written as

$$d\mu_1^v = \overline{V}_1 dP \qquad \ldots(ix)$$

For a dilute solution we can assume that $\overline{V}_1 = V_1^0$ (molar volume of the solvent.)

$\therefore \qquad d\mu_1^v = V_1^0 dP \qquad \ldots(x)$

Assuming ideal behaviour for the solvent vapour present in equilibrium with the solution, we have

$$V_1^0 = RT/P_1$$

Substituting this value in eqn. (x), we get

$$d\mu_1^v = RT\frac{dP_1}{P_1} \qquad \ldots(xi)$$

But at equilibrium, as already mentioned in eqn. (ii),

$$d\mu_1^l = d\mu_1^v$$

Substituting the values from Eqn. (v) and (xi) in the above equation, we have

$$RT\frac{dx_1}{x_1} = RT\frac{dP_1}{P_1}$$

or
$$\frac{dx_1}{x_1} = \frac{dP_1}{P_1} \qquad \ldots(xii)$$

If x_1 is the mole fraction of the solvent and x_2 is the mole fraction of the solute in the solution, then

$$x_1 + x_2 = 1$$

or $\qquad dx_1 + dx_2 = 0$

or $\qquad dx_1 = -dx_2 \qquad \ldots(xiii)$

From eqn. (xii) and (xiii)

$$\frac{dP}{P_1} = -\frac{dx_2}{1-x_2} \qquad \ldots(xiv)$$

When $x_2 = 0$, $P_1 = p_1°$ i.e. vapour pressure of the pure solvent.

and when $x_2 = x_2$, $P_1 = p_1$ i.e. vapour pressure of the solvent in the solution which is equal to the vapour pressure of the solution. Integrating eq. (xiv) between the above mentioned limits, we get

$$\int_{p_1°}^{p_1} \frac{dP}{P_1} = -\int_{x_2=0}^{x_2=x_2} \frac{dx_2}{1-x_2}$$

or $\qquad \ln(p_1/p_1°) = \ln(1-x_2)$ or $p_1/p_1° = 1 - x_2 = x_1 \qquad \ldots(xv)$

or $\qquad \boldsymbol{p_1 = x_1 p_1°} \qquad \ldots(xvi)$

which is the expression for Raoult's Law which states that *the vapour pressure of the solvent in the solution is equal to the product of the mole fraction and vapour pressure of the pure solvent*. We can also obtain the equation in terms of relative lowering of vapour pressure from eq. (xv).

$$p_1/p_1° = 1 - x_2$$

or $\qquad 1 - \dfrac{p_1}{p°} = x_2$ or $\dfrac{p° - p_1}{p°} = x_2$

or $\qquad \dfrac{p° - p_s}{p°} = x_2 \qquad (\because p_1 = p_s)$

i.e., *relative lowering of vapour pressure of a solution containing a non-volatile solute is equal to the mole fraction of the solute in the solution.*

4.9 OSMOSIS PHENOMENON

Before coming to the phenomenon of osmosis, let us understand clearly about semipermeable membrane.

The membrane which allows the flow of solvent molecules but not the solute molecules through it is called a semipermeable membrane. Examples are parchment, collodion, animal membranes etc. In nature the plant cell are protected by a semipermeable membrane. Chemically, we can get a semipermeable membrane of cupric ferrocyanide within the walls of a porous pot.

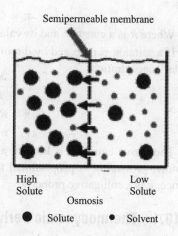

Osmosis and osmotic pressure: A solute tends to dissolve in a solvent as the most predominant randomness factor favours such a tendency. But when a pure solvent is separated from its solution by a semipermeable membrane, the molecules of the solvent diffuse through the semipermeable membrane into the solution. This is called **osmosis**. As a result of this transference of solvent to solution, the level of solvent decreases while that of solution increases. It may be demonstrated by taking solvent and solution in two compartments of a box separated by a semipermeable membrane as shown in Fig. 4.3 (a)

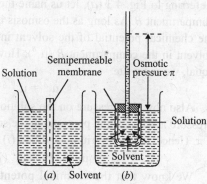

Fig. 4.3: *Osmosis and Osmotic pressure*

As more and more of the solvent molecules pass through the membrane into the solution, the concentration of the solution falls gradually. After some time the hydrostatic pressure exerted by the solution column prevents the flow of more of solvent molecules and *osmosis* stops. i.e., there is no further rise in the level of solution in the column [Fig. 4.3 (b)]. Hence, osmotic pressure is defined as follows:

Osmotic pressure is the equilibrium hydrostatic pressure exerted by the solution column which just prevents the flow of solvent molecules into the solution through a semipermeable membrane.

Osmosis is a phenomenon by which the pure solvent molecules tend to diffuse through a semipermeable membrane into the solution.

4.10 VAN'T HOFF RELATION BETWEEN THE OSMOTIC PRESSURE OF A SOLUTION AND MOLECULAR MASS OF THE SOLUTE

In dilute solutions the behaviour of solute molecules is similar to that of molecules of a gas. Osmotic pressure (π) of a solution is found to be directly proportional to the molar concentration C of solution and its absolute temperature T. Thus:

$$\pi \propto C \text{ and } \pi \propto T$$

$$\therefore \quad \pi \propto C.T \text{ or } \pi = R.C.T$$

Where R is a constant and its value is found to be same as that of gas constant.

If a solution is prepared by dissolving n_2 moles of the solute in V litres of the solution, the molar concentration

$$C = \frac{n_2}{V} \text{ moles per litre}$$

$$\therefore \quad \pi = RT$$

or $\quad \pi V = n_2 RT$

The above equation is known, as *Van't Hoff equation of dilute solution* and shows that osmotic pressure π is proportional to the molar concentration of the solute in the solution. Hence, it is a colligative property.

4.10.1 Thermodynamic Derivation of the Expression for Osmotic Pressure

Referring to Fig. 4.3 (*a*), let us name the solvent side as compartment A and solution side as compartment B. As long as the osmosis is taking place and equilibrium has not been attained, the chemical potential of the solvent in the compartment $A(\mu_1^A)$ is greater than that of the solvent in the compartment B (μ_1^B). However when equilibrium is attained, the two become equal, we can write

$$\mu_1^A = \mu_1^B \qquad \ldots(i)$$

Also now the pressure on the solution side is greater than on the solvent side by osmotic pressure π. Thus, if the pressure on the solvent side is P, then on the solution side, it will be $P + \pi$. Hence, the equilibrium condition (*i*) can be written as

$$\mu_1^A(T, P) = \mu_1^B(T, P + \pi) \qquad \ldots(ii)$$

We know that the chemical potential of the solvent in the solution is a function of temperature, pressure and composition, we can write

$$\mu_1 = f(T, P, x_1) \qquad \ldots(iii)$$

where x_1 represents the mole fraction of the solvent in the solution. (Subscript 1 stands for solvent)

For small changes in T, P and x_1, the change in the value of μ_1 will be given by

$$d\mu_1 = \left(\frac{\partial \mu_1}{\partial T}\right)_{P, x_1} dT + \left(\frac{\partial \mu_1}{\partial P}\right)_{P, x_1} dP + \left(\frac{\partial \mu_1}{\partial x_1}\right)_{T, P} dx_1$$

At constant T, dT becomes zero, therefore, this equation becomes

$$d\mu_1 = \left(\frac{\partial \mu_1}{\partial P}\right)_{T, x_1} dP + \left(\frac{\partial \mu_1}{\partial x_1}\right)_{T, P} dx_1 \qquad \ldots(iv)$$

We have the thermodynamic relation,

$$\left(\frac{\partial \mu_1}{\partial P}\right)_{T, x_1} = \overline{V_1} \text{ (called partial molar volume)}$$

$$= V_1^\circ \text{ (molar volume of the solvent if the solution is dilute)} \qquad \ldots(v)$$

There is another standard thermodynamic relation

$$\mu_1 = \mu_1^\circ + RT \ln x_1, \text{ which on differentiation can be written as}$$

$$\left(\frac{\partial \mu_1}{\partial x_1}\right)_{T, P} = \frac{RT}{x_1} \qquad \ldots(vi)$$

Solutions and Colligative Properties

Substituting the values from equation (v) and (vi) in eqn. (iv), we get

$$d\mu_1 = V^\circ_1 \, dP + \frac{RT}{x_1} dx_1$$

$$d\mu_1 = V_1^\circ \, dP + RT d \ln x_1 \qquad \ldots(vii)$$

When $x_1 = 1$, $P = P$ and when $x_1 = x_1$, $P = P + \pi$

Integrating both sides of eqn. (vii) between the above limits, we have,

$$\int_{T,P}^{T,P+\pi} d\mu_1 = V_1^\circ \int_P^{P+\pi} dP + RT \int_{x_1=1}^{x_1=x_1} d \ln x_1$$

or $\mu_1(T, P + \pi) - \mu_1(T, P) = V_1^\circ(P + \pi - P) + RT \ln x_1$ ...(viii)

From equations (ii) and (viii)

$$RT \ln x_1 = -V_1^\circ \pi \qquad [\because \mu_1(T, P + \pi) - \mu_1(T, P) = 0] \qquad \ldots(ix)$$

For a very dilute solution $x_1 \ll 1$. Hence, we can write

$$\ln x_1 = \ln(1 - x_2) = -x_2$$

$$[\because \ln(1-x_2) = -x_2 - \frac{1}{2}x_2^2 - \frac{1}{3}x_2^3 \ldots \text{ neglecting } x_2^2 \text{ and higher terms}]$$

$\therefore$ Eqn. (ix) becomes

$$-x_2 RT = -V_1^\circ \pi \quad \text{or} \quad \pi V_1^\circ = x_2 RT \qquad \ldots(x)$$

But $$x_2 = \frac{n_2}{n_1 + n_2} \simeq \frac{n_2}{n_1} \text{ for a very dilute solution.}$$

Hence, Eq (x) becomes: $\pi V_1^\circ = \frac{n_2}{V} RT$

or $$n_1 \pi V_1^\circ = n_2 RT \qquad \ldots(xi)$$

Further, for total volume of the solution, we can write the equation

$$V = n_1 V_1^\circ + n_2 V_2^\circ$$

For a very dilute solution, $n_2 \simeq 0$ so that

$$V = n_1 V_1^\circ \qquad \ldots(xii)$$

Substituting this value in eqn. (xi), we get

$$\pi V = n_2 RT \quad \text{or} \quad \pi = \frac{n_2}{V} RT$$

or $$\pi = \mathbf{CRT} \qquad \ldots(xiii)$$

where $C = n_2/V$ is the molar concentration of the solute in solution (in mol L^{-1}). Eqn. (xiii) is known as van't Hoff equation for dilute solutions.

Determination of Molecular Mass from Osmotic Pressure

If w gram of the solute are dissolved in V liters of the solution and m is the molecular of the solute, then

$$n = w/m$$

Substituting this value in equation $\pi V = nRT$, we get

$$\pi V = \frac{w}{m} \times R.T \text{ or } m = \frac{wR.T}{\pi V}$$

Then molecular mass m of the solute can be calculated from this equation if osmotic pressure π of the solution is known. The value of constant R is taken as 0.0821 litre-atmosphere per degree per mole when π is expressed in atmosphere and T in degree Kelvin.

4.10.2 Relation Between the Lowering of Vapour Pressure and Osmotic Pressure of a Solution

The relation between the lowering of vapour pressure and the mole fraction of the solute is

$$\frac{p^o - p_s}{p^o} = \frac{w}{m} \times \frac{M}{W} \quad \text{(for a dilute solution)} \quad \ldots(i)$$

or

$$\frac{p^o - p_s}{p^o} = n \times \frac{M}{W} \quad \ldots(ii)$$

where n is the no. of moles of the solute. p^o and p_s represent the vapour pressures of the solvent and solution respectively.

The relation between the osmotic pressure and the no. of moles of the solute is given by

$$\pi = \frac{nRT}{V} \quad \ldots(iii)$$

From equation (ii)

$$n = \frac{p^o - p_s}{p^o} \times \frac{W}{M} \quad \ldots(iv)$$

From equation (iii)

$$n = \frac{\pi V}{RT} \quad \ldots(v)$$

Equating R.H.S of equations (iv) and (v)

$$\frac{p^o - p_s}{p_s} \cdot \frac{W}{M} = \frac{\pi V}{RT} \quad \text{or} \quad \frac{p^o - p_s}{p_s} = \frac{M}{W} \cdot \frac{\pi V}{RT}$$

or

$$\frac{p^o - p_s}{p_s} = \frac{M\pi}{(W/V).RT} \quad \text{or} \quad \frac{p^o - p_s}{p^o} = \frac{M\pi}{\rho.RT}$$

where M = Mol. mass of the solvent, ρ = Density of the solution

R = Solution (gas) constant, T = Temperature.

4.10.3 Derivation of Raoults law from the Relationship Between Relative Lowering of Vapour Pressure and Osmotic Pressure

The relationship between the relative lowering vapour pressure and osmotic pressure, as derived in the previous section can be written as

$$\frac{p^o - p_s}{p^o} = \frac{M\pi}{\rho RT} \quad \ldots(i)$$

If the solution contains n_2 moles of the solute dissolved in V litres of the solvent, then according to van't Hoff equation for dilute solutions,

$$\pi V = n_2 RT \quad \text{or} \quad \pi = \frac{n_2 RT}{V} \quad \ldots(ii)$$

The volume V of the solvent can be expressed in terms of moles (n_1) as

$$n_1 = \frac{V \times \rho}{M} \quad \text{or} \quad V = \frac{Mn_1}{\rho} \quad \ldots(iii)$$

Substituting this value of V in equation (ii), we get

$$\pi = n_2 RT \times \frac{\rho}{Mn_1} \qquad ...(iv)$$

Substituting this value in equation (i), we get

$$\frac{p^o - p_s}{p^o} = \frac{M}{\rho RT} \times n_2 RT \times \frac{\rho}{Mn_1} = \frac{n_2}{n_1}$$

This is a modified form of Raoult's law when the solution is dilute, i.e., $n_2 < n_1$ so that in the denominator n_2 is neglected in comparison to n_1.

4.10.4 Interesting Experiments to Demonstrate the Phenomenon of Osmosis

Two interesting experiments to demonstrate the phenomenon of osmosis are being described here.

1. The egg experiment

The outer hard shell of two eggs of the same size is removed by dissolving in dilute hydrochloric acid. One of these is placed in distilled water and the other in saturated salt solution. After a few hours, it will be noticed that the egg placed in water swells and the one in salt solution shrinks. In the first case, water diffuses through the skin (a semipermeable membrane) into the egg which swells. In the second case, the concentration of the salt solution being higher than the material, the egg shrinks (Fig. 4.4).

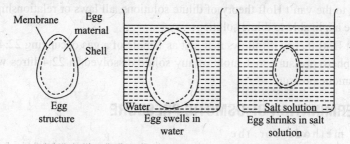

Fig. 4.4: Demonstration of osmosis by egg experiment

2. Silica garden

Crystals of many salts *e.g.*, ferrous sulphate, nickel chloride, cobalt nitrate and ferric chloride are placed in a solution of water glass (sodium silicate). The layers of metallic silicates formed on the surface of crystals by double decomposition are semipermeable. The water from outside enters through these membranes which burst and form what we call a Silica Garden. It makes an interesting sight.

4.11 VAN'T HOFF THEORY OF DILUTE SOLUTION TAKING THE EXAMPLE OF OSMOSIS

Van't Hoff noted the striking resemblance between the behaviour of dilute solutions and gases. Dilute solutions obeyed laws analogous to the gas laws. To explain it van't Hoff visualised that gases consist of molecules moving in vacant space (or vacuum), while in the solutions the solute particles are moving in the solvent. The exact analogy between solutions and gases is illustrated with Fig. 4.5.

As shown in Fig. 4.5 (a), the pure solvent flows into the solution by osmosis across the semipermeable membrane. The solute molecules striking the membrane cause osmotic pressure and the sliding membrane is moved towards the solvent chamber. In case of a gas [Fig. 4.5 (b)], the gas molecules strike the piston and produce pressure that pushes it towards the empty chamber. Here it is the vacuum which moves into the gas. This demonstrates clearly that there is close similarity between a gas and a dilute solution.

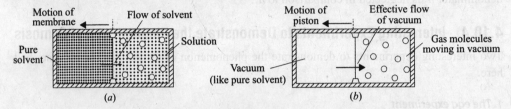

Fig. 4.5: The analogy between osmotic pressure and gas pressure

Thinking on these lines, van't Hoff propounded his theory of dilute solution. The **van't Hoff Theory of Dilute Solutions** states that:

A substance in solution behaves exactly like a gas and the osmotic pressure of a dilute solution is equal to the pressure which the solute would exert if it were a gas at the same temperature occupying the same volume as the solution.

According to the van't Hoff theory of dilute solutions, all laws or relationships obeyed by gases would be applicable to dilute solutions.

From van't Hoff theory if follows that just as 1 mole of a gas occupying 22.4 liters at 0°C exert 1 atmosphere pressure so 1 mole of any solute dissolved in 22.4 litres would exert 1 atmosphere osmotic pressure.

4.12 DETERMINATION OF OSMOTIC PRESSURE

Prominent methods for the determination of osmotic pressure are described as under:

1. Pfeffer's method. The apparatus used by Peffor consists of semipermeable membrane of copper ferrocyanide, supported on the walls of a porous pot. A T-shaped wide tube is fitted into the porous pot as shown in Fig. 4.6. The side tube S of the T-tube is connected to a closed manometer containing mercury and nitrogen. A simple tube R is fitted at the top of the T-tube, through which

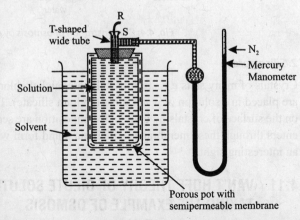

Fig. 4.6: Pfeffer's apparatus

the experimental solution is added into the porous pot till the porous pot and space above the mercury in the left limb of the manometer are completely filled. After filling, the tube t

is sealed off. The pot is then placed in the pure solvent maintained at constant temperature. As a result of osmosis, the solvent passes into the solution. The highest pressure developed is recorded on the manometer.

2. Morse and Frazer's Method. In this method, the solvent was taken in the porous pot (having semipermeable membrane in its walls) and the solution was kept outside the porous pot in the vessel made of bronze and connected to a manometer at the top. A tube T open at

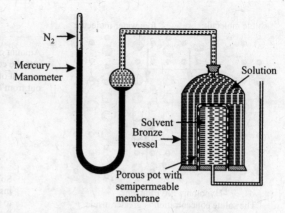

Fig. 4.7: Morse and Frazer's apparatus

both the ends was fixed to the porous pot, as shown in Fig. 4.7, to keep the porous pot filled with the solvent. Morse and Frazer prepared the semi-permeable membrane of better quality by electrolytic method to measure osmotic pressures upto 300 atmospheres.

3. Barkeley and Hartley method. The apparatus used for the determination of osmotic pressure by the above method is shown in Fig. 4.8.

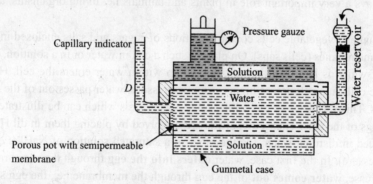

Fig. 4.8: Barkeley and Hartley methods for osmotic pressure determination

It consists of a porous pot containing potassium ferrocyanide deposited on its walles and fitted into a metallic cylinder. A piston and a pressure gauge is attached to the metallic cylinder which is filled with the solution whose osmotic pressure is to be determined. The porous pot is fitted with a capillary tube and water reservoir on the opposite sides. Water from the porous pot moves towards the solution in the cylinder through the semipermeable membrane. As a result the level of water tends to fall down. External pressure is applied on the piston to such an extent that the water level in the capillary tube does not change. The magnitude of pressure applied can be read from the pressure gauge. This pressure is equal to the osmotic pressure.

4.12.1 Isotonic Solutions

Solutions of equimolecular concentrations at the same absolute temperature have the same osmotic pressure. *Such solutions which have the same osmotic pressure at the same temperature are called isotonic solutions.*

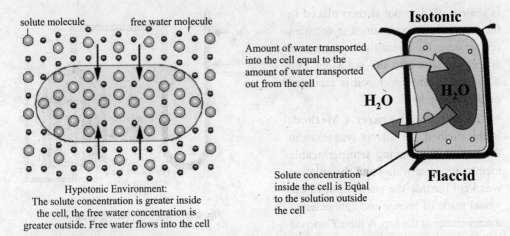

(a) Hypotonic solutions (b) Isotonic solutions

Hypotonic solutions and isotonic solutions play a vital role in biological systems.

4.12.2 Biological Importance of Osmosis

Osmosis plays a very important role in plants and animals i.e. living organisms, as described below:

1. Animal and vegetable cells contain solutions of sugars and salts enclosed in their semi-permeable membranes (cells saps). On placing such a cell in water or in a solution, the osmotic pressure of which is less than that of the cell sap within water enters the cell. However, on placing, the cells in a solution of higher osmotic pressure, water passes out of the cell and the cell shrinks. This shrinkage of the cell is called plasmolysis which can be illustrated by taking two hen eggs of the size. Their outer shells are dissolved by placing them in dil HCl, one egg is then placed in distilled water and the other in a saturated solution of NaCl (*i.e.*, of higher osmotic pressure). In the first case, water **enters into the egg** through the membrane while in the second case, **water comes out** of the egg through the membrane i.e., the egg **swells** in the first **case and shrinks** in the second case.

Using solutions of varying concentrations and placing plants or animals cells in them, it is possible to find out the concentration when plasmolysis just stops. Such a solution is then said to be **isotonic** (*i.e.*, having same osmotic pressure) with the cell solution. This process is thus helpful in measuring osmotic pressure of cell saps.

It is found that the red blood corpuscles and a 0.91% sol. of NaCl are isotonic.

2. Plant cell at their roots contain root hair, which are in contact with the soil, since the osmotic pressure inside the cell is higher, water from the soil flows into the cell by osmosis.

3. Damage done to the plant by use of excess of fertilizer is due to higher osmotic pressure of the fertilizer solution than that of the cell and makes the organs stretched and fully expanded.

4. Plant movements such as opening and closing of flowers, opening and closing of leaves etc. are also regulated by osmosis.

Solved Examples—Osmotic Pressure

Example 10: Osmotic pressure of solution containing 7 grams of dissolved protein per 100 cm³ of a solution is 25 mm of Hg at body temperature (310 K). Calculate the molar mass of the protein ($R = 0.08205$ litre atm deg^{-1} mol^{-1}).

Solution: We know

$$\pi = \frac{n}{V} \times RT$$

Given $\pi = 25$ mm $= \frac{25}{760}$ atmosphere ($\because$ 760 mm = 1 atmosphere)

$R = 0.08205$ L atmosphere K^{-1} mol^{-1}, $T = 310$ K

$$V = 100 \text{ cm}^3 = \frac{10}{1000} = \frac{1}{10} \text{ litres}$$

$\therefore$ n, the number of moles of the solute $= \dfrac{\pi \times V}{RT}$ ...(i)

Also $\quad n = \dfrac{\text{Mass of solute}}{\text{Molecular mass of solute}}$ ...(ii)

From (i) and (ii)

$\dfrac{\text{Mass of solute}}{\text{Molecular mass of solute}} = \dfrac{\pi \times V}{RT}$. Substituting the values, we have

$$\text{Molecular mass of solute} = \frac{\text{Weight of solute} \times R \times T}{\pi \times V}$$

$$= \frac{7 \times 0.08205 \times 310}{\dfrac{25}{760} \times \dfrac{1}{10}}$$

or molecular mass of the solute $= \dfrac{7 \times 0.08205 \times 310 \times 760 \times 10}{25} = 54126.7$

$\therefore$ Molecular mass of the solute = **54126.7**

Example 11: A solution of sucrose (Molecular mass 342) is prepared by dissolving 68.4 g of it per litre of the solution. What its osmatic pressure at 300 K? $R = 0.082$ litre atm K^{-1} mol^{-1}?

Solution: We know, $\quad \pi = \dfrac{n}{V} \times RT$

Given, molecular mass of the solute = 342, Mass of the solute = 68.4 g

Volume V of the solution = 1 litre, $T = 300$ K, Osmotic pressure $\pi = ?$

$R = 0.082$ litre atm K^{-1} mol^{-1}

n, the no. of moles of solute $= \dfrac{\text{Mass of solute}}{\text{Moleculare mass of solute}}$ or $n = \dfrac{68.4}{342}$

Substituting the values in the equation

$$\pi = \frac{n}{V} RT \text{ or } \pi = \frac{68.4}{342} \times \frac{0.082 \times 300}{1} = \mathbf{4.22 \text{ atm}}$$

$\therefore$ Osmotic pressure = **4.22 atm**

Example 12: Calculate the osmotic pressure of a 5% solution of cane sugar at 288 K. $R = 0.082$ lit atm K^{-1} mol^{-1}

Solution: Molecular mass of the cane sugar $(C_{12}H_{22}O_{11}) = 342$. The solution is 5%

∴ Mass of sugar per litre = 50 g

∴ n, the no. of moles $= \dfrac{\text{Mass of solute}}{\text{Molecular mass of solute}} = \dfrac{50}{342}$

$V = 1$ litre, $T = 288$ K

Substituting the values in the relation, we have

$$\pi = \dfrac{n}{V}.RT$$

or $\pi = \dfrac{50}{342} \times \dfrac{0.082 \times 288}{1} = 3.45$ atm

∴ Osmotic pressure = **3.45 atm**

Example 13: At 298 K, 100 cm³ of a solution, containing 3.002 g of an unidentified solute, exhibits an osmotic pressure of 2.55 atmosphere. What would be the molecular mass of the solute?

Solution: $\pi V = nRT$ where n is the no. moles of the solute

Also $nV = \dfrac{w}{M}RT$ where w is the mass and M the molecular mass of the solute

or $M = \dfrac{wRT}{\pi V}$, substituting the values, we have

$M = \dfrac{3.002 \times 0.0821 \times 298}{2.55 \times 0.1}$ ($V = 100$ cm³ = 0.1 litre)

or $M = $ **288 a.m.u.**

Problems for Practice

1. Calculate the value of the constant R in litre atmosphere from the observation that solution containing 34.2 g of cane-sugar in one litre of water has an osmotic pressure of 2.405 atm at 20°C.   **[Ans. 0.0821]**
2. Calculate the osmotic pressure at 25°C of a solution containing one gram of glucose $(C_6H_{12}O_6)$ and one gram of sucrose $(C_{12}H_{22}O_{11})$ in 100 g of water. If it were not known that the solute was a mixture of glucose and sucrose, what would be the molecular mass of the solute corresponding to the calculated osmotic pressure.   **[Ans. 0.2074 atm, 235.8]**
3. A solution of glucose containing 18 g/litre had an osmotic pressure of 2.40 atm at 27°C. Calculate the molecular mass of glucose ($R = 0.082$ litre atm).   **[Ans. 183.5]**
4. Calculate osmotic pressure of a solution containing 5 g of glucose in 100 ml of its solution at 17° C. ($R = 0.0821$ litre atm/mol/degree.)   **[Ans. 6.61 atm]**
5. A 6% solution of sucrose $(C_{12}H_{22}O_{11})$ is isotonic with a 3% solution of an unknown organic substance. Calculate the molecular mass of the unknown substance.   **[Ans. 171]**
6. Calculate the osmotic pressure of solution obtained by mixing one litre of 7.5% solution of substance A (mol. wt = 75) and two litres of 3% solution of a substance B (mol. wt = 60) at 18°C.   **[Ans. 7.954 atm]**
7. A 4% solution of cane-sugar gave an osmotic pressure of 208.0 cm of Hg at 15°C. Find its molecular mass.   **[Ans. 345.2]**
8. Calculate the concentration of solution of glucose which is isotonic at the same temperature with a solution of urea containing 6.2 g/litre.   **[Ans. 18.6 g/litre]**

Solved Example—Relationship between Osmotic Pressure and Lowering of Vapour Pressure

Example 14: The vapour pressure of a solution containing 2.47 g of ethyl benzoate in 100 g of benzene (mol. wt 78 and density 0.8149 g/ml) was found to be 742.6 mm of Hg at 80° C while that of pure benzene at the same temperature, it is 751.86 mm of Hg. Calculate the osmotic pressure of the solution.

Solution: It is given that

$$p_s = 742.6 \text{ mm}, p° = 751.86 \text{ mm}, M = 78$$
$$\rho = 0.8149 \text{ g/ml} = 814.9 \text{/g litre}$$
$$T = 80 + 273 = 353 \, K$$

Taking $R = 0.0821$ litre atmosphere per degree per mol and substituting these values in the formula

$$\frac{p° - p_s}{p_s} = \frac{M\pi}{\rho RT} \qquad \text{(Refer to Section 2.7.2)}$$

$$\frac{751.86 - 742.60}{751.86} = \frac{78 \times P}{814.9 \times 0.0821 \times 353}$$

Osmotic pressure, $\pi = \dfrac{9.26}{751.86} \times \dfrac{814.9 \times 0.0821 \times 353}{78} = \mathbf{3.73 \text{ atm}}$

Problem for Practice

1. The vapour pressure of a solution of urea is 736.2 mm at 100°C. What is the osmotic pressure of this solution at 15°C?   [**Ans.** 41.1atm]

4.13 THEORIES OF OSMOSIS

1. Molecular sieves theory

According to this theory, the membrane contains lots of fine pores and acts as a sort of molecular sieves. **Smaller solvent molecules can pass through the pores but the larger solute molecules cannot.** Solvent molecules flow from a region of higher solute concentration to one of lower concentration across such a membrane (Fig. 4.9). But we observe that some membranes can act as sieves even when the solute molecules are smaller than the solvent molecules. This theory does not provide a satisfactory answer to this.

Recently it has been shown that the pores or capillaries between the protein molecules constituting an animals membrane are lined with

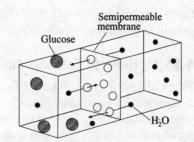

Fig. 4.9: A semipermeable membrane can separate particles on the basis of size. It allows the passage of small water molecules in both directions. But it prevents the passage of glucose molecules which are larger than water molecules

polar groups ($-COO^-$, $-NH_3^+$, $-S^{2-}$, etc.). Therefore, the membrane acts not simply as a sieve but also regulates the passage of solute molecules by electrostatic or 'chemical interactions'. In this way even solute molecules smaller than solvent molecules can be held back by the membrane.

2. Membrane solution theory

Membrane proteins bearing functional groups such as –COOH, –OH, –NH$_2$, etc., dissolve water molecules by hydrogen bonding or chemical intersection. Thus, membrane dissolves water from the pure water (solvent) forming what may be called '*membrane solution*'. The dissolved water flows into the solution across the membrane to equalise concentrations. In this way water molecules pass through the membrane while solute molecules being insoluble in the membrane do not.

3. Vapour pressure theory

It suggests that a semipermeable membrane has many fine holes or capillaries. The walls of these capillaries are not wetted by water (solvent) or solution.

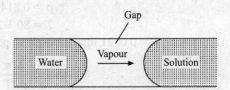

Fig. 4.10: *Water vapours diffuse into solution across the gap in a capillary of the membrane*

Thus neither solution nor water can enter the capillaries. Therefore, each capillary will have in it solution at one end and water at the other, separated by a small gap (Fig. 4.10). Since the vapour pressure of a solution is lower than that of the pure solvent, the diffusion of vapour will occur across the gap from water side to solution side. This results in the transfer of water into the solution.

4. Membrane bombardment theory

This theory suggest that osmosis results from an unequal bombardment pressure caused by solvent molecules on the two sides of the semipermeable membrane. On one side we have only solvent molecules while on the other side there are solute molecules occupying some of the surface area. Thus, there are fewer bombardments per unit area of surface on the solution side than on the solvent side. Hence the solvent molecules will diffuse more slowly through the membrane on the solution side than on the solvent side. The net result causes a flow of the solvent from the pure solvent to the solution across the membrane.

4.13.1 Reverse Osmosis and its Applications

When a solution is separated from pure water by a semipermeable membrane, osmosis of water occurs from water to solution. This osmosis can be stopped by applying pressure equal to or more than osmotic pressure, on the solution (Fig. 4.11). If pressure greater than osmotic pressure is applied, osmosis is made to proceed in the reverse direction to ordinary osmosis i.e., from solution to water.

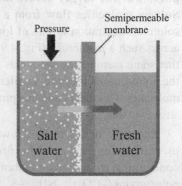

The osmosis taking place from solution to pure water by application of pressure greater than osmotic pressure, on the solution, is termed *Reverse Osmosis*.

This technology is used in the commercial production of water purifiers these days (R.O. Water purifiers).

Representation of principle of reverse osmosis

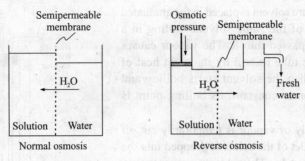

Fig. 4.11: Reverse osmosis versus ordinary osmosis

4.13.2 Desalination of Sea Water by Hollow-fibre Reverse Osmosis

Reverse osmosis is used for the desalination of sea water for getting fresh drinking water. This is done with the help of hollow fibres (nylon or cellulose acetate) whose wall acts as semipermeable membrane. A hollow-fibre reverse osmosis unit is shown in Fig. 4.12.

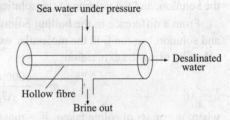

Fig. 4.12: Desalination of sea water by reverse osmosis in a hollow fibre unit

Water is introduced under pressure around the hollow fibres. The fresh water is obtained from the inside of the fibre. In actual practice, each unit contains more than three million fibres bundled together, each fibre is of about the diameter of a human hair.

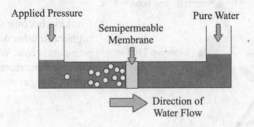

Design of a commercial R.O. Plant

4.14 ELEVATION IN BOILING POINT

Determination of boiling point elevation

Two methods that are generally employed to measure boiling point elevation are described below:

1. Landsberger-Walker Method

Apparatus. The apparatus used in this method is shown in Fig. 4.13 and consists of : (*i*) An *inner* tube with a hole in its side and graduated in ml; (*ii*) A boiling flask which sends solvent vapour into the graduated tube through a bulb with several holes (rose bulb), (*iii*) An *outer tube* which receives hot solvent vapour issuing from the side-hole of the inner tube; (*iv*) A *thermometer* reading to 0.1K, dipping in solvent or solution in the inner tube.

Procedure. Pure solvent is placed in the graduated tube and vapour of the same solvent boiling in a separate flask is passed into it. The vapour causes the solvent in the tube to boil by its latent heat of condensation. When the solvent starts boiling and temperature becomes constant, its boiling point is recorded.

Now the supply of vapour is temporarily cut off and a weighed pellet of the solute is dropped into the solvent in the inner tube. The solvent vapour is again passed through until the boiling point of the solution is reached and this is recorded. The solvent vapour is then cut off, thermometer and rosehead raised out of the solution, and the volume of the solution read.

From a difference in the boiling points of solvent and solution, we can find the molecular weight of the solute by using the expression

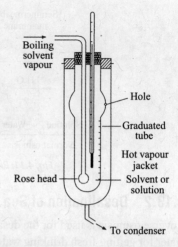

Fig. 4.13: Landsberger-Walker apparatus

$$m = \frac{1000 \times K_b \times w}{\Delta T_b \times W}$$

where w = mass of solute taken, W = mass of solvent which is given by the volume of solvent (or solution) measured in ml multiplied by the density of the solvent at its boiling point.

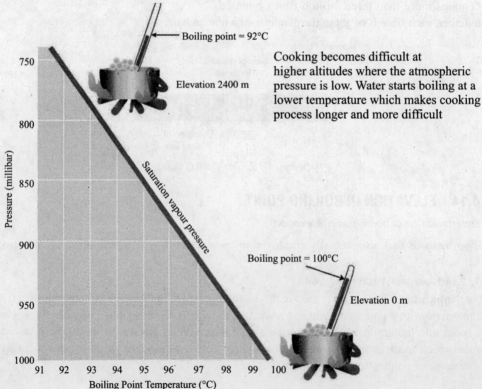

Cooking becomes difficult at higher altitudes where the atmospheric pressure is low. Water starts boiling at a lower temperature which makes cooking process longer and more difficult

2. Cottrell's Method

Apparatus. It consists of : (*i*) a graduated *boiling tube* containing solvent or solution; (*ii*) a reflux condenser which returns the vapourised solvent to the boiling tube; (*iii*) a thermometer reading to 0.01 K, enclosed in a glass hood; (*iv*) A small inverted funnel with a narrow stem which branches into three jets projecting at the thermometer bulb. Fig. 4.14 (*b*) showing all the four components.

Beckmann Thermometer [Fig. 4.14 (*a*)]. It is a *differential thermometer*. It is designed to measure small changes in temperature and not the temperature itself. It has a large bulb at the bottom of a fine capillary tube. The scale is calibrated from 0 to 6 K and subdivided into 0.01 K. The unique feature of this thermometer is the small reservoir of mercury at the top. The amount of mercury in this reservoir can be decreased or increased by tapping the thermometer gently. In this way the thermometer is adjusted so that the level of mercury thread will show up at the middle of the scale when the instrument is placed in the boiling (or freezing) solvent.

Procedure. The apparatus is set up as shown in [Fig. 4.14 (*b*)]. Solvent is placed in the boiling tube with a porcelain piece lying in it. It is heated on a small flame. As the solution starts boiling, solvent vapour arising from the porcelain piece pump the boiling liquid into the narrow stem. Thus, a mixture of solvent vapour and boiling liquid is continuously sprayed around the thermometer bulb. The temperature soon becomes constant and the boiling point of the pure solvent is recorded.

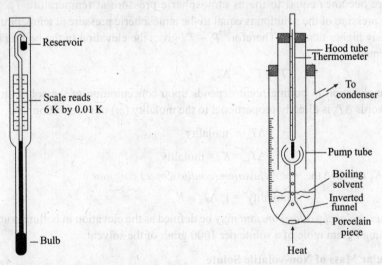

Fig. 4.14 (a): Beckmann thermometer reading to 0.01 K *Fig. 4.14 (b): Cottrell's apparatus*

Now a weighed amount of the solute is added to the solvent and the boiling point of the solution noted as the temperature becomes steady. Also, the volume of the solution in the boiling tube is noted. The difference of the boiling temperatures of the solvent and solute gives the elevation of boiling point. While calculating the molecular weight of the solute, the volume of solution is converted into mass by multiplying with density of solvent at its boiling point.

4.15 RELATION BETWEEN ELEVATION IN BOILING POINT OF THE SOLUTION AND THE MOLECULAR WEIGHT OF THE SOLUTE

The boiling point of a liquid is the temperature at which its vapour pressure becomes equal to the atmospheric pressure. Since, at any temperature, the vapour pressure of a solution of a non-volatile solute is always lower than that of the pure solvent, the boiling point of a solution is always higher than that of the pure solvent.

This fact can be illustrated in Fig. 4.15. The upper curve represents the pressure-temperature relationship of the pure solvent. The lower curve represents the vapour pressure-temperature relationship of the dilute solution of a known concentration. It is evident that the vapour pressure of the solution is less than that of the pure

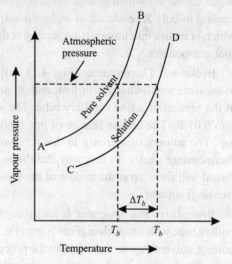

Fig. 4.15: Vapour pressure of pure solvent of solution

solvent at each temperature. From Fig. 4.15 it is clear that for the pure solvent the vapour pressure becomes equal to the as atmospheric pressure at temperature $T_b^°$. Similarly the vapour pressure of the solution is equal to the atmospheric pressure at temperature T_b, which is obviously higher than $T°_b$. Therefore $T_b - T°_b$ gives the elevation in the boiling point which is represented as ΔT_b.

or $$T_b - T°_b = \Delta T_b$$

The elevation in boiling point depends upon concentration of the solute in a solution. In other words ΔT_b is directly proportional to the molality (m) of the solution i.e.,

$$\Delta T_b \propto \text{molality} \quad \ldots(i)$$

or $$\Delta T_b = K_b \times \text{molality}$$

where K_b is called the *molal elevation (ebullioscopic) constant*.

If $\quad$ molality = 1, $\Delta T_b = K_b$

Thus, *molal elevation constant* may be defined as the elevation in boiling point of a solution containing 1 gram mole of a solute per 1000 gram of the solvent.

Molecular Mass of Non-volatile Solute

By definition,

$$\text{molality} = \frac{1000 \times w_2}{w_1 M_2} \quad \ldots(ii)$$

∴ eqn. (i) becomes

$$\Delta T_b = \frac{1000 \times K_b \times w}{\Delta T_b \times W} \quad \ldots(iii)$$

or

$$M_2 = \frac{1000 \times K_b \times w_2}{w_1 \times \Delta T_b} \quad \ldots(iv)$$

where ΔT_b = Elevation in boiling point.

K_b = Molal elevation (Ebullioscopic) constant of the solvent.

M_2 = Molecular mass of the solute

w_2 = Mass of the solute in grams

w_1 = Mass of the solvent in grams.

4.15.1 Thermodynamic Derivation of a Relationship Between Elevation in Boiling Point and Molecular Weight of a Non-volatile Solute

Fig. 4.16 gives the vapour pressure-temperature curves for the solvent and solution. The horizontal dotted line drawn corresponding the external pressure cuts the solvent and the solution curves at the points E and F which correspond to temperatures T_0 and T_s respectively. Then by definition, T_0 is the boiling point of the pure solvent and T_s that of the solution. Obviously $T_s > T_0$. The increase $(T_s - T_0)$ is called the elevation in boiling point and is usually represented by ΔT_b. Thus

$$\Delta T_b = T_s - T_0 \qquad ...(i)$$

The relationship between the elevation in boiling point and the concentration of the solution can be obtained by applying Clausius-Clapeyron equation and Raoult's law to the different conditions shown in Fig. 4.16.

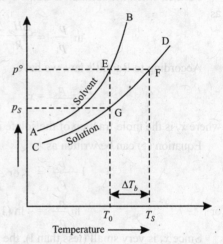

Fig. 4.16: Relationship between elevation in boiling point and lowering of vapour pressure

Clausius-Clapeyron equation which is applicable to a phase equilibrium, in the integrated form is given by

$$\ln \frac{p_2}{p_1} = \frac{\Delta H}{R} \left(\frac{1}{T_1} - \frac{1}{T_2} \right)$$

where p_1 is the vapour pressure at temperature T_1, p_2 is the vapour pressure at temperature T_2 and ΔH is the molar latent heat of transition.

Referring to Fig. 4.16 points G and F lie on the same solution curve. Suppose the vapour pressure of the pure solvent and the solution at temperature T_0 are p_0 and p_s respectively. Then from Fig. 2.22, we have

(a) Corresponding to point G, temperature = T_0, vapour pressure = p_s.

(b) Corresponding to point F, temperature = T_s, vapour pressure = p^o.

Hence for the equilibrium, solution $\rightleftharpoons$ vapour, according to Clausius-Clapeyron equation, we have (for the points G and F)

$$\ln \frac{p^o}{p_s} = \frac{\Delta H_v}{R} \left(\frac{1}{T_0} - \frac{1}{T_s} \right) \qquad ...(ii)$$

where ΔH_v is the latent heat of vaporisation of one mole of the solvent from the solution. When the solution is dilute, ΔH_v is nearly equal to the latent heat of evaporation of the pure solvent.

Equation (ii) can be written as

$$\ln \frac{p^o}{p_s} = \frac{\Delta H_v}{R}\left(\frac{T_s - T_0}{T_0 T_s}\right)$$

or

$$\ln \frac{p^o}{p_s} = \frac{\Delta H_v}{R} \frac{\Delta T_v}{T_b T_s} \qquad [\because T_s - T_0 = \Delta T_b] \quad ...(iii)$$

Further, when the solution is dilute T_s is nearly equal to T_0. Hence equation (iii) can written as

$$\ln \frac{p^o}{p_s} = \frac{\Delta H_v}{R} \frac{\Delta T_b}{T_0^2} \qquad ...(iv)$$

According to Raoult's law, we have

$$\frac{p^o - p_s}{P^o} = x_2 \qquad ...(v)$$

where x_2 is the mole fraction of the solute in the solution.

Equation (v) can be written as

$$1 - \frac{p_s}{p^o} = x_2 \text{ or } \frac{p_s}{p^o} = 1 - x_2 \text{ or } \ln \frac{p_s}{p^o} = \ln(1 - x_2)$$

or

$$\ln \frac{p_s}{p^o} = -\ln(1 - x_2) \qquad ...(vi)$$

Since x_2 is very small (less than 1), the expansion of $\ln(1 - x_2)$ as an infinite series is given by

$$\ln(1 - x_2) = -x_2 - \frac{1}{2}x_2^2 - \frac{1}{3}x_2^3 ...$$

Neglecting x_2^2, x_2^3 etc. (as x_2 is much smaller than 1), we have

$$\ln(1 - x_2) \simeq -x_2 \text{ or } -\ln(1 - x_2) \simeq -x_2 \qquad ...(vii)$$

From eq. (vi) and (vii) $\ln \frac{p_s}{p^o} = x_2$ ...(viii)

Substituting this value in equation (iv)

$$x_2 = \frac{\Delta H_v}{R} \frac{\Delta T_b}{T_0^2} \qquad ...(ix)$$

or

$$\Delta T_b = \frac{R T_0^2 x_2}{\Delta H_v} \qquad ...(x)$$

However the common practice in the studies on elevation in boiling point is not to express the concentrations in terms of mole fractions but in terms of *moles of the solute per 1000 g of the solvent i.e. in terms of molality, m*. Hence equation (x) is further modified as follows:

If n_2 moles of the solute are dissolved in n_1 moles of the solvent, the mole fraction of the solute (x_2) in the solution will be given by

$$x_2 = \frac{n_2}{n_1 + n_2}$$

As the solution is supposed to be dilute $n_2 \ll n_1$ so that in the denominator n_2 can be neglected in comparison to n_1. Hence for the dilute solution, above equation can be written as

$$x_2 = \frac{n_2}{n_1}$$

Further, if w_2 g of the solute is dissolved in w_1 g of the solvent and M_2 and M_1 are the molecular weights of the solute and solvent respectively, then

$$n_2 = \frac{w_2}{M_2} \text{ and } n_1 = \frac{w_1}{M_1}$$

Hence $$x_2 = \frac{n_2}{n_1} = \frac{w_2/M_2}{w_1/M_1} = \frac{w_2 \times M_1}{w_1 \times M_2}$$

Putting this value of x_2 in equation (x), we get

$$\Delta T_b = \frac{RT_0^2}{\Delta H_v} \frac{w_2 M_1}{w_1 M_2} = \frac{RT_0^2}{\Delta H_v/M_1} \frac{w_2}{w_1 M_2} \qquad \text{...(xi)}$$

Putting $\dfrac{\Delta H_v}{M_2} = l_v$, latent heat of vaporisation per gram of the solvent, equation (xi) becomes

$$\Delta T_b = \frac{RT_0^2}{l_v} \frac{w_2}{w_1 M_2} \qquad \text{..(xii)}$$

Let us say molality of the solution is m. Thus m moles of the solute are dissolved in 1000 g of the solvent, we can write

$$\frac{w_2}{M_2} = m \text{ and } w_1 = 1000$$

Putting these values in equation (xii), we get

$$\Delta T_b = \frac{RT_0^2}{l_v} \frac{m}{1000} \qquad \text{...(xiii)}$$

or $$\Delta T_b = \frac{RT_0^2}{1000 l_v} m \qquad \text{...(xiv)}$$

For a given solvent, the quantity $\dfrac{RT_0^2}{1000 l_v}$ is a constant quantity because l_v, T_0 and R are constant. It is represented by K_b and is called **molal elevation constant** or **ebullioscopic constant**.

i.e. $$K_b = \frac{RT_0^2}{1000 l_v} \qquad \text{...(xv)}$$

Hence equation (xiv) can be written as

$$\Delta T_b = K_b \cdot m \qquad \text{....(xvi)}$$

If $$m = 1, \Delta T_b = K_b$$

Thus molal elevation constant may be defined as the elevation in boiling point when the molality of the solution is unity.

Solved Examples— Elevation in Boiling Point

Example 15: A solution of 12.5 g of urea in 170 g of water gave boiling point elevation of 0.63 K. Calculate the molar mass of urea. $K_b = 0.52$ K kg mol^{-1}.

Solution: From the given data

Mass of the solute, $w_2 = 12.5$ g, Mass of the solvent $w_1 = 170$ g
Elevation of boiling point $\quad\quad\quad\quad \Delta T_b = 0.63$ K
Elevation constant $\quad\quad\quad\quad\quad\quad K_b = 0.52$ K kg mol^{-1}
Let the molecular mass of solute (urea) $\quad = M_2$

Calculation of molality

$$170 \text{ grams of water contain urea} = 12.5 \text{ grams}$$

$$\therefore \quad 1000 \text{ grams of water contain urea} = \frac{1000 \times 12.5}{170} \text{ g,} = \frac{1000 \times 12.5}{170 \times M_2} \text{ mole}$$

$$\therefore \quad \text{Molality (no. of mole in 1000 g of solvent)} = \frac{1000 \times 12.5}{170 \times M_2}$$

We know, $\quad\quad\quad\quad\quad\quad\quad\quad\quad \Delta T_b = K_b \times \text{molality}$

Substituting the value in the above relation,

We have, $0.63 = 0.52 \times \left(\dfrac{1000 \times 12.5}{170 \times M_2}\right)$ or $M_2 = \dfrac{0.52 \times 1000 \times 12.5}{170 \times 0.63} = 60.69$ a.m.u.

Example 16: A solution prepared from 0.3 g of an unknown non-volatile solute in 30.0 g of CCl_4 boils at 350.392 K. Calculate the molecular mass of the solute. The boiling point of CCl_4 and its K_b values are 350.0 K and 5.03 respectively.

Solution: From the given data

Mass of the solute, $\quad w_2 = 0.3$ g, Mass of the solvent $w_1 = 30.0$ g
Elevation of boiling point $= \Delta T_b = 350.392 - 350.0 = 0.392$ K
Elevation constant $\quad K_b = 5.03$ K kg mol^{-1}

Calculation of molality

30 g of CCl_4 contain $\quad\quad\quad\quad\quad\quad = 0.3$ g of solute

$$\therefore \quad 1000 \text{ g of CCl}_4 \text{ contain} = \frac{1000 \times 0.3}{30} \text{ g} = \frac{1000 \times 0.3}{30 \times M_2} \text{ moles}$$

$$\therefore \quad \text{Molality (no. of moles in 1000 g of CCl}_4) = \frac{1000 \times 0.3}{30 \times M_2}$$

Substituting the values in the relation

$\Delta T_b = K_b \times \text{molality} \quad$ or $\quad 0.92 = 5.03 \times \dfrac{1000 \times 0.3}{30 \times M_2}$

or $M_2 = \dfrac{5.03 \times 1000 \times 0.3}{30 \times 0.392} = 128.3 \quad \therefore$ Molecular mass = **128.3**

Example 17: Find the b.p. of a solution containing 0.36 g of glucose ($C_6H_{12}O_6$) dissolved in 100 g of water ($K_b = 0.52$ K/m).

Solution: Mass of glucose (w_2) = 0.36 g

Mass of water (w_1) = 100 g
Mol. Mass of glucose (M_2) = 180
Molal elevation constant for water (K_b) = 0.52
Substituting the values in the relation

$$\Delta T_b = \frac{1000 \times K_b \times w_2}{w_1 \times M_2} \text{ or } \Delta T_b = \frac{1000 \times 0.52 \times 0.36}{100 \times 180}$$

Elevation in b.p. = 0.0104, B. P. of pure water = 373 K
Hence b.p. of the solution = 373 + 0.0104 = 373.0104 K

Example 18: 10 g of a non-volatile solute when dissolved in 100g of benzene raises its b.p. by 1°. What is the molecular mass of the solute (K_b for benzene = 2.53 K mol^{-1})?

Solution: In this problem
Mass of the solute (w_2) = 10 g, Mass of the solvent (w_1) = 100 g
Elevation in b.p. (ΔT_b) = 1°, K_b = 2.53
Substituting the values in the equation

$$M_2 = \frac{1000 \times K_b \times w_2}{w_1 \times \Delta T_b} = \frac{1000 \times 2.53 \times 10}{100 \times 1} = 253$$

Example 19: A solution containing 0.5126 g of naphthalene (mol mass 128) in 50 g of carbon tetrachloride yields a b.p. elevation of 0.402°C while a solution of 0.6216 g of an unknown solute in the same mass of the solvent gives a b.p. elevation of 0.647°C. Find the molecular mass of the unknown solute.

Solution 1: Determination of K_b from the first data

$w_2 = 0.5126$, $M_2 = 128$, $w_1 = 50$

$$K_b = \frac{M_2 \times w_1 \times \Delta T_b}{1000 \times w_2} = \frac{128 \times 50 \times 0.402}{1000 \times 0.5126} = = 5.02$$

2. Mol. mass of the unknown solute

$w_2 = 0.6216$ g, $w_1 = 50$ g, $\Delta T_b = 0.647$, $K_b = 5.02$

The value of K_b remains the same because the solvent is the same. Substituting the values in the equation

$$M_2 = \frac{1000 \times K_b \times w_2}{w_1 \times \Delta T_b} = \frac{1000 \times 5.20 \times 0.6216}{50 \times 0.647} = 96.46$$

Problem for Practice

1. What is the molecular mass of a non-volatile organic compound if the addition of 1.0 g of it in 50.0 g of benzene raises the boiling point of benzene by 0.30°C? K_b for benzene is 2.53°C per 1000 g of benzene.   **[Ans. 170]**

2. A solution containing 36g of solute dissolved in one litre of water gave an osmotic pressure of 6.75 atmosphere at 27°C. The molal elevation constant of water is 0.52 K kg mol^{-1}. Calculate the boiling point of the solution.   **[Ans. 100.1425°C]**

3. Calculate the molal boiling point constant for chloroform (M = 119.4) from the fact than its boiling point is 61.2°C and its latent heat of vaporizatin is 59.0 cal/g.   **[Ans. 3.79°C]**

4. A solution of 3.795 g sulphur in 100 g carbon disulphide (boiling point 46.30°C, ΔH_v = 6400 cal/mol) boils at 46.66°C. What is the formula of sulphur molecule in the solution? **[Ans. S_8]**

[Hint. ΔH_v = 6400 cal/mol, $\therefore l_v = \dfrac{6400}{76}$ cal/g because mol mass of CS_2 = 76. First calculate K_b and then M_2.

We get M_2 = 255.2. Hence atomicity of molecule = $\dfrac{255.2}{32}$ = 8]

5. When 1.80g of a non-volatile compound are dissolved in 25.0 g of acetone, the solution boils at 56.86°C while pure acetone boils at 56.38°C under the same atmospheric pressure. Calculate the molecular mass of the compound. The molal elevation constant for acetone is 1.72°K kg mol^{-1}. **[Ans. 258]**

6. What elevation in boiling point of alcohol is to be expected when 5 g of urea (mol. mass = 60) are dissolved in 75 g of it? The molal elevation constant for alcohol is 1.15°C per molality. **[Ans. 1.28°C]**

4.15.2 Relationship between Elevation in Boiling Point and Relative Lowering of Vapour Pressure

We have the following relation between elevation in boiling point and molecular mass

$$\Delta T_b = \dfrac{1000 \times K_b \times w_2}{w_1 M_2} \qquad ...(i)$$

Also $\qquad \dfrac{w_2}{M_2} = n_2 \quad ...(ii) \quad$ and $\quad \dfrac{w_1}{M_1} = n_1 \qquad ...(iii)$

Substituting the values of w_2 and w_1 for eq. (ii) and (iii) in (i), we get

we get $\qquad \Delta T_b = \dfrac{1000 \times K_b \times n_2}{n_1 M_1} = \dfrac{1000 K_b}{M_1} \dfrac{n_2}{n_1} \qquad ...(iv)$

By Raoult's law, for dilute solutions, we have

$$\dfrac{\Delta p}{p^o} = \dfrac{n_2}{n_1}$$

where $\Delta p = p^o - p_s$ is the lowering of vapour pressure. From eqs (iv) and (v) we get

$$\Delta T_b = \dfrac{1000 K_b}{M_1} \dfrac{\Delta p}{p^o}$$

4.15.3 Relationship between Elevation in Boiling Point and Osmotic Pressure

The equation for the osmotic pressure for n_2 moles of the solute dissolved in V litres of solution is

$$\pi V = n_2 RT \quad \text{or} \quad n_2 = \dfrac{\pi v}{RT} \qquad ...(i)$$

The equation for the elevation in boiling point is

$$\Delta T_b = \dfrac{1000 K_b w_2}{w_1 M_2} \qquad ...(ii)$$

or $\qquad \Delta T_b = \dfrac{1000 K_b w_2/M_2}{w_1} = \dfrac{1000 K_b n_2}{w_1} \qquad ...(iii)$

Substituting the value of n_2 from eq (*i*) in eq. (*iii*)

$$\Delta T_b = \frac{1000 K_b}{w_1} \times \frac{\pi V}{RT} = \frac{1000 K_b \pi}{(w_1/V)RT} \qquad ...(iv)$$

As the solution is dilute, Volume of the solution (V) $\simeq$ Volume of the solvent.

∴ If d is the density of the solvent, $\dfrac{w_1}{V} = d$

Substituting this value in equation (*iv*), we get

$$\Delta T_b = \frac{1000 K_b \pi}{dRT}$$

4.16 DEPRESSION IN FREEZING POINT

Methods for the determination of depression in freezing point.

1. Beckmann's Method

Apparatus. It consists of : (*i*) *A freezing tube* with a side-arm to contain the solvent or solution, while the solute can be introduced through the side-arm; (*ii*) An outer tube into which is fixed the freezing tube, the space in between providing an air jacket which ensures a slower and more uniform rate of cooling; (*iii*) A *large jar* containing a freezing mixture e.g., ice and salt with a stirrer (Fig. 4.17).

Procedure. About 20 g of the solvent is taken in the freezing point tube and the apparatus set up as shown in Fig. 4.17 so that the bulb of the thermometer is completely immersed in the solvent. Determine the freezing point of the solvent by directly cooling the freezing-point tube in the cooling bath.

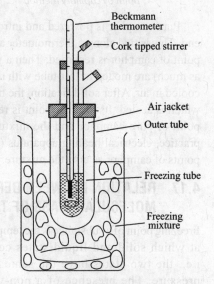

Fig. 4.17: Beckmann's freezing-point apparatus

The freezing point of the solvent having been accurately determined, the solvent is remelted by removing the tube from the bath, and a weight amount (0.1–0.2 g) of the solute is introduced through the side tube. Now the freezing point of the solution is determined in the same way as that of the solvent. A further quantity of solute may then be added and another reading taken. Knowing the depression of the freezing point, the molecular mass of the solute can be determined by using the expression.

$$m = \frac{1000 \times K_f \times w}{\Delta T \times W}$$

2. Rast's Camphor Method (Cryoscopic method)

This method is used for determination of molecular weights of solutes which are soluble in molten camphor. The freezing point depressions are so large that an ordinary thermometer can also be used.

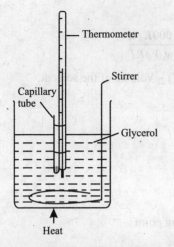
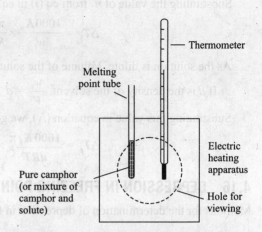

Fig. 4.18: *Determination of depression of melting point by capillary method*

Fig. 4.19: *Determination of depression of melting point by electrical apparatus*

Pure camphor is powered and introduced into a capillary tube which is sealed at the upper end. This is tied along a thermometer and heated in a glycerol bath (see Fig. 4.18). The melting point of camphor is recorded. Then a weighed amount of solute and camphor (about 10 times as much) are melted in test-tube with the open end sealed. The solution of solute in camphor is cooled in air. After solidification, the mixture is powdered and introduced into a capillary tube which is sealed. Its melting point is recorded described before. The difference of the melting point of pure camphor and the mixture, gives the depression of freezing point. In modern practice, electrical heating apparatus (Fig. 4.19) is used for a quick determination of melting points of camphor as also the mixture.

4.17 RELATION BETWEEN DEPRESSION IN FREEZING POINT AND MOLECULAR MASS OF THE SOLUTE

Freezing point of a substance is the temperature at which solid and liquid states co-exists i.e., the two states have the same vapour pressure. The presence of a non-volatile solute lowers the vapour pressure of the solution. Thus liquid and solid states will have equal vapour pressure at a much lower temperature. Hence there is a depression in the freezing point. In other words, the freezing point of a solution is lower than that of the pure solvent as is clear from the Fig. 4.20.

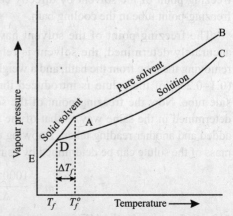

Fig. 4.20: *Plot of Vapour pressure vs. temperature*

The curve AB represents the vapour pressure curve of pure solvent. At point A, solvent co-exists with the liquid state. The temperature corresponding to point A, gives the freezing point of solvent. Similarly curve CD represents the vapour pressure curve of the solution in which point D corresponds to two states of the solvent existing simultaneously in the solution. Thus temperature corresponding to point D is the freezing point of solution which is clearly lower than the freezing point of the pure solvent. Hence there is a depression in the freezing point given by ΔT_f,

$$\Delta T_f = T^o_f - T_f$$

It has been found that depression in the freezing point is proportional to molal concentration of the solute in the solution i.e., molality as given below.

$$\Delta T_f = K_f \times \text{molality}$$

where K_f is constant known as *molal depression constant or cryoscopic constant of the solvent*.

If molality = 1 then $\quad \Delta T_f = K_f$

Hence *molal depression constant may be defined as the depression in freezing point when one mole of the solute is dissolved in 1000 grams of the solvent.*

Molecular mass of non-volatile solute

$$\text{Since molality} = \frac{1000 \times w}{W \times m}$$

$$\Delta T_f = \frac{K_f \times 1000 \times w}{W \times m} \quad \text{or} \quad m = \frac{1000 \times K_f \times w}{W \times \Delta T_f}$$

Here
$\quad \Delta T_f$ = depression in freezing point
$\quad K_f$ = Molal depression constant or cryoscopic constant
$\quad w$ = mass of the solute in grams
$\quad m$ = molecular mass of the solute
$\quad W$ = mass of the solvent in grams

4.17.1 Thermodynamic Derivation of a Relationship between the Freezing Point and Molecular Weight of a Non-volatile Solute

The depression in freezing point of a solution can be explained on the basis of lowering of vapour pressure of the solution. If vapour pressure are plotted against temperature the curve AB is obtained for the liquid solvent as shown in Fig. 4.21. At the point B, the liquid solvent starts solidifying and hence the liquid solvent and the solid solvent are in equilibrium with each other. The temperature corresponding to the point B is thus the freezing point of the pure solvent (T_0). When whole of the liquid solvent has been solidified, there is a sharp change in the vapour pressure-temperature curve beyond the point B, as shown by the curve BC for the solid solvent. Vapour pressure temperature curve for the solution lies below the vapour pressure curve of the liquid solvent, as represented by the

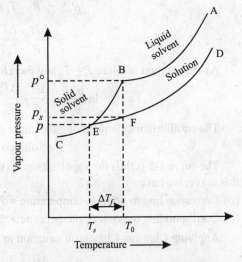

Fig. 4.21: *Relationship between depression in freezing point and lowering of vapour pressure*

curve DE in Fig. 4.21 As the cooling of the solution is continued, at the point E, the separation of the solid solvent starts. Thus the point E represents the freezing point of the solution (T_s). From Fig. 4.21, it is evident that $T_s < T_0$. The decrease $(T_0 - T_s)$ is called the depression in freezing point and is usually represented by ΔT_f. Thus

$$\Delta T_f = T_0 - T_s \qquad ...(i)$$

The relationship between the depression in freezing point and the concentration of the solution can be derived by applying Clausius-Clapeyron equation and Raoult's law to the curves shown in Fig. 4.21.

Suppose the vapour pressures corresponding to the point B, E and F are $p°$, p and p_s respectively.

The equilibrium existing along the curve BC is

$$\text{Solid} \rightleftharpoons \text{Vapour}$$

The curve BC is, therefore, the sublimation curve of the solid solvent. For the points B and E lying on this curve, we have

(a) Corresponding to point B, temperature = T_0, vapour pressure = $p°$

(b) Corresponding to point E, temperature = T_s, vapour pressure = p.

Applying Clausius-Clapeyron equation corresponding to points B and E, we have

$$\ln \frac{p°}{p} = \frac{\Delta H_s}{R} \cdot \left(\frac{1}{T_s} - \frac{1}{T_0} \right) \qquad ...(ii)$$

Rudolf Clausius (1822–1888)
Rudolf Clausius was a German physicist and mathematician and is considered as one of the central founder of thermodynamics. He deduced the Clausius-Clapeyron equation from thermodynamics. He also gave the first mathematical version of concept of entropy. His wife died in childbirth in 1875 leaving him to look after their six children. He continued to teach but had less time for research thereafter.

where ΔH_s is the molar latent heat of sublimation of the solid solvent.

Equation (ii) can be written as

$$\ln \frac{p°}{p} = \frac{\Delta H_s}{R} \cdot \frac{T_0 - T_s}{T_s T_0}$$

$$= \frac{\Delta H_s}{R} \cdot \frac{\Delta T_f}{T_0 T_f} \qquad ...(iii)$$

As the solution is dilute $T_s \simeq T_0$ and therefore equation (iii) becomes

$$\ln \frac{p°}{p} = \frac{\Delta H_s}{R} \cdot \frac{\Delta T_f}{T_0^2} \qquad ...(iv)$$

The equilibrium existing along the curve DE is

$$\text{Solution} \rightleftharpoons \text{Vapour}$$

The curve DE is thus the vaporisation curve of the solution. For the points E and F lying on this curve, we have

(a) Corresponding to point E, temperature = T_s, vapour pressure = $p°$
(b) Corresponding to point F, temperature = T_0, vapour pressure = p_s

Applying Clausius-Clapeyron equation to these conditions, we have

$$\ln \frac{p_s}{p} = \frac{\Delta H_v}{R} \left(\frac{1}{T_s} - \frac{1}{T_0} \right) = \frac{\Delta H_v}{R} \cdot \frac{T_0 - T_s}{T_s \cdot T_0}$$

$$= \frac{\Delta H_v}{R} \frac{\Delta T_f}{T_0^2} \quad \text{(Taking } T_s \simeq T_0) \qquad \ldots(v)$$

where ΔH_v is the latent heat of vaporisation of one mole of the solvent from the solution and is nearly equal to the molar latent of vaporisation of the pure solvent for a dilute solution.

In order to get a relationship for the equilibrium

$$\text{Solid} \rightleftharpoons \text{Liquid}$$

subtract equation (v), from equation (iv), we get

$$\ln \frac{p^o}{p} - \ln \frac{p_s}{p} = \frac{\Delta H_s}{R} \frac{\Delta T_f}{T_0^2} - \frac{\Delta H_v}{R} \frac{\Delta T_f}{T_0^2}$$

or $(\ln p^o - \ln p) - (\ln p_s - \ln p^o) = \dfrac{\Delta T_f}{RT_0^2} (\Delta H_s - \Delta H_v)$

$$\ln p^o - \ln p_s = \frac{\Delta T_f}{RT_0^2} \Delta H_f \quad \text{or} \quad \ln \frac{p^o}{p_s} = \frac{\Delta T_f}{RT_0^2} \Delta H_f \qquad \ldots(vi)$$

where $\Delta H_s - \Delta H_v = \Delta H_f$ is the molar latent heat of fusion of the solid solvent.

The left hand term of equation (vi) can be expressed in terms of mole fraction by applying Raoult's law,

$$\frac{p^o - p_s}{p^o} = x_2 \qquad \ldots(vii)$$

where x_2 is the mole fraction of the solute in the solution.

Equation (vii) can be written as

$$1 - \frac{p_s}{p^o} = x_2 \quad \text{or} \quad \frac{p_s}{p^o} = 1 - x_2 \quad \text{or} \quad \ln \frac{p_s}{p^o} = \ln(1 - x_2)$$

or $\ln \dfrac{p_s}{p^o} = -\ln(1 - x_2) \qquad \ldots(viii)$

The expansion of $\ln(1 - x_2)$ as an infinite series is given by

$$\ln(1 - x_2) = -x_2 - \frac{1}{2}x_2^2 - \frac{1}{3}x_2^3 \ldots$$

Neglecting x_2^2, x_2^3 etc. (because x_2 is very small, higher powers will be still smaller), we have $\ln(1 - x_2) = -x_2$. Hence equation (viii) becomes

$$\ln \frac{p^o}{p_s} = x_2$$

Substituting this value in equation (vi), we get

$$x_2 = \frac{\Delta T_f}{RT_0^2} \cdot \Delta H_f \quad \text{or} \quad \Delta T_f = \frac{RT_0^2}{\Delta H_f} \cdot x_2 \qquad \ldots(ix)$$

This equation gives the relationship between depression in freezing point and the mole fraction x_2 of the solute in the solution.

To obtain the equation in terms of molality instead of mole fraction x_2, we have

$$x_2 = \frac{n_2}{n_1 + n_2}$$

where n_2 = number of moles of the solute in the solution
and n_1 = number of moles of the solvent in the solution.

For dilute solution $n_2 \ll n_1$ so that on neglecting n_2 in comparison to n_1, we have

$$x_2 = \frac{n_2}{n_1} = \frac{w_2/M_2}{w_1/M_1} = \frac{w_2/M_1}{w_1/M_2} \qquad \ldots(x)$$

where w_2 = mass of the solute dissolved, M_2 = molecular mass of the solute
w_1 = mass of the solvent, M_1 = molecular mass of the solvent.

Putting the value of x_2 from equation (x) in equation (ix), we get

$$\Delta T_f = \frac{RT_0^2}{\Delta H_f} \frac{w_2 M_1}{w_1 M_2} \qquad \ldots(xi)$$

$$= \frac{RT_0^2}{\Delta H_f/M_1} \frac{w_2}{w_1 M_2} \qquad \ldots(xii)$$

$$= \frac{RT_0^2}{l_f} \frac{w_2}{w_1 M_2}$$

where $l_f = \dfrac{\Delta H_f}{M_1}$ is the latent heat of fusion *per gram* for the solid solvent.

If m is the molality of the solution *i.e.* m moles of the solute are dissolved in 1000 g of the solvent, then

$$\frac{w_2}{M_2} = m \text{ and } w_1 = 1000 \text{ g}$$

Putting these values in equation (xii), we get

$$\Delta T_f = \frac{RT_0^2}{l_f} \frac{m}{1000} = \frac{RT_0^2}{l_f}.m \qquad \ldots(xiii)$$

Since T_0 and l_f for a given solvent are constant, the quantity $\dfrac{RT_0^2}{1000 l_f}$ in equation (xiii) must be a constant quantity. It is usually represented by K_f and is called **molal depression constant or cryoscopic constant**. Thus

$$K_f = \frac{RT_0^2}{1000 l_f} \qquad \ldots(xiv)$$

Substituting this value in equation (xiii), we get

$$\Delta T_f = K_f m \qquad \ldots(xv)$$

If $m = 1$, $\Delta T_f = K_f$

Thus *molal depression constant may be defined as the depression in freezing point which takes place when the molality of the solution is unity.*

4.17.2 Relations between Depression in Freezing Point and Lowering of Vapour Pressure and Osmotic Pressure

Just as in Section 4.15.2 and Section 4.15.3 we derived the relations between elevation in boiling point with lowering of vapour pressure and osmotic pressure respectively, following the same procedure, we can deduce the relationships between depression in freezing point and lowering of vapour pressure and osmotic pressure. This is because the expression for elevation in boiling point is similar to that of depression in freezing point.

(a) Relationship between depression in freezing point and lowering of vapour pressure

$$\Delta T_f = \frac{1000 K_f}{M_1} \cdot \frac{\Delta p}{p^o}$$

The different terms used in the above equation have usual meaning.

(b) Relationship between depression in freezing point and osmotic pressure.

$$\Delta T_f = \frac{1000 K_f \pi}{\rho RT}$$

The different terms used in the above equation have usual meaning.

Solved Examples—Depression in Freezing Point

Example 20: Calculate the freezing point of a solution containing 0.520 g of glucose ($C_6H_{12}O_6$) in 80.2 gram of water. For water $K_f = 1.86$ K kg mol^{-1}.

Solution: Given

Mass of the solute (glucose), $w_2 = 0.520$ g, Mass of solvent water, $w_1 = 80.2$ g

Molar depression constant $K_f = 1.86$ K kg mol^{-1}

Molecular mass of the solute $C_6H_{12}O_6$, $M_2 = 180$

Calculation of molality

80.2 g of water contain glucose = 0.520 g

∴ 1000 g of water, contain glucose = $\frac{1000 \times 0.520}{80.2}$ ∴ Molality = $\frac{1000 \times 0.520}{80.2 \times 180}$

Applying the relation $\Delta T_f = K_f \times$ molality $= 1.86 \times \frac{1000 \times 0.520}{80.2 \times 180}$

∴ Freezing point = 273 − 0.067 = 272.933K (∵ Freezing point of water = 273K)

Example 21: A solution of 1.25 g of a certain non-electrolyte in 20 g of water freezes at 271.95 K. Calculate the molecular mass of the solute.   ($K_f = 1.86$ K kg mol^{-1})

Solution: Here we are given that:

Mass of the solute $w_2 = 1.25$ g, Mass of solvent $w_1 = 20$ g

Molar depression constant $K_f = 1.86$ K kg mol^{-1}

Depression in freezing point $\Delta T_f = 273 − 271.95 = 1.05$ K

Calculation of molality

20 g of water contain solute = 1.25 g

∴ 1000 g of water contain solute = $\frac{1000 \times 1.25}{20}$

Let the molecular mass of the solute = M_2

∴ Molality = $\frac{1000 \times 1.25}{20 \times M_2}$

Substituting the values in the relation: $\Delta T_f = K_f \times m$, we get

$$1.05 = 1.86 \times \frac{1000 \times 1.25}{20 \times M_2} \text{ or } M_2 = \frac{1.86 \times 1000 \times 1.25}{20 \times 1.05} = 110.71$$

Molecular mass of the solute (M_2) = **110.71**

Example 22: Find (*i*) the boiling point and (*ii*) the freezing point of a solution containing 0.52 g glucose ($C_6H_{12}O_6$) dissolved in 30.2 g of water. For water K_b = 0.52 K kg mol^{-1}; K_f = 1.86 K kg mol^{-1}.

Solution: Given

Molecular mass of glucose $\qquad M_2 = 180$

Mass of solute w_2 = 0.52 g, Mass of solvent w_1 = 80.2 g

Calculation of molality

80.2 g of the solvent contain solute $\quad = 0.52$ g

$\therefore$ 1000 g of the solvent contain solute $\quad = \dfrac{1000 \times 0.52}{80.2}$ g

$\therefore \qquad\qquad\qquad$ Molality $= \dfrac{1000 \times 0.52}{80.2 \times 180}$

We know, $\qquad\qquad\qquad \Delta T_b = K_b \times$ molality

$\qquad\qquad\qquad\qquad = 0.52 \times \dfrac{1000 \times 0.52}{80.2 \times 180}$ g $= 0.018$ K

$\therefore$ Boiling point $\qquad\qquad = 373 + 0.018 = 373.018$ K

$\therefore$ Boiling point of water $\qquad = 373$ K

$\therefore$ Boiling point of solution $\qquad = 373.018$ K

For freezing point, $\qquad\qquad \Delta T_f = K_f \times$ molality

$\qquad\qquad\qquad\qquad = 1.86 \times \dfrac{1000 \times 0.52}{80.2 \times 180} = 0.067$

Freezing point of $H_2O \qquad = 273$ K

$\therefore$ Freezing point of solution $\quad = 273 - 0.067 = \mathbf{272.933}$ **K**

Example 23: The normal freezing point of nitrobenzene $C_6H_5NO_2$ is 278.82 K. A 0.25 molal solution of a certain solute in nitrobenzene causes a freezing point depression of 2 degree. Calculate the value of K_f for nitrobenzene.

Solution: Apply the relation: $\qquad \Delta T_f = K_f \cdot m$

$\qquad\qquad\qquad\qquad\qquad \Delta T_f = 2, m = 0.25$

Substituting the values in the first equation, we have

$\qquad\qquad 2 = K_f \times 0.25 \quad$ or $\quad K_f = \dfrac{2}{0.25} = 8$ K kg mol^{-1}

Example 24: A sample of camphor used in the Rast method of determination molecular masses had a melting point of 176.5°C. The melting point of a solution containing 0.522 g camphor and 0.0386 g of an unknown substance was 158.5°C. Find the molecular mass of the substance. K_f of camphor per kg is 37.7.

Solution: Apply the expression

$$M_2 = \dfrac{1000 \times K_f \times w_2}{\Delta T \times w_1}$$

In the present case, we have

$\Delta T = 176.5 - 158.5 = 18 \quad$ or $\quad K_f = 37.7, w_2 = 0.0386$ g, $w_1 = 0.522$ g

Substituting these values, we have

$$M_2 = \dfrac{1000 \times 37.7 \times 0.0386}{18 \times 0.522} = 154.8$$

Example 25: A solution of sucrose (molar mass = 342 g/mole) is prepared by dissolving 68.4 g in 1000 g of water. What is the
(i) Vapour pressure of solution at 293 K
(ii) Osmotic pressure at 293 K
(iii) Boiling point of the solution
(iv) Freezing point of the solution

The vapour pressure of water at 293 K is 17.5 mm, K_b = 0.52, K_f = 1.86.

Solution: (i) Mole fraction of sucrose $X_B = \dfrac{68.4/342}{68.4/342 + 1000/18} = \dfrac{9}{2509}$

$$\dfrac{p_A^o - p_A}{p_A} = X_B$$

Vapour pressure of water $\quad p_A^o = 17.5$ mm

Substituting the values $\quad \dfrac{17.5 - p_A}{17.5} = \dfrac{9}{2509}$

or $\quad 17.5 \times 2509 - 2509\, p_A = 9 \times 17.5 \quad$ or $\quad 2509\, p_A = 43750$

or $\quad p_A = 17.43$ mm

(ii) $\quad \pi V = nRT,\ V = 1$ litre, $n = \dfrac{68.4}{342}$

$\pi \times 1 = \dfrac{68.4}{342} \times 0.0821 \times 293$

or $\quad \pi = 4.8$ atmospheres

(iii) $\quad \Delta T_b = K_b \times m \quad$ where m is the molality

Molality $(m) = \dfrac{68.4}{342}$, $K_b = 0.52$ m. Substituting the values, we have

$$\Delta T_b = 0.52 \times \dfrac{68.4}{342} = 0.104$$

Boiling point of the solution = 373 + 0.104 = 373.104

(iv) $m = \dfrac{68.4}{342}$, $K_f = 1.86$. Substituting the values in the eq. $\Delta T_f = k_f\, m$

$\Delta T_f = 1.86 \times \dfrac{68.4}{342} = 0.372 \quad \therefore$ Freezing point of the solution = 273 − 0.372 = **272.628 K**

Problems for Practice

1. Naphthalene (F.P. 80.1°C) has a molal freezing point constant of 6.82°C m^{-1}. A solution of 3.2 g in 100 g of naphthalene freezes at a temperature of 0.863°C less than pure naphthalene. What is the molecular formula of sulphur in naphthalene? **[Ans. S_8]**

2. The latent heat of fusion of ice is 79 .7 cal/g. Calculate the molal depression constant of water. **[Ans. 1.86 K kg mol^{-1}]**

3. In winter season, ethylene glycol is added to water so that water may not freeze. Assuming ethylene glycol to be non-volatile, calculate the minimum amount of ethylene glycol that must be added to 6.0 kg of water to prevent it from freezing at −0.30°C. The molal depression constant of water is 1.86 kg mol^{-1}. **[Ans. 60 g]**

4. The freezing point of pure benzene is 5.40°C. A solution containing 1 g of a solute dissolved in 50 g of benzene freezes at 4.40°C. Calculate the molecular mass of the solute. The molecular depression constant of benzene with reference to 100 g is 50°C.

[**Ans.** 100]

5. An aqueous solution contains 5% by mass of urea and 10% by mass of glucose. What will be its freezing point? Molal depression constant of water is 1.86°C/m. [**Ans.** –3.039°C]

6. An aqueous solution freezes at –0.2°C. What is the molality of the solution? Determine also
   (*i*) elevation in the boiling point
   (*ii*) lowering of vapour pressure at 25°C given that K_f = 1.86 kg mol^{-1} and K_b = 0.512° kg mol^{-1} and the vapour pressure of water at 25°C is 23.756 mm.

[**Ans.** Molality = 0.1075, ΔT_b = 0.055°C, Lowering of vapour pressure = 0.046 m]

Applying Chemsitry to Life

Both boiling point elevation and freezing point depression have practical uses. For example, solutions of water and ethylene glycol are used as coolants in automobile engines because the boiling point of such a solution is higher than 100°C, the normal boiling point of water.

In winter, in cold countries, salts like NaCl and $CaCl_2$ are sprinkled on the roads to melt ice or to keep ice from forming on roads or sidewalks. This is because the solution obtained by NaCl or $CaCl_2$ in water has a lower freezing point than pure water, so the formation of ice is inhibited.

4.18 CAUSE OF ABNORMAL MOLECULAR MASSES OF SOLUTES IN SOLUTIONS

In certain situations, we find that the value of colligative property measured is greater or smaller than the value expected. Also since molecular mass is inversely related to the observed colligative property, the molecular mass calculated from the observed value of colligative property by applying the relevant relation, comes out to be different from the calculated molecular mass. We express it by saying that the solute is showing abnormal molecular mass.

Abnormal molecular mass of solute in solution is due to *association* or *dissociation* of molecules in the solution. These are explained as below.

(*a*) **Association:** If the molecules of a solute undergo association in solution, there will be decrease in the number of species. As a result there will be proportionate decrease in the value of each colligative property. The experimental value of molecular mass of the solute will be higher in such a case. This is because molecular mass is inversely proportional to the value of any colligative property.

For example acetic acid and benzoic acid both associate in benzene. Similarly, chloroacetic acid associates in naphthalene. The molecular mass of solute in such cases is higher than the molecular mass obtained from their molecular formulae. Thus the molecular mass of acetic acid (CH_3COOH) in benzene, as determined from freezing point depression is 118 instead of 60.

(b) **Dissociation:** Inorganic acids, bases and salts dissociate or ionise in solution. As a result of this, number of effective particles increases and therefore, the value of colligative property is also increased. Therefore, the experimental value of molecular mass of the solute will be lower in such a case. This is because molecular mass is inversely proportional to the value of any colligative property as already discussed.

Evidently the observed molecular mass will be higher in case of association and lower in case of dissociation.

4.19 VAN'T HOFF FACTOR

To account for abnormal cases, van't Hoff introduced a factor i **known as the van't Hoff factor**

$$i = \frac{\text{Observed colligative property}}{\text{Calculated (normal) coligative property}}$$

Since colligative properties vary inversely as the molecular mass of the solute, it follows that

$$i = \frac{\text{Calculated (normal) molecular mass}}{\text{Observed molecular mass}}$$

van't Hoff (1852–1911)
van't Hoff was a Dutch physical and organic chemist and first winner of the Nobel Prize in Chemistry. He is known for his discoveries in chemical kinetics, chemical equilibrium, osmotic pressure and stereochemistry of organic compounds. van't Hoff's work in these areas helped strengthen the discipline of physical chemistry as it is today.

Relation between degree of association and van't Hoff factor

Let us consider the association of n molecules of a solute A to give one molecule of A_n

$$nA = A_n$$

Let the degree of association be α. If we start with 1 mole of A, the number of moles that associate is α and the number of moles that remain unchanged is $1 - \alpha$.

α moles on association will give α/n moles

Total number of moles at equilibrium = $\frac{\alpha}{n} + (1 - \alpha)$

van't Hoff factor is the ratio of observed colligative property to calculated colligative property. And colligative property is proportional to the number of moles of the solute. Hence

$$\text{van't Hoff factor } (i) = \frac{1 - \alpha + \frac{\alpha}{n}}{1}$$

Relation between degree of dissociation and van't Hoff factor

Let us consider the dissociation of molecule to give n molecules or ions in solution

$$A \rightarrow nB$$

Let us start with 1 mole of A and let α be the degree of dissociation.
At equilibrium
No. of moles of undissociated substance = $1 - \alpha$

No of moles of dissociated substance = $n\alpha$ (one molecule dissociates to give n molecules)
Total number of moles of solute at equilibrium = $1 - \alpha + n\alpha$.

Hence van't Hoff factor $(i) = \dfrac{1 - \alpha + n\alpha}{1}$

Solved Examples—Abnormal Molecular Masses and Van't Hoff Factor

Example 26: 0.1 M solution of KNO_3 has an osmotic pressure of 4.5 atmosphere at 300 K. Calculate the apparent degree of dissociation of the salt.

Solution: $\pi_{obs} = 4.5$ atm, $C = 0.1$ moles/litre

$n = 2$ because one KNO_3 molecule dissociates to give two ions, K^+ and NO^-_3, $T = 300$

$$\pi = C.R.T. = 0.1 \times 0.082 \times 300 = 2.46$$

$$i = \dfrac{\text{Observed osmotic pressure}}{\text{Calculated osmotic pressure}} = \dfrac{4.5}{2.46} = 1.83 \quad ...(i)$$

Also $\quad i = \dfrac{1-\alpha+n\alpha}{1} \quad$ or $\quad i = 1 + \alpha(n-1)$ or $\alpha(n-1) = i - 1$

or $\quad \alpha = \dfrac{i-1}{n-1} \quad ...(ii)$

Substitute the value of i from (1) in (2)

$$\alpha = \dfrac{1.83 - 1}{2 - 1} \quad \text{or} \quad \alpha = 0.83 = 83\%$$

Example 27: Calculate the osmotic pressure of 20% anhydrous calcium chloride solution at 273 K, assuming that the solution is completely dissociated. ($R = 0.082$ lit. atm K^{-1} mol^{-1}).

Solution: Molecular mass of $CaCl_2 = 40 + 71 = 111$. The solution is 20%

∴ Mass of $CaCl_2$ per litre = 200 g, $V = 1$ litre, $T = 273$ K

$$n = \dfrac{\text{Mass of solute}}{\text{Molecular mass of solute}} = \dfrac{200}{111}$$

Substituting the values in the relation

$$\pi = \dfrac{n}{V} \times R.T. \quad \text{or} \quad \pi = \dfrac{200}{111} \times \dfrac{0.082 \times 273}{1} = 40.335 \text{ atm}$$

Since $CaCl_2$ dissolves to give 3 particles for each molecule on complete dissociation ($CaCl_2 \rightarrow Ca^{++} + 2Cl^-$)

∴ Observed osmotic pressure = Normal osmotic pressure × 3
$$= 40.335 \times 3 = 121.00 \text{ atm}$$

∴ Observed osmotic pressure = **121.00 atm.**

Example 28: 2.0 g of benzoic acid dissolved in 25.0 g of benzene shows a depression in freezing point equal to 1.62 K. Molal depression constant (K_f) of benzene is 4.9 K kg mol^{-1}. What is the percentage association of the acid?

Solution: Mass of the solute $w_2 = 2.0$ g, Mass of solvent $w_1 = 25.0$ g
Observed $\Delta T_f = 1.62$ K, $K_f = 4.9$ kg mol^{-1}.

∴ Observed molecular mass of benzoic acid (solute)

$$M_2 = \frac{1000 \times K_f \times w_2}{\Delta T_f \times w_1} = \frac{1000 \times 4.9 \times 2}{1.62 \times 25.0} = 242$$

Calculated molecular mass of benzoic acid (C_6H_5COOH) (By adding atomic masses)
$$= 72 + 5 + 12 + 32 + 1 = 122$$

van't Hoff factor, $\quad i = \dfrac{\text{Calculated mol. mass}}{\text{Observed mol. mass}} = \dfrac{122}{242} = 0.504$

If α is the degree of association of benzoic acid, then

$$2C_6H_5COOH \rightleftharpoons (C_6H_5COOH)_2$$

Initial moles	1	0
After association	$1-\alpha$	$\dfrac{\alpha}{2}$

∴ Total no. of moles after association $= 1 - \alpha + \dfrac{\alpha}{2} = 1 - \dfrac{\alpha}{2}$

$$i = \frac{1 - \dfrac{\alpha}{2}}{1} = 0.504 \quad \text{or} \quad 1 - \dfrac{\alpha}{2} = 0.504$$

$\alpha = (1 - 0.504) \times 2 = 0.496 \times 2 = \mathbf{0.992}$

Per cent association $= 0.992 \times 100 = 99.2$

Problems for Practice

1. 0.5 g KCl (mol. mass 74.5) was dissolved in 100 g of water and the solution originally at 20°C froze at – 0.24°C. Calculate the percentage ionization of the salt. K_f for 1000 g water = 1.86°C. **[Ans. 92%]**

2. Phenol associates in benzene to a certain extent to form a dimer. A solution containing 2×10^{-2} kg of phenol in 1.0 g of benzene has its freezing point depressed by 0.60 K. Calculate the fraction of phenol that has dimerised (K_f for benzene is 5.12 K kg mol^{-1}). **[Ans. 0.733 or 73.3%]**

3. A solution containing 2.2965 g of benzoic acid, C_6H_5COOH, in 20.27 g of benzene froze at a temperature 0.317°C below the freezing point of solvent. The freezing point of pure solvent is 5.5°C and its latent heat of fusion is 30.1 cal/g. Calculate
   (a) the apparent molecular mass of benzoic acid, and
   (b) its degree of association assuming that in this solution it forms double molecules.
   [Ans. 184.2, 67.4%]

4. A certain number of grams of a given substance in 100 g of benzene lower the freezing point by 1.28°C. The same weight of solute in 100 g of water gives a freezing point of – 1.395°. If the substance has normal molecular mass in benzene and is completely dissociated in water, into how many ions does a molecule of this substance dissociate when placed in water? **[Ans. 3]**

5. 0.3015 g of silver nitrate when dissolved in 28.40 g of water depressed the freezing point by 0.212°. To what extent is silver nitrate dissociated? (K_f for water = 1.85°C mol^{-1}). **[Ans. 83.5%]**

Solved Conceptual Examples

Example 29: Out of one molar and one molal aqueous solutions, which one is more concentrated and why?

Solution: 1 M aqueous solution is more concentrated than 1 m aqueous solution. 1 M solution contains 1 moles of the solute in 1000 cc of the solution. This means that the solvent (water) present is less than 1000 cc because 1 mole of the solute is also present in the same volume. Hence 1000 g of the solvent contains more than one mole of the solute.

Example 30: How are the mole fractions of the components in the liquid phase and in the vapour phase related to each other? What result follows from it and what is it called?

Solution:
$$\frac{y_B}{y_A} = \frac{x_B}{x_A} \times \frac{p_B^o}{p_A^o}$$

Thus if B is more volatile than A, $p_B^o > p_A^o$ or $\frac{p_B^o}{p_A^o} > 1$ then $y_B/y_A > x_B/x_A$. This means that the vapour phase is richer in B than the liquid phase from which it vaporises. This is known as **Konowaloff's rule.**

Example 31: Taking a suitable example, explain why some solutions show negative deviations.

Solution: Negative deviations are observed on mixing A and B if $A - B$ attractions are stronger than $A - A$ and $B - B$ attractions. For example, when chloroform and acetone are mixed, new forces of attraction due to hydrogen bonding, start operating between them, as shown below

$$\underset{\text{Chloroform}}{Cl_3C-H}\cdots\cdots\underset{\text{Acetone}}{O=C(CH_3)_2}$$

Thus, the molecules have a smaller tendency to escape into the vapour phase.

Example 32: Out of the various methods of expressing concentration of a solution, which one are preferred over the other and why?

Solution: Molality, mole fraction and mass fraction are preferred over molarity, normality etc. because the former involves masses of the solutes and solvent which do not change with temperature. Molarity and normality involves volumes of the solution which change with temperature.

Example 33: Taking a suitable example, explain why some solutions show positive deviations.

Solution: Solutions show positive deviations if on mixing the two components A and B, the $A - B$ attractions are weaker than $A - A$ and $B - B$ attractions. For example, when chloroform is added to ethanol, the solution shows positive deviations. This is because there is hydrogen bonding in ethanol molecules which can be represented as

$$\overset{\delta+}{H}-\overset{\delta-}{\underset{C_2H_5}{O}}\cdots\cdots\overset{\delta+}{H}-\overset{\delta-}{\underset{C_2H_5}{O}}\cdots\cdots\overset{\delta+}{H}-\overset{\delta-}{\underset{C_2H_5}{O}}\cdots\cdots$$

The chloroform molecules get in between the molecules of ethanol, thereby reducing the attractions between them.

Example 34: **How can you justify that osmotic pressure is a colligative property?**

Solution: Osmotic pressure, $p = \dfrac{n}{V} RT$. Thus it depends only on the number of moles of the solute dissolved in a definite volume of the solution and there is no factor involving the nature of the solute. Hence it is a colligative property.

Example 35: **Of the following properties, which are colligative?**

(*i*) **Refractive index** (*ii*) **Vapour pressure** (*iii*) **Boiling point** (*iv*) **Depression in freezing point** (*v*) **Relative lowering of vapour pressure.**

Solution: Depression in freezing point and relative lowering of vapour pressure.

Example 36: **What is meant by the statement "the osmotic pressure of a solution is 5.0 atmospheres?**

Solution: This means that when the solution is separated from the solvent by a semipermeable membrane, there is a net flow of the solvent molecules from solvent to the solution through the semi-permeable membrane and that the pressure of the hydrostatic column set up is 5.0 atmosphere.

Example 37: **How is relative lowering of vapour pressure related to osmotic pressure?**

Solution: $\dfrac{p^o - p_s}{p^o} = \dfrac{M\pi}{\rho RT}$. Hence relative lowering of V.P. $\propto$ Osmotic pressure.

Example 38: **How is elevation in boiling point related to (*i*) relative lowering of vapour pressure (*ii*) osmotic pressure.**

Solution: The two relations are given as under

(i) $\Delta T_b = \dfrac{1000 K_b}{M_1} \dfrac{\Delta p}{p^o}$ i.e. $\Delta T_b \propto \dfrac{\Delta p}{p^o}$ (ii) $\Delta T_b = \dfrac{1000 K_b}{dRT} \pi$ i.e. $\Delta T_b \propto \pi$

Example 39: **Molecular weight of benzoic acid was determined by freezing point measurements using (i) its solution in water (ii) its solution in benzene. Will the results differ or not? Give reasons for your answer.**

Solution: The results will differ. In water, benzoic acid dissociates. Thus, number of particles increases. As a result observed value of ΔT_f will be more than expected value. Hence, observed molecular mass will be less than the actual value. In benzene, benzoic acid will associate. The situation is just the reverse. No. of particles decreases. Hence observed ΔT_f will be less or molecular mass will be more.

Example 40: **Justify the statement that "dilute solutions behave in the same manner as gases" or "there exists an exact analogy between the dissolved state of the substance and its gaseous state."**

Solution: Dilute solutions obey the same equation as ideal gas equation *i.e.* $PV = nRT$. Thus osmotic pressure of a dilute solution is equal to pressure which the solute would exert if it were a gas at the same temperature and occupied the same volume as the solution.

Example 41: **What are isotonic solutions? How are their molar concentrations related to each other?**

Solution: Isotonic solutions are the solutions which have the same osmotic pressure. As $\pi = CRT$, at constant T, osmotic pressures will be equal when they have same molar concentration.

Example 42: Explain why equimolar solutions of NaCl and cane sugar do not have the same osmotic pressure.

Solution: NaCl is an electrolyte which dissociates to give Na^+ and Cl^- ions. As a result, number of particles increases and so observed osmotic pressure is greater than expected. Canesugar (sucrose) is a non-electrolyte. It does not undergo any association or dissociation. Hence the expected osmotic pressure is observed.

Example 43: Which colligative property is used for finding the molecular masses of polymers and why?

Solution: Polymers give very small values of colligative properties due to their large molecular masses. Δp, ΔT_b, ΔT_f etc. are too small to be measured while osmotic pressure can be measured to acceptable level of accuracy.

Example 44: A solution of 4M HCl is expected to have osmotic pressure 197 atmosphere. How is that bottles containing this solution in the laboratory do not break?

Solution: Osmotic pressure has a meaning only when a solution is separated from the solvent by a semipermeable membrane. Thus, bottles containing this solution in the laboratory do not break.

Example 45: Arrange the following in order of ascending values of their osmotic pressure

(*i*) 0.1 M Na_3PO_4 solution

(*ii*) 0.1 M sugar solution

(*iii*) 0.1 M $BaCl_2$ solution

(*iv*) 0.1 M KCl solution

Solution: 0.1 M Na_3PO_4 solution = 0.3 moles of Na^+ and 0.1 moles of PO_4^{3-} = Total 0.4 mole

0.1 M Sugar solution = 0.1 M Sucrose molecules = Total 0.1 mole

0.1 M $BaCl_2$ solution = 0.1 mole of Ba^{2+} and 0.2 mole of Cl^- = Total 0.3 moles

0.1 M KCl solution = 0.1 mole of K^+ + 0.1 mole of Cl^- = Total 0.2 mole. All these values are per litre of the solution.

Greater the concentration of particles in the solution, higher the osmotic pressure. Hence the values in the ascending order are 0.1 M sugar solution < 0.1 M KCl sol < 0.1 M $BaCl_2$ sol < 0.1 M Na_3PO_4 solution.

Example 46: Account for the following:

(*i*) **Camphor is used as a solvent in the Rast method.**

(*ii*) **Benzoic acid in benzene shows less osmotic pressure than expected.**

(*iii*) **Boiling point of 0.1 m NaCl is greater than 0.1 m glucose solution.**

Solution: (*i*) Camphor is used because it has high molal depression constant. We get a higher value of ΔT_b which helps in obtaining accurate results for molecular mass.

(*ii*) Benzoic acid undergoes association in benzene.

(*iii*) 0.1 m NaCl dissociates to form 0.2 m concentration of particles (Na^+ and Cl^- ions) whereas glucose remains as such.

Example 47: Tell whether the osmotic pressure of M/10 solution of glucose to be same as that of M/10 solution of sodium chloride.

Solution: No. colligative property of a solution depends upon the number of moles of the compound or ions present per litre of the solution. M/10 solution of sodium chloride will contain twice the number of particles, as compared to M/10 glucose solution (one sodium chloride molecule gets dissociated into two ions). Consequently, M/10 sodium chloride will show nearly twice the osmotic pressure.

Example 48: How and why does vapour pressure of a liquid depends upon the temperature?

Solution: Vapour pressure of a liquid increases with the increase of temperature. With the increase of temperature, the kinetic energy of the molecules on the surface increases (K.E = $\frac{3}{2}$RT). More and more molecules leave the surface of the liquid and are converted into vapours, thereby raising the vapour pressure.

Key Terms

- Raoult's law
- Lowering of vapour pressure.
- Henry's law
- Depression in freezing point

Evaluate Yourself

Multiple Choice Questions

1. A liquid boils when its vapour pressure becomes equal to
   (a) one atmospheric pressure
   (b) zero
   (c) very high
   (d) very low

2. The elevation in boiling point is given by the formula
   $$\Delta T = K_b \times \frac{w}{m} \times \frac{1}{W}$$
   where K_b is called
   (a) boiling point constant
   (b) ebullioscopic constant
   (c) molal elevation constant
   (d) all of these

3. The depression in freezing point is measured by using the formula $\Delta T = K_f \times \frac{w}{m} \times \frac{1}{W}$ where K_f is called
   (a) molal depression constant
   (b) freezing point depression constant
   (c) cryoscopic constant
   (d) all of these

4. The ratio of the colligative effect produced by an electrolyte solution to the corresponding effect for the same concentration of a non-electrolyte solution is known as

(a) degree of dissociation
(b) degree of association
(c) activity coefficient
(d) van't Hoff factor

5. Molal elevation constant is the boiling point elevation when .... of the solute is dissolved in one kg of the solvent.
   (a) one gram
   (b) one kg
   (c) one mole
   (d) none of these

6. Freezing point depression is measured by
   (a) Beckmann's method
   (b) Rast's camphor method
   (c) both
   (d) none of these

7. The law of the relative lowering or vapour pressure was given by
   (a) van't Hoff
   (b) Ostwald
   (c) Raoult
   (d) Henry

8. Which of the following is a colligative property?
   (a) Molar refractivity
   (b) Optical rotation
   (c) Depression in freezing point
   (d) Viscosity

9. The study of depression in freezing point of a solution is called
   (a) osmotic pressure
   (b) ebullioscopy
   (c) cryoscopy
   (d) none of these

10. 36 g of glucose (molecular mass 180) is present in 500 g of water, the molality of the solution is
    (a) 0.2
    (b) 0.4
    (c) 0.8
    (d) 1.0

Short Answer Questions

1. State and explain Raoult's law (a) for volatile solute (b) for non-volatile solutes.
2. Define ideal and non-ideal solutions in as many ways as you can. Give at least two examples of each of them.
3. For an ideal solution, derive relationship between the mole fraction of a component in the liquid phase to that in the vapour phase. What do you conclude if out of the two components A and B, B is more volatile?

4. What are colligative properties? How can you say that 'Relative lowering of vapour pressure' is a colligative property?
5. Write the relationship between Relative lowering of vapour pressure and Osmotic pressure. Deduce Raoult's law from it.

Long Answer Questions

1. What are ideal and non-ideal solutions? Derive expressions for free energy change, entropy change, enthalpy change and volume change when n_1 moles of pure component 1 are mixed with n_2 moles of pure component 2 to form an ideal solution.
2. (*a*) Explain how non-ideal solutions show deviations from ideal behaviour or Raoult's law. Draw vapour pressure-composition diagrams and boiling point-composition diagrams for each of them.
   (*b*) What are Azeotropes? Briefly explain the types of azeotropes.
3. Briefly explain at least five different methods of expressing the concentration of a solution. Which out of these are preferred and why?
4. State and explain Raoult's law for volatile solutes as well as for non-volatile solutes. Derive the following:
   (*i*) Expression for total vapour pressure in terms of mole fractions of the components in the vapour phase.
   (*ii*) Relationship between the mole fractions of the components in the liquid phase and those in the vapour phase.
5. Briefly explain the following curves for ideal solutions.
   (*i*) Vapour pressure versus mole fraction of the component in the solution.
   (*ii*) Vapour pressure versus mole fraction of the component in the vapour phase.
6. What are 'colligative properties'? Briefly explain how lowering of vapour pressure is used in the calculation of molecular masses of solutes? Explain one method for the experimental determination of lowering of vapour pressure.
7. Why do we observe abnormal molecular masses of the solutes in certain cases when determined by studying colligative properties? What is van't Hoff factor? How is it used in the determination of degree of association and degree of dissociation of a solute?
8. Why there is a depression in freezing point when a non-volatile solute is dissolved in a solvent? Derive thermodynamically the relation between depression in freezing point and molecular mass of the solute.
9. Derive the following relationship between osmotic pressure and lowering of vapour pressure. Deduce Raoult's law from it.
10. Define osmotic pressure. How is it measured experimentally by Berkeley and Hartley's method? What are the advantages of this method over the other methods?
11. Briefly describe the following methods for the determination of depression in freezing point.
    (*i*) Beckmann's method
    (*ii*) Rast's method
12. Derive thermodynamically the expression for the Relative lowering of vapour pressure or Derive Raoult's law thermodynamically.

13. Define activity and activity coefficient. Describe vapour pressure method for their determination.
14. Derive the relationship between
    (*i*) Depression in freezing point and lowering of vapour pressure.
    (*ii*) Depression in freezing point and osmotic pressure.
15. Derive the relationship between
    (*i*) Elevation in boiling point and relative lowering of vapour pressure.
    (*ii*) Elevation in boiling point and osmotic pressure.

Answers
Multiple Choice Questions

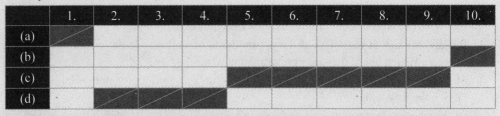

Suggested Readings

1. Engel, T. & Reid, P. Physical Chemistry, Prentice-Hall.

Section III

Laboratory Work

- Calibration of a Thermometer
- Purification of Organic Compounds by Crystallization
- Determination of Melting and Boiling Points
- Paper Chromatography
- Thin Layer Chromatography
- Thermochemistry

Chapter 1
Calibration of a Thermometer

Modern Chemistry Laboratory

Just as your watches can move fast or slow, our laboratory thermometers can also show higher or lower temperatures than the actual. A thermometer can develop fault by, for example, dipping in extremely hot and extremely cold substances, by accidental shocks or rough handling. It therefore becomes necessary to calibrate them frequently. If this is not done, the wrong observations from the thermometer can lead it to wrong results. Thermometers can be calibrated by immersing them in ice-cold water (0°C) or boiling water (100°C). We shall describe here the method of ice-cold water.

Experiment 1: Calibrate a laboratory thermometer.

Requirement: Thermometer 0°-100°C or 0°-350°C, iron stand clamp, beaker, ice, glass stirrer.

Procedure:
1. Take an iron stand with a clamp.
2. Suspend the thermometer from the clamp as shown in Figure 1.1.
3. Take some ice-cubes in a beaker and add a little cold water to it to make a uniform mixture.
4. Place the thermometer in the beaker and stir ice-water mixture with a glass rod.
5. Adjust the height of the thermometer such that its mercury bulb completely dips into the ice.
6. Keep stirring ice-water mixture for about two minutes.
7. Note down the temperature on the thermometer. This temperature is to be taken as 0°C. If the mercury thread in the thermometer is at a higher or lower (minus) level, that much correction is to be applied to the temperature.

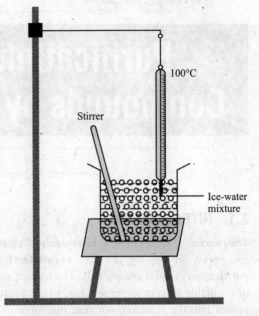

Fig. 1.1: Calibration of laboratory thermometer

Note: Suppose the mercury thread is at 2°, then 2°C is to be taken as 0°C. If in an observation, you note down the temperature as 45°C, then the real temperature is 45 – 2 = 43°C and so on.

Viva-Voce Questions with Answers

1. 1. How does a thermometer start showing a fault in showing correct temperature?
Ans. It could be due to dipping in very hot or very cold substance. It could also be due to rough handling or due to accidental shock.
2. What is meant by calibration of a thermometer?
Ans. Calibration means to find out the figure that should be added or subtracted from the observed temperature to obtain the correct temperature.
3. Which standard substances are taken to calibrate a thermometer?
Ans. Water at 100°C or ice at 0°C.
4. What are different types of available thermometers?
Ans. 1. With the range of 0–100°C for low boiling liquids.
   2. With the range of 0—350°C for solids and high boiling liquids.

Chapter 2
Purification of Organic Compounds by Crystallization

2.1 INTRODUCTION

The process of separation of the crystals of a compound from its saturated solution in a suitable solvent on the cooling is called **crystallisation**. It is the technique generally used for the purification of solid compounds. The impure solid substance is dissolved in minimum amount of a suitable solvent preferably hot, to get a saturated solution. The soluble organic compound passes into the solution, whereas insoluble impurities are left undissolved. These insoluble impurities are removed by filtration as residue from the hot solution. On cooling the hot filtrate, the pure organic compound is deposited as fine crystals which are separated from mother liquor by filtration and dried.

2.2 SELECTION OF THE SOLVENT

The selection of the solvent for crystallization is very important as the efficiency of the process depends upon it. A *suitable solvent is one which dissolves the organic compound completely on heating but on cooling yields back maximum amount as crystals.* So the organic compound should be insoluble or only partly soluble in cold but completely soluble in hot solvent. If the solvent dissolves the organic compound completely at room temperature, then it is not a good solvent. Many organic compounds are insoluble in water even on heating. Water is not used as a solvent in such cases. Some of the other solvents are alcohol, benzene, petroleum ether, acetone, etc. Sometimes if the solubility of organic compound in a solvent is very high at room temperature, two solvents mixed together in an appropriate ratio can be used. One of the solvents used in such a mixture should not dissolve the organic compound. A mixture of alocohol and water is found to be suitable in many cases.

Various steps involved in the process are described below:
1. **Preparation of solution.** A sufficient quantity of the selected solvent is added to the impure organic substance taken in a round bottom or conical flask. The contents are heated to prepare a saturated solution (preferably by using an air condenser or water condenser as most of the organic solvents are inflammable and volatile).
2. **Filtration.** The saturated solution is filtered hot so that organic compound does not crystallize during filtration. To avoid cooling of solution, the filtration can be hastened considerably by using a fluted filter paper or hot water funnel (Fig. 2.1). The hot filtrate is collected in a beaker or china dish.

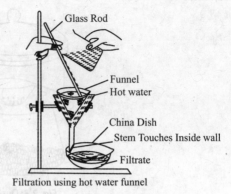

Filtration using fluted filter paper   Filtration using hot water funnel

Fig. 2.1: Filtration

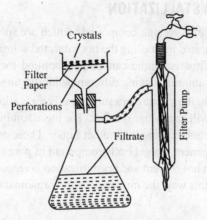

Fig. 2.2: Filtration using a Buchner funnel under reduced pressure

If the organic compound crystallizes during filtration, hot solvent can be passed through funnel again so as to dissolve the crystals deposited in the funnel.

3. **Cooling the filtrate.** The beaker or china dish containing the hot filtrate is placed over ice-water and cooled without disturbing. The pure solid organic compound separates as crystals. Scraching the walls of the vessel with a glass rod or addition of a few crystals of pure compound may facilitate the crystallization.

4. **Separation and drying of crystals.** When a sufficient amount of crystals is obtained, the crystals is separated from mother liquor by Buchner funnel under reduced pressure (Fig. 2.2) using a suction pump. The crystals are washed with small quantity of solvent and then dried by pressing strongly between thick filter paper pads or by heating in steam or electric oven.

5. **Decolorisation using animal charcoal.** Sometimes we get coloured crystals due to presence of some coloured impurities. These can be removed by dissolving the compound in the solvent and boiling with a small quantity of finely powdered animal charcoal for a while and then filtering it hot. The animal charcoal absorbs the coloured impurities and the filtrate obtained is colourless from which pure and colourless or white crystals are obtained.

6. Finally place the crystals in a clean china dish in a desiccator containing anhydrous calcium chloride to protect them from moisture as shown in Fig. 2.3.

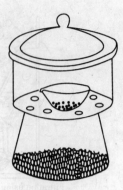

Fig. 2.3: Desiccator

2.3 FRACTIONAL CRYSTALLIZATION

A mixture of two or more solid organic compounds which are soluble in same solvent but to different extent, can be separated by cooling the hot saturated solution and the process is called fractional crystallization. This technique can also be applied for the purification of a solid compound containing solid impurity having different solubility in same solvent.

A hot saturated solution of the mixture in a suitable solvent is allowed to cool. The compound having the lower solubility will crystallize out first. The more soluble compound will crystallize out on further concentrating and cooling the mother liquor. These crops of crystals may contain some crystals of other component. To get each compound in pure state each crop of crystals is dissolved separately in fresh hot solvent and crystallization is repeated till pure crystals of each compound are obtained. In this way, the mixture of solid organic compounds get separated into different pure components.

Experiment 1: Recrystalllize an impure sample of acetanilide from water.

Requirement:

Acetanilide sample:	4 g
Charcoal powder:	0.5 g
250 ml Beaker:	1 No.
250 ml conical flask:	1 No.
Flutted filter paper:	_____
Suction pump with:	
Buchner funnel	

Procedure:

1. Add 4 gm acetanilide sample to about 50 ml water in a conical flask and heat nearly to boiling on an electric hot plate. Give constant movement to the flask to dissolve the solid into water.
2. If the solution is not colourless, add about 0.5 g of animla charcoal (or ordinary charcoal) powder and continue heating with continuous stirring till the colouring matter has been removed.
3. Filter the boiling solution using fluted filter paper or a hot water funnel. Collect the filtrate in a 250 ml beaker and allow to cool for about 30 minutes.

4. Filter the product under suction through a Buchner funnel. Wasth the crystals twice with water.
5. Remove the Buchner funnel from the filration pump and invert it on two sheets of filter paper. Press the product within the folds of filter paper. Allow the crystals to dry in air or place them in an oven at about 70°C.
6. Determine the malting point of the crystals. Pure acetanilide melts at 114°C

   Yield = ... g

   Melting point = ... °C

Experiment 2: Recrystallize an impure sample of naphthalene from alcohol.

Requirement:

Naphthalene:	5 g
Rectified spirit:	25 ml

Procedure:
1. Add 5 g of the commercial sample of naphthalene into 25 ml of rectified spirit taken in a 100 ml conical flask.
2. Add 2-3 small pieces of porous porcelain and fit a reflux condenser.
3. Heat the mixture on a water both or in an electric heating mantle until the solvent starts boiling and all naphthalene has dissolved.
4. If the solution is not colourless, remove the flask from the heat source, remove the condenser and quickly add 1 g of charcoal powder. Shake the flask and fix the condenser back.
5. Heat the flask for 5 minutes more. Extinguish the flame and switch off the electric heating mantle. *This is done as a percaution because the solvent alcohol is inflammable.*
6. Filter the hot solution through a fluted filter paper supported in a funnel with a small and wide neck, collect the filtrate in a beaker and allow to cool for 30 minutes. Reject the residue.
7. Pure naphthalene crystals will appear. Separate the crystals from the mother liquor by filtration at the pump using Bucher funnel. Give washings to the crystals twice with cooled alcohol.
8. Press the crystals with a flat glass stopper. Remove the funnel from the pump and invert it on a filter paper. Dry the crystals by pressing within the folds of filter paper 2 or 3 times. Allow the crystals to dry in air.
9. Determine the melting point of the product. Pure naphthalene melts at 80°C

   Yield = ... g

   Melting point = ... °C

Viva-Voce Questions with Answers

1. Name some methods employed for the purification of organic substances ?

Ans. Crystallization, fractional crystallization, distillation, fractional distillation, steam distillation, vacuum distillation, chromatography.

2. What are crystals ?

Ans. Solid particles, having well defined geometries and shape which separate out on slow and undistributed cooling of a hot saturated solution of the solid are called crystals.

3. How do you remove the coloured impurities present in the crystals ?

Ans. By heating the solution of the coloured crystals with animal charcoal, the coloured impurities are retained by charcoal.

4. What is the function of a hot water funnel in the process of filtration ?

Ans. It keeps the solution hot and thus helps to avert the premature crystallization on the filter paper in the funnel.

5. What is meant by mother liquor ?

Ans. The solution left behind after the removal of crystals is called mother liquor.

6. What is a saturated solution ?

Ans. A solution which contains the maximum amount of the solute that can be dissolved in a given amount of the solvent, at a given temperature, is called a saturated solution.

7. Define crystallization.

Ans. It is the process of obtaining pure crystals of a compound from impure sample.

Chapter 3
Determination of Melting and Boiling Points

3.1 DETERMINATION OF MELTING POINT

After the functional groups in the compound have been identified, the melting point of the solid compound or boiling point of the liquid compound are determined. These m.p. or b.p. are then compared with the tables of m.p. or b.p. of compound to arrive at the compound itself.

Melting point. *Melting point of a substance is defined as the temperature at which the solid and liquid **co-exist**,* i.e. *the temperature at which the solid and liquid forms of a substance are in **equilibrium** with each other at atmospheric pressure.*

Melting Point and Utility

Determination of m.p. has the following utilities:

(*i*) **Criterion of Purity.** Melting point of a substance helps to establish the purity of a crustalline solid. A pure solid has a definite and sharp melting point i.e. it starts melting at a certain temperature and melts completely within 1° of that temperature. Melting points of perfectly pure organic solids are known. Thus, if the observed melting point of a substance tallies with the known standard melting point, the substance is taken to be pure. It should be noted that the presence of **impurities** generally **lowers** the melting point of a solid.

(*ii*) **Identity of a substance.** Determination of melting point is useful in checking and confirming the identity of the given unknown organic compound.

Experiment 1: Determine the melting point of the given organic solid

Requirements: Pyrex beaker or Thiele's tube, a liquid bath (H_2SO_4, liquid paraffin or water) thermometer, capillary tubes, cork, burner, tripod stand, wire gauze, organic solid.

Procedure: The following steps are involved:

1. *Powdering the substance.* Take the organic compound on a clean porcelain plate. Powder it with the help of a spatula (Fig. 3.1)
2. *Filling of capillary tube with the organic compound.* Take a fine capillary tube about 6 cm long. Seal one end of the capillary by heating it gently in the flame. Slide the open end of the capillary tube into powdered organic compound and fill a small amount of the compound into the capillary tube. Tap the sealed end of the capillary tube gently against the porous plate, so that the powdered organic compound comes to the bottom of the sealed end. Repeat this procedure till the compound fills about 0.5 cm of the capillary tube. Prepare two or three capillaries likewise.

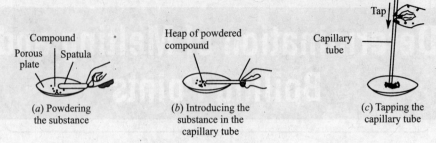

Fig. 3.1: Filling a capillary tube

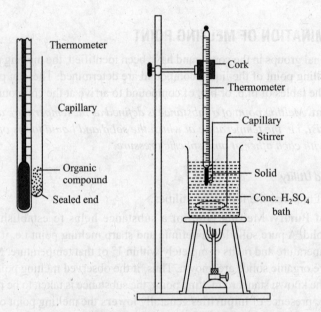

Fig. 3.2: Melting point apparatus

3. *Setting up of the apparatus.* Take a small pyrex beaker. Fill it half with conc. sulphuric acid or liquid paraffin. If the compound has a melting point much lower than 100°C, water may be taken in place of conc. H_2SO_4 or liquid paraffin in the beaker. Place the beaker on a white gauze kept over a triped stand. Fit up the centigrade thermometer as shown in the diagram. Wet the sealed end of the capillary tube with the acid and slip it along the side of the thermometer. The capillary will stick to the thermometer by capillary action due to surface tension of the bath liquid. The open end of the capillary should be well outside the acid layer (see Fig. 3.2).

4. *Determination of melting point.* Now start heating the beaker on a burner and stir the acid gently with a stirrer to keep the temperature uniform. Above 60-70° C, stirring may be stopped because by then convection currents are set up in the liquid and the bath is heated uniformly. Continue heating and watch the substance and the temperature closely. When the substance shows the sign of melting remove the burner.

As soon as the substance melts and becomes transparent, note the temperature.

Cool the acid bath and note the temperature at which the melted substance resolidifies. The mean of the two temperatures gives the correct m.p. of the compound.

Observations

Let the temp. at which the compound melts = t_1 °C

Temp. at which it resolidifies = t_2 °C

Correct melting point = $\dfrac{t_1 + t_2}{2}$ °C

Note: A Thiele's tube which is safer and more convenient may be used in place of the beaker as shown in Fig. 3.3. This convection currents set up in the tube maintain a uniform temperature throughout and thus eliminate the need of a stirrer.

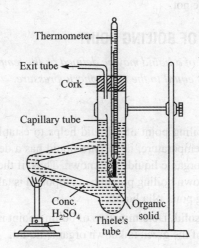

Fig. 3.3: *Thiele's tube for finding melting range*

Useful Hints

1. If the sulphuric acid gets darkened, add a few crystals of KNO_3 or $NaNO_3$.
2. For cooling the hot sulphuric acid, mix with it some fresh sulphuric acid.
3. A pure substance has always a sharp melting point. In case a substance melts in the range, it is impure.
4. Water bath may be used for substances melting below 100°C.

The melting points of some organic solids are listed in Table 3.1.

TABLE 3.1 *Melting Points of Some Organic solids*

Compound	M.P. °C	Compound	M.P. °C
Chloral hydrate	57	Acetanilide	114
Naphthalene	80	Iodoform	119
Acetamide	62	Benzoic acid	121
m-Dinitrobenzene	90	Benzamide	128
Oxalic acid	101	Urea	132

Effect of Impurities on the Melting Point

Impurities always lower down the melting point of a compound. If we find that a sample of a compound is giving lower melting point than found in literature, we can infer that the compound is not pure and there are impurities present in the sample. It has to be crystallised again and again till it gives the literature value of the melting point.

Mixed Melting Point

One of the procedure that is followed in the laboratory to check the purity of a compound is the determination of *mixed melting point*. Here, we mix the sample under investigation or the sample that we have prepared with an equal amount of pure sample. We do a perfect mixing and then determine the melting point. If this mixture gives the melting point as given in literature, our sample is pure, otherwise not.

3.2 DETERMINATION OF BOILING POINT

Boiling Point. *Boiling point of a liquid may be defined as the temperature at which the vapour pressure of a liquid becomes **equal** to the atmospheric pressure.*

Boiling Point and Its Utility

(*i*) **Criterion of purity.** Boiling point of a liquid helps to establish its purity. A pure liquid boils at a certain fixed temperature, i.e. a pure liquid has a definite boiling point. Boiling points of perfectly pure organic liquids are known. Thus, if the observed boiling point of a liquid tallies with the known boiling point, the compound is taken to be pure. Impurities, if present, **raise** the boiling point of a liquid.

(*ii*) **Identity of the organic solid.** Determination of boiling point is also useful in checking and confirming the identity of the given unknown organic liquid.

Experiment 2: Determine the boiling point of the given organic solid.

Requirements: Pyrex glass beaker or Thiele's tube, a small test tube, capillary tube, thermometer, wire gauze, tripod stand, stirrer, liquid paraffin or conc. sulphuric acid and organic liquid.

Procedure: For analytical work, generally a small amount of the organic liquid is provided and its boiling point is determined by **capillary tube method** or **Siwoloboff's method** which involves the following steps:

1. *Fixing the small tube (fusion tube) to the thermometer.* Take a small quantity of the liquid in the small tube so that it is filled to one-third. Take a capillary tube 5-6 cm. in length. Seal it at about 1 cm. from one end of it by bringing into a flame in order to have **constriction.** Drop the capillary tube in the test tube so that its sealed point dips in the liquid. Fix the tube containing the liquid to the thermometer with a rubber band such that the liquid in the tube is in level with the bulb of the thermometer. The rubber band should remain outside the acid bath.
2. *Adjusting the thermometer in an acid bath.* The thermometer along with the test tube is placed in a suitable bath (sulphuric acid or liquid paraffin) provided with a stirrer (Fig. 3.4).
3. *Heating the acid bath.* Heat the acid bath gradually and stir the bath slowly. Watch the capillary. At first a bubble or two will be seen escaping at the lower end of the capillary

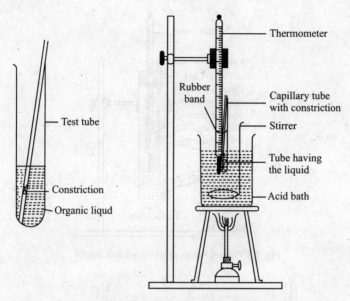

Fig. 3.4: Boiling point apparatus

tube. Finally, when the liquid reaches its boiling point, a rapid and continuous stream of bubbles escapes from it. Note this temperature and remove the burner. Continue stirring. When the stream of air bubbles stops, note again the temperature. The mean of the two temperature gives the boiling point of the liquid.

Observations

Temp. when the continuous stream of bubbles comes out = t_1 °C

Temp. when the stream of bubbles stops = t_2 °C

Correct boiling point = $\dfrac{t_1 + t_2}{2}$ °C

A Thiele's tube which is more safe and handy may also be used in place of the beaker as shown in Fig. 3.5. No stirring is required here as the convection current set-up maintain uniform temperature throughout the acid bath.

Useful Hints

1. If you find that the liquid has risen high in the capillary tube it means the sealing has not been properly done. In such a case, reject the capillary tube and ue a new one.
2. The sealed point (constriction) of the capillary tube should not be out of the liquid.
3. Boiling points obtained by this method are usually a little lower than the standard boiling points because the atmospheric pressure at which the experiment is performed is generally less than the standard pressure of 760 mm.

We can also determine the boiling point of a liquid by distillation. We use a set-up of the apparatus as shown in Fig. 3.6.

In the distillation set up, we have a distillation flask fitted with a thermometer and a water condenser as shown in Fig. 3.4. We take liquid in the distillation flask, fit a cork alongwith the

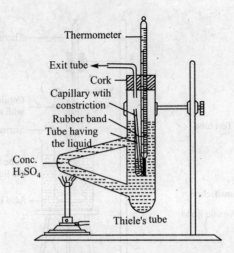

Fig. 3.5: Boiling Points of some organic liquids

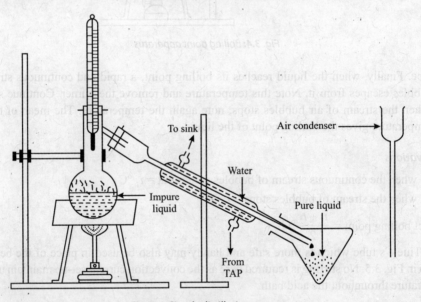

Fig. 3.6: Simple distillation process

TABLE 3.2	Boiling Points of Some Organic Liquids		
Compound	B.Pt. °C	Compound	B. Pt. °C
Acetone	56	Toluene	110.6
Chloroform	61	Acetic acid	118
Methyl alcohol	65	Chlorobenzene	132
Carbon tetrachloride	77	Benzaldehyde	179
Ethanol	78	Aniline	184
Benzene	80		
Cyclohexane	81.4		

thermometer. The delivery tube of the flask goes into the water condenser. A receiver is placed near the other end of water condenser. If the liquid boils at a temperature more than 200°C, we can also use an air condenser in place of water condenser.

Heating is started. Vapours of the liquid rise, enter the condenser and the condensed liquid is collected in the receiver.

Initially, the temperature will rise but will become constant as the boiling point of the liquid has reached. This constant temprature is the boiling point of the liquid.

Viva-Voce Questions with Answers

1. Define melting point.
Ans. The melting point of a substance is defined as the temperature at which the solid and the liquid phase co-exist under atmospheric pressure.
2. Define boiling point.
Ans. The boiling point of a liquid is defined as the temperature at which the vapour pressure of the liquid becomes equal to the atmospheric pressure.
3. What is the effect of impurities on the melting point of a substance?
Ans. The substance will melt over a range of temperature which is lower than its normal melting point.
4. What do you mean by 'sharp melting point'?
Ans. It means that the whole of solid melts at the same temperature.
5. What is the utility of stirring while heating the acid bath during the determination of the melting point?
Ans. It serves to maintain a uniform temperature of the acid.
6. Why is stirring not required in case of a Thiele's tube?
Ans. Because a uniform temperature is automatically maintained by convection currents.
7. What is the effect of impurities on the boiling point of a liquid?
Ans. The liquid will boil over a range of temperature which is higher than its boiling point.
8. Is boiling point as good a criterion for the purity of a liquid as the melting point is for a solid?
Ans. No. This is due to the fact that the effect of impurities is less apparent in case of boiling point. A liquid. In general, boils over a range of a few degrees and varies widely with pressure changes. On the other hand, the effect of change of pressure on the melting point of a solid is almost negligible.
9. How does hydrogen bonding alter the boiling point of a compound?
Ans. The formation of hydrogen bond causes the compound to boil at a higher temperature than its molecular weight would suggest.

Chapter 4

Paper Chromatography

4.1 INTRODUCTION

In paper chromatography, we take a strip of filter paper about 2.5 cm wide and about 15 cm long. Solution of the mixture under examination is prepared in a suitable solvent. A horizontal line is drawn with a pencil at a distance of about 2 cm from the bottom. A drop of the solution is applied with the help of a capillary tube, at the centre of the pencil line (Fig. 4.1)

The strip is then held vertically in a cylinder containing a liquid called as *developer* or *eluent* in such a way that this liquid level lies below the spot applied on the paper. The liquid rises up due to capillary action and along with the liquid the components of the mixture also rise. Due to different adsorption properties of the components, they travel with different speeds and after some time we find as many spots as the number of components in the mixture.

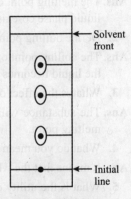

Fig. 4.1: Filter-paper strip for chromatography

The components are easily identified if they are coloured. In case of colourless or white components, the filter paper strip is subjected to *visualisation* i.e. it is treated with a *visualizing agent*. The white or colourless components become visible and can be identified. For separating and identifying different types of substances we use different developers and visualizing agents.

The strip is taken out when the eluent has risen about 15 cm. A pencil mark is put at the level upto which the liquid has risen. The central points of the spots are marked. The distances moved by the various spots and the liquid from the initial position are measured with a scale. R_f (Retention factor) value of a component is given by

$$R_f = \frac{\text{Distance moved by a component}}{\text{Distance moved by the liquid}}$$

The cylinder used for carrying out paper chromatography is shown in Fig. 4.2.

The procedure given above refers to **ascending paper chromatography** i.e. the developer rises upwards from the bottom due to capillary action.

In **descending paper chromatography**, the apparatus used is slightly different as shown in Fig. 4.3 and 4.4.

Here the developer is taken in a small reservoir hung at the top. The upper end of the strip dips in the reservoir and hangs vertically

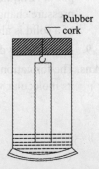

Fig. 4.2: Cylinder for paper chromatography

Paper Chromatography

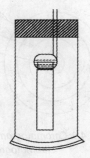

Fig. 4.3: Trough for descending paper chromatography

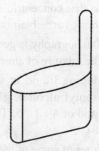

Fig. 4.4: Arrangement for descending paper chromatography

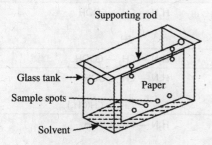

Fig. 4.5: Ascending technique

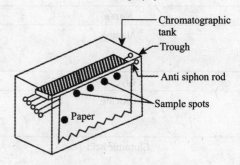

Fig. 4.6: Descending technique

down. The developer travels downwards again by capillary action and the various components move alongwith the developer.

Other types of chrmatographic apparatus to perform paper chromatography are shown below in Fig. 4.5 and 4.6. These are rectangular glass boxes with covers and provision to hold the paper vertically Straight.

4.2 CIRCULAR PAPER CHROMATOGRAPHY OR RADIAL CHROMATOGRAPHY

In this technique, a circular sheet of Whatman filter paper No. 1 is taken. A wick is made out of it as shown in the diagram (Fig. 4.7).

A spot of the solution of the mixture is applied at the centre of the filter paper, with the help of a capillary tube. It is then supported horizontally on a petri dish containing the eluent in such a way that the wick dips in the eluent (Fig. 4.8). The whole thing is covered with a bigger petri dish. The eluent rises up and makes the constituents of the mixture to move radially along the

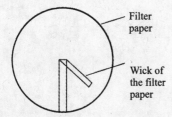

Fig. 4.7: Making wick in the circular filter paper

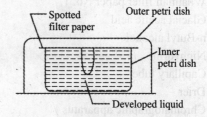

Fig. 4.8: Arrangement of apparatus for radial chromatography

filter paper, so that concentric rings corresponding to different substances are obtained (Fig. 4.9).

Paper chromatography is generally used to identify and separate the mixture of amino acids.

The developing liquid in such experiments is a mixture of n-butyl alcohol, glacial acetic acid and water in the ratio of 4 : 1 : 5. The visualising agent is ninhydrin.

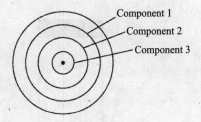

Fig. 4.9: Chromatogram of radial paper chromatography

The structures of some of the amino acids are given below:

Name	Structure
Alanine	$H_2N - CH - COOH$ $\quad\quad\quad\;\;|$ $\quad\quad\quad\; CH_3$
Glycine	$H_2N - CH_2 - COOH$
Leucine	$H_2N - CH - COOH$ $\quad\quad\quad\;\;|$ $\quad\quad\quad CH_2 - CH(CH_3)_2$
Glutamic acid	$H_2N - CH - COOH$ $\quad\quad\quad\;\;|$ $\quad\quad\quad CH_2 - CH_2 - COOH$
Aspartic acid	$H_2N - CH - COOH$ $\quad\quad\quad\;\;|$ $\quad\quad\quad CH_2 - COOH$

Experiment 1: Separate a mixture of phenylalanine and glycine and calculate thier R_f values by ascending paper chromatography.

Theory: R_f values of the amino acids phenylalanine and glycine are determined by paper chromatography. The mixture is also taken simultaneously. The components of the mixture get separated into two spots, one corresponding to phenylalanine and the other corresponding to glycine.

The R_f values of the spots obtained are calculated. They are matched with the Rf values of phenylalanine and glycine. Thus, the separation and identification of the components of the mixture is possible.

Requirement:
1. Whatman filter paper No. 1.
2. Glacial acetic acid
3. n-Butyl alcohol
4. Ninhydrin
5. Capillary tube
6. Drier
7. Chromatographic apparatus
8. Phenylalanine
9. Glycine

Procedure:

1. Cut strips of about 20 cm x 3 cm from Whatman filter paper No. 1.
2. Prepare concentrated solutions (about 2 ml) of phenyl alanine, glycine and a mixture of the two in water.
3. Draw a horizontal line with pencil about 2 cm from the bottom.
4. Apply thin spots of the three solutions on the pencil line by means of a capillary tube. The spots should be well separated from one another. Remember the positions of various components.
5. Dry the spots with the help of an air drier.
6. Lower the filter paper strip into chromatography jar in such a way that it does not touch the sides of the jar and hangs vertically straight (Fig. 4.10).
7. Take the developing liquid i.e. mixture of n-butanol. acetic acid and water in the ratio of 4 :1: 5 in such quantity in the jar that the liquid does not touch the spots directly.
8. Allow the developing liquid to rise till it has travelled a distance of about 15 cm.
9. Take out the filter strip from the jar. Mark the level upto which the liquid has risen. Dry the strip with an air drier.
10. Spray ninhydrin solution on the strip. The spots of phenylalanine and glycine will become distinct. The mixture wiil get separated into two spots (Fig. 4.11).

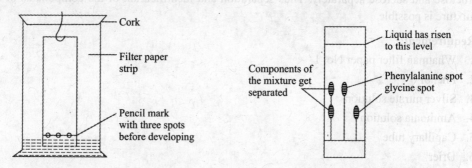

Fig. 4.10: Chromatography jar   *Fig. 4.11: Chromatogram*

Calculations:

The centres of various spots are marked with the pencil.

Let the distance moved by glycine = d_1 cm

Distance moved by phenylalanine = d_2 cm

Distance moved by the lower spot from the mixture = d_3 cm

Distance moved by the upper spot from the mixture = d_4 cm

Distance moved by developing liquid = d_5 cm

R_f value of glycine = $\dfrac{d_1}{d_5}$

R_f value of phenylalane = $\dfrac{d_2}{d_5}$

R_f value of lower spot from the mixture $= \dfrac{d_3}{d_5}$

R_f value of the upper spot from the mixture $= \dfrac{d_4}{d_5}$

It will be found that R_f value of the upper spot from the mixture will be equal to the R_f value of phenyl alanine. Thus this spot corresponds to phynyl alanine. Also, it is found that R_f value of the lower spot from the mixture is equal to the R_f value of glycine. Thus this spot corresponds to glycine. In this way indentification of the components of the mixture becomes possible.

Experiment 2: Separate a mixture of two amino acids by horizontal paper chromatography.

Follow the procedure as given in Art. 4.2 (Figs. 4.7 to 4.9)

Experiment 3: Separate a mixture of fructose and sucrose and calculate R_f values by paper chromatography.

Theory: The R_f values of the sugars fructose and sucrose are determined by performing paper chromatography. The mixture of the two is also taken separately. Components of the mixture get separated into two spots, one corresponding to fructose and the other corresponding to sucrose.

R_f values of the spots obtained are calculated. They are matched with the R_f values of fructose and sucrose separately. Thus separation and identification of the components of the mixture is possible.

Requirement:
1. Whatman filter paper No. 1
2. Phenol
3. Silver nitrate solution
4. Ammonia solution
5. Capillary tube
6. Drier
7. Chromatographic apparatus
8. Fructose
9. Sucrose

Procedure:
1. Cut out strips of 20 cm × 3 cm from Whatman filter paper No. 1.
2. Prepare concentrated solutions (about 2 ml) of fructose and sucrose and a mixture of the two in water.
3. Draw a horizontal line with pencil about 2 cm from the bottom of filter paper strip.
4. Apply thin spots of the three solutions on the pencil line by means of a capillary tube. Remember the positions of different components. The spots should be well separated from one another.
5. Dry the spots with the help of an air drier
6. Take the developing liquid, a 5% solution of phenol in the chromatography jar as in the previous experiment. The level of developing liquid in the jar should be about 1 cm.

7. Lower the filter paper strip with dried spots of sugars in such a way that it does not touch the sides of the jar and hangs vertically down as shown in Fig. 4.10 in the previous experiment.
8. Allow the developing liquid to rise till it has travelled a distance of about 12-15 cm.
9. Take out the filter paper strip from the jar. Mark the level with a pencil upto which the liquid has risen. Dry the strip with an air drier.
10. Spray ammoniacal silver nitrate solution on the strip.
11. The spots of fructose and sucrose will become visible. The spot containing the mixture will get separated into two spots. The upper spot corresponds to sucrose while the lower spot corresponds to fructose.
12. Note: Ammoniacal silver nitrate solution may be prepared by adding ammonia solution to a solution of silver nitrate taken in test tube. A precipitate is obtained. Add ammonia solution in small lots and shaking of the test tube till the precipitate dissolves again and a clear solution is obtained.

Calculations

Centres of different spots are marked with a pencil.
Let the distance moved by fructose (lower spot) = d_1 cm
And the distance moved by sucrose (upper spot) = d_2 cm
Distance moved by lower spot from the mixture = d_3 cm
Distance moved by upper spot from the mixture = d_4 cm
Distance moved by developing liquid = d_5 cm

R_f value of fructose = $\dfrac{d_1}{d_5}$

R_f value of sucrose = $\dfrac{d_2}{d_5}$

R_f value of the lower spot from the mixture = $\dfrac{d_3}{d_5}$

R_f value from the upper spot from the mixture = $\dfrac{d_4}{d_5}$

It will be found that R_f value of the lower spot from the mixture is equal to the R_f value of the spot for fructose while R_f value of upper spot from the mixture is equal to R_f value of the spot for sucrose. Thus, the components of the mixture have been identify. In general, R_f values for the sugars increase in the order:

fructose < lactose < glucose < maltose < sucrose

Viva-Voce Questions with Answers

1. **How do we carry out paper chromatography?**
Ans. In paper chromatography, we take a strip of filter paper about 2.5 cm wide and 15 cm long. A horizontal line is drawn with a pencil about 2 cm from the bottom. A drop of the solution is applied on this line and the paper is lowered in a jar containing the developing liquid.

2. How do we visualise the components on the filter paper if they are colourless?
Ans. We use a visualizing agent in order to see the components.
3. What is meant by R_f value of a component of a mixture?
Ans. R_f value of component is given by

$$Rf = \frac{\text{Distance moved by a component}}{\text{Distance moved by the liquid}}$$

It is a characteristic property of a substance.
4. Which developing agent is used for the separation of a mixture of amino acids?
Ans. We use a mixture of n-butyl alcohol, glacial acetic acid and water in ratio of 4:1:5.
5. Which visualizing agent is used in the identification of components of a mixture of amino acids?
Ans. Ninhydrin is used as a visualizing agent.
6. Draw the structures of glutamic acid and aspartic acid
Ans. Glutamic acid

$$H_2N - CH - COOH$$
$$|$$
$$CH_2-CH_2 - COOH$$

Aspartic acid

$$H_2N - CH - COOH$$
$$|$$
$$CH_2-COOH$$

7. Which developing liquid is used in the separation of a mixture of sugars?
Ans. Phenol-water mixture.
8. Which visualizing substance is used for the identification of a mixture of sugars?
Ans. Ammoniacal silver nitrate solution is used as visualizing agent.

Chapter 5
Thin Layer Chromatography

5.1 INTRODUCTION

Chromatography is a modern and versatile techniques used for separation and identification of organic compounds. The extent to which organic compounds get adsorbed on a solid surface depends widely on the structure of compound and nature of solid surface. This difference in adsorption tendencies is used for the separation and identification of organic compounds.

The separation of components present in a mixture is achieved by allowing them to pass in a moving solvent phase against a stationary adsorbent phase. Depending upon the extent of adsorption to the stationary phase, each component has a different rate of passage and these rates will be slower than that of the mobile solvent phase. In this way, the components will be separated into different zones or bands called **chromatogram**. The technique is known as **chromatography**.

There are various types of chromatography depending upon the physical state of phases involved. If the stationary adsorbent phase is a solid and mobile solvent phase a liquid, then it is called liquid-solid chromatography. The extent of adsorption of an organic compound depends on the polarity of its molecule, the activity of adsorbent and the polarity of mobile phase. Usually, a more polar organic molecule is adsorbed more strongly on a solid surface. The adsorptive power or the activity of adsorbent depends upon the type of material and mode of its preparation. Most widely used adsorbents are silica gel and alumina. Some other adsorbents are cellulose, starch, sugar, magnesium silicate, activated charcoal, etc. Choice of proper adsorbent depends on the nature of the compounds to be chromatographed. As the liquid phase crosses over the stationary adsorbent phase, the molecules of organic compounds experience a competition of attraction between moving solvent molecules and the stationary solid so that their rate of movement will be slower than the solvent and will differ for each of the components of the mixture. The mobile liquid phase is also known as **eluent** and the process as **elution**.

Three important chromatographic techniques that are employed in a chemistry laboratory are:
(*i*) **Thin-layer chromatography**
(*ii*) **Paper chromatography**
(*iii*) **Column chromatography**

Only thin-layer chromatography is discussed in this chapter in detail.

5.2 THIN-LAYER CHROMATOGRAPHY (TLC)

It is one of the various types of adsorption chromatography. In thin layer chromatography a thin layer of adsorbent on a smooth surface is used as stationary phase and chromatogram is developed by upside capillary movement of the solvent through thin layer of adsorbent.

Thin layer chromatography has the following advantages over other types:
(i) *Wide choice of stationary phase is possible.*
  (ii) *Ensures rapid separation of even minute quantities of the components.*
  (iii) *Early recovery and visualisation of separated components can be achieved.*
  (iv) *Spots of solution remain compact on the plates.*

Basic principle of thin layer chromatography and column chromatography is the same. Thin layer chromatography however, differs from the column chromatography in the following respects:
(i) Unlike column chromatography, TLC is carried out with the mobile phase ascending the thin layer of adsorbent by capillary action rather than descending through the column due to gravity.
(ii) Whereas in column chromatography, the stationary phase is packed in a column, in TLC it is coated on a glass plate as a uniform thin layer.

5.3 EXPERIMENTAL PROCEDURE FOR TLC

In thin layer chromatography, a glass plate is coated with an extremely thin layer of an adsorbent such as alumina or silica mixed with some binding material such as calcium sulphate. A drop of the solution to be analysed is applied as a small spot on one end of the plate and dried. The plate is then placed in a jar containing a small quantity of some suitable solvent known as developer or eluent. As a result of the capillary action, the solvent rises up the plate and the components of the mixture move up to varying heights depending upon the extent of their adsorption on the surface. As the solvent approaches the top end, the plate is removed from the jar and the position of the solvent front and of the components are marked. If the components are not coloured, the developed and dried plate, known as chromatogram, is placed in a jar containing some iodine crystals. Brown spots, corresponding to the positions of the components, appear on the chromatogram.

R_f, *retention factor* is given by the following relation.

$$R_f = \frac{\text{Distance moved up by the component}}{\text{Distance moved up by the solvent}}$$

Different substances have characteristic R_f values which, in turn, depend upon a number of factors such as nature of substance, nature of solvent, temperature, thickness of the layer, etc. It may be mentioned here that since the solvent always travels a larger distance, R_f value is always less than 1.

5.4 R_{st} VALUES

In order to eliminate most of these factors, the chromatogram of the pure substance is also developed along with that of the given substance (sample) and that of the pure substance is called R_{st} value *i.e.,*

$$R_{st} = \frac{R_f \text{ of the given substance}}{R_f \text{ of the standard (pure substance)}}$$

Unlike R_f value, R_{st} values may be greater, less than or equal to 1.

5.5 ADSORBENTS AND ELUENTS

Most commonly employed adsorbents in TLC are silica gel and alumina whereas the commonly used binder is plaster of Paris or gypsum. However various other adsorbents can also be employed. These are listed below in the increasing order of their adsorptive power:

1. Sucrose, starch
2. Talc
3. Sodium carbonate
4. Calcium carbonate
5. Calcium phosphate
6. Magnesium carbonate
7. Magnesia
8. Lime
9. Activated silicic acid (silica gel)
10. Activated magnesium silicate
11. Activated alumina
12. Fuller's earth.

Increasing Adsorptive power

Various solvents used as eluting agents are listed below in the order of their increasing eluting power or polarity:

1. Petroleum, ether
2. Carbon tetrachloride
3. Cyclohexane
4. Benzene
5. Diethyl ether
6. Chloroform
7. Acetone (anhydrous)
8. Ethyl acetate
9. Ethanol
10. Methanol
11. Pyridine
12. Glacial acetic acid.

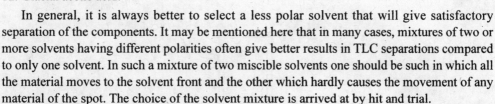
Increasing Polarity

In general, it is always better to select a less polar solvent that will give satisfactory separation of the components. It may be mentioned here that in many cases, mixtures of two or more solvents having different polarities often give better results in TLC separations compared to only one solvent. In such a mixture of two miscible solvents one should be such in which all the material moves to the solvent front and the other which hardly causes the movement of any material of the spot. The choice of the solvent mixture is arrived at by hit and trial.

Experiment 1: To determine the number of components present in the given organic mixture by thin layer chromatography.

Requirement: Glass plates, wide mouthed bottle with air-tight lids, two small glass jars with lids, pair of tongs, silica gel, chloroform, iodine crystals, experimental substance.

Procedure: The experiment is carried out in the following steps:

1. Preparation of a slurry of the adsorbent.

Take about 100 ml of chloroform in a wide-mouthed bottle having an air tight stopper. Now add about 30 g of silica gel in small instalments with a swirling motion, keeping the bottle tightly stoppered.

A slurry of alumina or cellulose if required can be obtained similarly by adding about 30 g of alumina or cellulose in rectified spirit or acetone respectively.

2. Preparation of the chromatographic plate.

(*i*) Take a glass plate or a microscopic slide and clean it thoroughly with distilled water. Dry it.

(*ii*) Shake the slurry bottle giving it a swirling motion. Now remove the stopper. Holding the two strips together with a pair of tongs obtain a uniform coating of the slurry on one side of each of the slides by alternately dipping and removing the slides quickly. Do not dip top one cm of the slides. Allow the slurry to drain. Holding only the top edges, separate apart the slides and place on a filter paper, with coated sides upwards, for drying. Stopper the slurry bottle again (Fig. 5.1).

(*iii*) Scrap off the excess adsorbent from the edges of the plate (Fig. 5.2).

3. Spotting of the chromatographic plate.

(*i*) Prepare a very dilute solution of the experimental mixture in some volatile solvent (say, benzene).

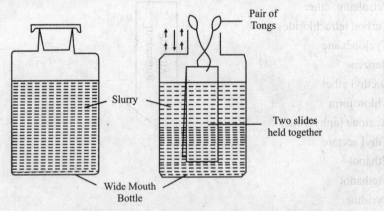

Fig. 5.1: *Preparation of chromatographic plates*

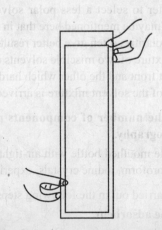

Fig. 5.2: *Removing the excess adsorbent from the edges*

Fig. 5.3: Applying the samples on the chromatographic plate

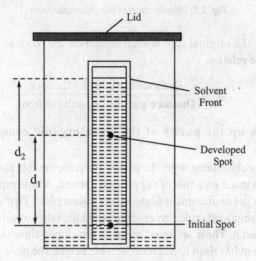

Fig. 5.4: Developing the chromatogram

(*ii*) With the help of a capillary tube apply a small spot of the above solution on the chromatographic plate about 1 cm from one end (Fig. 5.3) and let the spots dry.

4. Development of the chromatogram

(*i*) Place the slide (chromatographic plate) in the jar containing about 5 ml of solvent taking care that the spotted portions do not dip in the solvent, (Fig. 5.4).

(*ii*) When the solvent front has moved up a little below the top, remove the slide. Mark the solvent front with the marking pencil and also the spots (if coloured) and let them dry in air.

5. Visualisation of the chromatogram

If the spots are not coloured, place the slide in another jar having a few crystals of iodine & cover the jar with the lid within a few minutes, dark brown spots appear on the chromatogram. (Fig. 5.5).

6. Determining the number of components in the given mixture

Count the number of spots developed on the chromatogram. This gives the number of components present in the given mixture. Also measure the distance between the centres of

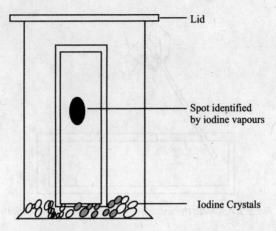

Fig. 5.5: Visualization of the chromatogram

each developed spot & the original spot & then determine the R_f values of the components of mixture by applying the relation.

$$R_f = \frac{\text{Distance travelled by the spot}}{\text{Distance travelled by solvent front}}$$

Experiment 2: Check up the purity of the given organic compound by thin layer chromatography.

Theory. In thin layer chromatography the purity of a compound is checked up by comparing its R_f (Retention factor) value with that of the pure compound. A uniform thin layer of a suitable absorbent such as silica gel or alumina is obtained on a glass plate. Two small spots, one of very dilute solution of the compound under investigation and the other of dilute solution of the pure compound are applied on it. These are dried and then developed simultaneously with a suitable solvent. When the solvent has risen to a certain suitable height, the plate is removed, the solvent front and spots are marked and then dried carefully. If the spots are not coloured, these are visualised with the help of I_2 crystals. The R_f values for both the compounds are determined separately by applying the relation.

$$R_f = \frac{\text{Distance moved up by the spot}}{\text{Distance moved up by the solvent front}}$$

If it comes out to be the same in both the cases then the given compound is pure.

Requirement: Glass plates, wide-mouthed bottle with air-tight stopper, two small glass jars with lids, a pair of tongs, a thin capillary tube, filter paper sheet, silica gel, chloroform, acetone, benzene, iodine crystals, the experimental and the pure compounds.

Procedure: The experiment is carried out in the following steps:

1. Preparation of a slurry of the adsorbent. Proceed as in experiment 1.
2. Preparation of the chromatographic plate. Proceed as in experiment 1.
3. Development and visualization are also done as in experiment 1.

Experiment 3: Separate a mixture of *o*- and *p*-nitrophenol by thin layer chromatography (TLC).

Requirement: Glass plates, wide mouth bottles with air-tight lids, two small glass jars with lids, pair of tougs, silica gel, chloroform toluene.

Procedure:

1. **Preparation of a slurry of the adsorbent (silica gel)**
   Proceed as explained in Experiment 1.
2. **Preparation of chromatographic plates**
   Proceed as in Experiment 1.
3. **Spotting of chromatographic plates**
   (i) Dissolve 0.5 g of the mixture of o- and p-nitrophenol in 2 ml of ethanol.
   (ii) With the help of a capillary tube, apply a small drop of the solution on the chromatographic plate about 1 cm from one end and let the spot dry.
4. **Elution**
   Toluene will be used as eluting agent.
5. **Development of the chromatogram**
   (i) Place the slide in the jar containing about 5 mL of toluene, taking care that the spotted portion does not dip into toluene.
   (ii) When the solvent front has moved up a little below the top, remove the slide. Mark the solvent front (Fig. 5.6).

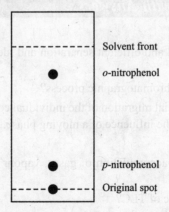

Fig. 5.6: TLC chromatogram of o- and p-nitrophenol

6. It will be observed that the original yellow spot will separate into two yellow spots corresponding to the two constituents of the mixture.
7. The upper spot corresponds to *o*-nitrophenol and the lower spot corresponds to *p*-nitrophenol.
8. Measure the distance travelled by the solvent from the orginal spot.
9. Measure the distance travelled by *o*- and *p*-nitrophenol spots from the original spot with the help of a scale and record in the table.

Solvent	Distance travelled by (in cm)		R_f value of	
	o-nitrophenol	p-nitrophenol	o-nitrophenol	p-nitrophenol

$$R_f \text{ value of } o\text{-nitrophenol} = \frac{\text{Distance travelleed by } o\text{-nitrophenol}}{\text{Distance travelled by solvent}}$$

$$R_f \text{ value of } p\text{-nitrophenol} = \frac{\text{Distance travelleed by } p\text{-nitrophenol}}{\text{Distance travelled by solvent}}$$

Viva-Voce Questions with Answers

1. **What is chromatography?**
Ans. It is a technique for rapid and efficient separation and identification of components of a mixture.

2. **What is the principle of chromatographic process?**
Ans. It is based on the differential migration of the individual components of a mixture through a stationary phase under the influence of a moving phase.

3. **What is adsorption?**
Ans. It is the phenomenon of accumulation of gases, vapour or liquids by a solid or liquid surface or interface.

4. **What is the basic principle of TLC?**
Ans. It is based on the fact that the components of a mixture get adsorbed on a given adsorbent to different extents.

5. **What types of solvents are employed in chromatography?**
Ans. Solvents having low viscosities are employed in general. This is due to the fact that the rate of flow of a solvent varies inversely as its viscosity.

6. **What is meant by the term 'development'?**
Ans. It is the process by which the chromatographic plate, after applying the sample, is placed in a suitable solvent and different constituents get separated.

7. **What are moving and stationary phase in TLC?**

Ans. The stationary phase is a thin layer of the adsorbent on a glass plate and moving phase is a suitable developing solvent.

8. **What is a chromatogram?**

Ans. The developed and dried chromatographic plate is called a chromatogram.

9. **What is meant by 'R_f value'?**

Ans. R_f (retention factor) of a substance is defined as the ratio of the distance moved up by the solute from the point of its application to the distance moved up by the solvent from the same point.

10. **On what factors does R_f value depend?**

Ans. It depends upon a number of factors. Some of these are quality of the adsorbent material, activation grade of the adsorbent used, thickness of the adsorbent layer, quality of the solvent, concentration of the substance applied, presence of impurities, etc.

11. **Mention some of the adsorbents employed in TLC.**

Ans. Starch, calcium carbonate, calcium oxide, activated silica gel, activated alumina, Fuller's earth.

12. **What is visualization?**

Ans. If the components of the mixture are not coloured, their positions are not visible on the chromatogram. They may then be located with the help of some suitable reagents such as iodine. This process is called visualization.

13. **What are the different procedures for visualizing the white spots of APC?**

Ans. (*i*) By spraying fluorescene solution on the slide and viewing under a UV lamp.

(*ii*) Subjecting them to I_2 vapours in a closed jar.

Chapter 6

Thermochemistry

6.1 ENTHALPY OF SOLUTION

When a solid is dissolved in water or any other suitable solvent, some amount of heat is evolved or absorbed. This is called enthalpy or *heat of dissolution* or *heat of solution*. Enthalpy of solution may be defined as the amount of heat change accompanying the dissolution of one mole of the substance in large excess of water so that further dilution does not produce heat effects.

$$\text{Solid} + \text{Solvent} \longrightarrow \text{Solution} ; \Delta H$$

ΔH is the heat of solution.

If we try to look into the causes that produce heat change during the dissolution process, we will notice that on adding water to a solid (particularly an electrolyte), the ionisation takes place. Ions are separated from each other in solution. This requires energy which is drawn from the surroundings. Hence solution becomes colder *i.e.*, on dissolving a solid into a solvent, the temperature falls.

But there is another factor, hydration of ions, that accompanies the dissolution process. The cations and the anions produced have a tendency to get hydrated *i.e.*, to have a coating of water around it. This is an exothermic process unlike the ionisation phenomenon which is endothermic. The net heat of solution observed will, therefore, be dependent upon the heating and cooling effects of the two processes.

For non-electrolytic solids which do not ionise, lattice energy of the crystals determines the heat of solution.

For majority of solids, heat of reaction ΔH is positive *i.e.*, heat is absorbed and cooling is produced. For some solids, particularly the anhydrous solids, heat of reaction is negative *i.e.*, heat is evolved. $BaCl_2$ (anhydrous) for example when dissolved in water evolves heat.

The thermochemical equations for the dissolution of KCl and $CuSO_4$ in water may be represented as

$$KCl (s) + aq \longrightarrow KCl (aq), \Delta H = +18.6 \text{ kJ}$$
$$Cu SO_4 (s) + aq \longrightarrow CuSO_4 (aq), \Delta H = -66.5 \text{ kJ}$$

It is interesting to note that salts like copper sulphate, calcium chloride, etc. when present in the hydrated state (as $CuSO_4.5H_2O$, $CaCl_2.6H_2O$, etc.) dissolve with absorption of heat. For example,

$$CuSo_4.5H_2O + aq \longrightarrow CuSO_4 (aq), \Delta H = +11.7 \text{ kJ}$$

It can therefore be generalized that the process of dissolution is usually endothermic for
(i) salts which do not form hydrates like NaCl, KCl, KNO$_3$, etc.
(ii) hydrated salts like CuSO$_4$.5H$_2$O, CaCl$_2$.6H$_2$O, etc.

How to Calculate Enthalpy or Heat of Hydration of a Salt?

Enthalpy of hydration of a salt (which forms hydrates) can be calculated by determining the enthalpy of solution (or dissolution) of the anhydrous and hydrated samples of the salt. We illustrate this by taking the example of copper sulphate. Let us say the enthalpy of solution of anhydrous and hydrated samples of copper sulphate are ΔH_1 and ΔH_2 respectively. Then, the thermochemical equations for the two processes are

$$CuSO_4 (s) + aq \longrightarrow CuSO_4 (aq), \Delta H_1 \quad \ldots(i)$$
$$CuSO_4.5H_2O (s) + aq \longrightarrow CuSO_4 (aq), \Delta H_2 \quad \ldots(ii)$$

We want to find out the enthalpy of the following reaction (enthalpy of hydration)

$$CuSO_4 + 5H_2O (l) \longrightarrow CuSO_4.5H_2O (s), \Delta H \quad \ldots(iii)$$

This can be obtained by performing (i) – (ii) i.e., by subtracting eq. (ii) from eq. (i)

$$CuSO_4 (s) + aq \longrightarrow CuSO_4.5H_2O (s), \Delta H_1 - \Delta H_2$$

or $\quad CuSO_4 (s) + 5H_2O (l) \longrightarrow CuSO_4.5H_2O (s), \Delta H_1 - \Delta H_2$

(Please note water may be added as per the difference in the molecular formulas of the reactants and molecules without making change in the heat of reaction)

∴ Enthalpy of hydration of copper sulphate = $\Delta H_1 - \Delta H_2$

It may be noted that ΔH_1 comes out to be negative while ΔH_2 is positive. These signs (+ or –) are to be considered while computing the heat of hydration of copper sulphate.

Enthalpy of solution for some of the substances are given below in Table 6.1.

TABLE 6.1

Solid	ΔH in k Joules per mole
KCl	+ 18.60
BaCl$_2$.2H$_2$O	+ 20.63
BaCl$_2$	– 11.30
NH$_4$Cl	+ 16.23
KNO$_3$	+ 35.65
CuSO$_4$ Anhyd.	– 66.50
CuSO$_4$.5 H$_2$O	+ 11.70

6.2 ENTHALPY OF NEUTRALIZATION

Acids and bases react with each other forming neutral salts and water. This reaction is called neutralization. Heat change accompanying the neutralization processes is called *enthalpy of neutralization*.

Enthalpy of neutralization may be defined as the heat evolved when one gram equivalent of a strong acid (or base) is neutralized by excess of base (or acid) in dilute solution. For strong acid and strong base, enthalpy of neutralization is –56.4 kJ. Reaction between a strong acid and a strong base is written as

$$H_3O^+ + OH^- \longrightarrow 2H_2O; \Delta H = -56.4 \text{ kJ}$$

The above value is valid for any acid or base which ionises completely to give hydrogen and hydroxyl ions. Hydrogen ions normally get hydrated to give hydronium ions and hence they are written as such in the neutralization reaction. In fact a neutralization process is the interaction between hydrogen and hydroxyl ions. Strong acids like H_2SO_4, HCl or HNO_3 are completely ionised in solution to give H^+ ions. Similarly strong bases like NaOH, KOH and $Ba(OH)_2$ are completely ionised to give OH^- ions. Hence any one of the above acids with one of the above bases will give heat of neutralization equal to –56.4 kJ. Since neutralization of the acid and base is accompanied by neutralization of positive and negative charges, there is an evolution of heat in every acid-base neutralization.

Experiment 1: **Determine the integral enthalpy of the given solid in water at room temperature.**

Theory: The enthalpy of dissolution of the solid is determined calorimetrically. The experiment is carried out in a copper calorimeter insulated properly from the surroundings. A polythene bottle can give approximate results (Fig. 6.1). It is necessary to know the heat capacity or water equivalent of the calorimeter before starting the experiment on heat of solution. Water equivalent of the calorimeter is determined by taking a known volume of water at room temperature in the calorimeter and adding a known volume of hot water of known temperature and noting down the temperature on mixing. Heat capacity of the calorimeter can, then, be calculated.

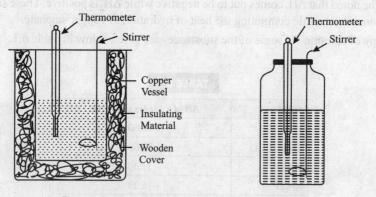

Fig. 6.1: Copper calorimeter and polythene calorimeter

Calorimeter is emptied. It is filled again with a known volume of water at room temperature. A known weight of the solid of known molecular mass is added to it, stirred thoroughly and the change in temperature of the calorimeter is noted. If w is the mass of solid with molecular mass m and Q joules is the heat evolved on mixing the solid with water, then

$$\text{Heat of solution} = \frac{Q}{w} \times m \text{ Joules}$$

Note: There are two kinds of enthalpy of solution, *integral enthalpy* and *differential enthalpy*. For details refer to the chapter on chemical thermodynamics. The enthalpy of solution that we shall measure in this experiment is integral enthalpy because a specific quantity of solute is dissolved in specific quantity of solvent.

Apparatus: Calorimeter, thermometer with 0.1°C calibration, stirrer.

Material: 8 g of the powdered solid. (KNO_3 or NH_4Cl)

Procedure:

(i) *Determination of heat capacity of the calorimeter.*

Take 100 ml water in the calorimeter. Wait for 5 minutes till the temperature of water becomes constant. Note down this temperature. Heat some water in a beaker on a flame, to about 20°C above the room temperature. Measure 100 ml of this hot water in a measuring cylinder. Note down the exact temperature of water. Add hot water to the calorimeter. Shake, note the highest temperature reached in the thermometer.

(ii) *Determination of integral enthalpy*

Powder 8.0 g of the given solid. Take 200 ml of water in the calorimeter. Note down the temperature. Add the powdered solid to the calorimeter. Replace the lid on the calorimeter. Stir with a glass rod, keeping an eye on the thermometer. Note down the maximum change in temperature that takes place.

(iii) Record the observations and calculate the water equivalent of the calorimeter and heat of solution as given below:

Observations

(a) *For heat capacity of the calorimeter:*

Let the temperature of cold water = t_1°C

Temperature of hot water = t_2°C

Temperature after mixing = t_3°C

(b) *For enthalpy of solution:*

Temperature of water taken = t_1°C

Temperature after mixing solid = t_4°C

Calculation of heat capacity

We shall apply the principle of heat gained = heat lost.

On adding 100 ml hot water to 100 ml cold water in the calorimeter

(i) heat is gained by the calorimeter and 100 ml cold water,

(ii) heat is lost by 100 ml hot water.

If W is the water equivalent of the calorimeter, then

$$\text{Heat gained} = (100 + W) \times (t_3 - t_1) \text{ calorie}$$

It is assumed here that 100 ml water weigh 100 g.

$$\text{Heat lost} = 100 \times (t_2 - t_3) \text{ calorie}$$

As heat gained = heat lost

$$(100 + W)(t_3 - t_1) = 100(t_2 - t_3)$$

or

$$100 + W = \frac{100(t_2 - t_3)}{t_3 - t_1}$$

or

$$W = \left[\frac{100(t_2 - t_3)}{t_3 - t_1}\right] - 100$$

Calculation of enthalpy of solution

When the solid is added to water and stirred with a glass rod, there is a fall of temp. (except when an anhydrous salt is used).

$$\text{Heat lost} = (200 + 8 + W) \times (t_1 - t_4) \text{ calorie}$$

Heat lost by dissolving 8 g solid $= (200 + 8 + W) \times (t_1 - t_4)$ calorie.

Heat lost by dissolving M g of the solid (M = Mol. wt.)

$$= \frac{(208 + W) \times (t_1 - t_4)}{8} \times M \text{ calorie}.$$

Convert the heat into joules by multiplying the above value by 1.184.

Experiment 2: Determine the enthalpy of neutralization of a strong acid (hydrochloric acid) and a strong base (sodium hydroxide solution)

Theory: Neutralisation process basically involves the reaction:

$$H_3O^+ + OH^- \longrightarrow 2H_2O + \text{Heat}$$

Every neutralization process is accompanied by evolution of heat as there is a neutralization of positive and negative charges. A neutralization between 1 g equivalent of our strong acid with a strong base will produce the same amount of heat *i.e.,* 57 kJ because the amounts of hydrogen and hydroxyl ions will be the same. However, if we take the strong acid and weak base or a weak acid and strong base or weak acid and weak base, the heats of neutralization will differ.

We perform the experiment in a calorimeter (copper calorimeter or polythene bottle). Its heat capacity is determined. Then we take 100 ml N/2 acid in the calorimeter and add an equivalent volume of the base and note down the rise of temperature. Finally the enthalpy of neutralisation is calculated.

Apparatus: A calorimeter, a thermometer (0.1°C) a stirrer.

Material: N/2 HCl, N/2 NaOH.

Procedure:

1. Determine the heat capacity or water equivalent of the calorimeter as described in experiment 1.
2. Take N/2 HCl and N/2 NaOH in different beakers. Let them attain the room temperatures. Sometimes, the temperatures of the two solutions do not attain the same value.
3. Measure 100 ml of acid and transfer to the calorimeter. Note down the temperature.
4. Note down the temperature of the alkali, measure 100 ml of it and transfer to the calorimeter.
5. Replace the lid immediately. Stir with the stirrer for complete mixing, keeping an eye on the mercury level in the thermometer. Note down the highest temperature reached.
6. Calculate the enthalpy of neutralization as given below:

Observations

For heat capacity of calorimeter:

Let the temperature of cold water $= t_1°C$

Temperature of hot water $= t_2°C$

Temperature after mixing $= t_3°C$

For enthalpy of neutralization:

Let the temperature of the acid = t_4°C
Temperature of the alkali = t_5°C
Temperature after mixing = t_6°C

Calculations

Heat capacity of the calorimeter (W) as explained in the previous experiment.

$$= \left[\frac{100(t_2 - t_3)}{t_3 - t_1}\right] - 100$$

If the temperature of acid and alkali before mixing are different, the mean temperature will be taken as the temperature before mixing. Heat evolved is calculated as under:

$$\text{Heat evolved} = (200 + W)\left[t_6 - \left(\frac{t_5 + t_5}{2}\right)\right] \text{ calories}$$

Where W is heat capacity of the calorimeter.

Above heat is evolved by neutralising 100 ml of $\frac{N}{2}$ acid *i.e.*, $\frac{1}{20}$th of the gram equivalent of the acid with a base. Hence heat evolved in neutralizing 1 g equivalent of the acid *i.e.*, enthalpy of neutralization

$$= 20(200 + W)\left[t_6 - \left(\frac{t_4 + t_5}{2}\right)\right]$$

Result: Enthalpy of neutralisation of HCl with NaOH

= .... kcal/g eq.
= .... kJ/g eq.
(1 kcal = 4.184 kJ).

Basicity/proticity of a polyprotic acid by thermochemical Method

Consider that we are carrying out a neutralisation of HCl with NaOH

HCl + NaOH ⟶ NaCl + H_2O + Heat
1 Mole   1 Mole

1 mole of NaOH is required to neutralise one mole of HCl. And the temperature of the mixture rises as neutralisation process is exothermic. We take 100 mL of 1 M HCl and 100 mL of 1 M NaOH to carry out the reaction and note down the highest temperature reached. Now, if you add 100 mL of more of 1 M NaOH to the above mixture, will there be a further rise in temperature? No, because complete neutralisation has already taken place.

Now consider a diabasic acid, like oxalic acid.

COOH   COONa
| + 2NaOH ⟶ | + $2H_2O$
COOH   COONa
1 Mole   2 Moles

Here 2 moles of NaOH are required to neutralise 1 mole of oxalic acid.

If you add 100 mL of 1 M NaOH to 100 mL of 1 M oxalic acid, temperature will rise and neutralisation will take place. But this is not complete neutralisation because only one of the —COOH group has been neutralised, one more —COOH group is still left. So if you now add one more lot of 1 M NaOH to this mixture, again the temperature will rise because neutralisation of second —COOH group is taking place. This can be represented as under:

$$\begin{array}{c}\text{COOH}\\|\\\text{COOH}\end{array} + \text{NaOH} \xrightarrow{\text{1st lot}} \begin{array}{c}\text{COONa}\\|\\\text{COOH}\end{array} + \text{NaOH} \xrightarrow{\text{2nd lot}} \begin{array}{c}\text{COONa}\\|\\\text{COONa}\end{array}$$

If you add third lot of 100 mL 1 M NaOH to this final mixture and stir, will there be further rise in temperaure? No, because complete neutralisation has taken. Thus, number of lots of NaOH solution that were added which resulted in the rise of temperature gives the basicity/proticity of the acid.

We can extend this treatment to a tribasic acid like citric acid which may represented as below for simplicity.

$$R\!\!-\!\!\begin{array}{l}\text{—COOH}\\\text{—COOH}\\\text{—COOH}\end{array}$$

$$R\!\!-\!\!\begin{array}{l}\text{—COOH}\\\text{—COOH}\\\text{—COOH}\end{array} + 3\text{NaOH} \longrightarrow R\!\!-\!\!\begin{array}{l}\text{—COONa}\\\text{—COONa}\\\text{—COONa}\end{array} + 3\text{H}_2\text{O}$$

1 Mole   3 Moles   1 Mole

In this case, 3 moles of NaOH will be required for complete neutralisation of 1 mole of citric acid. In other words, if we are taking 100 mL of 1 M citric acid, we shall find that these successive addition of 100 mL lots of 1 M NaOH will result is rise in temperature. After that, if we add 4th lot of 1 M NaOH will not show any rise in temperature if the neutralisation has already been completed.

To carry out such experiments, we have to keep on adding 100 mL of 1 M NaOH till there is no rise in temperature. Number of lots of NaOH that resulted in rise in temperature given the basicity of the acid. This is the underlying principle of this experiment

The enthalpy of neutralisation of any step of the reaction will be calculated from the initial and final temperature of that step. It may be noted that heat of neutralisation of different steps is not the same due to a umber of factors.

Experiment 3: Determination the basicity of a polyprotic acid (oxalic acid say) by thermochemical method with the help of a graph of temperature versus time. Also calculate the enthalpy of neutralisation of the first step.

Theory: As explained above

Reaction:

$$\begin{array}{c}\text{COOH}\\|\\\text{COOH}\end{array} + 2\text{NaOH} \longrightarrow \begin{array}{c}\text{COONa}\\|\\\text{COONa}\end{array} + 2\text{H}_2\text{O}$$

1 Mole   2 Moles

Apparatus: A calorimeter, thermometer (0.1°C), a stirrer, measuring cylinder

Materials: 1 M oxalic acid solution, 1 M NaOH solution

Procedure:

1. Determine the heat capacity (or water equivalent) of the calorimeter as described in experiment 1.
2. Take 1 M oxalic acid and 1 M NaOH solution in separate beakers. Let them attain room temperature. Sometimes, the temperature of two different solutions does not attain in the same value.
3. Measure 100 mL of the acid and transfer it to the calorimeter. Note down the temperature.
4. Note down the temperature of the alkali solution, measure 100 mL of it with a measuring cylinder and transfer to the calorimeter.
5. Replace the lid immediately. Stir with the stirrer for complete mixing, keeping an eye on the mercury level in the thermometer. Note down the highest temperature reached.
6. Once again note down the temperature and immediately add 100 mL of 1 M NaOH solution to the above mixture, stir and note down the maximum temperature attained.
7. If there has been a rise of temperature in step 6, once again note down the temperature and immediately add 100 mL of 1 M NaOH solution to the above mixture in the calorimeter, stir and note down the maximum temperature attained.
8. Repeat as required.

Observations

For heat capacity of the calorimeter
As in Experiment 1
For enthalpy of neutralisation
1st step
Temperature of oxalic acid solution = t_4
Temperature of NaOH solution = t_5
Highest temperature on mixing = t_6
2nd step
Initial temperature = t_7
Temperature on mixing second lot of NaOH = t_8

Calculations

Heat capacity of the calorimeter as explained in experiment 1

$$W = \left[\frac{100(t_2 - t_3)}{t_3 - t_1}\right] - 100$$

If the temperature of acid and alkali before mixing are different, the mean temperature will be taken as the temperature before mixing. Enthalpy of neutralisation of the first step is calculated as under:

$$\text{Heat evolved} = (200 \times W) \times \left[t_6 - \left(\frac{t_4 + t_5}{2}\right)\right] \text{ calories}$$

where W is the heat capacity of the calorimeter.

Above heat is evolved by neutralising 100 mL of 1 M acid and 100 mL of 1 M NaOH

100 mL of 1 M NaOH = 0.1 mole of NaOH

or 0.1 gm eq of NaOH

As per definition, enthalpy of neutralisation is the heat evolved when 1 gm equivalent of the alkali neutralises 1 gm equivalent of the acid.

The enthalpy of neutralisation of the 1st step

$$= (200 + W) \left[t_6 - \left(\frac{t_4 + t_5}{2} \right) \right] \times 10 \text{ calories}$$

$$= \text{....... k cal/gm eq}$$

$$= \text{....... k J/gm eq} \qquad \qquad (1 \text{ K cal} = 4.184 \text{ kJ})$$

Graph between Temperature and Time (use a graph paper)

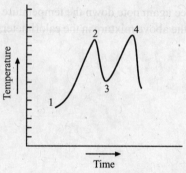

First step
1 = Temperature before mixing
2 = Maximum temperature obtained after mixing
Second step
3 = Temperature before mixing
4 = Temperature after mixing

Basicity = No. of crests in the graph
= 2

Experiment 4: Determine the enthalpy of hydration of copper sulphate ($CuSO_4$).

Theory: Enthalpy of hydration of $CuSO_4$ can be obtained by determining the enthalpy of solution of anhydrous copper sulphate ($CuSO_4$), ΔH_1 and that of hydrated copper sulphate ($CuSO_4 \cdot 5H_2O$) ΔH_2. The difference between the two values provides the enthalpy of hydration of copper sulphate

$$\Delta H_{Hydration} = \Delta H_1 - \Delta H_2$$

Apparatus: A calorimeter, a thermometer (0.1°C), a stirrer.

Materials: Hydrated copper sulphate

Procedure:

1. Determine the heat capacity or water equivalent of the calorimeter as described in experiment 1.
2. Take about 20 g of powdered copper sulphate in a china dish. Place it in an oven at 250°C for about one hour. Blue hydrated copper sulphate will turn into white anhydrous copper sulphate.
3. Alternatively, heat about 20 g of powdered hydrated copper sulphate in a china dish at low flame till a white powder is obtained. Remove the china dish from the flame and place it in a desiccator to bring it to room temperature. This white powder is to be protected from atmospheric moisture, otherwise it will convert back to blue hydrated copper sulphate.

4. When cooled, weigh 8 g of white powder in a weighing bottle (with a lid) speedily, taking care that white powder is exposed to atmospheric moisture for a minimum time.
5. Take 200 mL water in a calorimeter. Note down its temperature. Add the weighed white powder to it quickly. Replace the lid on the calorimeter. Stir with a glass rod, keeping an eye on the thermometer. Note down the maximum change in temperature that takes place.
6. Repeat the experiment with 8 g of powdered hydrated copper sulphate (blue) in the same way as with white powder (step 5).
7. Record the observations and calculate the heat capacity of the calorimeter and enthalpy of solution as given below.

Observations

For heat capacity of calorimeter

Temperature of cold water = t_1°C
Temperature of hot water = t_2°C
Temperature on mixing = t_3°C

For enthalpy of solution

Temperature of water taken = t_4°C
Temperature after mixing anhyd. $CuSO_4$ = t_5°C
Temperature after mixing hydrated $CuSO_4$ = t_6°C

Calculations

Heat of hydration. For heat capacity of calorimeter proceed as in experiment 1

I. Heat change when 8 g of anhydrous copper sulphate is dissolved = $(200 + 8 + W) \times (t_4 - t_5)$ calories

Heat change when 159.5 g (1 mole) of anhydrous copper sulphate is dissolved

$$= \frac{(208 + W)(t_4 - t_5)}{8} \times 159.5 \text{ calories}$$

Thus $\quad \Delta H_1 = \dfrac{(208 + W)(t_4 - t_5)}{8} \times 159.5 \text{ calories}$

II. Heat change when 8 g of hydrated copper sulphate is dissolved = $(200 + 8 + W) \times (t_4 - t_6)$ calories

Heat change when 249.5 g (1 mole) of hydrated copper sulphate is dissolved

$$= \frac{(208 + W)(t_4 - t_6)}{8} \times 249.5 \text{ calories}$$

Thus $\quad \Delta H_2 = \dfrac{(208 + W)(t_4 - t_6)}{8} \times 249.5 \text{ calories}$

Enthalpy of hydration of copper sulphate = $\Delta H_1 - \Delta H_2$

$$= \frac{(208 + W)(t_4 - t_5)}{8} \times 159.5 - \frac{(208 + W)(t_4 - t_6)}{8} \times 249.5$$

$$= \frac{(208 + W)}{8} [(t_4 - t_5) \times 159.5 - (t_4 - t_6) \times 249.51] \text{ cal}$$

It may be kept in mind that heat of solution of anhydrous copper sulphate ΔH_1 comes out to be negative (heat is evolted) and heat of solution of hydrated copper sulphate ΔH_2 comes out to be positive (heat is absorbed). These signs (+ or –) are to be taken into consideration while computing heat of hydration ΔH_{hyd}.

Multiply by 4.184 to convert calories into joules.

The expected value of heat of hydration of copper sulphate is –78.2 kJ.

Experiment 5: Determine the enthalpy of ionisation of acetic acid

Theory: We determine the enthalpy of ionization of acetic acid by performing a titration with a strong alkali solution say NaOH solution. Enthalpy of neutralization of a strong acid and a strong base is 57 kJ/g. eq. However with acetic acid and sodium hydroxide, the enthalpy of neutralization comes out to be smaller. This is because acetic acid is a week electrolyte and some energy is required to ionize it. We determine the enthalpy of neutralization of acetic acid with sodium hydroxide. The difference between the enthalpy of neutralization of a strong acid and strong base and that of acetic acid and sodium hydroxide gives the enthalpy of ionization of acetic acid.

We perform the experiment in a calorimeter. Its heat capacity (water equivalent) is first determined. The reaction involved is:

$$CH_3COOH + NaOH \longrightarrow \underset{\text{Sod. acetate}}{CH_3COONa} + H_2O$$

Apparatus: A calorimeter, a thermometer (0.1°C), a stirrer

Material: N/2 CH_3COO, N/2 NaOH

Procedure:
1. Determine the heat capacity of the calorimeter as described in experiment 1.
2. Take N/2 CH_3COOH and N/2 NaOH in different beakers. Let them attain room temperature. Sometimes, the temperatures in the two beakers do not attain the some value. Take the mean temperature in that case.
3. Measure 100 ml of the acid and therefore it to the calorimeter. Note the temperature.
4. Note down the temperature of the alkali, measure 100 ml of it and transfer it to the calorimeter.
5. Replace the lid immediately. Stir with a stirrer (glass rod) for complete mixing, keeping an eye on the mercury level in the thermometer. Note down the highest temperature reached.
6. Calculate the enthalpy of ionization as given below:

Observations

For heat capacity of calorimeter
    Let   The temperature of cold water = $t_1°$ C
                  Temperature of hot water = $t_2°$ C
                  Temperature after mixing = $t_3°$ C

For enthalpy of ionization
    Let   The temperature of the acid = $t_4°$ C
                  Temperature of the alkali = $t_5°$ C
                  Temperature after mixing = $t_6°$ C

Calculationos

Heat capacity of the calorimeter, explained in the previous experiment

$$= \left[\frac{100(t_2 - t_3)}{t_3 - t_1}\right] - 100$$

If the temperature of the acid and alkali solutions before mixing are different, the mean temperature will be taken as the temperature before mixing. Heat evolved is calculated as under:

$$\text{Heat evolved} = (200 + W) \times \left[t_6 - \left(\frac{t_4 + t_5}{2}\right)\right] \text{calories}$$

where W is the heat capacity of the calorimeter.

The above heat is evolved by neutralizing 100 ml of N/2 acetic acid i.e. $\frac{1}{20}$ th gram equivalent of the acid with a base. Hence heat evolved in neutralizing 1 g equivalent of the acid i.e. enthalpy of neutralization

$$= 20 \times (200 + W) \times \left[t_6 - \left(\frac{t_4 + t_5}{2}\right)\right] \text{Cal/g-eq}$$

Convert this value first into joules and then into kJ. (1 cal = 4.184 J)

Let the enthalpy of neutralization = x kJ

We know the enthalpy of neutralization of a strong acid and strong base is 57 kJ/g. eq

Since, the acid (CH_3COOH) in this case is not a strong acid, some amount of energy is consumed in ionizing the acid into ions. *The heat required to produce ionization in one g-eq (or one mole) of the acid in order to complete the neutralization process is called enthalpy of ionization.* Hence

Enthalpy of ionization = 57 kJ – observed enthalpy of neutralization
$$= 57 \text{ kJ} - x \text{ kJ}$$
$$= (57 - x) \text{ kJ}$$

Experiment 6: Determine the solubility of benzoic acid at different temperatures and to determine ΔH of the dissolution process

Theory: Varying amounts of benzoic acid are taken in four test tubes. 20 ml of water is added to each tube. The four tubes are placed in a beaker containing water and the beaker is heated on a flame (Fig. 6.2). A stirrer is placed in the test tube containing the least amount of benzoic acid. The contents of this test tube are stirred. The temperature is noted at which at benzoic acid just completely

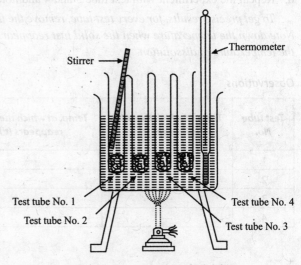

***Fig. 6.2:** Determination of solubility*

dissolves. The stirrer is now placed in the test tube containing next higher amount of benzoic acid. Here also the temperature at which the solid just dissolves is noted. The same observations are repeated with the other test tubes. The amounts of benzoic acid are converted into no. of moles/litre of the solvent. These values represent the solubilities. A curve is plotted between the solubility and temperature. With the help of this curve, we can find out the solubility of benzoic acid at any desired temperature.

Let S_1 and S_2 be the solubilities of benzoic acid at two different temperatures $t_1°C$ and $t_2°C$. van't Hoff's equation can be applied to the solubilities as under:

$$\log S_2 - \log S_1 = \frac{\Delta H (T_2 - T_1)}{2.303 R \times T_1 \times T_2}$$

Substituting the values of S_1, S_2, T_1, T_2 and R, ΔH can be evaluated. It may be noted that in the above equation, S represents the solubility in gm mole per litre of the solvent as against the *normal definition of no. of grams of the solute dissolved in 100 g of the solvent*.

Apparatus: A 500 ml pyrex beaker, thermometer, 4 boiling tubes.

Material: Water benzoic acid.

Procedure:
1. Take four boiling tubes of 40-50 ml capacity. Label them as 1, 2, 3 and 4.
2. Weigh 0.1 g of benzoic acid on a watch glass and transfer it to test tube No. 1. Similarly take 0.15 g, 0.2 g and 0.3 g benzoic acid in test tubes no. 2, 3, 4 respectively.
3. Add 20 ml water to each tube and place them in a beaker containing water.
4. Place the stirrer in test tube 1. Stir the contents keeping an eye on the temperature in the thermometer. Note down the temperature as soon as benzoic acid in test tube 1 just dissolves.
5. Remove the stirrer from test tube 1 and palce it in test tube 2. Stir the contents and note down the temperature at which benzoic acid in test tube 2 just dissolves.
6. Repeat the experiment with test tube 3 and 4 and note down the observations.

To get precise results, for every test tube, remove the flame when the solid has just dissolved. Note down the temperature when the solid just reappears. Mean of the two temperatures gives the temperature of dissolution.

Observations

Test tube No.	Temp at which the solid dissolves (t)	Temp. at which the solid reappears (t')	Mean temp. of dissolution $\frac{(t + t')}{2}$
1.	—	—	t_1
2.	—	—	t_2
3.	—	—	t_1
4.	—	—	t_1

S. No.	Test tube No.	Amount of benzoic acid (g)	Solubility (moles/litre)	Temp. at which the solid dissolves (°C)
1.	1	1.1	$\frac{0.1}{122} \times 50$	t_1
2.	2	0.15	$\frac{0.15}{122} \times 50$	t_2
3.	3	0.2	$\frac{0.2}{122} \times 50$	t_3
4.	4	0.3	$\frac{0.3}{122} \times 50$	t_4

Calculations

Draw a graph between solubility taken along y-axis and temperature long x-axis (Fig. 6.3). The curve is of the type as shown.

Determine the solubilities of benzoic acid at 40°C (313 K) and 50°C (323 K) from the graph.

Let the solubility at 313 K be $= S_{313}$
andsolubility at 323 K be $= S_{323}$
Applying van't Hoff's equation, we have

$$\log S_{323} - \log S_{313} = \frac{\Delta H (323 - 313)}{2.303 \times 313 \times 323 \times 8.314}$$

$(R = 8.314 \text{ J})$

Compute the value of ΔH, it will be in joules

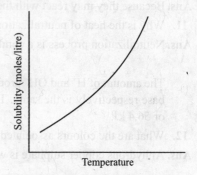

Fig. 6.3: Graph between the solubility of benzoic acid and temperature

Viva-Voce Questions with Answers

1. Define heat of reaction?

Ans. The heat of reaction is defined as the amount of heat change accompanying the conversion of the various reactants, as indicated by a balanced chemical equation into products.

2. Define heat of solution?

Ans. Heat of solution (dissolution) of a substance, at a given temperature, is defined as the amount of heat change accompanying the dissolution of 1 mole of it in such a large excess of the solvent that further dilution produces no heat change at that temperature.

3. Define neutralization.

Ans. It is the phenomenon of chemical combination between an acid and a base to form salt and water.

4. Define heat of neutralization.

Ans. It is the amount of heat evolved when 1 gram equivalent of an acid is neutralized by 1 gram equivalent of a base in dilute solutions.

5. Is heat of neutralization of a strong acid and a strong base constant?
Ans. Yes, it is.
6. Why do the heat changes taking place in a reaction vary with change in temperature?
Ans. On account of the difference in the specific heats of various substances involved in the reaction.
7. Is neutralization an exothermic reaction or endothermic?
Ans. It is an exothermic reaction.
8. What is meant by radiation error?
Ans. Heat lost due to radiation of heat leads to an error referred to as radiation error.
9. Why is stirring necessary in calorimetric experiments?
Ans. In order to have a uniform temperature.
10. Why are metallic calorimeters not used in measurement of heat of neutralization?
Ans. Because they may react with the acid or the base involved.
11. Why is the heat of neutralization of any strong acid with any strong base the same?
Ans. Neutralization process is a combination between H^+ and OH^- ions to produce water.
$$H^+ + OH^- \longrightarrow H_2O$$
The amount of H^+ and OH^- produced per gm equivalent of any strong acid and any strong base respectively is the same. Therefore, heat of neutralization is the same *i.e.*, 13.7 kcal or 56.4 kJ.
12. What are the colours associated with anhydrous and hydrated copper sulphate?
Ans. Anhydrous copper sulphate is white while the hydrated compound is blue in colour.